GERARD J. TORTORA
Biology Coordinator, Bergen Community College, Paramus, New

Laboratory Exercises in HUMAN ANATOMY with Cat Dissections

FOURTH EDITION

Prentice Hall
Upper Saddle River, NJ 07458

Executive Editor: David Kendric Brake
Acquisitions Editor: Linda Schreiber
Assistant Vice President and Director of ESM Production: David W. Riccardi
Special Projects Manager: Barbara A. Murray
Editorial Production/Design: Publication Services
Text Illustrations: Mary Dersh; Publication Services
Cover Design: Joseph Sengotta
Cover Illustrations: Mary Dersh; Publication Services
Photo Researchers: Jan Fisher; Stuart Kenter, Associates

This book was set in 10/12 Palatino by Publication Services, Inc.,
and was printed and bound by Von Hoffmann Press.
The cover was printed by Von Hoffmann Press.

Published by Prentice-Hall, Inc.
Simon & Schuster/A Viacom Company

Printed in the United States of America

ISBN 0-13-268251-6

10 9 8 7 6 5 4 3 2 1

Prentice Hall International (UK) Limited, *London*
Prentice Hall of Australia Pty. Limited, *Sydney*
Prentice Hall Canada Inc., *Toronto*
Prentice Hall Hispanoamericana, S.A., *Mexico*
Prentice Hall of India Private Limited, *New Delhi*
Prentice Hall of Japan, Inc., *Tokyo*
Simon & Schuster Asia Private Limited, *Singapore*
Editora Prentice Hall do Brazil, Ltda., *Rio de Janeiro*

Preface

Laboratory Exercises in Human Anatomy with Cat Dissections, Fourth Edition, has been written to guide students in the laboratory study of introductory anatomy. The manual was written to accompany most of the leading anatomy textbooks.

COMPREHENSIVENESS

This manual examines virtually every structure of the human body that is typically studied in an introductory anatomy course. Because of its detail, the need for supplemental handouts is minimized; the manual is a strong teaching device in itself.

USE OF THE SCIENTIFIC METHOD

Anatomy (the science of structure) cannot be understood without the practical experience of laboratory work. The exercises in this manual challenge students to understand the way scientists work by asking them to make microscopic examinations and evaluations of cells and tissues, to record data, to make gross examinations of organs and systems, to dissect, and to interpret and apply the results of this work.

ILLUSTRATIONS

The manual contains a large number and variety of illustrations. The illustrations of the body systems of the human have been carefully drawn to exhibit structures that are essential to students' understanding of anatomy. Numerous photographs, photomicrographs, and scanning electron micrographs are presented to show students how the structures of the body actually look. We feel that this laboratory manual has better and more complete illustrations than any other anatomy manual.

IMPORTANT FEATURES

Among the key features of this manual are (1) dissection of the white rat, and selected mammalian organs; (2) emphasis on the study of anatomy through histology; (3) lists of appropriate terms accompanying drawings and photographs to be labeled; (4) inclusion of numerous scanning electron micrographs and specimen photos; (5) a separate exercise on surface anatomy; (6) phonetic pronunciations and derivations for the vast majority of anatomical terms; (7) laboratory report questions at the end of each exercise that can be filled in, removed, and turned in for grading if the instructor so desires; (8) three appendixes dealing with units of measurement, a periodic table of elements, and eponyms used in the laboratory manual; and (9) emphasis on laboratory safety throughout the manual.

NEW TO FOURTH EDITION

Numerous changes have been made in the fourth edition of this manual in response to suggestions from instructors and students. We have added some new line drawings, cadaver photographs, photomicrographs, and phonetic pronunciations and derivations. The principal additions to various exercises are as follows:

Exercise 2, "Introduction to the Human Body," has a new section dealing with organs in various body cavities.

In Exercise 4, "Tissues," there is a new exercise dealing with the identification of epithelial tissues.

In Exercise 6, "Bone Tissue," the chemistry of bone is now discussed before the structure of a long bone. There is also a new section on the epiphyseal plate and one dealing with fractures.

In Exercise 7, "Bones," the black and white bone photos have been replaced by new, clearer illustrations.

In Exercise 8, "Articulations," several photographs have been replaced.

In Exercise 10, "Skeletal Muscles," there is a new section on the arrangement of muscle fascicles and a new exercise on naming skeletal muscles.

In Exercise 12, "Nervous Tissue," there is a new introduction and a new section on neuronal circuits.

In Exercise 13, "Nervous System," there are new illustrations on plexuses and a new section on somatic sensory and motor pathways. Phonetic pronunciations have been added to *all* the nerves of the plexuses.

In Exercise 14, "General and Special Senses," the discussion of somatic sensory and motor pathways has been expanded and the details of sensory and motor tracts have been moved to Exercise 13.

In Exercise 15, "Endocrine System," all hormone functions have been updated and the histology of the testes and ovaries has been revised.

Several new illustrations have been added to Exercise 17, "The Heart." In addition, there is a new section on the cardiac conduction system.

In Exercise 18, "Blood Vessels," phonetic pronunciations have been added to *all* blood vessel names and the discussions of hepatic portal, pulmonary, and fetal circulation have been expanded.

In Exercise 20, "Respiratory System," the introduction is new and the section dealing with the nose has been revised. New art has also been added.

In Exercise 21, "Digestive System," the sections dealing with the salivary glands and stomach histology have been revised and a new illustration has been added.

In Exercise 23, "Reproductive Systems," a new introduction has been added and the discussions of the testes, ovaries, and female reproductive cycle have been revised.

The discussions of spermatogenesis, oogenesis, and the embryonic period have been expanded and revised in Exercise 24, "Development."

CHANGES IN TERMINOLOGY

In recent years, the use of eponyms for anatomical terms has been minimized or eliminated. Anatomical eponyms are terms named after various individuals. Examples include Fallopian tube (after Gabriello Fallopio) and Eustachian tube (after Bartolommeo Eustachio).

Anatomical eponyms are often vague and nondescriptive and do not necessarily mean that the person whose name is applied contributed anything very original. For these reasons, we have also decided to minimize their use. However, because some still prevail, we have provided eponyms, in parentheses, after the first reference in each chapter to the more acceptable synonym. Thus, you will expect to see terms such as ***uterine (Fallopian) tube*** or ***auditory (Eustachian) tube.*** See Appendix C.

INSTRUCTOR'S GUIDE

A complimentary instructor's guide by Gerard J. Tortora to accompany the manual is available from the publisher. This comprehensive guide contains: (1) a listing of materials needed to complete each exercise, (2) suggested audiovisual materials, (3) answers to illustrations and questions within the exercises, and (4) answers to laboratory report questions.

Gerard J. Tortora
Biology Coordinator
Science and Health, S229
Bergen Community College
400 Paramus Road
Paramus, NJ 07652

Contents

Laboratory Safety*

In 1989, The Centers for Disease Control and Prevention (CDC) published "Guidelines for Prevention of Transmission of Human Immunodeficiency Virus and Hepatitis B Virus to Health-Care and Public-Safety Workers" (*MMWR*, vol. 36, No. 6S). The CDC guidelines recommend precautions to protect health care and public safety workers from exposure to human immunodeficiency virus (HIV), the causative agent of acquired immunodeficiency syndrome (AIDS), and hepatitis B virus (HBV), the causative agent of hepatitis B. These guidelines are presented to reaffirm the basic principles involved in the transmission of not only the AIDS and hepatitis B viruses, but also any disease-producing organism.

Based on the CDC guidelines for health care workers, as well as on other standard additional laboratory precautions and procedures, the following list has been developed for your safety in the laboratory. Although specific cautions and warnings concerning laboratory safety are indicated throughout the manual, read the following ***before*** performing any experiments.

A. GENERAL SAFETY PRECAUTIONS AND PROCEDURES

1. Arrive on time. Laboratory directions and procedures are given at the beginning of the laboratory period.
2. Read all experiments before you come to class to be sure that you understand all the procedures and safety precautions. Ask the instructor about any procedure you do not understand exactly. Do not improvise any procedure.
3. Protective eyewear and laboratory coats or aprons must be worn by all students performing or observing experiments.
4. Do not perform any unauthorized experiments.
5. Do not bring any unnecessary items to the laboratory and do not place any personal items (pocketbooks, bookbags, coats, umbrellas, etc.) on the laboratory table or at your feet.
6. Make sure each apparatus is supported and squarely on the table.
7. Tie back long hair to prevent it from becoming a laboratory fire hazard.
8. Never remove equipment, chemicals, biological materials, or any other materials from the laboratory.
9. Do not operate any equipment until you are instructed in its proper use. If you are unsure of the procedures, ask the instructor.
10. Dispose of chemicals, biological materials, used apparatus, and waste materials according to your instructor's directions. Not all liquids are to be disposed of in the sink.
11. Some exercises in the laboratory manual are designed to induce some degree of cardiovascular stress. Students should not participate in these exercises if they are pregnant or have hypertension or any other known or suspected condition that might compromise health. Before you perform any of these exercises, check with your physician.
12. Do not put anything in your mouth while in the laboratory. Never eat, drink, taste chemicals, lick labels, smoke, or store food in the laboratory.
13. Your instructor will show you the location of emergency equipment such as fire extinguishers, fire blankets, and first-aid kits as well as eyewash stations. Memorize their locations and know how to use them.
14. Wash your hands before leaving the laboratory. Because bar soaps can become contaminated, liquid or powdered soaps should be used. Before leaving the laboratory, remove any protective clothing, such as laboratory coats or aprons, gloves, and eyewear.

*The authors and publisher urge consultation with each instructor's institutional policies concerning laboratory safety and first-aid procedures.

B. PRECAUTIONS RELATED TO WORKING WITH BLOOD, BLOOD PRODUCTS, OR OTHER BODY FLUIDS

1. Work only with *your own* body fluids, such as blood, saliva, urine, tears, and other secretions and excretions; blood from a clinical laboratory that has been tested and certified as noninfectious; or blood from a mammal (other than a human).
2. Wear gloves when touching another person's blood or other body fluids.
3. Wear safety goggles when working with another person's blood.
4. Wear a mask and protective eyewear or a face shield during procedures that are likely to generate droplets of blood or other body fluids.
5. Wear a gown or an apron during procedures that are likely to generate splashes of blood or other body fluids.
6. Wash your hands immediately and thoroughly if contaminated with blood or other body fluids. Hands can be rapidly disinfected by using (1) a phenol disinfectant-detergent for 20 to 30 seconds (sec) and then rinsing with water, or (2) alcohol (50 to 70%) for 20 to 30 sec, followed by a soap scrub of 10 to 15 sec and rinsing with water.
7. Spills of blood, urine, or other body fluids onto bench tops can be disinfected by flooding them with a disinfectant-detergent. The spill should be covered with disinfectant for 20 minutes (min) before being cleaned up.
8. Potentially infectious wastes, including human body secretions and fluids, and objects such as slides, syringes, bandages, gloves, and cotton balls contaminated with those substances, should be placed in an autoclave container. Sharp objects (including broken glass) should be placed in a puncture-proof sharps container. Contaminated glassware should be placed in a container of disinfectant and autoclaved before it is washed.
9. Use only single-use, disposable lancets, and needles. Never recap, bend, or break the lancet once it has been used. Place used lancets, needles, and other sharp instruments in a *fresh* 1 : 10 dilution of household bleach (sodium hypochlorite) or other disinfectant such as phenols (Amphyl), aldehydes (glutaraldehyde, 1%), and 70% ethyl alcohol and then dispose of the instruments in a puncture-proof container. These disinfectants disrupt the envelope of HIV and HBV. The fresh household bleach solution or other disinfectant should be prepared for *each* laboratory session.
10. All reusable instruments, such as hemocytometers, well slides, and reusable pipettes, should be disinfected with a *fresh* 1 : 10 solution of household bleach or other disinfectant and thoroughly washed with soap and hot water. The fresh household bleach solution or other disinfectant should be prepared for *each* laboratory session.
11. A laboratory disinfectant should be used to clean laboratory surfaces *before* and after procedures, and should be available for quick cleanup of any blood spills.
12. Mouth pipetting should never be done. Use mechanical pipetting devices for manipulating all liquids in the laboratory.
13. All procedures and manipulations that have a high potential for creating aerosols or infectious droplets (such as centrifuging, sonicating, and blending) should be performed carefully. In such instances, a biological safety cabinet or other primary containment device is required.

C. PRECAUTIONS RELATED TO WORKING WITH REAGENTS

1. Use extreme care when working with reagents. Should any reagents make contact with your eyes, flush with water for 15 min; or, if they make contact with your skin, flush with water for 5 min. Notify your instructor immediately should a reagent make contact with your eyes or skin, and seek immediate medical attention.
2. Report all accidents to your instructor, no matter how minor they may appear.
3. When you are working with chemicals or preserved specimens, the room should be well ventilated. Avoid breathing fumes for any extended period of time.
4. Never point the opening of a test tube containing a reacting mixture (especially when heating it) toward yourself or another person.

5. Exercise care in noting the odor of fumes. Use "wafting" if you are directed to note an odor. Your instructor will demonstrate this procedure.
6. Do not force glass tubing or a thermometer into rubber stoppers. Lubricate the tubing and introduce it gradually and gently into the stopper. Protect your hands with toweling when inserting the tubing or thermometer into the stopper.
7. Never heat a flammable liquid over or near an open flame.
8. Use only glassware marked Pyrex or Kimax. Other glassware may shatter when heated. Handle hot glassware with test-tube holders.
9. If you have to dilute an acid, always add acid (AAA) to water.
10. When shaking a test tube or bottle to mix its contents, do not use your fingers as a stopper.
11. Read the label on a chemical twice before using it.
12. Replace caps or stoppers on bottles immediately after using them. Return spatulas to their correct place immediately after using them and do not mix them up.
13. Mouth pipetting should never be done. Use mechanical pipetting devices for manipulating all liquids in the laboratory.

D. PRECAUTIONS RELATED TO DISSECTION

1. When you are working with chemicals or preserved specimens, the room should be well ventilated. Avoid breathing fumes for any extended period of time.
2. Wear rubber gloves when dissecting.
3. To reduce the irritating effects of chemical preservatives to your skin, eyes, and nose, soak or wrap your specimen in a substance such as "Biostat." If this is not available, hold your specimen under running water for several minutes to wash away excess preservative and dilute what remains.
4. When dissecting, there is always the possibility of skin cuts or punctures from dissecting equipment or the specimens themselves, such as the teeth or claws of an animal. Should you sustain a cut or puncture in this manner, wash your hands with disinfectant soap, notify your instructor, and seek immediate medical attention to decrease the possibility of infection. A first-aid kit should be readily available for your use.
5. When cleaning dissecting instruments, always hold the sharp edges away from you.
6. Dispose of any damaged or worn-out dissecting equipment in an appropriate container supplied by your instructor.

SELECTED LABORATORY SAFETY SIGNS/LABELS

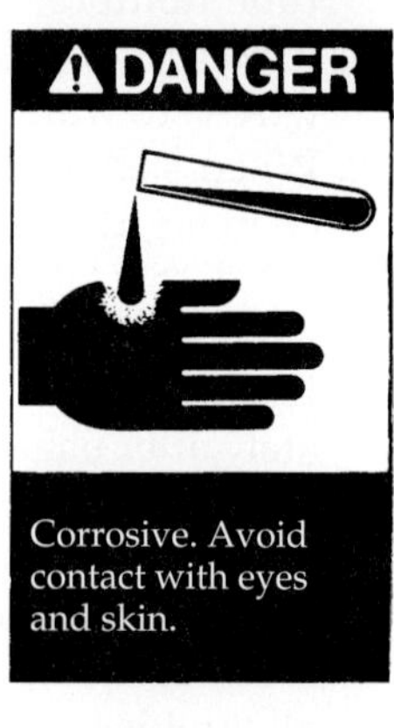

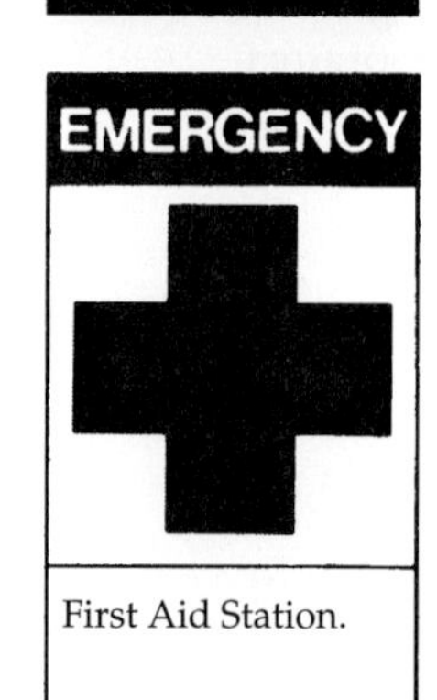

COMMONLY USED LABORATORY EQUIPMENT

Pronunciation Key

A unique feature of this revised manual is the phonetic pronunciations given for many anatomical terms. The pronunciations are given in parentheses immediately after the particular term is introduced. The following key explains the essential features of the pronunciations.

1. The syllable with the strongest accent appears in capital letters; for example, bilateral (bī-LAT-er-al) and diagnosis (dī-ag-NŌ-sis).
2. A secondary accent is denoted by a single quote mark ('); for example, constitution (kon'-sti-TOO-shun) and physiology (fiz'-ē-OL-ō-jē). Additional secondary accents are also noted by a single quotation mark; for example, decarboxylation (dē'-kar-bok'-si-LĀ-shun).
3. Vowels marked with a line above the letter are pronounced with the long sound, as in the following common words:

   ā as in *māke* — ī as in *īvy*
   ē as in *bē* — ō as in *pōle*

4. Unmarked vowels are pronounced with the short sound, as in the following words:

   e as in *bet* — *o* as in *not*
   i as in *sip* — *u* as in *bud*

5. Other phonetic symbols are used to indicate the following sounds:

   a as in *above* — *yoo* as in *cute*
   oo as in *soon* — *oy* as in *oil*

Microscopy

1

Note: *Before you begin any laboratory exercises in this manual, please read the section on LABORATORY SAFETY on page (xi).*

One of the most important instruments that you will use in your anatomy course is a compound light microscope. In this instrument, the lenses are arranged so that images of objects too small to be seen with the naked eye can become highly magnified; that is, apparent size can be increased, and their minute details can be revealed. Before you actually learn the parts of a compound light microscope and how to use it properly, discussion of some of the principles employed in light microscopy (mī-KROS-kō-pē) will be helpful. Later in this exercise, some of the principles used in electron microscopy will also be discussed.

A. COMPOUND LIGHT MICROSCOPE

A ***compound light microscope*** uses two sets of lenses, ocular and objective, and employs light as its source of illumination. Magnification is achieved as follows. Light rays from an illuminator are passed through a condenser, which directs the light rays through the specimen under observation; from here, light rays pass into the objective lens, the magnifying lens that is closest to the specimen; the image of the specimen then forms on a prism and is magnified again by the ocular lens.

A general principle of microscopy is that the shorter the wavelength of light used in the instrument, the greater the resolution. ***Resolution (resolving power),*** is the ability of the lenses to distinguish fine detail and structure, that is, to distinguish between two points as separate objects. As an example, a microscope with a resolving power of 0.3 micrometers (mī-KROM-e-ters), symbolized μm, is capable of distinguishing two points as separate objects if they are at least 0.3 μm apart. 1 μm = 0.000001 or 10^{-6} m. (See Appendix A.) The light used in a compound light microscope has a relatively long wavelength and cannot resolve structures smaller than 0.3 μm. This fact, as well as practical considerations, means that even the best compound light microscopes can magnify images only about 2000 times.

A ***photomicrograph*** (fō-tō-MĪ-krō'-graf), a photograph of a specimen taken through a compound light microscope, is shown in Figure 4.1. In later exercises you will be asked to examine photomicrographs of various specimens of the body before you actually view them yourself through the microscope.

1. Parts of the Microscope

Carefully carry the microscope from the cabinet to your desk by placing one hand around the arm and the other hand firmly under the base. Gently place it on your desk, directly in front of you, with the arm facing you. Locate the following parts of the microscope and, as you read about each part, label Figure 1.1 by placing the correct numbers in the spaces next to the list of terms that accompanies the figure.

1. ***Base*** The bottom portion on which the microscope rests.
2. ***Body tube*** The portion that receives the ocular.
3. ***Arm*** The angular or curved part of the frame.
4. ***Inclination joint*** A movable hinge in some microscopes that allows the instrument to be tilted to a comfortable viewing position.
5. ***Stage*** A platform on which microscope slides or other objects to be studied are placed. The opening in the center, called the ***stage opening,*** allows light to pass from below through the specimen being examined. Some microscopes have a ***mechanical stage.*** An adjustor knob below the stage moves the stage forward and backward and from side to side. With a mechanical stage, the slide and the stage move simultaneously. A mechanical stage permits a smooth, precise movement of a slide. Sometimes a mechanical stage is fitted with calibrations that permit the numerical "mapping" of a specimen on a slide.

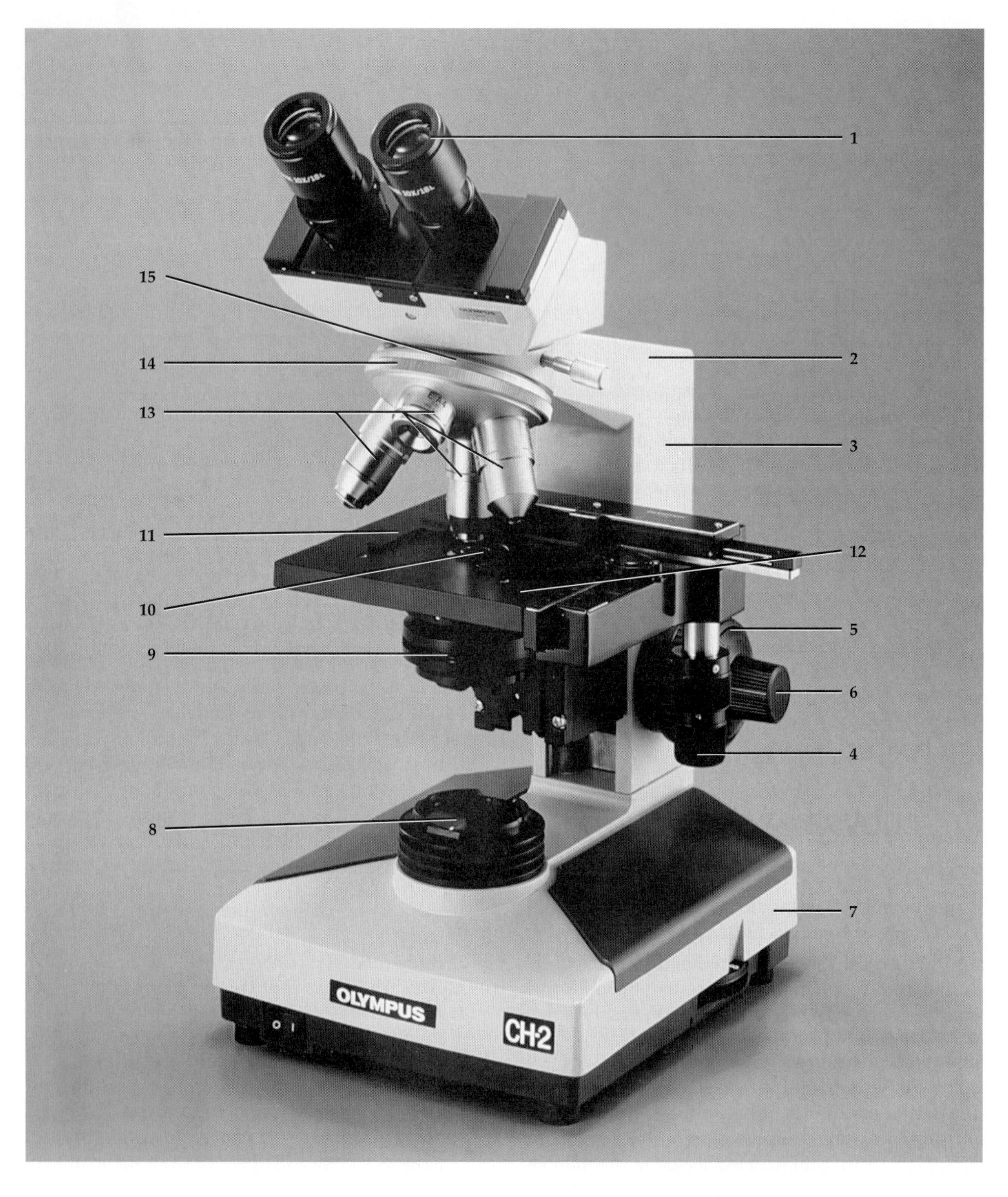

___ Arm	___ Diaphragm	___ Ocular
___ Base	___ Fine adjustment knob	___ Revolving nosepiece
___ Body tube	___ Mechanical stage knob	___ Stage
___ Coarse adjustment knob	___ Nosepiece	___ Stage clip of mechanical stage
___ Condenser	___ Objectives	___ Substage lamp

FIGURE 1.1 Olympus CH-2 microscope

6. ***Stage (spring) clips*** Two clips mounted on the stage that hold slides securely in place.
7. ***Substage lamp*** The source of illumination for some light microscopes with a built-in lamp.
8. ***Mirror*** A feature found in some microscopes below the stage. The mirror directs light from its source through the stage opening and through the lenses. If the light source is built-in, a mirror is not necessary.
9. ***Condenser*** A lens located beneath the stage opening that concentrates the light beam on the specimen.
10. ***Condenser adjustment knob*** A knob that functions to raise and lower the condenser. In its highest position, it allows full illumination and thus can be used to adjust illumination.
11. ***Diaphragm*** (DĪ-a-fram) A device located below the condenser that regulates light intensity passing through the condenser and lenses to the observer's eyes. Such regulation is needed because transparent or very thin specimens cannot be seen in bright light. One of two types of diaphragms is usually used. An ***iris diaphragm,*** as found in cameras, is a series of sliding leaves that vary the size of the opening and thus the amount of light entering the lenses. The leaves are moved by a ***diaphragm lever*** to regulate the diameter of a central opening. A ***disc diaphragm*** consists of a plate with a graded series of holes, any of which can be rotated into position.
12. ***Coarse adjustment knob*** A usually larger knob that raises and lowers the body tube (or stage) to bring a specimen into general view.
13. ***Fine adjustment knob*** A usually smaller knob found below or external to the coarse adjustment knob and used for fine or final focusing. Some microscopes have both coarse and fine adjustment knobs combined into one.
14. ***Nosepiece*** A plate, usually circular, at the bottom of the body tube.
15. ***Revolving nosepiece*** The lower, movable part of the nosepiece that contains the various objective lenses.
16. ***Scanning objective*** A lens, marked 5× on most microscopes (× means the same as "times"); it is the shortest objective and is not present on all microscopes.
17. ***Low-power objective*** A lens, marked 10× on most microscopes; it is the next longer objective.
18. ***High-power objective*** A lens, marked 43× or 45× on most microscopes; also called a ***high-dry objective;*** it is an even longer objective.
19. ***Oil-immersion objective*** A lens, marked 100× on most microscopes and distinguished by an etched colored circle (special instructions for this objective are discussed later); it is the longest objective.
20. ***Ocular (eyepiece)*** A removable lens at the top of the body tube, marked 10× on most microscopes. An ocular is sometimes fitted with a pointer or measuring scale.

2. Rules of Microscopy

You must observe certain basic rules at all times to obtain maximum efficiency and provide proper care for your microscope.

1. Keep all parts of the microscope clean, especially the lenses of the ocular, objectives, condenser, and also the mirror. *You should use the special lens paper that is provided and never use paper towels or cloths, because these tend to scratch the delicate glass surfaces.* When using lens paper, use the same area on the paper only once. As you wipe the lens, change the position of the paper as you go.
2. Do not permit the objectives to get wet, especially when observing a ***wet mount***. You must use a ***cover slip*** when you examine a wet mount or the image becomes distorted.
3. Consult your instructor if any mechanical or optical difficulties rise. *Do not try to solve these problems yourself.*
4. Keep *both* eyes open at all times while observing objects through the microscope. This is difficult at first, but with practice becomes natural. This important technique will help you to draw and observe microscopic specimens without moving your head. Only your eyes will move.
5. Always use either the scanning or low-power objective first to locate an object; then, if necessary, switch to a higher power.
6. If you are using the high-power or oil-immersion objectives, *never focus using the coarse adjustment knob.* The distance between these objectives and the slide, called ***working distance,*** is very small and you may break the cover slip and the slide and scratch the lens.
7. Some microscopes have a stage that moves while focusing, others have a body tube that moves while focusing Be sure you are familiar with which type you are using. Look at your microscope from the side and using the scanning or low power objective, and gently turn the coarse adjustment knob. Which moves? The stage or the body tube? *Never focus downward* if the microscope's body tube moves when focusing. *Never focus upward* if the microscope's stage

moves when focusing. By observing from one side you can see that the objectives do not make contact with the cover slip or slide.

8. Make sure that you raise the body tube before placing a slide on the stage or before removing a slide.

3. Setting Up the Microscope

PROCEDURE

1. Place the microscope on the table with the ocular toward you and with the back of the base at least 1 inch (in.) from the edge of the table.
2. Position yourself and the microscope so that you can look into the ocular comfortably.
3. Wipe the objectives, the top lens of the ocular, the condenser, and the mirror with lens paper. Clean the most delicate and the least dirty lens first. Apply xylol or ethanol to the lens paper only to remove grease and oil from the lenses and microscope slides.
4. Position the low-power objective in line with the body tube. When it is in its proper position, it will click. Lower the body tube using the coarse adjustment knob until the bottom of the lens is approximately 1/4 in. from the stage.
5. Admit the maximum amount of light by opening the diaphragm, if it is an iris diaphragm, or turning the disc to its largest opening, if it is a disc diaphragm.
6. Place your eye to the ocular, and adjust the light. When a uniform circle (the ***microscopic field***) appears without any shadows, the microscope is ready for use.

4. Using the Microscope

PROCEDURE

1. Using the coarse adjustment knob, raise the body tube to its highest fixed position.
2. Make a temporary mount using a single letter of newsprint, or use a slide that has been specially prepared with a letter, usually the letter "e." If you prepare such a slide, cut a single letter—"a," "b," or "e"—from the smallest print available and place this letter in the correct position to be read with the naked eye. Your instructor will provide directions for preparing the slide.
3. Place the slide on the stage, making sure that the letter is centered over the stage opening, directly over the condenser. Secure the slide in place with the stage clips.
4. Align the low-power objective with the body tube.
5. Lower the body tube or raise the stage as far as it will go *while you watch it from the side*, taking care not to touch the slide. The tube should reach an automatic stop that prevents the low-power objective from hitting the slide.
6. While looking through the ocular, turn the coarse adjustment knob counterclockwise, raising the body tube. Or, turn the coarse adjustment knob clockwise, lowering the stage. When focusing, always *raise* the body tube or *lower* the stage. Watch for the object to suddenly appear in the microscopic field. If it is in proper focus, the low-power objective is about 1/2 in. above the slide. When focusing, always *raise* the body tube.
7. Use the fine adjustment knob to complete the focusing; you will usually use a counterclockwise motion once again.
8. Compare the position of the letter as originally seen with the naked eye to its appearance under the microscope.
   Has the position of the letter been changed?

   ____

9. While looking at the slide through the ocular, move the slide by using your thumbs, or, if the microscope is equipped with them, the mechanical stage knobs. This exercise teaches you to move your specimen in various directions quickly and efficiently.
   In which direction does the letter move when you move the slide to the left?

   ____

   This procedure, called "scanning" a slide, will be useful for examining living objects and for centering specimens so you can observe them easily.

   Make a drawing of the letter as it appears under low power in the microscopic field in the space below on the left side.

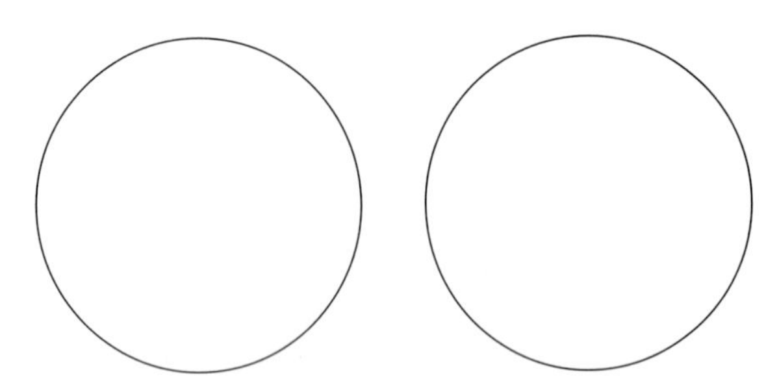

Drawing of letter as seen under low power

Drawing of letter as seen under high power

10. Change your magnification from low to high power by carrying out the following steps:

a. Place the letter in the center of the field under low power. Centering is important because you are now focusing on a smaller area of the microscopic field. As you will see, *microscopic field size decreases with higher magnifications.*
b. Make sure the illumination is at its maximum. Illumination must be increased at higher magnifications because the amount of light entering the lens decreases as the size of the objective lens increases.
c. The letter should be in focus, and if the microscope is ***parfocal*** (meaning that when clear focus has been attained using any objective at random, revolving the nosepiece results in a change in magnification but leaves the specimen still in focus), the high-power objective can be switched into line with the body tube without changing focus. If it is not completely in focus after switching the lens, a slight turn of the fine adjustment knob will focus it.
d. If your microscope is not parfocal, observe the stage from one side and carefully switch the high-power objective in line with the body tube.
e. While still observing from the side and using the coarse adjustment knob, *carefully* lower the objective or raise the stage until the objective almost touches the slide.
f. Look through the ocular and focus up slowly. Finish focusing by turning the fine adjustment knob.
g. If your microscope has an oil-immersion objective, you must follow special procedures. Place a drop of special ***immersion oil*** directly over the letter on the microscope slide, and lower the oil-immersion objective until it just contacts the oil. If your microscope is parfocal, you do not have to raise or lower the objectives. For example, if you are using the high-power objective and the specimen is in focus, just switch the high-power objective out of line with the body tube. Then add the oil and switch the oil-immersion objective into position; the specimen should be in focus. The same holds true when you switch from low power to high power. The special light-transmitting properties of the oil are such that light is refracted (bent) toward the specimen, permitting the use of powerful objectives in a relatively narrow field of vision. This objective is extremely close to the slide being examined, so when it is in position, take precautions *never to focus downward* while you are looking through the ocular. Whenever you finish using immersion oil, be sure to saturate a piece of lens paper with xylol or alcohol and clean the oil-immersion objective and the slide if it is to be used again.

Is as much of the letter visible under high power as under low power? Explain. ________

____

Make a drawing of the letter as it appears under high power in the microscopic field in the space next to the drawing you made for low power following step 9 on page 4.

11. Now select a prepared slide of three different-colored threads. Examination will show that a specimen mounted on a slide has depth as well as length and width. At lower magnification the amount of depth of the specimen that is clearly in focus, the depth of field, is greater than that at higher magnification. You must focus at different depths to determine the position (depth) of each thread.

After you make your observation under low power and high power, answer the following questions about the location of the different threads:

What color is at the bottom, closest to the slide? ________________________

On top, closest to the cover slip? __________

____

In the middle? ______________________

____

12. Your instructor might want you to prepare a wet mount as part of your introduction to microscopy. If so, the directions are given in Exercise 4, A.2.f, on page 44.

13. When you are finished using the microscope

a. Remove the slide from the stage.
b. Clean all lenses with lens paper.
c. Align the mechanical stage so that it does not protrude.
d. Leave the scanning or low-power objective in place.
e. Lower the body tube or raise the stage as far as it will go.

f. Wrap the cord according to your instructor's directions.

g. Replace the dust cover or place the microscope in a cabinet.

5. Magnification

The total magnification of your microscope is calculated by multiplying the magnification of the ocular by the magnification of the objective used. Example: An ocular of 10× used with an objective of 5× gives a total magnification of 50×. Calculate the total magnification of each of the objectives on your microscope:

1. Ocular ______ × ______ Objective = ______

2. Ocular ______ × ______ Objective = ______

3. Ocular ______ × ______ Objective = ______

4. Ocular ______ × ______ Objective = ______

B. ELECTRON MICROSCOPE

Examination of specimens smaller than 0.3 μm requires an ***electron microscope***, an instrument with a much greater resolving power than a compound light microscope. An electron microscope uses a beam of electrons instead of light. Electrons travel in waves just as light does. Instead of glass lenses, however, magnets are used in an electron microscope to focus a beam of electrons through a vacuum tube onto a specimen. Since the wavelength of electrons is about 1/100,000 that of visible light, the resolving power of very sophisticated transmission electron microscopes (described shortly) is close to 10 nanometers (nm). 1 nm = 0.000,000,001 m or 10^{-9} m. (See Appendix A.) Most electron microscopes have a working resolving power of about 100 nm and can magnify images up to 200,000×.

Two types of electron microscopes are available. In a ***transmission electron microscope***, a finely-focused beam of electrons passes through a specimen, usually ultra-thin sections of material. The beam is then refocused, and the image reflects what the specimen has done to the transmitted electron beam. The photograph produced from an image generated in this manner is called a ***transmission electron micrograph***. This method is extremely valuable in providing details of the interior of specimens at different layers but does not give a three-dimensional effect. Such an effect can be obtained with a ***scanning electron microscope***. With this instrument, a finely- focused beam of electrons is directed over the specimen and then reflected from the surface of the specimen onto a televisionlike screen or photographic plate. The photograph produced from an image generated in this manner is called a ***scanning electron micrograph*** (see Figure 5.2). Scanning electron micrographs commonly magnify specimens up to 10,000× and are especially useful in studying surface features of specimens.

Note: A section of *LABORATORY REPORT QUESTIONS* is located at the end of each exercise. These questions can be answered by the student and handed in for grading at the discretion of the instructor. Even if the instructor does not require you to answer these questions, we recommend that you do so anyway to check your understanding.

Some exercises also have a section of ***LABORATORY REPORT RESULTS***, in which students can record results of laboratory exercises, in addition to laboratory report questions. As with the laboratory report questions, the laboratory report results are located at the end of selected exercises and can be handed in as the instructor directs. Instructions in the manual tell students when and where to record laboratory results.

ANSWER THE LABORATORY REPORT QUESTIONS AT THE END OF THE EXERCISE.

LABORATORY REPORT QUESTIONS

Microscopy 1

Student ______________________ **Date** ______________________

Laboratory Section ______________________ **Score/Grade** ______________________

PART 1. Multiple Choice

_______ **1.** The amount of light entering a microscope may be adjusted by regulating the (a) ocular (b) diaphragm (c) fine adjustment knob (d) nosepiece

_______ **2.** If the ocular on a microscope is marked 10× and the low-power objective is marked 15×, the total magnification is (a) 50× (b) 25× (c) 150× (d) 1500×

_______ **3.** The size of the light beam that passes through a microscope is regulated by the (a) revolving nosepiece (b) coarse adjustment knob (c) ocular (d) condenser

_______ **4.** Parfocal means that (a) the microscope employs only one lens (b) final focusing can be done only with the fine adjustment knob (c) changing objectives by revolving the nosepiece will still keep the specimen in focus (d) the highest magnification attainable is 1000×

_______ **5.** Which of these is *not* true when changing magnification from low power to high power? (a) the specimen should be centered (b) illumination should be decreased (c) the specimen should be in clear focus (d) the high-power objective should be in line with the body tube

_______ **6.** The ability of a microscope to distinguish between two points as separate objects is called (a) parfocal focusing (b) working distance (c) diffraction (d) resolution

PART 2. Completion

7. The advantage of using immersion oil is that it has special ______________________ properties that permit the use of a powerful objective in a narrow field of vision.

8. The uniform circle of light that appears when one looks into the ocular is called the

______________________.

9. In determining the position (depth) of the colored threads, the ______________________ (red, blue, yellow) colored thread was in the middle.

10. If you move your slide to the right, the specimen moves to the ______________________ as you are viewing it microscopically.

11. After switching from low power to high power, ______________________ (more or less) of the specimen will be visible.

12. Microscopic field size ______________________ (increases or decreases) with higher magnifications.

13. The wavelength of electrons is about ______________________ (what proportion?) that of visible white light.

14. A photograph of a specimen taken through a compound light microscope is called a(n)

______________________.

15. The type of microscope that provides three-dimensional images of nonliving specimens is the

______________________ microscope.

PART 3. Matching

_______	**16.** Ocular	A.	Platform on which slide is placed
_______	**17.** Stage	B.	Mounting for objectives
_______	**18.** Arm	C.	Lens below stage opening
_______	**19.** Condenser	D.	Brings specimen into sharp focus
_______	**20.** Revolving nosepiece	E.	Eyepiece
_______	**21.** Low-power objective	F.	An objective usually marked 43× or 45×
_______	**22.** Fine adjustment knob	G.	An objective usually marked 10×
_______	**23.** Diaphragm	H.	Angular or curved part of frame
_______	**24.** Coarse adjustment knob	I.	Brings specimen into general focus
_______	**25.** High-power objective	J.	Regulates light intensity

Introduction to the Human Body

2

In this exercise, you will be introduced to the organization of the human body through a study of the principal subdivisions of anatomy, levels of structural organization, principal body systems, the anatomical position, regional names, directional terms, planes of the body, body cavities, abdominopelvic regions, and abdominopelvic quadrants.

A. SUBDIVISIONS OF ANATOMY

Whereas ***anatomy*** (a-NAT-o-mē; *anatome* = to cut up) refers to the study of *structure* and the relationships among structures, ***physiology*** (fiz-ē-OL-ō-jē) deals with the *functions* of body parts, that is, how they work. Each structure of the body is designed to carry out a particular function.

Following are selected subdivisions of anatomy. In the spaces provided, define each term.

Surface anatomy ________________________

____

____

____

Gross (macroscopic) anatomy ________________

____

____

____

Systemic anatomy ________________________

____

____

____

Regional anatomy ________________________

____

____

____

Radiographic (rā′-dē-ō-GRAF-ik) ***anatomy*** ______

____

____

____

Developmental anatomy ____________________

____

____

____

Embryology (em′-brē-OL-ō-jē) ________________

____

____

____

Histology (hiss′-TOL-ō-jē) __________________

____

____

____

Cytology (sī-TOL-ō-jē) ____________________

____

____

____

Pathological (path′-ō-LOJ-i-kal) ***anatomy*** ______

____

____

____

B. LEVELS OF STRUCTURAL ORGANIZATION

The human body is composed of several levels of structural organization associated with one another in various ways:

1. ***Chemical level*** Composed of all atoms and molecules necessary to maintain life.
2. ***Cellular level*** Consists of cells, the structural and functional units of the body.
3. ***Tissue level*** Formed by tissues, groups of cells (and their intercellular material) that usually arise from common ancestor cells and work together to perform a particular function.
4. ***Organ level*** Consists of organs, structures composed of two or more different tissues, having specific functions and usually having recognizable shapes.
5. ***System level*** Formed by systems, associations of organs that have a common function. The systems together constitute an ***organism.***

C. SYSTEMS OF THE BODY

Using an anatomy textbook, torso, wall chart, and any other materials that might be available to you, identify the principal organs that compose the following body systems.[1] In the spaces that follow, indicate the organs and functions of the systems.

Integumentary

Organs ______________________________

____

____

____

Functions ______________________________

____

____

____

[1]You will probably need other sources, plus any aids the instructor might provide, to label many of the figures and answer some questions in this manual. You are encouraged to use other sources as you find necessary.

Skeletal

Organs ______________________________

____

____

____

Functions ______________________________

____

____

____

Muscular

Organs ______________________________

____

____

____

Functions ______________________________

____

____

____

Nervous

Organs ______________________________

____

____

____

Functions ______________________________

____

____

____

Endocrine

Organs ______________________________

____

____

____

Functions ______________________________

____

____

____

Cardiovascular

Organs ______________________________

____

____

____

Functions ______________________________

____

____

____

Lymphatic and Immune

Organs ______________________________

____

____

____

Functions ______________________________

____

____

____

Respiratory

Organs ______________________________

____

____

____

Functions ______________________________

____

____

____

Digestive

Organs ______________________________

____

____

____

Functions ______________________________

____

____

____

Urinary

Organs ______________________________

____

____

____

Functions ______________________________

____

____

____

Reproductive

Organs ______________________________

____

____

____

Functions ______________________________

____

____

____

D. ANATOMICAL POSITION AND REGIONAL NAMES

Figure 2.1 shows anterior and posterior views of a subject in the ***anatomical position.*** The subject is standing erect (upright position) and facing the observer with the head and eyes facing forward, the upper limbs (extremities) are at the sides, the palms are facing forward, and the feet are flat on the floor and facing forward. The figure also shows the common names for various regions of the body. When you as the observer make reference to the left and right sides of the subject you are studying, this refers to the *subject's* left and right sides. In the spaces next to the list of terms in Figure 2.1 write the number of each common term next to each corresponding anatomical term. For example, the skull (29) is cranial, so write the number *29* next to the term *Cranial.*

E. EXTERNAL FEATURES OF THE BODY

Referring to your textbook and human models, identify the following external features of the body:

1. ***Head (cephalic region*** or ***caput)*** This is divided into the ***cranium*** (brain case) and ***face.***
2. ***Neck (collum)*** This region consists of an anterior ***cervix,*** two lateral surfaces, and a posterior ***nucha*** (NOO-ka), or the nape of the neck.
3. ***Trunk*** This region is divided into the ***back*** (dorsum), ***chest*** (thorax), ***abdomen*** (venter), and ***pelvis.***
4. ***Upper limb (extremity)*** This consists of the ***armpit*** (axilla), ***shoulder*** (acromial region or omos), ***arm*** (brachium), ***elbow*** (cubitus), ***forearm*** (antebrachium), and ***hand*** (manus). The hand, in turn, consists of the ***wrist*** (carpus), ***palm*** (metacarpus), and ***fingers*** (digits). Individual bones of a digit (finger or toe) are called ***phalanges. Phalanx*** is singular.
5. ***Lower limb (extremity)*** This consists of the ***buttocks*** (gluteal region), ***thigh*** (femoral region), ***knee*** (genu), ***leg*** (crus), and ***foot*** (pes). The foot includes the ***ankle*** (tarsus), ***sole*** (metatarsus), and ***toes*** (digits). The ***groin*** is the point of attachment between the lower limbs and the trunk.

F. DIRECTIONAL TERMS

To explain exactly where a structure of the body is located, it is a standard procedure to use ***directional terms.*** Such terms are very precise and avoid the use of unnecessary words. Commonly used directional terms for humans are as follows:

1. ***Superior*** (soo′-PEER-ē-or) ***(cephalic*** or ***cranial)*** Toward the head or the upper part of a structure; generally refers to structures in the trunk.
2. ***Inferior*** (in′-FEER-¯e-or) ***(caudal)*** Away from the head or toward the lower part of a structure; generally refers to structures in the trunk.
3. ***Anterior*** (an-TEER-ē-or) ***(ventral)*** Nearer to or at the front surface of the body. In the ***prone position,*** the body lies anterior side down; in the ***supine position,*** the body lies anterior side up.
4. ***Posterior*** (pos-TEER-ē-or) ***(dorsal)*** Nearer to or at the back or backbone surface of the body.
5. ***Medial*** (MĒ-dē-al) Nearer the midline of the body or a structure. The ***midline*** is an imaginary vertical line that divides the body into equal left and right sides.
6. ***Lateral*** (LAT-er-al) Farther from the midline of the body or a structure.
7. ***Intermediate*** (in′-ter-MĒ-dē-at) Between two structures.
8. ***Ipsilateral*** (ip-si-LAT-er-al) On the same side of the midline.
9. ***Contralateral*** (CON-tra-lat-er-al) On the opposite side of the midline.

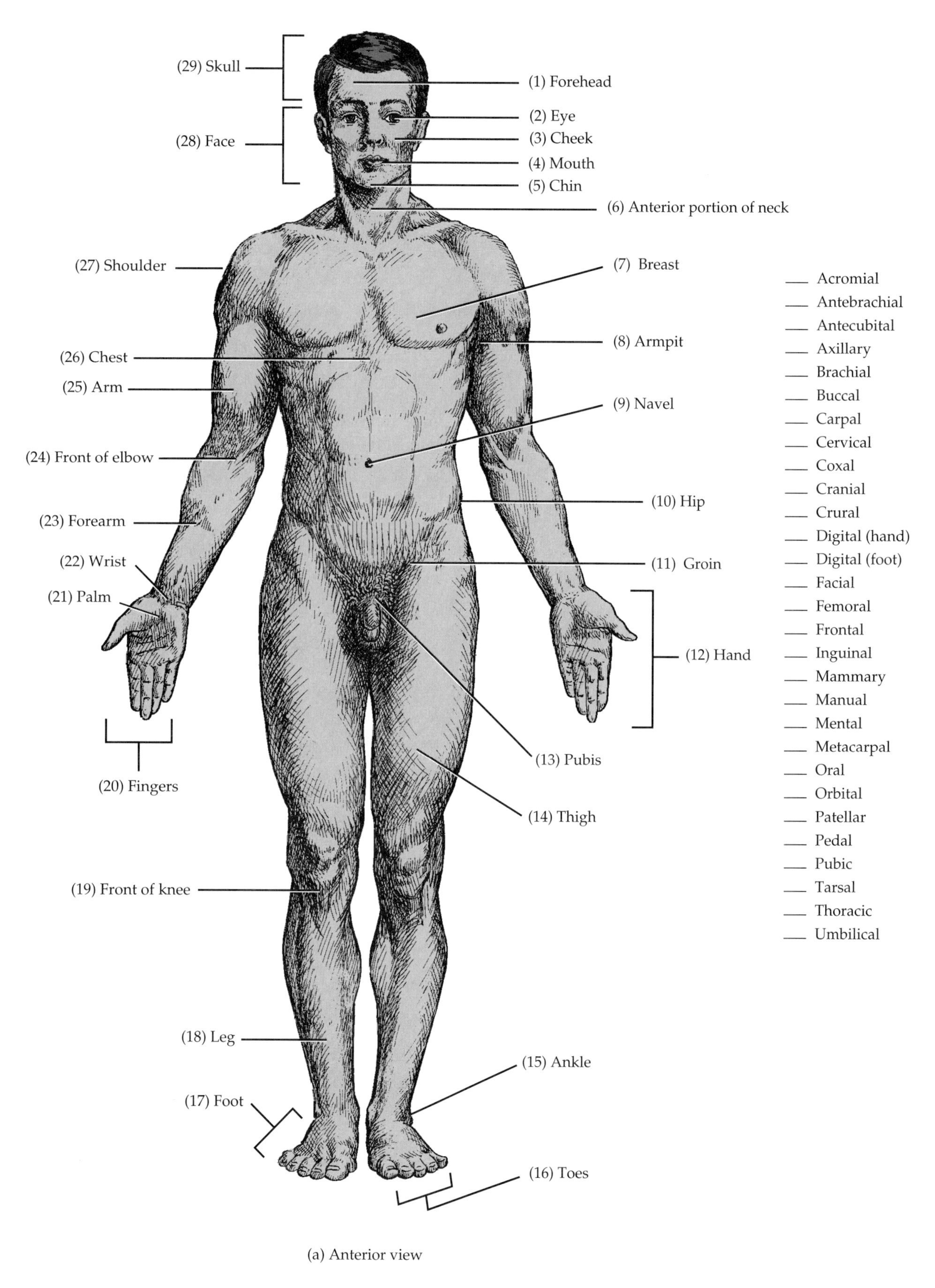

FIGURE 2.1 The anatomical position.

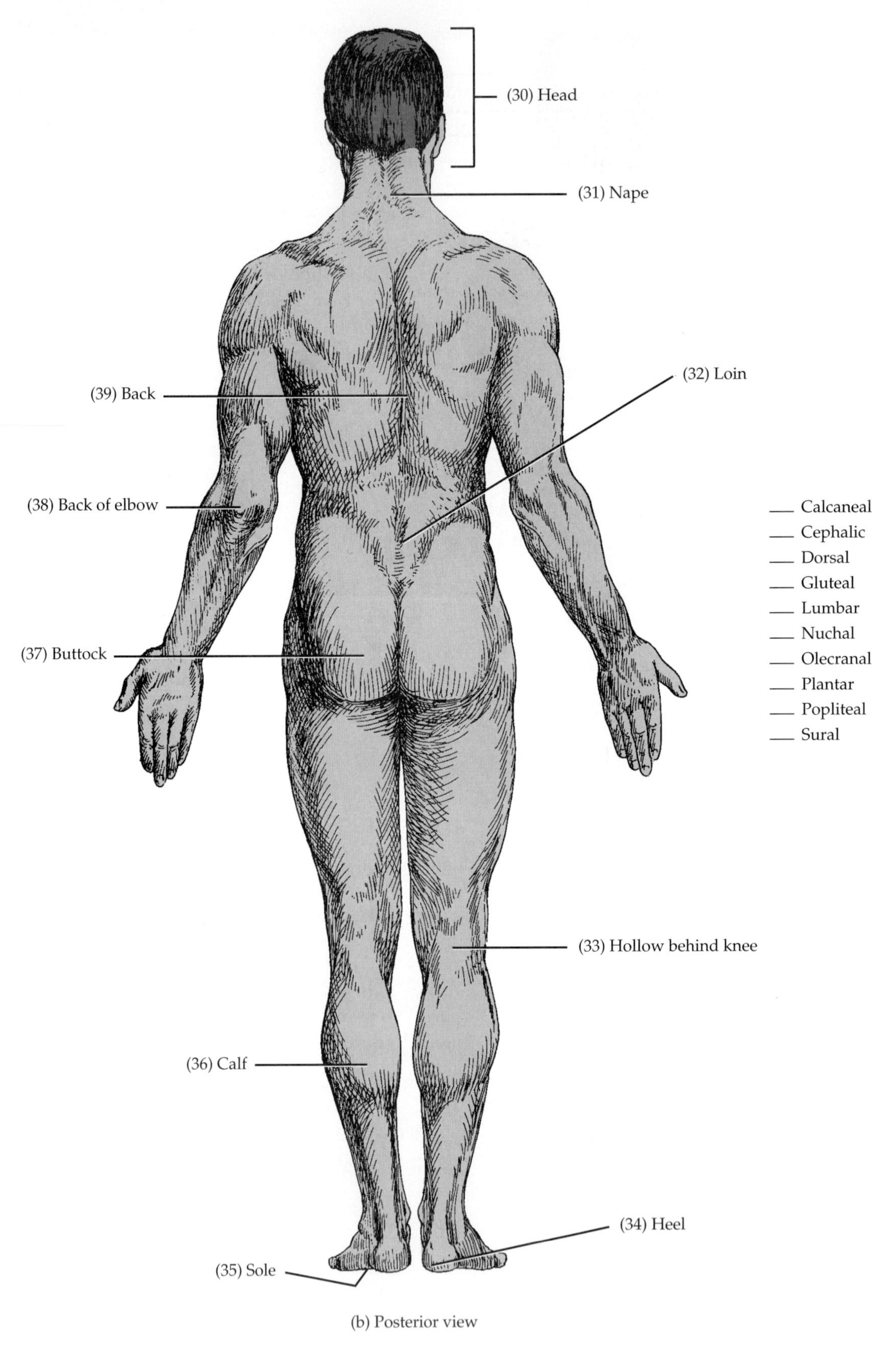

(b) Posterior view

FIGURE 2.1 ***(Continued)*** The anatomical position.

10. ***Proximal*** (PROK-si-mal) Nearer the attachment of a limb to the trunk; nearer to the point of origin.
11. ***Distal*** (DIS-tal) Farther from the attachment of a limb to the trunk; farther from the point of origin.
12. ***Superficial*** (soo′-per-FISH-al) ***(external)*** Toward or on the surface of the body.
13. ***Deep*** (DĒP) ***(internal)*** Away from the surface of the body.

Using a torso and an articulated skeleton, and consulting with your instructor as necessary, describe the location of the following by inserting the proper directional term.

1. The ulna is on the ______________ side of the forearm.
2. The lungs are ______________ to the heart.
3. The heart is ______________ to the liver.
4. The muscles of the arm are ______________ to the skin of the arm.
5. The sternum is ______________ to the heart.
6. The humerus is ______________ to the radius.
7. The stomach is ______________ to the lungs.
8. The muscles of the thoracic wall are ______________ to the viscera in the thoracic cavity.
9. The esophagus is ______________ to the trachea.
10. The phalanges are ______________ to the carpals.
11. The ring finger is ______________ between the little (medial) and middle (lateral) fingers.
12. The ascending colon of the large intestine and the gallbladder are ______________.
13. The ascending and descending colons of the large intestine are ______________.

G. PLANES OF THE BODY

The structural plan of the human body may be described with respect to ***planes*** (imaginary flat surfaces) passing through it. Planes are frequently used to show the anatomical relationship of several structures in a region to one another.

Commonly used planes are as follows:

1. ***Midsagittal*** (mid-SAJ-it-tal; *sagittalis* = arrow) or ***median*** A vertical plane that passes through the midline of the body and divides the body or an organ into *equal* right and left sides.
2. ***Parasagittal*** (par-a-SAJ-it-tal; *para* = near) A vertical plane that does not pass through the midline of the body and divides the body or an organ into *unequal* right and left sides.
3. ***Frontal*** (***coronal;*** kō-RŌ-nal) A vertical plane that divides the body or an organ into anterior (front) and posterior (back) portions.
4. ***Transverse*** (***horizontal*** or ***cross-sectional***) A plane that divides the body or an organ into superior (top) and inferior (bottom) portions.
5. ***Oblique*** (ō-BLĒK) A plane that passes through the body or an organ at an angle between the transverse plane and either the midsagittal, parasagittal, or frontal plane.

Refer to Figure 2.2 and label the planes shown.

H. BODY CAVITIES

Confined spaces within the body that contain various internal organs are called ***body cavities.*** One way of organizing the principal body cavities follows:

Dorsal body cavity
- *Cranial* (KRĀ-nē-al) *cavity*
- *Vertebral* (VER-te-bral) or *spinal cavity*

Ventral body cavity
- *Thoracic* (thor-AS-ik) *cavity*
  - Right pleural (PLOOR-al) cavity
  - Left pleural cavity
  - Pericardial (per′-i-KAR-dē-al) cavity
- *Abdominopelvic cavity*
  - Abdominal cavity
  - Pelvic cavity

The mass of tissue in the thoracic cavity between the coverings (pleurae) of the lungs and extending from the sternum (breast bone) to the backbone is called the ***mediastinum*** (mē′-dē-as-TĪ-num). It contains all structures in the thoracic cavity, except the lungs themselves. Included are the heart, thymus gland, esophagus, trachea, and many large blood and lymphatic vessels.

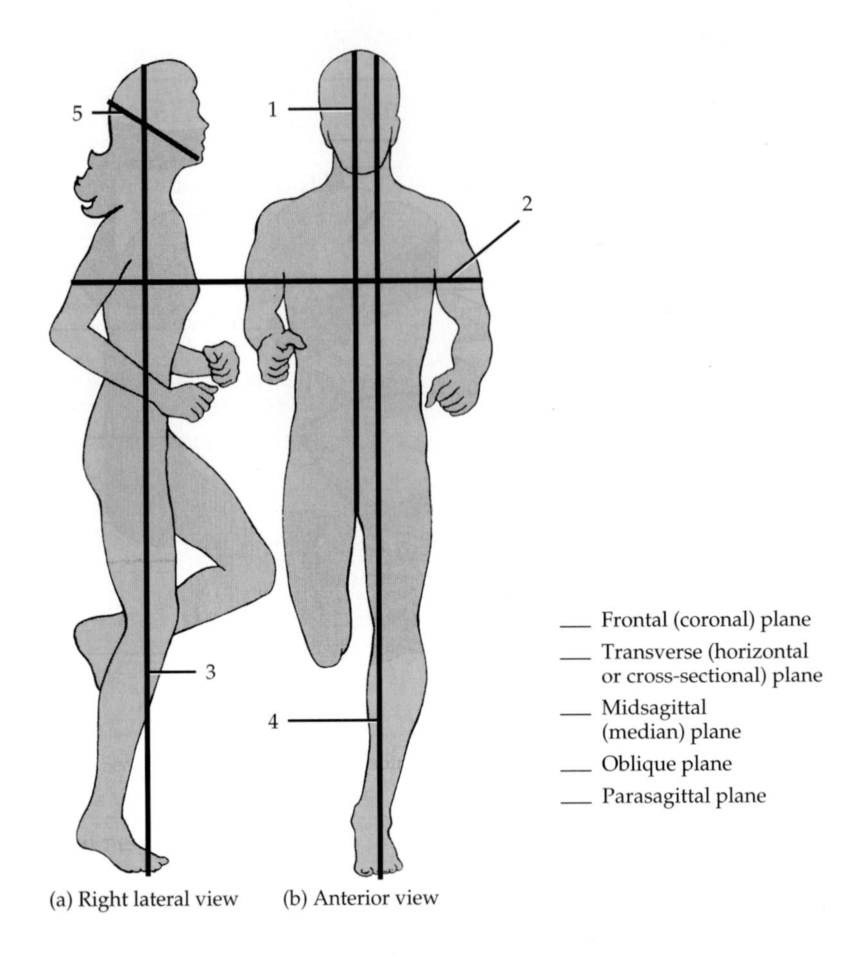

___ Frontal (coronal) plane
___ Transverse (horizontal or cross-sectional) plane
___ Midsagittal (median) plane
___ Oblique plane
___ Parasagittal plane

FIGURE 2.2 Planes of the body.

Label the body cavities shown in Figure 2.3. Then examine a torso or wall chart, or both, and determine which organs lie within each cavity.

Using *T* (for thoracic), *A* (for abdominal), and *P* (for pelvic), indicate which organs are found in their respective cavities.

1. ____ Urinary bladder
2. ____ Stomach
3. ____ Spleen
4. ____ Lungs
5. ____ Liver
6. ____ Internal reproductive organs
7. ____ Pancreas
8. ____ Heart
9. ____ Gallbladder
10. ____ Small portion of large intestine

I. ABDOMINOPELVIC REGIONS

To describe the location of viscera more easily, the abdominopelvic cavity may be divided into ***nine regions*** by using four imaginary lines: (1) an upper horizontal ***subcostal*** (sub-KOS-tal) ***line*** that passes just below the bottom of the rib cage through the pylorus (lower portion) of the stomach, (2) a lower horizontal line, the ***transtubercular*** (trans-too-BER-kyoo′-lar) ***line,*** just below the top surfaces of the hipbones, (3) a ***right midclavicular*** (mid-kla-VIK-yoo′-lar) ***line*** drawn through the midpoint of the right clavicle slightly medial to the right nipple, and (4) a ***left midclavicular line*** drawn through the midpoint of the left clavicle slightly medial to the left nipple.

The four imaginary lines divide the abdominopelvic cavity into the following nine regions: (1) ***umbilical*** (um-BIL-i-kul) ***region,*** which is centrally located; (2) ***left lumbar*** (*lumbus* = loin) ***region,*** to the left of the umbilical region; (3) ***right lumbar,*** to the right of the umbilical region; (4) ***epigastric*** (ep-i-GAS-trik; *epi* = above; *gaster* = stomach) ***region,*** directly above the umbilical region; (5) ***left hypochondriac*** (hī′-pō-KON-drē-ak; *hypo* = under; *chondro* = cartilage) ***region,*** to the left of the epigastric

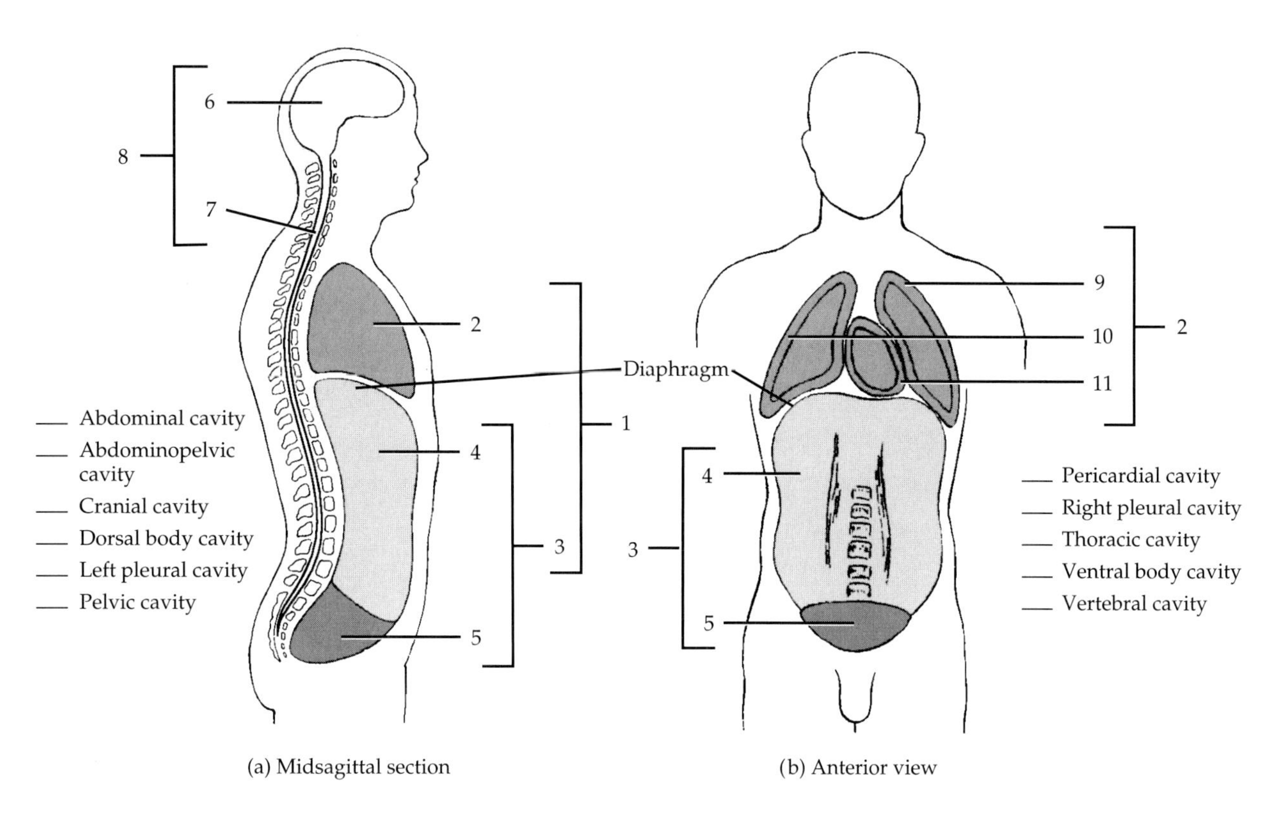

FIGURE 2.3 Body cavities.

region; (6) ***right hypochondriac region,*** to the right of the epigastric region; (7) ***hypogastric (pubic) region,*** directly below the umbilical region; (8) ***left iliac*** (IL-ē-ak; *iliacus* = superior part of hipbone) or ***inguinal region,*** to the left of the hypogastric (pubic) region; and (9) ***right iliac (inguinal) region,*** to the right of the hypogastric (pubic) region.

Label Figure 2.4 by indicating the names of the four imaginary lines and the nine abdominopelvic regions.

Examine a torso and determine which organs or parts of organs lie within each of the nine abdominopelvic regions.

In the space provided, list several organs or parts of organs found in the following abdominopelvic regions:

Right hypochondriac ______________________

____

____

Epigastric ______________________

____

____

Left hypochondriac ______________________

____

____

Right lumbar ______________________

____

____

Umbilical ______________________

____

____

Left lumbar ______________________

____

____

Right iliac ______________________

____

____

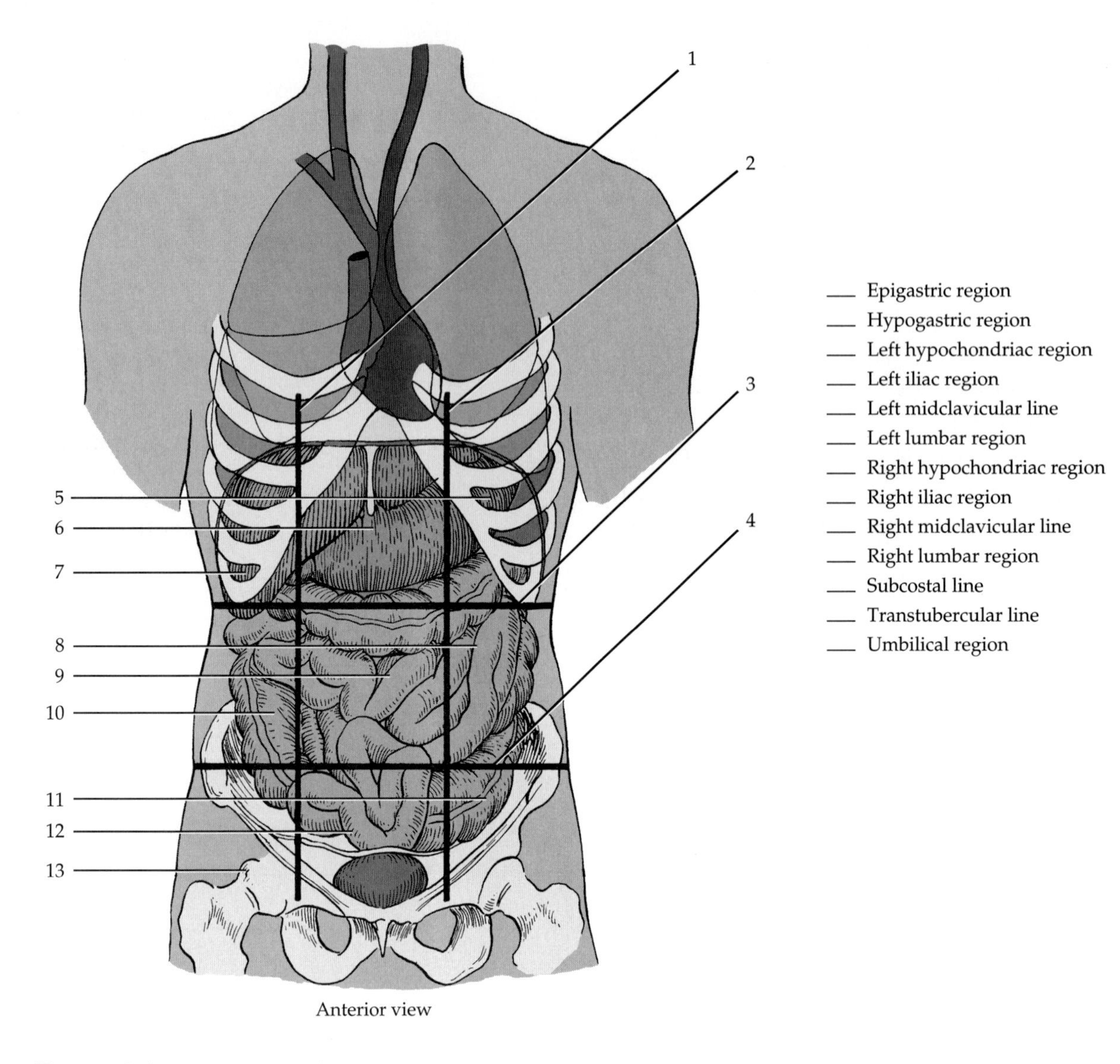

FIGURE 2.4 Abdominopelvic regions.

Hypogastric ________________________________

____

____

____

Left iliac ________________________________

____

____

____

J. ABDOMINOPELVIC QUADRANTS

Another way to divide the abdominopelvic cavity is into ***quadrants*** by passing one horizontal line and one vertical line through the umbilicus (navel). The two lines thus divide the abdominopelvic cavity into a ***right upper quadrant (RUQ), left upper quadrant (LUQ), right lower quadrant (RLQ),*** and ***left lower quadrant (LLQ).*** Quadrant names are frequently used by health care professionals for locating the site of an abdominopelvic pain, tumor, or other abnormality.

Examine a torso or wall chart, or both, and determine which organs or parts of organs lie within each of the abdominopelvic quadrants.

In the space provided, list several organs or parts of organs found in the following abdominopelvic quadrants:

Right upper quadrant ______________________

____

____

____

Left upper quadrant ______________________

____

____

____

Right lower quadrant ______________________

____

____

____

Left lower quadrant ______________________

____

____

____

K. DISSECTION OF WHITE RAT

Now that you have some idea of the names of the various body systems and the principal organs that comprise each, you can actually observe some of these organs by dissecting a white rat. ***Dissect*** means "to separate." This dissection gives you an excellent opportunity to see the different sizes, shapes, locations, and relationships of organs and to compare the different textures and external features of organs. In addition, this exercise will introduce you to the general procedure for dissection before you dissect in later exercises.

CAUTION! *Please reread Section D, "Precautions Related to Dissection" at the beginning of the laboratory manual on page xi before you begin your dissection.*

PROCEDURE

1. Place the rat on its backbone on a wax dissecting pan (tray). Using dissecting pins, anchor each of the four limbs to the wax (Figure 2.5a).
2. To expose the contents of the thoracic, abdominal, and pelvic cavities, you will have to first make a midline incision. This is done by lifting the abdominal skin with a forceps to separate the skin from the underlying connective tissue and muscles. While lifting the abdominal skin, cut through it with scissors and make an incision that extends from the lower jaw to the anus (Figure 2.5a).
3. Now make four lateral incisions that extend from the midline incision into the four limbs (Figure 2.5a).
4. Peel the skin back and pin the flaps to the wax to expose the superficial muscles (Figure 2.5b).
5. Next, lift the abdominal muscles with a forceps and cut through the muscle layer, being careful not to damage any underlying organs. Keep the scissors parallel to the rat's backbone. Extend this incision from the anus to a point just below the bottom of the rib cage (Figure 2.5b). Make two lateral incisions just below the rib cage and fold back the muscle flaps to expose the abdominal and pelvic viscera (Figure 2.5b).
6. To expose the thoracic viscera, cut through the ribs on either side of the sternum. This incision should extend from the diaphragm to the neck (Figure 2.5b). The ***diaphragm*** is the thin muscular partition that separates the thoracic from the abdominal cavity. Again make lateral incisions in the chest wall so that you can lift the ribs to view the thoracic contents.

1. Examination of Thoracic Viscera

You will first examine the thoracic viscera (large internal organs). As you dissect and observe the various structures, palpate (feel with the hand) them so that you can compare their texture. Use Figure 2.6 on page 21 as a guide.

a. ***Thymus gland*** An irregular mass of glandular tissue superior to the heart and superficial to the trachea. Push the thymus gland aside or remove it.
b. ***Heart*** A structure located in the midline, deep to the thymus gland and between the lungs. The sac covering the heart is the ***pericardium,*** which may be removed. The large vein that returns blood from the lower regions of the body is the ***inferior vena cava;***

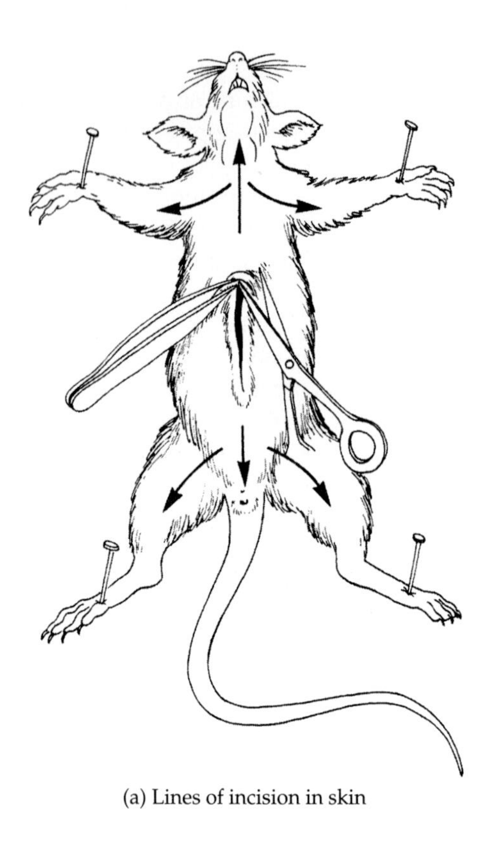

(a) Lines of incision in skin

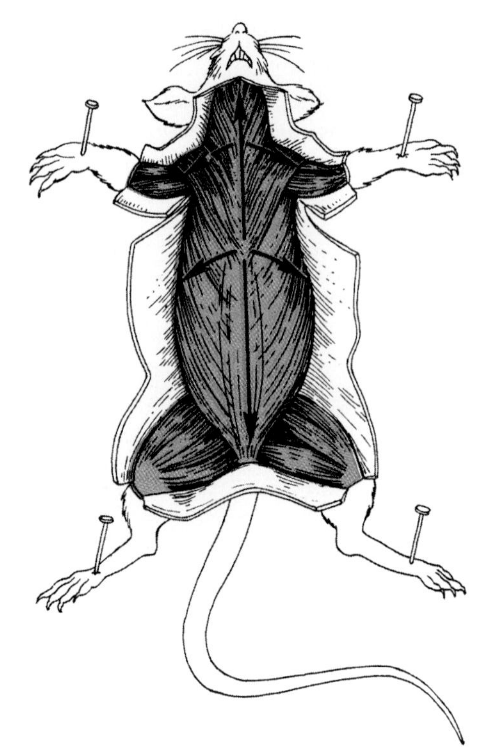

(b) Peeling back skin and lines of incision in muscles

FIGURE 2.5 Dissection procedure for exposing thoracic and abdominopelvic viscera of the white rat for examination.

the large vein that returns blood to the heart from the upper regions of the body is the ***superior vena cava.*** The large artery that carries blood from the heart to most parts of the body is the ***aorta.***

c. ***Lungs*** Reddish, spongy structures on either side of the heart. Note that the lungs are divided into regions called ***lobes.***
d. ***Trachea*** A tubelike passageway superior to the heart and deep to the thymus gland. Note that the wall of the trachea consists of rings of cartilage. Identify the ***larynx*** (voice box) at the superior end of the trachea and the ***thyroid gland,*** a bilobed structure on either side of the larynx. The lobes of the thyroid gland are connected by a band of thyroid tissue, the isthmus.
e. ***Bronchial tubes*** Trace the trachea inferiorly and note that it divides into bronchial tubes that enter the lungs and continue to divide within them.
f. ***Esophagus*** A muscular tube posterior to the trachea that transports food from the throat into the stomach. Trace the esophagus inferiorly to see where it passes through the diaphragm to join the stomach.

2. Examination of Abdominopelvic Viscera

You will now examine the principal viscera of the abdomen and pelvis. As you do so, again refer to Figure 2.6.

a. ***Stomach*** An organ located on the left side of the abdomen and in contact with the liver. The digestive organs are attached to the posterior abdominal wall by a membrane called the ***mesentery.*** Note the blood vessels in the mesentery.
b. ***Small intestine*** An extensively coiled tube that extends from the stomach to the first portion of the large intestine called the ***cecum.***
c. ***Large intestine*** A wider tube than the small intestine that begins at the cecum and ends at the rectum. The cecum is a large, saclike structure. In humans, the appendix arises from the cecum.

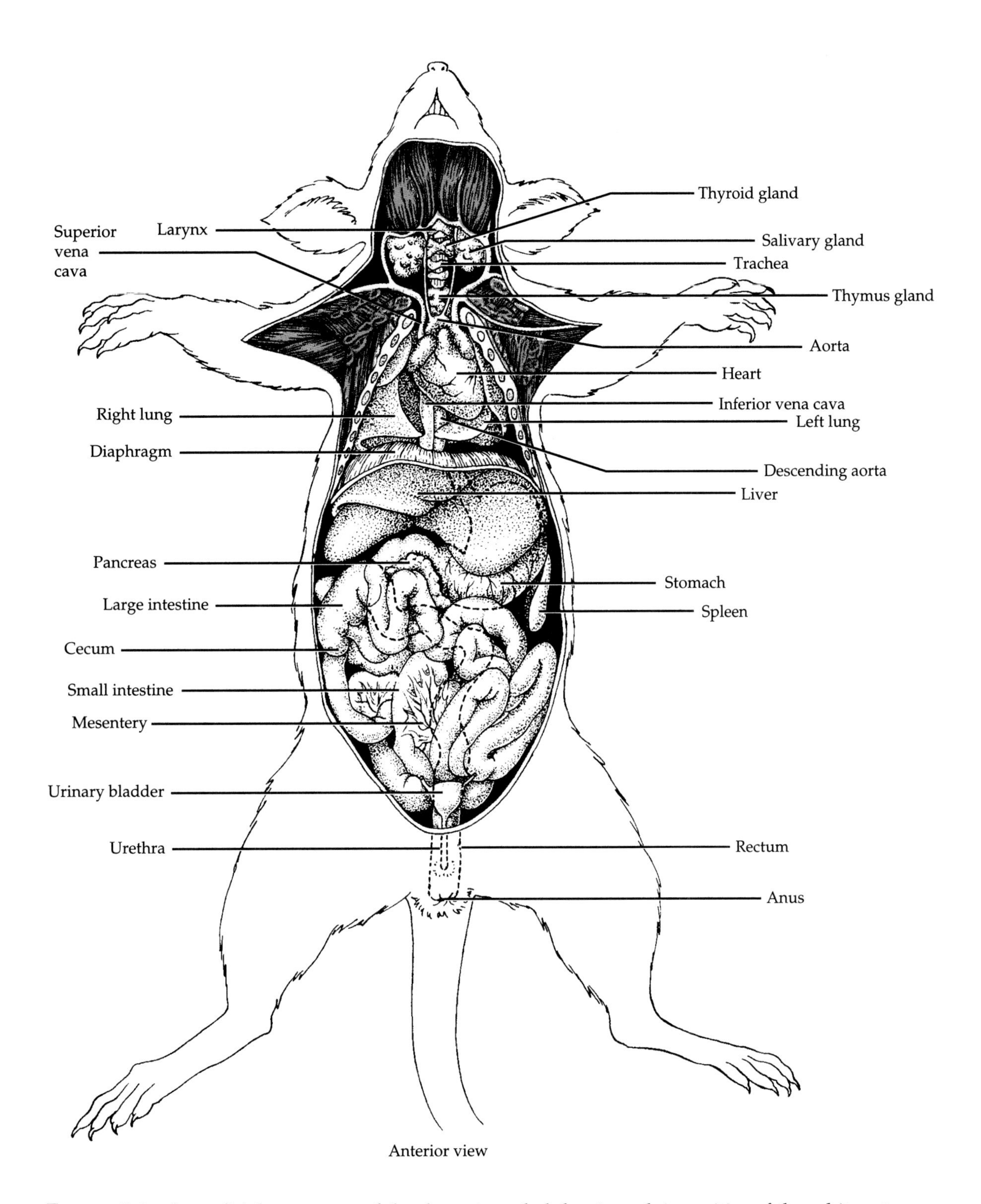

FIGURE 2.6 Superficial structures of the thoracic and abdominopelvic cavities of the white rat.

d. ***Rectum*** A muscular passageway, located on the midline in the pelvic cavity, that terminates in the anus.

e. ***Anus*** Terminal opening of the digestive system to the exterior.

f. ***Pancreas*** A pale gray, glandular organ posterior and inferior to the stomach.

g. ***Spleen*** A small, dark red organ lateral to the stomach.

h. ***Liver*** A large, brownish-red organ directly inferior to the diaphragm. The rat does not have a gallbladder, a structure associated with the liver. To locate the remaining viscera, either move the superficial viscera aside or remove them. Use Figure 2.7 as a guide.

i. ***Kidneys*** Bean-shaped organs embedded in fat and attached to the posterior abdominal wall on either side of the backbone. As will be explained later, the kidneys and a few other structures are behind the membrane that lines the abdomen ***(peritoneum).*** Such structures are referred to as ***retroperitoneal*** and are not actually within the abdominal cavity. See if you can find the ***abdominal aorta,*** the large artery located along the midline behind the inferior vena cava. Also, locate the ***renal arteries*** branching off the abdominal aorta to enter the kidneys.

j. ***Adrenal (suprarenal) glands*** Glandular structures. One is located on top of each kidney.

k. ***Ureters*** Tubes that extend from the medial surface of the kidneys inferiorly to the urinary bladder.

l. ***Urinary bladder*** A saclike structure in the pelvic cavity that stores urine.

m. ***Urethra*** A tube that extends from the urinary bladder to the exterior. Its opening to the exterior is called the ***urethral orifice.*** In male rats, the urethra extends through the penis; in female rats, the tube is separate from the reproductive tract.

If your specimen is female (no visible scrotum anterior to the anus), identify the following:

n. ***Ovaries*** Small, dark structures inferior to the kidneys.

o. ***Uterus*** An organ located near the urinary bladder consisting of two sides (horns) that join separately into the vagina.

p. ***Vagina*** A tube that leads from the uterus to the external vaginal opening, the ***vaginal orifice.*** This orifice is in front of the anus and behind the urethral orifice.

If your specimen is male, identify the following:

q. ***Scrotum*** Large sac anterior to the anus that contains the testes.

r. ***Testes*** Egg-shaped glands in the scrotum. Make a slit into the scrotum and carefully remove one testis. See if you can find a coiled duct attached to the testis ***(epididymis)*** and a duct that leads from the epididymis into the abdominal cavity ***(vas deferens).***

s. ***Penis*** Organ of copulation medial to the testes.

When you have finished your dissection, store or dispose of your specimen according to your instructor's directions. Wash your dissecting pan and dissecting instruments with laboratory detergent, dry them, and return them to their storage areas.

ANSWER THE LABORATORY REPORT QUESTIONS AT THE END OF THE EXERCISE.

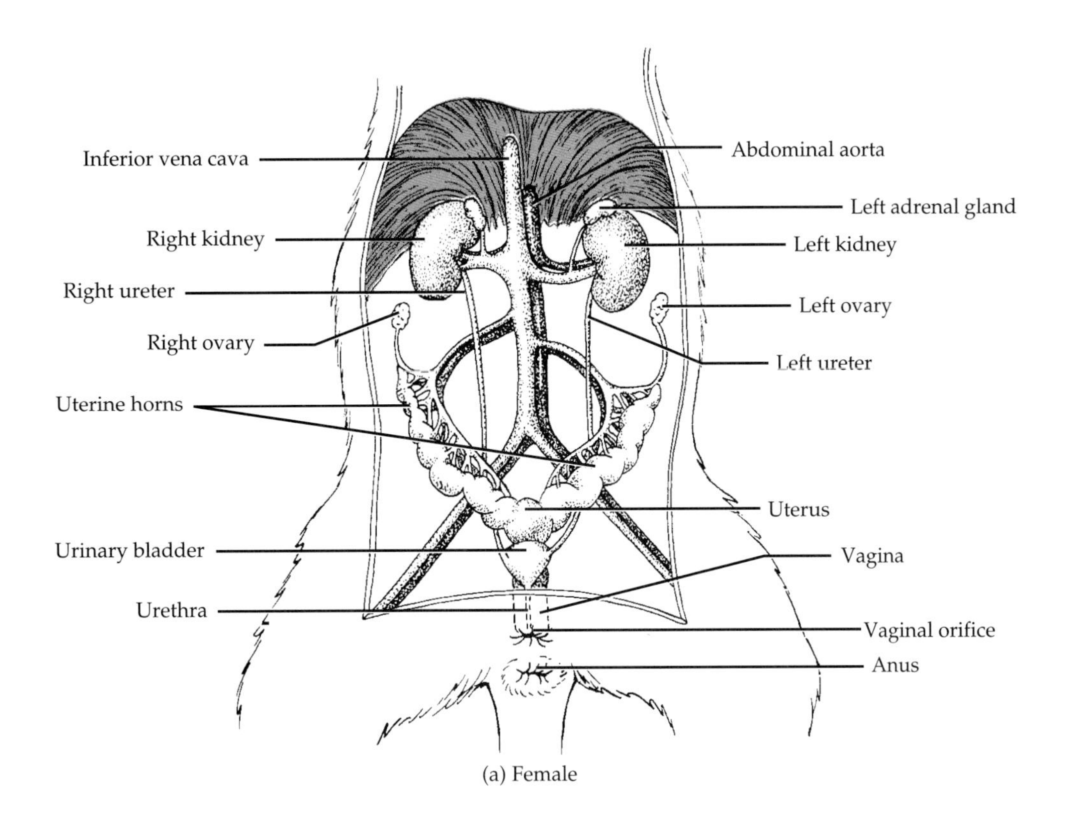

(a) Female

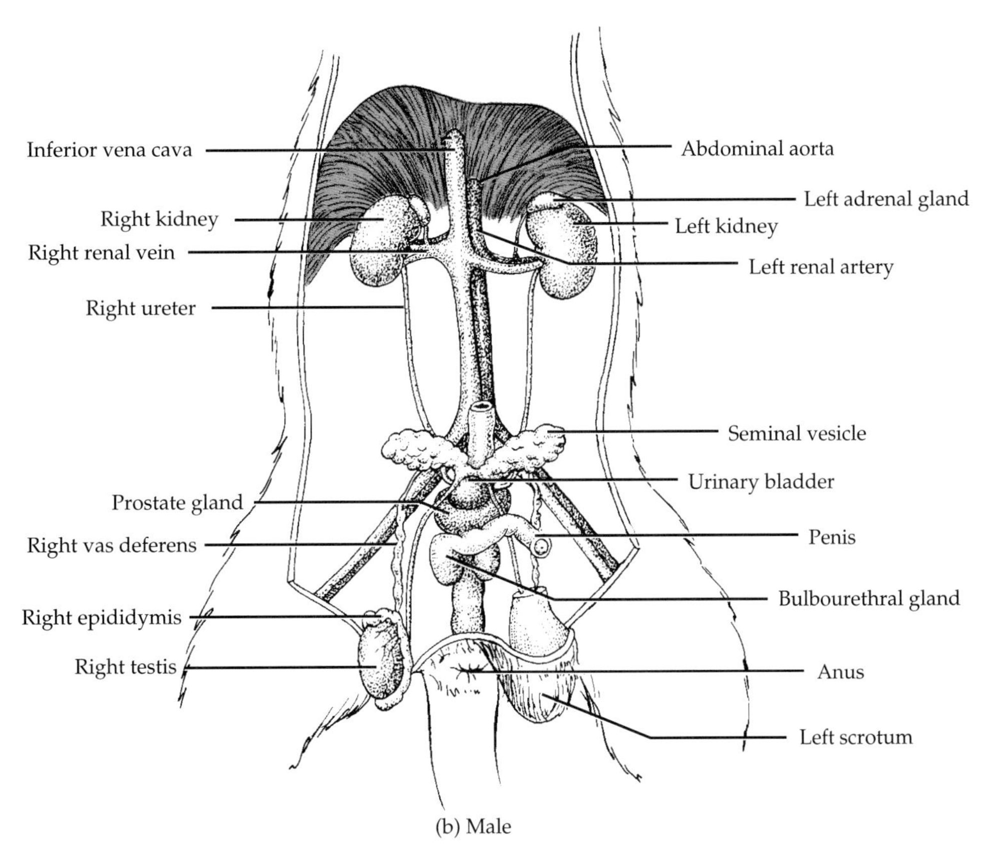

(b) Male

FIGURE 2.7 Deep structures of the abdominopelvic cavity of the white rat.

LABORATORY REPORT QUESTIONS

Introduction to the Human Body 2

Student ______________________________ **Date** ______________________________

Laboratory Section ______________________ **Score/Grade** ______________________

PART 1. Multiple Choice

_______ **1.** The directional term that best describes the eyes in relation to the nose is (a) distal (b) superficial (c) anterior (d) lateral

_______ **2.** Which does *not* belong with the others? (a) right pleural cavity (b) pericardial cavity (c) vertebral cavity (d) left pleural cavity

_______ **3.** Which plane divides the brain into an anterior and a posterior portion? (a) frontal (b) median (c) sagittal (d) transverse

_______ **4.** The urinary bladder lies in which region? (a) umbilical (b) hypogastric (c) epigastric (d) left iliac

_______ **5.** Which is *not* a characteristic of the anatomical position? (a) the subject is erect (b) the subject faces the observer (c) the palms face backward (d) the upper limbs are at the sides

_______ **6.** The abdominopelvic region that is bordered by all four imaginary lines is the (a) hypogastric (b) epigastric (c) left hypochondriac (d) umbilical

_______ **7.** Which directional term best describes the position of the phalanges with respect to the carpals? (a) lateral (b) distal (c) anterior (d) proximal

_______ **8.** The pancreas is found in which body cavity? (a) abdominal (b) pericardial (c) pelvic (d) vertebral

_______ **9.** The anatomical term for the leg is (a) brachial (b) tarsal (c) crural (d) sural

_______ **10.** In which abdominopelvic region is the spleen located? (a) left lumbar (b) right lumbar (c) epigastric (d) left hypochondriac

_______ **11.** Which of the following represents the most complex level of structural organization? (a) organ (b) cellular (c) tissue (d) chemical

_______ **12.** Which body system is concerned with support, protection, leverage, blood-cell production, and mineral storage? (a) cardiovascular (b) integumentary (c) skeletal (d) digestive

_______ **13.** The skin and structures derived from it, such as nails, hair, sweat glands, and oil glands, are components of which system? (a) respiratory (b) integumentary (c) muscular (d) skeletal

_______ **14.** Hormone-producing glands belong to which body system? (a) cardiovascular (b) lymphatic and immune (c) endocrine (d) digestive

_______ **15.** Which body system brings about movement, maintains posture, and produces heat? (a) skeletal (b) respiratory (c) reproductive (d) muscular

_______ **16.** Which abdominopelvic quadrant contains most of the liver? (a) RUQ (b) RLQ (c) LUQ (d) LLQ

_______ **17.** The physical and chemical breakdown of food for use by body cells and the elimination of solid wastes are accomplished by which body system? (a) respiratory (b) urinary (c) cardiovascular (d) digestive

PART 2. Completion

18. The tibia is ____________________ to the fibula.

19. The ovaries are found in the ____________________ body cavity.

20. The upper horizontal line that helps divide the abdominopelvic cavity into nine regions is the ____________________ line.

21. The anatomical term for the hollow behind the knee is ____________________.

22. A plane that divides the stomach into a superior and an inferior portion is a(n) ____________________ plane.

23. The wrist is described as ____________________ to the elbow.

24. The heart is located in the ____________________ cavity within the thoracic cavity.

25. The abdominopelvic region that contains the rectum is the ____________________ region.

26. A plane that divides the body into unequal left and right sides is the ____________________ plane.

27. The spinal cord is located within the ____________________ cavity.

28. The body system that removes carbon dioxide from body cells, delivers oxygen to body cells, helps maintain acid-base balance, helps protect against disease, helps regulate body temperature, and prevents hemorrhage by forming clots is the ____________________ system.

29. The ____________________ abdominopelvic quadrant contains the descending colon of the large intestine.

30. Which body system returns proteins and plasma to the cardiovascular system, transports lipids from the digestive system to the cardiovascular system, filters blood, protects against disease, and produces white blood cells? ____________________.

PART 3. Matching

_______ **31.** Right hypochondriac region
_______ **32.** Hypogastric region
_______ **33.** Left iliac region
_______ **34.** Right lumbar region
_______ **35.** Epigastric region
_______ **36.** Left hypochondriac region
_______ **37.** Right iliac region
_______ **38.** Umbilical region
_______ **39.** Left lumbar region

A. Junction of descending and sigmoid colons of large intestine
B. Descending colon of large intestine
C. Spleen
D. Most of right lobe of liver
E. Appendix
F. Ascending colon of large intestine
G. Middle of transverse colon of large intestine
H. Adrenal (suprarenal) glands
I. Sigmoid colon of large intestine

PART 4. Matching

______	**40.** Anterior	A.	Passes through iliac crests
______	**41.** Skull	B.	Contains spinal cord
______	**42.** Transtubercular line	C.	Nearer the midline
______	**43.** Armpit	D.	Thoracic
______	**44.** Umbilical region	E.	Cervical
______	**45.** Medial	F.	Axillary
______	**46.** Cranial cavity	G.	Contains the heart
______	**47.** Front of knee	H.	Cranial
______	**48.** Breast	I.	Antebrachial
______	**49.** Chest	J.	Gluteal
______	**50.** Buttock	K.	Mammary
______	**51.** Superior	L.	Sole
______	**52.** Groin	M.	Contains navel
______	**53.** Vertebral cavity	N.	Buccal
______	**54.** Cheek	O.	Farther from the attachment of a limb
______	**55.** Front of neck	P.	Patellar
______	**56.** Distal	Q.	Toward the head
______	**57.** Pericardial cavity	R.	Nearer to or at the front of the body
______	**58.** Forearm	S.	Oral
______	**59.** Plantar	T.	Contains brain
______	**60.** Mouth	U.	Inguinal

Cells

3

A *cell* is the basic living structural and functional unit of the body. The study of cells is called ***cytology*** (sī-TOL-ō-jē; *cyto* = cell; *logos* = study of). The different kinds of cells—blood, nerve, bone, muscle, epithelial, and others—perform specific functions and differ from one another in shape, size, and structure. You will start your study of cytology by learning the important components of a theoretical, generalized cell.

A. CELL PARTS

Refer to Figure 3.1, a generalized cell based on electron micrograph studies. With the aid of your textbook and any other items made available by your instructor, label the parts of the cell indicated. In the spaces that follow, describe the function of the cellular structures indicated:

1. ***Plasma (cell) membrane*** ______________________________

____

____

____

2. ***Cytoplasm*** (SĪ-tō-plazm') ______________________________

____

____

____

3. ***Nucleus*** (NOO-klē-us) ______________________________

____

____

____

4. ***Endoplasmic reticulum*** (en'-dō-PLAS-mik re-TIK-yoo-lum) or ***ER*** ______________________________

____

____

____

5. ***Ribosome*** (RĪ-bō-sōm) ______________________________

____

____

____

6. ***Golgi*** (GOL-jē) ***complex*** ______________________________

____

____

____

7. ***Mitochondrion*** (mī'-tō-KON-drē-on) ______________________________

____

____

____

8. ***Lysosome*** (LĪ-sō-sōm) ______________________________

____

____

____

9. ***Peroxisome*** (pe-ROKS-i-sōm) ______________________________

____

____

____

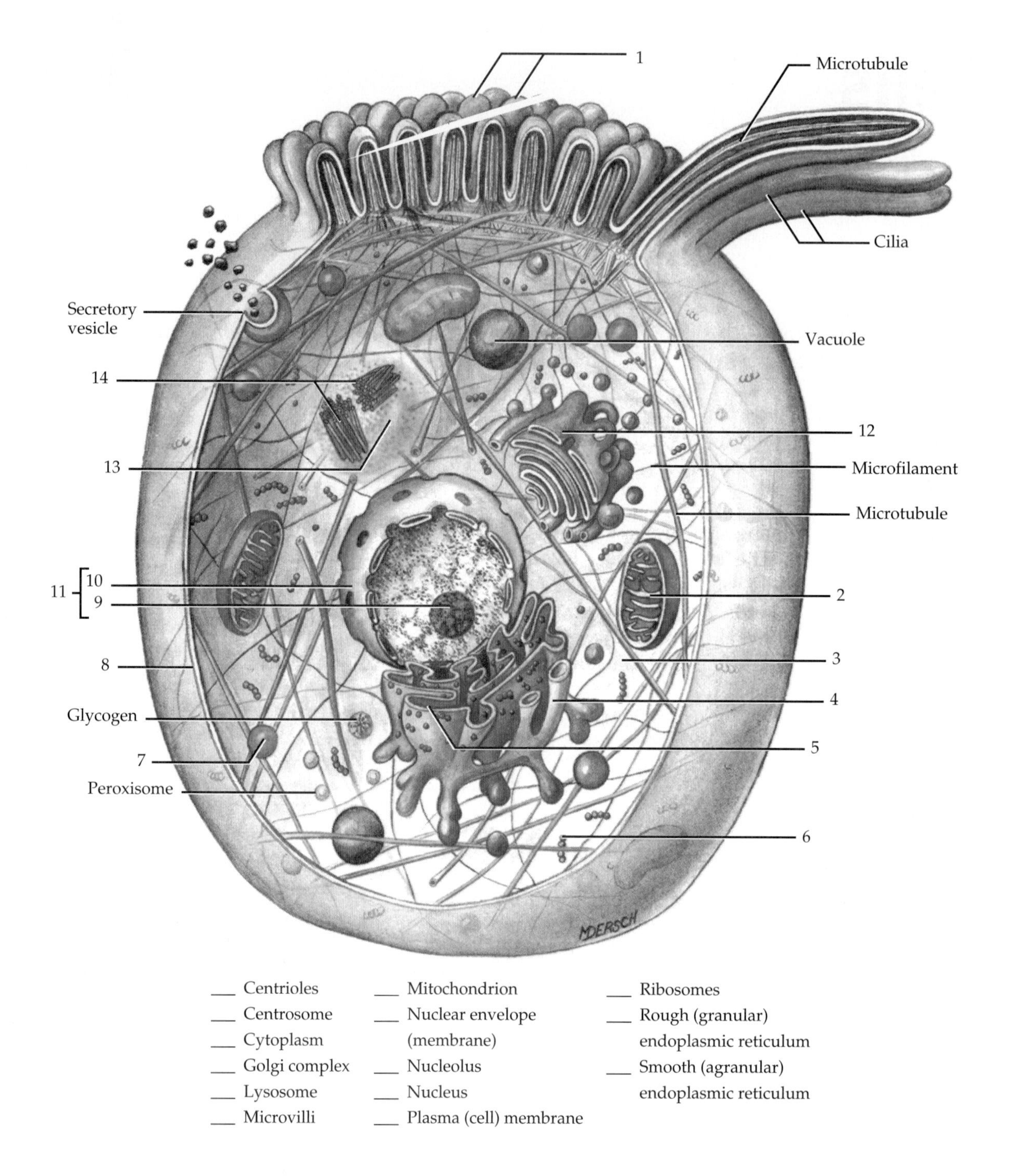

___ Centrioles
___ Centrosome
___ Cytoplasm
___ Golgi complex
___ Lysosome
___ Microvilli
___ Mitochondrion
___ Nuclear envelope (membrane)
___ Nucleolus
___ Nucleus
___ Plasma (cell) membrane
___ Ribosomes
___ Rough (granular) endoplasmic reticulum
___ Smooth (agranular) endoplasmic reticulum

FIGURE 3.1 Generalized animal cell.

10. ***Cytoskeleton*** ________________________________

____

____

____

11. ***Centrosome*** (SEN-trō-sōm') ____________________

____

____

____

12. ***Cilium*** (SIL-ē-um)________________

____

____

____

13. ***Flagellum*** (fla-JEL-um)________________

____

____

____

B. DIVERSITY OF CELLS

Now obtain prepared slides of the following types of cells and examine them under the magnifications suggested:

1. Ciliated columnar epithelial cells (high power)
2. Sperm cells (oil immersion)
3. Nerve cells (high power)
4. Muscle cells (high power)

After you have made your examination, draw an example of each of these kinds of cells in the spaces provided and under each cell indicate how each is adapted to its particular function.

Ciliated columnar epithelial cell

Sperm cell

Nerve cell

Muscle cell

C. CELL INCLUSIONS

Cell inclusions are a large and diverse group of mostly organic substances produced by cells that may appear or disappear at various times in the life of a cell. Using your textbook as a reference, indicate the function of the following cell inclusions:

1. ***Melanin***________________

____

____

____

2. ***Glycogen***________________

____

____

____

3. ***Triglycerides***________________

____

____

____

D. EXTRACELLULAR MATERIALS

Substances that lie outside the plasma membranes of body cells are referred to as ***extracellular materials.*** They include body fluids, such as interstitial fluid and plasma, which provide a medium for dissolving, mixing, and transporting substances. Extracellular materials also include special substances in which some cells are embedded.

Some extracellular materials are produced by certain cells and deposited outside their plasma membranes where they support cells, bind them together, and provide strength and elasticity. They have no definite shape and are referred to as ***amorphous.*** These include hyaluronic (hī-a-loo-RON-ik) acid and chondroitin (kon-DROY-tin) sulfate. Others are ***fibrous*** (threadlike). Examples include collagen, reticular, and elastic fibers.

Using your textbook as a reference, indicate the location and function for each of the following extracellular materials:

Hyaluronic acid

Location ______________________________

____

____

____

Function ______________________________

____

____

____

Chondroitin sulfate

Location ______________________________

____

____

____

Function ______________________________

____

____

____

Collagen fibers

Location ______________________________

____

____

____

Function ______________________________

____

____

____

Reticular fibers

Location ______________________________

____

____

____

Function ______________________________

____

____

____

Elastic fibers

Location ______________________________

____

____

____

Function ______________________________

____

____

____

E. CELL DIVISION

Cell division is the basic mechanism by which cells reproduce themselves. It consists of a nuclear division and a cytoplasmic division. Because nuclear division can be of two types, two kinds of cell division are recognized. In the first type, called ***somatic cell division,*** a single starting cell called a ***parent cell*** duplicates itself and the result is two identical cells called ***daughter cells.*** Somatic cell division consists of a nuclear division called ***mitosis*** (mī-TŌ-sis) and a cytoplasmic division called ***cytokinesis*** (sī-tō-ki-NĒ-sis; *cyto* = cell; *kinesis* = motion). It provides the body with a means of growth and of replacement of diseased or damaged cells (Figure 3.2). The second type of cell division is called ***reproductive cell division*** and is the mechanism by which sperm and eggs are produced (Exercise 24). Reproductive cell division consists of a nuclear division called ***meiosis*** and two cytoplasmic divisions (cytokinesis), and it results in the development of four nonidentical daughter cells.

In order to study somatic cell division, obtain a prepared slide of a whitefish blastula and examine it under high power.

A cell between divisions is said to be in ***interphase*** of the cell cycle. Interphase is the longest part of the cell cycle and is the period of time during which a cell carries on its physiological activities. One of the most important activities of interphase is the replication of DNA so that the two daughter cells that eventually form will each have the same kind and amount of DNA as the parent cell. In addition, the proteins needed to produce structures required for doubling all cellular components are manufactured. Scan your slide and find a cell in interphase. Such a parent cell is characterized by a clearly defined nuclear envelope. Within the nucleus, look for the nucleolus (or nucleoli) and ***chromatin,*** DNA that is associated with protein in the form of a granular substance. Also locate the centrioles.

Draw a labeled diagram of an interphase cell in the space provided.

Once a cell completes its interphase activities, mitosis begins. Mitosis is the distribution of two sets of chromosomes into two separate and equal nuclei after replication of the chromosomes of the parent cell, an event that takes place in the interphase preceding mitosis. Although a continuous process, mitosis is divided into four stages for purposes of study: prophase, metaphase, anaphase, and telophase.

1. ***Prophase***—The first stage of mitosis is called ***prophase*** (*pro* = before). During early prophase, the chromatin condenses and shortens into chromosomes. Because DNA replication took place during interphase, each prophase chromosome contains a pair of identical double-stranded DNA molecules called ***chromatids.*** Each chromatid pair is held together by a small spherical body called a ***centromere*** that is required for the proper segregation of chromosomes. Attached to the outside of each centromere is a protein complex known as the ***kinetochore*** (ki-NET-ō-kor), whose function will be described shortly. Later in prophase, the nucleolus (or nucleoli) disperses, and the nuclear envelope breaks up. In addition, each centrosome with its pair of centrioles, moves to opposite poles (ends) of the cell. As they do so, the centrosomes start to form the ***mitotic spindle,*** a football-shaped assembly of microtubules that are responsible for the movement of chromosomes.

   The lengthening of microtubules between centrosomes pushes the centrosomes to the poles of the cell so that the spindle extends from pole to pole. As the mitotic spindle continues to develop, three types of microtubules form: (a) ***nonkinetochore microtubules,*** which grow from centrosomes and extend inward, but do not bind to kinetochores; (b) ***kinetochore microtubules,*** which grow from centrosomes, extend inward, and attach to kinetochores; and (c) ***aster microtubules,*** which grow out of centrosomes but radiate outward from the mitotic spindle. Overall, the spindle is an attachment site for chromosomes. It also distributes chromosomes to opposite poles of the cell. Draw and label a cell in prophase in the space provided.

2. ***Metaphase***—During ***metaphase*** (*meta* = after), the second stage of mitosis, the kinetochore microtubules line up the centromeres of the chromatid pairs at the exact center of the mitotic spindle. This midpoint region is called the

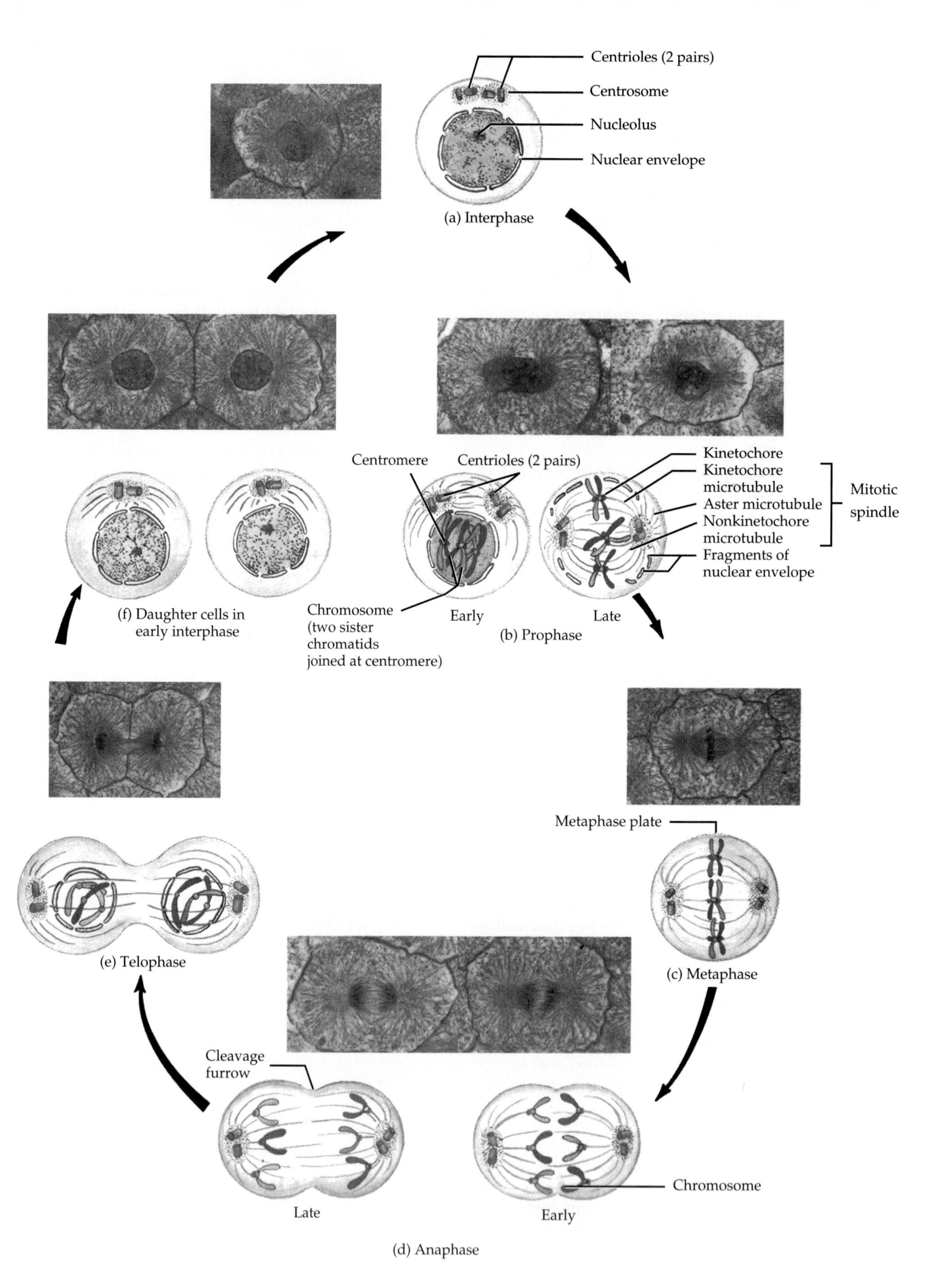

FIGURE 3.2 Cell division: mitosis and cytokinesis. Diagrams and photomicrographs of the various stages of cell division in whitefish eggs.

metaphase plate, or ***equatorial plane region.*** Draw and label a cell in metaphase in the space provided.

3. ***Anaphase***—The third stage of mitosis, ***anaphase*** (*ana* = upward), is characterized by the splitting and separation of the centromeres (and kinetochores) and the movement of the two sister chromatids of each pair toward opposite poles of the cell. Once separated, the sister chromatids are referred to as ***chromosomes.*** The movement of chromosomes is the result of the shortening of kinetochore microtubules and elongation of the nonkinetochore microtubules, processes that increase the distance between separated chromosomes. As the chromosomes move during anaphase, they appear V-shaped. Draw and label a cell in anaphase in the space provided.

4. ***Telophase***—The final stage of mitosis, ***telophase*** (*telo* = far or end), begins as soon as chromosomal movement stops. Telophase is essentially the opposite of prophase. During telophase, the identical sets of chromosomes at opposite poles of the cell uncoil and revert to their threadlike chromatin form; kinetochore microtubules disappear; nonkinetochore microtubules elongate even more; a new nuclear envelope reforms around each chromatin mass; new nucleoli reappear in the daughter nuclei; and eventually the mitotic spindle breaks up. Draw and label a cell in telophase in the space provided.

Cytokinesis begins during late anaphase and terminates during telophase with the formation of a ***cleavage furrow,*** a slight indentation of the plasma membrane that extends around the center of the cell. The furrow gradually deepens until opposite surfaces of the cell make contact and the cell is split in two. The result is two separated daughter cells, each with separate portions of cytoplasm and organelles and its own set of identical chromosomes.

Following cytokinesis, each daughter cell returns to interphase. Each cell in most tissues of the body eventually grows and undergoes mitosis and cytokinesis, and a new divisional cycle begins. Examine your telophase cell again and be sure that it contains a cleavage furrow.

Using high power and starting at 12 o'clock, move around the blastula and count the number of cells in interphase and in each mitotic phase. It will be easier to do this if you imagine lines dividing the blastula into quadrants. Count the interphase cells in each quadrant, then assign each dividing cell to a specific mitotic stage. It will be hard to assign some cells to a phase—e.g., to distinguish late anaphase from early telophase. If you cannot make a decision, assign one cell to the earlier phase in question and the next cell to the later phase.

Divide the number of cells in each stage by the total number of cells counted and multiply by 100 to determine the percent of the cells in each mitotic stage at a given point in time. Record your results below.

Percent of cells in interphase ______

Percent of cells in prophase ______

Percent of cells in metaphase ______

Percent of cells in anaphase ______

Percent of cells in telophase ______

ANSWER THE LABORATORY REPORT QUESTIONS AT THE END OF THE EXERCISE.

LABORATORY REPORT QUESTIONS

Cells 3

Student ______________________ Date ______________________

Laboratory Section ______________________ Score/Grade ______________________

PART 1. Multiple Choice

_______ 1. The portion of the cell that forms part of the mitotic spindle during division is the (a) endoplasmic reticulum (b) Golgi complex (c) cytoplasm (d) centrosome

_______ 2. A cell that carries on a great deal of digestion also contains a large number of (a) lysosomes (b) centrioles (c) mitochondria (d) nuclei

_______ 3. The area of the cell between the plasma membrane and nuclear envelope where chemical reactions occur is the (a) centrosome (b) vacuole (c) peroxisome (d) cytoplasm

_______ 4. The "powerhouses" of the cell where ATP is produced are the (a) ribosomes (b) mitochondria (c) centrioles (d) lysosomes

_______ 5. The sites of protein synthesis in the cell are (a) peroxisomes (b) flagella (c) ribosomes (d) centrosomes

_______ 6. A cell inclusion that is a pigment in skin and hair is (a) glycogen (b) melanin (c) mucus (d) collagen

_______ 7. Which extracellular material is found in ligaments and tendons? (a) elastic fibers (b) chondroitin sulfate (c) collagen fibers (d) mucus

_______ 8. The organelles that contain enzymes for the metabolism of hydrogen peroxide are (a) lysosomes (b) mitochondria (c) Golgi complexes (d) peroxisomes

_______ 9. The framework of cilia, flagella, centrioles, and the mitotic spindle is formed by (a) endoplasmic reticulum (b) collagen fibers (c) chondroitin sulfate (d) microtubules

_______ 10. A viscous fluidlike substance that binds cells together, lubricates joints, and maintains the shape of the eyeballs is (a) elastin (b) hyaluronic acid (c) mucus (d) plasmin

PART 2. Completion

11. The external boundary of the cell through which substances enter and exit is called the

______________________.

12. The cytoskeleton is formed by microtubules, intermediate filaments, and ______________________.

13. The portion of the cell that contains hereditary information is the ______________________.

14. The tail of a sperm cell is a long whiplash structure called a(n) ______________________.

15. Division of the cytoplasm is referred to as ____________________.

16. Lipid and protein secretion, carbohydrate synthesis, and assembly of glycoproteins are functions of the ____________________.

17. Storage of digestive enzymes is accomplished by the ____________________ of a cell.

18. Projections of cells that move substances along their surfaces are called ____________________.

19. The ____________________ provides a surface area for chemical reactions, a pathway for transporting molecules, and a storage area for synthesized molecules.

20. ____________________ is a cell inclusion that represents stored glucose in the liver and skeletal muscles.

21. A jellylike substance that supports cartilage, bone, heart valves, and the umbilical cord is ____________________.

22. The framework of many soft organs is formed by ____________________ fibers.

23. In an interphase cell, DNA is in the form of a granular substance called ____________________.

24. Distribution of chromosomes into separate and equal nuclei is referred to as ____________________.

PART 3. Matching

________	**25.** Anaphase	A. Mitotic spindle appears
________	**26.** Metaphase	B. Movement of chromosome sets to opposite poles of cell
________	**27.** Interphase	C. Centromeres line up on metaphase plate
________	**28.** Telophase	D. Formation of two identical nuclei
________	**29.** Prophase	E. Phase between divisions

Tissues

4

A *tissue* is a group of similar cells that usually have the same embryological origin and function together to perform a specific function. The study of tissues is called ***histology*** (hiss-TOL-ō-jē; *histio* = tissue; *logos* = study of). The various body tissues can be categorized into four principal kinds: (1) epithelial, (2) connective, (3) muscular, and (4) nervous. In this exercise you will examine the structure and functions of epithelial and connective tissues, except for bone or blood. Other tissues will be studied later as parts of the systems to which they belong.

A. EPITHELIAL TISSUE

Epithelial (ep'-i-THĒ-lē-al) ***tissue,*** or ***epithelium,*** may be divided into two types: (1) covering and lining and (2) glandular. Covering and lining epithelium forms the outer layer of the skin and some internal organs; forms the inner lining of blood vessels, ducts, body cavities, and many internal organs; and helps make up special sense organs for smell, hearing, vision, and touch. Glandular epithelium constitutes the secreting portion of glands.

1. Characteristics

Following are the general characteristics of epithelial tissue:

a. Epithelium consists largely or entirely of closely packed cells with little extracellular material between adjacent cells.
b. Epithelial cells are arranged in continuous sheets, in either single or multiple layers.
c. Epithelial cells have an ***apical*** (free) ***surface*** that is exposed to a body cavity, lining of an internal organ, or the exterior of the body and a ***basal surface*** that is attached to the basement membrane (described shortly).
d. Cell junctions are plentiful, providing secure attachments among the cells.
e. Epithelia are ***avascular*** (*a* = without; *vascular* = blood vessels). The vessels that supply nutrients and remove wastes are located in the adjacent connective tissue. The exchange of materials between epithelium and connective tissue is by diffusion.
f. Epithelia adhere firmly to nearby connective tissue, which holds the epithelium in position and prevents it from being torn. The attachment between the epithelium and the connective tissue is a thin extracellular layer called the ***basement membrane.*** It consists of two layers. The ***basal lamina*** contains collagen, laminin, and proteoglycans secreted by the epithelium. Cells in the connective tissue secrete the second layer, the ***reticular lamina,*** which contains reticular fibers, fibronectin, and glycoproteins. The basement membrane provides physical support for epithelium, provides for cell attachment, serves as a filter in the kidneys, and guides cell migration during development and tissue repair.
g. Epithelial tissue has a nerve supply.
h. Epithelial tissue is the only tissue that makes direct contact with the external environment.
i. Since epithelium is subject to a certain amount of wear and tear and injury, it has a high capacity for renewal (high mitotic rate).
j. Epithelia are diverse in origin. They are derived from all three primary germ layers (ectoderm, mesoderm, and endoderm).
k. Functions of epithelia include protection, filtration, lubrication, secretion, digestion, absorption, transportation, excretion, sensory reception, and reproduction.

2. Covering and Lining Epithelium

Before you start your microscopic examination of epithelial tissues, refer to Figure 4.1. Study the tissues carefully to familiarize yourself with their general structural characteristics. For each of the types of epithelium listed, obtain a prepared slide and, unless otherwise specified by your instructor, examine each under high power. In conjunction with your examination, consult a textbook of anatomy.

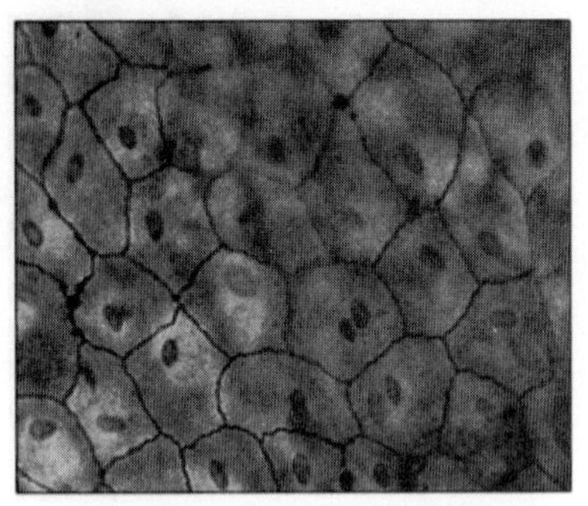

Surface view of mesothelial lining of peritoneal cavity (240×)

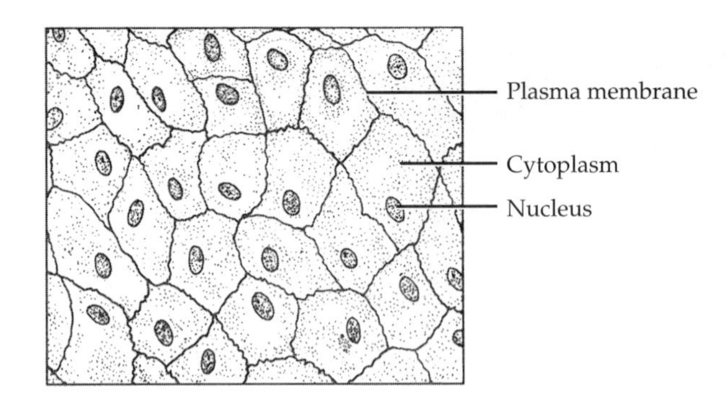

(a) Simple squamous epithelium

FIGURE 4.1 Epithelial tissues. Photomicrographs are unlabeled; the line drawings of the same tissues are labeled.

a. ***Simple squamous*** (SKWĀ-mus; *squama* = flat) ***epithelium*** This tissue consists of a single layer of flat cells and is highly adapted for diffusion, osmosis, and filtration because of its thinness. Simple squamous epithelium lines the air sacs (alveoli) of the lungs, glomerular (Bowman's) capsules (filtering units) of the kidneys, and inner surface of the tympanic membrane (eardrum) of the ear. Simple squamous epithelium that lines the heart, blood vessels, and lymphatic vessels and forms capillary walls is called ***endothelium*** (*endo* = within; *thelium* = covering). Simple squamous epithelium that forms the epithelial layer of a serous membrane is called ***mesothelium*** (*meso* = middle). Serous membranes line the thoracic and abdominopelvic cavities and cover viscera within the cavities. After you make your microscopic examination, draw several cells in the space that follows and label plasma membrane, cytoplasm, and nucleus.

Simple squamous epithelium

b. ***Simple cuboidal epithelium*** This tissue consists of a single layer of cube-shaped cells. When the tissue is sectioned at right angles to the surface, its cuboidal nature is obvious. Highly adapted for secretion and absorption, this tissue covers the surface of the ovaries; lines the smaller ducts of some glands and the anterior surface of the lens capsule of the eye; and forms the pigmented epithelium of the retina of the eye, part of the tubules of the kidneys, and the secreting units of other glands, such as the thyroid gland.

After you make your microscopic examination, draw several cells in the space that follows and label plasma membrane, cytoplasm, nucleus, basement membrane, and connective tissue layer.

Simple cuboidal epithelium

c. ***Simple columnar (nonciliated) epithelium*** This tissue consists of a single layer of columnar cells and, when sectioned at right angles,

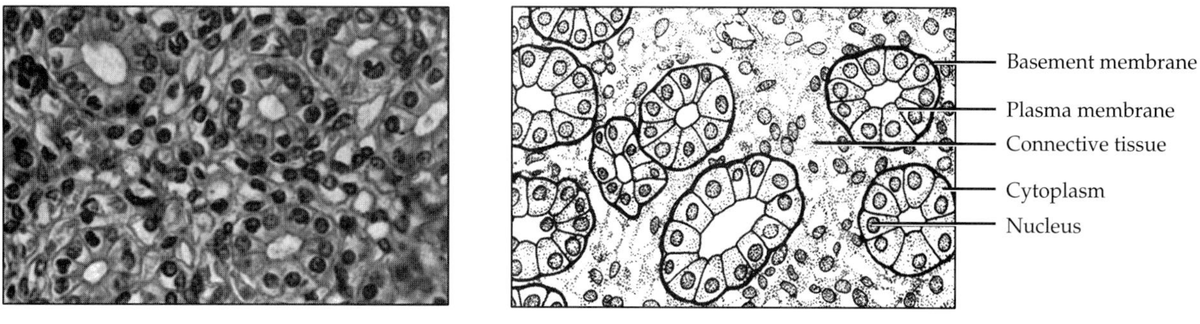

Sectional view of kidney tubules (400×)

(b) Simple cuboidal epithelium

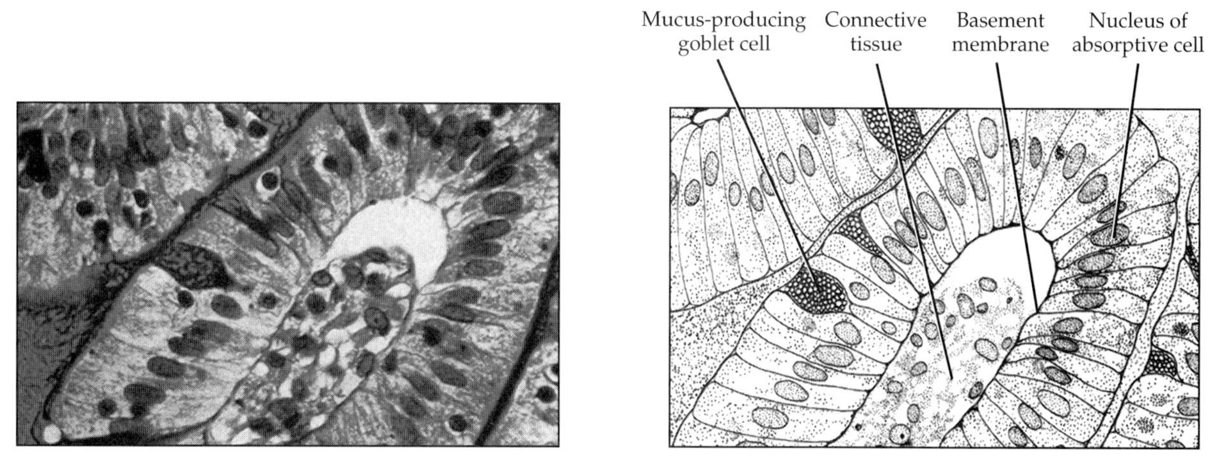

Sectional view of colonic glands (140×)

(c) Simple columnar (nonciliated) epithelium

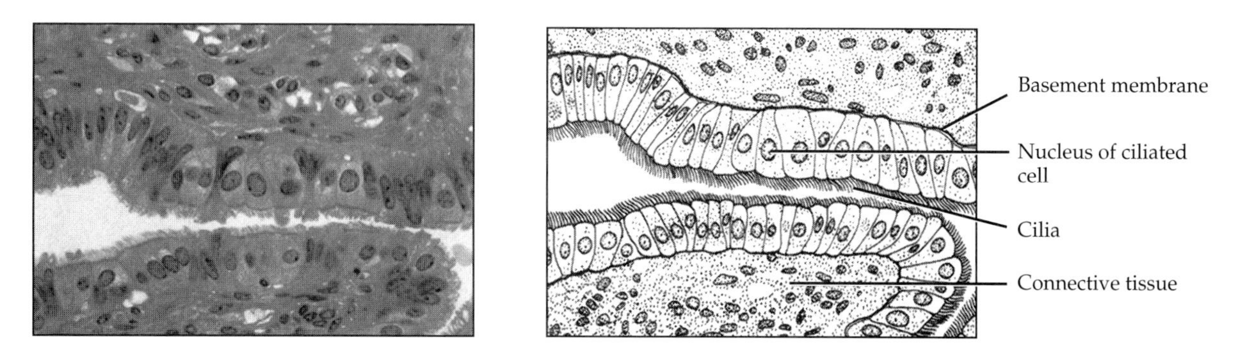

Sectional view of uterine (Fallopian) tube (100×)

(d) Simple columnar (ciliated) epithelium

FIGURE 4.1 ***(Continued)*** Epithelial tissues.

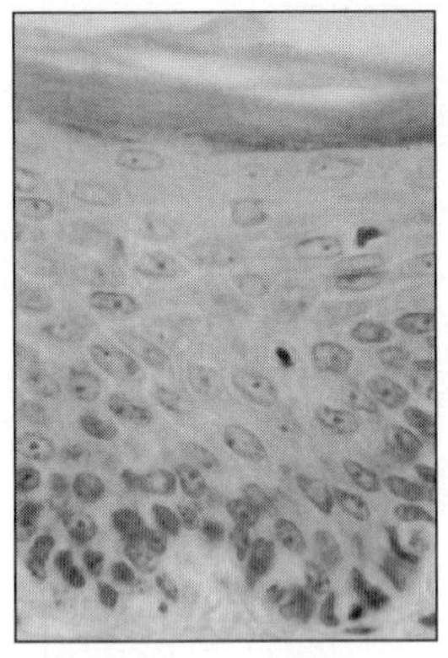

Sectional view of skin

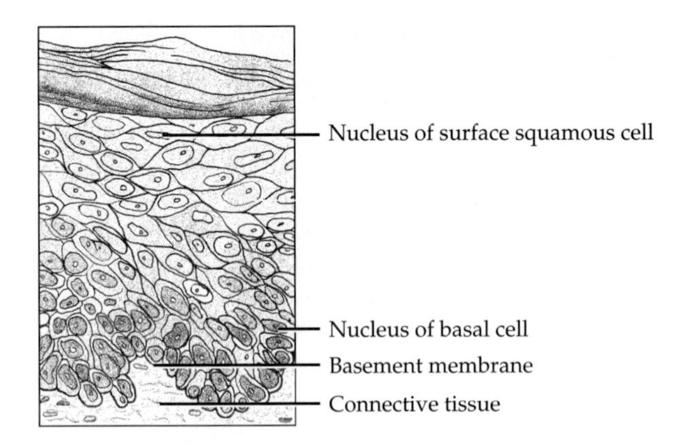

(e) Stratified squamous epithelium

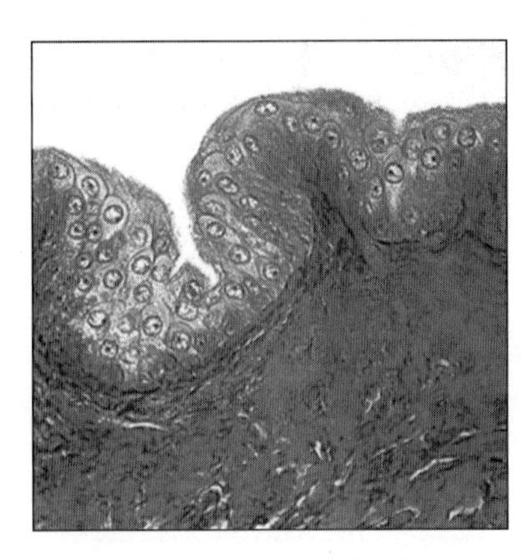

Sectional view of urinary bladder in relaxed state (100×)

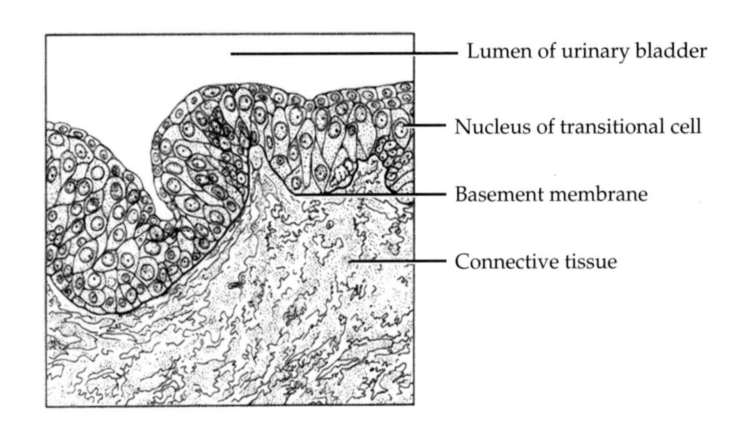

(f) Transitional epithelium

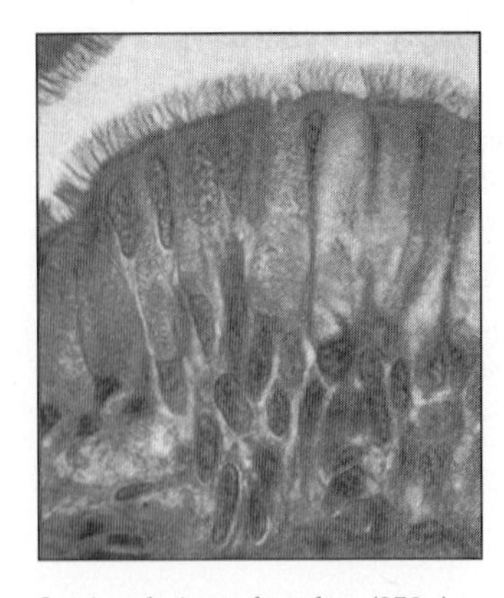

Sectional view of trachea (250×)

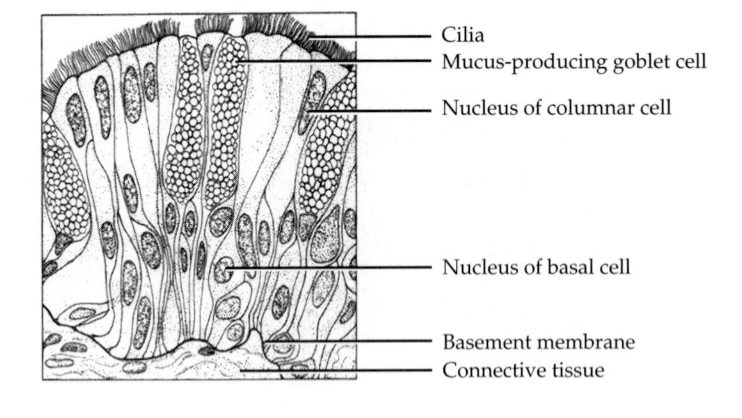

(g) Pseudostratified columnar epithelium

FIGURE 4.1 ***(Continued)*** Epithelial tissues.

these cells appear as rectangles. Adapted for secretion and absorption, this tissue lines the gastrointestinal tract from the stomach to the anus, gallbladder, and ducts of many glands. Some columnar cells are modified in that the plasma membranes are folded into microscopic fingerlike cytoplasmic projections called ***microvilli*** (*micro* = small; *villus* = tuft of hair) that increase the surface area for absorption. Other cells are ***goblet cells***, modified columnar cells that secrete and store mucus to protect the lining of the gastrointestinal tract. After you make your microscopic examination, draw several cells in the space that follows and label plasma membrane, cytoplasm, nucleus, goblet cell, absorptive cell, basement membrane, and connective tissue layer.

Simple columnar (nonciliated) epithelium

d. ***Simple columnar (ciliated) epithelium*** This type of epithelium consists of a single layer of columnar absorptive, goblet, and ciliated cells. ***Cilia*** (*cilia* = eyelashes) are hairlike processes that move substances over the surfaces of cells. Simple columnar (ciliated) epithelium lines some portions of the upper respiratory tract, uterine (Fallopian) tubes, uterus, some paranasal sinuses, and the central canal of the spinal cord. Mucus produced by goblet cells forms a thin film over the surface of the tissue, and movements of the cilia propel the mucus and the trapped substances over the surface of the tissue. After you make your microscopic examination, draw several cells in the space that follows and label plasma membrane, cytoplasm, nucleus, cilia, goblet cell, basement membrane, and connective tissue layer.

Simple columnar (ciliated) epithelium

e. ***Stratified squamous epithelium*** This tissue consists of several layers of cells and affords considerable protection against friction. The superficial cells are flat whereas cells of the deep layers vary in shape from cuboidal to columnar. The basal (bottom) cells continually multiply by cell division. As surface cells are sloughed off, new cells replace them from the basal layer. The surface cells of ***keratinized stratified squamous epithelium*** contain a waterproofing protein called ***keratin*** (*kerato* = horny) that also resists friction and bacterial invasion. The keratinized variety forms the outer layer of the skin. Surface cells of ***nonkeratinized stratified squamous epithelium*** do not contain keratin. The nonkeratinized variety lines wet surfaces such as the mouth, esophagus, vagina, and part of the epiglottis, and it covers the tongue. After you make your microscopic examination, draw several cells in the space that follows and label plasma membrane, cytoplasm, nucleus, squamous surface cells, basal cells, basement membrane, and connective tissue layer.

Stratified squamous epithelium

f. ***Stratified squamous epithelium (student prepared)*** Before examining the next slide, prepare a smear of cheek cells from the epithelial lining of the mouth. As noted previously, epithelium that lines the mouth is nonkeratinized stratified squamous epithelium. However, you will be examining surface cells only, and these will appear similar to simple squamous epithelium.

PROCEDURE

CAUTION! *Please reread Section B, "Precautions Related to Working with Blood, Blood Products, or Other Body Fluids" on page x at the beginning of the laboratory manual before you begin any of the following experiments. You should also read the experiments before you perform them to be sure that you understand all the procedures and safety precautions. When you finish this part of the exercise, place the reusable items in a fresh bleach solution and the discarded items in a biohazard container.*

a. Using the blunt end of a toothpick, *gently* scrape the lining of your cheek several times to collect some surface cells of the stratified squamous epithelium.
b. Now move the toothpick across a clean glass microscope slide until a thin layer of scrapings is left on the slide.
c. Allow the preparation to air dry.
d. Next, cover the smear with several drops of 1% methylene blue stain. After about 1 min, gently rinse the slide in cold tap water or distilled water to remove excess stain.
e. *Gently* blot the slide dry using a paper towel.
f. Examine the slide under low and high power. See if you can identify the plasma membrane, cytoplasm, nuclear membrane, and nucleoli. Some bacteria are commonly found on the slide and usually appear as very small rods or spheres.

g. ***Transitional epithelium*** This tissue resembles nonkeratinized stratified squamous epithelium, except that the superficial cells are larger and more rounded. When stretched, the surface cells are drawn out into squamouslike cells. This drawing out permits the tissue to stretch without the outer cells breaking apart from one another. The tissue lines parts of the urinary system that are subject to expansion from within, such as the urinary bladder, parts of the ureters, and urethra. After you have made your microscopic examination, draw several cells in the space that follows and label plasma membrane, cytoplasm, nucleus, surface cells, basement membrane, and connective tissue layer.

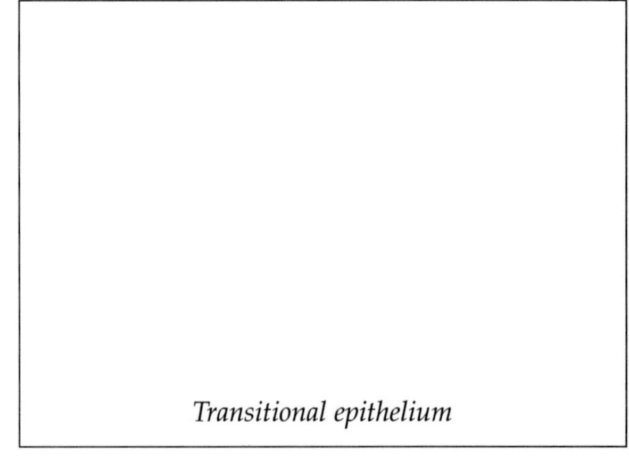

Transitional epithelium

h. ***Pseudostratified columnar epithelium*** Nuclei of cells in this tissue are at varying depths, and, although all the cells are attached to the basement membrane in a single layer, some do not reach the surface. This arrangement gives the impression of a multilayered tissue when sectioned, thus the name *pseudostratified.* This tissue lines large ducts of many glands, the epididymis, parts of the male urethra, and parts of the auditory (Eustachian) tubes; and a special variety that lines most of the upper respiratory tract is called pseudostratified ciliated columnar epithelium. After you have made your microscopic examination, draw and label basement membrane, a cell that reaches the surface, a cell that does not reach the surface, and nuclei of each cell.

Pseudostratified columnar epithelium

3. Glandular Epithelium

A ***gland*** may consist of a single epithelial cell or a group of highly specialized epithelial cells that secrete various substances. Glands that have no ducts (ductless), secrete hormones, and release their secretions into the blood are called ***endocrine glands.*** Examples include the pituitary gland and thyroid gland (Exercise 15). Glands that secrete their products into ducts are called ***exocrine glands***. Examples include sweat glands and salivary glands.

a. STRUCTURAL CLASSIFICATION OF EXOCRINE GLANDS

Based on the shape of the secretory portion and the degree of branching of the duct, exocrine glands can be structurally classified as follows:

1. ***Unicellular*** One-celled glands that secrete mucus. An example is the goblet cell (see Figure 4.1c). These cells line portions of the respiratory and digestive systems.
2. ***Multicellular*** Many-celled glands that occur in several different forms (Figure 4.2).
   Simple—Single, nonbranched duct.
   Tubular—Secretory portion is straight and tubular (intestinal glands).
   Branched tubular—Secretory portion is branched and tubular (gastric and uterine glands).
   Coiled tubular—Secretory portion is coiled (sudoriferous [soo-dor-IF-er-us], or sweat, glands).
   Acinar (AS-i-nar)—Secretory portion is flasklike (seminal vesicle glands).
   Branched acinar—Secretory portion is branched and flasklike (sebaceous [se-BĀ -shus], or oil, glands).
   Compound—Branched duct.
   Tubular—Secretory portion is tubular (bulbourethral or Cowper's glands, testes, liver).
   Acinar—Secretory portion is flasklike (sublingual and submandibular salivary glands).
   Tubuloacinar—Secretory portion is both tubular and flasklike (parotid salivary glands, pancreas).

Obtain a prepared slide of a representative of each of the types of multicellular exocrine glands just described. As you examine each slide, compare your observations to the diagrams of the glands in Figure 4.2.

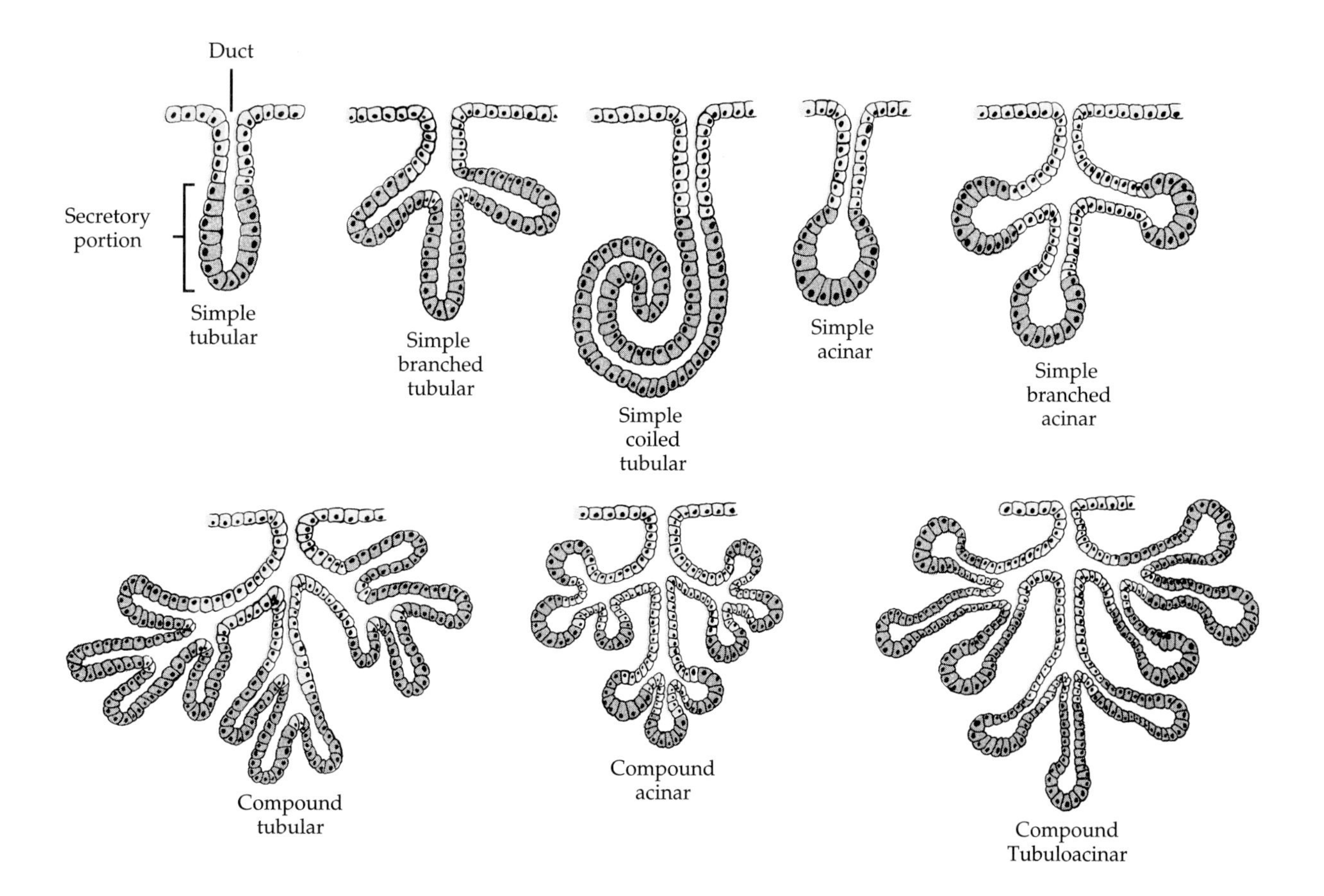

FIGURE 4.2 Structural classification of multicellular exocrine glands.

b. FUNCTIONAL CLASSIFICATION OF EXOCRINE GLANDS

Functional classification is based on whether a secretion is a product of a cell or consists of entire or partial glandular cells themselves. ***Holocrine glands,*** such as sebaceous (oil) glands, accumulate their secretory product in their cytoplasm. The cell then dies and is discharged with its contents as the glandular secretion; the discharged cell is replaced by a new one. ***Apocrine glands,*** such as the sudoriferous (sweat) glands in the axilla, accumulate secretory products at the outer margins of the secreting cells. The margins pinch off as the secretion and the remaining portions of the cells are repaired so that the process can be repeated. ***Merocrine (eccrine) glands,*** such as the pancreas and salivary glands, produce secretions that are simply formed by the secretory cells and then discharged into a duct.

B. CONNECTIVE TISSUE

Connective tissue, the most abundant tissue in the body, functions by protecting, supporting, and separating structures (e.g., skeletal muscles), and binding structures together.

1. Characteristics

Following are the general characteristics of connective tissue.

- **a.** Connective tissue consists of three basic elements: cells, ground substance, and fibers. Together, the ground substance (fluid, gel, or solid) and fibers, both of which are outside the cells, form the ***matrix.*** Unlike epithelial cells, connective tissue cells rarely touch one another; they are separated by a considerable amount of matrix.
- **b.** In contrast to epithelia, connective tissues do not usually occur on free surfaces, such as the surfaces of a body cavity or the external surface of the body.
- **c.** Except for cartilage, connective tissue, like epithelium, has a nerve supply.
- **d.** Unlike epithelium, connective tissue usually is highly vascular (has a rich blood supply). Exceptions include cartilage, which is avascular, and tendons, which have a scanty blood supply.
- **e.** The matrix of a connective tissue, which may be fluid, semifluid, gelatinous, fibrous, or calcified, is usually secreted by the connective tissue cells and adjacent cells and determines the tissue's qualities. In blood the matrix, which is not secreted by blood cells, is fluid. In cartilage it is firm but pliable. In bone it is considerably harder and not pliable.

2. Connective Tissue Cells

Following are some of the cells contained in various types of connective tissue. The specific tissues to which they belong will be described shortly.

- **a.** ***Fibroblasts*** (FĪ-brō-blasts; *fibro* = fiber) are large, flat, spindle-shaped cells with branching processes; they secrete the molecules that become the connective tissue fibers and ground substance of the matrix.
- **b.** ***Macrophages*** (MAK-rō-fā-jez; *macro* = large; *phagein* = to eat), or ***histiocytes,*** develop from ***monocytes,*** a type of white blood cell. Macrophages have an irregular shape with short branching projections and are capable of engulfing bacteria and cellular debris by phagocytosis. Thus, they provide a vital defense for the body.
- **c.** ***Plasma cells*** are small and either round or irregular in shape. They develop from a type of white blood cell called a ***B lymphocyte (B cell).*** Plasma cells secrete specific antibodies and, accordingly, provide a defense mechanism through immunity.
- **d.** ***Mast cells*** are abundant alongside blood vessels. They produce histamine, a chemical that dilates small blood vessels and increases their permeability during inflammation.
- **e.** Other cells in connective tissue include ***adipocytes (fat cells)*** and ***white blood cells (leukocytes).***

3. Connective Tissue Ground Substance

The ***ground substance*** is amorphous, meaning that it has no specific shape. Fibroblasts produce the ground substance and deposit it in the space between the cells.

Several examples of ground substance are as follows. ***Hyaluronic*** (hī-a-loo-RON-ik) acid is a viscous, slippery substance that binds cells together, lubricates joints, and helps maintain the shape of the eyeballs. It also appears to play a role in helping phagocytes migrate through connective tissue during development and wound repair. ***Chondroitin*** (kon-DROY-tin) ***sulfate*** is a jellylike substance that provides support and adhesiveness in cartilage, bone, the skin, and blood vessels. The skin, tendons, blood vessels, and heart valves contain ***dermatan sulfate,*** while bone, cartilage, and the cornea of the eye contain ***keratan sulfate. Adhesion proteins*** (fibronectin, laminin, collagen, and fibrinogen) interact with plasma membrane receptors to anchor cells in position.

The ground substance supports cells and binds them together and provides a medium through which substances are exchanged between the blood and cells. Until recently, the ground substance was thought to function mainly as an inert scaffolding to support tissues. Now it is clear that the ground substance is quite active in functions such as influencing development, migration, proliferation, shape, and even metabolic functions.

4. Connective Tissue Fibers

Fibers in the matrix are secreted by fibroblasts and provide strength and support for tissues. Three types of fibers are embedded in the matrix between the cells of connective tissue: collagen, elastic, and reticular fibers.

a. ***Collagen*** (*kolla* = glue) ***fibers,*** of which there are at least five different types, are very tough and resistant to a pulling force, yet allow some flexibility in the tissue because they are not taut. These fibers often occur in bundles made up of many minute fibrils lying parallel to one another. The bundle arrangement affords great strength. Chemically, collagen fibers consist of the protein ***collagen.*** This is the most abundant protein in your body, representing about 25% of the total protein. Collagen fibers are found in most types of connective tissues, especially bone, cartilage, tendons, and ligaments.

b. ***Elastic fibers*** are smaller than collagen fibers and freely branch and rejoin one another. They consist of a protein called ***elastin.*** Like collagen fibers, elastic fibers provide strength. In addition, they can be stretched 150% of their relaxed length without breaking. Elastic fibers are plentiful in the skin, blood vessels, and lungs.

c. ***Reticular*** (*rete* = net) ***fibers*** consisting of the protein collagen and a coating of glycoprotein, provide support in the walls of blood vessels and form a network around fat cells, nerve fibers, and skeletal and smooth-muscle cells. They are much thinner than collagen fibers and form branching networks. Like collagen fibers, reticular fibers provide support and strength and also form the ***stroma*** (framework) of many soft-tissue organs, such as the spleen and lymph nodes. These fibers also help form the basement membrane.

5. Types

Before you start your microscopic examination of connective tissues, refer to Figure 4.3. Study the tissues carefully to familiarize yourself with their general structural characteristics. For each type of connective tissue listed, obtain a prepared slide and, unless otherwise specified by your instructor, examine each under high power.

Here, we will concentrate on various types of ***mature connective tissues,*** meaning connective tissues that are present in the newborn and that do not change afterward. The types of mature connective tissue are ***loose connective tissue, dense connective tissue, cartilage, bone,*** and ***blood.***

a. LOOSE CONNECTIVE TISSUE

In this general type of connective tissue, the fibers are *loosely* woven and there are many cells.

1. ***Areolar*** (a-RĒ-ō-lar; *areola* = small space) ***connective tissue*** This is one of the most widely distributed connective tissues in the body. It contains at one time or another all cells normally found in connective tissue, including fibroblasts, macrophages, plasma cells, mast cells, adipocytes, and a few white blood cells. All three types of fibers—collagen, elastic, and reticular—are present and randomly arranged. The fluid, semifluid, or gelatinous ground substance contains hyaluronic acid, chondroitin sulfate, dermatan sulfate, and keratan sulfate. Areolar connective tissue is present in many mucous membranes, the superficial region of the dermis of the skin, around blood vessels, nerves, and organs, and, together with adipose tissue, forms the ***subcutaneous*** (sub'-kyoo-TĀ-nē-us) ***layer*** or ***superficial fascia*** (FASH-ē-a). This layer is located between the skin and underlying tissues. After you make your microscopic examination, draw a small area of the tissue in the space that follows on page 51 and label the elastic fibers, collagen fibers, fibroblasts, and mast cells.

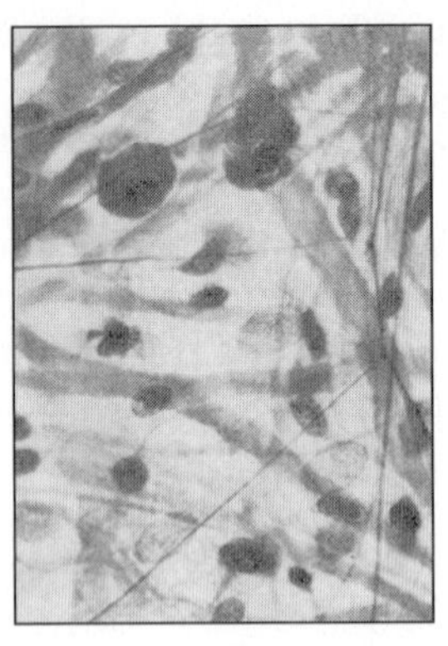

Surface view of subcutaneous tissue (160×)

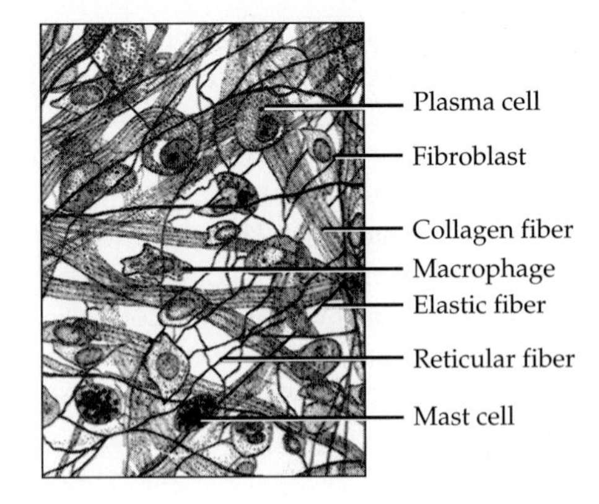

(a) Areolar connective tissue

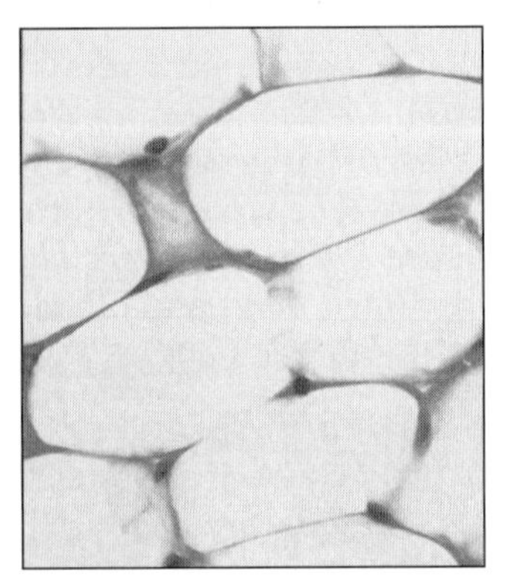

Sectional view of white fat of pancreas (1,600×)

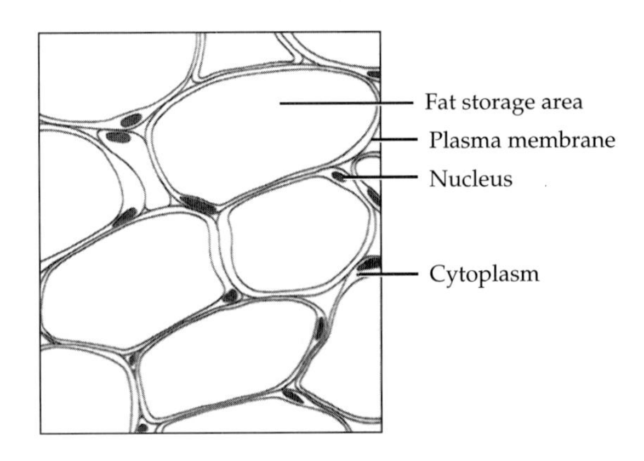

(b) Adipose tissue

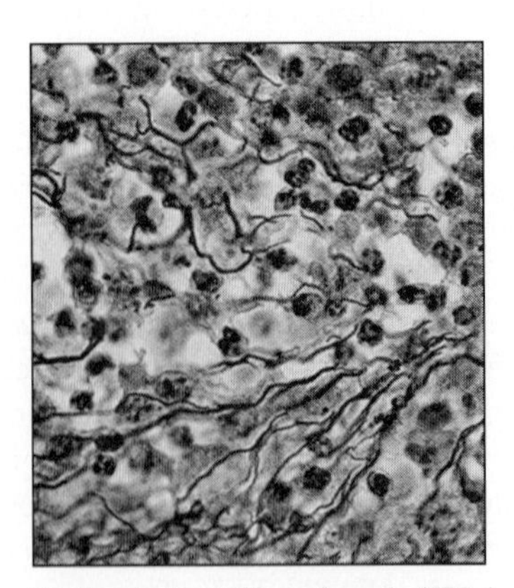

Sectional view of lymph node (250×)

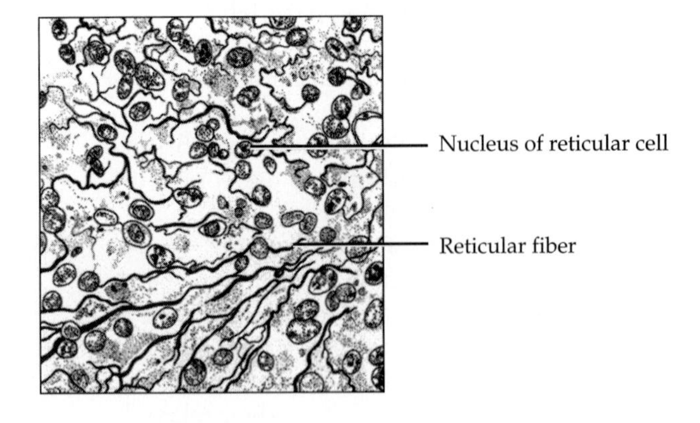

(c) Reticular connective tissue

FIGURE 4.3 Connective tissues. Photomicrographs are unlabeled; the line drawings of the same tissues are labeled.

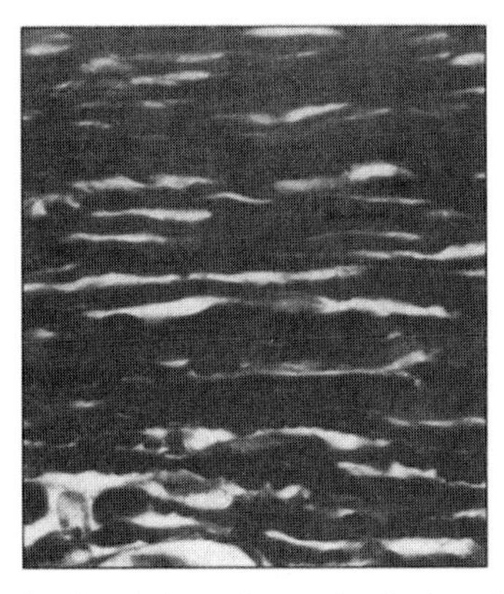

Sectional view of capsule of adrenal gland (250×)

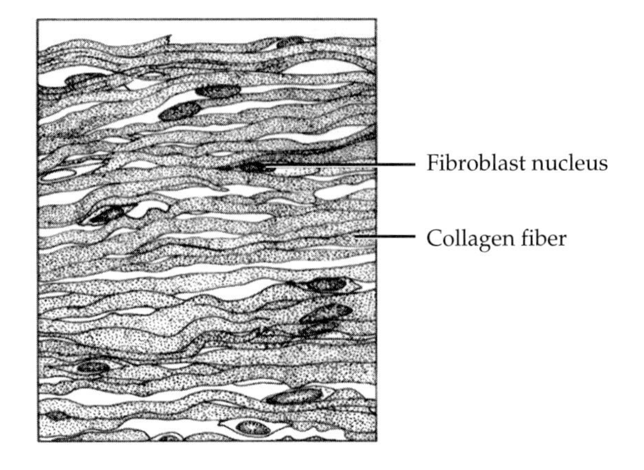

(d) Dense regular connective tissue

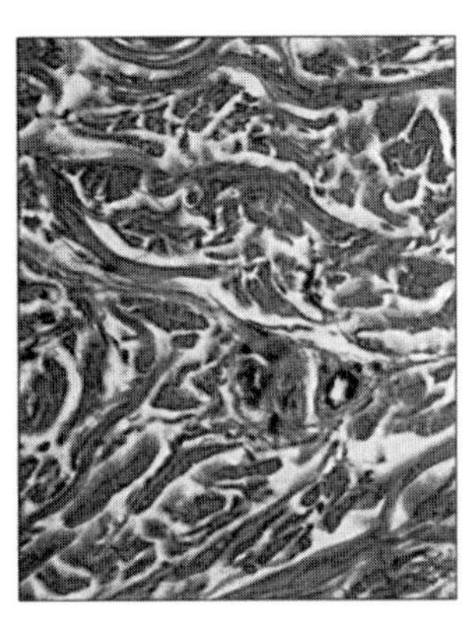

Sectional view of dermis of skin (275×)

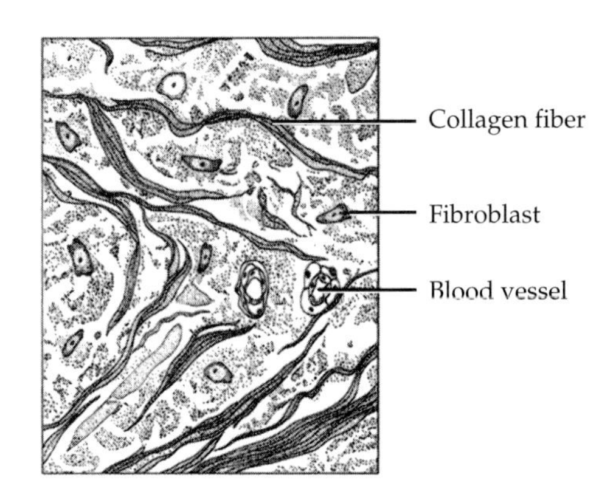

(e) Dense irregular connective tissue

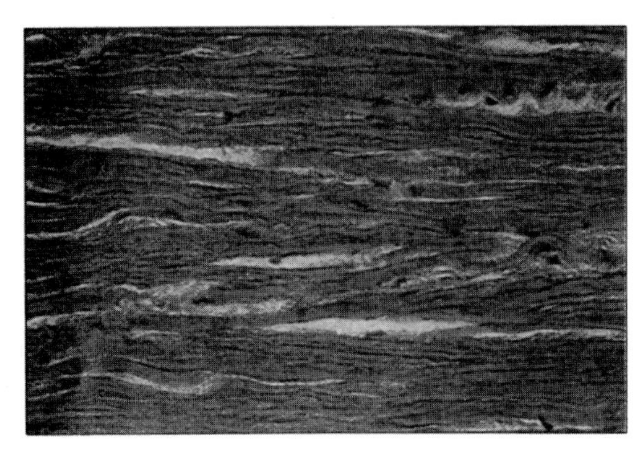

Sectional view of ligamentum nuchae (400×)

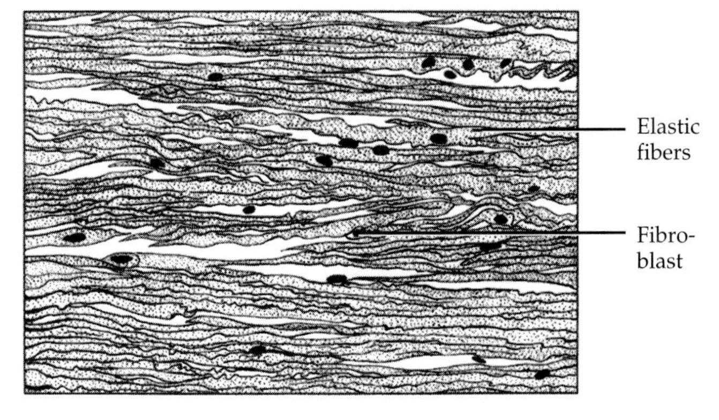

(f) Elastic connective tissue

FIGURE 4.3 ***(Continued)*** Connective tissues.

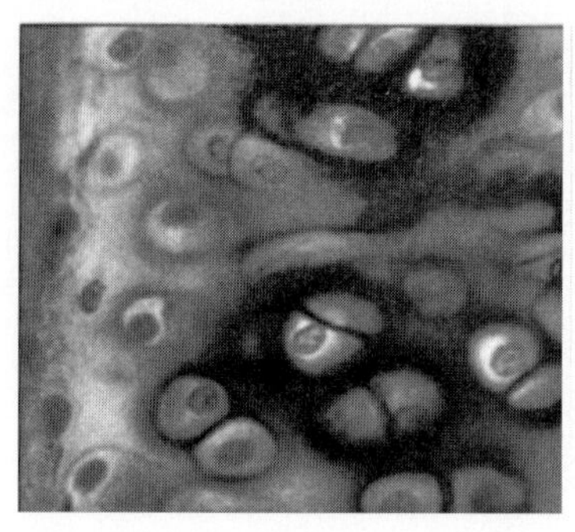

Sectional view of hyaline cartilage from trachea (160×)

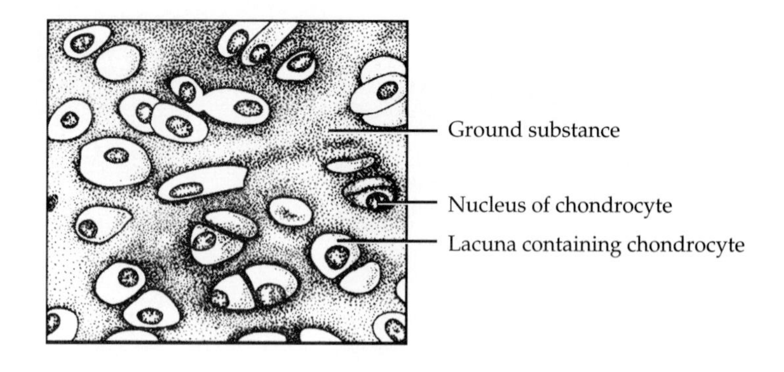

(g) Hyaline cartilage

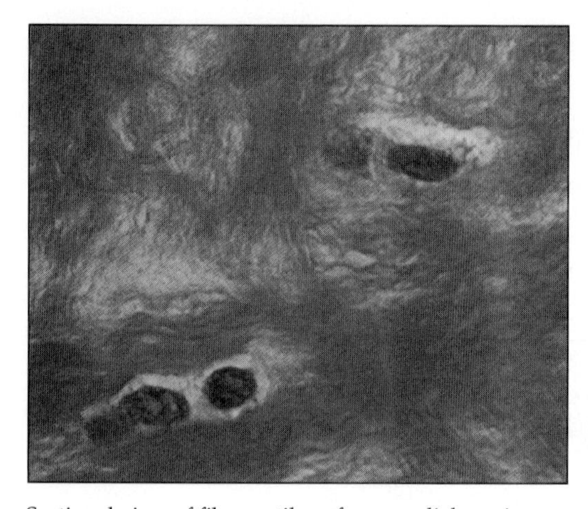

Sectional view of fibrocartilage from medial meniscus of knee (315×)

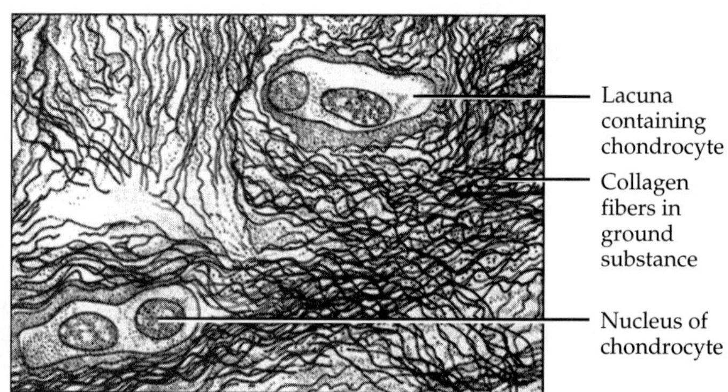

(h) Fibrocartilage

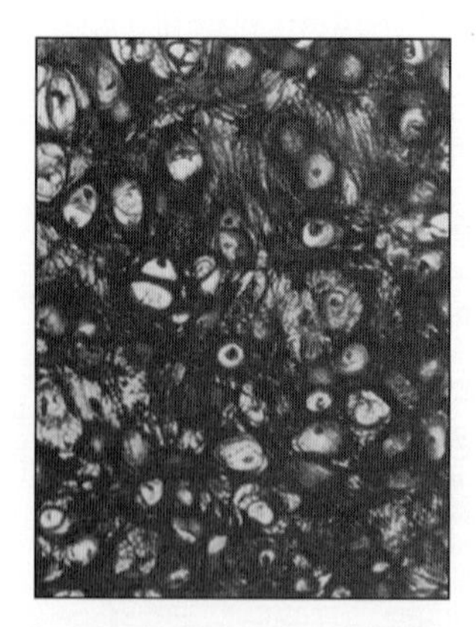

Sectional view of elastic cartilage from auricle (pinna) of external ear (175×)

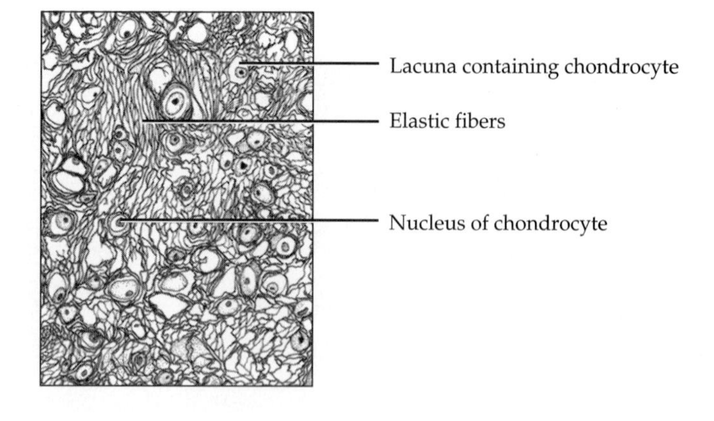

(i) Elastic cartilage

FIGURE 4.3 ***(Continued)*** Connective tissues.

Areolar connective tissue

Reticular connective tissue

2. ***Adipose tissue*** This is fat tissue in which cells derived from fibroblasts, called ***adipocytes*** (*adeps* = fat), are modified for triglyceride (fat) storage. The cytoplasm and nuclei of the cells are pushed to the edge. The tissue is found wherever areolar connective tissue is located and around the kidneys and heart, in the yellow bone marrow of long bones, and behind the eyeball. It provides insulation, energy reserve, support, and protection. After your microscopic examination, draw several cells in the space that follows and label the fat storage area, cytoplasm, nucleus, and plasma membrane.

Adipose tissue

3. ***Reticular connective tissue*** This tissue consists of fine interlacing reticular fibers in which reticular cells are interspersed between the fibers. It provides the stroma (framework) of certain organs and helps bind certain cells together. It is found in the liver, spleen, and lymph nodes; in a portion of the basement membrane; and in red bone marrow. After you make your microscopic examination, draw a sample of the tissue in the space that follows and label the reticular fibers and cells of the organ.

b. DENSE CONNECTIVE TISSUE

In this general type of connective tissue, the fibers are more numerous, *thicker*, and *densely* packed and there are fewer cells than loose connective tissue.

1. ***Dense regular connective tissue*** In this tissue, bundles of collagen fibers have a *regular* (orderly), parallel arrangement that confers great strength. The tissue structure withstands pulling in one direction. Fibroblasts, which produce the fibers and ground substance, appear in rows between the fibers. The tissue is silvery white, tough, yet somewhat pliable. Because of its great strength, it is the principal component of ***tendons***, which attach muscles to bones; ***aponeuroses*** (ap'-ō-noo-RŌ-sēz), which are sheetlike tendons connecting one muscle with another or with bone; and most ***ligaments*** (collagen ligaments), which hold bones together at joints. After you make your microscopic examination, draw a sample of the tissue in the space that follows and label the collagen fibers and fibroblasts.

Dense regular connective tissue

2. ***Dense irregular connective tissue*** This tissue contains collagen fibers that are *irregularly* arranged (without regular orientation) and is found in parts of the body where tensions are exerted in various directions. The tissue usually occurs in sheets. It forms some fasciae, the reticular (deeper) region of the dermis of the skin, the pericardium of the heart, the periosteum of bone, the perichondrium of cartilage, joint capsules, heart valves, and the membrane (fibrous) capsules around organs, such as the kidneys, liver, testes, and lymph nodes. After you make your microscopic examination, draw a sample of the tissue in the space that follows and label the collagen fibers.

Dense irregular connective tissue

3. ***Elastic connective tissue*** This tissue has a predominance of freely branching elastic fibers. These fibers give the unstained tissue a yellowish color. Fibroblasts are present in the spaces between fibers. Elastic connective tissue can be stretched and will snap back into shape (elasticity). It is a component of the walls of elastic arteries, the trachea, bronchial tubes of the lungs, and the lungs themselves. Elastic connective tissue provides stretch and strength, allowing structures to perform their functions efficiently. Yellow elastic ligaments, as contrasted with collagen ligaments, are composed mostly of elastic fibers; they form the ligamenta flava of the vertebrae (ligaments between successive vertebrae), the suspensory ligament of the penis, and the true vocal cords. After you make your microscopic examination, draw a sample of the tissue in the space that follows and label the elastic fibers and fibroblasts.

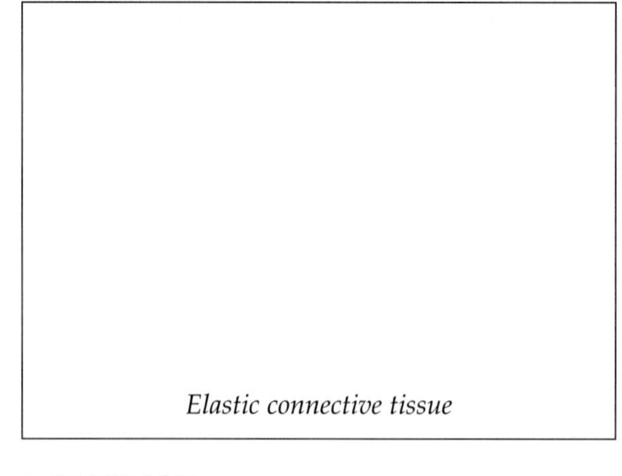
Elastic connective tissue

C. CARTILAGE

Cartilage is capable of enduring considerably more stress than the tissues just discussed. Unlike other connective tissues, cartilage has no blood vessels or nerves, except for those in the perichondrium (membranous covering). Cartilage consists of a dense network of collagen fibers and elastic fibers firmly embedded in chondroitin sulfate, a rubbery component of the ground substance. Whereas the strength of cartilage is due to its collagen fibers, its resilience (ability to assume its original shape after deformation) is due to chondroitin sulfate.

The cells of mature cartilage, called ***chondrocytes*** (KON-drō-sīts; *chondros* = cartilage), occur singly or in groups within spaces called ***lacunae*** (la-KOO-nē; *lacuna* = little lake) in the matrix. The surface of cartilage, except for fibrocartilage and some hyaline cartilage, is surrounded by dense irregular connective tissue called the ***perichondrium*** (per'-i-KON-drē-um; *peri* = around). Three kinds of cartilage are recognized: hyaline cartilage, fibrocartilage, and elastic cartilage.

1. ***Hyaline cartilage*** This cartilage, also called *gristle,* contains a resilient gel as its ground substance and appears in the body as a bluish-white, shiny substance. The fine collagen fibers, although present, are not visible with ordinary staining techniques, and the prominent chondrocytes are found in lacunae. Hyaline cartilage is the most abundant cartilage in the body. It is found at joints over the ends of the long bones (articular cartilage) and at the anterior ends of the ribs (costal cartilage). Hyaline cartilage also helps to support the nose, larynx, trachea, bronchi, and bronchial tubes leading to the lungs. Most of the embryonic skeleton consists of hyaline cartilage, which gradually becomes calcified and develops into bone.

Hyaline cartilage affords flexibility and support and, at joints, reduces friction and absorbs shock. After you make your microscopic examination, draw a sample of the tissue in the space that follows and label the perichondrium, chondrocytes, lacunae, and ground substance.

Hyaline cartilage

2. ***Fibrocartilage*** Chondrocytes are scattered among clearly visible bundles of collagen fibers within the matrix of this type of cartilage. Fibrocartilage forms the pubic symphysis, the point where the hipbones fuse anteriorly at the midline. It is also found in the intervertebral discs between vertebrae, and the menisci (cartilage pads) of the knee. This tissue combines strength and rigidity. After you make your microscopic examination, draw a sample of the tissue in the space that follows and label the chondrocytes, lacunae, ground substance, and collagen fibers.

Fibrocartilage

3. ***Elastic cartilage*** In this tissue, chondrocytes are located in a threadlike network of elastic fibers within the matrix. Elastic cartilage provides strength and elasticity and maintains the shape of organs—the epiglottis of the larynx, the external part of the ear (auricle), and the auditory (Eustachian) tubes. After you make your microscopic examination, draw a sample of the tissue in the space that follows and label the perichondrium, chondrocytes, lacunae, ground substance, and elastic fibers.

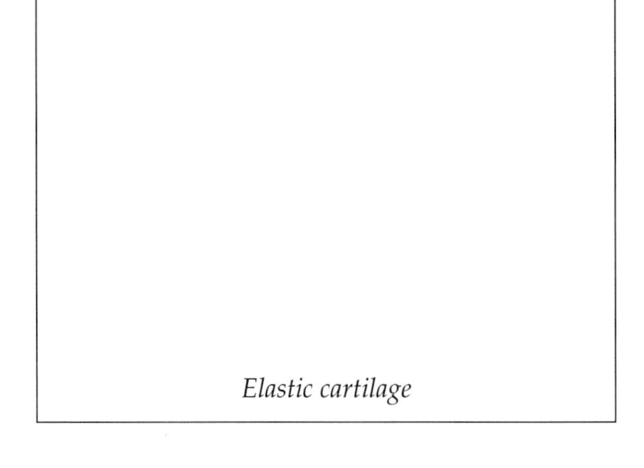

Elastic cartilage

C. MEMBRANES

The combination of an epithelial layer and an underlying layer of connective tissue constitutes an ***epithelial membrane***. Examples are mucous, serous, and cutaneous membranes (skin). Another kind of membrane, a synovial membrane, has no epithelium. It contains only connective tissue. ***Mucous membranes***, also called the ***mucosa***, line body cavities that open directly to the exterior, such as the gastrointestinal, respiratory, urinary, and reproductive tracts. The surface tissue of a mucous membrane consists of epithelium and has a variety of functions, depending on location. Accordingly, the epithelial layer secretes mucus but may also secrete enzymes, filter dust, and have a protective and absorbent action. The underlying connective tissue layer of a mucous membrane, called the *lamina propria* (LAM-i-na PRŌ-prē-a), binds the epithelial layer in place, protects underlying tissues, provides the epithelium with nutrients and oxygen and removes wastes, and holds blood vessels in place.

Serous (*serous* = watery) ***membranes***, also called the ***serosa***, line body cavities that do not open to the exterior and cover organs that lie within the cavities. Serous membranes consist of a surface layer of mesothelium (simple squamous epithelium) and an underlying layer of areolar connective tissue. The mesothelium secretes a lubricating fluid. Serous membranes consist of two layers. The layer attached to the cavity wall is called the ***parietal*** (pa-RĪ-e-tal; *paries* = wall) ***layer;*** the layer that covers the organs in the cavity is called the ***visceral*** (*viscus* = body organ) ***layer***. Examples of serous membranes are the pleurae, pericardium, and peritoneum.

The ***cutaneous membrane,*** or skin, is the principal component of the integumentary system, which will be considered in the next exercise.

Synovial (sin-Ō-vē-al) ***membranes*** line joint cavities. They do not contain epithelium but rather consist of areolar connective tissue, adipose tissue, and elastic fibers. Synovial membranes produce synovial fluid, which lubricates the ends of bones as they move at joints and nourishes the articular cartilage around the ends of bones.

ANSWER THE LABORATORY REPORT QUESTIONS AT THE END OF THE EXERCISE.

LABORATORY REPORT QUESTIONS

Tissues 4

Student ______________________ Date ______________________

Laboratory Section ______________________ Score/Grade ______________________

PART 1. Multiple Choice

_______ **1.** In parts of the body such as the urinary bladder, where considerable distension (stretching) occurs, you can expect to find which epithelial tissue? (a) pseudostratified columnar (b) cuboidal (c) columnar (d) transitional

_______ **2.** Stratified epithelium is usually found in areas of the body where the principal activity is (a) filtration (b) absorption (c) protection (d) diffusion

_______ **3.** Ciliated epithelium destroyed by disease would cause malfunction in which system? (a) digestive (b) respiratory (c) skeletal (d) cardiovascular

_______ **4.** The tissue that provides the skin with resistance to wear and tear and serves to waterproof it is (a) keratinized stratified squamous (b) pseudostratified columnar (c) transitional (d) simple columnar

_______ **5.** The connective tissue cell that would most likely increase its activity during an infection is the (a) melanocyte (b) macrophage (c) adipocyte (d) fibroblast

_______ **6.** Torn ligaments would involve damage to which tissue? (a) dense regular (b) reticular (c) elastic (d) areolar

_______ **7.** Simple squamous tissue that lines the heart, blood vessels, and lymphatic vessels is called (a) transitional (b) adipose (c) endothelium (d) mesothelium

_______ **8.** Microvilli and goblet cells are associated with which tissue? (a) hyaline cartilage (b) simple columnar nonciliated (c) transitional (d) stratified squamous

_______ **9.** Superficial fascia contains which tissue? (a) elastic (b) reticular (c) fibrocartilage (d) areolar connective tissue

_______ **10.** Which tissue forms articular cartilage and costal cartilage? (a) fibrocartilage (b) elastic cartilage (c) adipose (d) hyaline cartilage

_______ **11.** Because the sublingual gland contains a branched duct and flasklike secretory portions, it is classified as (a) simple coiled tubular (b) compound acinar (c) simple acinar (d) compound tubular

_______ **12.** Which glands discharge an entire dead cell and its contents as their secretory products? (a) merocrine (b) apocrine (c) endocrine (d) holocrine

_______ **13.** Membranes that line cavities that open directly to the exterior are called (a) synovial (b) serous (c) mucous (d) cutaneous

_______ **14.** Which statement about connective tissue is false? (a) Cells are always very closely packed together. (b) Connective tissue always has an abundant blood supply. (c) Matrix is always present in large amounts. (d) It is the most abundant tissue in the body.

_______ **15.** A group of similar cells that has a similar embryological origin and operates together to perform a specialized activity is called a(n) (a) organ (b) tissue (c) system (d) organ system

_______ **16.** Which statement best describes covering and lining epithelium? (a) It is always arranged in a single layer of cells. (b) It contains large amounts of intercellular substance. (c) It has an abundant blood supply. (d) Its free surface is exposed to the exterior of the body or to the interior of a hollow structure.

_______ **17.** Which statement best describes connective tissue? (a) usually contains a large amount of matrix (b) always arranged in a single layer of cells (c) primarily concerned with secretion (d) usually lines a body cavity

_______ **18.** A gland (a) is either exocrine or endocrine (b) may be single celled or multicellular (c) consists of epithelial tissue (d) is described by all of the preceding statements.

_______ **19.** Which of the following statements is not correct? (a) Simple squamous epithelium lines blood vessels. (b) Endothelium is composed of cuboidal cells. (c) Ciliated epithelium is found in the respiratory system. (d) Transitional epithelium is found in the urinary bladder.

PART 2. Completion

20. Cells found in epithelium that secrete mucus are called _______________ cells.

21. A type of epithelium that appears to consist of several layers but actually contains only one layer of cells is _______________.

22. The cell in connective tissue that forms new fibers is the _______________.

23. Histamine, a substance that dilates small blood vessels during inflammation, is secreted by _______________ cells.

24. Cartilage cells found in lacunae are called _______________.

25. The simple squamous epithelium of a serous membrane that covers viscera is called _______________.

26. The tissue that provides insulation, support, protection, and serves as a food reserve is _______________.

27. _______________ tissue forms the stroma of organs such as the liver and spleen.

28. The cartilage that provides support for the larynx and external ear is _______________.

29. The ground substance that helps lubricate joints and binds cells together is _______________.

30. Ductless glands that secrete hormones are called _______________ glands.

31. Multicellular exocrine glands that contain branching ducts are classified as _______________ glands.

32. The mammary glands are classified as _______________ glands because their secretory products are the pinched-off margins of cells.

33. _______________ membranes consist of parietal and visceral layers and line cavities that do not open to the exterior.

34. If the secretory portion of a gland is flasklike, it is classified as a(n)____________________gland.

35. Membranes that line joint cavities are called____________________membranes.

36. An example of a simple branched acinar gland is a(n)____________________gland.

37. The structure that attaches epithelium to underlying connective tissue is called

    the____________________.

PART 3. Matching

________ 38. Lines the inner surface of the stomach and intestine

________ 39. Lines urinary tract, as in urinary bladder, permitting distention

________ 40. Lines mouth; present on outer surface of skin

________ 41. Single layer of cube-shaped cells; found in kidney tubules and ducts of some glands

________ 42. Lines air sacs of lungs where thin cells are required for diffusion of gases into blood

________ 43. Not a true stratified tissue; all cells on basement membrane, but some do not reach surface

________ 44. Derived from lymphocyte, gives rise to antibodies and so is helpful in defense

________ 45. Phagocytic cell; engulfs bacteria and cleans up debris; important during infection

________ 46. Believed to form collagen and elastic fibers in injured tissue

________ 47. Abundant along walls of blood vessels; produces histamine, which dilates blood vessels

________ 48. Contains lacunae and chondrocytes

________ 49. Forms fasciae and dermis of skin

________ 50. Stores fat and provides insulation

A. Transitional epithelium
B. Fibroblast
C. Pseudostratified columnar epithelium
D. Dense irregular connective tissue
E. Simple columnar epithelium
F. Macrophage
G. Stratified squamous epithelium
H. Adipose
I. Simple cuboidal epithelium
J. Plasma cell
K. Simple squamous epithelium
L. Mast cell
M. Cartilage

Integumentary System

5

The skin and the organs derived from it (hair, nails, and glands), and several specialized receptors constitute the ***integumentary*** (in-teg-yoo-MEN-tar-ē; *integumentum* = covering) ***system,*** which you will study in this exercise. An ***organ*** is an aggregation of tissues of definite form and usually recognizable shape that performs a specific function; a ***system*** is a group of organs that operate together to perform specialized functions.

A. SKIN

The ***skin*** is one of the largest organs of the body in terms of surface area, occupying a surface area of about 2 square meters (2 m^2) (22 ft^2). Among the functions performed by the skin are regulation of body temperature; protection of underlying tissues from physical abrasion, microorganisms, dehydration, and ultraviolet (UV) radiation; excretion of water and salts and several organic compounds; synthesis of vitamin D in the presence of sunlight; reception of stimuli for touch, pressure, pain, and temperature change sensations; serving as a blood reservoir; and immunity.

The skin consists of an outer, thinner ***epidermis*** (*epi* = above), which is avascular, and an inner, thicker ***dermis*** (*derm* = skin), which is vascular. Below the dermis is the ***subcutaneous layer (superficial fascia*** or ***hypodermis)*** that attaches the skin to underlying tissues and organs.

1. Epidermis

The epidermis consists of four principal kinds of cells. ***Keratinocytes*** (ker-a-TIN-ō-sīts; *kerato* = horny) are the most numerous cells and they undergo keratinization, that is, newly formed cells produced in the basal layers are pushed up to the surface and in the process synthesize a protective chemical, keratin. ***Melanocytes*** (MEL-a-nō-sīts; *melan* = black) are pigment cells that impart color to the skin. The third type of cell in the epidermis is called a ***Langerhans*** (LANG-er-hans) ***cell.*** These cells are a small population of cells that arise from red bone marrow and migrate to the epidermis and other stratified squamous epithelial tissue in the body. They are sensitive to UV radiation and lie above the basal layer of keratinocytes. Langerhans cells interact with white blood cells called ***helper T cells*** to assist in the immune response. ***Merkel cells*** are found in the bottom layer of the epidermis. Their bases are in contact with flattened portions of the terminations of sensory nerves ***(Merkel discs)*** and function as receptors for touch. At this point, we will concentrate only on keratinocytes.

Obtain a prepared slide of human skin and carefully examine the epidermis. Identify the following layers from the outside inward:

a. ***Stratum corneum*** (*corneum* = horny) 25 to 30 rows of flat, dead cells that are filled with ***keratin;*** these cells are continuously shed and replaced by cells from deeper strata.
b. ***Stratum lucidum*** (*lucidus* = clear) Several rows of clear, flat cells; more apparent in the thick skin of the palms and soles.
c. ***Stratum granulosum*** (*granulum* = little grain) 3 to 5 rows of flat cells that contain ***keratohyalin*** (ker'-a-tō-HĪ-a-lin) and ***lamellar granules*** which produce a lipid waterproof sealant.
d. ***Stratum spinosum*** (*spinosum* = thornlike) 8 to 10 rows of polyhedral (many-sided) cells.
e. ***Stratum basale*** (*basale* = base) Single layer of cuboidal to columnar cells that constantly undergo division. In hairless skin, this layer contains tactile (Merkel) discs, receptors sensitive to touch. Also called the ***stratum germinativum.***

Label the epidermal layers in Figure 5.1.

2. Dermis

The dermis is divided into two regions and is composed of connective tissue containing collagen and elastic fibers and a number of other structures. The

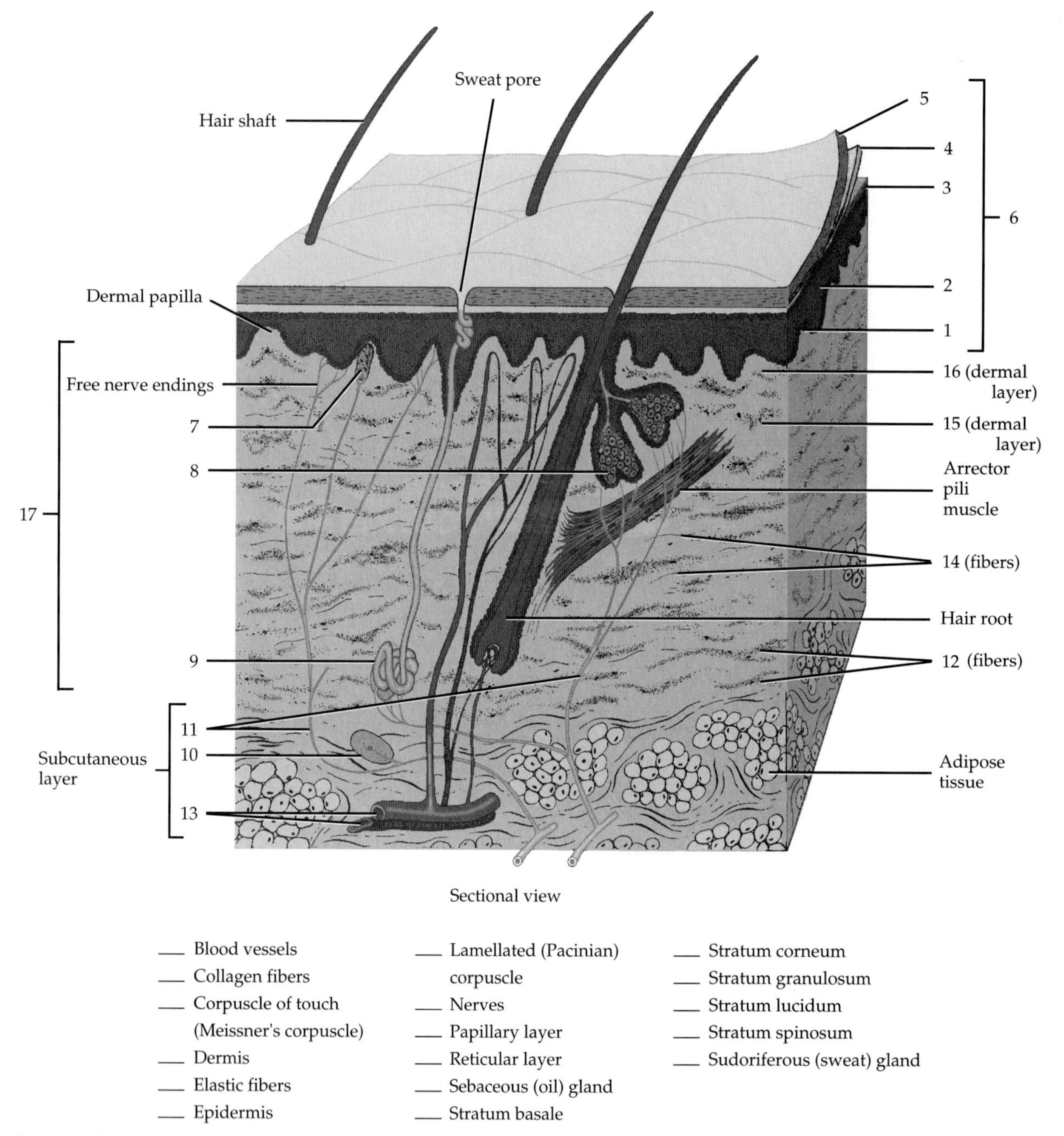

___ Blood vessels
___ Collagen fibers
___ Corpuscle of touch (Meissner's corpuscle)
___ Dermis
___ Elastic fibers
___ Epidermis
___ Lamellated (Pacinian) corpuscle
___ Nerves
___ Papillary layer
___ Reticular layer
___ Sebaceous (oil) gland
___ Stratum basale
___ Stratum corneum
___ Stratum granulosum
___ Stratum lucidum
___ Stratum spinosum
___ Sudoriferous (sweat) gland

FIGURE 5.1 Structure of the skin.

superficial region of the dermis ***(papillary layer)*** is areolar connective tissue containing fine elastic fibers. This layer contains fingerlike projections, the ***dermal papillae*** (pa-PIL-ē; *papilla* = nipple). Some papillae enclose blood capillaries; others contain ***corpuscles of touch (Meissner's corpuscles),*** nerve endings sensitive to touch. The deeper region of the dermis ***(reticular layer)*** consists of dense, irregular connective tissue with interlacing bundles of larger collagen and some coarse elastic fibers. Spaces between the fibers may be occupied by ***hair follicles, sebaceous (oil) glands, bundles of smooth muscle (arrector pili muscle), sudoriferous (sweat) glands, blood vessels,*** and ***nerves.***

The reticular layer of the dermis is attached to the underlying structures (bones and muscles) by the subcutaneous layer. This layer also contains nerve endings sensitive to pressure called ***lamellated*** or ***Pacinian*** (pa-SIN-ē-an) ***corpuscles.***

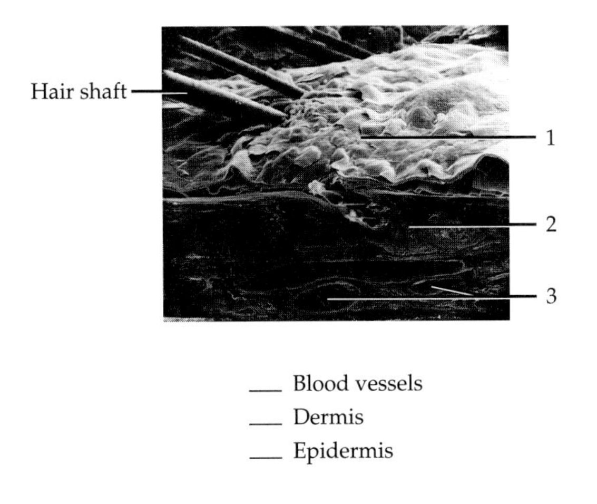

FIGURE 5.2 Scanning electron micrograph of the skin and several hairs at a magnification of 260×. (Reproduced by permission from R. G. Kessel and R. H. Kardon, *Tissues and Organs: A Text Atlas of Scanning Electron Microscopy*, W. H. Freeman and 1979.)

Carefully examine the dermis and subcutaneous layer on your microscope slide. Label the following structures in Figure 5.1: papillary layer, reticular layer, corpuscle of touch, blood vessels, nerves, elastic fibers, collagen fibers, sebaceous gland, sudoriferous gland, and lamellated corpuscle. Also label the epidermis, dermis, and blood vessels in Figure 5.2.

If a model of the skin and subcutaneous layer is available, examine it to see the three-dimensional relationship of the structures to one another.

3. Skin Color

The color of skin results from (1) ***hemoglobin in red blood cells in capillaries*** of the dermis (beneath the epidermis); (2) ***carotene*** (KAR-o-tēn; *keraton* = carrot), a yellow-orange pigment in the stratum corneum of the epidermis and fatty areas of the dermis and subcutaneous layer; and (3) ***melanin*** (MEL-a-nin), a pale yellow to black pigment found primarily in the melanocytes in the stratum basale and spinosum of the epidermis. Whereas hemoglobin in red blood cells in capillaries imparts a pink color to Caucasian skin, carotene imparts a yellowish color to skin. Because the number of ***melanocytes*** (MEL-a-nō-sīts), or melanin-producing cells, is about the same in all races, most differences in skin color are due to the amount of melanin that the melanocytes synthesize and disperse. Exposure to ultraviolet (UV) radiation increases melanin synthesis, resulting in darkening (tanning) of the skin to protect the body against further UV radiation.

An inherited inability of a person of any race to produce melanin results in ***albinism*** (AL-bi-nizm). The pigment is absent from the hair and eyes as well as from the skin, and the individual is referred to as an ***albino.*** In some people, melanin tends to form in patches called ***freckles.*** Others inherit patches of skin that lack pigment, a condition called ***vitiligo*** (vit-i-LĪ-gō).

B. HAIR

Hairs (pili) develop from the epidermis and are variously distributed over the body. Each hair is composed of columns of dead, keratinized cells and consists of a ***shaft,*** most of which is visible above the surface of the skin, and a ***root,*** the portion below the surface that penetrates deep into the dermis and even into the subcutaneous layer.

The shaft of a coarse hair consists of the following parts:

1. ***Medulla*** Inner region composed of several rows of polyhedral cells containing pigment and air spaces. Poorly developed or not present in fine hairs.
2. ***Cortex*** Middle layer; contains several rows of dark cells surrounding the medulla; contains pigment in dark hair and mostly air in white hair.
3. ***Cuticle of the hair*** Outermost layer; consists of a single layer of flat, keratinized cells arranged like shingles on a house.

The root of a hair also contains a medulla, cortex, and cuticle of the hair along with the following associated parts:

1. ***Hair follicle*** Structure surrounding the root that consists of an external root sheath and an internal root sheath. These epidermally derived layers are surrounded by a dermal layer of connective tissue.
2. ***External root sheath*** Downward continuation of the epidermis.
3. ***Internal root sheath*** Cellular tubular sheath that separates the hair from the external root sheath; consists of (a) the ***cuticle of the internal root sheath,*** an inner single layer of flattened cells with atrophied nuclei, (b) ***granular (Huxley's) layer,*** a middle layer of one to three rows of cells with flattened nuclei, and (c) ***pallid (Henle's) layer,*** an outer single layer of cuboidal cells with flattened nuclei.
4. ***Bulb*** Enlarged, onion-shaped structure at the base of the hair follicle.

5. ***Papilla of the hair*** Dermal indentation into the bulb; contains areolar connective tissue and blood vessels to nourish the hair.
6. ***Matrix*** Region of cells at the base of the bulb derived from the stratum basale and that divides to produce new hair.
7. ***Arrector*** (*arrector* = to raise) ***pili muscle*** Bundle of smooth muscle extending from the superficial dermis of the skin to the side of the hair follicle; its contraction, under the influence of fright or cold, causes the hair to move into a vertical position, producing "goose bumps."
8. ***Hair root plexus*** Nerve endings around each hair follicle that are sensitive to touch and respond when the hair shaft is moved.

Obtain a prepared slide of a transverse (cross) section and a longitudinal section of a hair root and identify as many parts as you can. Using your textbook as a reference, label Figure 5.3. Also label the parts of a hair shown in Figure 5.4.

C. GLANDS

Sebaceous (se-BĀ-shus; *sebo* = grease) or ***oil glands,*** with few exceptions, are connected to hair follicles (see Figures 5.1 and 5.4). They are simple, branched acinar glands and secrete an oily substance called ***sebum*** (SĒ-bum), a mixture of fats, cholesterol, proteins, and salts. Sebaceous glands

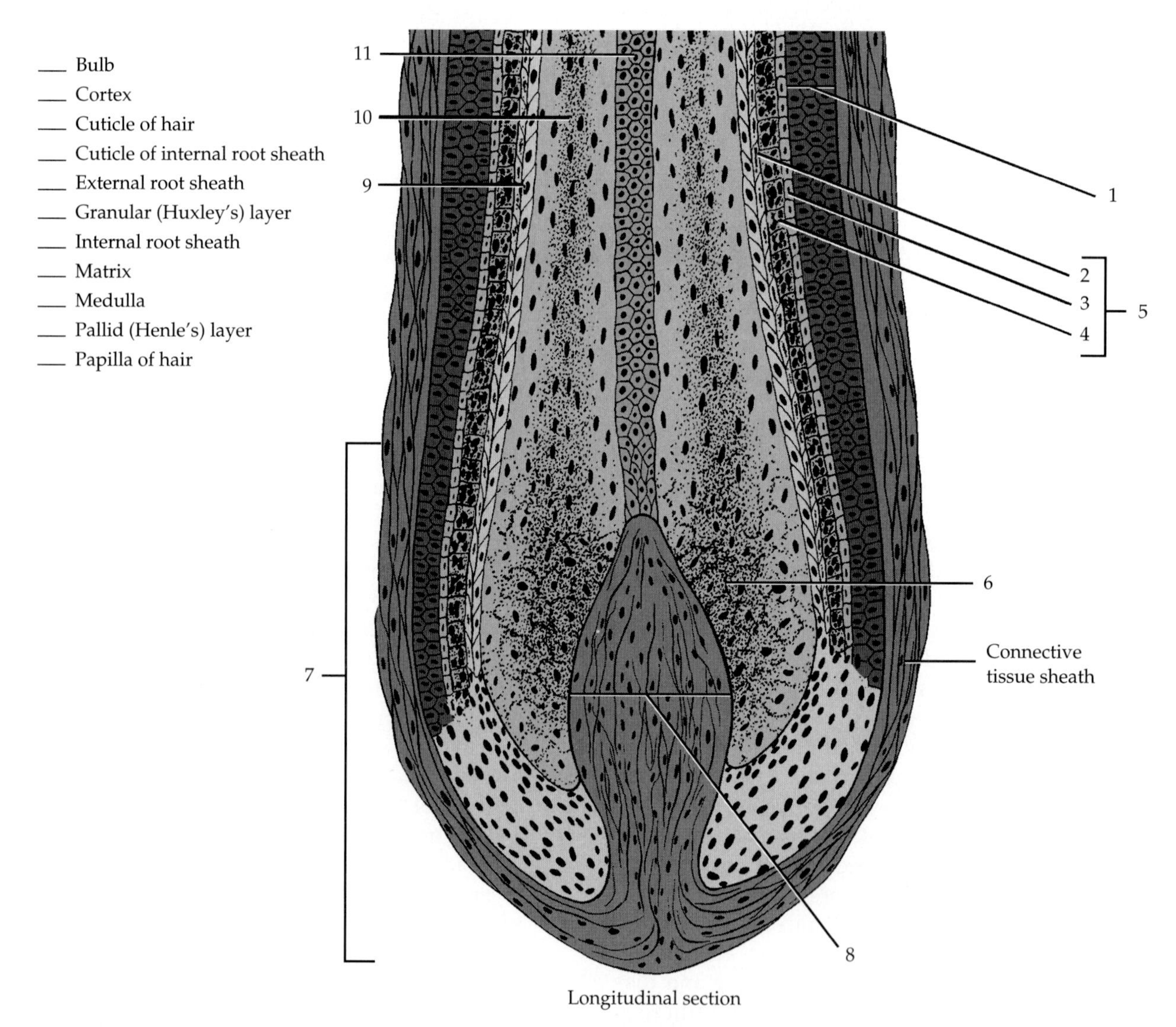

FIGURE 5.3 Hair root.

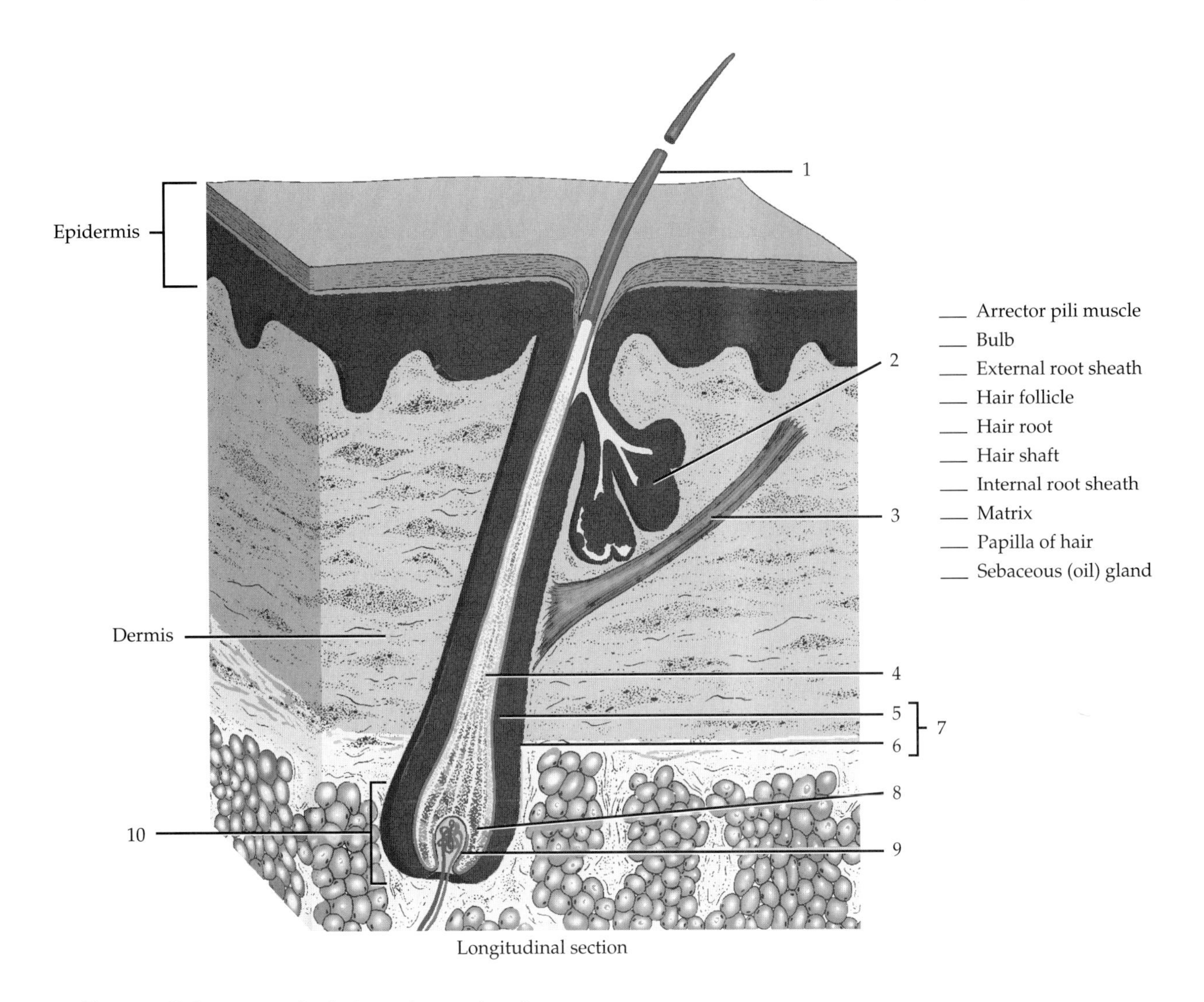

FIGURE 5.4 Parts of a hair and associated structures.

are absent in the skin of the palms and soles but are numerous in the skin of the face, neck, upper chest, and breasts. Sebum helps prevent hair from drying and forms a protective film over the skin that prevents excessive evaporation and keeps the skin soft and pliable.

Sudoriferous (soo'-dor-IF-er-us; *sudor* = sweat; *ferre* = to bear) or ***sweat glands*** are separated into two principal types on the basis of structure, location, and secretion. ***Apocrine sweat glands*** are simple, branched tubular glands found primarily in the skin of the axilla (armpit), pubic region, and areolae (pigmented areas) of the breasts. Their secretory portion is located in the dermis or subcutaneous layer; the excretory duct opens into hair follicles. Apocrine sweat glands begin to function at puberty and produce a more viscous secretion than the other type of sweat gland. ***Eccrine sweat glands*** are simple, coiled tubular glands found throughout the skin, except for the margins of the lips, nail beds of the fingers and toes, glans penis, glans clitoris, and eardrums. They are most numerous in the skin of the palms and soles. The secretory portion of these glands is in the subcutaneous layer; the excretory duct projects upward and terminates at a pore at the surface of the epidermis (see Figure 5.1). Eccrine sweat glands function throughout life and produce a more watery secretion than the apocrine glands. Sudoriferous glands produce ***perspiration,*** a mixture of water, salt, urea, uric acid, amino acids, ammonia, sugar, lactic acid, and ascorbic acid. The evaporation of perspiration helps to maintain normal body temperature.

Ceruminous (se-ROO-mi-nus; *cera* = wax) ***glands*** are modified sudoriferous glands in the external auditory meatus. They are simple, coiled tubular glands. The combined secretion of ceruminous and sudoriferous glands is called ***cerumen*** (earwax). Cerumen, together with hairs in the external auditory meatus, provides a sticky barrier that prevents foreign bodies from reaching the eardrum.

D. NAILS

Nails are plates of tightly packed, hard, keratinized epidermal cells that form a clear covering over the dorsal surfaces of the terminal portions of the fingers and toes. Each nail consists of the following parts:

1. ***Nail body*** Portion that is visible.
2. ***Free edge*** Part that may project beyond the distal end of the digit.
3. ***Nail root*** Portion hidden in nail groove (see item 7).
4. ***Lunula*** (LOO-nyoo-la; *lunula* = little moon) Whitish semilunar area at proximal end of body.
5. ***Nail fold*** Fold of skin that extends around the proximal end and lateral borders of the nail.
6. ***Nail bed*** Strata basale and spinosum of the epidermis beneath the nail.
7. ***Nail groove*** Furrow between the nail fold and nail bed.
8. ***Eponychium*** (ep'-ō-NIK-ē-um) Cuticle; a narrow band of epidermis.
9. ***Hyponychium*** Thickened area of stratum corneum below the free edge of the nail.
10. ***Nail matrix*** Epithelium of the proximal part of the nail bed; division of the cells brings about growth of nails.

Using your textbook as a reference, label the parts of a nail shown in Figure 5.5. Also, identify the parts that are visible on your own nails.

ANSWER THE LABORATORY REPORT QUESTIONS AT THE END OF THE EXERCISE.

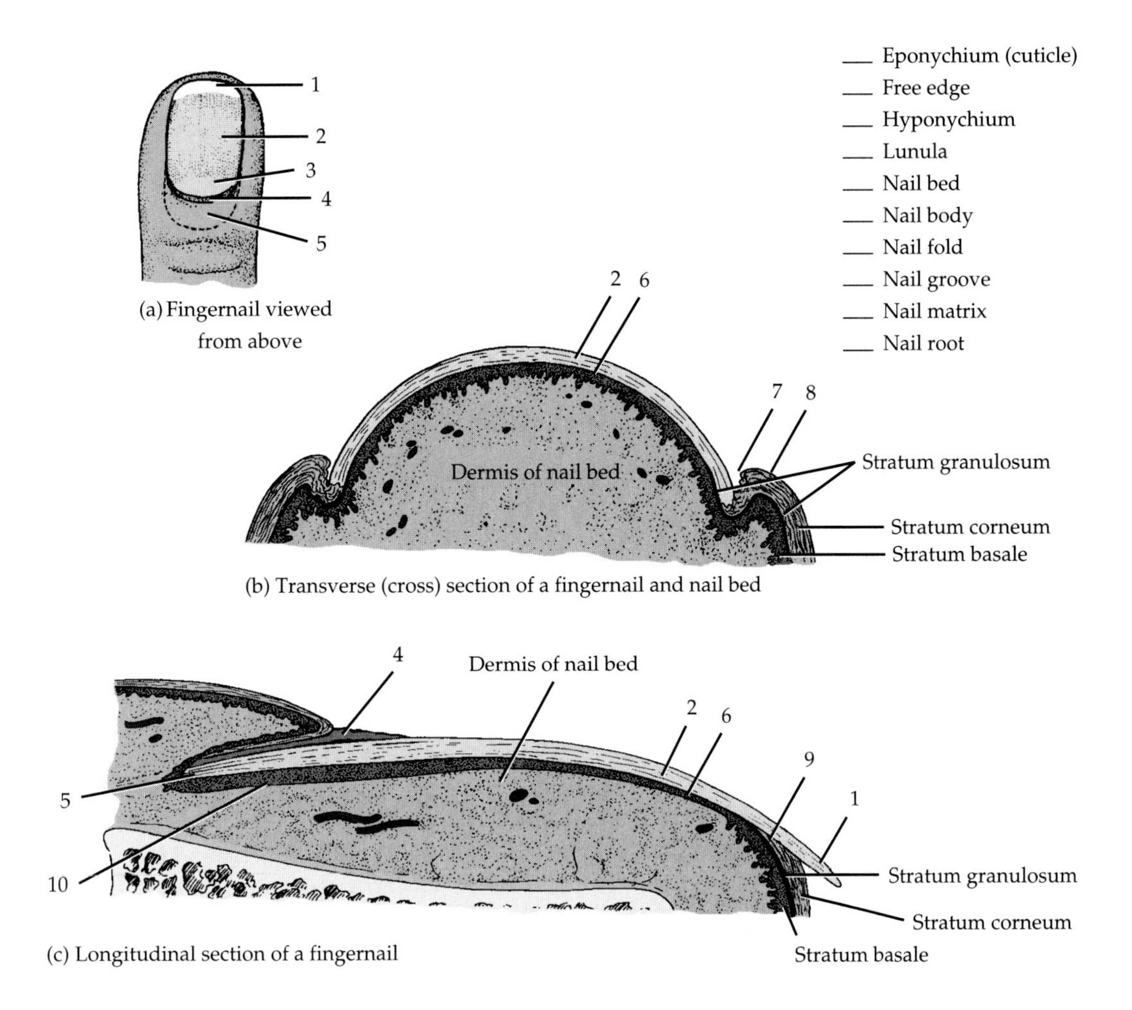

FIGURE 5.5 Structure of nails.

LABORATORY REPORT QUESTIONS

Integumentary System 5

Student ______________________ **Date** ______________________

Laboratory Section ______________________ **Score/Grade** ______________________

PART 1. Multiple Choice

_______ **1.** The waterproofing quality of skin is due to the presence of (a) melanin (b) carotene (c) lamellar granules (d) receptors

_______ **2.** Sebaceous glands (a) produce a watery solution called sweat (b) produce an oily substance that prevents excessive water evaporation from the skin (c) are associated with mucous membranes (d) are part of the subcutaneous layer

_______ **3.** Which of the following is the proper sequence of layering of the epidermis, going from the free surface toward the underlying tissues? (a) basale, spinosum, granulosum, corneum (b) spinosum, basale, granulosum, corneum (c) corneum, lucidum, granulosum, spinosum, basale (d) corneum, granulosum, lucidum, spinosum

_______ **4.** Skin color is *not* determined by the presence or absence of (a) melanin (b) carotene (c) keratin (d) hemoglobin in red blood cells in capillaries in the dermis

_______ **5.** Destruction of what part of a single hair would result in its inability to grow? (a) sebaceous gland (b) arrector pili muscle (c) matrix (d) bulb

_______ **6.** One would expect to find relatively few, if any, sebaceous glands in the skin of the (a) palms (b) face (c) neck (d) upper chest

_______ **7.** Which of the following sequences, from outside to inside, is correct? (a) epidermis, reticular layer, papillary layer, subcutaneous layer (b) epidermis, subcutaneous layer, reticular layer, papillary layer (c) epidermis, reticular layer, subcutaneous layer, papillary layer (d) epidermis, papillary layer, reticular layer, subcutaneous layer

_______ **8.** The attached visible portion of a nail is called the (a) nail bed (b) nail root (c) nail fold (d) nail body

_______ **9.** Nerve endings sensitive to touch are called (a) corpuscles of touch (Meissner's corpuscles) (b) papillae (c) lamellated (Pacinian) corpuscles (d) follicles

_______ **10.** The cuticle of a nail is referred to as the (a) matrix (b) eponychium (c) hyponychium (d) fold

_______ **11.** One would *not* expect to find sudoriferous glands associated with the (a) forehead (b) axilla (c) palms (d) nail beds

_______ **12.** Fingerlike projections of the dermis that contain loops of capillaries and receptors are called (a) dermal papillae (b) nodules (c) polyps (d) pili

_______ **13.** Which of the following statements about the function of skin is *not* true? (a) it helps control body temperature (b) it prevents excessive water loss (c) it synthesizes several compounds (d) it absorbs water and salts

_______ **14.** Which is *not* part of the internal root sheath? (a) granular (Huxley's) layer (b) cortex (c) pallid (Henle's) layer (d) cuticle of the internal root sheath

_______ **15.** Growth in the length of nails is the result of the activity of the (a) eponychium (b) nail matrix (c) hyponychium (d) nail fold

PART 2. Completion

16. A group of tissues that performs a definite function is called a(n) ____________________.

17. The outer, thinner layer of the skin is known as the ____________________.

18. The skin is attached to underlying tissues and organs by the ____________________.

19. A group of organs that operate together to perform a specialized function is called a(n) ____________________.

20. The epidermal layer that is more apparent in the palms and soles is the stratum ____________________.

21. The epidermal layers that produce new cells are the stratum spinosum and stratum ____________________.

22. The smooth muscle attached to a hair follicle is called the ____________________ muscle.

23. An inherited inability to produce melanin is called ____________________.

24. Nerve endings sensitive to deep pressure are referred to as ____________________ corpuscles.

25. The inner region of a hair shaft and root is the ____________________.

26. The portion of a hair containing areolar connective tissue and blood vessels is the ____________________.

27. Modified sweat glands that line the external auditory meatus are called ____________________ glands.

28. The whitish semilunar area at the proximal end of the nail body is referred to as the ____________________.

29. The secretory product of sudoriferous glands is called ____________________.

30. Melanin is synthesized in cells called ____________________.

Bone Tissue

6

Structurally, the ***skeletal system*** consists of two types of connective tissue: cartilage and bone. The microscopic structure of cartilage has been discussed in Exercise 4. In this exercise the gross structure of a typical bone and the histology of ***bone (osseous) tissue*** will be studied. ***Osteology*** (os-tē-OL-ō-jē; *osteo* = bone; *logos* = study of) is the study of bone structure and the treatment of bone disorders.

A. FUNCTIONS OF BONE

The skeletal system has the following basic functions:

1. ***Support*** It provides a supporting framework for the soft tissues, maintaining the body's shape and posture and provides points of attachment for many skeletal muscles.
2. ***Protection*** It protects delicate structures such as the brain, spinal cord, heart, lungs, major blood vessels in the chest, and pelvic viscera.
3. ***Assistance in movement*** When skeletal muscles contract, they pull on bones to help produce body movements.
4. ***Mineral homeostasis*** Bone tissue stores several minerals, especially calcium and phosphorus, which are important in muscle contraction and nerve activity, among other functions. On demand, bone releases minerals into the blood to maintain critical mineral balances and for distribution to other parts of the body.
5. ***Blood cell production*** or ***hemopoiesis*** (hē'-mō-poy-Ē-sis) ***Red bone marrow*** in certain parts of bones consists of primitive blood cells in immature stages, fat cells, and macrophages. It is responsible for producing red blood cells, white blood cells, and platelets.
6. ***Storage of energy*** Lipids stored in cells of yellow bone marrow are an important source of a chemical energy reserve. ***Yellow bone marrow*** consists of mostly adipose cells and a few scattered blood cells.

B. CHEMISTRY OF BONE

Unlike other connective tissues, the matrix of bone is very hard. This hardness results from the presence of mineral salts, mainly ***tricalcium phosphate*** ($Ca_3(PO_4)_2 \cdot [OH]_2$), called ***hydroxyapatite,*** and some calcium carbonate ($CaCO_3$). Mineral salts compose about 50% of the weight of bone. Despite its hardness, bone is also flexible, a characteristic that enables it to resist various forces. The flexibility of bone comes from organic substances in its matrix, especially collagen fibers. Organic materials compose about 25% of the weight of bone. The remaining 25% of the bone matrix is water.

PROCEDURE

1. Obtain a bone that has been baked. How does this compare to an untreated one? ______________

   ____

   ____

   ____

   What substances does baking remove from the bone (inorganic or organic)? ______________

   ____

   ____

   ____

2. Now obtain a bone that has already been soaked in nitric acid by your instructor. How does this bone compare to an untreated one?

   ____

   ____

   ____

   ____

   What substances does nitric acid treatment remove from the bone (inorganic or organic)?

   ____

   ____

   ____

   ____

C. GROSS STRUCTURE OF A LONG BONE

Examine the external features of a fresh long bone and locate the following structures:

1. ***Diaphysis*** (dī-AF-i-sis; *dia* = through) Elongated shaft of a bone between the epiphyses.
2. ***Epiphysis*** (e-PIF-i-sis; *epi* = above; *physis* = growth) End or extremity of a bone; the epiphyses are referred to as proximal and distal.
3. ***Metaphysis*** (me-TAF-i-sis; *meta* = after or beyond) In mature bone, the region where the diaphysis joins the epiphysis; in growing bone, the region that includes the epiphyseal plate where calcified cartilage is replaced by bone as the bone lengthens.
4. ***Articular cartilage*** Thin layer of hyaline cartilage covering the ends of the bone where joints are formed. It reduces friction and absorbs shock at freely movable joints.
5. ***Periosteum*** (per'-ē-OS-tē-um; *peri* = around; *osteo* = bone) Connective tissue membrane covering the surface of the bone, except for areas covered by articular cartilage. The outer ***fibrous layer*** is composed of dense, irregular connective tissue and contains blood vessels, lymphatic vessels, nerves that pass into the bone; the inner ***osteogenic*** (os'-te-ō-JEN-ik) ***layer*** contains elastic fibers, blood vessels, ***osteoprogenitor*** (os'-tē-ō-prō-JEN-i-tor, *pro* = precursor, *gen* = to produce) ***cells*** (stem cells derived from mesenchyme that differentiate into osteoblasts), and ***osteoblasts*** (OS-tē-ō-blasts; *blast* = germ or bud), cells that secrete the organic components and mineral salts involved in bone formation. The periosteum functions in bone growth, nutrition, and repair, and as an attachment site for tendons and ligaments.
6. ***Medullary*** (MED-yoo-lar-ē; *medulla* = central part) or ***marrow cavity*** Cavity within the diaphysis that contains yellow bone marrow in adults.
7. ***Endosteum*** (end-OS-tē-um; *endo* = within) Membrane that lines the medullary cavity and contains osteoprogenitor cells and ***osteoclasts*** (OS-tē-ō-clasts; *clast* = to break), cells that remove bone by destroying the matrix, a process called ***resorption***.

Label the parts of a long bone indicated in Figure 6.1.

D. HISTOLOGY OF BONE

Bone tissue, like other connective tissues, contains a large amount of matrix that surrounds widely separated cells. As noted earlier, this matrix consists of abundant mineral salts, collagen fibers, and water. The cells include osteoprogenitor cells, osteoblasts, osteoclasts, and ***osteocytes*** (OS-tē-ō-sīts; *cyte* = cell), mature bone cells that maintain daily activities of bone tissue.

Depending on the size and distribution of spaces between its hard components, bone tissue may be categorized as spongy (cancellous) or compact (dense). ***Spongy (cancellous) bone tissue,*** which contains many large spaces, composes most of the bone tissue of short, flat, and irregularly shaped bones and most of the epiphyses of long bones. The spaces within the spongy bone tissue of some bones contain red bone marrow. ***Compact (dense) bone tissue,*** which contains few spaces, forms the external layer of all bones of the body and the bulk of the diaphyses of long bones.

Label the spongy and compact bone tissue in Figure 6.1.

Spongy bone tissue is composed of concentric layers of hardened matrix, called ***lamellae*** (la-MEL-ē), that are arranged in an irregular latticework of thin columns of bone called ***trabeculae*** (tra-BEK-yoo-lē). See Figure 6.2a.

Compact bone tissue is composed of microscopic units called ***osteons*** (***Haversian systems***).

Obtain a prepared slide of compact bone tissue in which several osteons (Haversian systems) are shown in transverse (cross) section. Observe under high power. Look for the following structures:

1. ***Central (Haversian) canal*** Circular canal in the center of an osteon (Haversian system) that runs longitudinally through the bone; the canal contains blood vessels, lymphatic vessels, and nerves.
2. ***Concentric lamellae*** Concentric layers of calcified matrix.
3. ***Lacunae*** (la-KOO-nē; *lacuna* = little lake) Spaces or cavities between lamellae that contain osteocytes.
4. ***Canaliculi*** (kan'-a-LIK-yoo-lē; *canaliculi* = small canal) Minute canals that radiate in all directions from the lacunae and interconnect with each other; contain slender processes of osteocytes; canaliculi provide routes so that

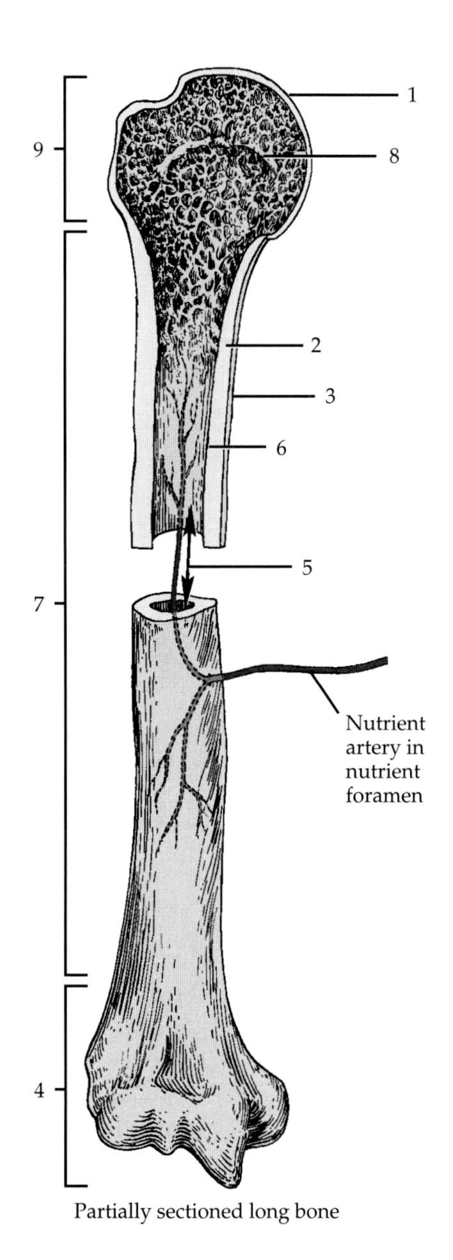

___ Articular cartilage
___ Compact bone tissue
___ Diaphysis
___ Distal epiphysis
___ Endosteum
___ Medullary (marrow) cavity
___ Periosteum
___ Proximal epiphysis
___ Spongy bone tissue

FIGURE 6.1 Parts of a long bone.

nutrients can reach osteocytes and wastes can be removed from them.

5. ***Osteocyte*** Mature bone cell located within a lacuna.
6. ***Osteon (Haversian system)*** Microscopic structural unit of compact bone made up of central (Haversian) canal plus its surrounding lamellae, lacunae, canaliculi, and osteocytes.

Label the parts of the osteon (Haversian system) shown in Figure 6.2.

Now obtain a prepared slide of a longitudinal section of compact bone tissue and examine under high power. Locate the following:

1. ***Perforating (Volkmann's) canals*** Canals that extend obliquely or horizontally inward from the periosteum and contain blood vessels, lymphatic vessels, and nerves; they extend into central (Haversian) canals and medullary cavity.
2. ***Endosteum***
3. ***Medullary (marrow) cavity***
4. ***Concentric lamellae***
5. ***Lacunae***
6. ***Canaliculi***
7. ***Osteocytes***

Label the parts indicated in the microscopic view of bone in Figure 6.2.

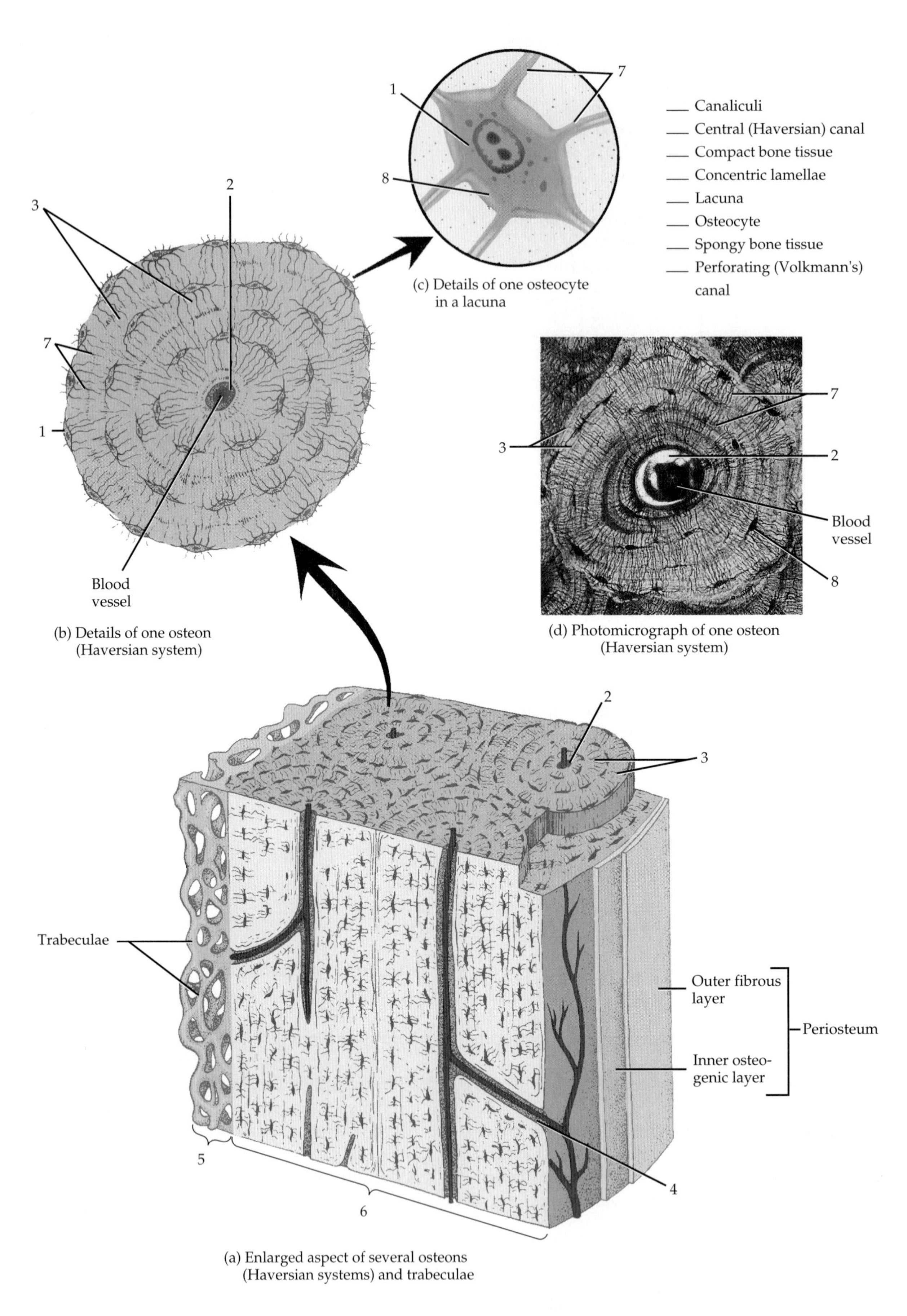

FIGURE 6.2 Histology of bone.

E. BONE FORMATION: OSSIFICATION

The process by which bone forms is called ***ossification*** (os'-i-fi-KĀ-shun; *facere* = to make). The "skeleton" of a human embryo is composed of fibrous connective tissue membranes formed by embryonic connective tissue (mesenchyme) or hyaline cartilage that are loosely shaped like bones. They provide the supporting structures for ossification. Ossification begins around the sixth or seventh week of embryonic life and continues throughout adulthood. Bone formation follows one of two patterns.

1. ***Intramembranous*** (in'-tra-MEM-bra-nus; *intra* = within; *membranous* = membrane) ***ossification*** refers to the formation of bone directly on or within loose fibrous connective tissue membranes. Such bones form *directly* from mesenchyme without first going through a cartilage stage. The flat bones of the skull and mandible (lower jawbone) form by this process. As you will see later, the fontanels ("soft spots") of an infant's skull, which are composed of loose fibrous connective tissue membranes, are also eventually replaced by bone through intramembranous ossification.
2. ***Endochondral*** (en'-dō-KON-dral; *endo* = within; *chondro* = cartilage) ***ossification*** refers to the formation of bone in hyaline cartilage. In this process, mesenchyme is transformed into chondroblasts which produce a hyaline cartilage matrix that is gradually replaced by bone. Most bones of the body form by this process.

These two kinds of ossification do *not* lead to differences in the gross structure of mature bones. They are simply different methods of bone formation. Both mechanisms involve the replacement of a preexisting connective tissue with bone.

The first stage in the development of bone is the migration of embryonic mesenchymal cells into the area where bone formation is about to begin. These cells increase in number and size and become osteoprogenitor cells. In some skeletal structures where capillaries are lacking, they become chondroblasts; in others where capillaries are present, they become osteoblasts. The **chondroblasts** are responsible for cartilage formation. Osteoblasts form bone tissue by intramembranous or endochondral ossification.

F. BONE GROWTH

During childhood, bones throughout the body grow in diameter by appositional growth (deposition of matrix on the surface) and long bones lengthen by interstitial growth (the addition of bone material at the epiphyseal plate). Growth in length of bones normally ceases by age 25, although bones may continue to thicken.

1. Growth in Length

To understand how a bone grows in length, you will need to know some of the details of the structure of the epiphyseal plate.

The ***epiphyseal*** (ep'-i-FIZ-ē-al; *epiphyein* = to grow upon) ***plate*** is a layer of hyaline cartilage in a growing bone that consists of four zones. The ***zone of resting cartilage*** is near the epiphysis and consists of small, scattered chondrocytes. The cells do not function in bone growth (thus the term "resting"); they anchor the epiphyseal plate to the bone of the epiphysis.

The ***zone of proliferating cartilage*** consists of slightly larger chondrocytes arranged like stacks of coins. Chondrocytes divide to replace those that die at the diaphyseal surface of the epiphyseal plate.

The zone of ***hypertrophic*** (hī-per-TRŌF-ik) or ***maturing cartilage*** consists of even larger chondrocytes that are also arranged in columns. The lengthwise expansion of the epiphyseal plate is the result of cell divisions in the zone of proliferating cartilage and maturation of the cells in the zone of hypertrophic cartilage.

The ***zone of calcified cartilage*** is only a few cells thick and consists mostly of dead cells because the matrix around them has calcified. The calcified matrix is taken up by osteoclasts, and the area is invaded by osteoblasts and capillaries from the bone in the diaphysis. These cells lay down bone on the calcified cartilage that persists. As a result, the diaphyseal border of the epiphyseal plate is firmly cemented to the bone of the diaphysis.

Label the various zones in the epiphyseal plate in Figure 6.3.

The activity of the epiphyseal plate is the only mechanism by which the diaphysis can increase in length. Unlike cartilage, which can grow by both interstitial and appositional growth, bone can grow in diameter only by appositional growth. Eventually, the epiphyseal cartilage cells stop dividing and bone

Epiphyseal side

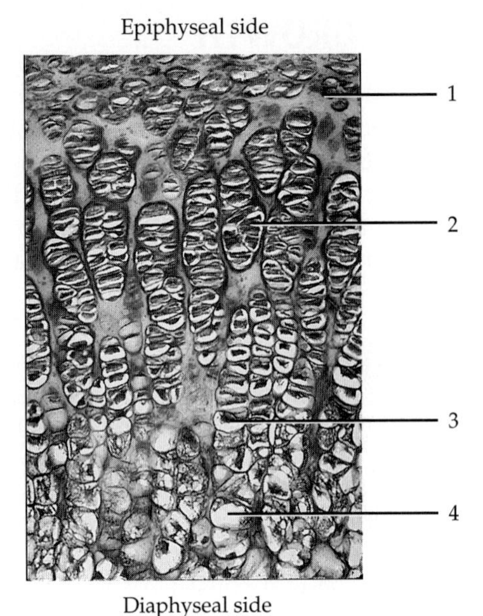

Diaphyseal side

Photomicrograph of epiphyseal plate

___ Zone of calcified cartilage
___ Zone of hypertrophic cartilage
___ Zone of proliferating cartilage
___ Zone of resting cartilage

FIGURE 6.3 Histology of epiphyseal plate.

replaces the cartilage. The newly formed bony structure is called the ***epiphyseal line***, a remnant of the once active epiphyseal plate. With the appearance of the epiphyseal line, bone stops growing in length. In general, lengthwise growth in bones in females is completed before that in males.

2. Growth in Thickness

Enlargement of bone thickness or diameter is by appositional growth and occurs as follows. First, the bone lining the medullary cavity is destroyed by osteoclasts in the endosteum so that the cavity increases in diameter. At the same time, osteoblasts from the periosteum add new bone tissue to the outer surface. Initially, diaphyseal and epiphyseal ossification produce only spongy bone. Later, the outer region of spongy bone is reorganized into compact bone.

G. FRACTURES

A ***fracture*** is any break in a bone. Usually, the fractured ends of a bone can be reduced (aligned to their normal positions) by manipulation without surgery. This procedure of setting a fracture is called ***closed reduction.*** In other cases, the fracture must be exposed by surgery before the break is rejoined. This procedure is known as ***open reduction.***

Using your textbook as a reference, define the fractures listed in Table 6.1 and label the fractures indicated in Figure 6.4 on page 76.

H. TYPES OF BONES

The 206 named bones of the body may be classified by shape into four principal types:

1. ***Long*** Have greater length than width, consist of a diaphysis and a variable number of epiphyses, and have a medullary cavity; contain more compact than spongy bone tissue. Example: humerus.
2. ***Short*** Somewhat cube-shaped, and contain more spongy than compact bone tissue. Example: wrist bones.
3. ***Flat*** Generally thin and flat and composed of two more-or-less parallel plates of compact bone tissue enclosing a layer of spongy bone tissue. Example: sternum.
4. ***Irregular*** Very complex shapes; cannot be grouped into any of the three categories just described. Example: vertebrae.

Other bones that are recognized, although not considered in the structural classification, include the following:

1. ***Sutural*** (SOO-chur-al) Small bones between certain cranial bones; variable in number.
2. ***Sesamoid*** *Sesamoid* means "resembling a grain of sesame." Small bones found in tendons; variable in number; the only constant sesamoid bones are the paired kneecaps.

Examine the disarticulated skeleton, Beauchene (disarticulated) skull, and articulated skeleton and find several examples of long, short, flat, and irregular bones. List examples of each type you find.

1. *Long* ______________________________

____

____

____

2. *Short* ______________________________

____

____

____

Table 6.1
Summary of selected fractures

Type of fracture	Definition
Partial	
Complete	
Closed (simple)	
Open (compound)	
Comminuted (KOM-i-nyoo'-ted)	
Greenstick	
Spiral	
Transverse	
Impacted	
Displaced	
Nondisplaced	
Stress	
Pathologic	
Pott's	
Colles' (KOL-ez)	

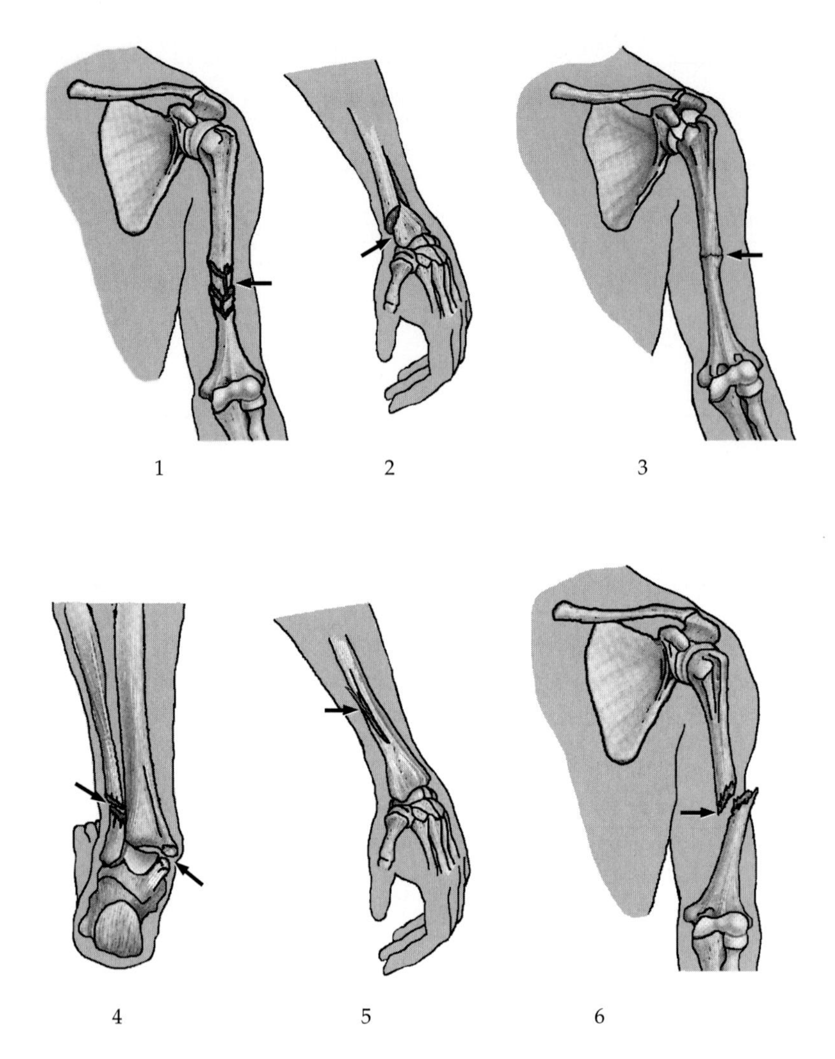

___ Colles' fracture
___ Comminuted fracture
___ Greenstick fracture
___ Impacted fracture
___ Open fracture
___ Pott's fracture

FIGURE 6.4 Types of fractures.

3. *Flat* ______________________________

____

____

____

4. *Irregular* ______________________________

____

____

____

I. BONE SURFACE MARKINGS

The surfaces of bones contain various structural features that have specific functions. These features are called ***bone surface markings*** and are listed in Table 6.2. Knowledge of the bone surface markings will be very useful when you learn the bones of the body in Exercise 7.

Next to each marking listed in Table 6.2, write its definition, using your textbook as a reference.

ANSWER THE LABORATORY REPORT QUESTIONS AT THE END OF THE EXERCISE.

Table 6.2
Bone surface markings

Marking	Description
DEPRESSIONS AND OPENINGS	
Fissure (FISH-ur)	
Foramen (fō-RĀ-men; *foramen* = hole)	
Meatus (mē-Ā-tus; *meatus* = canal)	
Paranasal sinus (*sin* = cavity)	
Sulcus (*sulcus* = ditchlike groove)	
Fossa (*fossa* = basinlike depression)	
PROCESSES (PROJECTIONS) THAT FORM JOINTS	
Condyle (KON-dīl; *condulus* = knucklelike process)	
Head	
Facet	
PROCESSES (PROJECTIONS) TO WHICH TENDONS, LIGAMENTS, AND OTHER CONNECTIVE TISSUES ATTACH	
Tubercle (TOO-ber-kul; *tube* = knob)	
Tuberosity	
Trochanter (trō-KAN-ter)	
Crest	
Line	
Spinous process (spine)	
Epicondyle (*epi* = above)	

LABORATORY REPORT QUESTIONS

Bone Tissue 6

Student ______________________ **Date** ______________________

Laboratory Section ______________________ **Score/Grade** ______________________

PART 1. Completion

1. Small clusters of bones between certain cranial bones are referred to as ______________ bones.
2. The technical name for a mature bone cell is a(n) ______________.
3. Canals that extend obliquely inward or horizontally from the bone surface and contain blood vessels and lymphatic vessels are ______________ canals.
4. The end, or extremity, of a bone is referred to as the ______________.
5. Cube-shaped bones that contain more spongy bone tissue than compact bone tissue are known as ______________ bones.
6. The cavity within the shaft of a bone that contains yellow bone marrow in the adult is the ______________ cavity.
7. The thin layer of hyaline cartilage covering the end of a bone where joints are formed is called ______________ cartilage.
8. Minute canals that connect lacunae are called ______________.
9. The covering around the surface of a bone, except for the areas covered by cartilage, is the ______________.
10. The shaft of a bone is referred to as the ______________.
11. The ______________ are concentric layers of calcified matrix.
12. The membrane that lines the medullary cavity and contains osteoblasts and a few osteoclasts is the ______________.
13. The technical name for bone tissue is ______________ tissue.
14. In a mature bone, the region where the shaft joins the extremity is called the ______________.
15. The microscopic structural unit of compact bone tissue is called a(n) ______________.
16. The hardness of bone is primarily due to the mineral salt ______________.
17. The process by which bone is formed is called ______________.

18. The zone of ______________________ is closest to the diaphysis of the bone.

19. Growth in diameter of bones occurs by ______________________ growth.

20. The destruction of matrix by osteoclasts is called ______________________.

21. Most bones of the body form by which type of ossification? ______________________.

22. The lengthwise expansion of the epiphyseal plate is partly the result of cell divisions in the zone of ______________________.

23. Any break in a bone is called a ______________________.

24. The term ______________________ refers to the irregular latticework of thin columns of spongy bone.

25. On the basis of shape, vertebrae are classified as ______________________ bones.

Bones

7

The 206 named bones of the adult skeleton are grouped into two divisions: axial and appendicular. The ***axial skeleton*** consists of bones that compose the axis of the body. The longitudinal axis is an imaginary straight line that runs through the center of gravity of the body, through the head, and down to the space between the feet. The ***appendicular skeleton*** consists of the bones of the limbs or extremities (upper and lower) and the girdles (pectoral and pelvic), which connect the limbs to the axial skeleton.

In this exercise you will study the names and locations of bones and their markings by examining various regions of the skeleton:

Region	Number of bones
Axial skeleton	
Skull	
Cranium	8
Face	14
Hyoid (above the larynx)	1
Auditory ossicles, 3 in each ear	6
Vertebral column	26
Thorax	
Sternum	1
Ribs	24
	80
Appendicular skeleton	
Pectoral (shoulder) girdles	
Clavicle	2
Scapula	2
Upper limbs (extremities)	
Humerus	2
Ulna	2
Radius	2
Carpals	16
Metacarpals	10
Phalanges	28
Pelvic (hip) girdle	
Hipbone (pelvic or coxal bone)	2
Appendicular skeleton (Continued)	
Lower limbs (extremities)	
Femur	2
Fibula	2
Tibia	2
Patella	2
Tarsals	14
Metatarsals	10
Phalanges	28
	126

A. BONES OF ADULT SKULL

The ***skull*** is composed of two sets of bones—cranial and facial. The 8 ***cranial*** (*cranium* = brain case) ***bones*** form the cranial cavity and enclose and protect the brain. They are 1 ***frontal,*** 2 ***parietals*** (pa-RĪ-i-tals), 2 ***temporals,*** 1 ***occipital*** (ok-SIP-i-tal), 1 ***sphenoid*** (SFĒ-noyd), and 1 ***ethmoid.*** The 14 ***facial bones*** form the face and include 2 ***nasals,*** 2 ***maxillae*** (mak-SIL-ē), 2 ***zygomatics,*** 1 ***mandible,*** 2 ***lacrimals*** (LAK-ri-mals), 2 ***palatines*** (PAL-a-tīns), 2 ***inferior conchae*** (KONG-kē), or ***turbinates,*** and 1 ***vomer.*** These bones are indicated by *arrows* in Figure 7.1. Using the illustrations in Figure 7.1 for reference, locate the cranial and facial bones on both a Beauchene (disarticulated) and an articulated skull.

Obtain a Beauchene skull and an articulated skull and observe them in anterior view. Using Figure 7.1a for reference, locate the parts indicated in the figure.

Turn the Beauchene skull and articulated skull so that you are looking at the right side. Using Figure 7.1b on page 83 for reference, locate the parts indicated in the figure. Note also the ***hyoid bone*** below the mandible. This is not a bone of the skull; it is noted here because of its proximity to the skull.

If a skull in median section is available, use Figure 7.1c on page 84 for reference and locate the parts indicated in the figure.

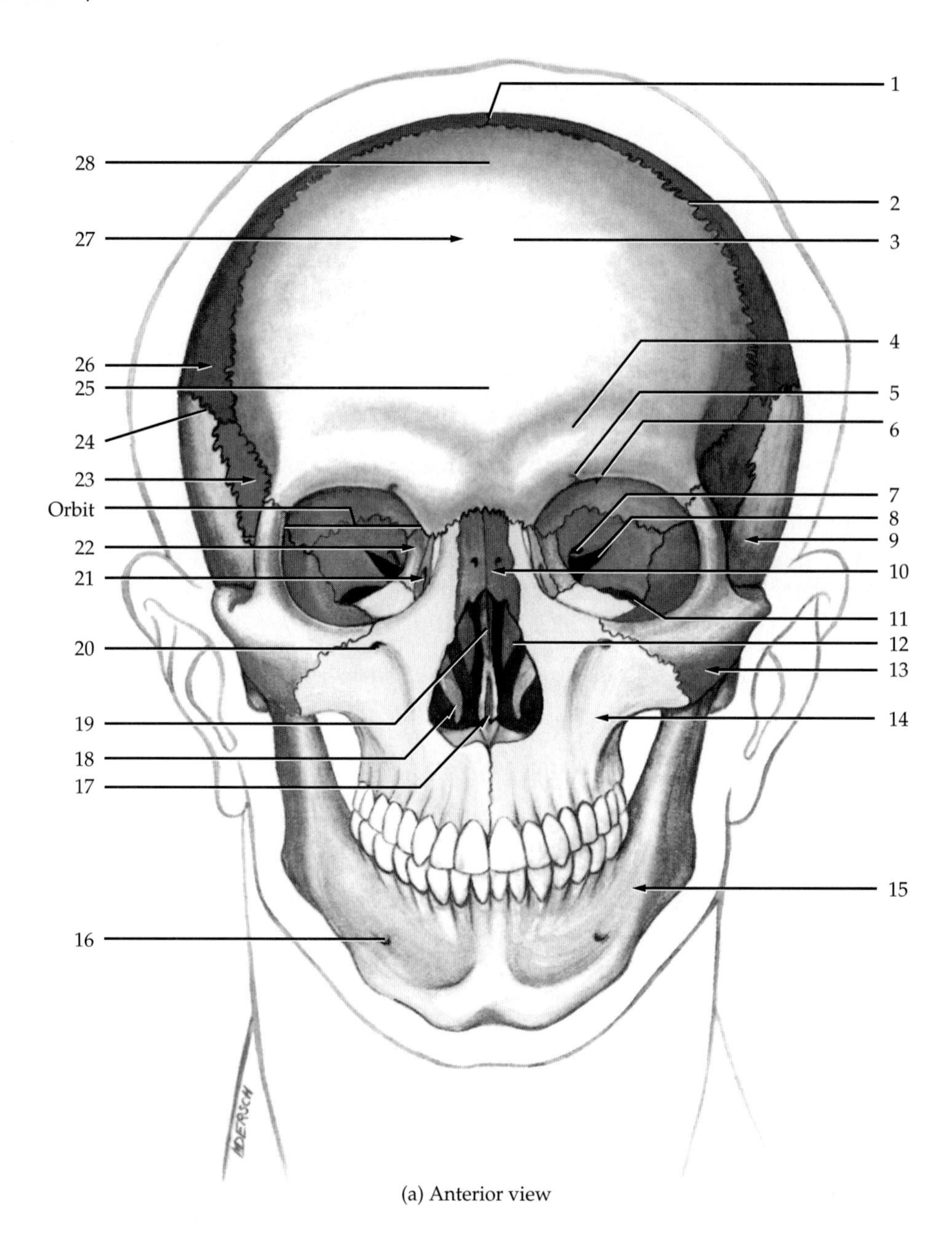

(a) Anterior view

1. Sagittal suture
2. Coronal suture
3. Frontal squama
4. Superciliary arch
5. Supraorbital foramen
6. Supraorbital margin
7. Optic foramen
8. Superior orbital fissure
9. Temporal bone
10. Nasal bone
11. Inferior orbital fissure
12. Middle nasal concha (turbinate)
13. Zygomatic bone
14. Maxilla
15. Mandible
16. Mental foramen
17. Vomer
18. Inferior nasal concha (turbinate)
19. Perpendicular plate
20. Infraorbital foramen
21. Lacrimal bone
22. Ethmoid bone
23. Sphenoid bone
24. Squamous suture
25. Glabella
26. Parietal bone
27. Frontal bone
28. Frontal eminence

FIGURE 7.1 Skull.

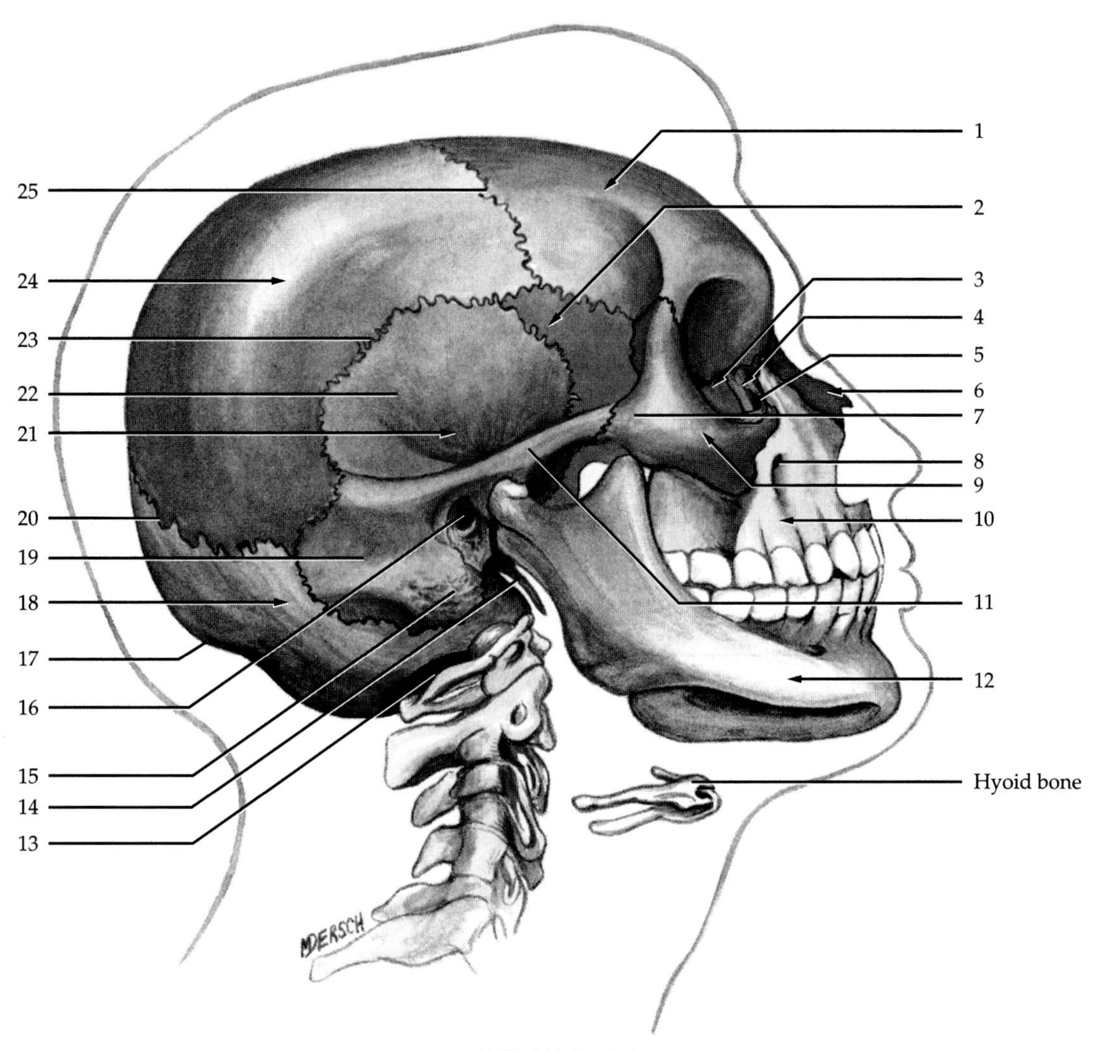

(b) Right lateral view

1. Frontal bone
2. Sphenoid bone
3. Ethmoid bone
4. Lacrimal bone
5. Lacrimal foramen
6. Nasal bone
7. Temporal process
8. Infraorbital foramen
9. Zygomatic bone
10. Maxilla
11. Zygomatic process
12. Mandible
13. Foramen magnum
14. Styloid process
15. Mastoid process
16. External auditory (acoustic) meatus
17. External occipital protuberance
18. Occipital bone
19. Mastoid portion
20. Lambdoid suture
21. Temporal bone
22. Temporal squama
23. Squamous suture
24. Parietal bone
25. Coronal suture

FIGURE 7.1 ***(Continued)*** Skull.

Take an articulated skull and turn it upside down so that you are looking at the inferior surface. Using Figure 7.1d on page 85 for reference, locate the parts indicated in the figure.

Obtain an articulated skull with a removable crown. Using Figure 7.1e on page 86 for reference, locate the parts indicated in the figure.

Examine the right orbit of an articulated skull. Using Figure 7.2 on page 87 for reference, locate the parts indicated in the figure.

Obtain a mandible and, using Figure 7.3 on page 87 for reference, identify the parts indicated in the figure.

Before you move on, refer to Table 7.1, "Summary of Foramina of the Skull" on page 88. Complete the

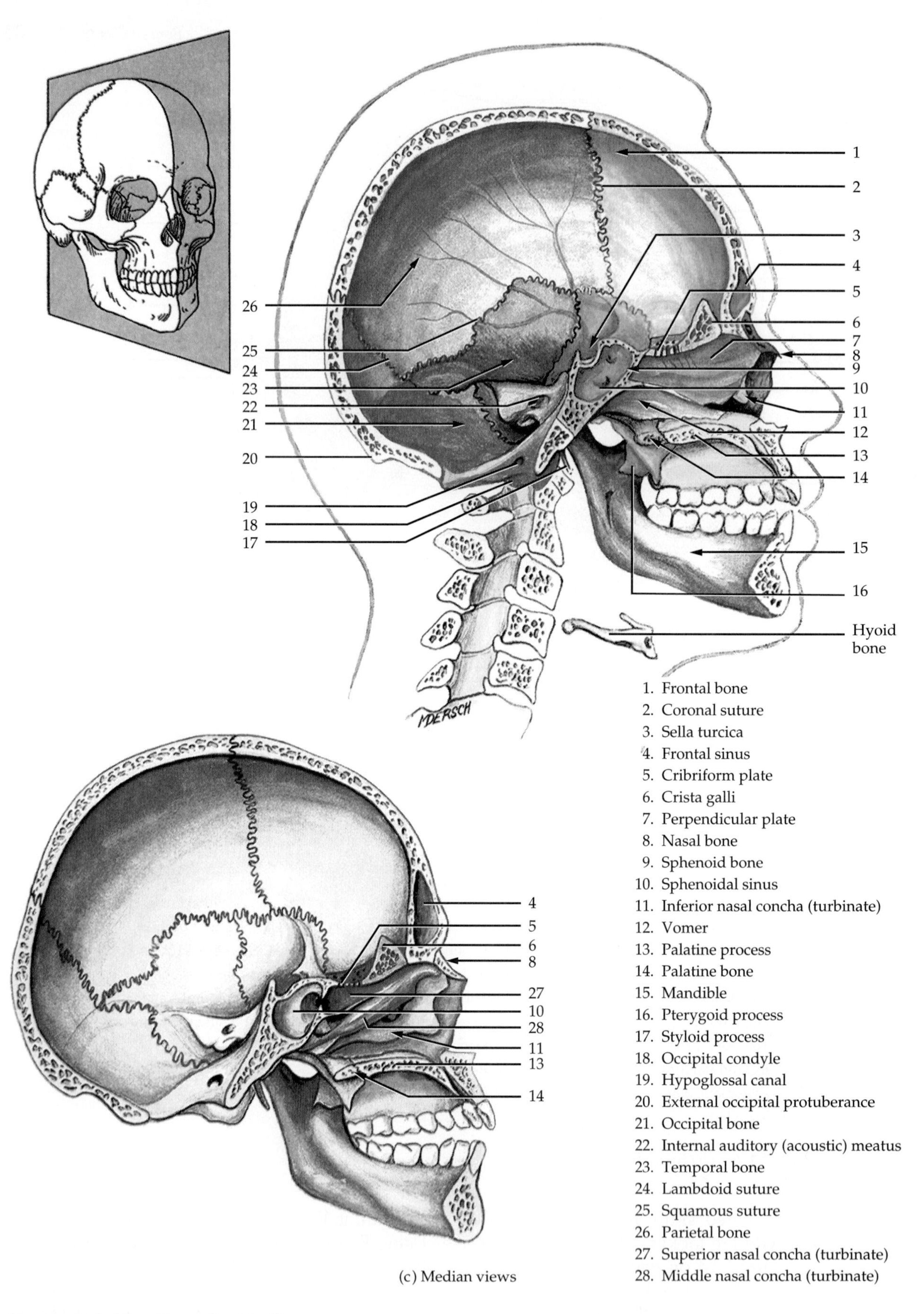

FIGURE 7.1 *(Continued)* Skull.

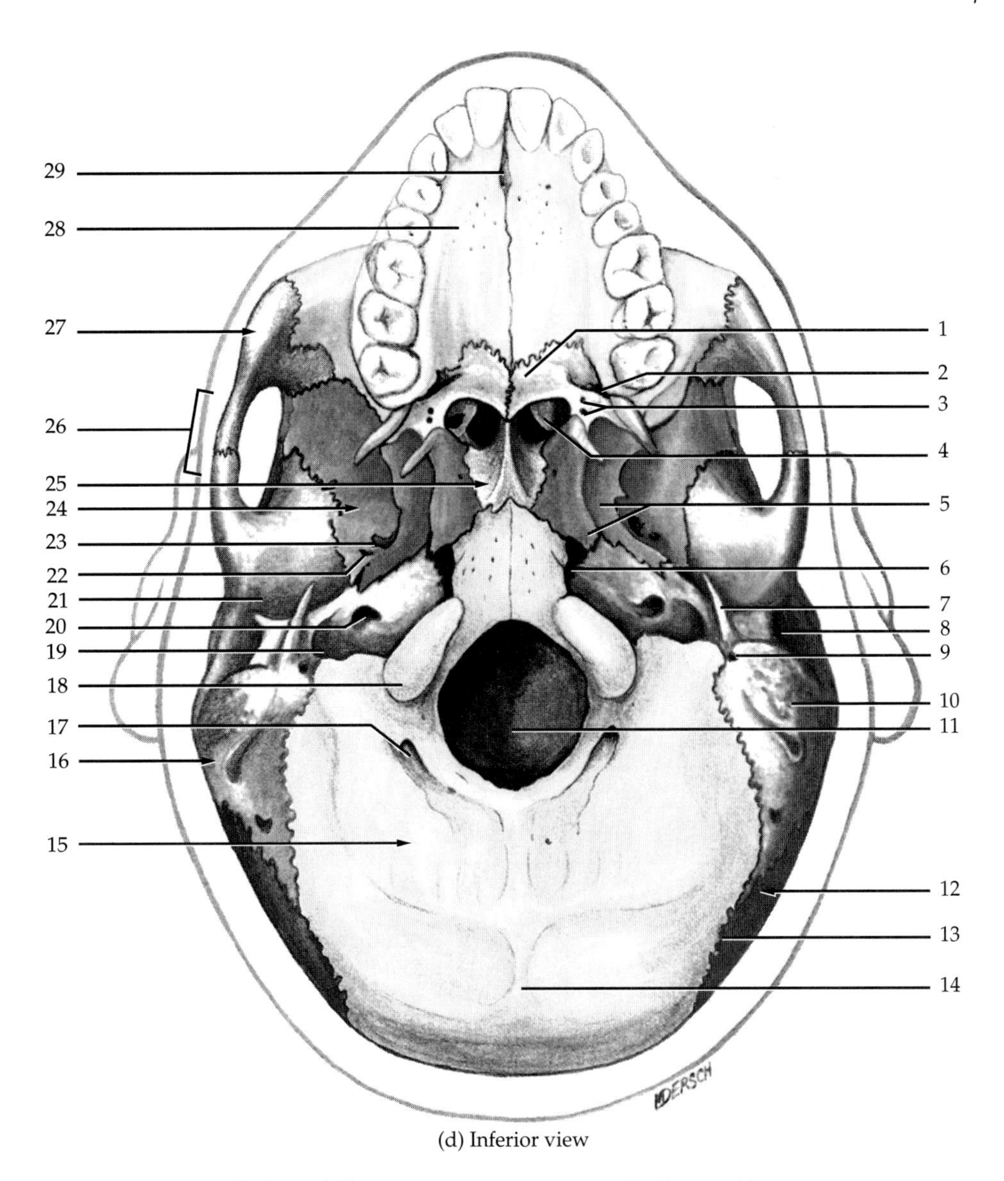

(d) Inferior view

1. Horizontal plate
2. Greater palatine foramen
3. Lesser palatine foramina
4. Middle nasal concha (turbinate)
5. Pterygoid process
6. Foramen lacerum
7. Styloid process
8. External auditory (acoustic) meatus
9. Stylomastoid foramen
10. Mastoid process
11. Foramen magnum
12. Parietal bone
13. Lambdoid suture
14. External occipital protuberance
15. Occipital bone
16. Temporal bone
17. Condylar canal
18. Occipital condyle
19. Jugular foramen
20. Carotid foramen
21. Mandibular fossa
22. Foramen spinosum
23. Foramen ovale
24. Sphenoid bone
25. Vomer
26. Zygomatic arch
27. Zygomatic bone
28. Palatine process
29. Incisive foramen

FIGURE 7.1 ***(Continued)*** Skull.

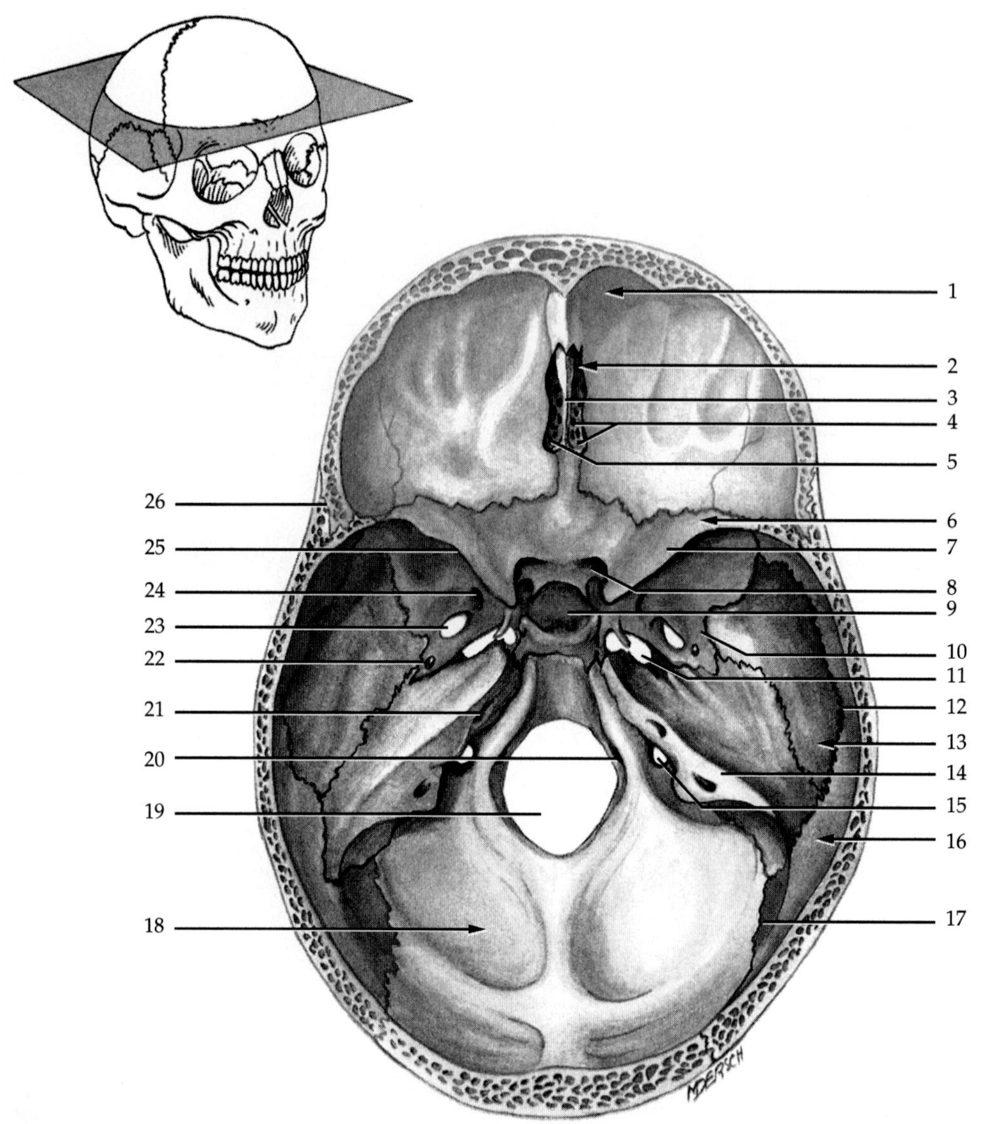

(e) Superior view of floor of cranium

1. Frontal bone
2. Ethmoid bone
3. Crista galli
4. Olfactory foramina
5. Cribriform plate
6. Sphenoid bone
7. Lesser wing
8. Optic foramen
9. Sella turcica
10. Greater wing
11. Foramen lacerum
12. Squamous suture
13. Temporal bone
14. Petrous portion
15. Jugular foramen
16. Parietal bone
17. Lambdoid suture
18. Occipital bone
19. Foramen magnum
20. Hypoglossal canal
21. Internal auditory (acoustic) meatus
22. Foramen spinosum
23. Foramen ovale
24. Foramen rotundum
25. Superior orbital fissure
26. Coronal suture

FIGURE 7.1 *(Continued)* Skull.

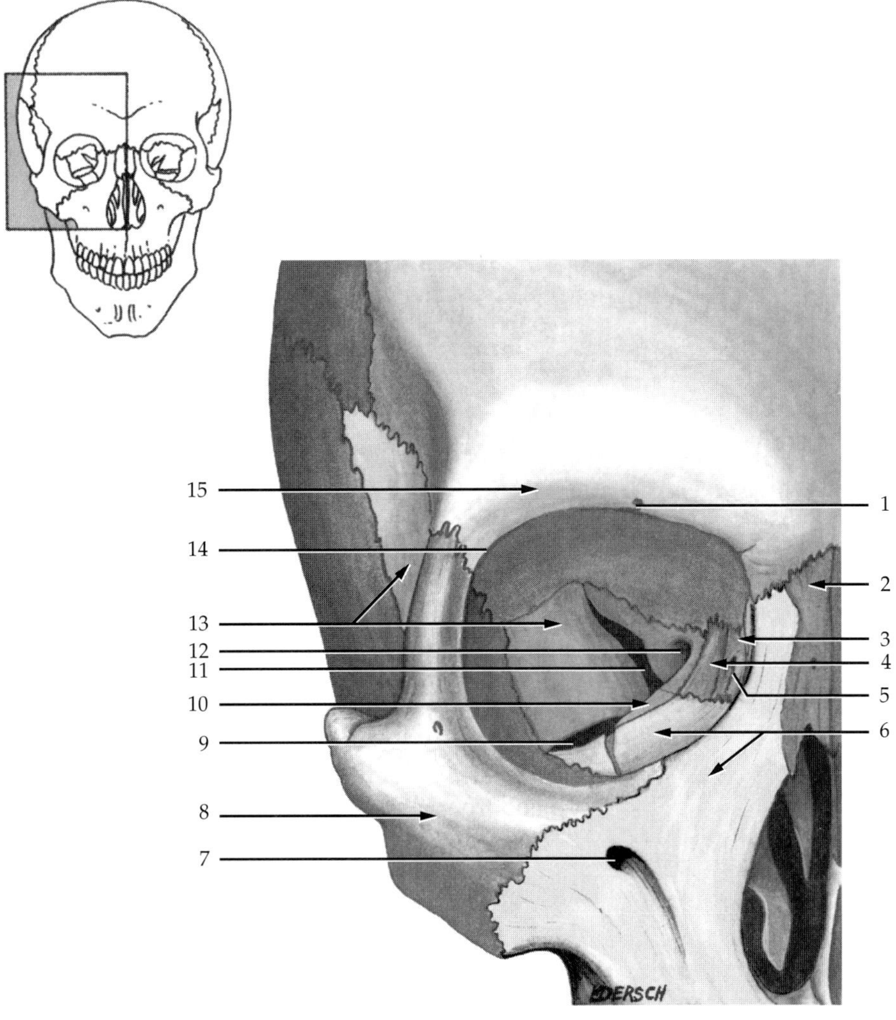

1. Supraorbital foramen
2. Nasal bone
3. Lacrimal bone
4. Ethmoid bone
5. Lacrimal foramen
6. Maxilla
7. Infraorbital foramen
8. Zygomatic bone
9. Inferior orbital fissure
10. Palatine bone
11. Superior orbital fissure
12. Optic foramen
13. Sphenoid bone
14. Supraorbital margin
15. Frontal bone

FIGURE 7.2 Right orbit.

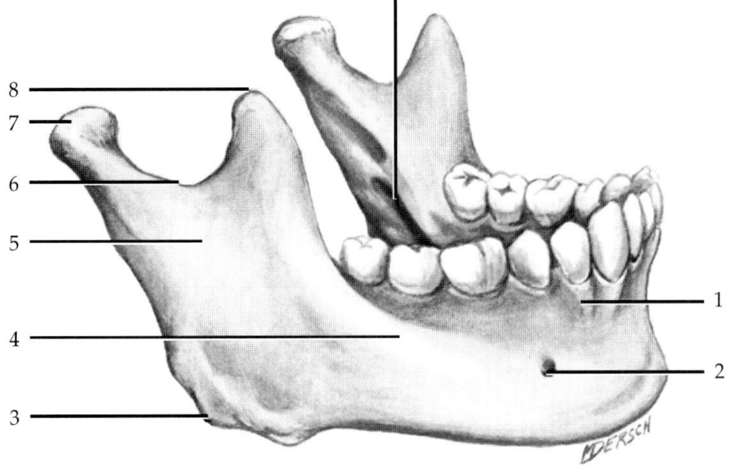

1. Alveolar process
2. Mental foramen
3. Angle
4. Body
5. Ramus
6. Mandibular notch
7. Condylar process
8. Coronoid process
9. Mandibular foramen

Right lateral view

FIGURE 7.3 Mandible.

Table 7.1
Summary of foramina of the skull

Foramen	Structures passing through
Carotid (relating to carotid artery in neck)	
Greater palatine (*palatum* = palate)	
Hypoglossal (*hypo* = under; *glossus* = tongue)	
Incisive (*incisive* = pertaining to incisor teeth)	
Inferior orbital (*inferior* = below; *orbital* = orbit)	
Infraorbital (*infra* = below)	
Jugular (*jugular* = pertaining to jugular vein)	
Lacerum (*lacerum* = lacerated)	
Lacrimal (*lacrima* = pertaining to tears)	
Lesser palatine (*palatum* = palate)	
Magnum (*magnum* = large)	
Mandibular (*mandere* = to chew)	
Mastoid (*mastoid* = breast-shaped)	
Mental (*mentum* = chin)	
Olfactory (*olfacere* = to smell)	
Optic (*optikas* = eye)	
Ovale (*ovale* = oval)	
Rotundum (*rotundum* = round opening)	
Spinosum (*spinosum* = resembling a spine)	
Stylomastoid (*stylo* = stake or pole)	
Superior orbital (*superior* = above)	
Supraorbital (*supra* = above)	
Zygomaticofacial (*zygoma* = cheekbone)	

table by indicating the structures that pass through the foramina listed.

B. SUTURES OF SKULL

A ***suture*** (SOO-chur; *sutura* = seam) is an immovable joint found only between skull bones. The four prominent sutures are the ***coronal*** (*corona* = crown), ***sagittal*** (*sagitta* = arrow), ***lambdoid*** (LAM-doyd), and ***squamous*** (SKWĀ-mos; *squama* = flat). Using Figures 7.1 and 7.4 for reference, locate these sutures on an articulated skull.

C. FONTANELS OF SKULL

At birth, the skull bones are separated by fibrous connective-tissue membrane-filled spaces called ***fontanels*** (fon'-ta-NELZ; *fontanelle* = little fountain). They (1) enable the fetal skull to compress as it passes through the birth canal, (2) permit rapid growth of the brain during infancy, (3) facilitate determination of the degree of brain development by their state of closure, (4) serve as landmarks (anterior fontanel) for withdrawal of blood from the superior sagittal sinus, and (5) aid in determining the position of the fetal head prior to birth. The principal fontanels are ***anterior (frontal), posterior (occipital), anterolateral (sphenoidal),*** and ***posterolateral (mastoid).*** Using Figure 7.4 for reference, locate the fontanels on the skull of a newborn infant.

D. PARANASAL SINUSES OF SKULL

A ***paranasal*** (*para* = beside) ***sinus,*** or simply ***sinus,*** is a cavity in a bone located near the nasal cavity. Paired paranasal sinuses are found in the maxillae and the frontal, sphenoid, and ethmoid bones. Locate the paranasal sinuses on the Beauchene skull or other demonstration models that may be available. Label the paranasal sinuses shown in Figure 7.5 on page 91.

E. VERTEBRAL COLUMN

The ***vertebral column*** (***backbone*** or ***spine***) is composed of a series of bones called ***vertebrae.*** The vertebrae of the adult column are distributed as follows: 7 ***cervical*** (SER-vi-kal) (neck), 12 ***thoracic*** (thō-RAS-ik) (chest), 5 ***lumbar*** (lower back), 5 ***sacral*** (fused into one bone), the ***sacrum*** (SĀ-krum) (between the hipbones), and usually 4 ***coccygeal*** (kok-SIJ-ē-al) (fused into one bone, the ***coccyx*** [KOK-six] forming the tail of the column). Locate each of these regions on the articulated skeleton. Label the same regions in Figure 7.6, the anterior view.

Examine the vertebral column on the articulated skeleton and identify the cervical, thoracic, lumbar, and sacral (sacrococcygeal) curves. Label the curves in Figure 7.6 on page 91, the right lateral view.

F. VERTEBRAE

A typical ***vertebra*** consists of the following portions:

1. ***Body*** Thick, disc-shaped anterior portion.
2. ***Vertebral (neural) arch*** Posterior extension from the body that surrounds the spinal cord and consists of the following parts:
   a. ***Pedicles*** (PED-i-kuls; *pediculus* = little feet) Two short, thick processes that project posteriorly; each has a superior and inferior notch (***vertebral notch***) and, when successive vertebrae are fitted together, the adjoining notches form an ***intervertebral foramen*** through which a spinal nerve and blood vessels pass.
   b. ***Laminae*** (LAM-i-nē; *lamina* = thin layer) Flat portions that form the posterior wall of the vertebral arch.
   c. ***Vertebral foramen*** Opening through which the spinal cord passes; when all the vertebrae are fitted together, the foramina form a canal, the ***vertebral canal.***
3. ***Processes*** Seven processes arise from the vertebral arch:
   a. Two ***transverse processes*** Lateral extensions where the laminae and pedicles join.
   b. One ***spinous process*** (***spine***) Posterior projection of the lamina.
   c. Two ***superior articular processes*** Articulate with the vertebra above. Their top surfaces, called ***superior articular facets*** (*facet* = little face), articulate with the vertebra above.
   d. Two ***inferior articular processes*** Articulate with the vertebra below. Their bottom surfaces, called ***inferior articular facets,*** articulate with the vertebra below.

Obtain a thoracic vertebra and locate each part just described. Now label the vertebra in Figure 7.7a on page 93. You should be able to distinguish the general parts on all the different vertebrae that contain them.

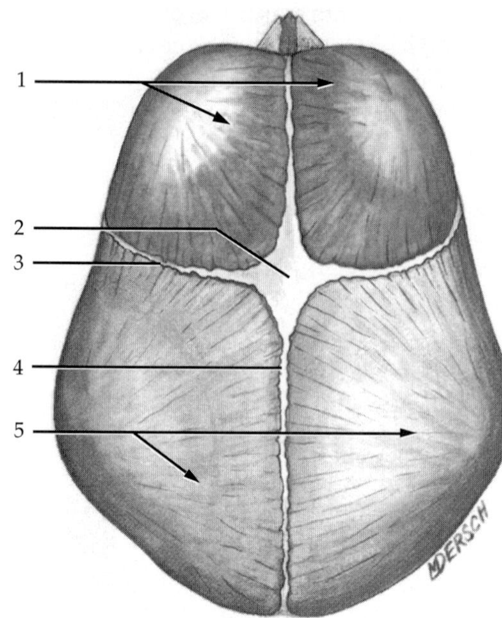

(a) Superior view

1. Frontal bones
2. Anterior (frontal) fontanel
3. Coronal suture
4. Sagittal suture
5. Parietal bones

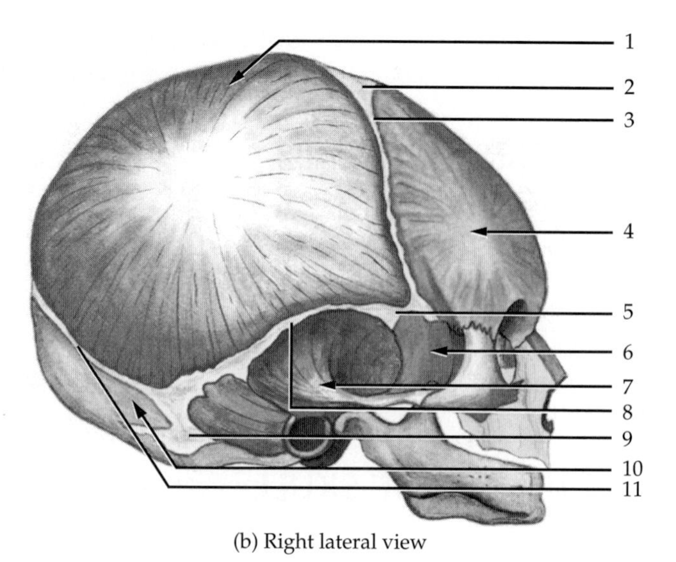

(b) Right lateral view

1. Parietal bone
2. Anterior (frontal) fontanel
3. Coronal suture
4. Frontal bone
5. Anterolateral (sphenoid) fontanel
6. Sphenoid bone
7. Temporal bone
8. Squamous suture
9. Posterolateral (mastoid) fontanel
10. Occipital bone
11. Lambdoid suture

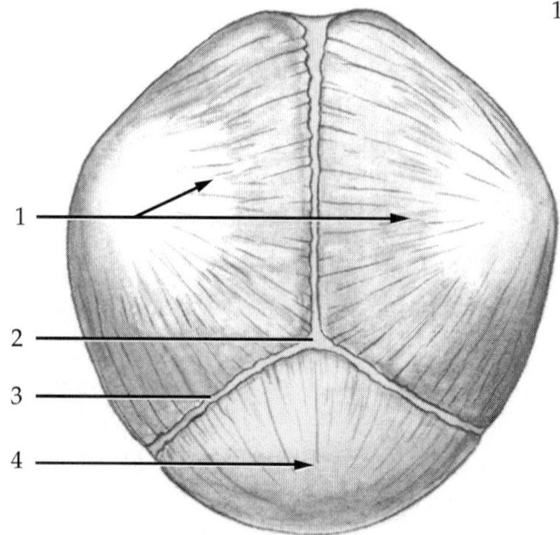

(c) Posterior view

1. Parietal bones
2. Posterior (occipital) fontanel
3. Lambdoid suture
4. Occipital bone

FIGURE 7.4 Fontanels.

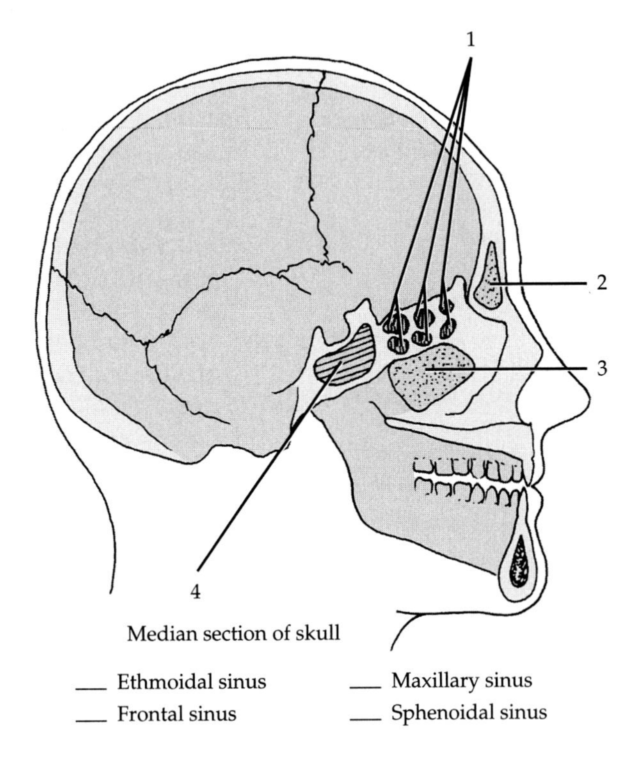

FIGURE 7.5 Paranasal sinuses.

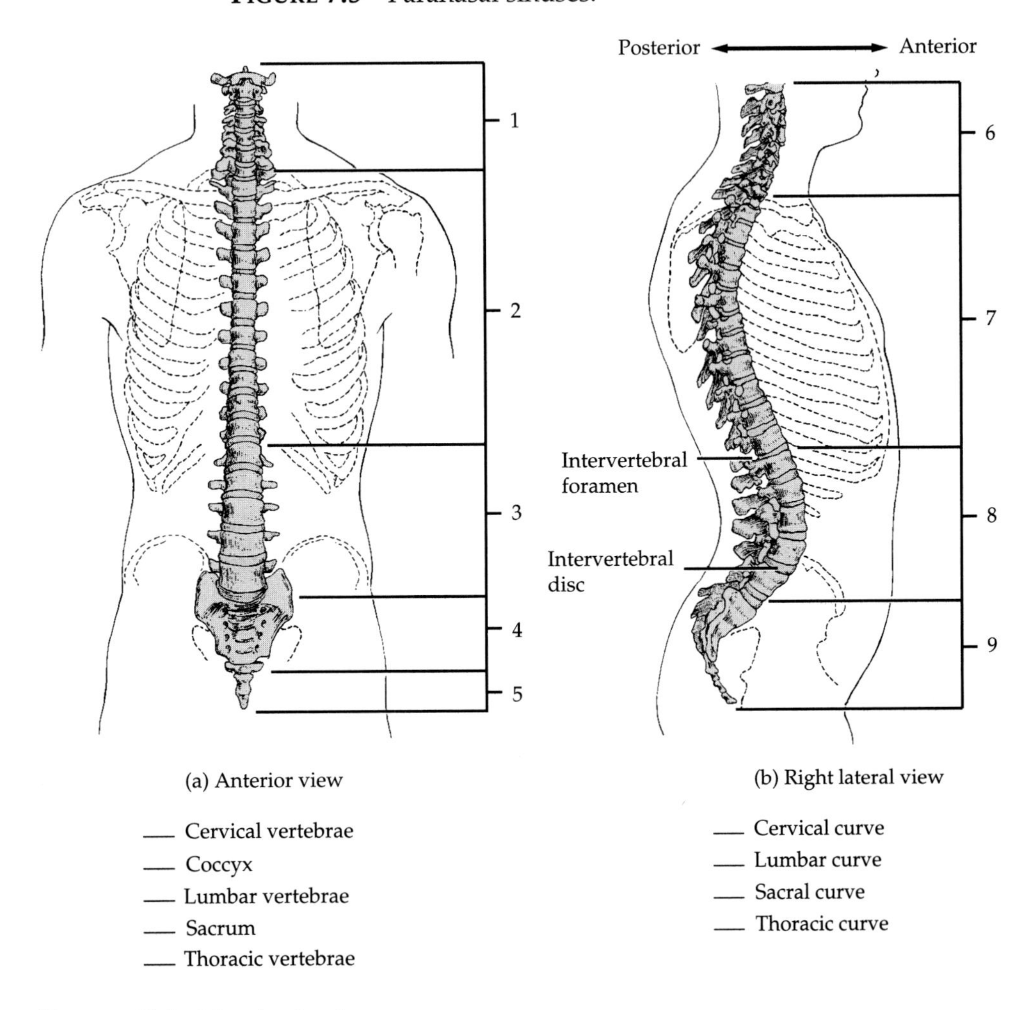

___ Cervical vertebrae
___ Coccyx
___ Lumbar vertebrae
___ Sacrum
___ Thoracic vertebrae

___ Cervical curve
___ Lumbar curve
___ Sacral curve
___ Thoracic curve

FIGURE 7.6 Vertebral column.

Although vertebrae have the same basic design, those of a given region have special distinguishing features. Obtain examples of the following vertebrae and identify their distinguishing features.

1. ***Cervical vertebrae***
   a. ***Atlas (C1)*** First cervical vertebra (Figure 7.7b)
      Transverse foramen Opening in transverse process through which an artery, a vein, and a branch of a spinal nerve pass.
      Anterior arch Anterior wall of vertebral foramen.
      Posterior arch Posterior wall of vertebral foramen.
      Lateral mass Side wall of vertebral foramen.

Label the other indicated parts.

   b. ***Axis (C2)*** Second cervical vertebra (Figure 7.7c)
      Dens (*dens* = tooth) Superior projection of body that articulates with atlas.

Label the other indicated parts.

   c. ***Cervicals 3 through 6*** (Figure 7.7d on page 94)
      Bifid spinous process (C3-C6) Cleft in spinous processes of cervical vertebrae 2 through 6.

Label the other indicated parts.

   d. ***Vertebra prominens (C7)*** Seventh cervical vertebra; contains a nonbifid and long spinous process.
2. ***Thoracic vertebrae (T1-T12)*** (Figure 7.7e on page 94)
   a. ***Facets*** For articulation with the tubercle and head of a rib; found on body and transverse processes.
   b. ***Spinous processes*** Usually long, pointed, downward projections.

Label the other indicated parts.

3. ***Lumbar vertebrae (L1-L5)*** (Figure 7.7f on page 94)
   a. ***Spinous processes*** Broad, blunt.
   b. ***Superior articular processes*** Directed medially, not superiorly.
   c. ***Inferior articular processes*** Directed laterally, not inferiorly.

Label the other indicated parts.

4. ***Sacrum*** (Figure 7.7g and h on page 95). Formed by the fusion of five sacral vertebrae.
   a. ***Transverse lines*** Areas where bodies of sacral vertebrae are joined.
   b. ***Anterior (pelvic) sacral foramina*** Four pairs of foramina that communicate with dorsal sacral foramina; passages for blood vessels and nerves.
   c. ***Median sacral crest*** Spinous processes of fused sacral vertebrae.
   d. ***Lateral sacral crest*** Transverse processes of fused sacral vertebrae.
   e. ***Posterior (dorsal) sacral foramina*** Four pairs of foramina that communicate with pelvic foramina; passages for blood vessels and nerves.
   f. ***Sacral canal*** Continuation of vertebral canal.
   g. ***Sacral hiatus*** (hi-Ā-tus) Inferior entrance to sacral canal where laminae of S5, and sometimes S4, fail to meet.
   h. ***Sacral promontory*** Superior, anterior projecting border.
   i. ***Auricular surface*** Articulates with ilium of hipbone.
5. ***Coccyx*** (Figure 7.7g and h) Formed by the fusion of usually four coccygeal vertebrae.

Label the coccyx and the parts of the sacrum in Figure 7.7g and h.

G. STERNUM AND RIBS

The skeleton of the ***thorax*** consists of the ***sternum, costal cartilages, ribs,*** and bodies of the ***thoracic vertebrae.***

Examine the articulated skeleton and disarticulated bones and identify the following:

1. ***Sternum (breastbone)*** Flat bone in midline of anterior thorax.
   a. ***Manubrium*** (ma-NOO-brē-um; *manubrium* = handle-like) Superior portion.
   b. ***Body*** Middle, largest portion.
   c. ***Sternal angle*** Junction of the manubrium and body.
   d. ***Xiphoid*** (ZI-foyd; *xipho* = sword-like) ***process*** Inferior, smallest portion.
   e. ***Suprasternal notch*** Depression on the superior surface of the manubrium that may be felt.
   f. ***Clavicular notches*** Articular surfaces lateral to suprasternal notches for articulation with the clavicles.

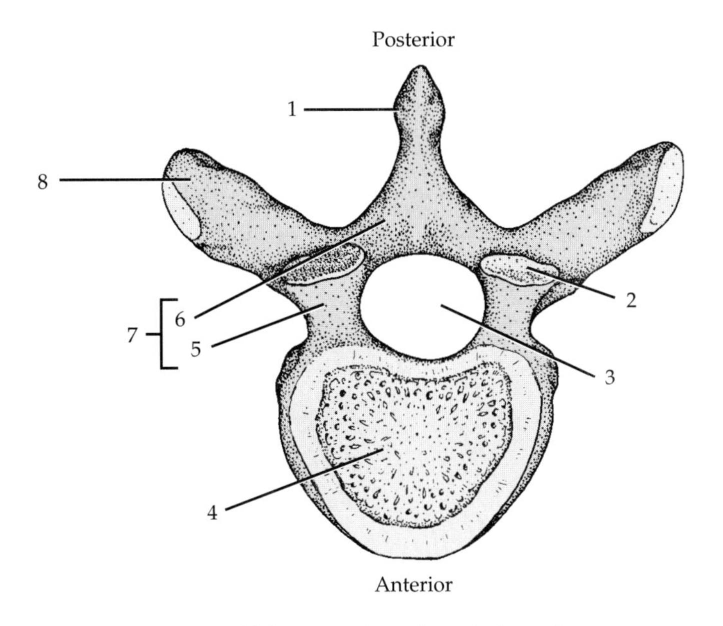

(a) Superior view of a typical vertebra

___ Body
___ Lamina
___ Pedicle
___ Spinous process
___ Superior articular facet
___ Transverse process
___ Vertebral arch
___ Vertebral foramen

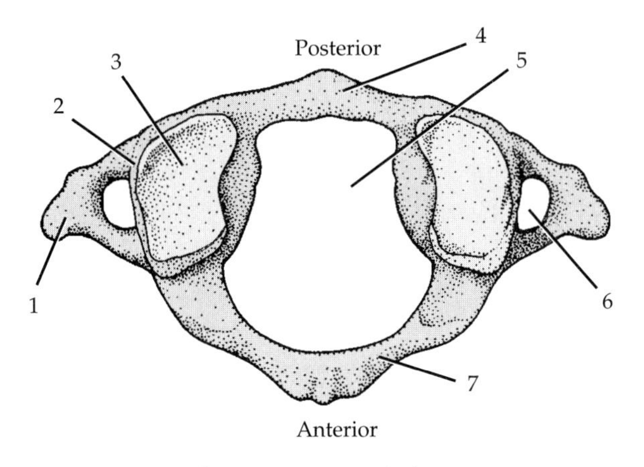

(b) Superior view of atlas

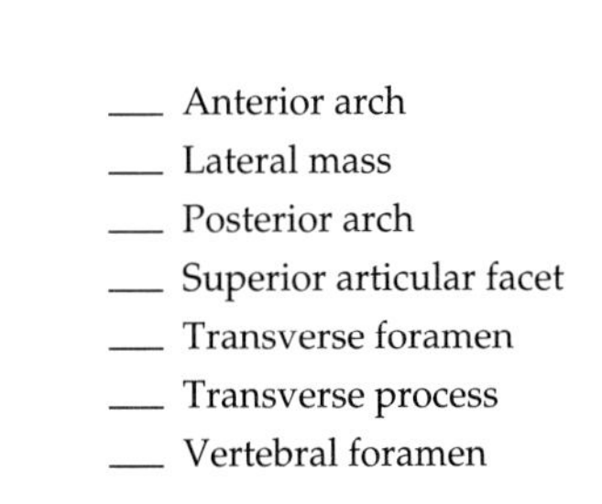
___ Anterior arch
___ Lateral mass
___ Posterior arch
___ Superior articular facet
___ Transverse foramen
___ Transverse process
___ Vertebral foramen

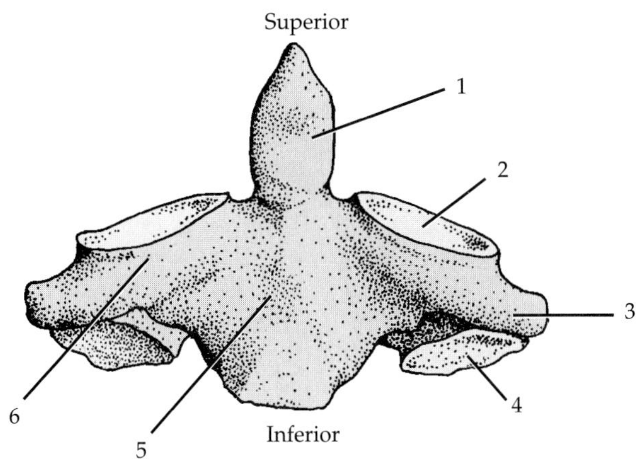

(c) Anterior view of the axis

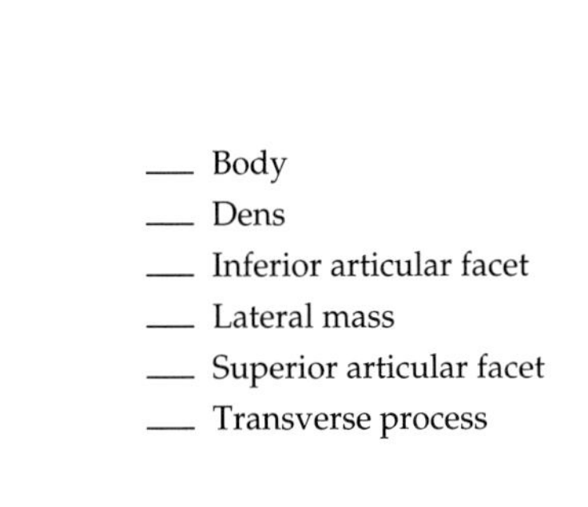
___ Body
___ Dens
___ Inferior articular facet
___ Lateral mass
___ Superior articular facet
___ Transverse process

FIGURE 7.7 Vertebrae.

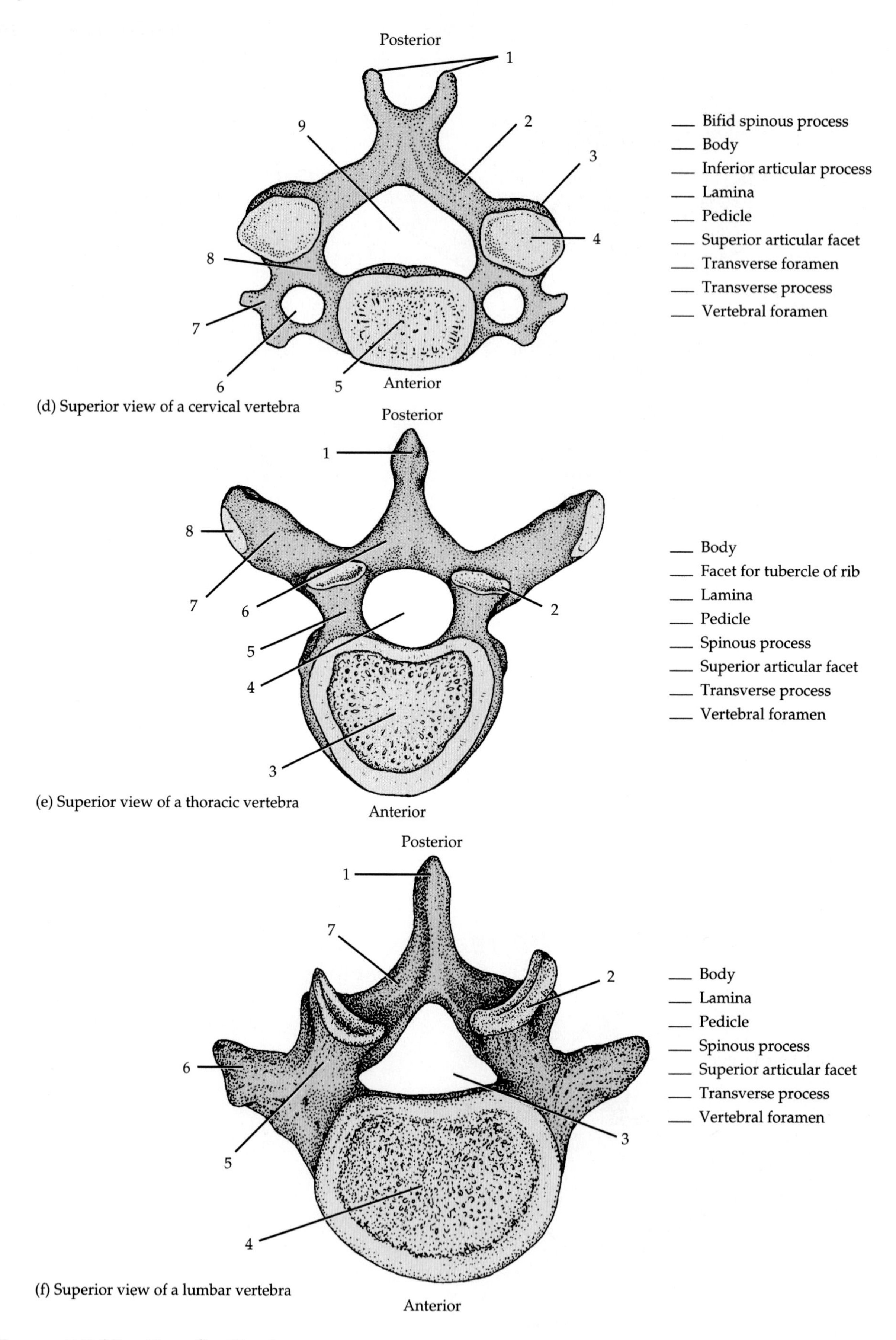

(d) Superior view of a cervical vertebra

(e) Superior view of a thoracic vertebra

(f) Superior view of a lumbar vertebra

FIGURE 7.7 *(Continued)* Vertebrae.

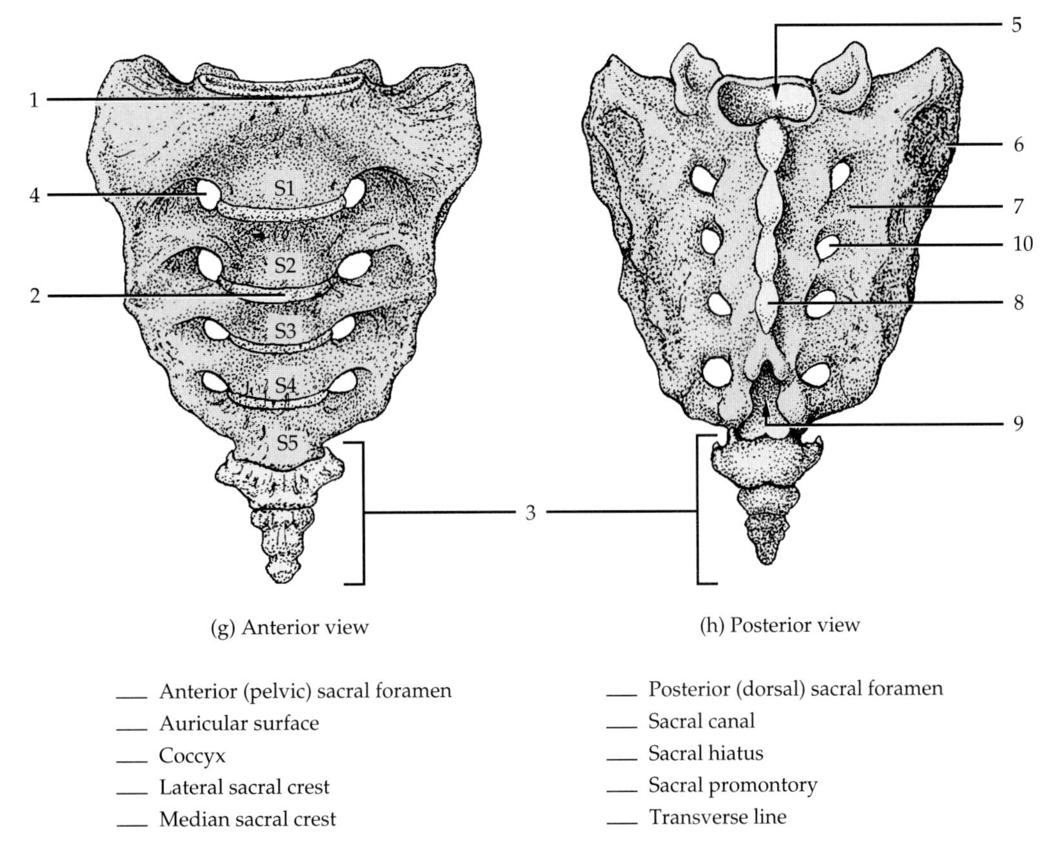

FIGURE 7.7 ***(Continued)*** Vertebrae. Sacrum and coccyx.

Label the parts of the sternum in Figure 7.8a.

2. ***Costal*** (*costa* = rib) ***cartilage*** Strip of hyaline cartilage that attaches a rib to the sternum.
3. ***Ribs*** Parts of a typical rib (third through ninth) include
   a. ***Body*** Shaft, main part of rib.
   b. ***Head*** Posterior projection.
   c. ***Neck*** Constricted portion behind head.
   d. ***Tubercle*** (TOO-ber-kul) Knoblike elevation just below neck; consists of a ***nonarticular part*** that affords attachment for a ligament and an ***articular part*** that articulates with an inferior vertebra.
   e. ***Costal groove*** Depression on the inner surface containing blood vessels and a nerve.
   f. ***Superior facet*** Articulates with facet on superior vertebra.
   g. ***Inferior facet*** Articulates with facet on inferior vertebra.

Label the parts of a rib in Figure 7.8b.

H. PECTORAL (SHOULDER) GIRDLES

Each ***pectoral*** (PEK-tō-ral) or ***shoulder girdle*** consists of two bones—***clavicle*** (collar bone) and ***scapula*** (shoulder blade). Its purpose is to attach the bones of the upper limb to the axial skeleton.

Examine the articulated skeleton and disarticulated bones and identify the following:

1. ***Clavicle*** (KLAV-i-kul; *clavus* = key) Slender bone with a double curvature; lies horizontally in superior and anterior part of the thorax.
   a. ***Sternal extremity*** Rounded, medial end that articulates with manubrium of sternum.
   b. ***Acromial*** (a-KRŌ-mē-al) ***extremity*** Broad, flat, lateral end that articulates with the acromion of the scapula.
   c. ***Conoid tubercle*** (TOO-ber-kul; *konos* = cone) Projection on the inferior, lateral surface for attachment of a ligament.

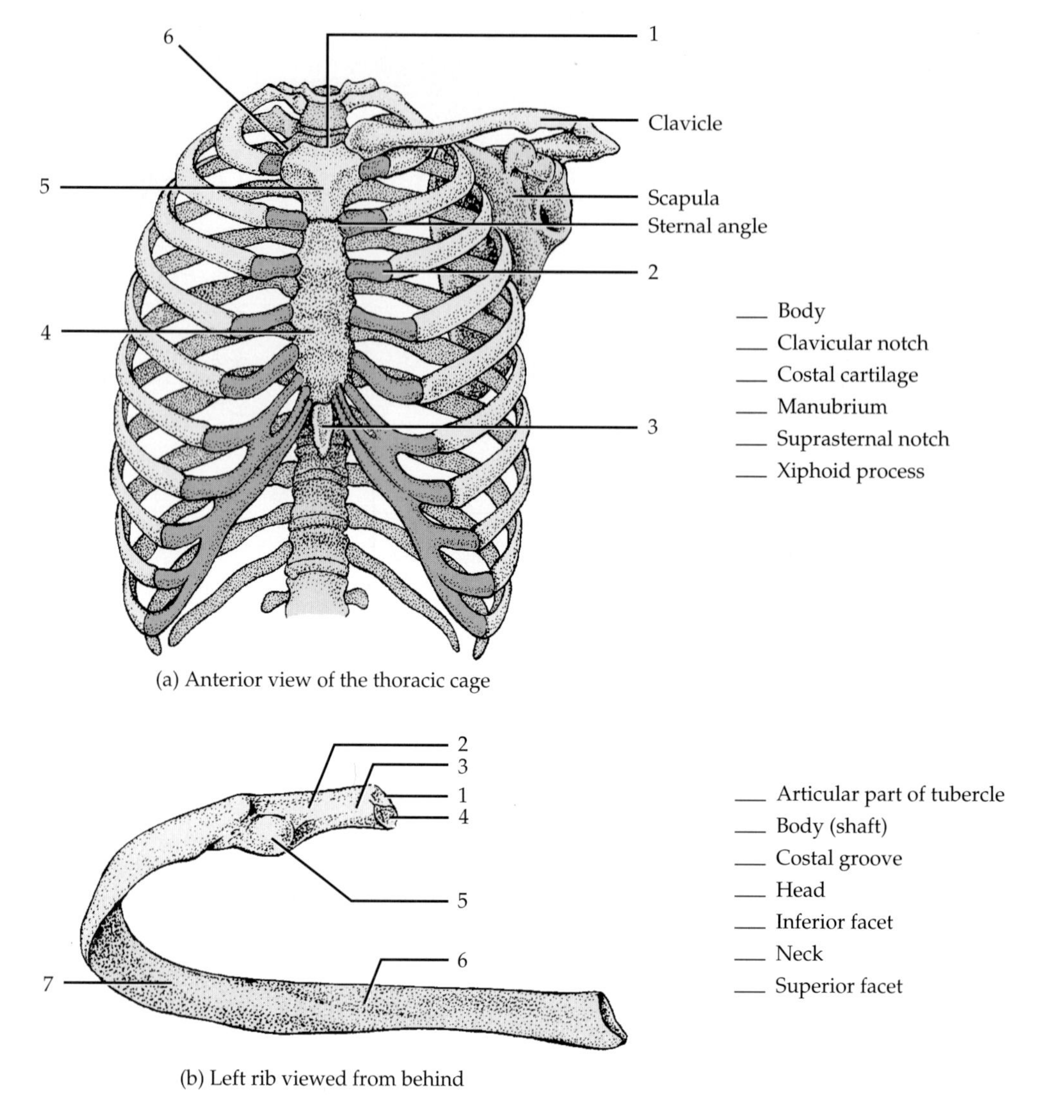

FIGURE 7.8 Bones of the thorax.

Label the parts of the clavicle in Figure 7.9a.

2. ***Scapula*** (SCAP-yoo-la) Large, flat triangular bone in dorsal thorax between the levels of ribs 2 through 7.
   a. ***Body*** Flattened, triangular portion.
   b. ***Spine*** Ridge across posterior surface.
   c. ***Acromion*** (a-KRŌ-mē-on; *acro* = top or summit) Flattened, expanded process of spine.
   d. ***Medial (vertebral) border*** Edge of body near vertebral column.
   e. ***Lateral (axillary) border*** Edge of body near arm.
   f. ***Inferior angle*** Bottom of body where medial and lateral borders join.
   g. ***Glenoid cavity*** Depression below acromion that articulates with head of humerus to form the shoulder joint.
   h. ***Coracoid*** (KOR-a-koyd; *korakodes* = like a crow's beak) ***process*** Projection at lateral end of superior border.
   i. ***Supraspinous*** (soo'-pra-SPĪ-nus) ***fossa*** Surface for muscle attachment above spine.
   j. ***Infraspinous fossa*** Surface for muscle attachment below spine.
   k. ***Superior border*** Superior edge of the body.
   l. ***Superior angle*** Top of body where superior and medial borders join.

Label the parts of the scapula in Figure 7.9b, c, and d.

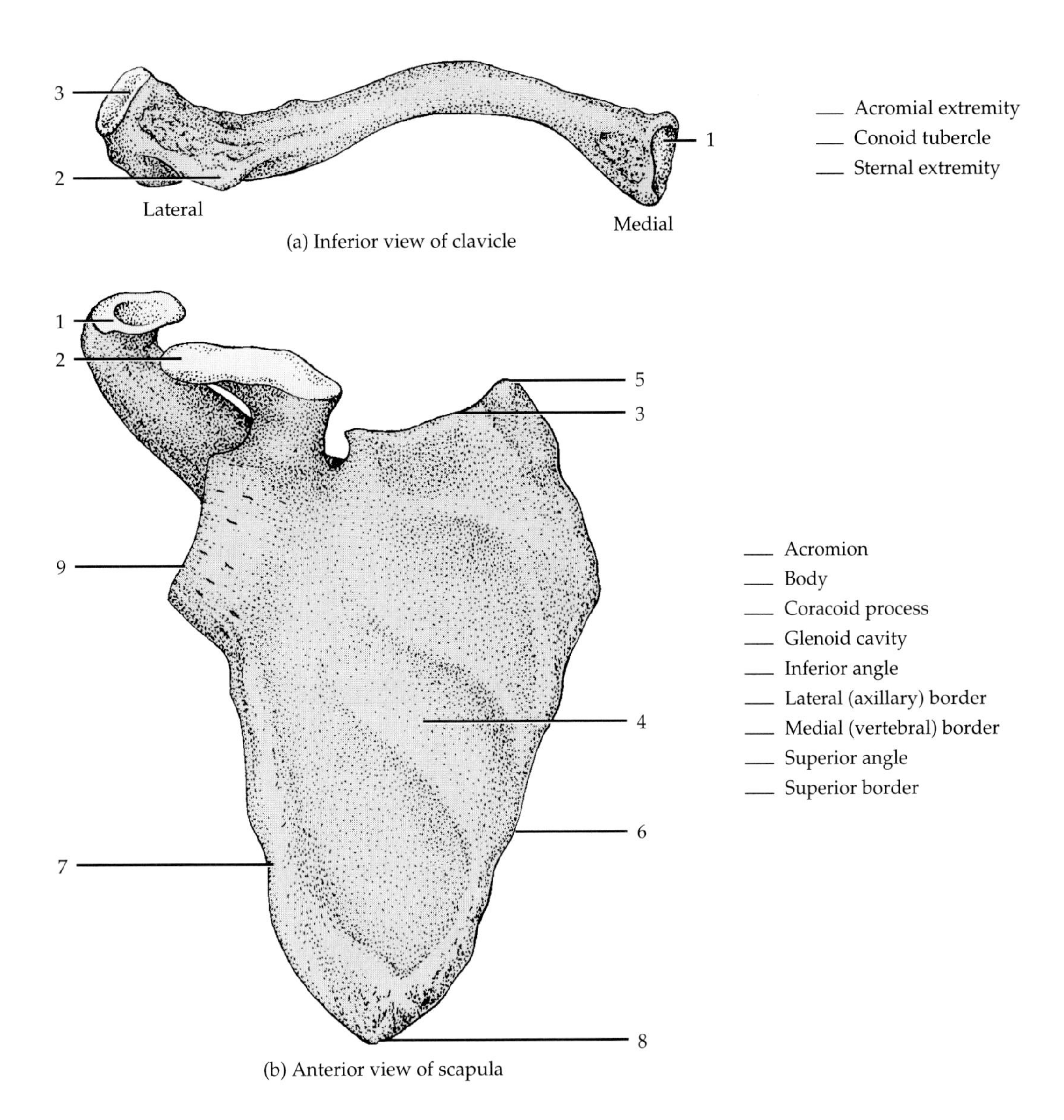

FIGURE 7.9 Pectoral (shoulder) girdle.

I. UPPER LIMBS

The skeleton of the ***upper limbs*** consists of a humerus in each arm, an ulna and radius in each forearm, carpals in each wrist, metacarpals in each palm, and phalanges in the fingers.

Examine the articulated skeleton and disarticulated bones and identify the following:

1. ***Humerus*** (HYOO-mer-us) Arm bone; longest and largest bone of the upper limb.
   a. ***Head*** Articulates with the glenoid cavity of scapula.
   b. ***Anatomical neck*** Oblique groove below the head; the former site of the epiphyseal plate.
   c. ***Greater tubercle*** Lateral projection below the anatomical neck.
   d. ***Lesser tubercle*** Anterior projection.
   e. ***Intertubercular sulcus (bicipital groove)*** Between the tubercles.
   f. ***Surgical neck*** Constricted portion below the tubercles.
   g. ***Body*** Shaft.
   h. ***Deltoid tuberosity*** Roughened, V-shaped area about midway down the lateral surface of shaft.
   i. ***Capitulum*** (ka-PIT-yoo-lum) Rounded knob that articulates with head of radius.
   j. ***Radial fossa*** Anterior lateral depression that receives head of radius when forearm is flexed.
   k. ***Trochlea*** (TRŌK-lē-a) Projection that articulates with the ulna.
   l. ***Coronoid*** (KOR-ō-noyd; *korne* = crown-shaped) ***fossa*** Anterior medial depression that receives part of the ulna when the forearm is flexed.

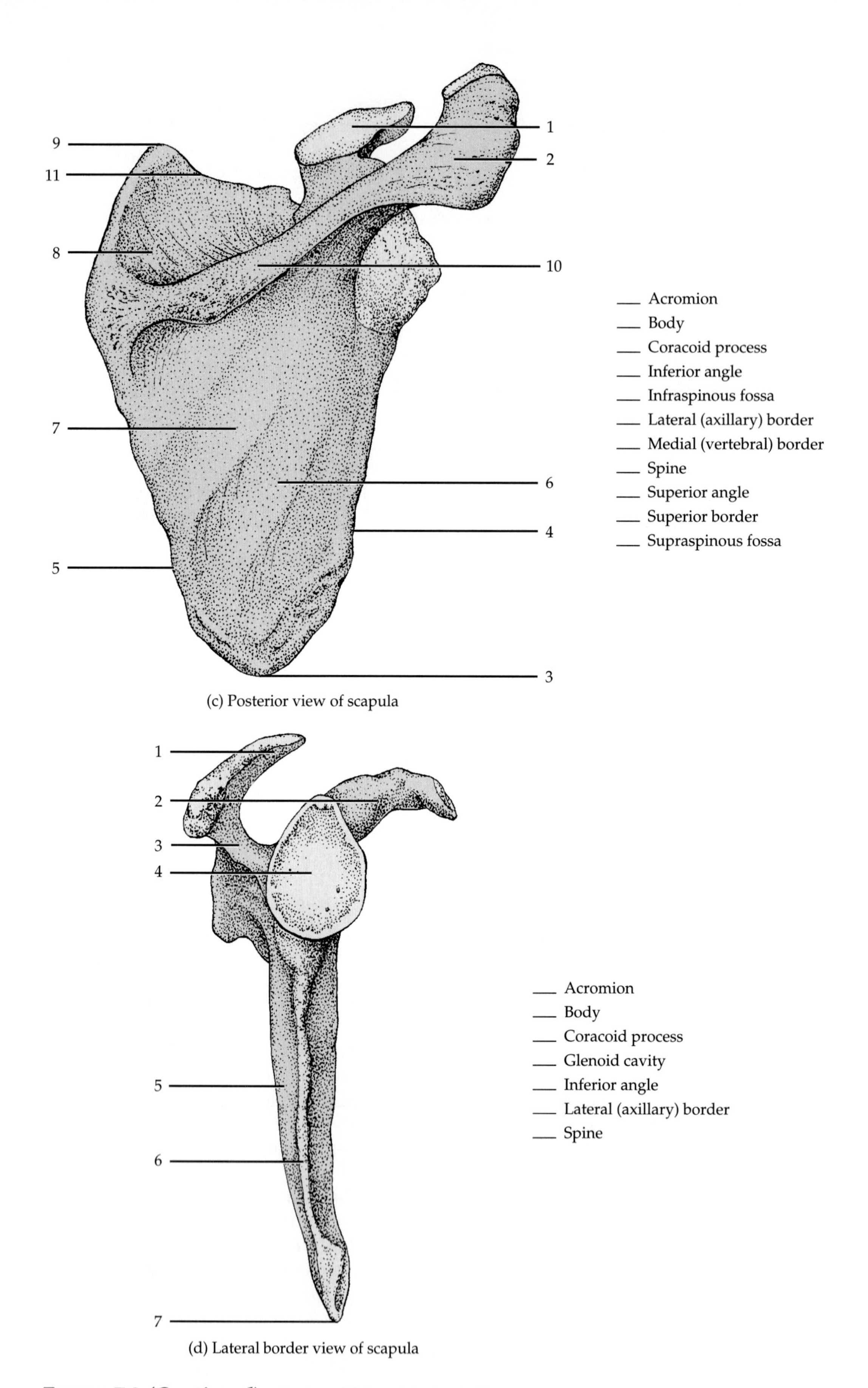

(c) Posterior view of scapula

(d) Lateral border view of scapula

FIGURE 7.9 ***(Continued)*** Pectoral (shoulder) girdle.

m. ***Olecranon*** (ō-LEK-ra-non) ***fossa*** Posterior depression that receives the olecranon of the ulna when the forearm is extended.
n. ***Medial epicondyle*** Projection on medial side of distal end.
o. ***Lateral epicondyle*** Projection on lateral side of distal end.

Label the parts of the humerus in Figure 7.10a and b.

2. *Ulna* Medial bone of forearm.
a. ***Olecranon (olecranon process)*** Prominence of elbow at proximal end.
b. ***Coronoid process*** Anterior projection that, with olecranon, receives trochlea of humerus.
c. ***Trochlear (semilunar) notch*** Curved area between olecranon and coronoid process into which trochlea of humerus fits.

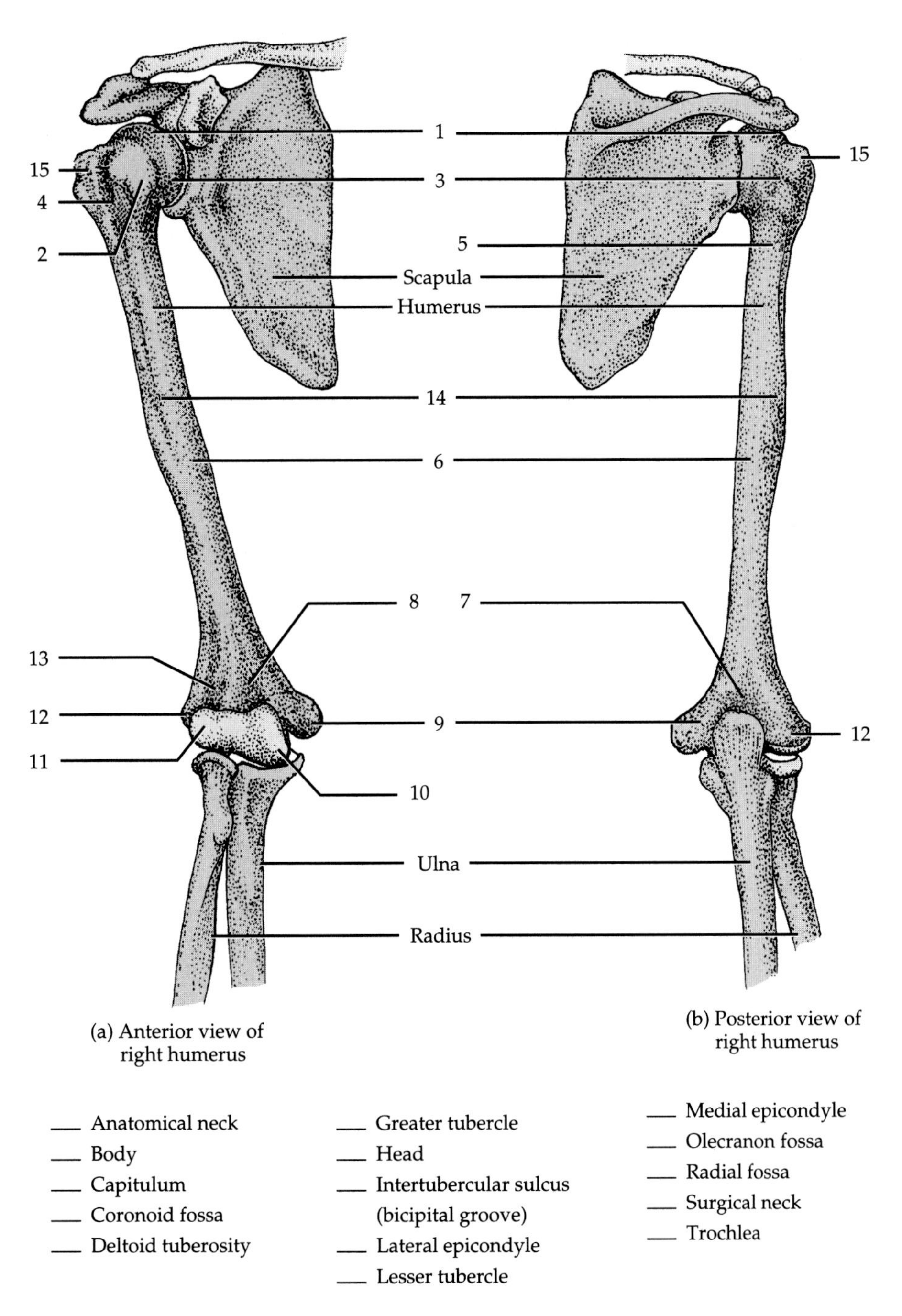

___ Anatomical neck
___ Body
___ Capitulum
___ Coronoid fossa
___ Deltoid tuberosity
___ Greater tubercle
___ Head
___ Intertubercular sulcus (bicipital groove)
___ Lateral epicondyle
___ Lesser tubercle
___ Medial epicondyle
___ Olecranon fossa
___ Radial fossa
___ Surgical neck
___ Trochlea

FIGURE 7.10 Bones of the upper limb.

d. ***Radial notch*** Depression lateral and inferior to trochlear notch that receives the head of the radius.
e. ***Head*** Rounded portion at distal end.
f. ***Styloid*** (*stylo* = pillar) ***process*** Projection on posterior side of distal end.

Label the parts of the ulna in Figure 7.10c and d.

3. ***Radius*** Lateral bone of forearm.
   a. ***Head*** Disc-shaped process at proximal end.
   b. ***Radial tuberosity*** Medial projection for insertion of the biceps brachii muscle.
   c. ***Styloid process*** Projection on lateral side of distal end.
   d. ***Ulnar notch*** Medial, concave depression for articulation with ulna.

Label the parts of the radius in Figure 7.10e.

4. ***Carpus*** Wrist, consists of eight small bones, called ***carpals,*** united by ligaments.
   a. ***Proximal row*** From medial to lateral are called ***pisiform, triquetrum, lunate,*** and ***scaphoid.***
   b. ***Distal row*** From medial to lateral are called ***hamate, capitate, trapezoid,*** and ***trapezium.***
5. ***Metacarpus*** (*meta* = after or beyond) Five bones in the palm, numbered as follows beginning with thumb side: I, II, III, IV, and V metacarpals.
6. ***Phalanges*** (fa-LAN-jēz; *phalanx* = closely knit row) Bones of the fingers; two in each thumb (proximal and distal) and three in each finger (proximal, middle, and distal). The singular of phalanges is ***phalanx.***

Label the parts of the carpus, metacarpus, and phalanges in Figure 7.10f on page 102.

J. PELVIC (HIP) GIRDLE

The ***pelvic (hip) girdle*** consists of the two hipbones or ***coxal*** (KOK-sal) ***bones.*** It provides a strong and stable support for the lower limbs on which the weight of the body is carried and attaches the lower limbs to the axial skeleton.

Examine the articulated skeleton and disarticulated bones and identify the following parts of the hipbone:

1. ***Ilium*** Superior flattened portion.
   a. ***Iliac crest*** Superior border of ilium.
   b. ***Anterior superior iliac spine*** Anterior projection of iliac crest.
   c. ***Anterior inferior iliac spine*** Projection under anterior superior iliac spine.
   d. ***Posterior superior iliac spine*** Posterior projection of iliac crest.
   e. ***Posterior inferior iliac spine*** Projection below posterior superior iliac spine.
   f. ***Greater sciatic*** (sī-AT-ik) ***notch*** Concavity under posterior inferior iliac spine.
   g. ***Iliac fossa*** Medial concavity for attachment of iliacus muscle.
   h. ***Iliac tuberosity*** Point of attachment for sacroiliac ligament posterior to iliac fossa.
   i. ***Auricular*** (*auricula* = little ear) ***surface*** Point of articulation with sacrum.
2. ***Ischium*** (IS-kē-um) Lower, posterior portion.
   a. ***Ischial spine*** Posterior projection of ischium.
   b. ***Lesser sciatic notch*** Concavity under ischial spine.
   c. ***Ischial tuberosity*** Roughened projection.
   d. ***Ramus*** Portion of ischium that joins the pubis and surrounds the ***obturator*** (OB-too-rā-ter) ***foramen.***
3. ***Pubis*** Anterior, inferior portion.
   a. ***Superior ramus*** Upper portion of pubis.
   b. ***Inferior ramus*** Lower portion of pubis.
   c. ***Pubic symphysis*** (SIM-fi-sis) Joint between left and right hipbones.
4. ***Acetabulum*** (as'-e-TAB-yoo-lum) Socket that receives the head of the femur to form the hip joint; two-fifths of the acetabulum is formed by the ilium, two-fifths is formed by the ischium, and one-fifth is formed by the pubis.

Label Figure 7.11 on page 103.

Again, examine the articulated skeleton. This time compare the male and female pelvis. The ***pelvis*** consists of the two hipbones, sacrum, and coccyx. Identify the following:

1. ***Greater (false) pelvis*** Expanded portion situated above the brim of the pelvis; bounded laterally by the ilia (plural of ilium) and posteriorly by the upper sacrum.
2. ***Lesser (true) pelvis*** Below and behind the brim of the pelvis; constructed of parts of the ilium, pubis, sacrum, and coccyx; contains an opening above, the ***pelvic inlet,*** and an opening below, the ***pelvic outlet.***

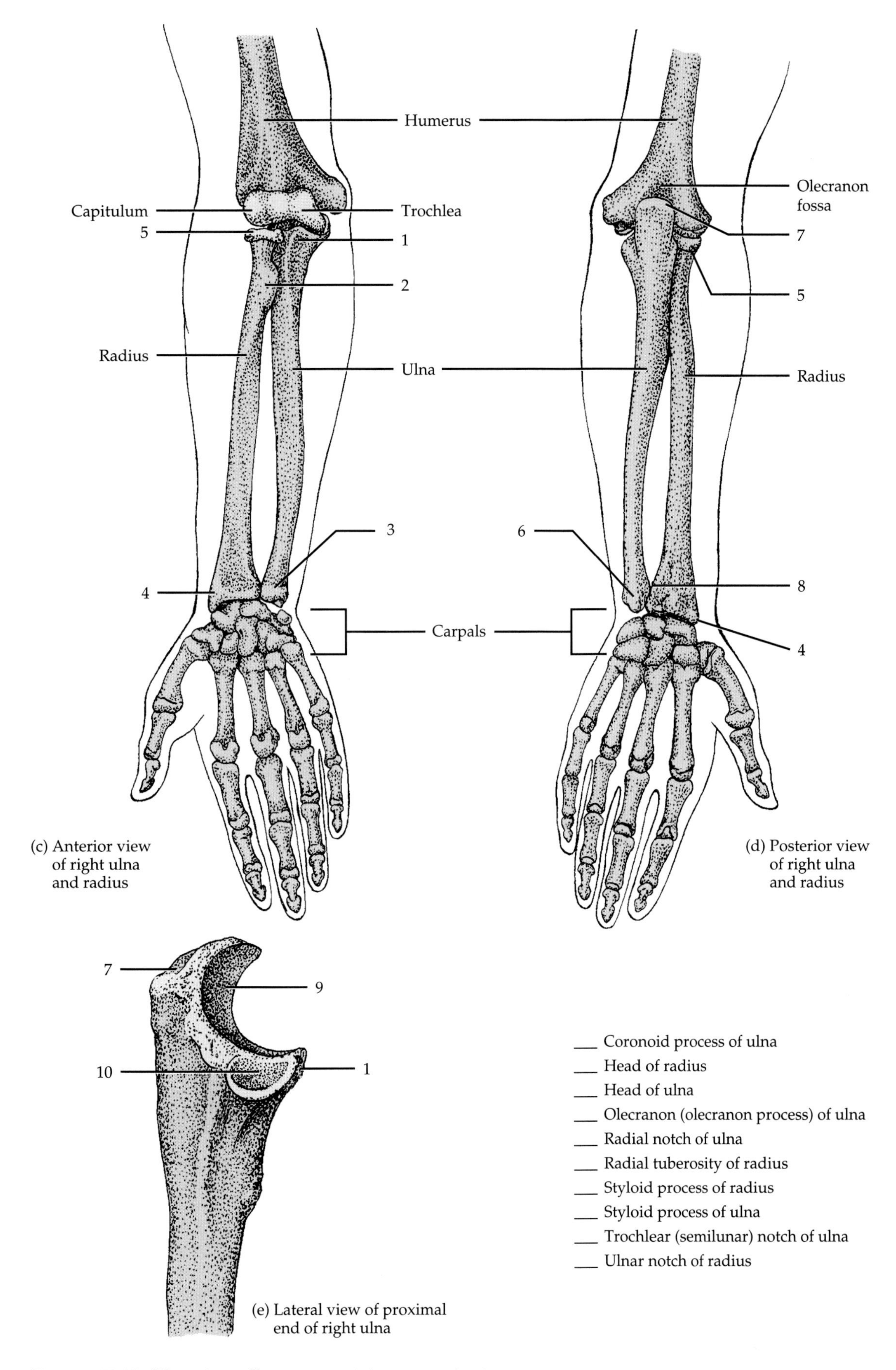

FIGURE 7.10 ***(Continued)*** Bones of the upper limb.

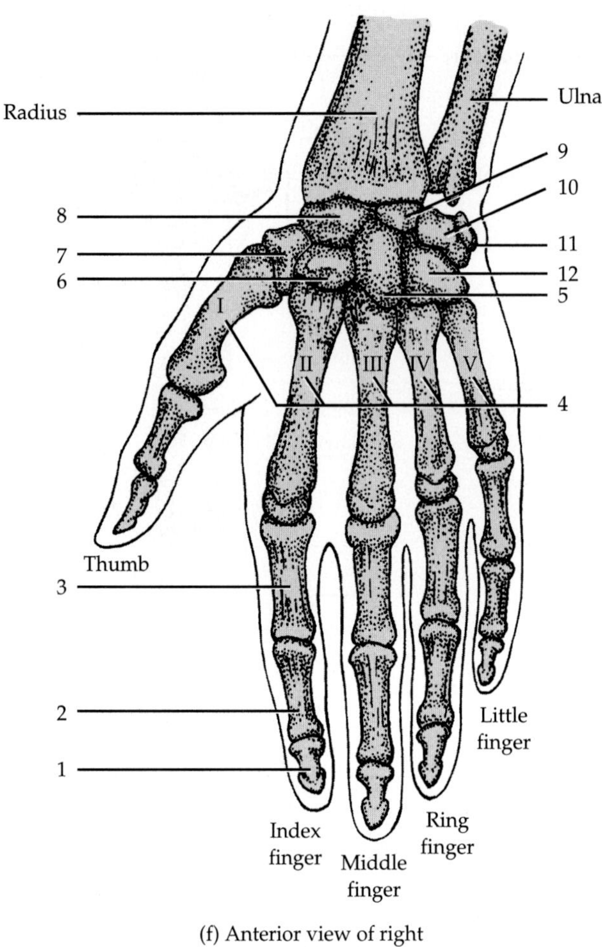

(f) Anterior view of right wrist and hand

___ Capitate
___ Distal phalanx
___ Hamate
___ Lunate
___ Metacarpal
___ Middle phalanx
___ Pisiform
___ Proximal phalanx
___ Scaphoid
___ Trapezium
___ Trapezoid
___ Triquetrum

FIGURE 7.10 *(Continued)* Bones of the upper limb.

K. LOWER LIMBS

The bones of the ***lower limbs*** consist of a femur in each thigh, a patella in front of each knee joint, a tibia and fibula in each leg, tarsals in each ankle, metatarsals in each foot, and phalanges in the toes.

Examine the articulated skeleton and disarticulated bones and identify the following:

1. ***Femur*** Thigh bone; longest and heaviest bone in the body.
   a. ***Head*** Rounded projection at proximal end that articulates with acetabulum of hipbone.
   b. ***Neck*** Constricted portion below head.
   c. ***Greater trochanter*** (trō-KAN-ter) Prominence on lateral side.
   d. ***Lesser trochanter*** Prominence on posteromedial side.
   e. ***Intertrochanteric line*** Ridge on anterior surface.
   f. ***Intertrochanteric crest*** Ridge on posterior surface.
   g. ***Linea aspera*** (LIN-ē-a AS-per-a) Vertical ridge on posterior surface.
   h. ***Medial condyle*** Medial posterior projection on distal end that articulates with tibia.
   i. ***Lateral condyle*** Lateral posterior projection on distal end that articulates with tibia.
   j. ***Intercondylar*** (in'-ter-KON-di-lar) ***fossa*** Depressed area between condyles on posterior surface.
   k. ***Medial epicondyle*** Projection above medial condyle.
   l. ***Lateral epicondyle*** Projection above lateral condyle.
   m. ***Patellar surface*** Between condyles on the anterior surface.

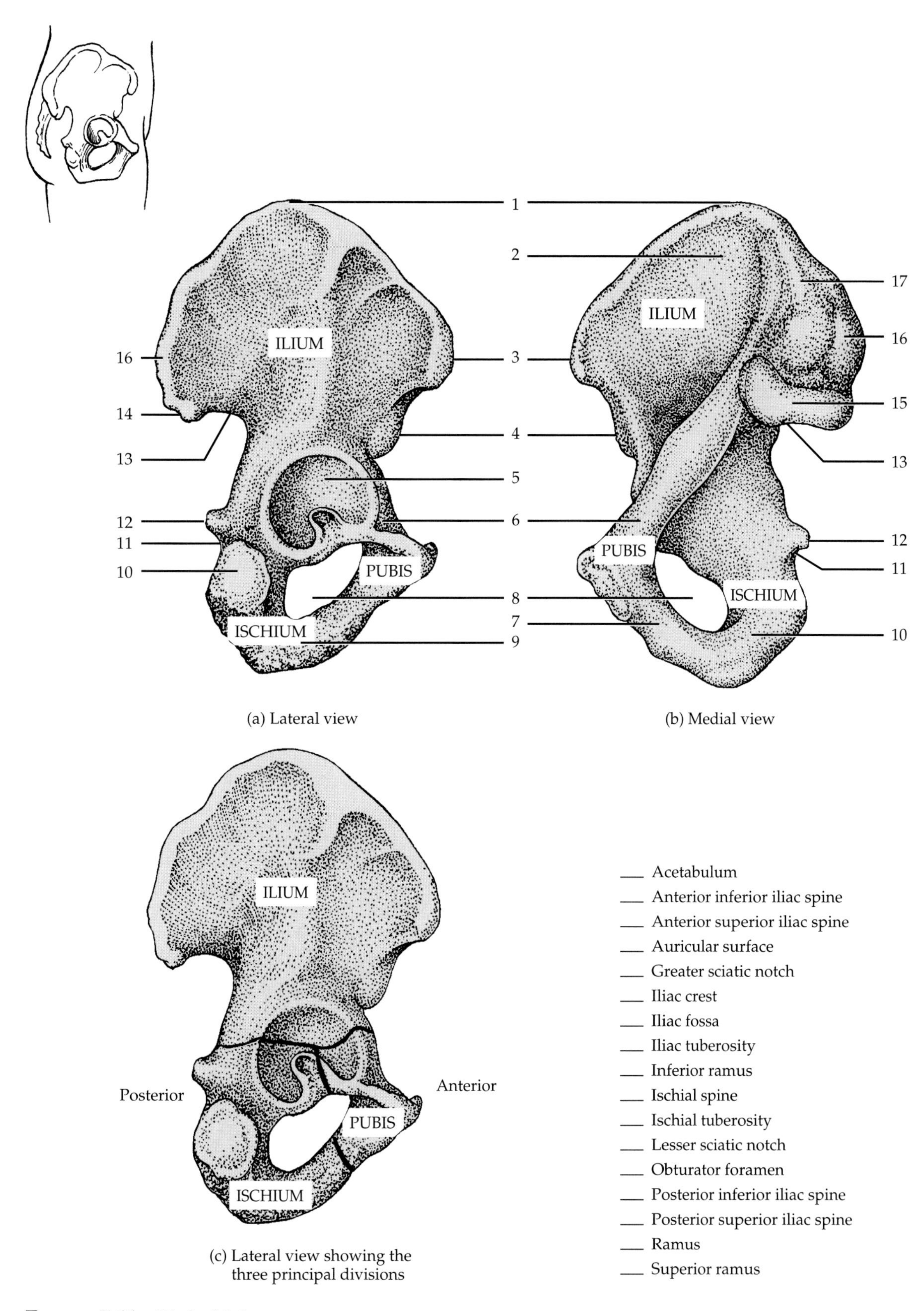

FIGURE 7.11 Right hipbone.

Label the parts of the femur in Figure 7.12a and b.

2. ***Patella*** (*patella* = small plate) Kneecap in front of knee joint; develops in tendon of quadriceps femoris muscle.
   a. ***Base*** Broad superior portion.
   b. ***Apex*** Pointed inferior portion.
   c. ***Articular facets*** Articulating surfaces on posterior surface for medial and lateral condyles of femur.

Label the parts of the patella in Figure 7.12c and d on page 106.

3. ***Tibia*** Shinbone; medial bone of leg.
   a. ***Lateral condyle*** Articulates with lateral condyle of femur.
   b. ***Medial condyle*** Articulates with medial condyle of femur.
   c. ***Intercondylar eminence*** Upward projection between condyles.
   d. ***Tibial tuberosity*** Anterior projection for attachment to patellar ligament.
   e. ***Medial malleolus*** (mal-LĒ-ō-lus; *malleus* = little hammer) Distal projection that articulates with talus bone of ankle.
   f. ***Fibular notch*** Distal depression that articulates with the fibula.
4. ***Fibula*** Lateral bone of leg.
   a. ***Head*** Proximal projection that articulates with tibia.
   b. ***Lateral malleolus*** Projection at distal end that articulates with the talus bone of ankle.

Label the parts of the tibia and fibula in Figure 7.12e on page 106.

5. ***Tarsus*** Seven bones of the ankle called ***tarsals.***
   a. ***Posterior bones—talus*** (TĀ-lus) and ***calcaneus*** (kal-KĀ-nē-us) (heel bone).
   b. ***Anterior bones—cuboid, navicular (scaphoid),*** and three ***cuneiforms*** called the ***first (medial), second (intermediate),*** and ***third (lateral) cuneiforms.***
6. ***Metatarsus*** Consists of five bones of the foot called metatarsals, numbered as follows, beginning on the medial (great or big toe) side: I, II, III, IV, and V metatarsals.
7. ***Phalanges*** Bones of the toes, comparable to phalanges of fingers; two in each great toe (proximal and distal) and three in each small toe (proximal, middle, and distal).

Label the parts of the tarsus, metatarsus, and phalanges in Figure 7.12f on page 107.

L. ARTICULATED SKELETON

Now that you have studied all the bones of the body, label the entire articulated skeleton in Figure 7.13 on page 108.

M. SKELETAL SYSTEM OF CAT

The ***skeletal system*** of the cat, like that of the human, is conveniently divided into two principal divisions: axial skeleton and appendicular skeleton. The ***axial skeleton*** of the cat consists of the bones of the skull, vertebral column, ribs, sternum, and hyoid apparatus. The ***appendicular skeleton*** of the cat consists of the bones of the pectoral girdle, pelvic girdle, forelimbs, and hindlimbs.

Before you start your examination of the bones of the cat skeleton, it should be noted that the cat walks on its digits with the remainder of the palm and foot elevated. This is referred to as ***digitigrade locomotion.*** Humans, by contrast, walk on the entire sole. This is called ***plantigrade locomotion.***

Refer to Figure 7.14 on page 109, which shows the skeleton of the cat in lateral view. Be sure that you study the figure carefully so that you can identify the principal bones.

Obtain a mounted articulated skeleton and identify all the bones indicated in Figure 7.14. Compare the cat skeleton to a human articulated skeleton.

ANSWER THE LABORATORY REPORT QUESTIONS AT THE END OF THE EXERCISE.

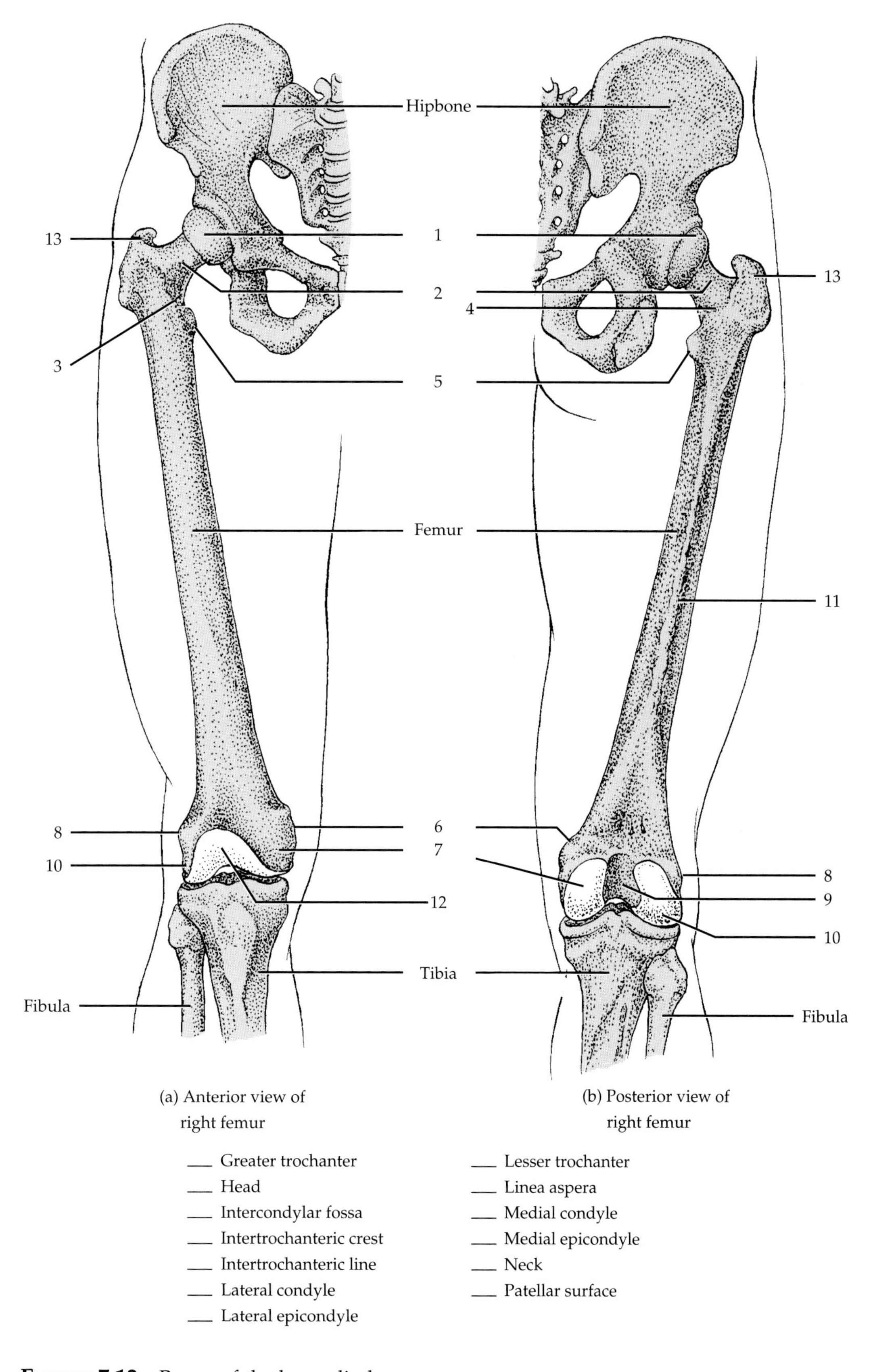

___ Greater trochanter
___ Head
___ Intercondylar fossa
___ Intertrochanteric crest
___ Intertrochanteric line
___ Lateral condyle
___ Lateral epicondyle
___ Lesser trochanter
___ Linea aspera
___ Medial condyle
___ Medial epicondyle
___ Neck
___ Patellar surface

FIGURE 7.12 Bones of the lower limb.

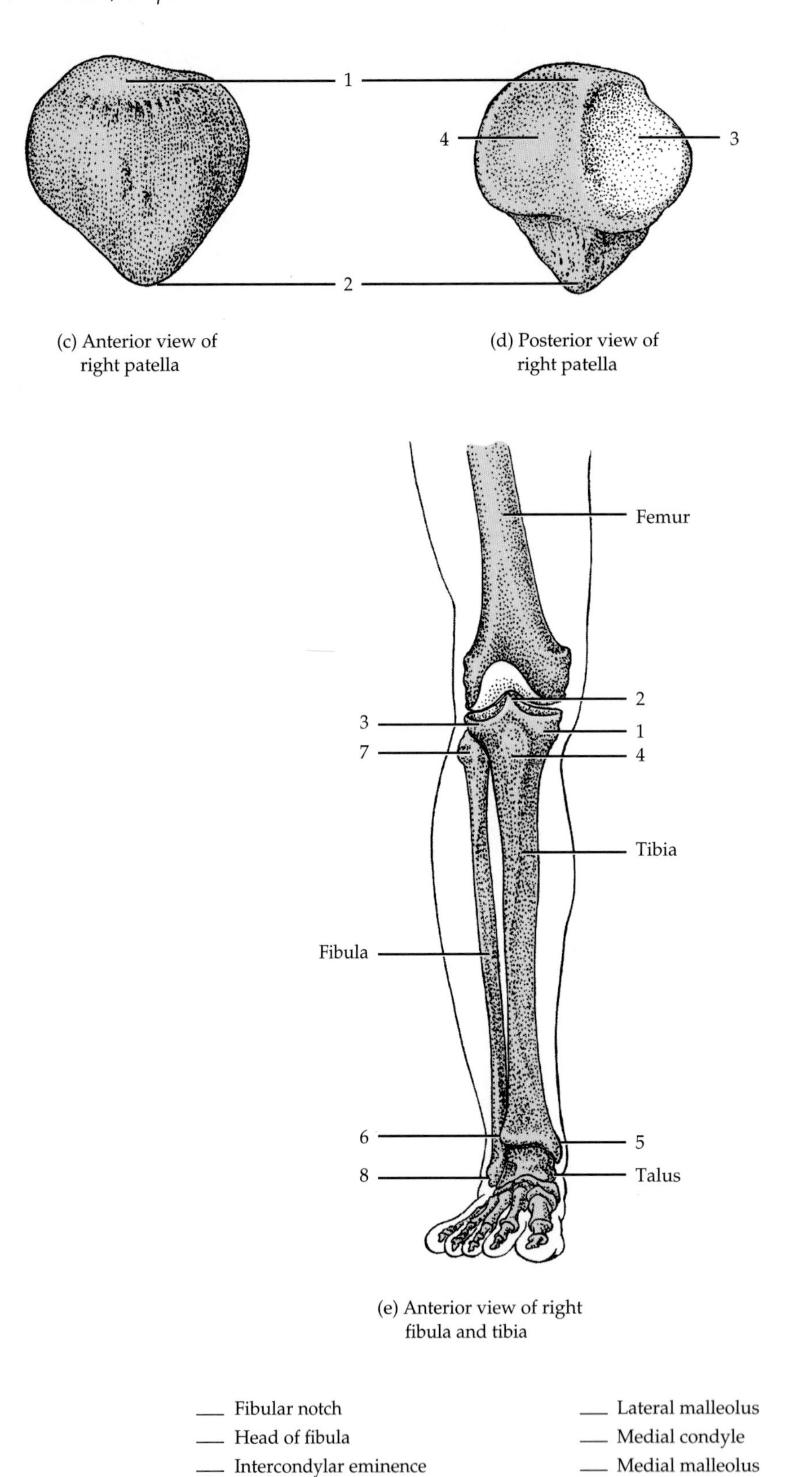

(c) Anterior view of right patella

(d) Posterior view of right patella

(e) Anterior view of right fibula and tibia

___ Apex
___ Articular facet for lateral femoral condyle
___ Articular facet for medial femoral condyle
___ Base

___ Fibular notch
___ Head of fibula
___ Intercondylar eminence
___ Lateral condyle
___ Lateral malleolus
___ Medial condyle
___ Medial malleolus
___ Tibial tuberosity

FIGURE 7.12 *(Continued)* Bones of the lower limb.

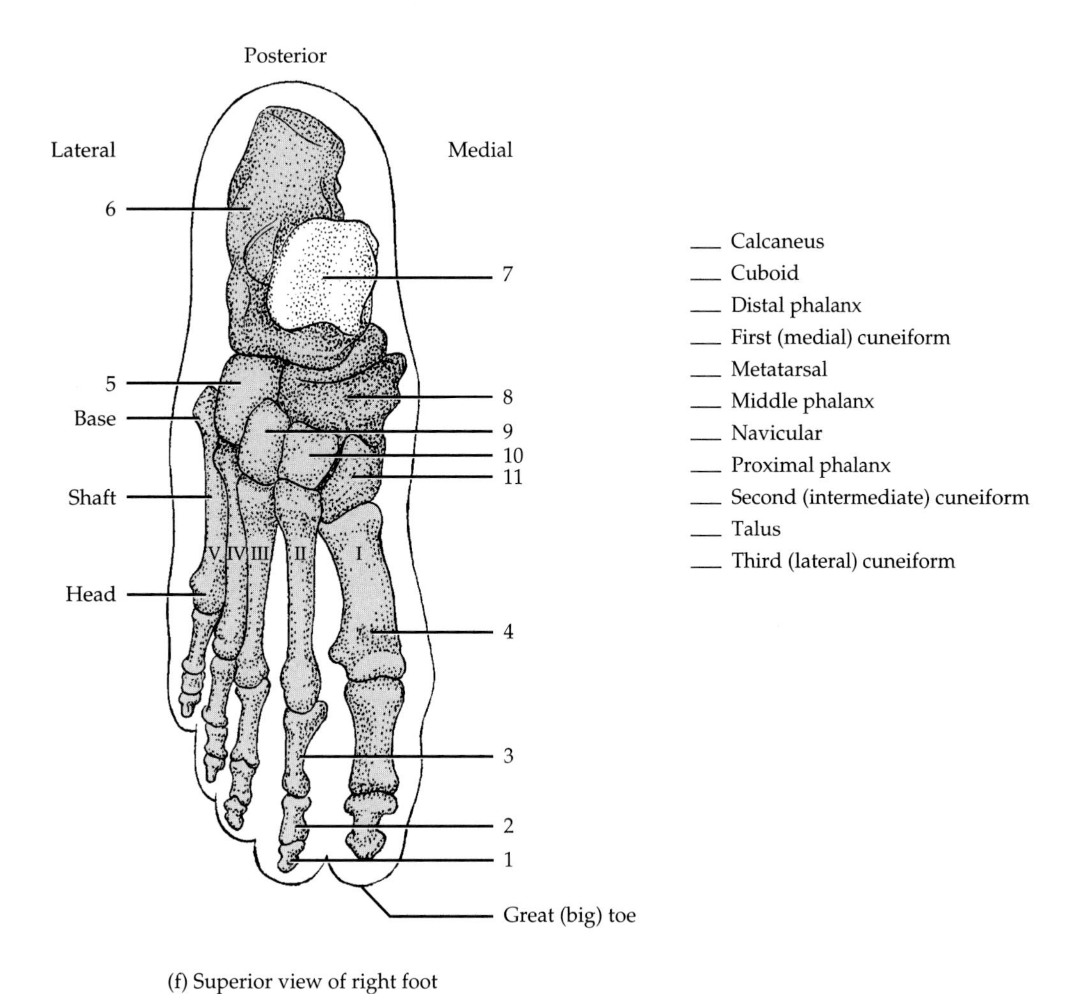

(f) Superior view of right foot

FIGURE 7.12 ***(Continued)*** Bones of the lower limb.

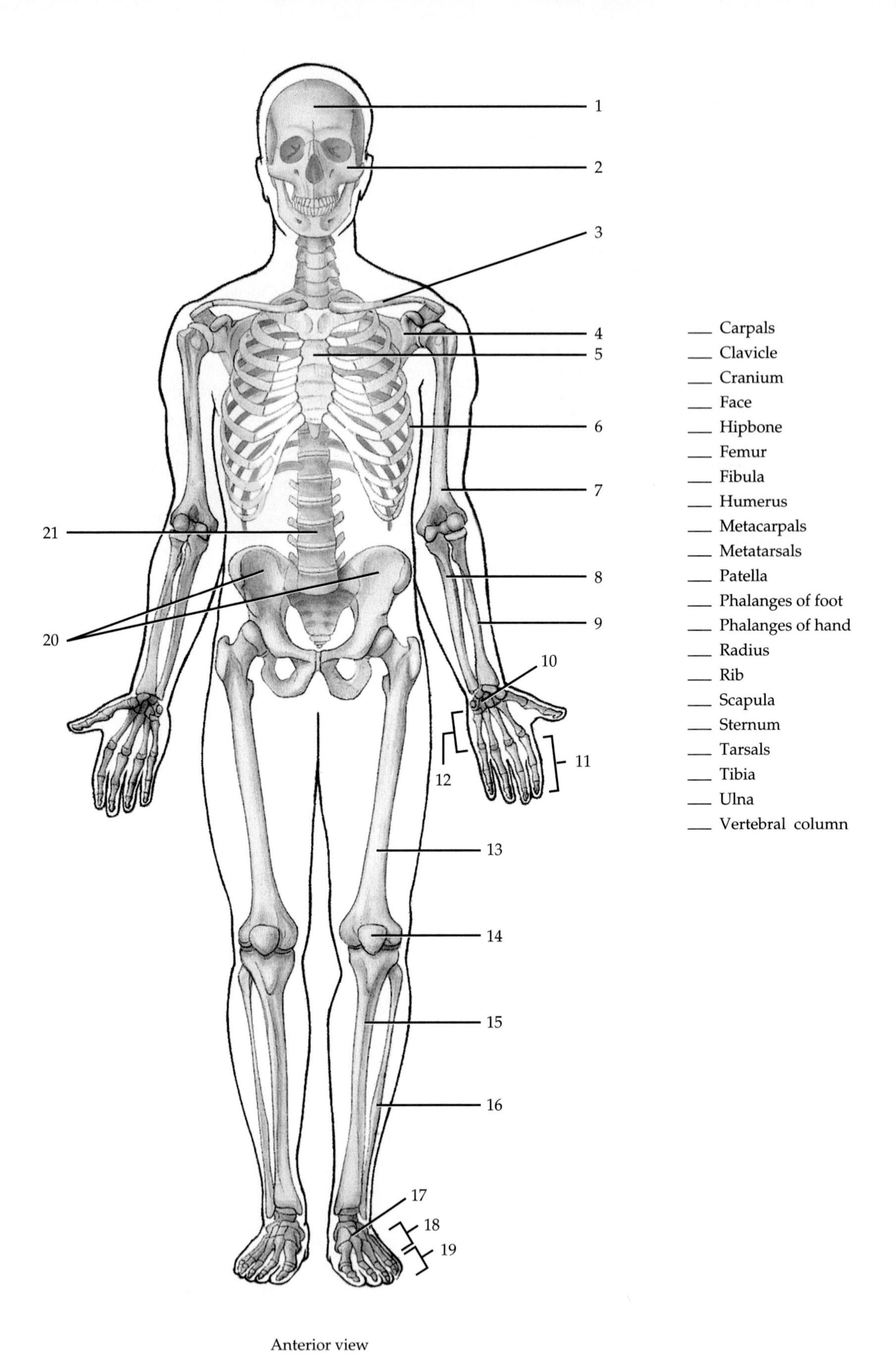

___ Carpals
___ Clavicle
___ Cranium
___ Face
___ Hipbone
___ Femur
___ Fibula
___ Humerus
___ Metacarpals
___ Metatarsals
___ Patella
___ Phalanges of foot
___ Phalanges of hand
___ Radius
___ Rib
___ Scapula
___ Sternum
___ Tarsals
___ Tibia
___ Ulna
___ Vertebral column

FIGURE 7.13 Entire skeleton.

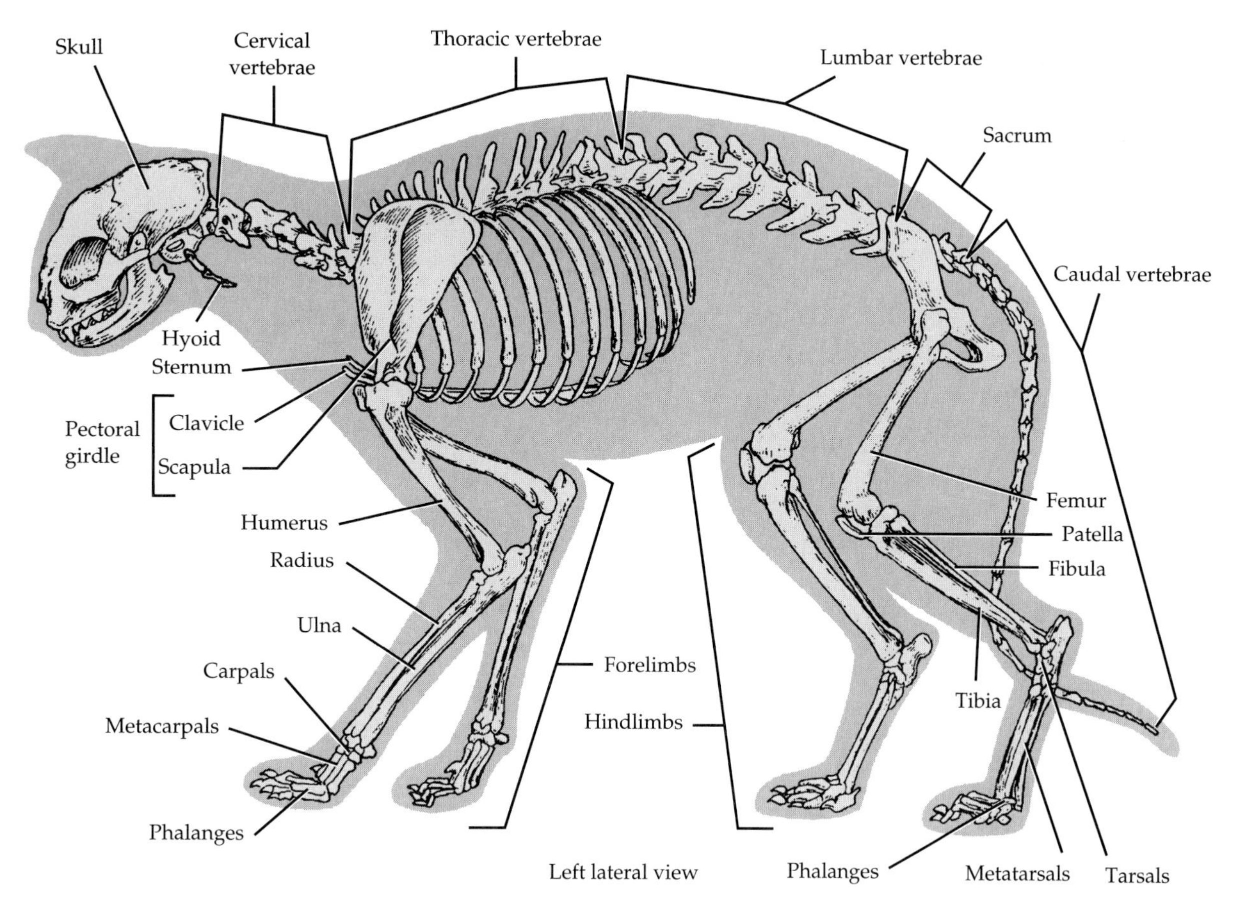

FIGURE 7.14 Skeleton of cat.

LABORATORY REPORT QUESTIONS

Bones 7

Student ______________________ Date ______________________

Laboratory Section ______________________ Score/Grade ______________________

PART 1. Multiple Choice

_______ 1. The suture between the parietal and temporal bones is the (a) lambdoid (b) coronal (c) squamous (d) sagittal

_______ 2. Which bone does *not* contain a paranasal sinus? (a) ethmoid (b) maxilla (c) sphenoid (d) sacrum

_______ 3. Which is the superior, concave curve in the vertebral column? (a) thoracic (b) lumbar (c) cervical (d) sacral

_______ 4. The fontanel between the parietal and occipital bones is the (a) anterolateral (b) anterior (c) posterior (d) posterolateral

_______ 5. Which is *not* a component of the upper limb? (a) radius (b) femur (c) carpus (d) humerus

_______ 6. All are components of the appendicular skeleton *except* the (a) humerus (b) occipital bone (c) calcaneus (d) triquetral

_______ 7. Which bone does *not* belong with the others? (a) occipital (b) frontal (c) parietal (d) mandible

_______ 8. Which region of the vertebral column is closer to the skull? (a) thoracic (b) lumbar (c) cervical (d) sacral

_______ 9. Of the following bones, the one that does *not* help form part of the orbit is the (a) sphenoid (b) frontal (c) occipital (d) lacrimal

_______ 10. Which bone does *not* form a border for a fontanel? (a) maxilla (b) temporal (c) occipital (d) parietal

PART 2. Identification

For each surface marking listed, identify the skull bone to which it belongs:

11. Glabella ______________________

12. Mastoid process ______________________

13. Sella turcica ______________________

14. Cribriform plate ______________________

15. Foramen magnum ______________________

16. Mental foramen ______________________

17. Infraorbital foramen ____________________

18. Crista galli ____________________

19. Foramen ovale ____________________

20. Horizontal plate ____________________

21. Optic foramen ____________________

22. Superior nasal concha ____________________

23. Zygomatic process ____________________

24. Styloid process ____________________

25. Mandibular fossa ____________________

PART 3. Matching

_______ 26. Iliac crest

_______ 27. Capitulum

_______ 28. Medial malleolus

_______ 29. Laminae

_______ 30. Vertebral foramen

_______ 31. Talus

_______ 32. Olecranon

_______ 33. Pisiform

_______ 34. Acromial extremity

_______ 35. Pubic symphysis

_______ 36. Hamate

_______ 37. Costal groove

_______ 38. Xiphoid process

_______ 39. Greater trochanter

_______ 40. Transverse lines

_______ 41. Radial tuberosity

_______ 42. Greater tubercle

_______ 43. Ischium

_______ 44. Glenoid cavity

_______ 45. Lateral malleolus

A. Inferior portion of sternum
B. Medial bone of distal carpals
C. Distal projection of tibia
D. Lateral end of clavicle
E. Points where bodies of sacral vertebrae join
F. Portion of rib that contains blood vessels
G. Prominence of elbow
H. Lateral projection of humerus
I. Medial projection for insertion of biceps brachii muscle
J. Lower posterior portion of hipbone
K. Articulates with head of radius
L. Prominence on lateral side of femur
M. Distal projection of fibula
N. Superior border of ilium
O. Articulates with head of humerus
P. Anterior joint between hipbones
Q. Opening through which spinal cord passes
R. Form posterior wall of vertebral arch
S. Medial bone of proximal carpals
T. Component of tarsus

Articulations

8

An ***articulation*** (ar-tik'-yoo-LĀ-shun), or ***joint***, is a point of contact between bones, cartilage and bones, or teeth and bones. When we say that one bone *articulates* with another, we mean that one bone forms a joint with another. The scientific study of joints is called ***arthrology*** (ar-THROL-ō-jē; *arthro* = joint; *logos* = study of).

Some joints permit no movement, others permit a slight degree of movement, and still others permit free movement. In this exercise you will study the structure and action of joints.

A. KINDS OF JOINTS

The joints of the body may be classified into several principal kinds on the basis of their structure and function (degree of movement they permit).

The structural classification of joints is based on the presence or absence of a space between articulating bones called a ***synovial*** or ***joint cavity*** and the type of connective tissue that binds the bones together. Structurally, a joint is classified as

1. ***Fibrous*** (FĪ-brus) There is no synovial cavity and the bones are held together by fibrous connective tissue.
2. ***Cartilaginous*** (kar-ti-LAJ-i-nus) There is no synovial cavity and the bones are held together by cartilage.
3. ***Synovial*** (si-NŌ-vē-al) There is a synovial cavity and the bones forming the joint are united by a surrounding articular capsule and frequently by accessory ligaments (described in detail later).

The functional classification of joints is as follows:

1. ***Synarthroses*** (sin'-ar-THRŌ-sēz; *syn* = together; *arthros* = joint) Immovable joints.
2. ***Amphiarthroses*** (am'-fē-ar-THRŌ-sēz; *amphi* = on both sides) Slightly movable joints.
3. ***Diarthroses*** (dī-ar-THRŌ-sēz; *diarthros* = movable joint) Freely movable joints.

We will discuss the joints of the body on the basis of their structural classification, referring to their functional classification as well.

B. FIBROUS JOINTS

Allow little or no movement; do not contain synovial cavity; articulating bones held together by fibrous connective tissue.

1. ***Sutures*** (SOO-cherz; *sutura* = seam) A fibrous joint composed of a thin layer of dense fibrous connective tissue that unites the bones of the skull. Example: coronal suture between the frontal and parietal bones.
2. ***Syndesmoses*** (sin'-dez-MŌ-sēz; *syndesmo* = band or ligament) A fibrous joint in which there is considerably more fibrous connective tissue than in a suture; the fibrous connective tissue forms an interosseous membrane or ligament that permits some flexibility and movement. Example: the distal articulation of the tibia and fibula.
3. ***Gomphoses*** (gom-FŌ-sēz; *gomphosis* = to bolt together) A type of fibrous joint in which a cone-shaped peg fits into a socket. The substance between the bones is the periodontal ligament. Example: articulations of the roots of the teeth with the alveoli (sockets) of the maxillae and mandible.

C. CARTILAGINOUS JOINTS

Allow little or no movement; do not contain synovial cavity; articulating bones held together by cartilage.

1. ***Synchondroses*** (sin'-kon-DRŌ-sēz; *syn* = together; *chondro* = cartilage) A cartilaginous

joint in which the connecting material is hyaline cartilage. Example: epiphyseal plate between the epiphysis and diaphysis of a growing bone.

2. ***Symphyses*** (SIM-fi-sēz; *symphysis* = growing together) A cartilaginous joint in which the connecting material is a broad, flat disc of fibrocartilage. Example: intervertebral discs between the bodies of vertebrae and the pubic symphysis between the anterior surfaces of the hipbones.

D. SYNOVIAL JOINTS

Synovial joints have a variety of shapes and permit several different types of movements. Following are the parts of a synovial joint.

1. Structure of a Synovial Joint

A distinguishing anatomical feature of a synovial joint is a space, called a ***synovial*** (si-NŌ-vē-al) or ***joint cavity*** (see Figure 8.1), that separates the articulating bones. Another characteristic of such joints is the presence of ***articular cartilage***. Articular cartilage (hyaline) covers the surfaces of the articulating bones but does not bind the bones together.

A sleevelike ***articular capsule*** surrounds a synovial joint, encloses the synovial cavity, and unites the articulating bones. The articular capsule is composed of two layers. The outer layer, the ***fibrous capsule***, usually consists of dense, irregular connective tissue. It attaches to the periosteum of the articulating bones at a variable distance from the edge of the articular cartilage. The flexibility of the fibrous capsule permits movement at a joint, whereas its great tensile strength resists dislocation. The fibers of some fibrous capsules are arranged in parallel bundles and are therefore highly adapted to resist recurrent strain. Such fibers are called ***ligaments*** (*ligare* = to bind) and are given special names. The strength of the ligaments is one of the principal factors in holding bone to bone. Synovial joints are freely movable joints (diarthroses) because of the synovial cavity and the arrangement of the articular capsule and accessory ligaments. The inner layer of the articular capsule is formed by a ***synovial membrane.*** The synovial membrane is composed of areolar connective tissue with elastic fibers and a variable amount of adipose tissue. It secretes ***synovial fluid (SF)***, which lubricates the joint and provides nourishment for the articular cartilage.

Many synovial joints also contain ***accessory ligaments***, called extracapsular ligaments and intracapsular ligaments. ***Extracapsular ligaments*** lie outside the articular capsule. An example is the fibular (lateral) collateral ligament of the knee joint (see Figure 8.4b). ***Intracapsular ligaments*** occur within the articular capsule but are excluded from the synovial cavity by folds of the synovial membrane. Examples are the cruciate ligaments of the knee joint (see Figure 8.4b).

Inside some synovial joints are pads of fibrocartilage that lie between the articular surfaces of the bones and are attached by their margins to the fibrous capsule. These pads are called ***articular discs (menisci).*** Singular is ***meniscus.*** See Figure 8.4b. The discs usually subdivide the synovial cavity into two separate spaces. Articular discs allow two bones of different shapes to fit tightly, modify the shape of the joint surfaces of the articulating bones, help to maintain the stability of the joint, and direct the flow of synovial fluid to areas of greatest friction.

Saclike structures called ***bursae*** (*bursa* = pouch) are strategically situated to alleviate friction in some spots (see Figure 8.4f). Bursae resemble joint capsules in that their walls consist of connective tissue lined by a synovial membrane. They are also filled with a fluid similar to synovial fluid. Bursae are located between the skin and bone in places where skin rubs over bone. They are also found between tendons and bones, muscles and bones, and ligaments and bones. Such fluid-filled sacs cushion the movement of one part of the body over another. Inflammation of a bursa is called ***bursitis.***

Label Figure 8.1, the principal parts of a synovial joint.

2. Movements at Synovial Joints

Movements at synovial joints may be classified as follows:

a. ***Gliding*** One surface moves back and forth and from side to side without any angular or rotary movement.
b. ***Angular*** Increase or decrease the angle between bones.
   1. ***Flexion*** Decrease in angle between articulating bones.
   2. ***Extension*** Increase in angle between articulating bones.
   3. ***Hyperextension*** Continuation of extension beyond the anatomical position.

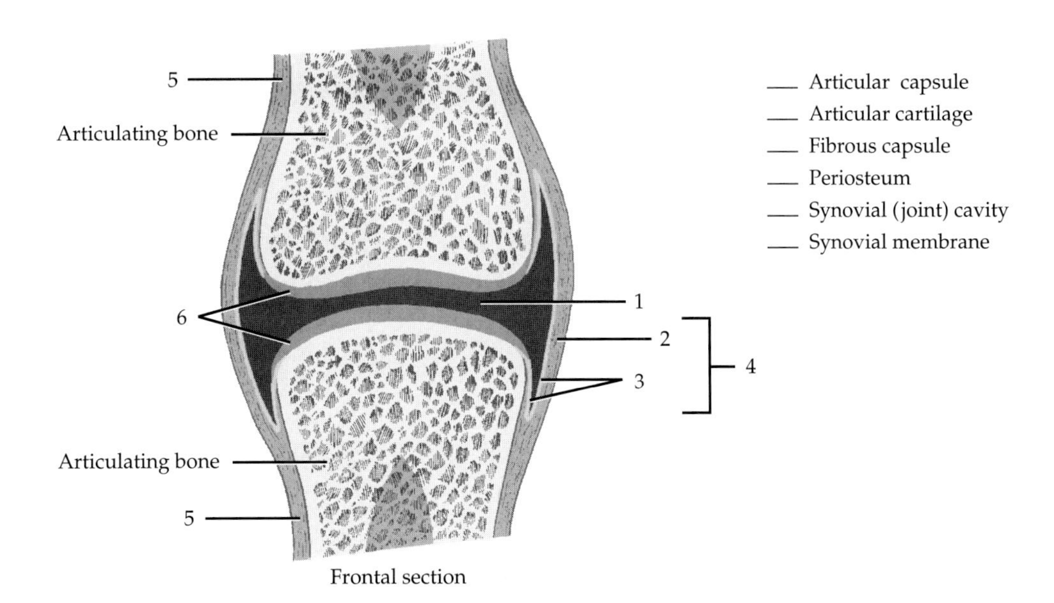

FIGURE 8.1 Parts of a synovial joint.

4. ***Abduction*** (= taking away) Movement of a bone away from midline.
5. ***Adduction*** (= to move toward) Movement of a bone toward midline.

c. ***Rotation*** Movement of a bone in a plane around its longitudinal axis.

d. ***Circumduction*** Movement in which the distal end of a bone moves in a circle while the proximal end remains relatively stable; bone outlines a cone in the air.

e. ***Special*** Found only at the joints indicated:
1. ***Inversion*** Movement of sole inward (medially).
2. ***Eversion*** Movement of sole outward (laterally).
3. ***Dorsiflexion*** Bending the foot in the direction of the dorsum (upper surface).
4. ***Plantar flexion*** Bending the foot in the direction of the plantar surface (sole).
5. ***Protraction*** Movement of mandible or shoulder girdle forward on a plane parallel to ground.
6. ***Retraction*** Movement of protracted part of the body backward on a plane parallel to ground.
7. ***Supination*** Movement of forearm in which palm is turned anteriorly or superiorly.
8. ***Pronation*** Movement of forearm in which palm is turned posteriorly or inferiorly.
9. ***Depression*** Movement in which part of body, such as the mandible or scapula, moves inferiorly.
10. ***Elevation*** Movement in which part of body moves superiorly.

Label the various movements illustrated in Figure 8.2.

3. Types of Synovial Joints

The principal types of synovial joints are as follows:

a. ***Gliding*** Articulating surfaces usually flat; permits gliding movement in two planes, side-to-side and back-and-forth. Example: between carpals, tarsals, sacrum and ilium, sternum and clavicle, scapula and clavicle, and articular processes of vertebrae.

b. ***Hinge*** Convex surface of one bone fits into the concave surface of another; movement in single plane (monoaxial or uniaxial movement), usually flexion and extension. Example: elbow, knee, ankle, interphalangeal joints.

c. ***Pivot*** Rounded or pointed surface of one bone articulates within a ring formed partly by another bone and partly by a ligament; primary movement is rotation; joint is monoaxial. Example: between atlas and axis and between proximal ends of radius and ulna.

d. ***Condyloid*** Oval-shaped condyle of one bone fits into an elliptical cavity of another bone; movement is biaxial, side-to-side, and back-and-forth. Example: between radius and carpals.

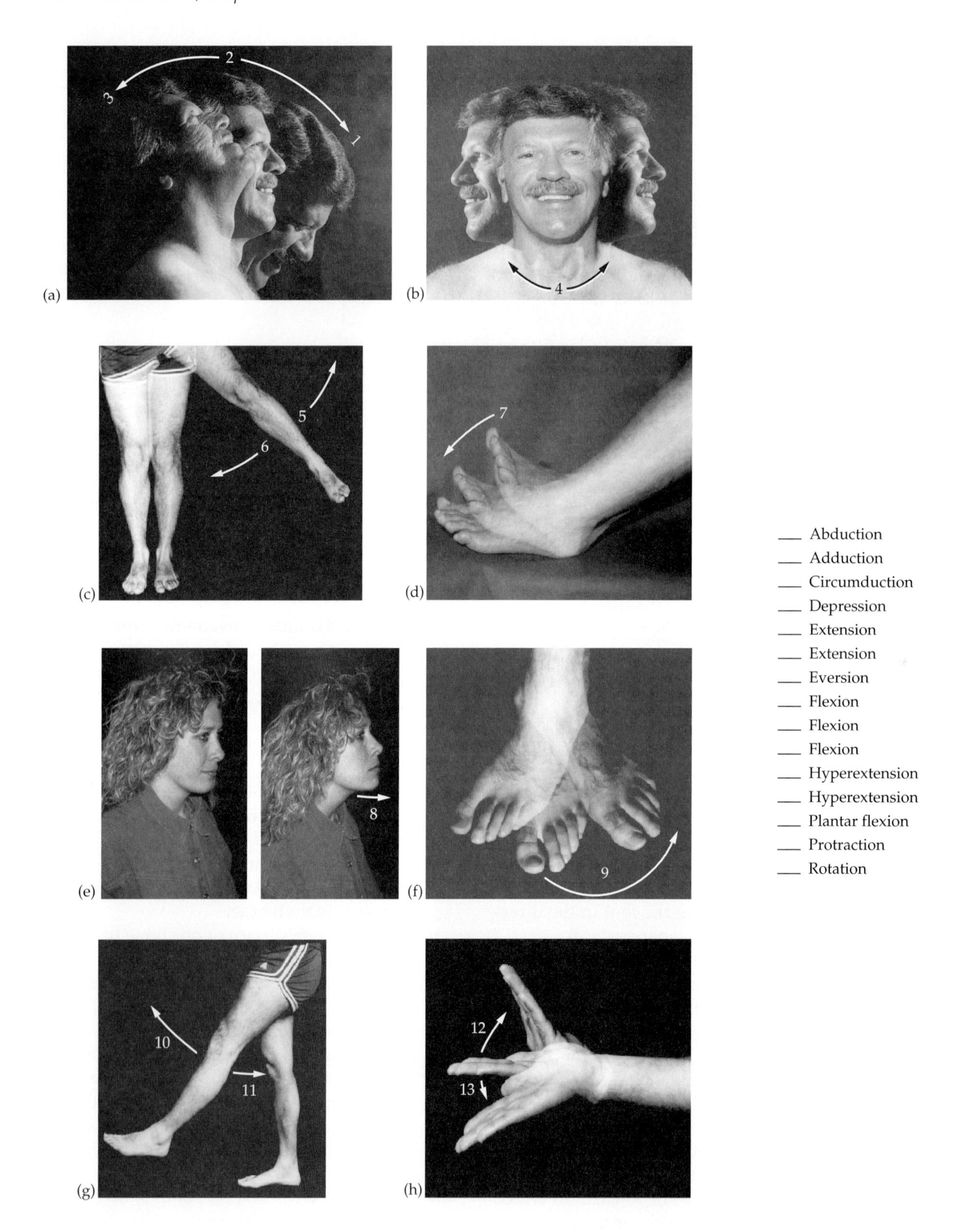

Figure 8.2 Movements at synovial joints.

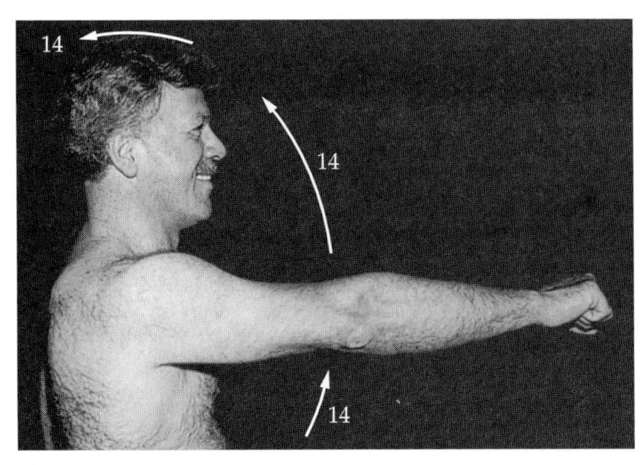

(i)

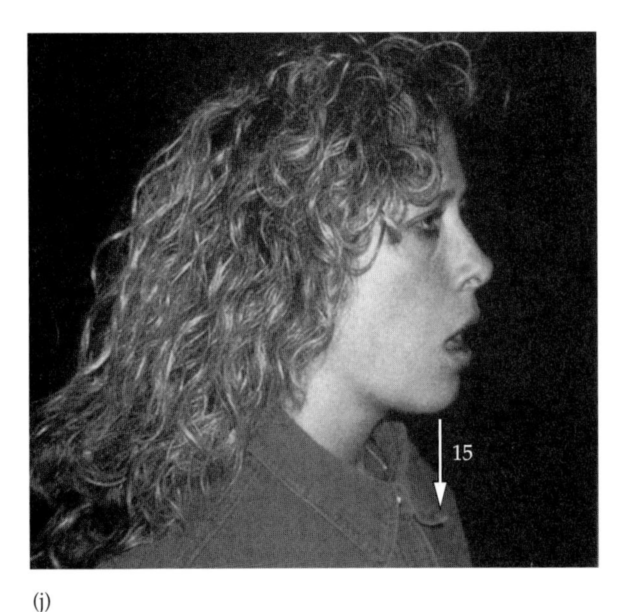

(j)

Figure 8.2 ***(Continued)*** Movements at synovial joints.

e. ***Saddle*** Articular surface of one bone is saddle-shaped and the articular surface of the other bone is shaped like a rider sitting in the saddle; movement similar to that of a condyloid joint. Example: between trapezium of carpus and metacarpal of thumb.
f. ***Ball-and-socket*** Ball-like surface of one bone fits into cuplike depression of another bone; movement is in three planes (triaxial movement), flexion-extension, abduction-adduction, and rotation. Example: shoulder and hip joints.

Examine the articulated skeleton and find as many examples as you can of the joints just described. As part of your examination, be sure to note the shapes of the articular surfaces and the movements possible at each joint.

Label Figure 8.3.

E. KNEE JOINT

The knee (tibiofemoral) joint is one of the largest joints in the body and illustrates the basic structure of a synovial joint and the limitations on its movement. Some of the structures associated with the knee joint are as follows:

1. ***Tendon of quadriceps femoris muscle*** Strengthens joint anteriorly and externally.
2. ***Gastrocnemius muscle*** Strengthens joint posteriorly and externally.
3. ***Patellar ligament*** Continuation of the tendon of the quadriceps femoris tendon below the patella that strengthens anterior portion of joint and prevents leg from being flexed too far backward.
4. ***Fibular (lateral) collateral ligament*** Between femur and fibula; strengthens the lateral side of the joint and prohibits side-to-side movement at the joint.
5. ***Tibial (medial) collateral ligament*** Between femur and tibia; strengthens the medial side of the joint and prohibits side-to-side movement at the joint.
6. ***Oblique popliteal ligament*** Starts in a tendon that lies over the tibia and runs upward and laterally to the lateral side of the femur; supports the posterior surface of the knee.
7. ***Anterior cruciate ligament (ACL)*** Passes posteriorly and laterally from the tibia and attaches to the femur; strengthens the joint internally and may help stabilize the knee during its movements.
8. ***Posterior cruciate ligament (PCL)*** Passes anteriorly and medially from the tibia and attaches to the femur; strengthens the joint internally and may help stabilize the knee during its movements.
9. ***Articular discs (menisci)*** Concentric wedge-shaped pieces of fibrocartilage between the femur and tibia; called ***lateral meniscus*** and the ***medial meniscus;*** help compensate for the irregular shapes of articulating bones and circulate synovial fluid.

Label the structures associated with the knee joint in Figure 8.4 on pages 119–121.

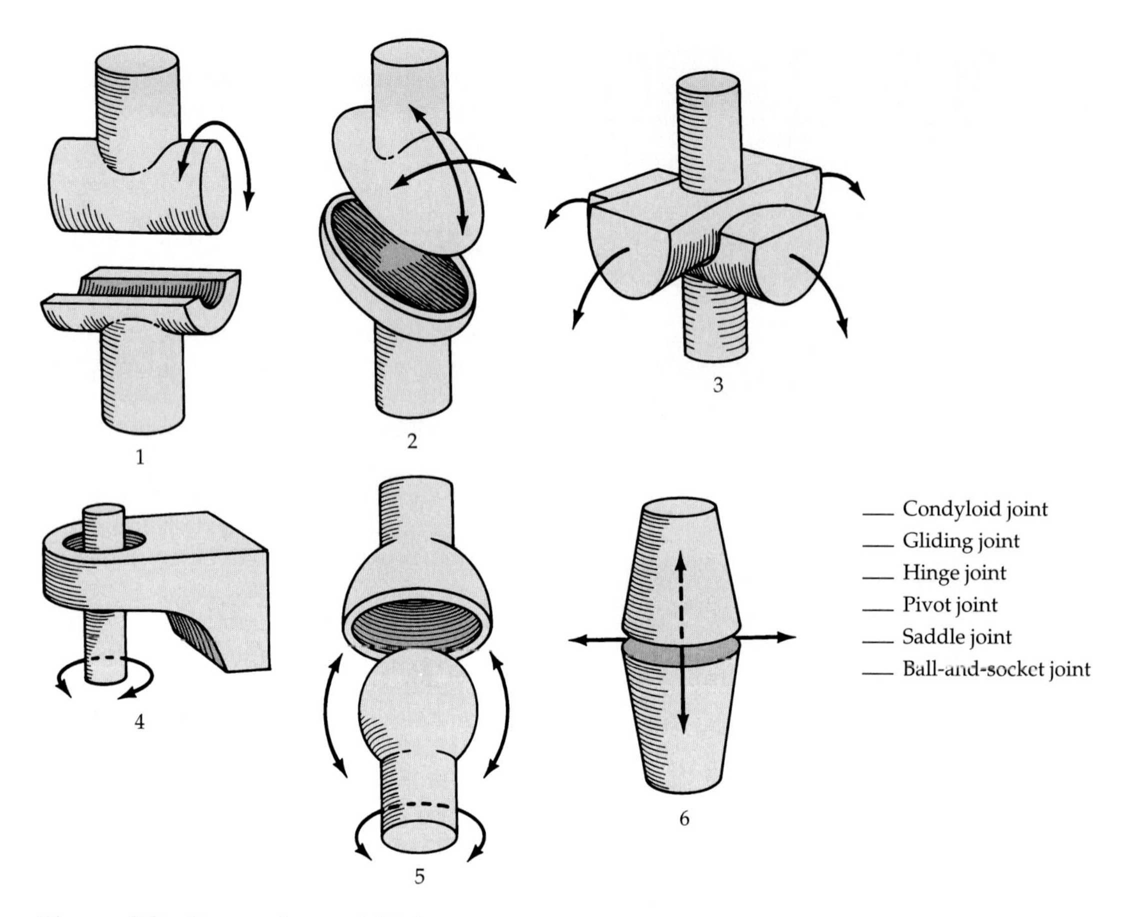

Figure 8.3 Types of synovial joints.

If a longitudinally sectioned knee joint of a cow or lamb is available, examine it and see how many structures you can identify.

ANSWER THE LABORATORY REPORT QUESTIONS AT THE END OF THE EXERCISE.

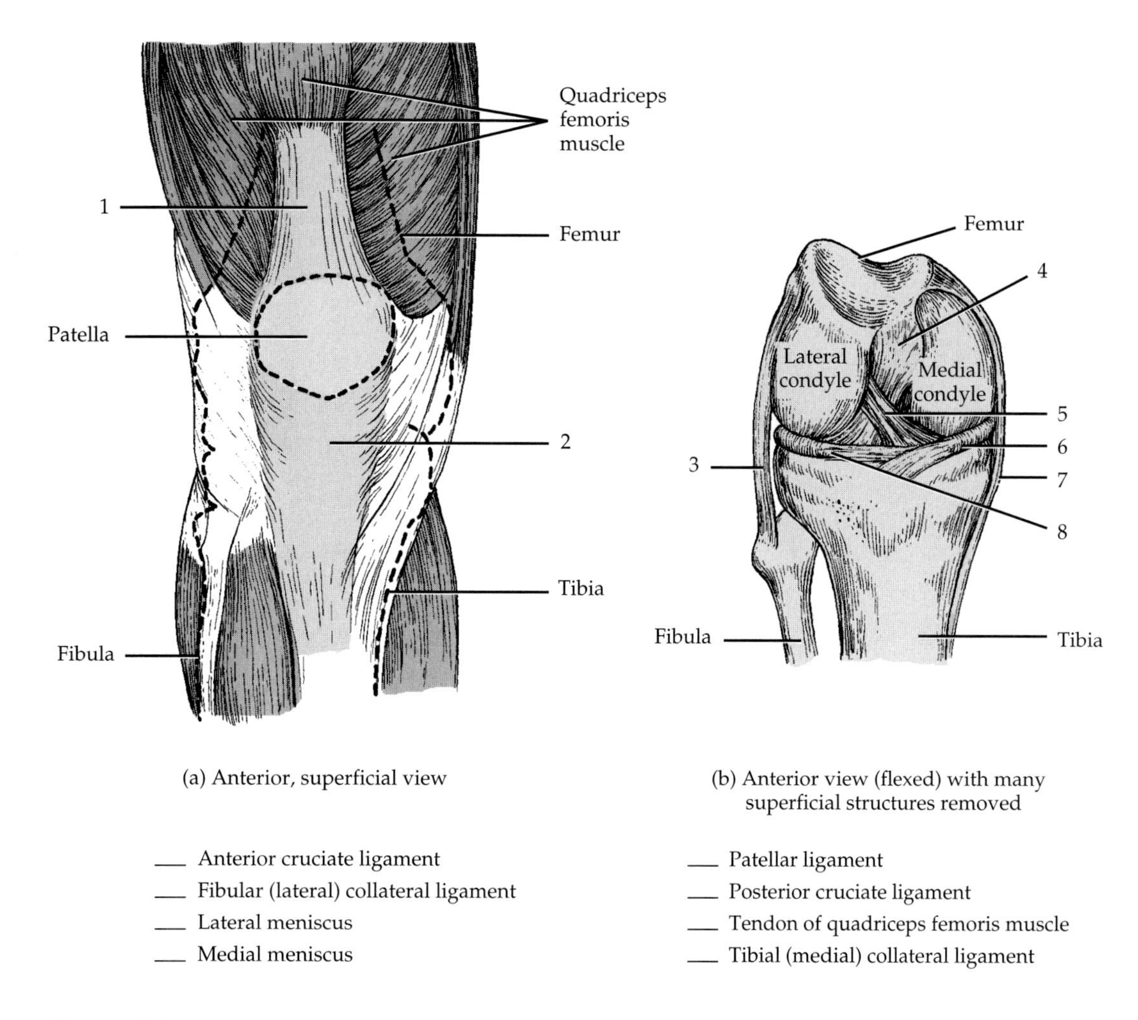

(a) Anterior, superficial view

(b) Anterior view (flexed) with many superficial structures removed

___ Anterior cruciate ligament
___ Fibular (lateral) collateral ligament
___ Lateral meniscus
___ Medial meniscus
___ Patellar ligament
___ Posterior cruciate ligament
___ Tendon of quadriceps femoris muscle
___ Tibial (medial) collateral ligament

Figure 8.4 Ligaments, tendons, bursae, and menisci of right knee joint.

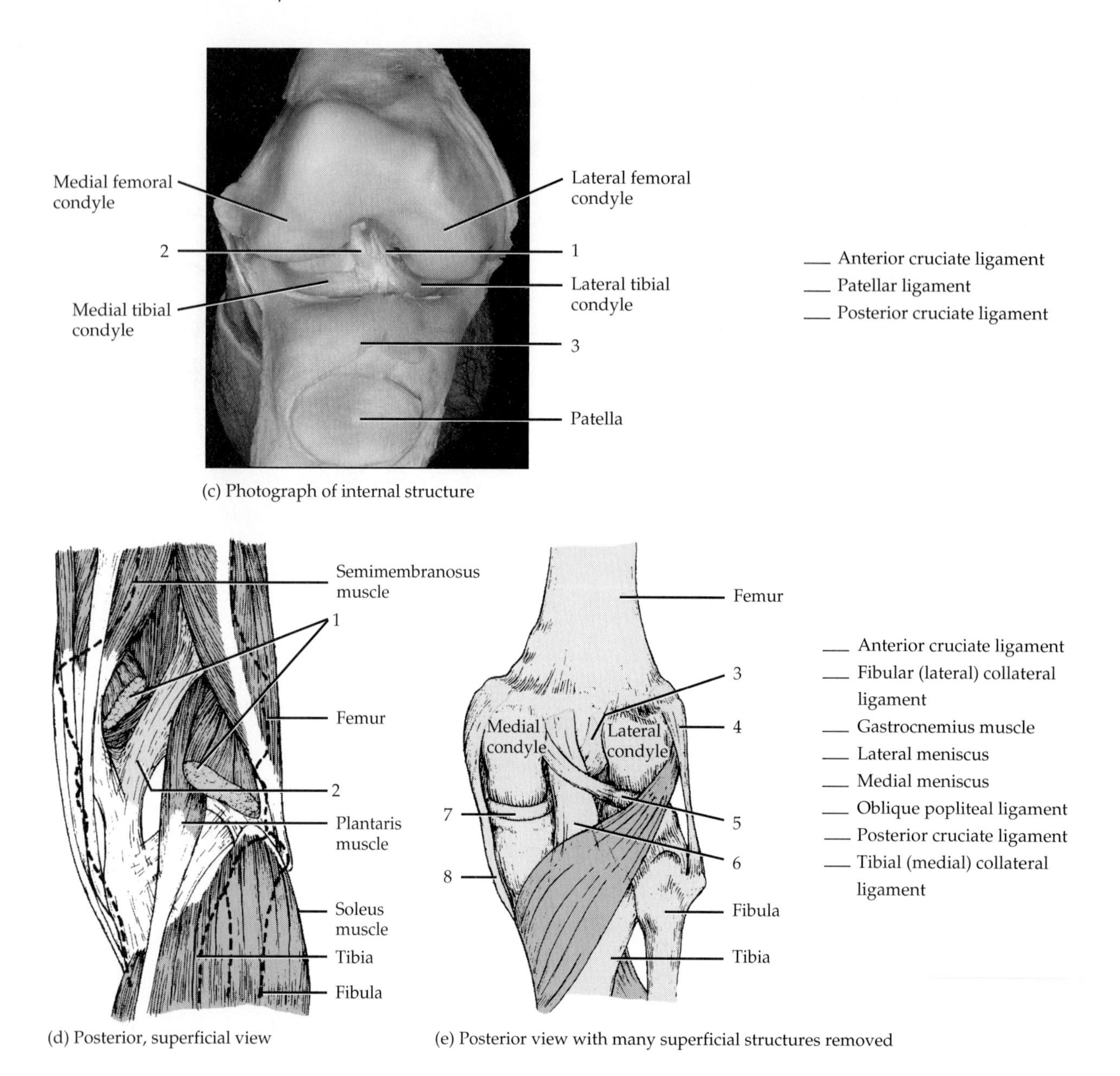

(c) Photograph of internal structure

(d) Posterior, superficial view

(e) Posterior view with many superficial structures removed

Figure 8.4 ***(Continued)*** Ligaments, tendons, bursae, and menisci of right knee joint.

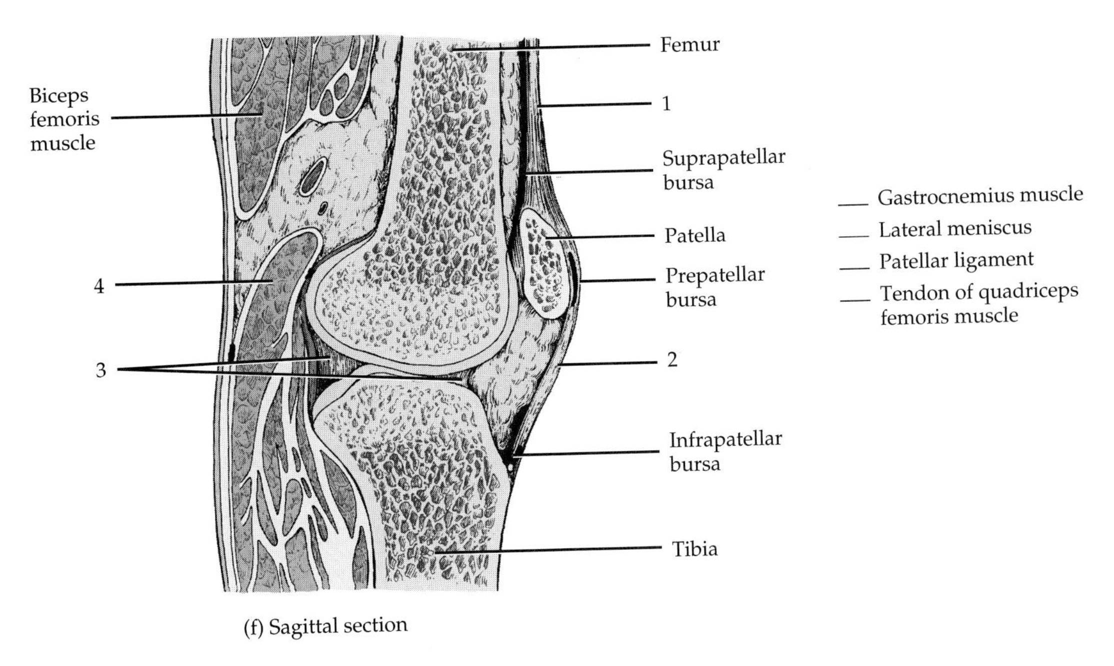

(f) Sagittal section

Figure 8.4 ***(Continued)*** Ligaments, tendons, bursae, and menisci of right knee joint.

LABORATORY REPORT QUESTIONS

Articulations 8

Student ______________________ Date ______________________

Laboratory Section ______________________ Score/Grade ______________________

PART 1. Multiple Choice

_______ **1.** A joint united by dense fibrous tissue that permits a slight degree of movement is a (a) suture (b) syndesmosis (c) symphysis (d) synchondrosis

_______ **2.** A joint that contains a broad flat disc of fibrocartilage is classified as a (a) ball-and-socket joint (b) suture (c) symphysis (d) gliding joint

_______ **3.** The following characteristics define what type of joint? Presence of a synovial cavity, articular cartilage, synovial membrane, and ligaments. (a) suture (b) synchondrosis (c) syndesmosis (d) hinge

_______ **4.** Which joints are slightly movable? (a) diarthroses (b) amphiarthroses (c) synovial (d) synarthroses

_______ **5.** Which type of joint is immovable? (a) synarthrosis (b) syndesmosis (c) symphysis (d) diarthrosis

_______ **6.** What type of joint provides triaxial movement? (a) hinge (b) ball-and-socket (c) saddle (d) condyloid

_______ **7.** Which ligament provides strength on the medial side of the knee joint? (a) oblique popliteal (b) posterior cruciate (c) fibular collateral (d) tibial collateral

_______ **8.** On the basis of structure, which joint is fibrous? (a) symphysis (b) synchondrosis (c) pivot (d) syndesmosis

_______ **9.** The elbow, knee, and interphalangeal joints are examples of which type of joint? (a) pivot (b) hinge (c) gliding (d) saddle

_______ **10.** Functionally, which joint provides the greatest degree of movement? (a) diarthrosis (b) synarthrosis (c) amphiarthrosis (d) syndesmosis

PART 2. Completion

11. The thin layer of hyaline cartilage on articulating surfaces of bones is called ______________________ cartilage.

12. The synovial membrane and fibrous capsule together form the ______________________ capsule.

13. Pads of fibrocartilage between the articular surfaces of bones that maintain stability of the joint are called ______________________.

14. Fluid-filled connective tissue sacs that cushion movements of one body part over another are referred to as ____________________.

15. The ____________________ ligament supports the back of the knee and helps to prevent hyper-extension.

PART 3. Matching

_______ **16.** Circumduction
_______ **17.** Adduction
_______ **18.** Flexion
_______ **19.** Pronation
_______ **20.** Elevation
_______ **21.** Protraction
_______ **22.** Rotation
_______ **23.** Plantar flexion
_______ **24.** Dorsiflexion
_______ **25.** Inversion

A. Decrease in the angle between articulating bones
B. Moving a part upward
C. Bending the foot in the direction of the upper surface
D. Forward movement parallel to the ground
E. Movement of the sole inward at the ankle joint
F. Movement toward the midline
G. Bending the foot in the direction of the sole
H. Turning the palm posteriorly
I. Movement of a bone around its own axis
J. Distal end of a bone moves in a circle while the proximal end remains relatively stable

Muscle Tissue

9

Muscle tissue constitutes 40 to 50% of the total body weight and is composed of fibers (cells) that are highly specialized with respect to four characteristics: (1) ***excitability (irritability),*** or ability to receive and respond to certain stimuli by producing electrical signals called action potentials (impulses); (2) ***contractility,*** or ability to contract (shorten and thicken); (3) ***extensibility (extension),*** or ability to stretch when pulled; and (4) ***elasticity,*** or ability to return to original shape after contraction or extension. Through contraction, muscle tissue performs three basic functions: motion, maintenance of posture, and heat production. In this exercise you will examine the histological structure of muscle tissue.

A. KINDS OF MUSCLE TISSUE

Histologically, three kinds of muscle tissue are recognized:

1. ***Skeletal muscle tissue*** Usually attached to bones; contains conspicuous striations (light and dark bands) when viewed microscopically; voluntary because it contracts under conscious control.
2. ***Cardiac muscle tissue*** Found only in the wall of the heart; striated when viewed microscopically; involuntary because it contracts usually without conscious control.
3. ***Smooth (visceral) muscle tissue*** Located in walls of viscera and blood vessels; referred to as nonstriated because it lacks striations when viewed microscopically; involuntary because it contracts usually without conscious control.

B. SKELETAL MUSCLE TISSUE

Examine a prepared slide of skeletal muscle tissue in longitudinal and transverse (cross) section under high power. Look for the following:

1. ***Sarcolemma*** (*sarco* = flesh; *lemma* = sheath) Plasma membrane of the muscle fiber.
2. ***Sarcoplasm*** Cytoplasm of the muscle fiber.
3. ***Nuclei*** Several in each muscle fiber lying close to sarcolemma.
4. ***Striations*** Alternating light and dark bands in each muscle fiber (described on page 126).
5. ***Epimysium*** (ep'-i-MĪZ-ē-um; *epi* = upon) Fibrous connective tissue that surrounds the entire skeletal muscle.
6. ***Perimysium*** (per'-i-MĪZ-ē-um; *peri* = around) Fibrous connective tissue surrounding a bundle (fascicle) of muscle fibers.
7. ***Endomysium*** (en'-dō-MĪZ-ē-um; *endo* = within) Fibrous connective tissue surrounding individual muscle fibers.

Refer to Figure 9.1 and label the structures indicated.

With the use of an electron microscope, additional details of skeletal muscle tissue may be noted. Among these are the following:

1. ***Mitochondria*** Organelles that have smooth outer membrane and folded inner membrane in which ATP is generated.
2. ***Sarcoplasmic reticulum*** (sar'-kō-PLAZ-mik re-TIK-yoo-lum), or ***SR*** Network of fluid-filled cisterns similar to the endoplasmic reticulum of nonmuscle cells; stores calcium ions in relaxed muscle fibers.
3. ***Transverse (T) tubules*** Tunnel-like infoldings of sarcolemma that run perpendicular to and connect with sarcoplasmic reticulum; open to outside of muscle fiber.
4. ***Triad*** Transverse tubule and the segments of sarcoplasmic reticulum on either side of it.
5. ***Myofibrils*** Threadlike structures that run lengthwise through a fiber and consist of ***thin filaments*** composed of the protein actin and ***thick filaments*** composed of the protein myosin.
6. ***Sarcomere*** (*meros* = part) Contractile unit of a muscle fiber; compartment within a muscle

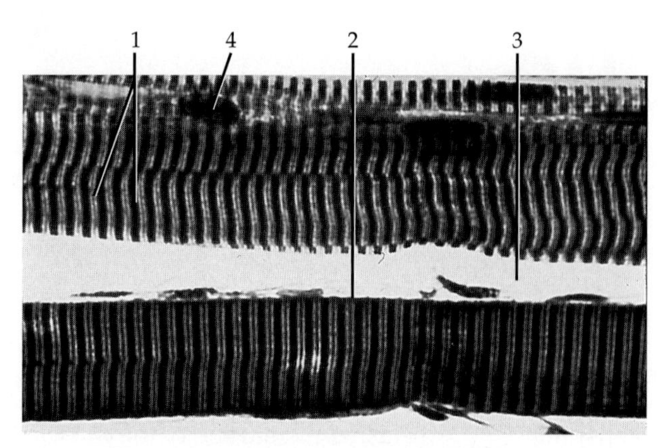

Photomicrograph of several muscle fibers in longitudinal section

FIGURE 9.1 Histology of skeletal muscle tissue.

fiber separated from other sarcomeres by dense material called ***Z discs (lines).***

7. ***A (anisotropic) band*** Dark region in a sarcomere that consists mostly of thick filaments and portions of thin filaments where they overlap the thick filaments.
8. ***I (isotropic) band*** Light region in a sarcomere composed of the rest of the thin filaments but no thick filaments. The combination of alternating dark A bands and light I bands gives the muscle fiber the striated (striped) appearance.
9. ***H zone*** Region in the center of the A band of a sarcomere consisting of thick filaments only.
10. ***M line*** Series of fine threads in the center of the H zone formed by proteins that connect adjacent thick filaments.

Figure 9.2 is a diagram of skeletal muscle tissue based on electron micrographic studies. Label the structures shown.

C. CARDIAC MUSCLE TISSUE

Examine a prepared slide of cardiac muscle tissue in longitudinal and transverse (cross) section under high power. Locate the following structures: ***sarcolemma, endomysium, nuclei, striations,*** and ***intercalated discs*** (transverse thickenings of the sarcolemma that separate individual fibers). Label Figure 9.3.

D. SMOOTH (VISCERAL) MUSCLE TISSUE

Examine a prepared slide of smooth muscle tissue in longitudinal and transverse (cross) section under high power. Locate and label the following structures in Figure 9.4 on page 128: ***sarcolemma, sarcoplasm, nucleus,*** and ***muscle fiber.***

E. SKELETAL MUSCLE CONTRACTION

Skeletal muscle contractions occur only if a stimulus of minimum intensity is applied to the muscle. Such a stimulus is called a ***threshold stimulus,*** and a nerve cell that delivers such a stimulus is known as a ***motor neuron.*** A motor neuron plus all the skeletal muscle fibers it stimulates is called a ***motor unit.*** Stimulation of a motor neuron produces a contraction in all of the muscle fibers of a particular motor unit.

Each motor neuron has a thread-like process, called an ***axon,*** that extends from the spinal cord to a group of skeletal muscle fibers. Close to its target skeletal muscle fibers, the axon enters the endomysium and branches into several ***axon terminals.*** The term ***neuromuscular junction (NMJ),*** or ***myoneural junction,*** is applied to the structure formed by the axon terminal and its associated skeletal muscle fiber (Figure 9.5 on page 129). The region of the muscle fiber sarcolemma adjacent to the axon terminal is called the ***motor end plate.***

The distal end of each axon terminal expands into clusters of bulb-like structures called ***synaptic end bulbs*** that contain many membrane-enclosed sacs called ***synaptic vesicles.*** Inside each vesicle are thousands of chemicals called ***neurotransmitters.*** Although many different neurotransmitters exist, the one released at the NMJ is ***acetylcholine*** (as′-ē-til-KŌ-lēn), abbreviated ***ACh.***

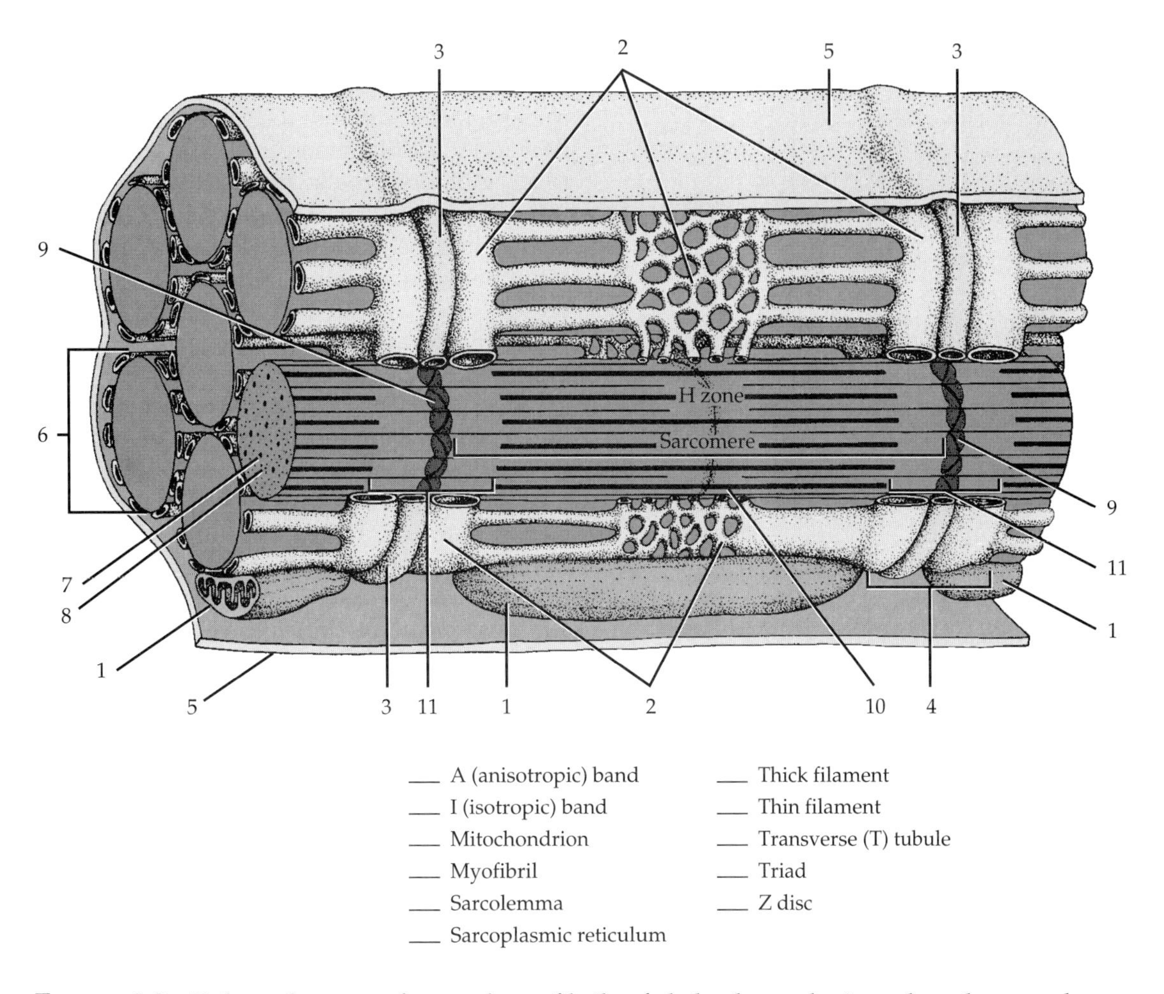

___ A (anisotropic) band
___ I (isotropic) band
___ Mitochondrion
___ Myofibril
___ Sarcolemma
___ Sarcoplasmic reticulum
___ Thick filament
___ Thin filament
___ Transverse (T) tubule
___ Triad
___ Z disc

FIGURE 9.2 Enlarged aspect of several myofibrils of skeletal muscle tissue based on an electron micrograph.

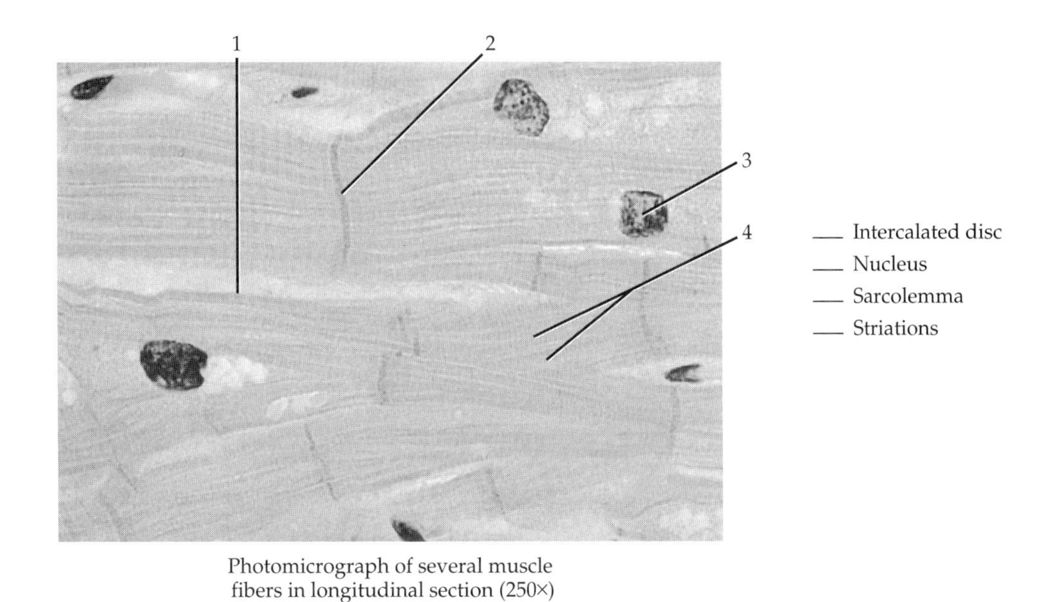

Photomicrograph of several muscle fibers in longitudinal section (250×)

___ Intercalated disc
___ Nucleus
___ Sarcolemma
___ Striations

FIGURE 9.3 Histology of cardiac muscle tissue.

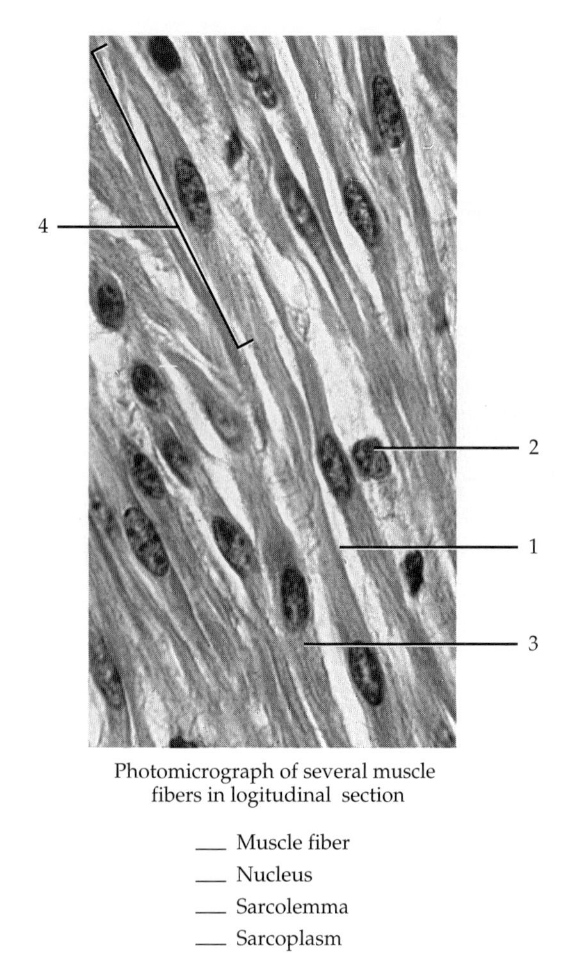

Photomicrograph of several muscle fibers in logitudinal section

___ Muscle fiber
___ Nucleus
___ Sarcolemma
___ Sarcoplasm

FIGURE 9.4 Histology of smooth (visceral) muscle tissue.

When an action potential (nerve impulse) reaches the synaptic end bulbs, it causes the synaptic vesicles to fuse with the sarcolemma of the skeletal muscle fiber, and the vesicles liberate ACh into a space between the axon terminal and sarcolemma called the ***synaptic cleft.*** ACh diffuses across the synaptic cleft and binds to receptors on the motor end plate. This binding of ACh to the motor end plate initiates a sequence of events that ultimately leads to the initiation of an action potential along the sarcolemma of the muscle fiber. From the sarcolemma, the action potential travels into the transverse (T) tubules to enter the interior of a muscle fiber and then to the sarcoplasmic reticulum which, in response, releases calcium ions (Ca^{2+}) into the sarcoplasm among the thin and thick filaments. Once released, Ca^{2+} binds to a regulatory protein in muscle associated with thin filaments called ***troponin.*** This causes a change in the shape of the thin filament, thus permitting myosin cross bridges to attach to receptors on the thin filament. This attachment causes ATP to be split by an enzyme called ***ATPase*** found on the thick filament. The energy released from the splitting of ATP causes the myosin cross bridges to move, pulling the thin filaments inward toward the H zone. This sliding of the thin filaments draws the Z discs toward each other, and the sarcomere shortens. The myofibrils thus contract and the whole muscle fiber shortens (Figure 9.6). This model of skeletal muscle contraction is called the ***sliding filament mechanism.***

Two changes permit a muscle fiber to relax after it has contracted. First, ACh is rapidly broken down by an enzyme called ***acetylcholinesterase (AChE).*** When action potentials cease in a motor neuron, release of ACh stops and AChE rapidly breaks down any ACh already present in the synaptic cleft. Second, active transport pumps rapidly remove Ca^{2+} from the sarcoplasm back into the sarcoplasmic reticulum. As the Ca^{2+} level in sarcoplasm decreases, the myosin cross bridges are prevented from attaching to receptors on the thin filaments, and the filaments slip back to their relaxed positions.

According to the ***all-or-none principle,*** a skeletal muscle fiber will contract to its fullest extent or not at all if a threshold or greater stimulus is applied; if a subthreshold stimulus is applied, no contraction will occur. In other words, muscle fibers do not contract partially. With respect to the entire skeletal muscle, however, the muscle can contract partially since at any given time some muscle fibers in the muscle are contracted, while others are relaxed.

ANSWER THE LABORATORY REPORT QUESTIONS AT THE END OF THE EXERCISE.

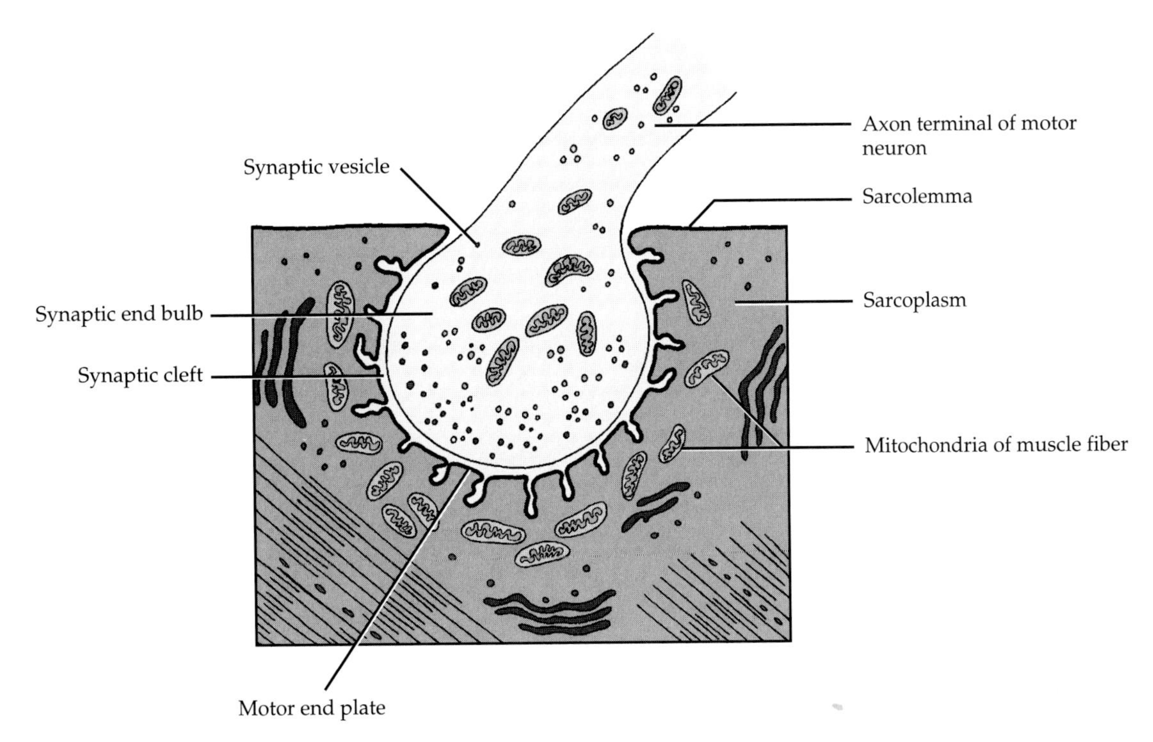

FIGURE 9.5 Neuromuscular junction.

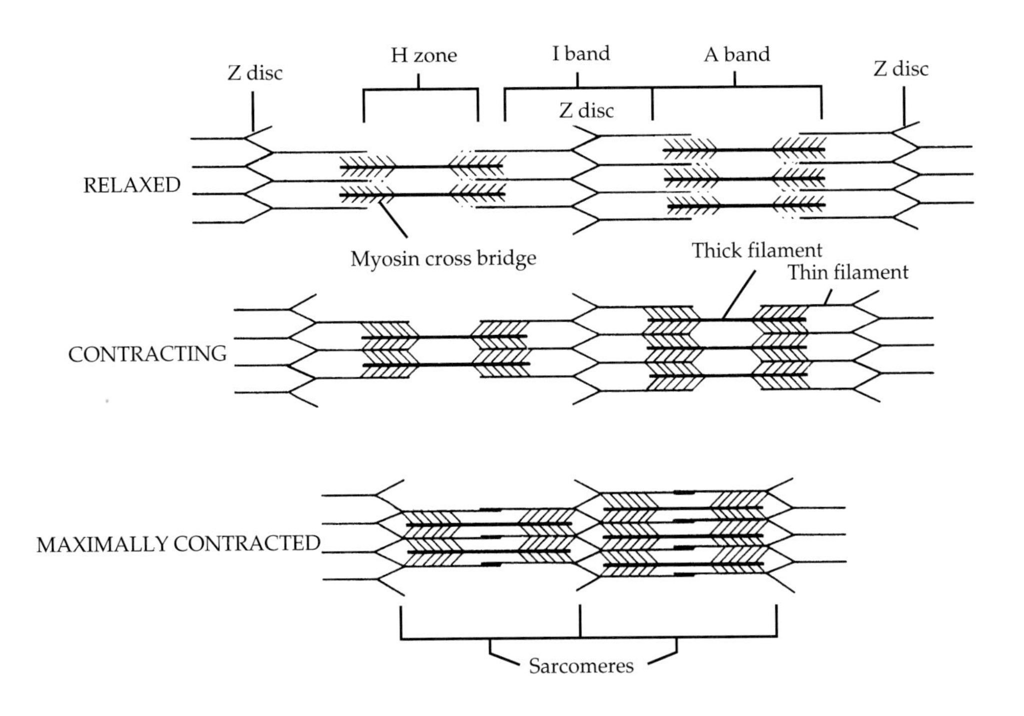

FIGURE 9.6 Sliding filament theory of a skeletal muscle contraction. The positions of the various parts of two sarcomeres in relaxed, contracting, and maximally contracted states are shown. Note the movement of thin filaments and relative size of H zone.

LABORATORY REPORT RESULTS

Muscle Tissue 9

Student ______________________________ **Date** ______________________________

Laboratory Section ______________________ **Score/Grade** ______________________

PART 1. Multiple Choice

_______ **1.** The ability of muscle tissue to return to its original shape after contraction or extension is called (a) excitability (b) elasticity (c) extension (d) tetanus

_______ **2.** Which of the following is striated and voluntary? (a) skeletal muscle tissue (b) cardiac muscle tissue (c) visceral muscle tissue (d) smooth muscle tissue

_______ **3.** The portion of a sarcomere composed of thin filaments only is the (a) H zone (b) A band (c) I band (d) Z disc

_______ **4.** The area of contact between a motor axon terminal and a muscle fiber sarcolemma (motor end plate) is called the (a) synapse (b) filament (c) transverse tubule (d) neuromuscular junction

_______ **5.** The connective tissue layer surrounding bundles of muscle fibers is called the (a) perimysium (b) endomysium (c) ectomysium (d) myomysium

_______ **6.** Which of the following is striated and involuntary? (a) smooth muscle tissue (b) skeletal muscle tissue (c) cardiac muscle tissue (d) visceral muscle tissue

_______ **7.** The portion of a sarcomere that consists mostly of thick filaments and portions of filaments where thin and thick filaments overlap is called the (a) H zone (b) triad (c) A band (d) I band

_______ **8.** Intercalated discs are characteristic of which type of muscle tissue? (a) cardiac muscle tissue (b) skeletal muscle tissue (c) smooth muscle tissue (d) visceral muscle tissue

_______ **9.** The space between an axon terminal and sarcolemma is called the (a) end plate (b) synaptic cleft (c) synaptic gutter (d) synaptic end bulb

PART 2. Completion

10. The ability of muscle tissue to receive and respond to stimuli is called ______________________.

11. Fibrous connective tissue located between muscle fibers is known as ______________________.

12. The sections of a muscle fiber separated by Z discs are called ______________________.

13. The plasma membrane surrounding a muscle fiber is called the ______________________.

14. The region in a sarcomere consisting of thick filaments only is known as the ______________________.

15. Muscle tissue that is nonstriated and involuntary is ______________________.

16. The ability of muscle tissue to stretch when pulled is called ____________________.
17. The phenomenon by which a muscle fiber contracts to its fullest extent or not at all is known as the

    ____________________.

18. A stimulus strong enough to initiate a nerve impulse is called a(n) ____________________ stimulus.
19. The neurotransmitter released at neuromuscular junctions is called ____________________.
20. The change in the shape of thin filaments that permits myosin cross bridges to attach to receptors on thin filaments is due to the binding of calcium ions to a regulatory protein called

    ____________________.

21. In relaxed muscle fibers, calcium ions are stored in the ____________________.
22. Whereas thin filaments are composed of the protein actin, thick filaments are composed of the protein ____________________.
23. The combination of alternating dark A bands and light ____________________ bands gives a muscle fiber its striated appearance.
24. A motor neuron and all the muscle fibers it innervates is called a(n) ____________________.
25. Neurotransmitters are stored in ____________________, which are located within synaptic end bulbs.

Skeletal Muscles

10

In this exercise you will learn the names, locations, and actions of the principal skeletal muscles of the body.

A. HOW SKELETAL MUSCLES PRODUCE MOVEMENT

1. Origin and Insertion

Skeletal muscles produce movements by exerting force on tendons, which in turn pull on bones or other structures, such as the skin. Most muscles cross at least one joint and are usually attached to the articulating bones that form the joint. When such a muscle contracts, it draws one articulating bone toward the other. The two articulating bones usually do not move equally in response to the contraction. One is held nearly in its original position because other muscles contract to pull it in the opposite direction or because its structure makes it less movable. Ordinarily, the attachment of a muscle tendon to the stationary bone is called the ***origin.*** The attachment of the other muscle tendon to the movable bone is the ***insertion.*** A good analogy is a spring on a door. The part of the spring attached to the door represents the insertion; the part attached to the frame is the origin. The fleshy portion of the muscle between the tendons of the origin and insertion is called the ***belly (gaster).*** The origin is usually proximal and the insertion distal, especially in the limbs. In addition, muscles that move a body part generally do not cover the moving part. For example, although contraction of the biceps brachii muscle moves the forearm, the belly of the muscle lies over the humerus.

2. Group Actions

Most movements require several skeletal muscles acting in groups rather than individually. Also, most skeletal muscles are arranged in opposing pairs at joints, that is, flexors-extensors, abductors-adductors, and so on. Consider flexing the forearm at the elbow, for example. A muscle that causes a desired action is referred to as the ***prime mover*** or ***agonist*** (*agogos* = leader). In this instance, the biceps brachii is the prime mover (see Figure 10.13a). Simultaneously with the contraction of the biceps brachii, another muscle, called the ***antagonist*** (*antiagonistes* = opponent), is stretching. In this movement, the triceps brachii serves as the antagonist (see Figure 10.13b). The antagonist has an action that is opposite to that of the prime mover; that is, the antagonist stretches and yields to the movement of the prime mover. You should not assume, however, that the biceps brachii is always the prime mover and the triceps brachii is always the antagonist. For example, when extending the forearm at the elbow, the triceps brachii serves as the prime mover, and the biceps brachii functions as the antagonist; their roles are reversed. Note that if the prime mover and antagonist contracted simultaneously with equal force, there would be no movement.

In addition to prime movers and antagonists, most movements also involve muscles called ***synergists*** (SIN-er-jists; *syn* = together; *ergon* = work), which assist the prime mover in performing a particular task. For example, flex your hand at the wrist and then make a fist. Note how difficult this is to do. Now extend your hand at the wrist and then make a fist. Note how much easier it is to clench your fist. In this case, the extensor muscles of the wrist act as synergists in cooperation with the flexor muscles of the fingers acting as prime movers. The extensor muscles of the fingers serve as antagonists (see Figure 10.14c, d).

Some muscles in a group also act as ***fixators,*** which stabilize the origin of the prime mover so that the prime mover can act more efficiently. Fixators steady the proximal end of a limb while movements occur at the distal end. For example, the scapula is a freely movable bone in the pectoral (shoulder) girdle that serves as an origin for several muscles that move the arm. However, for the scapula to serve as a firm origin for muscles that move the arm, it must be held steady. This is accomplished by fixator muscles that hold the scapula firmly against the back of the chest. In abduction of the arm, the deltoid muscle serves as the prime mover, whereas fixators (pectoralis minor, rhomboideus major, rhomboideus minor, trapezius, subclavius, and serratus anterior muscles) hold the scapula firmly (see Figure 10.11). These fixators stabilize the scapula that serves as the attachment site for the origin of the deltoid muscle while the insertion of the muscle pulls on the humerus to abduct the arm. Under different conditions and depending on the movement and which point is fixed, many muscles act, at various times, as prime movers, antagonists, synergists, or fixators.

B. ARRANGEMENT OF FASCICLES

Recall from Exercise 9 that skeletal muscle fibers (cells) are arranged within the muscle in bundles called ***fascicles (fasciculi).*** The muscle fibers are arranged in a parallel fashion within each bundle, but the arrangement of the fascicles with respect to the tendons may take several characteristic patterns.

Table 10.1 describes the major patterns of fascicles. Using your textbook as a guide, provide an example of each.

C. NAMING SKELETAL MUSCLES

Most of the almost 700 skeletal muscles of the body are named on the basis of one or more distinctive characteristics. If you understand these characteristics, you will find it much easier to learn and remember the names of individual muscles.

Table 10.2 describes the major characteristics that are used to name skeletal muscles. Using your textbook as a guide, provide an example for each.

D. CONNECTIVE TISSUE COMPONENTS

Skeletal muscles are protected, strengthened, and attached to other structures by several connective

TABLE 10.1
Arrangements of Fascicles

Arrangement	Description	Example
PARALLEL	Fascicles are parallel with longitudinal axis of muscle and terminate at either end in flat tendons.	
FUSIFORM	Fascicles are nearly parallel with longitudinal axis of muscle and terminate at either end in flat tendons, but muscle tapers toward tendons where the diameter is less than that of the belly.	
PENNATE	Fascicles are short in relation to muscle length and the tendon extends nearly the entire length of the muscle.	
Unipennate	Fascicles are arranged on only one side of tendon.	
Bipennate	Fascicles are arranged on both sides of a centrally positioned tendon.	
Multipennate	Fascicles attach obliquely from many directions to several tendons.	
CIRCULAR	Fascicles are arranged in a circular pattern and enclose an orifice (opening).	
TRIANGULAR	Fascicles attached to a broad tendon converge to give the muscle a triangular appearance.	

TABLE 10.2
Characteristics Used for Naming Skeletal Muscles

Characteristic	Description	Example
Direction of muscle fibers	Direction of muscle fibers relative to the midline of the body. **Rectus** means the fibers run straight; sometimes they run parallel to the midline. **Transverse** means the fibers run perpendicular to the midline. **Oblique** means the fibers run diagonally to the midline.	
Location	Structure near which a muscle is found.	
Size	Relative size of the muscle. **Maximus** means largest. **Minimus** means smallest. **Longus** means longest. **Brevis** means short.	
Number of origins	Number of tendons of origin. **Biceps** means two origins. **Triceps** means three origins. **Quadriceps** means four origins.	
Shape	Relative shape of the muscle. **Deltoid** means triangular. **Trapezius** means trapezoidal. **Serratus** means saw-toothed. **Rhomboideus** means rhomboid- or diamond-shaped.	
Origin and insertion	Sites where muscle originates and inserts.	
Action	Principal action of the muscle. **Flexor** (FLEK-sor): decreases the angle at a joint. **Extensor** (eks-TEN-sor): increases the angle at a joint. **Abductor** (ab-DUK-tor): moves a bone away from the midline. **Adductor** (ad-DUK-tor): moves a bone closer to the midline. **Levator** (le-VĀ-tor): produces a superior movement. **Depressor** (de-PRES-or): produces an inferior movement. **Supinator** (soo'-pi-NĀ-tor): turns the palm superiorly or anteriorly. **Pronator** (prō-NĀ-tor): turns the palm inferiorly or posteriorly. **Sphincter** (SFINGK-ter): decreases the size of an opening. **Tensor** (TEN-sor): makes a body part more rigid. **Rotator** (RŌ-tāt-or): moves a bone around its longitudinal axis.	

tissue components. For example, the entire muscle is usually wrapped with a dense, irregular connective tissue called the ***epimysium*** (ep'-i-MĪZ-ē-um; *epi* = upon). When the muscle is cut in transverse (cross) section, invaginations of the epimysium divide the muscle into fascicles. These invaginations of the epimysium are called the ***perimysium*** (per'-i-MĪZ-ē-um; *peri* = around). In turn, invaginations of the perimysium, called ***endomysium*** (en'-dō-MĪZ-ē-um; *endo* = within), penetrate into the interior of each fascicle and separate individual muscle fibers from one another. The epimysium, perimysium, and endomysium are all extensions of deep fascia and are all continuous with the connective tissue that attaches the muscle to another structure, such as bone or other muscle. All three elements may be extended beyond the muscle cells as a ***tendon*** (*tendere* = to stretch out)—a cord of connective tissue that attaches a muscle to the periosteum of bone. The connective tissue may also extend as a broad, flat band of tendons called an ***aponeurosis*** (*apo* = from; *neuron* = a tendon). Aponeuroses also attach to the coverings of a bone or another muscle. When a muscle contracts, the tendon and its corresponding bone or muscle are pulled toward the contracting muscle. In this way skeletal muscles produce movement.

In Figure 10.1, label the epimysium, perimysium, endomysium, fascicles, and muscle fibers.

E. PRINCIPAL SKELETAL MUSCLES

In the pages that follow, a series of tables has been provided for you to learn the principal skeletal muscles by region.[1] Use each table as follows:

1. Take each muscle, in sequence, and study the ***learning key*** that appears in parentheses after the name of the muscle. The learning key is a list of prefixes, suffixes, and definitions that explain the derivations of the muscles' names. It will help you to understand the reason for giving a muscle its name.
2. As you learn the name of each muscle, determine its origin, insertion, and action and write these in the spaces provided in the table. Consult your textbook as necessary.
3. Again, using your textbook as a guide, label the diagram referred to in the table.
4. Try to visualize what happens when the muscle contracts so that you will understand its action.
5. Do steps 1 through 4 for each muscle in the table. Before moving to the next table, examine a torso or chart of the skeletal system so that you can compare and approximate the positions of the muscles.
6. When possible, try to feel each muscle on your own body.

Refer to Tables 10.3 through 10.23 and Figures 10.2 through 10.20.

[1]A few of the muscles listed are not illustrated in the diagrams. Please consult your textbook to locate these muscles.

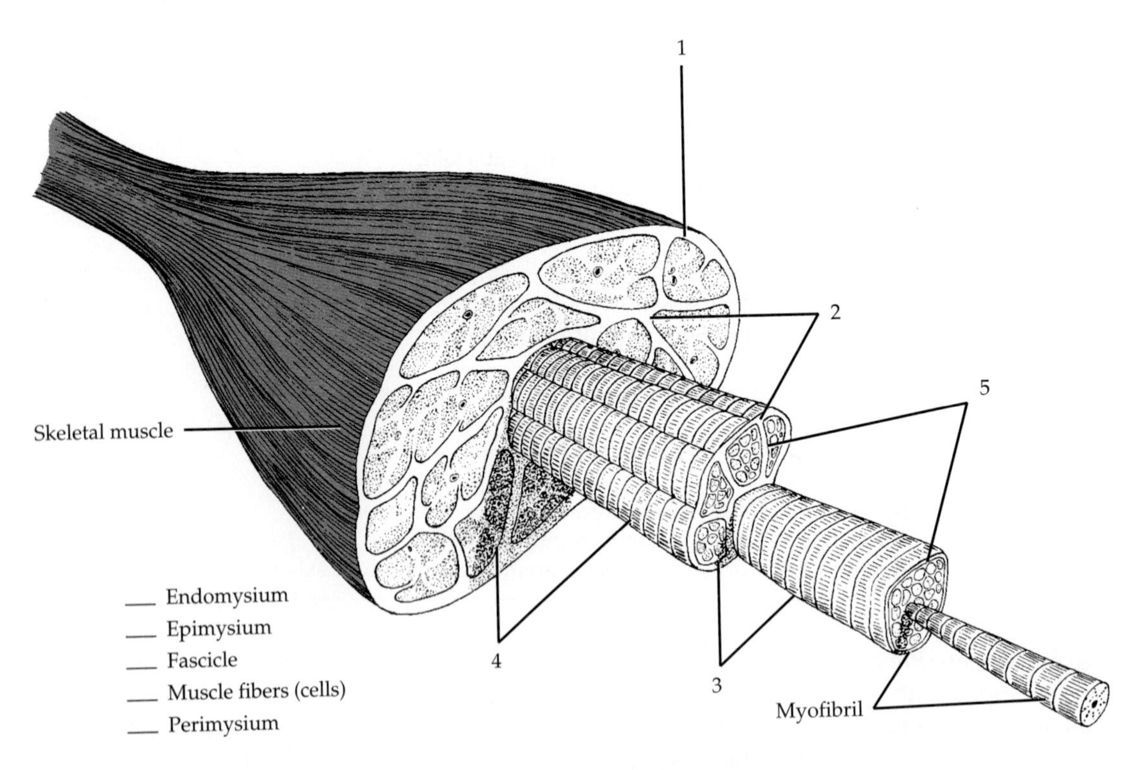

FIGURE 10.1 Connective tissue components of a skeletal muscle.

TABLE 10.3
Muscles of Facial Expression (After completing the table, label Figure 10.2.)

OVERVIEW: The muscles in this group provide humans with the ability to express a wide variety of emotions, including grief, surprise, fear, and happiness. The muscles themselves lie within the layers of superficial fascia. As a rule, they arise from the fascia or bones of the skull and insert into the skin. Because of their insertions, the muscles of facial expression move the skin rather than a joint when they contract.

Muscle	Origin	Insertion	Action
Epicranius (ep-i-KRĀ-nē-us; *epi* = over; *crani* = skull)	This muscle is divisible into two portions: the frontalis, over the frontal bone, and the occipitalis, over the occipital bone. The two muscles are united by a strong aponeurosis, the galea aponeurotica (epicranial aponeurosis), which covers the superior and lateral surfaces of the skull.		
Frontalis (fron-TA-lis; *front* = forehead)			
Occipitalis (ok-si'-pi-TA-lis; *occipito* = base of skull)			
Orbicularis oris (or-bi'-kyoo-LAR-is OR-is; *orb* = circular; *or* = mouth)			
Zygomaticus (zī-gō-MA-ti-kus) **major** (*zygomatic* = cheek bone; *major* = greater)			
Levator labii superioris (le-VĀ-ter LA-bē-ī soo-per'-ē-OR-is; *levator* = raises or elevates; *labii* = lip; *superioris* = upper)			
Depressor labii inferioris (de-PRE-ser LA-bē-ī in-fer'-ē-OR-is; *depressor* = depresses or lowers; *inferioris* = lower)			
Buccinator (BUK-si-nā'-tor; *bucc* = cheek)			
Mentalis (men-TA-lis; *mentum* = chin)			
Platysma (pla-TIZ-ma; *platy* = flat, broad)			
Risorius (ri-ZOR-ē-us; *risor* = laughter)			
Orbicularis oculi (or-bi'-kyoo-LAR-is O-kyoo-lī; *oculus* = eye)			
Corrugator supercilii (KOR-a-gā'-tor soo-per-SI-lē-ī; *corrugo* = to wrinkle; *supercilium* = eyebrow)			
Levator palpebrae superioris (le-VĀ-tor PAL-pe-brē soo-per'-ē-OR-is; *palpebrae* = eyelids) (See Figure 10.4)			

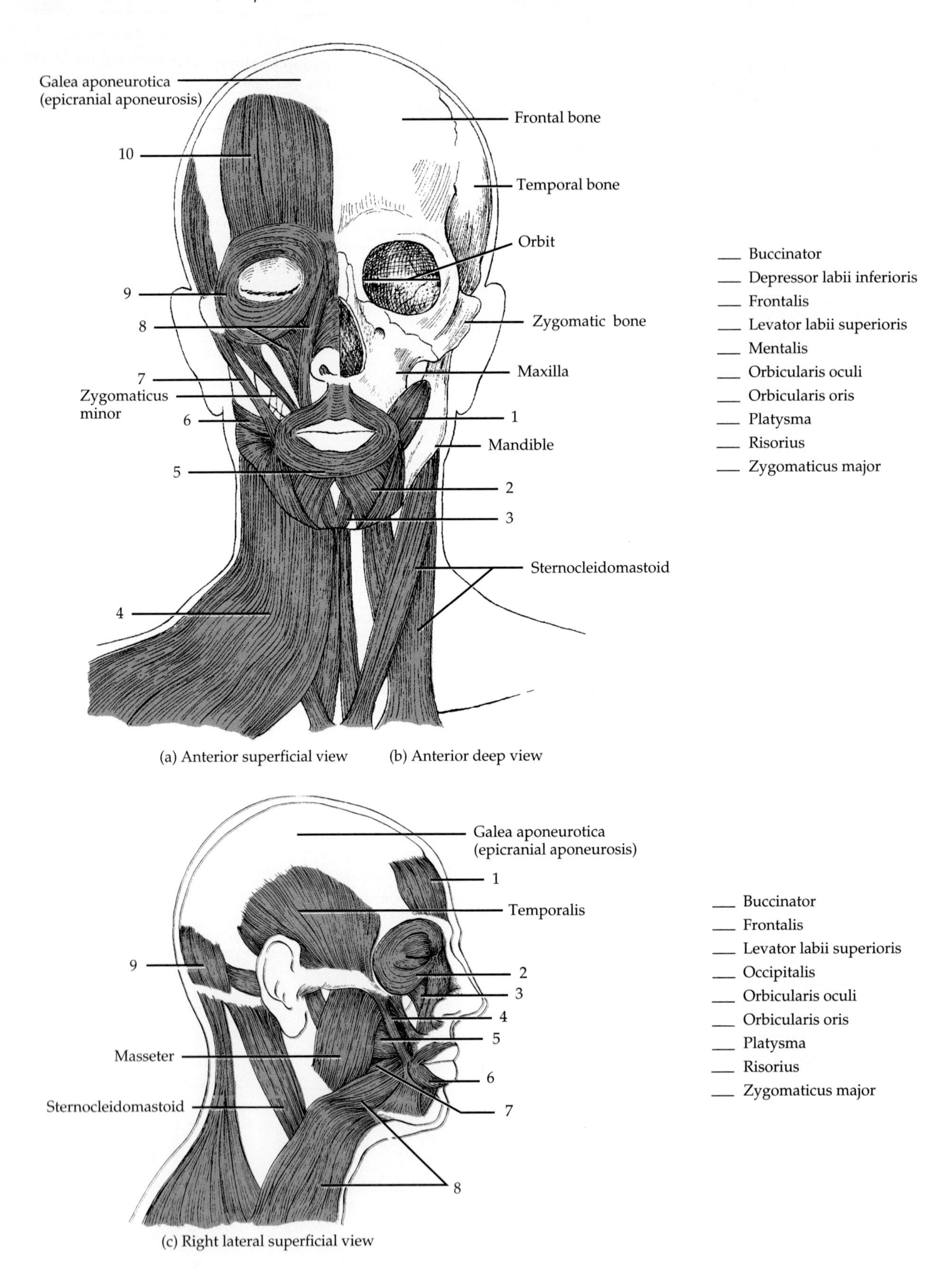

(a) Anterior superficial view (b) Anterior deep view

___ Buccinator
___ Depressor labii inferioris
___ Frontalis
___ Levator labii superioris
___ Mentalis
___ Orbicularis oculi
___ Orbicularis oris
___ Platysma
___ Risorius
___ Zygomaticus major

(c) Right lateral superficial view

___ Buccinator
___ Frontalis
___ Levator labii superioris
___ Occipitalis
___ Orbicularis oculi
___ Orbicularis oris
___ Platysma
___ Risorius
___ Zygomaticus major

FIGURE 10.2 Muscles of facial expression.

TABLE 10.4
Muscles That Move the Mandible (Lower Jaw) (After completing the table, label Figure 10.3.)

OVERVIEW: Muscles that move the mandible (lower jaw) are also known as muscles of mastication because they are involved in biting and chewing. These muscles also assist in speech.

Muscle	Origin	Insertion	Action
Masseter (MA-se-ter; *masseter* = chewer)			
Temporalis (tem'-por-A-lis; *tempora* = temples)			
Medial pterygoid (TER-i-goid; *medial* = closer to midline; *pterygoid* = like a wing)			
Lateral pterygoid (TER-i-goid; *lateral* = farther from midline)			

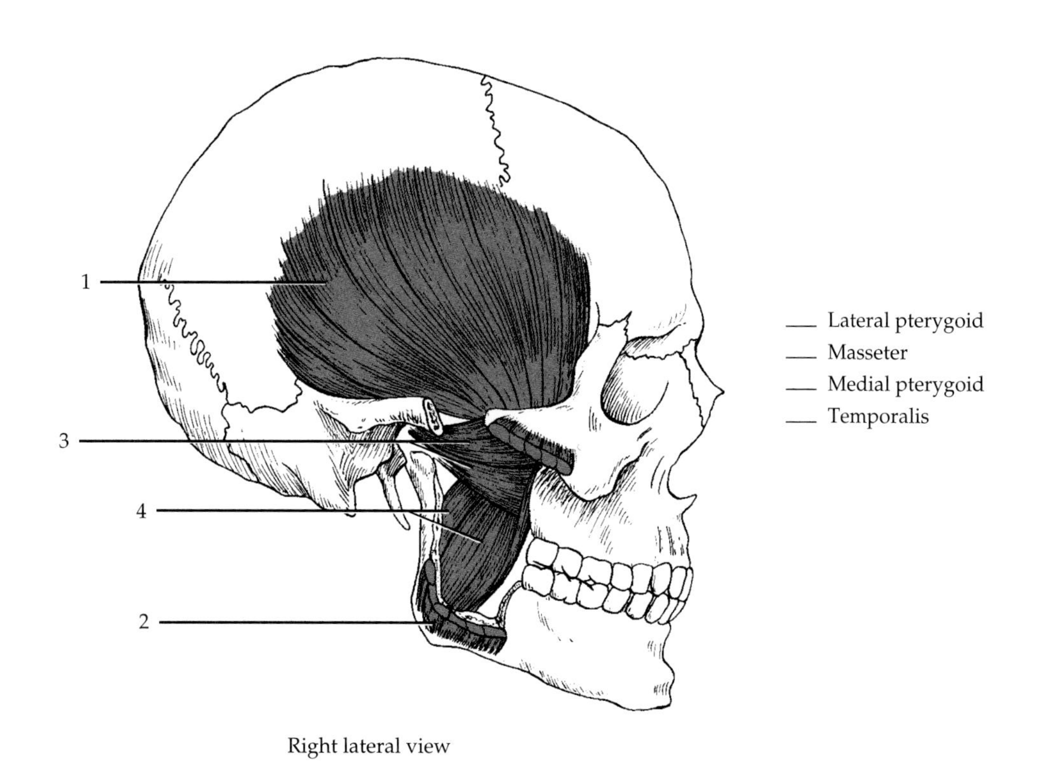

FIGURE 10.3 Muscles that move mandible (lower jaw).

TABLE 10.5
Muscles That Move the Eyeballs—The Extrinsic Muscles* (After completing the table, label Figure 10.4.)

OVERVIEW: Muscles associated with the eyeballs are of two principal types: extrinsic and intrinsic. ***Extrinsic muscles*** originate outside the eyeballs and are inserted on their outer surfaces (sclera). They move the eyeballs in various directions. ***Intrinsic muscles*** originate and insert entirely within the eyeballs. They move structures within the eyeballs.

Movements of the eyeballs are controlled by three pairs of extrinsic muscles. The two pairs of rectus muscles move the eyeballs in the direction indicated by their respective names—superior, inferior, lateral, and medial. The pair of oblique muscles—superior and inferior—rotate the eyeballs on their axes. The extrinsic muscles of the eyeballs are among the fastest contracting and most precisely controlled skeletal muscles of the body.

Muscle	Origin	Insertion	Action
Superior rectus (REK-tus; *superior* = above; *rectus* = in this case, muscle fibers running parallel to long axis of eyeball) **Inferior rectus** (REK-tus; *inferior* = below)			
Lateral rectus (REK-tus)			
Medial rectus (REK-tus)			
Superior oblique (ō-BLĒK; *oblique* = in this case, muscle fibers running diagonally to long axis of eyeball)			
Inferior oblique (ō-BLĒK)			

*Muscles situated on the outside of the eyeballs.

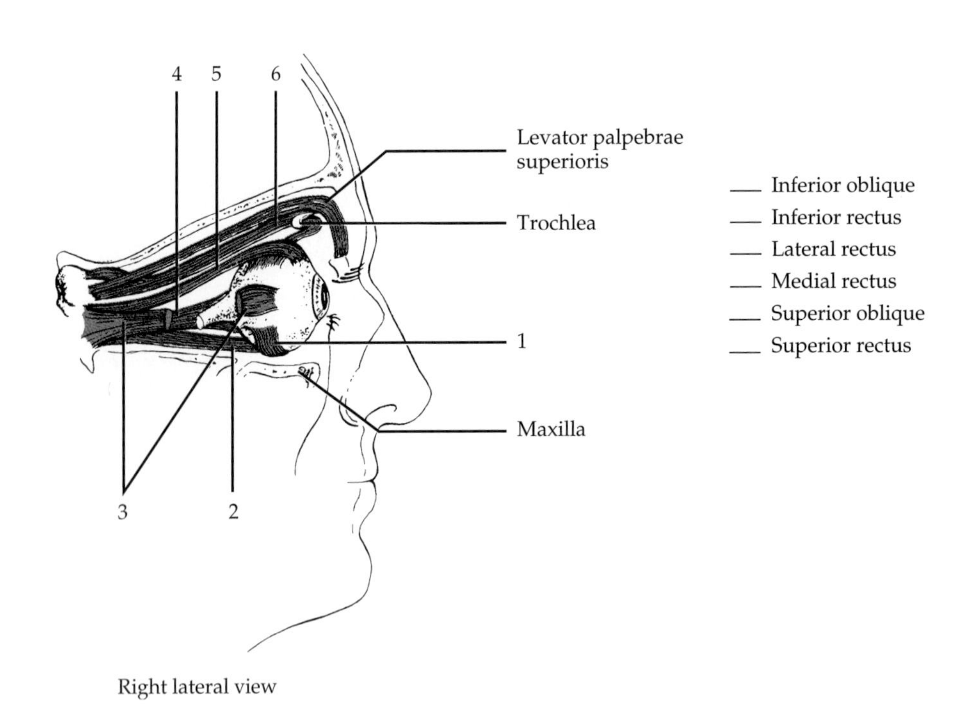

FIGURE 10.4 Extrinsic muscles of eyeballs.

TABLE 10.6
Muscles That Move the Tongue—The Extrinsic Muscles (After completing the table, label Figure 10.5.)

OVERVIEW: The tongue is divided into lateral halves by a median fibrous septum. The septum extends throughout the length of the tongue and is attached inferiorly to the hyoid bone. Like the muscles of the eyeballs, muscles of the tongue are of two principal types—extrinsic and intrinsic. ***Extrinsic muscles*** originate outside the tongue and insert into it. They move the entire tongue in various directions, such as anteriorly, posteriorly, and laterally. ***Intrinsic muscles*** originate and insert within the tongue. They alter the shape of the tongue rather than move the entire tongue. The extrinsic and intrinsic muscles of the tongue are arranged in both lateral halves of the tongue.

Muscle	Origin	Insertion	Action
Genioglossus (jē'-nē-ō-GLOS-us; *geneion* = chin *glossus* = tongue)			
Styloglossus (stī'-lō-GLOS-us; *stylo* = stake or pole)			
Palatoglossus (pal'-a-tō-GLOS-us; *palato* = palate)			
Hyoglossus (hī-ō-GLOS-us)			

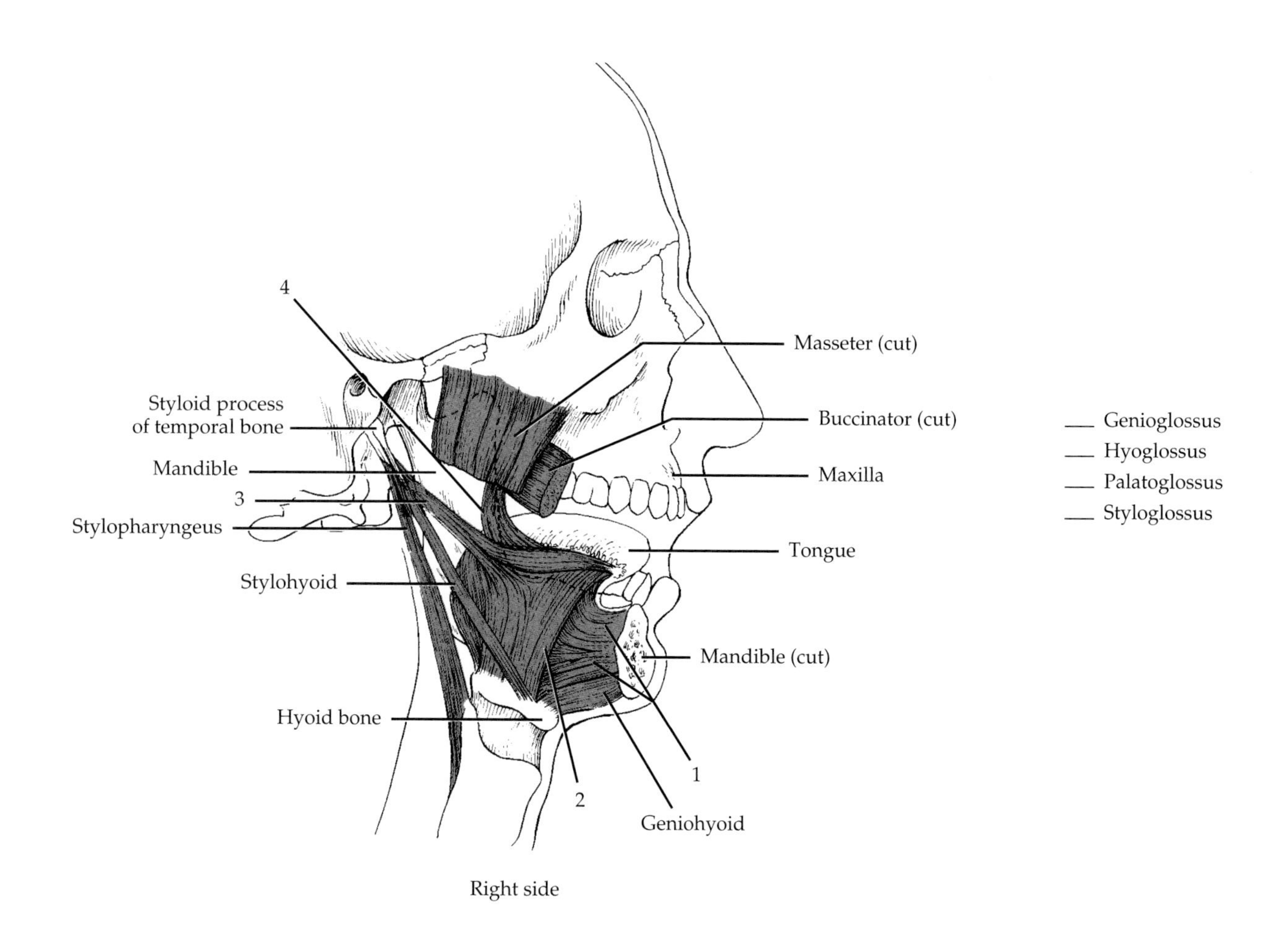

FIGURE 10.5 Muscles that move the tongue.

TABLE 10.7
Muscles of the Floor of the Oral Cavity (After completing the table, label Figure 10.6.)

OVERVIEW: As a group, these muscles are referred to as ***suprahyoid muscles.*** They lie superior to the hyoid bone and all insert into it. The diagstric muscle consists of an anterior belly and a posterior belly united by an intermediate tendon that is held in position by a fibrous group.

Muscle	Origin	Insertion	Action
Digastric (di-GAS-trik; *di* = two; *gaster* = belly)			
Stylohyoid (sti′-lo-HĪ-oid; *stylo* = stake or pole styloid process of temporal bone; *hyoedes* = U-shaped, pertaining to hyoid bone)			
Mylohyoid (mi′-lo-HĪ-oid)			
Geniohyoid (je′-ne-o-HĪ-oid; *geneion* = chin (See Figure 10.5)			

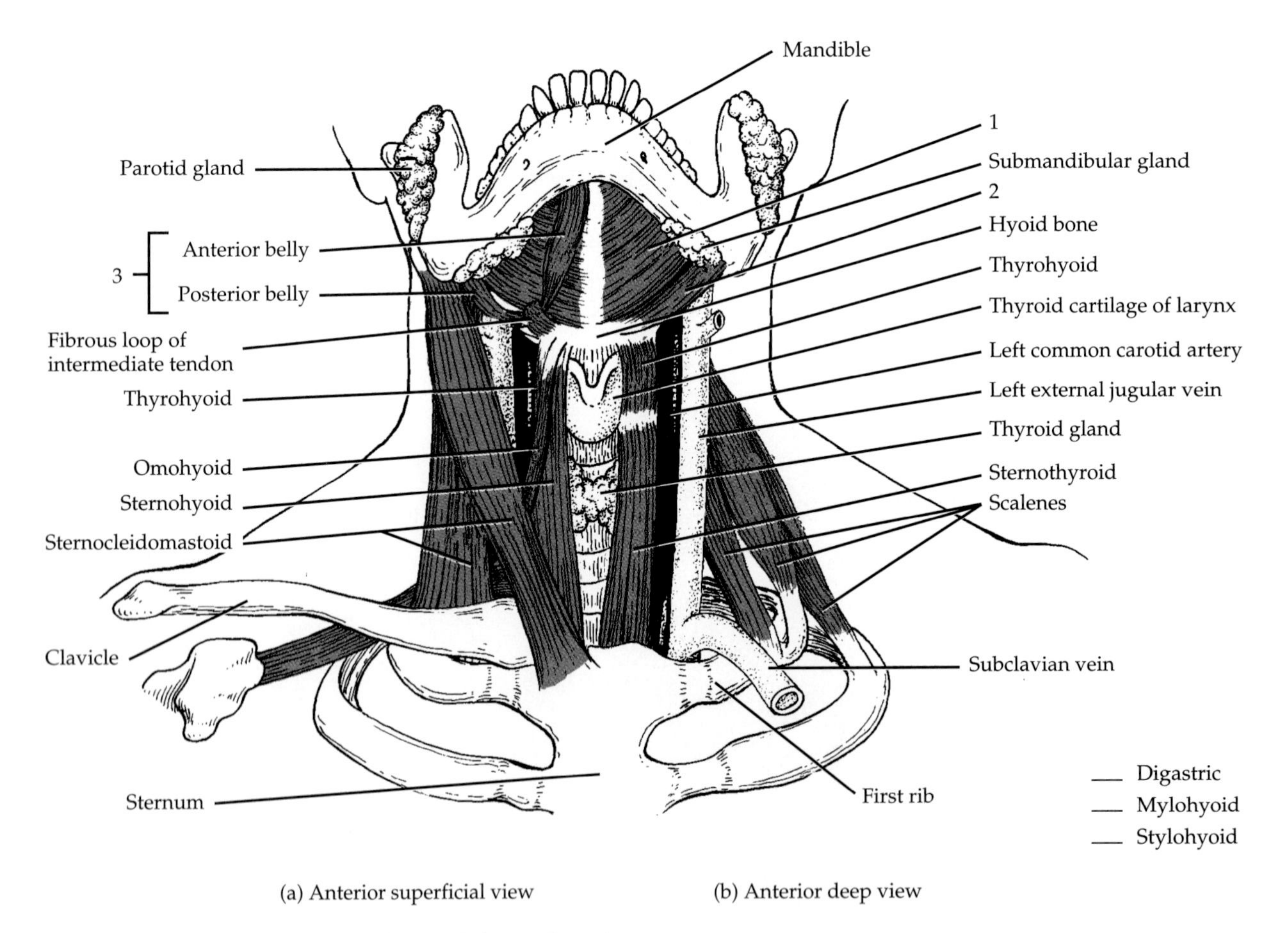

FIGURE 10.6 Muscles of the floor of the oral cavity.

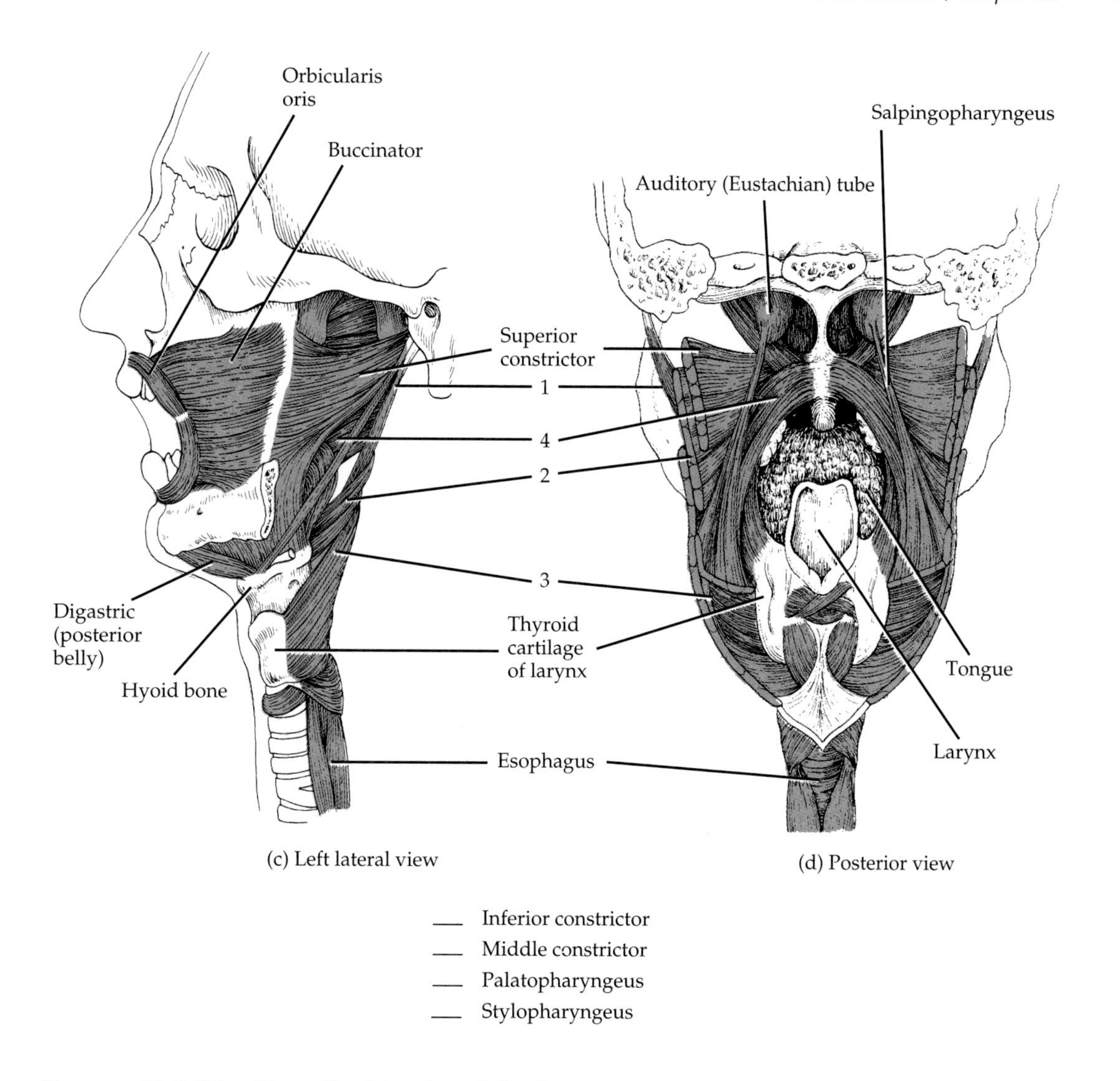

___ Inferior constrictor
___ Middle constrictor
___ Palatopharyngeus
___ Stylopharyngeus

FIGURE 10.6 ***(Continued)*** Muscles of the floor of the oral cavity.

TABLE 10.8
Muscles of the Larynx (Voice Box) (After completing the table, label Figure 10.7.)

OVERVIEW: The muscles of the larynx (voice box), like those of the eyeballs and tongue, are grouped into ***extrinsic*** and ***intrinsic muscles.*** The extrinsic muscles of the larynx marked (*) are together referred to as ***infrahyoid (strap) muscles*** because they lie inferior to the hyoid bone. The omohyoid muscle, like the diagstric muscle, is composed of two bellies and an intermediate tendon. In this case, however, the two bellies are referred to as superior and inferior, rather than anterior and posterior.

Muscle	Origin	Insertion	Action
EXTRINSIC **Omohyoid*** (ō′-mō-HĪ-oid; *omo* = relationship to shoulder; *hyoedes* = U-shaped)			
Sternohyoid* (ster′-nō-HĪ-oid; *sterno* = sternum)			
Sternothyroid* (ster′-nō-THĪ-roid; *thyro* = thyroid gland)			
Thyrohyoid* (thī′-rō-HĪ-oid)			
Stylopharyngeus (stī′-lō-fa-RIN-jē-us)			
Palatopharyngeus (pal′-a-tō-fa-RIN-jē-us)			
Inferior constrictor (kon-STRIK-tor)			
Middle constrictor (kon-STRIK-tor)			
INTRINSIC **Cricothyroid** (kri-kō-THĪ-roid; *crico* = cricoid cartilage of larynx)			
Posterior cricoarytenoid (kri-kō-ar′-i-TĒ-noid; *arytaina* = shaped like a jug)			

TABLE 10.8 (*Continued*)

Muscle	Origin	Insertion	Action
Lateral cricoarytenoid (kri′-kō-ar′-i-TĒ-noid)			
Arytenoid (ar′-i-TĒ-noid)			
Thyroarytenoid (thī′-rō-ar-i-TĒ-noid)			

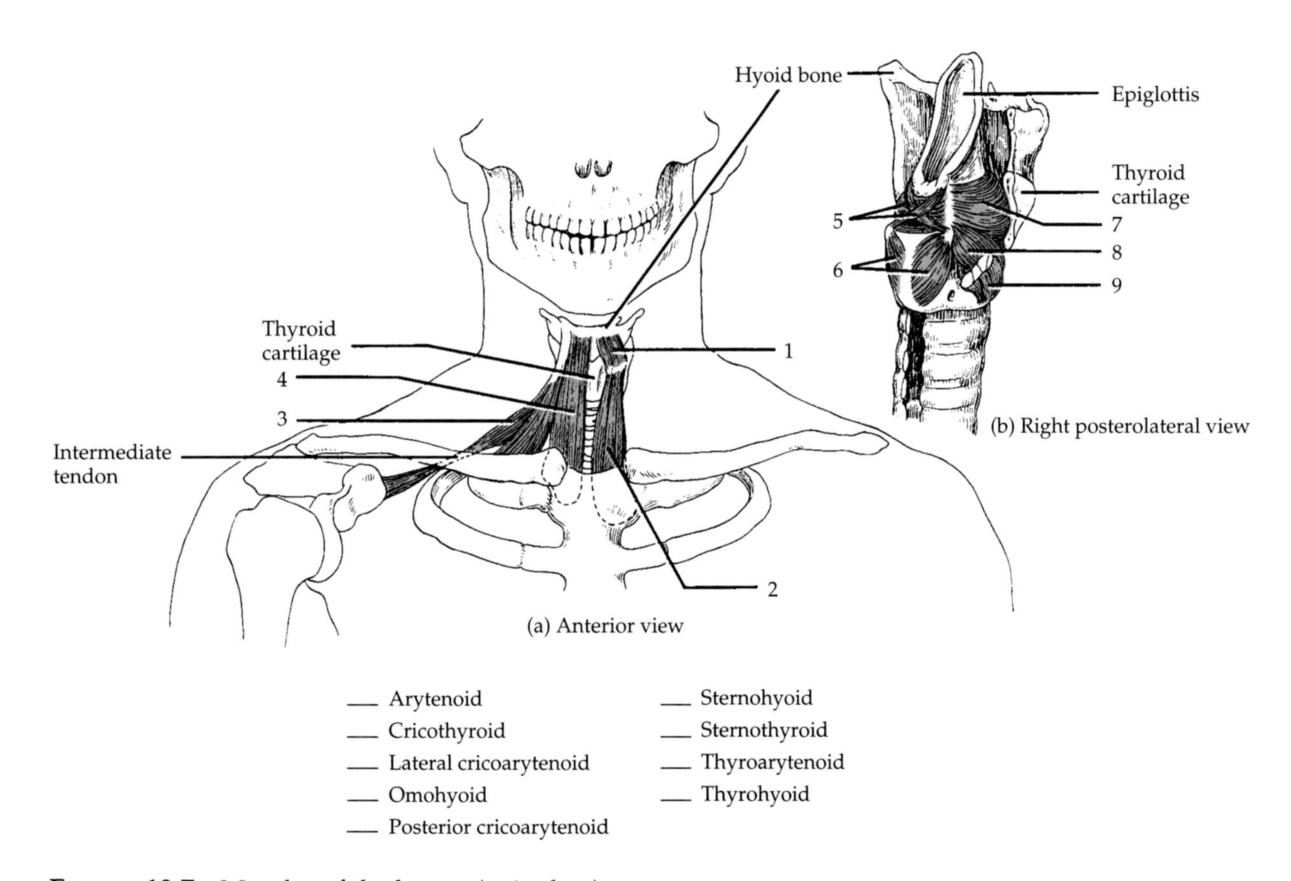

___ Arytenoid
___ Cricothyroid
___ Lateral cricoarytenoid
___ Omohyoid
___ Posterior cricoarytenoid
___ Sternohyoid
___ Sternothyroid
___ Thyroarytenoid
___ Thyrohyoid

FIGURE 10.7 Muscles of the larynx (voice box).

TABLE 10.9
Muscles That Move the Head

OVERVIEW: The cervical region is divided by the sternocleidomastoid muscle into two principal triangles—anterior and posterior. The ***anterior triangle*** is bordered superiorly by the mandible, inferiorly by the sternum, medially by the cervical midline, and laterally by the anterior border of the sternocleidomastoid muscle (see Figure 11.2). The ***posterior triangle*** is bordered inferiorly by the clavicle, anteriorly by the posterior border of the sternocleidomastoid muscle, and posteriorly by the anterior border of the trapezius muscle (see Figure 11.2). Subsidiary triangles exist within the two principal triangles.

Muscle	Origin	Insertion	Action
Sternocleidomastoid (ster'-nō-klī'-dō-MAS-toid; *sternum* = breastbone; *cleido* = clavicle; *mastoid* = mastoid process of temporal bone) (label this muscle in Figure 10.11)			
Semispinalis capitis (se'-mē-spi-NA-lis KAP-i-tis; *semi* = half; *spine* = spinous process; *caput* = head) (label this muscle in Figure 10.16)			
Splenius capitis (SPLĒ-nē-us KAP-i-tis; *splenion* = bandage) (label this muscle in Figure 10.16)			
Longissimus capitis (lon-JIS-i-mus KAP-i-tis; *longissimus* = longest) (label this muscle in Figure 10.16)			

TABLE 10.10
Muscles That Act on the Anterior Abdominal Wall (After completing the table, label Figure 10.8.)

OVERVIEW: The anterolateral abdominal wall is composed of skin, fascia, and four pairs of flat, sheetlike muscles: rectus abdominis, external oblique, internal oblique, and transversus abdominis. The anterior surfaces of the rectus abdominis muscles are interrupted by transverse fibrous bands of tissue called ***tendinous intersections,*** believed to be remnants of septa that separated myotomes during embryonic development. The aponeuroses of the external oblique, internal oblique, and transversus abdominis muscles meet at the midline to form the ***linea alba*** (white line), a tough fibrous band that extends from the xiphoid process of the sternum to the pubic symphysis of the hip bone. The inferior free border of the external oblique aponeurosis, plus some collagen fibers, forms the ***inguinal ligament,*** which runs from the anterior superior iliac spine to the pubic tubercle (see Figure 10.17). The ligament demarcates the thigh and body wall.

Just superior to the medial end of the inguinal ligament is a triangular slit in the aponeurosis referred to as the ***superficial inguinal ring,*** the outer opening of the ***inguinal canal.*** The canal contains the spermatic cord and ilioinguinal nerve in males and round ligament of the uterus and ilioinguinal nerve in females.

The posterior abdominal wall is formed by the lumbar vertebrae, parts of the ilia of the hipbones, psoas major muscle (described in Table 10.20), quadratus lumborum muscle, and iliacus muscle (also described in Table 10.20). Whereas the anterolateral abdominal wall is contractile and distensible, the posterior abdominal wall is bulky and stable by comparison.

Muscle	Origin	Insertion	Action
Rectus abdominis (REK-tus ab-do-MIN-is; *rectus* = fibers parallel to midline; *abdomino* = abdomen)			
External oblique (ō-BLĒK; *external* = closer to the surface; *oblique* = fibers diagonal to midline)			
Internal oblique (ō-BLĒK; *internal* = farther from the surface)			
Transversus abdominis (tranz-VER-sus ab-do-MIN-is; *transverse* = fibers perpendicular to midline)			
Quadratus lumborum (kwod-RĀ-tus lum-BOR-um; *quad* = four; *lumbo* = lumbar region) (See Figure 10.16a)			

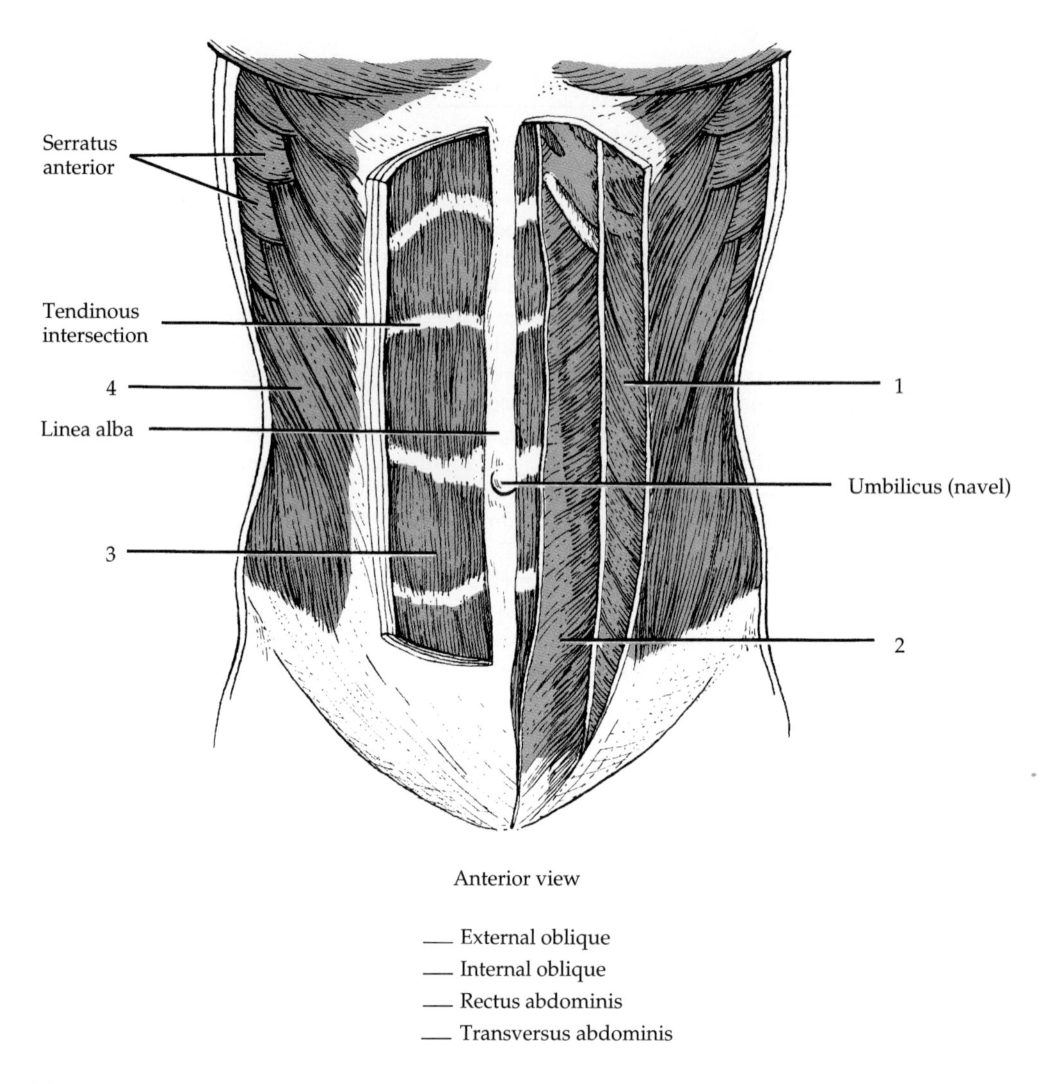

Anterior view

___ External oblique
___ Internal oblique
___ Rectus abdominis
___ Transversus abdominis

FIGURE 10.8 Muscles of anterior abdominal wall.

TABLE 10.11
Muscles Used in Breathing (After completing the table, label Figure 10.9.)

OVERVIEW: The muscles described here are attached to the ribs and by their contraction and relaxation alter the size of the thoracic cavity during breathing. Essentially, inspiration occurs when the thoracic cavity increases in size. Expiration occurs when the thoracic cavity decreases in size. The principal muscles of ***inspiration*** during normal breathing are the diaphragm and external intercostals. During forced inspiration, accessory muscles, such as the sternocleidomastoid, scalenes, and pectoralis minor are also used. The principal muscles of ***expiration*** during normal breathing are also the diaphragm and external intercostals. During forced expiration, accessory muscles, such as the internal intercostals and abdominal muscles (external oblique, internal oblique, transversus abdominis, and rectus abdominis) are also used.

The diaphragm is dome-shaped and has three major openings through which various structures pass between the thorax and abdomen. These structures include the aorta along with the thoracic duct and azygos vein, which pass through the ***aortic hiatus;*** the esophagus with accompanying vagus (X) nerves, which pass through the ***esophageal hiatus;*** and the inferior vena cava, which passes through the ***foramen for the vena cava.*** In a condition called a hiatal hernia, the stomach protrudes superiorly through the esophageal hiatus.

Muscle	Origin	Insertion	Action
Diaphragm (DĪ-a-fram; *dia* = across, between; *phragma* = wall)			
External intercostals (in'-ter-KOS-tals; *inter* = between; *costa* = rib)			
Internal intercostals (in'-ter-KOS-tals; *internal* = farther from surface)			

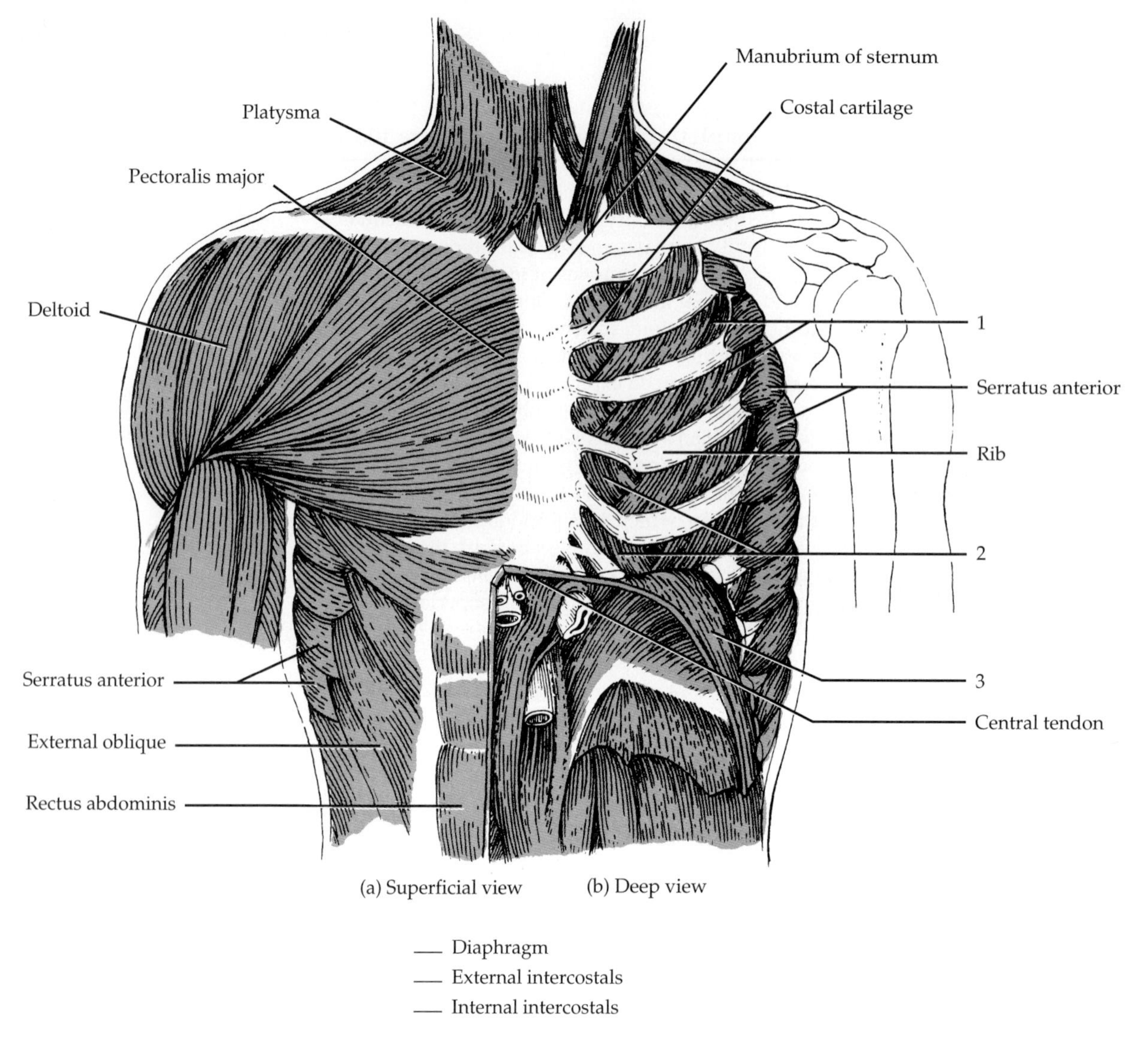

FIGURE 10.9 Muscles used in breathing.

TABLE 10.12
Muscles of the Pelvic Floor (After completing the table, label Figure 10.10.)

OVERVIEW: The muscles of the pelvic floor, together with the fascia covering their external and internal surfaces, are referred to as the ***pelvic diaphragm.*** This diaphragm is funnel-shaped and forms the floor of the abdominopelvic cavity. It is pierced by the anal canal and urethra in both sexes and also by the vagina in the female.

Muscle	Origin	Insertion	Action
Levator ani (le-VĀ-tor Ā-nē; *levator* = raises; *ani* = anus)	This muscle is divisible into two parts: the pubococcygeus muscle and the iliococcygeus muscle.		
Pubococcygeus (pu'-bo-kok-SIJ-ē-us; *pubo* = pubis; *coccygeus* = coccyx)			
Iliococcygeus (il'-ē-o-kok-SIJ-ē-us; *ilio* = ilium)			
Coccygeus* (kok-SIJ-ē-us)			

*Not illustrated

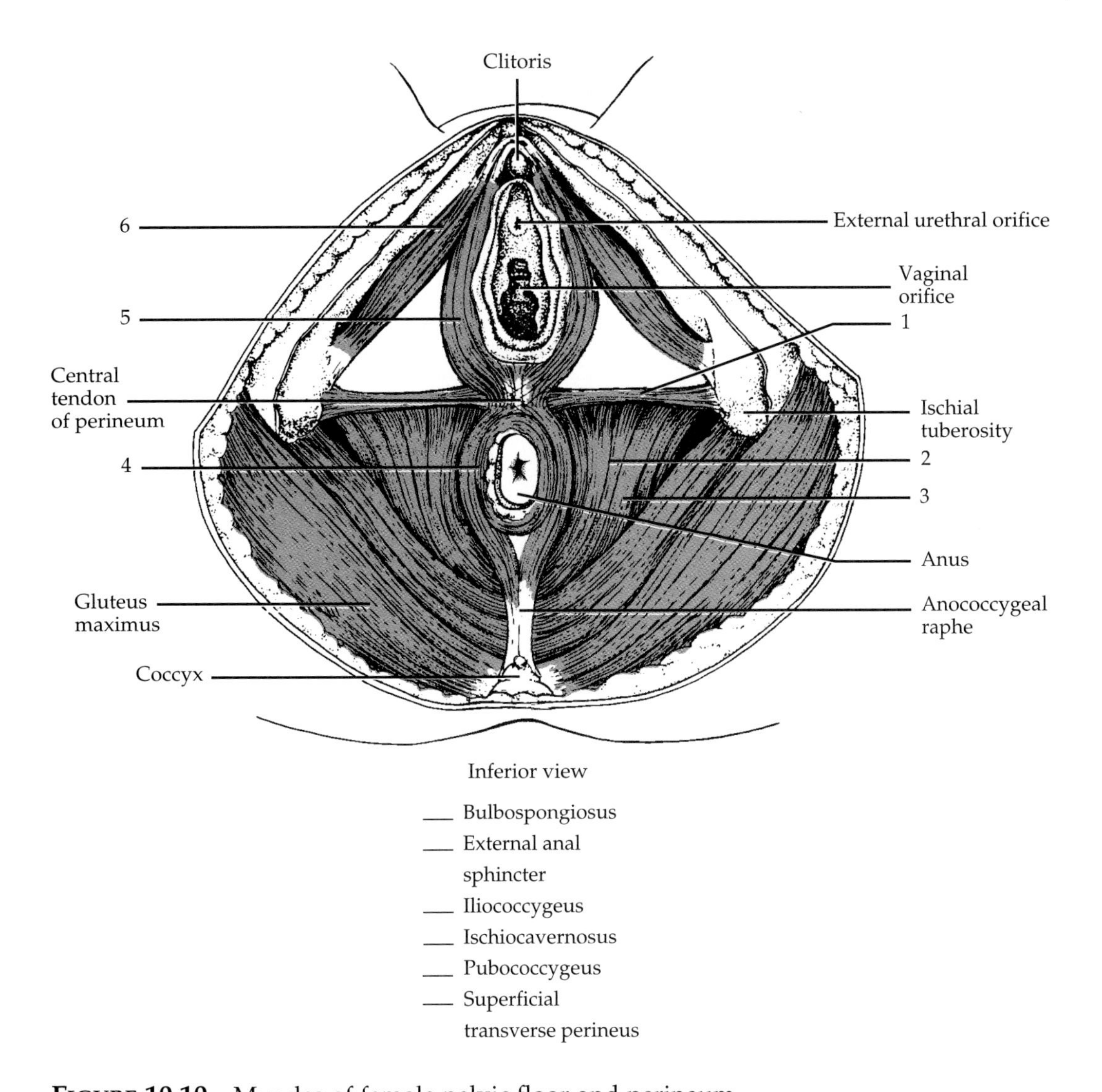

FIGURE 10.10 Muscles of female pelvic floor and perineum.

Table 10.13
Muscles of the Perineum (After completing the table, label Figure 10.10.)

OVERVIEW: The ***perineum*** is the entire outlet of the pelvis. It is a diamond-shaped area at the inferior end of the trunk between the thighs and buttocks. It is bordered anteriorly by the pubic symphysis, laterally by the ischial tuberosities, and posteriorly by the coccyx. A transverse line drawn between the ischial tuberosities divides the perineum into an anterior ***urogenital triangle*** that contains the external genitals and a posterior ***anal triangle*** that contains the anus.

The deep transverse perineus muscle, the urethral sphincter, and a fibrous membrane constitute the ***urogenital diaphragm.*** It surrounds the urogenital ducts and helps to strengthen the pelvic floor.

Muscle	Origin	Insertion	Action
Superficial transverse perineus (per-i-NĒ-us; *superficial* = near surface; *transverse* = across; *perineus* = perineum)			
Bulbospongiosus (bul'-bō-spon'-jē-Ō-sus; *bulbus* = bulb; *spongio* = sponge)			
Ischiocavernosus is'-kē-ō-ka'-ver-NŌ-sus; *ischion* = hip)			
Deep transverse perineus* (per-i-NĒ-us; *deep* = farther from surface)			
Urethral sphincter* (yoo-RĒ-thral SFINGK-ter; *urethral* = pertaining to urethra; *sphincter* = circular muscle that decreases the size of an opening)			
External anal (Ā-nal) **sphincter**			

*Not illustrated

TABLE 10.14
Muscles that Move the Pectoral (Shoulder) Girdle (After completing the table, label Figure 10.11.)

OVERVIEW: Muscles that move the pectoral (shoulder) girdle can be grouped into ***anterior*** and ***posterior*** muscles. The principal action of the muscles is to stabilize the scapula so that it can function as a stable point of origin for most of the muscles that move the humerus (arm).

Muscle	Origin	Insertion	Action
ANTERIOR MUSCLES **Subclavius** (sub-KLĀ-vē-us; *sub* = under; *clavius* = clavicle)			
Pectoralis (pek'-tor-A-lis) **minor** (*pectus* = breast, chest, thorax; *minor* = lesser)			
Serratus (ser-Ā-tus) **anterior** (*serratus* = sawtoothed; *anterior* = front)			
POSTERIOR MUSCLES **Trapezius** (tra-PĒ-zē-us; *trapezoides* = trapezoid-shaped)			
Levator scapulae (le-VĀ-tor SKA-pyoo-lē; *levator* = raises; *scapulae* = scapula)			
Rhomboideus (rom-BOID-ē-us) **major** (*rhomboides* = rhomboid- or diamond-shaped)			
Rhomboideus (rom-BOID-ē-us) **minor**			

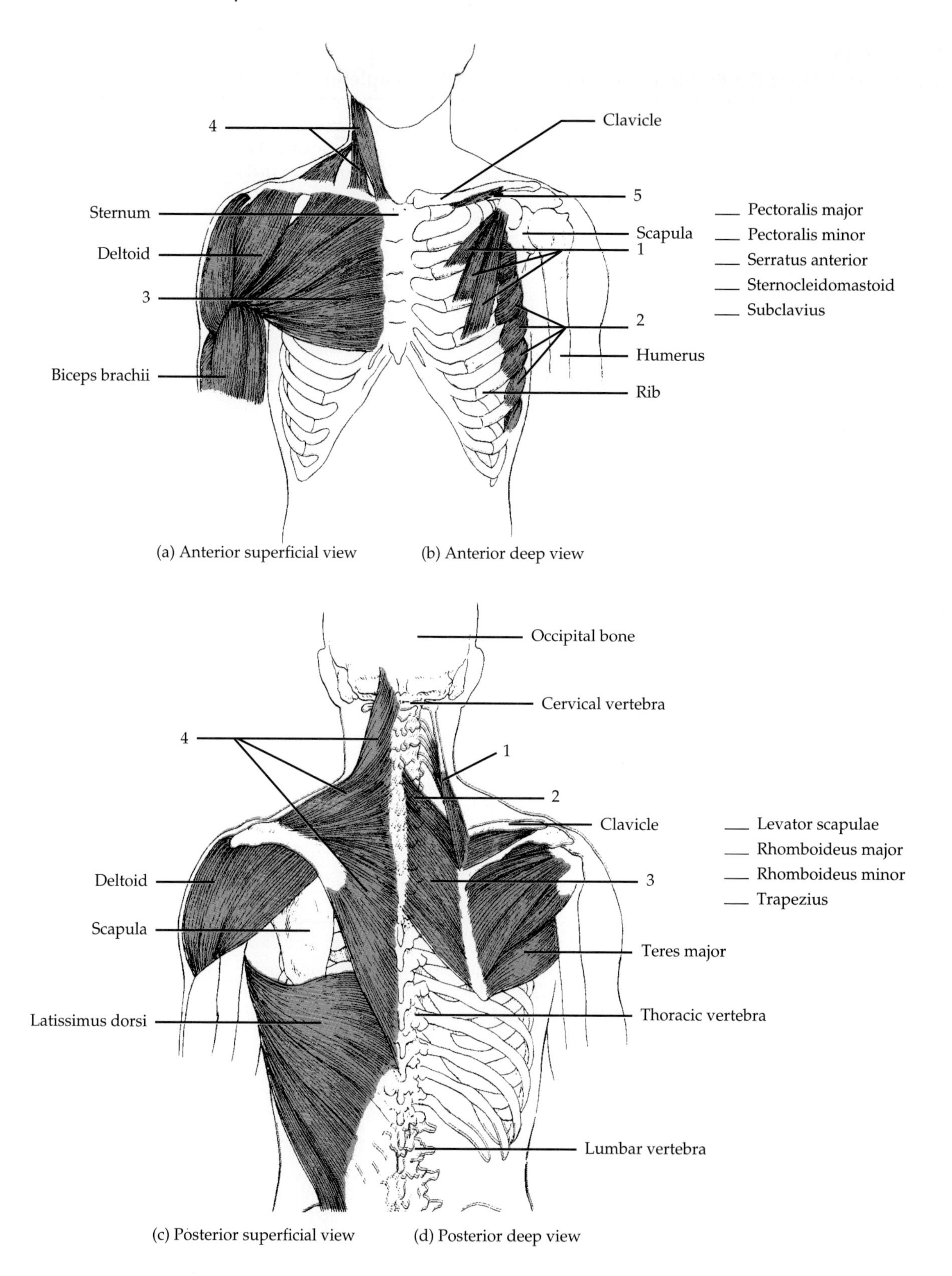

FIGURE 10.11 Muscles that move the pectoral (shoulder) girdle.

Table 10.15
Muscles that Move the Humerus (Arm) (After completing the table, label Figure 10.12.)

OVERVIEW: The muscles that move the humerus (arm) cross the shoulder joint. Of the nine muscles that cross the shoulder joint, only two of them (pectoralis major and latissimus dorsi) do not originate on the scapula. These two muscles are thus designated as ***axial muscles,*** since they originate on the axial skeleton. The remaining seven muscles, the ***scapular muscles,*** arise from the scapula.

The strength and stability of the shoulder joint are not provided by the shape of the articulating bones or its ligaments. Instead, four deep muscles of the shoulder—subscapularis, supraspinatus, infraspinatus, and teres minor—strengthen and stabilize the shoulder joint. The muscles join the scapula to the humerus. Their tendons are arranged to form a nearly complete circle around the joint. This arrangement is referred to as the ***rotator (musculotendinous) cuff*** and is a common site of injury to baseball pitchers, especially tearing of the supraspinatus muscle tendon. This tendon is especially predisposed to wear-and-tear changes because of its location between the head of the humerus and acromion of the scapula, which compresses the tendon during shoulder movements.

After you have studied the muscles in this table, arrange them according to the following actions: flexion, extension, abduction, adduction, medial rotation, and lateral rotation. (The same muscle can be used more than once.)

Muscle	Origin	Insertion	Action
AXIAL MUSCLES **Pectoralis** (pek'-tor-A-lis) **major** (label this muscle in Figure 10.11)			
Latissimus dorsi (la-TIS-i-mis DOR-sī; *dorsum* = back)			
SCAPULAR MUSCLES **Deltoid** (DEL-toyd; *delta* = triangular)			
Subscapularis (sub-scap'-yoo-LA-ris; *sub* = below; *scapularis* = scapula)			
Supraspinatus (soo'-pra-spi-NĀ-tus; *supra* = above; *spinatus* = spine of scapula)			
Infraspinatus (in'-fra-spi-NĀ-tus; *infra* = below)			
Teres (TE-rēz) **major** (*teres* = long and round)			
Teres (TE-rēz) **minor**			
Coracobrachialis (kor'-a-kō-BRĀ-kē-a'-lis; *coraco* = coracoid process)			

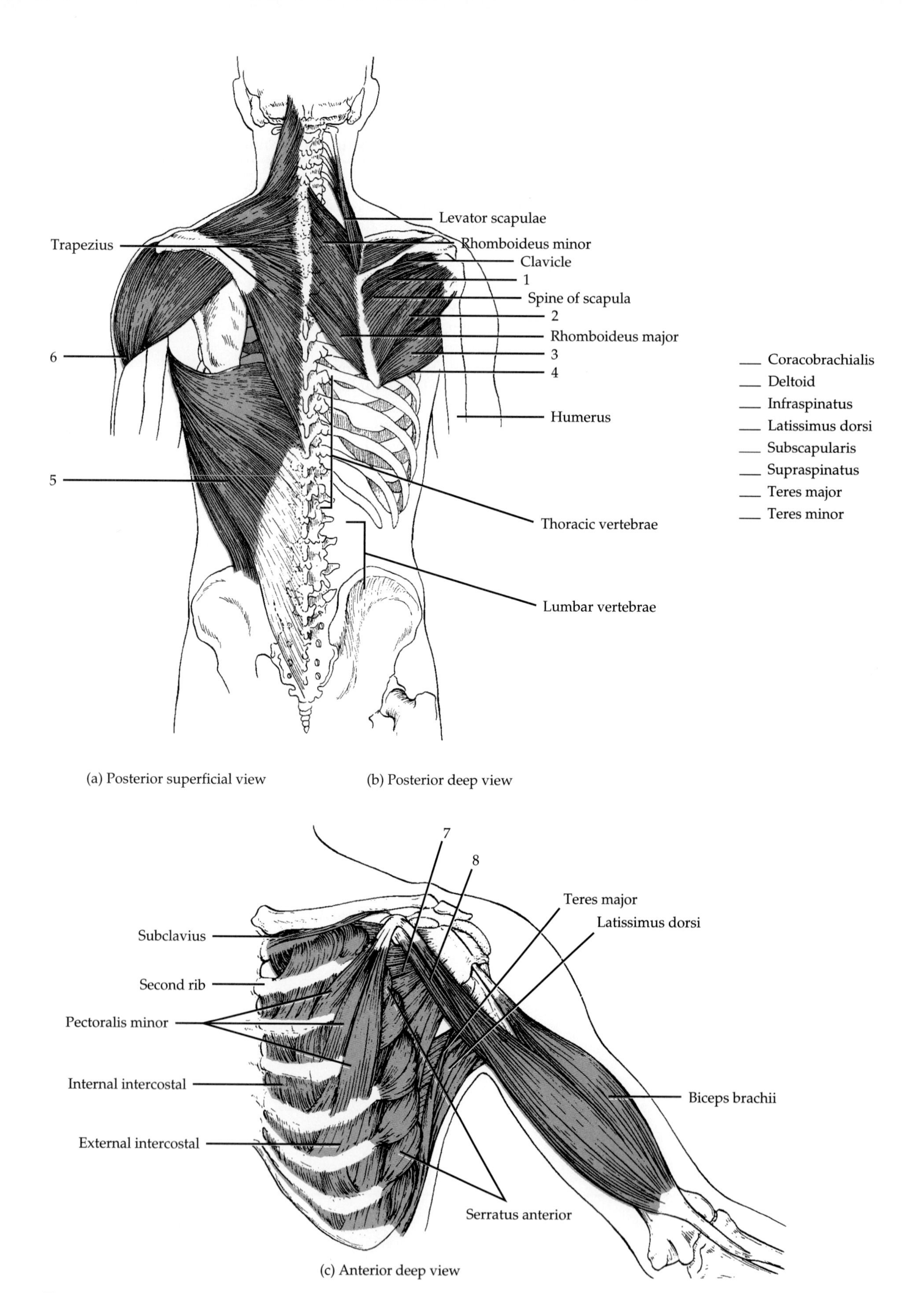

FIGURE 10.12 Muscles that move the humerus (arm).

TABLE 10.16
Muscles that Move the Radius and Ulna (Forearm) (After completing the table, label Figure 10.13.)

OVERVIEW: The muscles that move the radius and ulna (forearm) are divided into ***flexors*** and ***extensors.*** Recall that the elbow joint is a hinge joint, capable only of flexion and extension. Whereas the biceps brachii, brachialis, and brachioradialis are flexors of the forearm, the triceps brachii and anconeus are extensors. Other muscles that move the forearm are concerned with pronation and supination. (The biceps brachii muscle also permits supination of the forearm.)

Muscle	Origin	Insertion	Action
FLEXORS **Biceps brachii** (BĪ-ceps BRĀ-kē-ī; *biceps* = two heads of origin; *brachion* = arm)			
Brachialis (brā'-kē-A-lis)			
Brachioradialis (bra'-kē-ō-rā'-dē-A-lis; *radialis* = radius) (See also Figure 10.14a)			
EXTENSORS **Triceps brachii** (TRĪ-ceps BRĀ-kē-ī; *triceps* = three heads of origin)			
Anconeus (an-KŌ-nē-us; *anconeal* = pertaining to the elbow) (see Figure 10.14c)			
PRONATORS **Pronator teres** (PRŌ-na'-ter TE-rēz; (*pronation* = turning palm downward or posteriorly)			
Pronator quadratus (PRŌ-na'-ter kwod-RĀ-tus; *quadratus* = squared, four-sided) (see Figure 10.14a)			
SUPINATOR **Supinator** (sup-pi-nā-tor; *supination* = turning palm upward or anteriorly)			

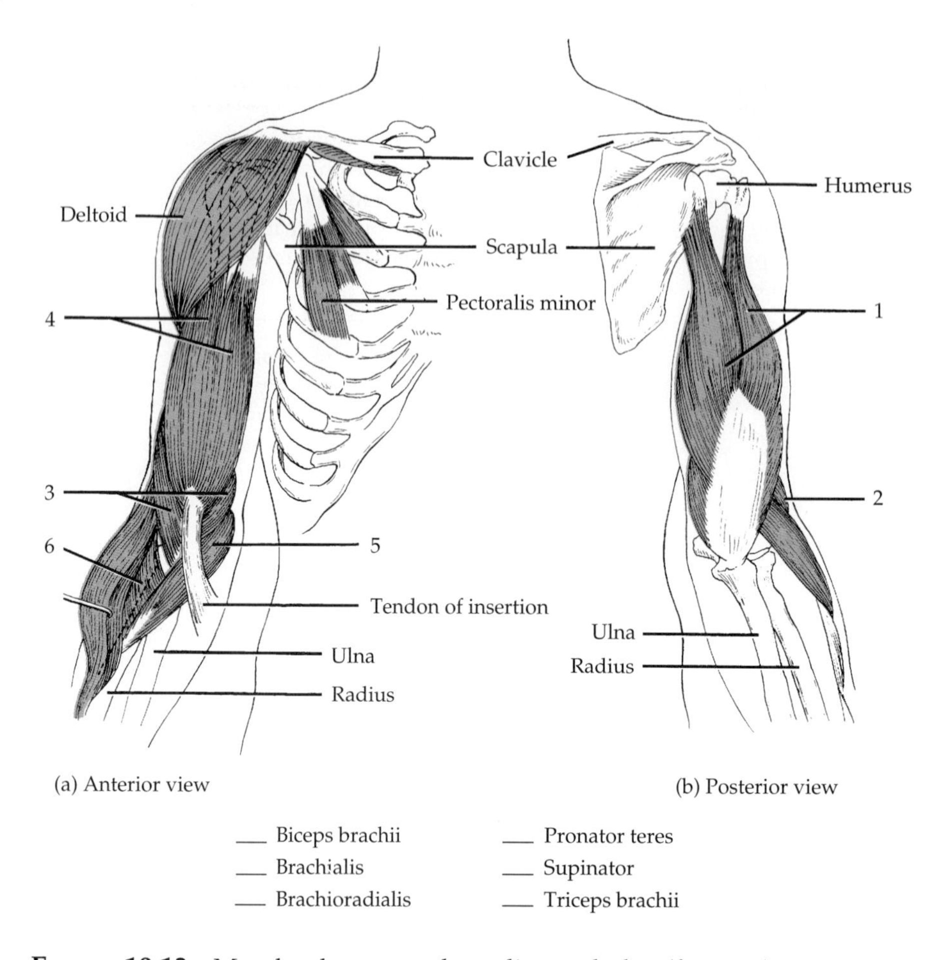

FIGURE 10.13 Muscles that move the radius and ulna (forearm).

TABLE 10.17
Muscles that Move the Wrist, Hand, and Fingers (Figure 10.14)

OVERVIEW: Muscles that move the wrist, hand, and fingers are many and varied. However, as you will see, their names usually give some indication of their origin, insertion, or action. On the basis of location and function, the muscles are divided into two groups—anterior and posterior compartments. The ***anterior compartment muscles*** function as flexors. They originate on the humerus and typically insert on the carpals, metacarpals, and phalanges. The bellies of these muscles form the bulk of the proximal forearm. The ***posterior compartment muscles*** function as extensors. These muscles arise on the humerus and insert on the metacarpals and phalanges. Each of the two principal groups is also divided into superficial and deep muscles.

The tendons of the muscles of the forearm that attach to the wrist or continue into the hand, along with blood vessels and nerves, are held close to bones by strong fascial structures. The tendons are also surrounded by tendon sheaths. At the wrist, the deep fascia is thickened into fibrous bands called ***retinacula*** (*retinere* = retain). The ***flexor retinaculum (transverse carpal ligament)*** is located over the palmar surface of the carpal bones. Through it pass the long flexor tendons of the digits and wrist and median nerve. The ***extensor retinaculum (dorsal carpal ligament)*** is located over the posterior surface of the carpal bones. Through it pass the extensor tendons of the wrist and digits.

After you have studied the muscles in the table, arrange them according to the following actions: flexion, extension, abduction, adduction, supination, and pronation. (The same muscles can be used more than once.)

Muscle	Origin	Insertion	Action
ANTERIOR GROUP (flexors) Superficial			
Flexor carpi radialis (FLEK-sor KAR-pē rā'-dē-A-lis; *flexor* = decreases angle at a joint; *carpus* = wrist; *radialis* = radius)			
Palmaris longus (pal-MA-ris LON-gus; *palma* = palm *longus* = long)			
Flexor carpi ulnaris (FLEK-sor KAR-pē ul-NAR-is; *ulnaris* = ulna)			
Flexor digitorum superficialis (FLEK-sor di'-ji-TOR-um soo'-per-fish'-ē-A-lis; *digit* = finger or toe; *superficialis* = closer to surface)			
Flexor digitorum profundus (FLEK-sor di'-ji-TOR-um pro-FUN-dus; *profundus* = deep)			
Flexor pollicis longus (FLEK-sor POL-li-kis LON-gus; *pollex* = thumb)			
POSTERIOR GROUP (extensors) Superficial			
Extensor carpi radialis longus (eks-TEN-sor KAR-pē rā'-dē-A-lis LON-gus; *extensor* = increases angle at a joint)			
Extensor carpi radialis brevis (eks-TEN-sor KAR-pē rā'-dē-A-lis BREV-is; *brevis* = short)			

Table 10.17 (*Continued*)

Muscle	Origin	Insertion	Action
Deep			
Extensor digitorum (eks-TEN-sor di′-ji-TOR-um)			
Extensor digiti minimi (eks-TEN-sor DIJ-i-tē MIN-i-mē; *digiti* = digit; *minimi* = finger)			
Extensor carpi ulnaris (eks-TEN-sor KAR-pē ul-NAR-is)			
Abductor pollicis longus (ab-DUK-tor POL-li-kis LON-gus; *abductor* = moves a part away from midline)			
Extensor pollicis brevis (eks-TEN-sor POL-li-kis BREV-is)			
Extensor pollicis longus (eks-TEN-sor POL-li-kis LON-gus)			
Extensor indicis (eks-TEN-sor IN-di-kis; *indicis* = index)			

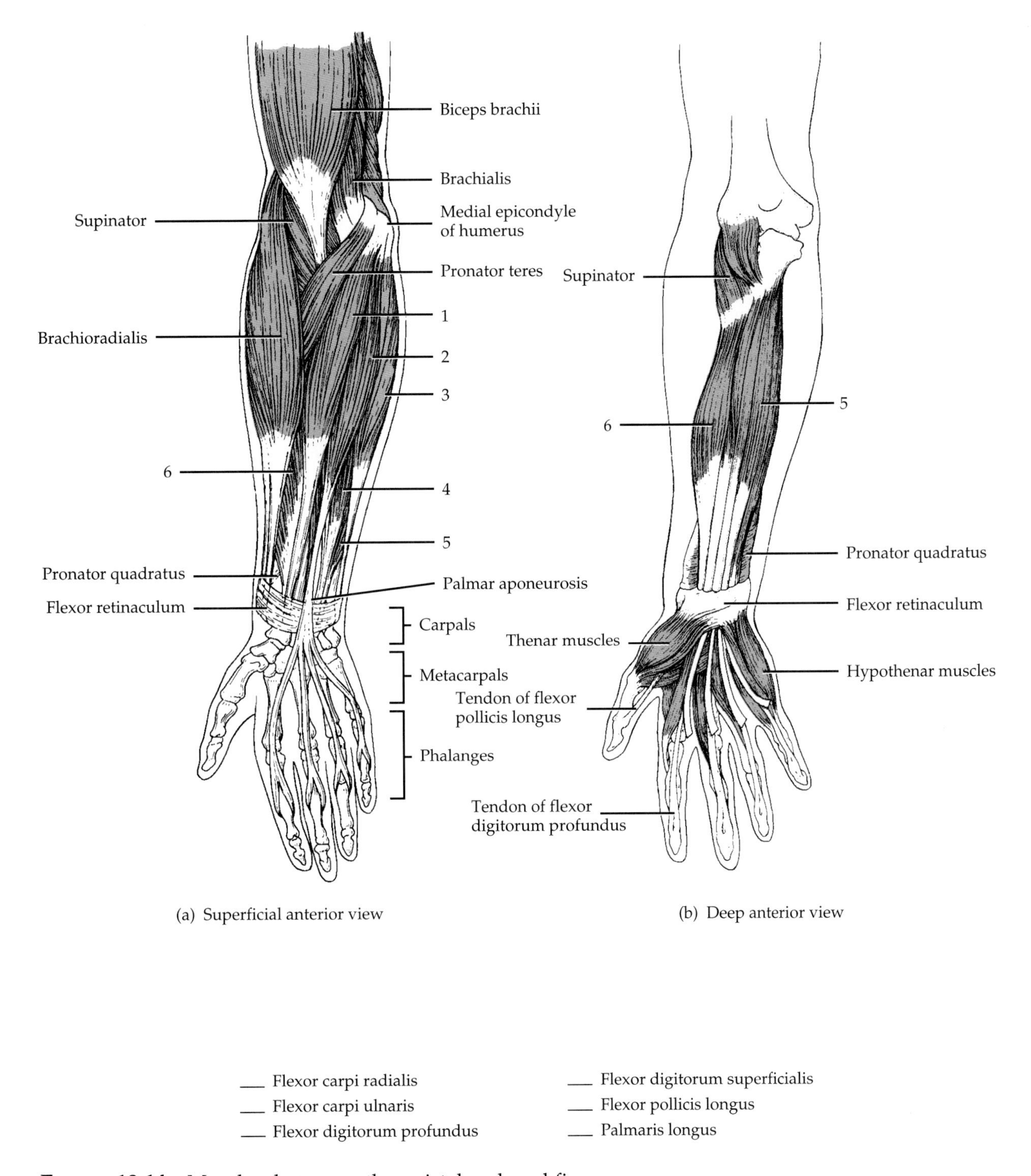

___ Flexor carpi radialis
___ Flexor carpi ulnaris
___ Flexor digitorum profundus
___ Flexor digitorum superficialis
___ Flexor pollicis longus
___ Palmaris longus

FIGURE 10.14 Muscles that move the wrist, hand, and fingers.

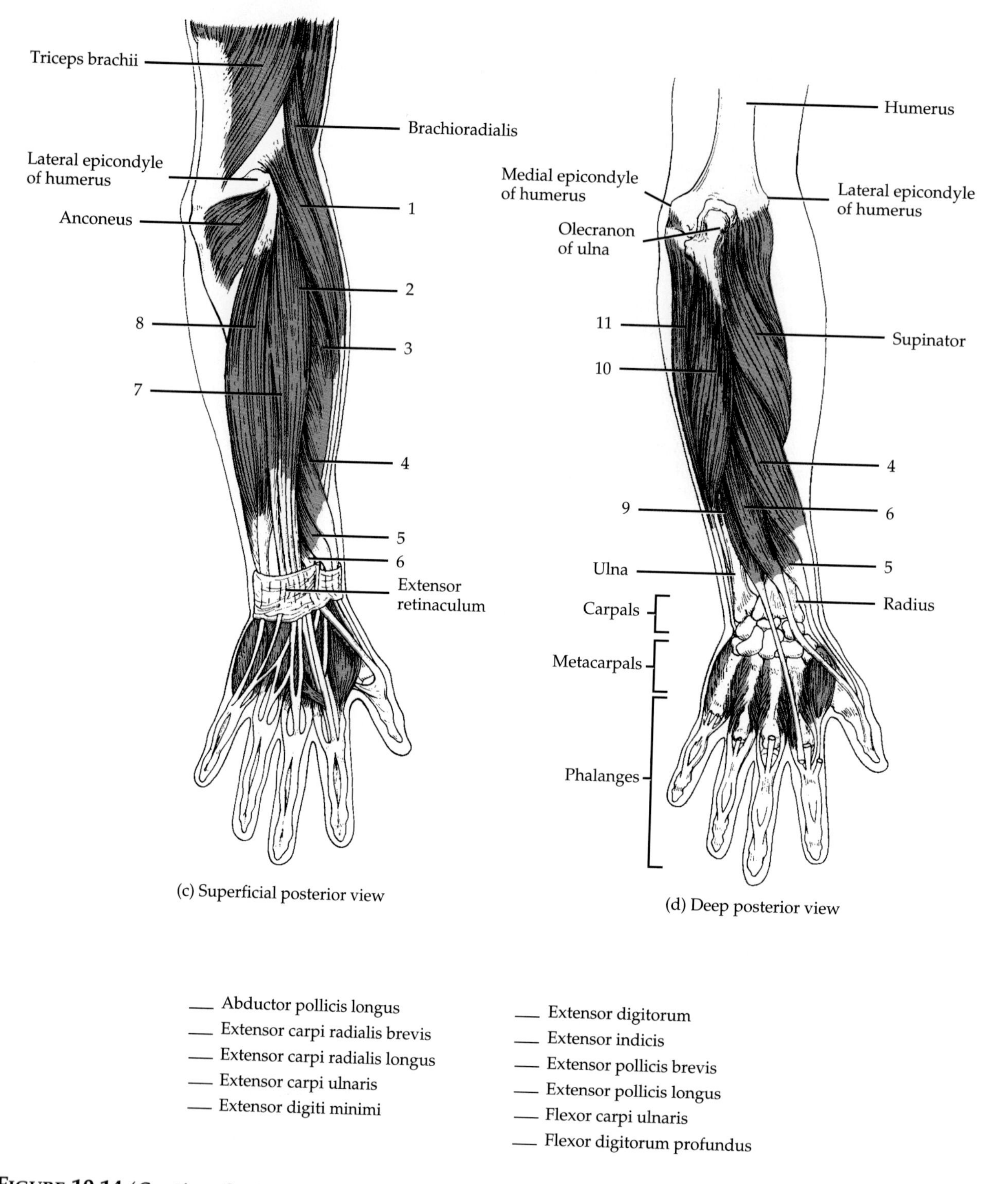

(c) Superficial posterior view

(d) Deep posterior view

— Abductor pollicis longus
— Extensor carpi radialis brevis
— Extensor carpi radialis longus
— Extensor carpi ulnaris
— Extensor digiti minimi
— Extensor digitorum
— Extensor indicis
— Extensor pollicis brevis
— Extensor pollicis longus
— Flexor carpi ulnaris
— Flexor digitorum profundus

FIGURE 10.14 (*Continued*) Muscles that move the wrist, hand, and fingers.

TABLE 10.18
Intrinsic Muscles of the Hand (After completing the table, label Figure 10.15.)

OVERVIEW: Several of the muscles discussed in Table 10.17 help to move the digits in various ways. In addition, there are muscles in the palmar surface of the hand called ***intrinsic muscles*** that also help to move the digits. Such muscles are so named because their origins and insertions are both within the hands. These muscles assist in the intricate and precise movements that are characteristic of the human hand.

The intrinsic muscles of the hand are divided into three principal groups—thenar, hypothenar, and intermediate. The four ***thenar*** (THĒ-nar) ***muscles***, shown in Figure 10.14b, act on the thumb and form the ***thenar eminence*** (see Figure 11.8). The four ***hypothenar*** (HĪ-pō-thē-nar) ***muscles***, shown in Figure 10.14b, act on the little finger and form the ***hypothenar eminence*** (see Figure 11.8). The eleven ***intermediate*** (midpalmar) ***muscles*** act on all the digits, except the thumb.

The functional importance of the hand is readily apparent when one considers that certain hand injuries can result in permanent disability. In fact, most of the dexterity of the hand depends on the movements of the thumb. The general activities of the hand are free motion, power grip (forcible movement of the fingers and thumb against the palm, as in squeezing), precision handling (a change in position of a handled object that requires exact control of finger and thumb positions, as in winding a watch or threading a needle), and pinch (compression between the thumb and index finger or between the thumb and first two fingers).

Movement of the thumb is very important in the precise activities of the hand. The five principal movements of the thumb, illustrated below, are flexion (movement of the thumb medially across the palm), extension (movement of the thumb laterally away from the palm), abduction (movement of the thumb in an anteroposterior plane away from the palm), adduction (movement of the thumb in an anteroposterior plane toward the palm), and opposition (movement of the thumb across the palm so that the tip of the thumb meets the tips of the fingers). Opposition is the single most distinctive digital movement that gives humans and other primates the ability to precisely grasp and manipulate objects.

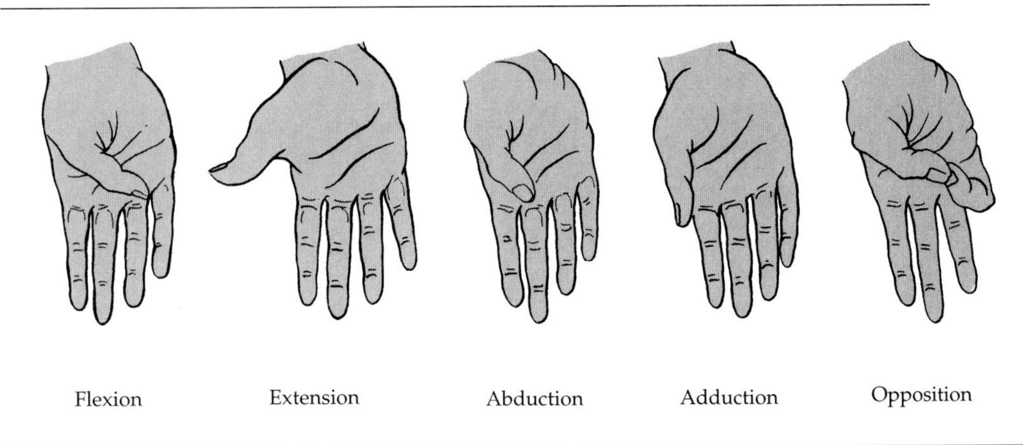

Muscle	Origin	Insertion	Action
THENAR (THĒ-nar) MUSCLES **Abductor pollicis brevis** (ab-DUK-tor POL-li-kis BREV-is); *abductor* = moves part away from midline; *pollex* = thumb; *brevis* = short)			

TABLE 10.18 *(Continued)*

Muscle	Origin	Insertion	Action
Opponens pollicis (o-PŌ-nez POL-li-kis; *opponens* = opposes)			
Flexor pollicis brevis (FLEK-sor POL-li-kis BREV-is *flexor* = decreases angle at joint)			
Adductor pollicis (ad-DUK-tor POL-li-kis; *adductor* = moves part toward midline)			
HYPOTHENAR (HĪ-pō-thē′-nar) MUSCLES **Palmaris brevis** (pal-MA-ris BREV-is; *palma* = palm)			
Abductor digiti minimi (ab-DUK-tor DIJ-i-tē MIN-i-mē; *digit* = finger or toe; *minimi* = little finger)			
Flexor digiti minimi brevis (FLEK-sor DIJ-i-tē MIN-i-mē BREV-is)			
Opponens digiti minimi (o-PŌ-nenz DIJ-i-tē MIN-i-mē)			
INTERMEDIATE (MIDPALMAR) MUSCLES **Lumbricals** (LUM-bri-kals; four muscles)			
Dorsal interossei (DOR-sal in′-ter-OS-ē-ī; four muscles; *dorsal* = back surface; *inter* = between; *ossei* = bones)			
Palmar interossei (PAL-mar in′-ter-OS-ē-ī; three muscles)			

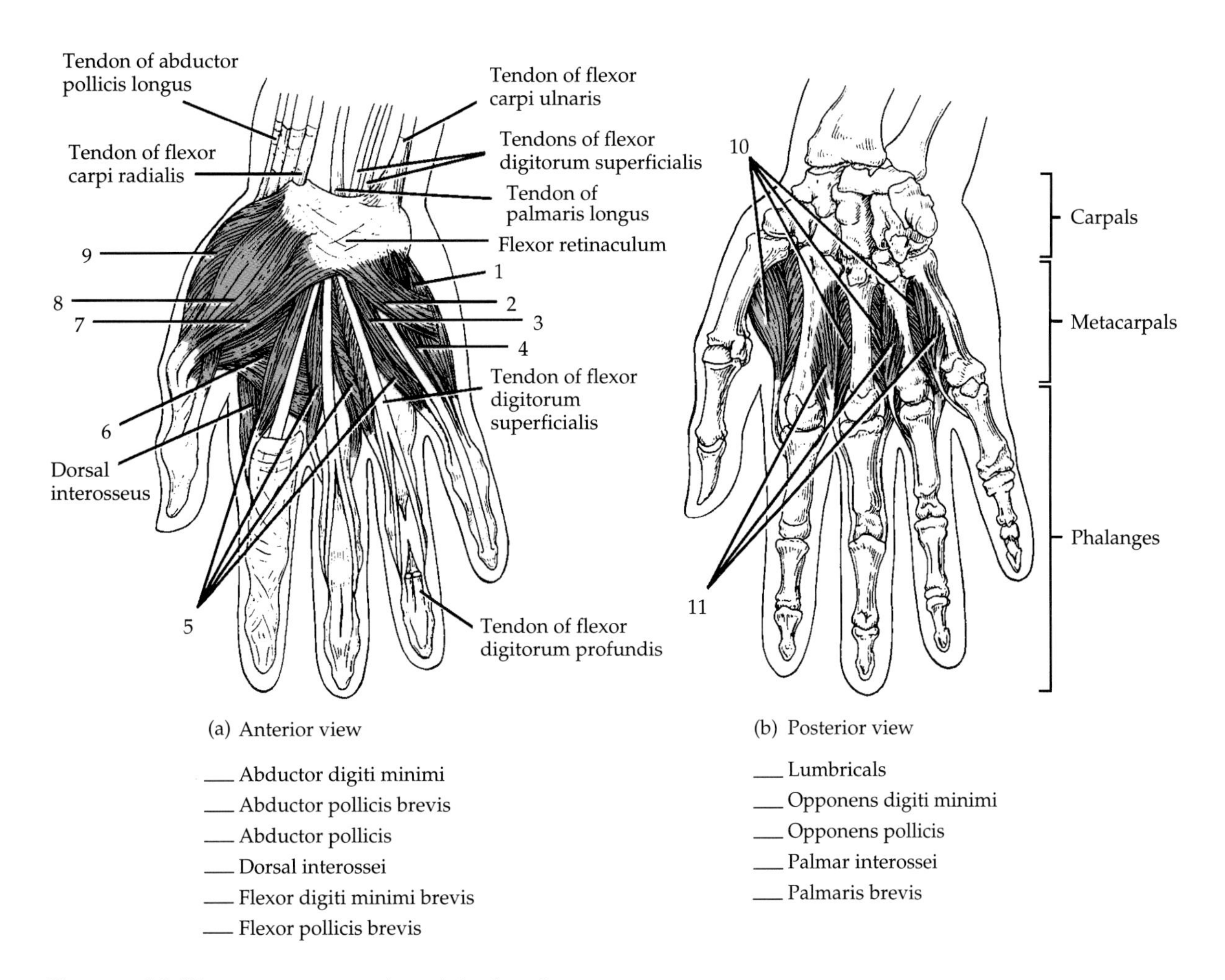

FIGURE 10.15 Intrinsic muscles of the hand.

TABLE 10.19
Muscles that Move the Vertebral Column (Backbone) (After completing the table, label Figure 10.16.)

OVERVIEW: The muscles that move the vertebral column (backbone) are quite complex because they have multiple origins and insertions and there is considerable overlapping among them. One way to group the muscles is on the basis of the general direction of the muscle bundles and their approximate lengths. For example, the ***splenius muscles*** arise from the midline and run laterally and superiorly to their insertions. The ***erector spinae (sacrospinalis) muscle*** arises from either the midline or more laterally, but usually runs almost longitudinally, with neither a marked lateral nor medial direction as it is traced superiorly. The ***transversospinalis muscles*** arise laterally, but run toward the midline as they are traced superiorly. Deep to these three muscles groups are small ***segmental muscles*** that run between spinous processes or transverse processes of vertebrae. Since the scalene muscles also assist in moving the vertebral column, they are included in this table.

Note in Table 10.10 that the rectus abdominis and quadratus lumborum muscles also assume a role in moving the vertebral column.

Muscle	Origin	Insertion	Action
SPLENIUS (SPLĒ-nē-us) MUSCLES			
Splenius capitis (SPLĒ-nē-us KAP-i-tis; *splenium* = bandage; *caput* = head)			
Splenius cervicis (SPLĒ-nē-us SER-vi-kis; *cervix* = neck)			
ERECTOR SPINAE (SACROSPINALIS)			
This is the largest muscular mass of the back and consists of three groupings—iliocostalis, longissimus, and spinalis. These groups, in turn, consist of a series of overlapping muscles. The iliocostalis group is laterally placed, the longissimus group is intermediate in placement, and the spinalis group is medially placed.			
Iliocostalis (Lateral) Group			
Iliocostalis cervicis (il′-ē-ō-kos-TAL-is SER-vi-kis; *ilium* = flank; *costa* = rib)			
Iliocostalis thoracis (il′-ē-ō-kos-TAL-is thō-RA-kis; *thorax* = chest)			
Iliocostalis lumborum (il′-ē-ō-kos-TAL-is lum-BOR-um)			
Longissimus (Intermediate) Group			
Longissimus capitis (lon-JIS-i-mus KAP-i-tis; *longissimus* = longest)			
Longissimus cervicis (long-JIS-i-mus SER-vi-kis)			
Longissimus thoracis (lon-JIS-i-mus thō-RA-kis)			

Table 10.19 *(Continued)*

Muscle	Origin	Insertion	Action
Spinalis (Medial) Group			
Spinalis capitis (spi-NA-lis KAP-i-tis; *spinalis* = vertebral column)			
Spinalis cervicis (spi-NA-lis SER-vi-kis)			
Spinalis thoracis (spi-NA-lis tho-RA-kis)			
TRANSVERSOSPINALIS (trans-ver′-sō-spi-NA-lis) MUSCLES **Semispinalis capitis** (sem′-ē-spi-NA-lis KAP-i-tis; *semi* = partially or one half)			
Semispinalis cervicis (sem′-ē-spi-NA-lis SER-vi-kis)			
Semispinalis thoracis (sem′-ē-spi-NA-lis tho-RA-kis)			
Multifidus (mul-TIF-i-dus; *mutli* = many; *fidere* = to split)			
Rotatores (rō-ta-TO-rēz; *rotate* = to turn)			
SEGMENTAL MUSCLES **Interspinales** (in-ter-SPĪ-na-lēz; *inter* = between)			
Intertransversarii (in′-ter-trans-vers-AR-ē-ī)			
SCALENE (SKĀ-lēn) MUSCLES **Anterior scalene** (SKĀ-lēn; *anterior* = front *skalenos* = uneven)			
Middle scalene (SKĀ-lēn)			
Posterior scalene (SKĀ-lēn; *posterior* = back)			

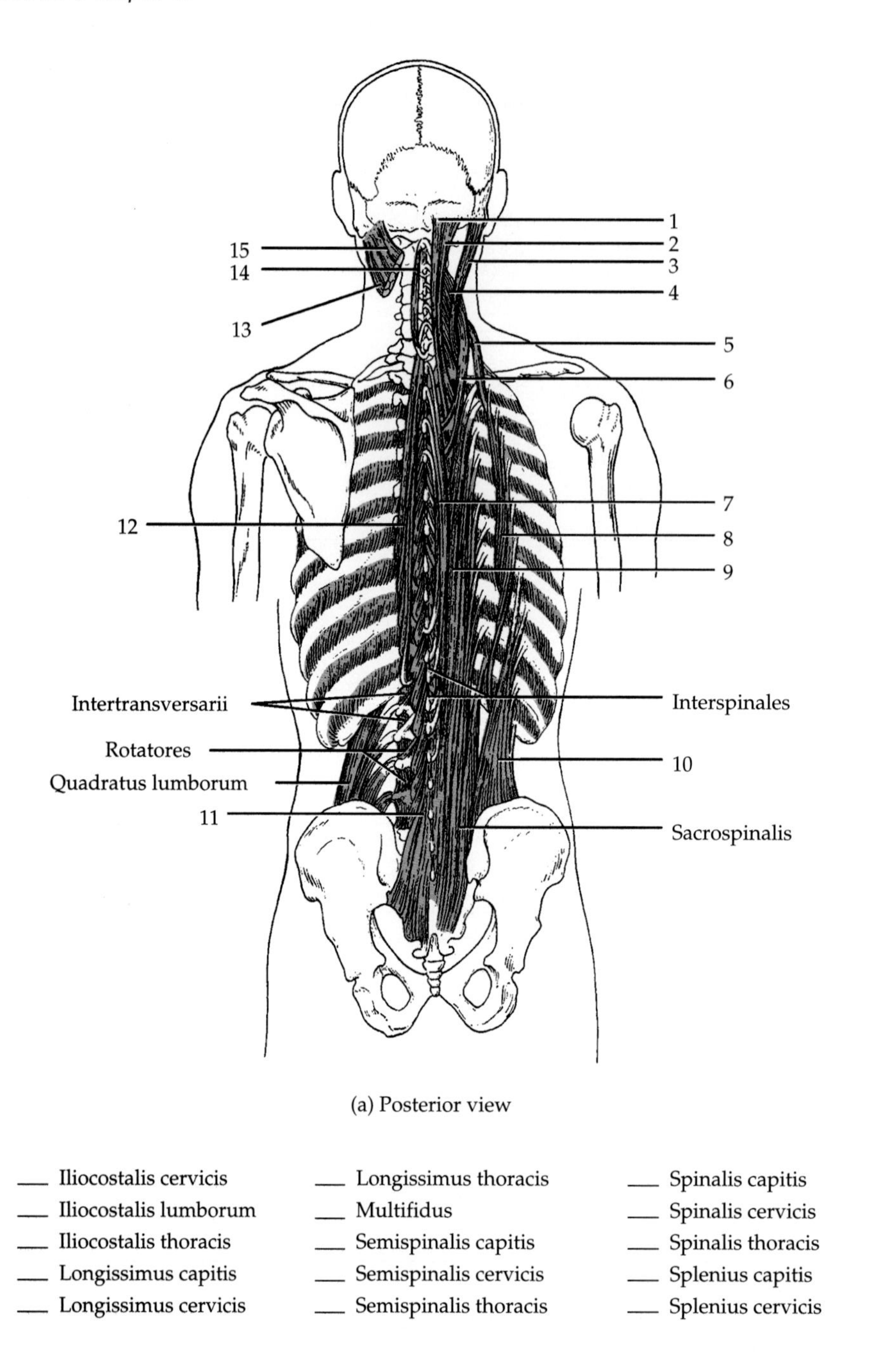

FIGURE 10.16 Muscles that move the vertebral column (backbone). Many superficial muscles have been removed or cut to show deep muscles.

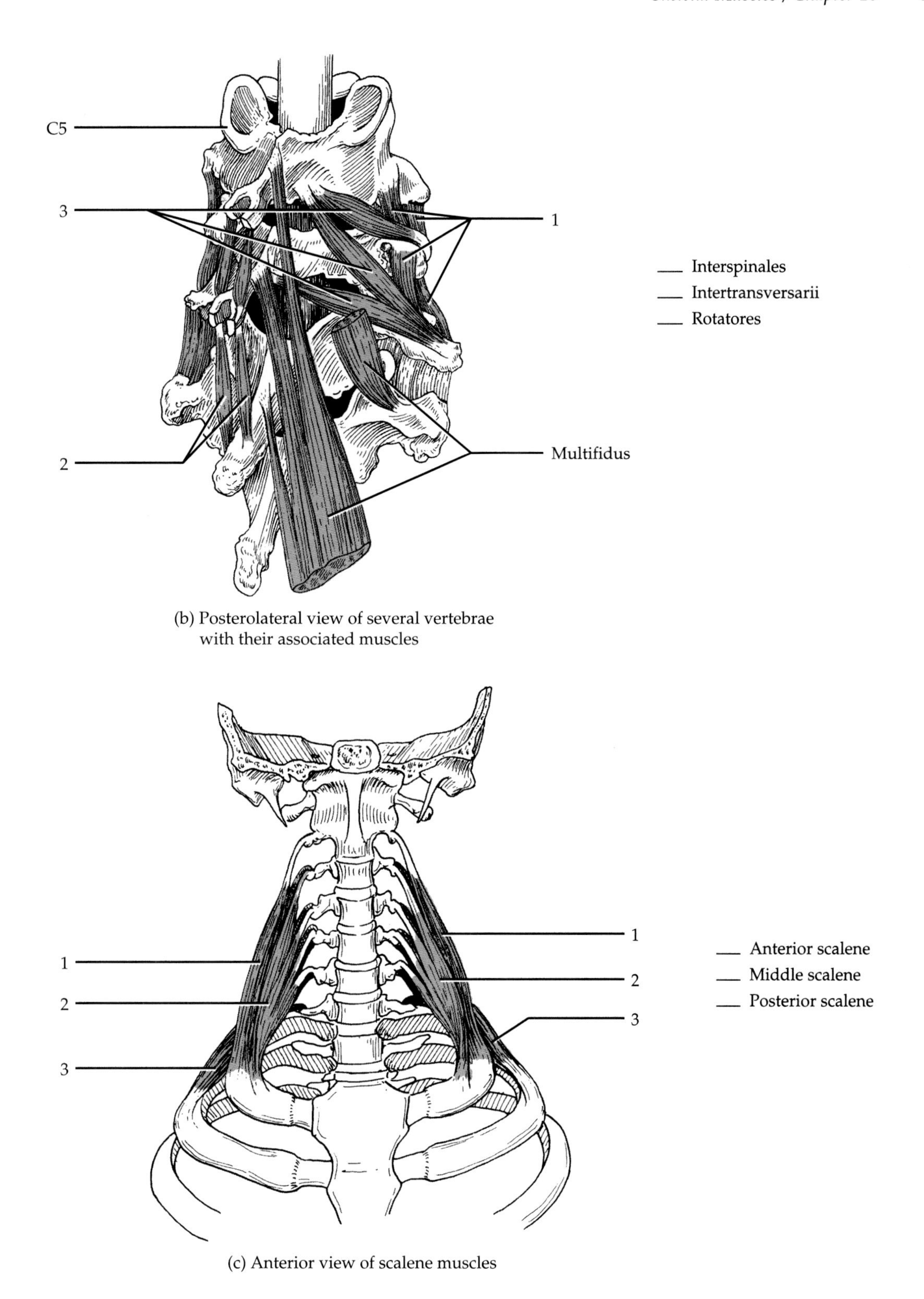

(b) Posterolateral view of several vertebrae with their associated muscles

(c) Anterior view of scalene muscles

FIGURE 10.16 (*Continued*) Muscles that move the vertebral column (backbone).

TABLE 10.20
Muscles That Move the Femur (Thigh) (After completing the table, label Figure 10.17.)

OVERVIEW: As you will see, muscles of the lower limbs are larger and more powerful than those of the upper limbs since lower limb muscles function in stability, locomotion, and maintenance of posture. Upper limb muscles are characterized by versatility of movement. In addition, muscles of the lower limbs frequently cross two joints and act equally on both.

The anterior muscles are the psoas major and iliacus, together referred to as the ***iliopsoas*** (il′-ē-ō-SŌ-as) ***muscle.*** The remaining muscles (except for the pectineus, adductors, and tensor fasciae latae) are posterior muscles. Technically, the pectineus and adductors are components of the medial compartment of the thigh, but they are included in this table because they act on the femur. The tensor fasciae latae muscle is laterally placed. The ***fascia lata*** is a deep fascia of the thigh that encircles the entire thigh. It is well developed laterally where, together with the tendons of the gluteus maximus and tensor fasciae latae muscles, it forms a structure called the ***iliotibial tract.*** The tract inserts into the lateral condyle of the tibia.

After you have studied the muscles in this table, arrange them according to the following actions: flexion, extension, abduction, adduction, medial rotation, and lateral rotation. (The same muscles can be used more than once.)

Muscle	Origin	Insertion	Action
Psoas (SŌ-as) **major** (*psoa* = muscle of loin)			
Iliacus (il'-ē-AK-us; *iliac* = ilium)			
Gluteus maximus (GLOO-tē-us MAK-si-mus; *glutos* = buttock; *maximus* = largest)			
Gluteus medius (GLOO-tē-us MĒ-dē-us; *media* = middle)			
Gluteus minimus (GLOO-tē-us MIN-i-mus; *minimus* = smallest)			

Tensor fasciae latae
(TEN-sor FA-shē-ē LĀ-tē;
tensor = makes tense;
fascia = band;
latus = wide)			

Table 10.20 (*Continued*)

Muscle	Origin	Insertion	Action
Piriformis (pir-i-FOR-mis; *pirum* = pear; *forma* = shape)			
Obturator internus* (OB-too-rā'-tor in-TER-nus; *obturator* = obturator foramen; *internus* = inside)			
Obturator externus (OB-too-rā'-tor ex-TER-nus; *externus* = outside)			
Superior gemellus (jem-EL-lus; *superior* = above; *gemellus* = twins)			
Inferior gemellus (jem-EL-lus; *inferior* = below)			
Quadratus femoris (kwod-RĀ-tus FEM-or-is; *quad* = four; *femoris* = femur)			
Adductor longus (LONG-us; *adductor* = moves part closer to midline; *longus* = long)			
Adductor brevis (BREV-is; *brevis* = short)			
Adductor magnus (MAG-nus; *magnus* = large)			
Pectineus (pek-TIN-ē-us; *pecten* = comb-shaped)			

*Not illustrated

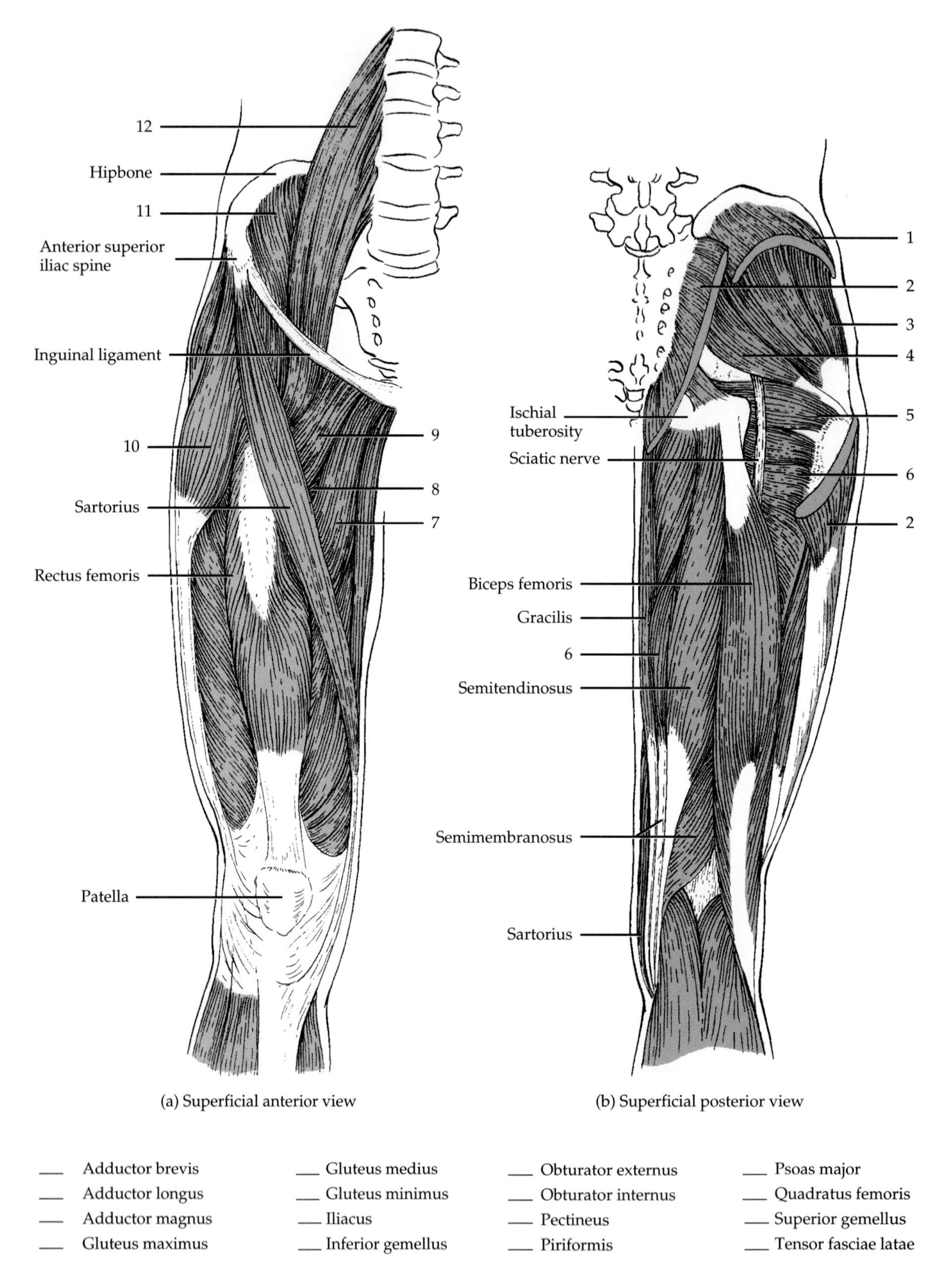

___ Adductor brevis
___ Adductor longus
___ Adductor magnus
___ Gluteus maximus
___ Gluteus medius
___ Gluteus minimus
___ Iliacus
___ Inferior gemellus
___ Obturator externus
___ Obturator internus
___ Pectineus
___ Piriformis
___ Psoas major
___ Quadratus femoris
___ Superior gemellus
___ Tensor fasciae latae

FIGURE 10.17 Muscles that move the femur (thigh).

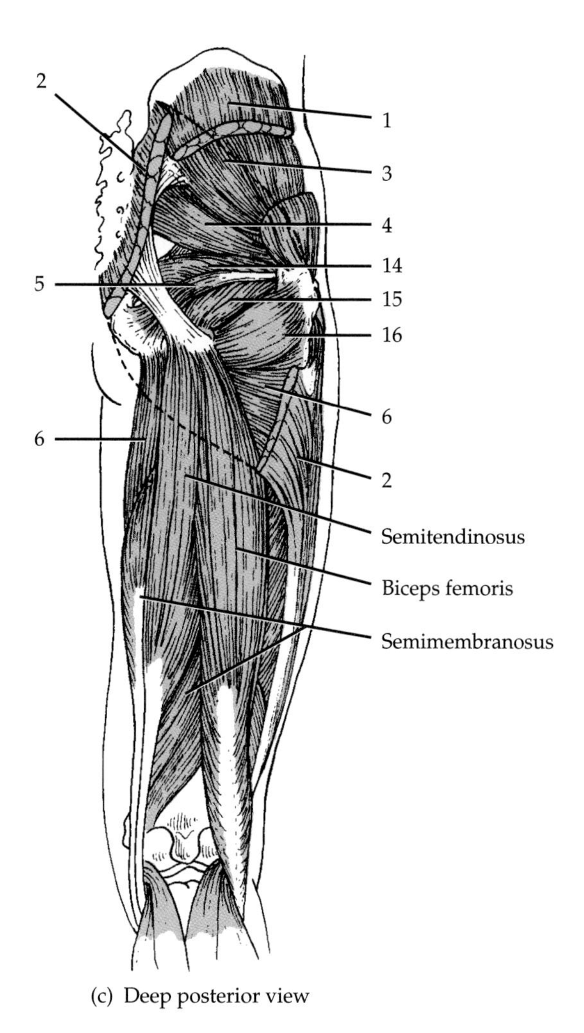

(c) Deep posterior view

___ Adductor magnus
___ Gluteus maximus
___ Gluteus medius
___ Gluteus minimus
___ Inferior gemellus
___ Obturator internus
___ Piriformis
___ Quadratus femoris
___ Superior gemellus

FIGURE 10.17 (*Continued*) Muscles that move the femur (thigh).

TABLE 10.21
Muscles that Act on the Femur (Thigh) and Tibia and Fibula (Leg) (After completing the table, label Figure 10.18.)

OVERVIEW: The muscles that act on the femur (thigh) and tibia and fibula (leg) are separated into compartments by deep fascia. The ***medial (adductor) compartment*** is so named because its muscles adduct the thigh. As noted earlier, the adductor magnus, adductor longus, adductor brevis, and pectineus muscles, components of the medial compartment, are included in Table 10.20 because they act on the femur. The gracilis, the other muscle in the medial compartment, not only adducts the thigh but also flexes the leg. For this reason, it is included in this table.

The ***anterior (extensor) compartment*** is so designated because its muscles act to extend the leg (some also flex the thigh). It is composed of the quadriceps femoris and sartorius muscles. The quadriceps femoris muscle is a composite muscle that includes four distinct parts, usually described as four separate muscles (rectus femoris, vastus lateralis, vastus intermedius, and vastus medialis). The common tendon for the four muscles is known as the ***quadriceps tendon,*** which attaches to the patella. The tendon continues inferiorly to the patella as the ***patellar ligament,*** which attaches to the tibial tuberosity.

The ***posterior (flexor) compartment*** is so named because its muscles flex the leg. Included are the hamstrings (biceps femoris, semitendinosus, and semimembranosus). The hamstrings are so named because their tendons are long and stringlike in the popliteal area. The ***popliteal fossa*** is a diamond-shaped space on the posterior aspect of the knee bordered laterally by the biceps femoris and medially by the semitendinosus and semimembranosus muscles (see also Figure 11.10b).

Muscle	Origin	Insertion	Action
MEDIAL (ADDUCTOR) COMPARTMENT **Adductor magnus** (MAG-nus)			
Adductor longus (LONG-us)			
Adductor brevis (BREV-is)			
Pectineus (pek-TIN-ē-us)			
Gracilis (gra-SIL-is; *gracilis* = slender)			
ANTERIOR (EXTENSOR) COMPARTMENT **Quadriceps femoris** (KWOD-ri-ceps FEM-or-is; *quadriceps* = four heads of origin; *femoris* = femur)			

TABLE 10.21 (*Continued*)

Muscle	Origin	Insertion	Action
Rectus femoris (REK-tus FEM-or-is; *rectus* = fibers parallel to midline)			
Vastus lateralis (VAS-tus lat'-er-A-lis; *vastus* = large; *lateralis* = lateral)			
Vastus medialis (VAS-tus mē-dē-A-lis; *medialis* = medial)			
Vastus intermedius (VAS-tus in'-ter-MĒ-dē-us; *intermedius* = middle)			
Sartorius (sar-TOR-ē-us; *sartor* = tailor; refers to cross-legged position of tailors)			
POSTERIOR (FLEXOR) COMPARTMENT **Hamstrings** **Biceps femoris** (BĪ-ceps FEM-or-is; *biceps* = two heads of origin)			
Semitendinosus (sem'-ē-TEN-di-nō-sus; *semi* = half; *tendo* = tendon)			
Semimembranosus (sem'-ē-MEM-bra-nō-sus; *membran* = membrane)			

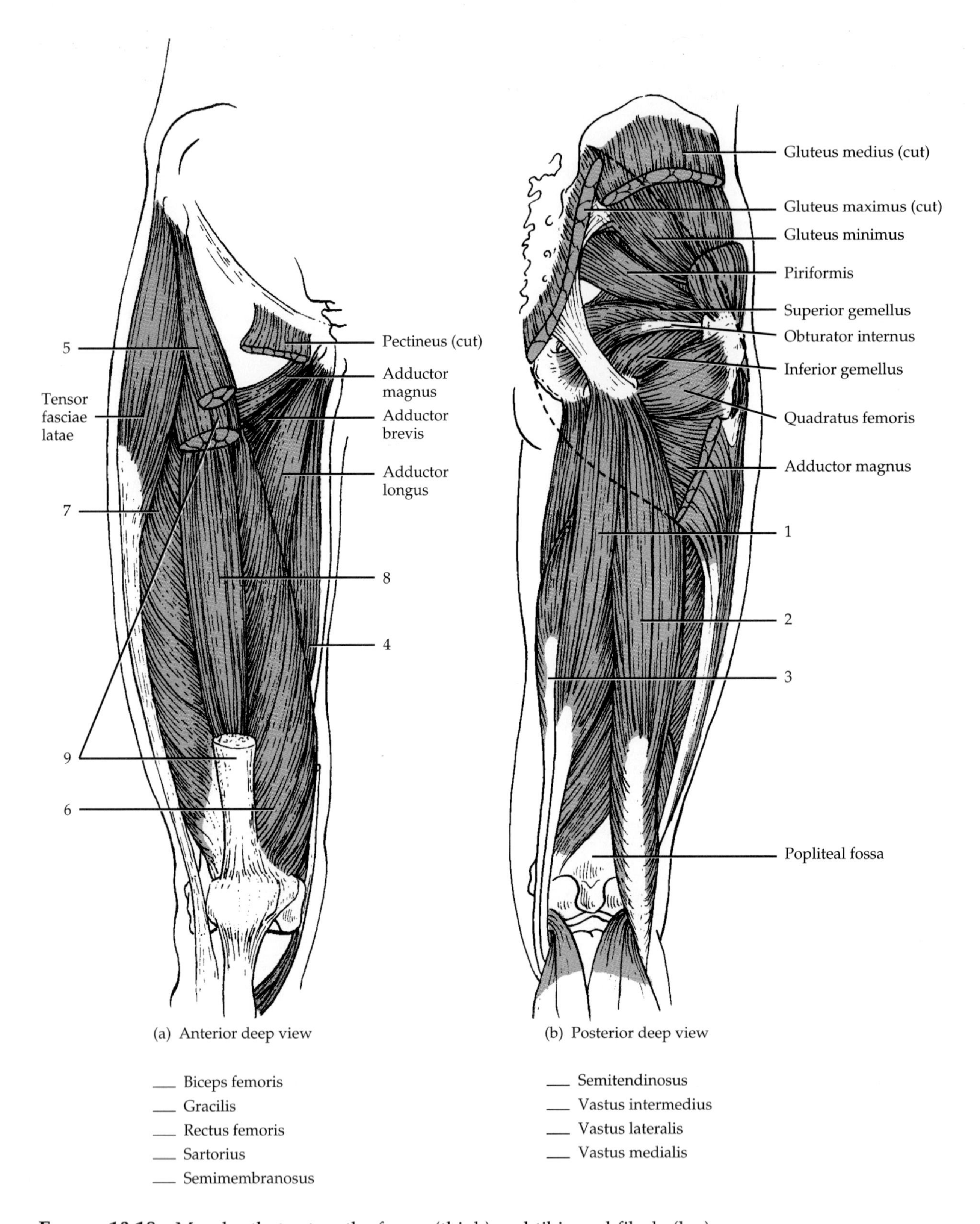

___ Biceps femoris
___ Gracilis
___ Rectus femoris
___ Sartorius
___ Semimembranosus

___ Semitendinosus
___ Vastus intermedius
___ Vastus lateralis
___ Vastus medialis

FIGURE 10.18 Muscles that act on the femur (thigh) and tibia and fibula (leg).

TABLE 10.22
Muscles that Move the Foot and Toes (After completing the table, label Figure 10.19.)

OVERVIEW: Muscles that move the foot and toes are located in the leg. The musculature of the leg, like that of the thigh, can be grouped into three compartments by deep fascia. The ***anterior compartment*** consists of muscles that dorsiflex the foot. In a situation analogous to the wrist, the tendons of the muscles of the anterior compartment are held firmly to the ankle by thickenings of deep fascia called the ***superior extensor retinaculum (transverse ligament of the ankle)*** and ***inferior extensor retinaculum (cruciate ligament of the ankle).***

The ***lateral*** (***peroneal***) ***compartment*** contains two muscles that plantar flex and evert the foot.

The ***posterior compartment*** consists of muscles that are divisible into superficial and deep groups. All three superficial muscles share a common tendon of insertion, the calcaneal (Achilles) tendon that inserts into the calcaneal bone of the ankle. The superficial muscles and most deep muscles are plantar flexors of the foot.

Muscle	Origin	Insertion	Action
ANTERIOR COMPARTMENT **Tibialis** (tib'-ē-A-lis) **anterior** (*tibialis* = tibia; *anterior* = front)			
Extensor hallucis longus (HAL-u-kis LON-gus; *extensor* = increases angle at joint; *hallucis* = hallux or great toe; *longus* = long)			
Extensor digitorum longus (di'-ji-TOR-um LON-gus)			
Peroneus tertius (per'-ō-NĒ-us TER-shus; *perone* = fibula; *tertius* = third)			
LATERAL (PERONEAL) COMPARTMENT **Peroneus longus** (per'-ō-NĒ-us LON-gus;)			
Peroneus brevis (per'-ō-NĒ-us BREV-is; *brevis* = short)			

TABLE 10.22 (*Continued*)

Muscle	Origin	Insertion	Action
POSTERIOR COMPARTMENT Superficial			
Gastrocnemius (gas'-trok-NĒ-mē-us; *gaster* = belly; *kneme* = leg)			
Soleus (SŌ-lē-us; *soleus* = sole)			
Plantaris (plan-TA-ris; *plantar* = sole)			
Deep			
Popliteus (pop-LIT-ē-us; *poples* = posterior surface of knee)			
Tibialis (tib'-ē-A-lis) **posterior** (*posterior* = back)			
Flexor digitorum longus (di'-ji-TOR-um LON-gus; *digitorum* = finger or toe)			
Flexor hallucis longus (HAL-u-kis LON-gus; *flexor* = decreases angle at joint)			

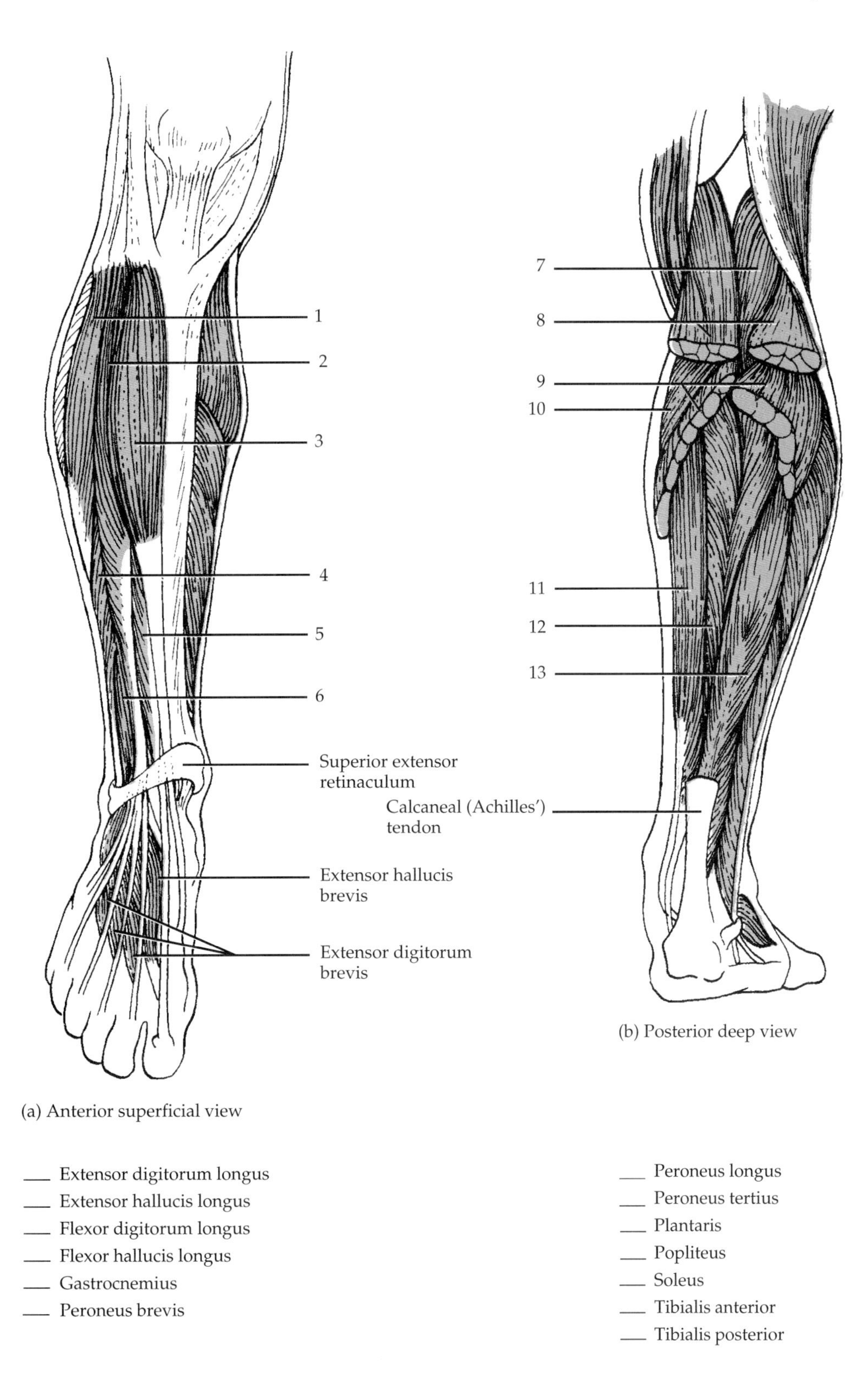

___ Extensor digitorum longus
___ Extensor hallucis longus
___ Flexor digitorum longus
___ Flexor hallucis longus
___ Gastrocnemius
___ Peroneus brevis

___ Peroneus longus
___ Peroneus tertius
___ Plantaris
___ Popliteus
___ Soleus
___ Tibialis anterior
___ Tibialis posterior

FIGURE 10.19 Muscles that move the foot and toes.

TABLE 10.23
Intrinsic Muscles of the Foot (After completing the table, label Figure 10.20.)

OVERVIEW: The intrinsic muscles of the foot are comparable to those in the hand. Where the muscles of the hand are specialized for precise and intricate movements, these of the foot are limited to support and locomotion. The deep fascia of the foot forms the ***plantar aponeurosis (fascia),*** that extends from the calcaneus to the phalanges. The aponeurosis supports the longitudinal arch of the foot and transmits the flexor tendons of the foot.

The intrinsic musculature of the foot is divided into two groups—dorsal and plantar. There is only one ***dorsal muscle.*** The ***plantar muscles*** are arranged in four layers, the most superficial layer being referred to as the first layer.

Muscle	Origin	Insertion	Action
DORSAL MUSCLE			
Extensor digitorum brevis (di'-ji-TOR-um BREV-is; *extensor* = increases angle at joint; *digit* = finger or toe; *brevis* = short) (Illustrated in Figure 10.19a)			
PLANTAR MUSCLES			
First (Superficial) Layer			
Abductor hallucis (HAL-a-kis; *abductor* = moves part away from midline; *hallucis* = hallux or great toe)			
Flexor digitorum brevis (di'-ji-TOR-um BREV-is; *flexor* = decreases angle at joint)			
Abductor digiti minimi (DIJ-i-tē; MIN-i-mē; *minimi* = small toe)			
Second Layer			
Quadratus plantae (quod-RĀ-tus PLAN-tē; *quad* = four *plantar* = sole of foot)			
Lumbricals (LUM-bri-kals; four muscles)			
Third layer			
Flexor hallucis brevis (HAL-a-kis BREV-is)			

TABLE 10.23 (*Continued*)

Muscle	Origin	Insertion	Action
Adductor hallucis (HAL-a-kis; *adduct* = moves part closer to midline)			
Flexor digiti minimi brevis (DIJ-i-tē MIN-i-mē BREV-is)			
Fourth (Deep) Layer			
Dorsal interossei (in′-ter-OS-ē-ī; four muscles; *dorsal* = back surface; *inter* = between; *ossei* = bone)			
Plantar interossei (in′-ter-OS-ē-ī)			

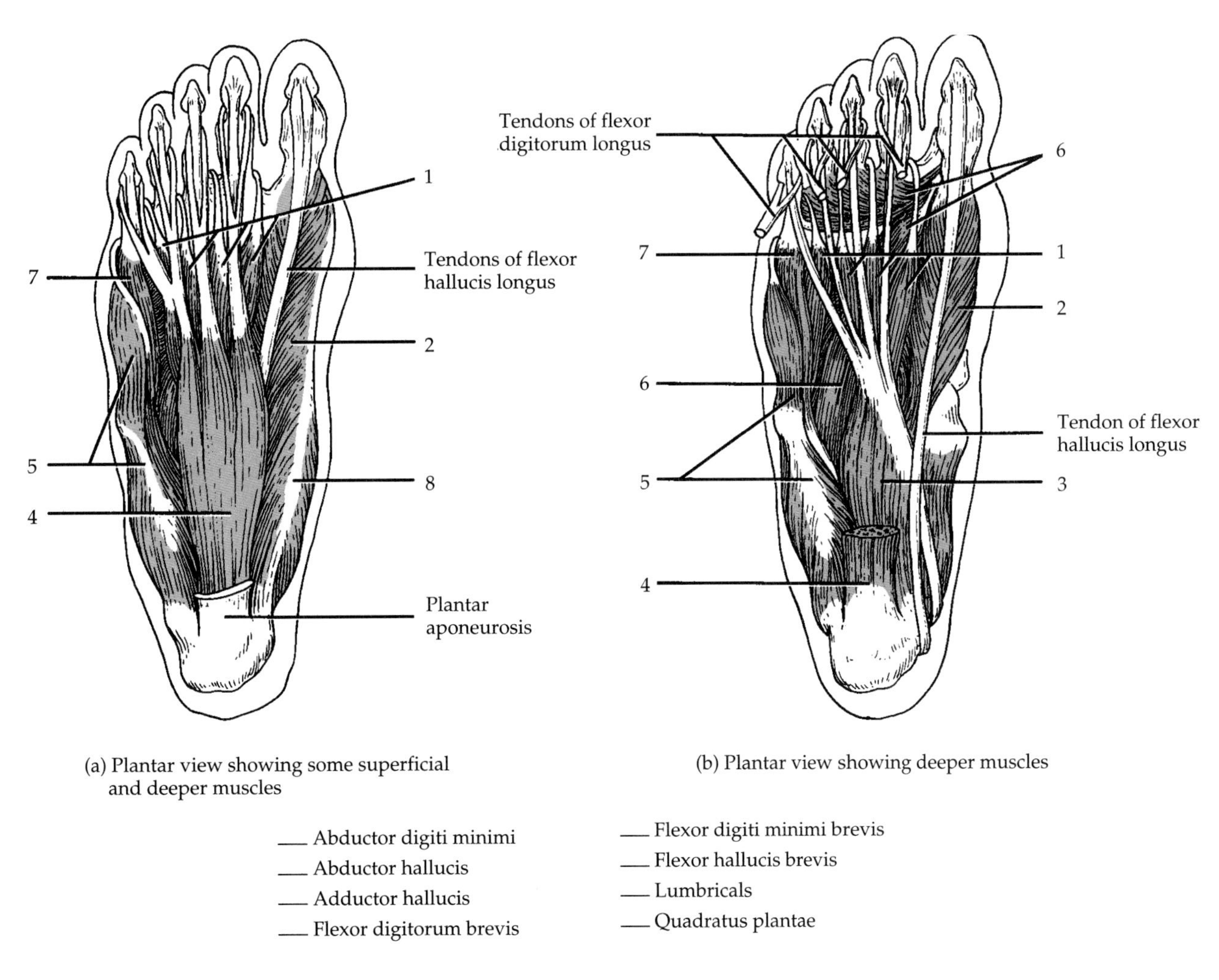

FIGURE 10.20 Intrinsic muscles of the foot.

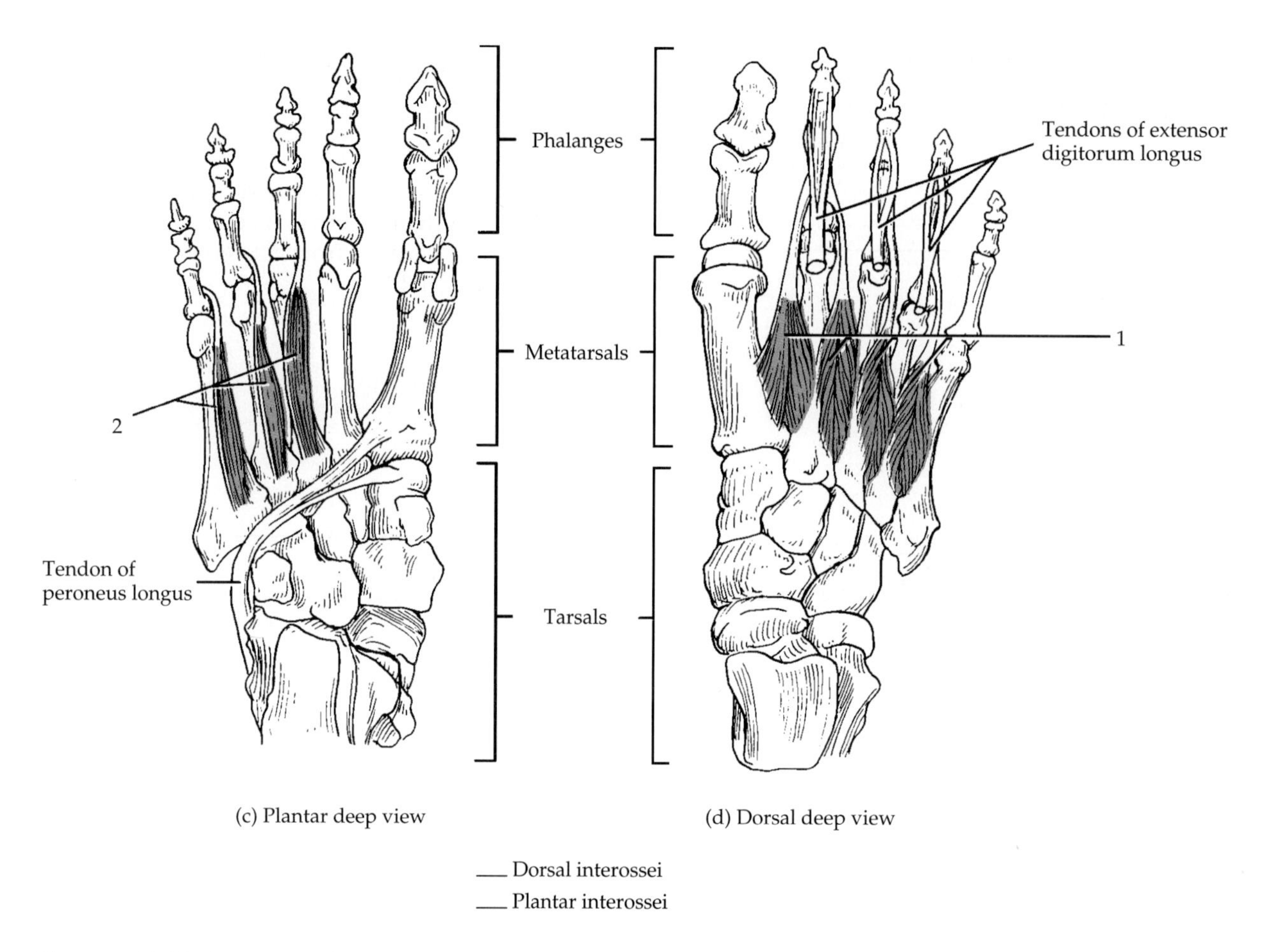

FIGURE 10.20 (*Continued*) Intrinsic muscles of the foot.

F. COMPOSITE MUSCULAR SYSTEM

Now that you have studied the muscles of the body by region, label the composite diagram shown in Figure 10.21.

G. DISSECTION OF CAT MUSCULAR SYSTEM

Assuming that you have completed your study of the human muscular system, you can now begin your examination of the cat musculature. The principal objective of dissecting the cat muscular system is to increase your understanding of the human muscular system. Do not be concerned that the cat and human muscular systems differ somewhat. Instead, concentrate on the similarities and, as you perform the dissection, try to relate your knowledge of the human muscular system to that of the cat.

CAUTION! *Please reread Section D, "Precautions Related to Dissection," at the beginning of the laboratory manual on p. xi before you begin your dissection.*

PROCEDURE

1. After obtaining a cat from your instructor, remove it from the plastic bag and place it on your dissecting tray ventral side up. Keep the liquid preservative in the plastic bag.
2. Determine the gender of your cat. A male has a ***scrotum*** ventral to the base of the tail. The ***prepuce*** appears as a slight elevation anterior to the scrotum. In contrast to the human male, the cat penis is usually retracted.
3. A female cat has a ***urogenital aperture,*** the external opening of the vagina and the urethra, just ventral to the anus. Both male and female cats have four or five ***nipples (teats)*** on either side of the midline, so their presence does not necessarily mean that your cat is female. Consult your instructor if you have difficulty determining the gender of your cat. Compare your cat with one of the opposite sex and learn to identify both sexes externally.
4. Before skinning your cat, review Section I in Exercise 2, in which you were introduced to the planes of the body. Identify the planes of the cat's body and compare their orientation to those in the human (Figures 2.3 and 10.22 on page 186). Also review Section G in Exercise 2 and locate, on your cat, the structures described therein for the human. You will note the following additional structures on the cat:
   a. ***Virissae*** Tactile whiskers located around the mouth and above the eyes.
   b. ***Claws*** Derived from the epidermis. They are retractable.
   c. ***Tori (foot pads*** or ***friction pads)*** Thickened areas of skin on the ventral surface of the paws. They cushion the body when the cat walks. Count the tori on the front and the back paws.

1. Skinning the Cat

PROCEDURE

Note: *Please read this entire section before starting.*

1. Grasp the skin at the ventral surface of the neck, pinch it up, and make a small longitudinal incision at the midline with the scissors. The incision, as well as all others you will make in skinning, should not be too deep because the skin is only about 1/8 in. thick and underlying muscles are easily damaged.
2. Now continue the incision upward along the midline to the lower lip and downward along the midline to the genital region. Do not remove the skin from the genital region.
3. If your specimen is female, the mammary glands may be removed. These appear as large masses along the ventral surface of the abdomen. Do not damage the underlying muscles as you remove the glands.
4. Next make incisions from the midline incision on the ventral side to the wrists and from the midline incision to the ankles. Cut all the way around the wrists and ankles.
5. Now you can begin to separate the skin from the underlying muscles to expose the muscles. This can be done by pulling the skin away from the body, using your fingers as much as possible. If you must use a scalpel, make sure that the sharp edge is directed away from the muscle and toward the skin.
6. Continue peeling the skin until only two points of attachment remain, one at the base of the tail and the other at the sides and top of the head.
7. Cut across the skin at the base of the tail to free the skin from the tail. The skin around the tail may be left in position.

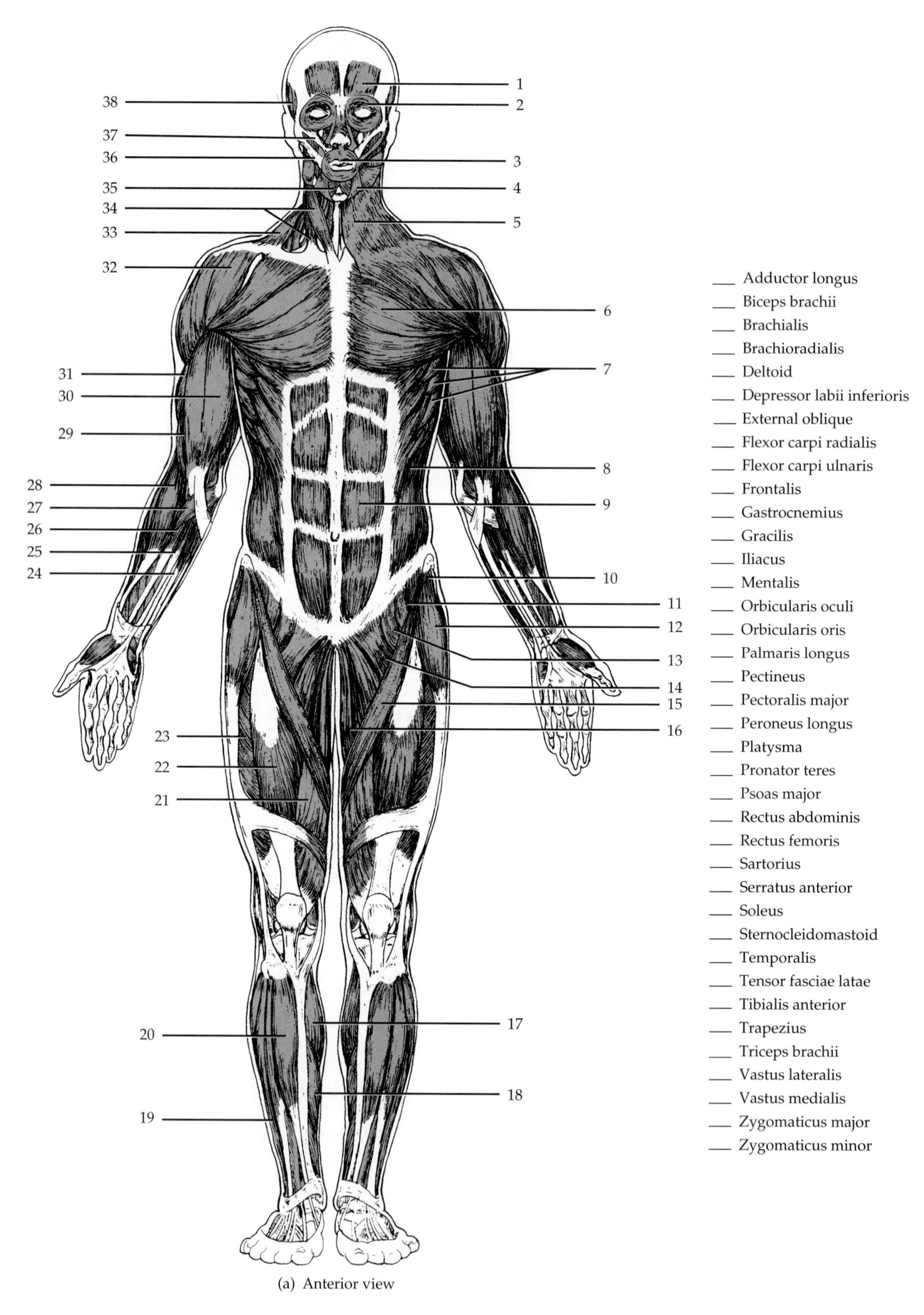

(a) Anterior view

FIGURE 10.21 Principal superficial muscles.

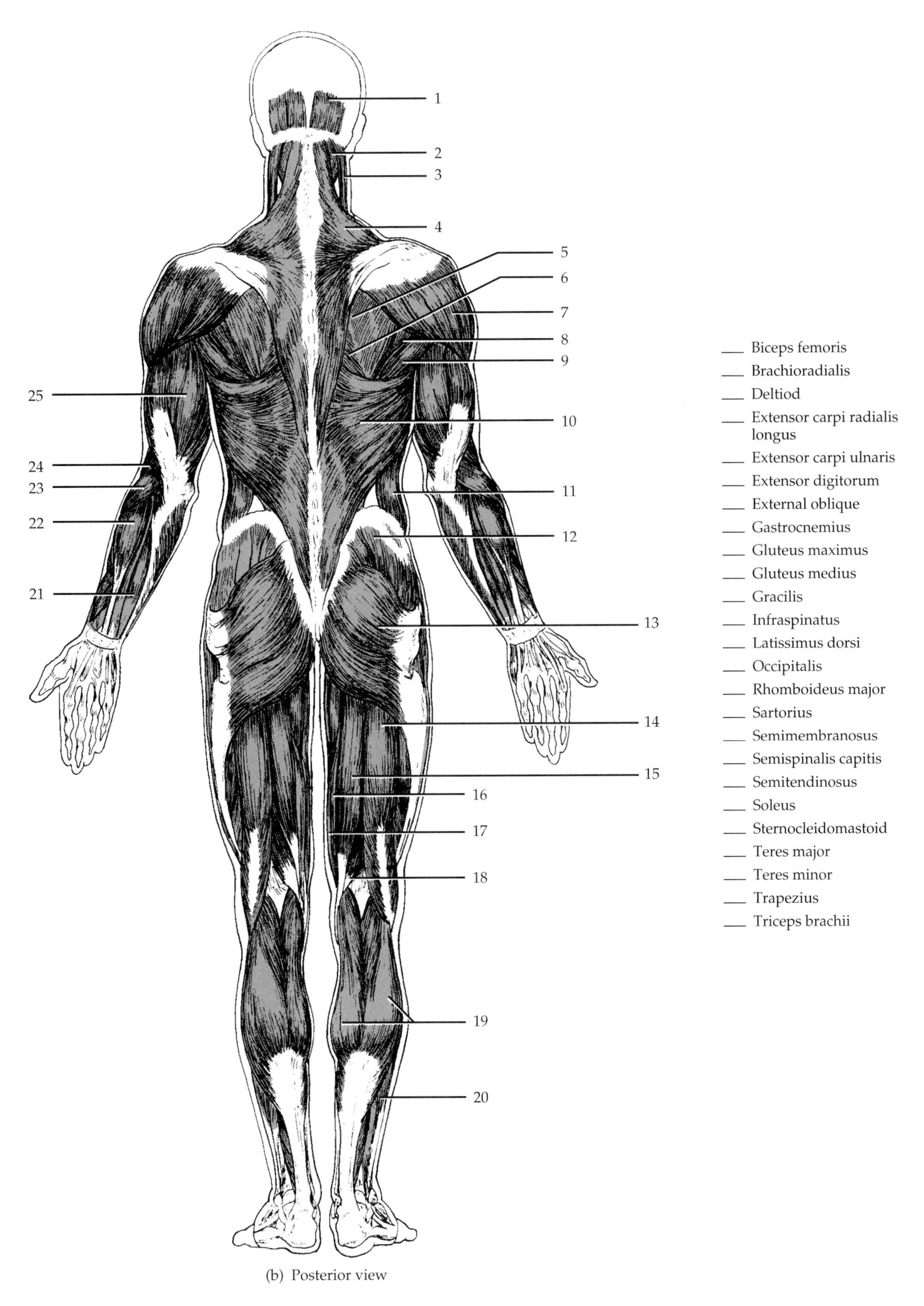

___ Biceps femoris
___ Brachioradialis
___ Deltiod
___ Extensor carpi radialis longus
___ Extensor carpi ulnaris
___ Extensor digitorum
___ External oblique
___ Gastrocnemius
___ Gluteus maximus
___ Gluteus medius
___ Gracilis
___ Infraspinatus
___ Latissimus dorsi
___ Occipitalis
___ Rhomboideus major
___ Sartorius
___ Semimembranosus
___ Semispinalis capitis
___ Semitendinosus
___ Soleus
___ Sternocleidomastoid
___ Teres major
___ Teres minor
___ Trapezius
___ Triceps brachii

FIGURE 10.21 (*Continued*) Principal superficial muscles.

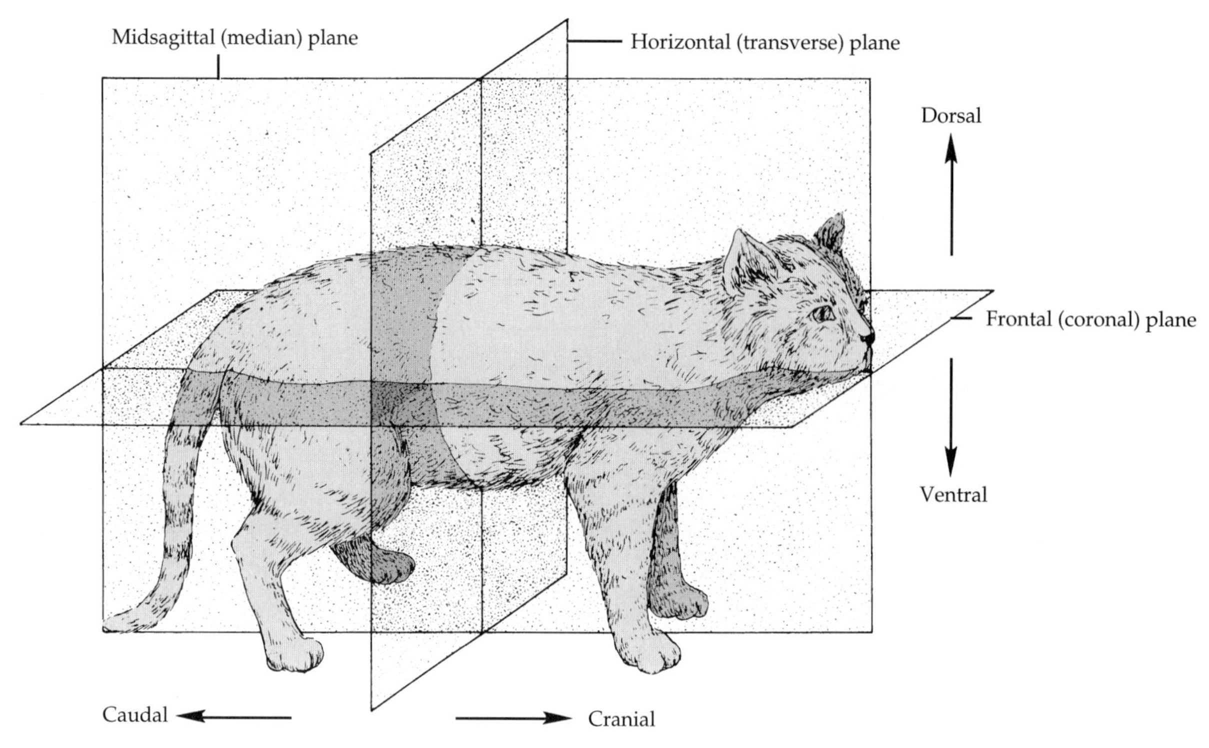

FIGURE 10.22 Anatomical planes and directional terms of the cat.

8. Make a final incision from the left side of the lower lip, up to the top of the left eye, across to the top of the right eye, and down to the right side of the lower lip.
9. Remove the skin from the head, cutting around the bases of the ears as you come to them.
10. As part of the skinning procedure, be careful of the following:

    a. While skinning the neck, do not damage the external jugular veins in the neck (see Figure 10.23a).
    b. While skinning the forelimbs, do not damage the cephalic veins (see Figure 18.22).
    c. While skinning the hindlimbs, do not damage the greater saphenous veins (see Figure 18.20).
    d. While skinning the dorsal surface of the abdomen, do not cut through the deep fascia (lumbodorsal fascia) along the dorsal midline (see Figure 10.25a).
    e. The edge of the latissimus dorsi muscle (see Figure 10.25a) has a tendency to tear.

11. Before continuing, find the ***cutaneous muscles,*** which are attached to the undersurface of the skin and are generally removed with the skin. One of these, the ***cutaneous maximus,*** is closely bound to the underside of the skin in the trunk area. Another is the ***platysma,*** which is closely bound to the skin on either side of the neck. The cutaneous muscles move the skin.[2]
12. Clean off as much fat as possible. Large fat deposits are usually found in the neck, axillary regions, and inguinal regions. As you pick away the fat with forceps, do not damage blood vessels or nerves. Also remove as much of the superficial fascia covering muscles as possible. This can be done with your forceps. If your specimen has been properly cleaned, the individual fibers in a muscle will be visible. Usually, the fibers of one muscle run in one direction, while those of another muscle run in another direction.
13. Save the skin and wrap it around the cat at the end of each laboratory period to prevent your specimen from drying out. Moist paper towels or an old cotton towel may also be used.

[2]If preskinned cats are used the instructor may wish to demonstrate the cutaneous maximus and the platysma.

2. Dissecting Skeletal Muscles

Keep in mind that *dissect* means to separate. Dissection of a skeletal muscle includes separating the body of the muscle from surrounding tissues, but points of attachment are left intact. These are the ***origin*** (attachment to a fixed bone) and the ***insertion*** (attachment to a movable bone). When a muscle contracts, the insertion moves toward the origin. You can pull the insertion toward the origin to observe the action of the muscle. Most muscles are attached to bones by bands of white fibrous connective tissue called ***tendons.*** Some muscles are attached by white, flattened, sheetlike tendons called ***aponeuroses.***

Skeletal muscles are distinguished from each other by normal lines of cleavage. These lines consist of deep fascia and may be exposed by slightly pulling adjacent muscles apart. The muscles can be separated from each other by picking the fascia out with your forceps. If, after removing the fascia, the muscles separate as distinct units, your procedure is correct. If, instead, the separated muscles have a ragged appearance, you have damaged a muscle by tearing it into two portions.

Sometimes superficial muscles must be severed in order to examine deep muscles. When you cut a muscle, ***transect*** it (i.e., cut it at right angles to the fibers at about the center of the belly of the muscle) using a scissors. Then ***reflect*** the cut ends (i.e., pull them back, one toward the origin and one toward the insertion). As you reflect the ends, carefully break the underlying fascia with your scalpel to separate the superficial muscle from the deep muscle. Transection keeps the origin and insertion of the superficial muscle intact and preserves the original shape and position of the superficial muscle.

3. Dissection, Transection, and Reflection of a Single Skeletal Muscle

Before you start to identify skeletal muscles, you should dissect a single muscle to familiarize yourself with the procedure and the terminology. The sample muscle that has been selected is the pectoantebrachialis (see Figure 10.23a).

PROCEDURE

1. Place the cat on its dorsal surface and find the pectoantebrachialis and pectoralis major muscles. Separate the muscles from each other.
2. Dissect the body of the pectoantebrachialis muscle toward its origin and insertion. Its origin is the manubrium of the sternum and it inserts on the fascia on the proximal end of the forearm. Leave the origin and insertion intact.
3. By drawing the insertion toward the origin you can note the action of the muscle—adduction of the arm.
4. Transect the pectoantebrachialis muscle and reflect the cut ends to expose the underlying pectoralis major. As you reflect, break the fascia under the pectoantebrachialis muscle.

4. Identification of Skeletal Muscles

You are now ready to begin your identification of selected skeletal muscles of the cat. The muscles you will study are grouped according to region. Each muscle discussed is shown in the corresponding figure. Locate each muscle discussed and try to determine its origin and insertion. Then dissect it and pull its insertion toward its origin to observe its action. If you are told that a particular muscle is a deep muscle, you will also have to transect and reflect the superficial muscle to locate and study the deep muscle.

As you dissect, transect, and reflect muscles of various regions, work only on either the left or right side of the cat. You will use the other side later on to study blood vessels and nerves.

a. SUPERFICIAL CHEST MUSCLES: PECTORAL GROUP (FIGURE 10.23a)

The ***pectoralis*** muscle group covers the ventral surface of the chest. The large, triangular muscle group arises from the sternum and inserts primarily on the humerus. On each side, the pectoralis group is divided into four muscles.

1. ***Pectoantebrachialis*** A narrow ribbonlike muscle superficial to the pectoralis major muscle. It consists of a narrow band ($1/2$ in.) of fibers that run from the sternum to the forearm. Dissect, transect, and reflect.
2. ***Pectoralis major*** A muscle 2 to 3 in. wide that lies immediately deep to the pectoantebrachialis and the clavobrachialis. Its fibers run roughly parallel to those of the pectoantebrachialis. Separate the pectoralis major from the pectoralis minor and clavobrachialis, transect, and reflect.

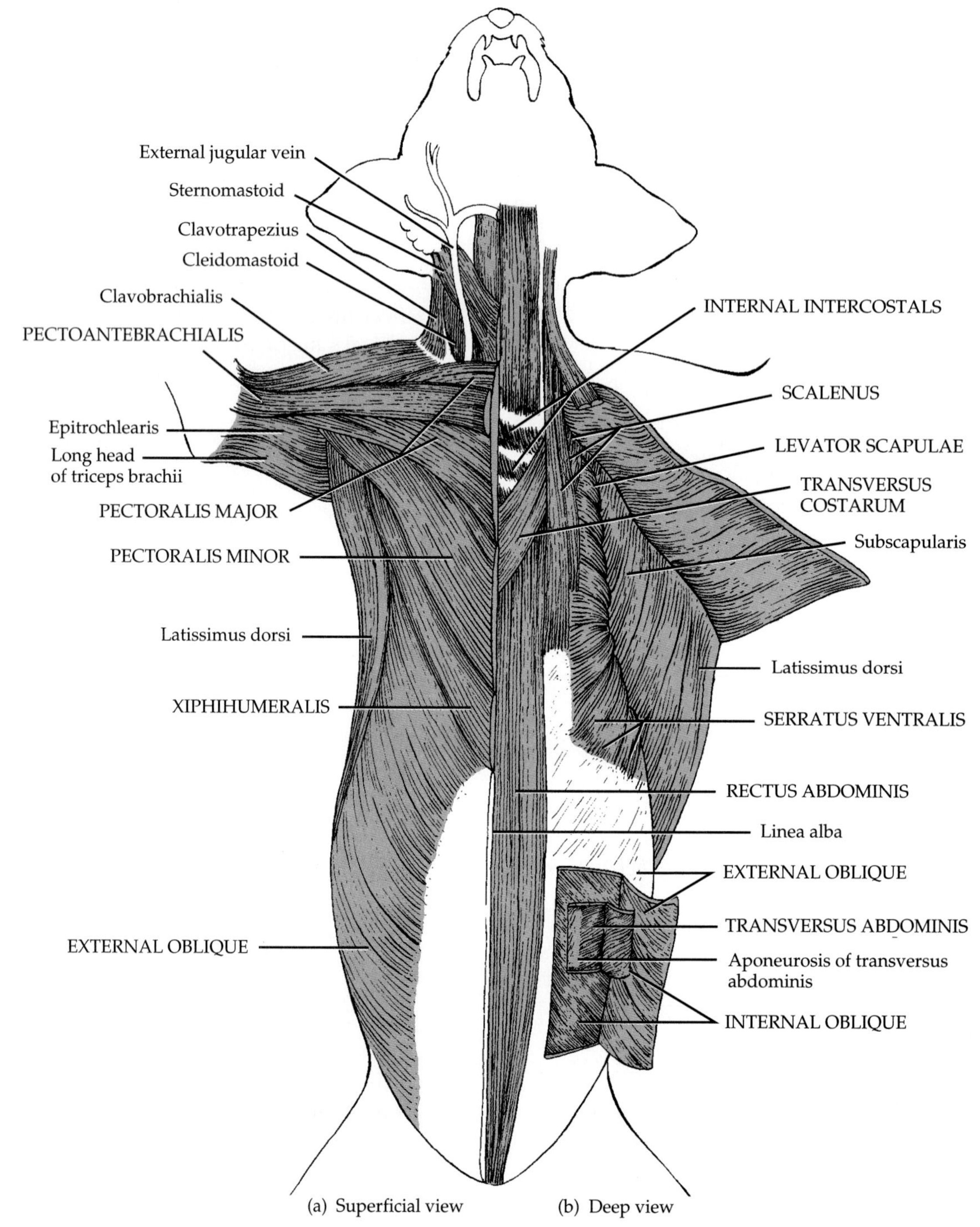

FIGURE 10.23 Ventral view of chest and abdominal muscles.

3. ***Pectoralis minor*** A muscle lying dorsal and caudal to the pectoralis major. Dissect, transect, and reflect.
4. ***Xiphihumeralis*** The most caudad of the pectoral group. Its fibers run parallel to pectoralis minor and eventually pass deep to the pectoralis minor. Dissect, transect, and reflect.

In humans, the pectoral group has only two muscles—the pectoralis major and minor.

b. DEEP CHEST MUSCLES (FIGURE 10.23b)

1. *Serratus ventralis* Fan-shaped muscle extending obliquely upward from the ribs to the shoulder blade. Dissect.
2. *Levator scapulae* Continuous with the serratus ventralis at its cranial border. Dissect.
3. *Transversus costarum* Small, thin muscle that covers the cranial end of the rectus abdominis muscle. Dissect.
4. *Scalenus* Complex muscle that extends longitudinally along the ventrolateral thorax and neck. Its three main parts are called the scalenus dorsalis, scalenus medius, and scalenus ventralis. The scalenus medius is easiest to recognize. Dissect.
5. *External intercostals* Series of muscles[3] in the intercostal spaces. Their fibers run in the same direction (caudoventrally) as those of the external oblique muscle (next section). Dissect, transect, and reflect.
6. *Internal intercostals* Series of muscles deep to the external intercostals. Their fibers run in the same direction (caudodorsally) as those of the internal oblique muscle (next section).

c. MUSCLES OF THE ABDOMINAL WALL (FIGURE 10.23a AND b)

Before examining the muscles of the abdominal wall, locate the ***linea alba,*** a midventral fascial line formed by fusion of the aponeuroses of the ventrolateral abdominal muscles. The muscles of the abdominal wall include the following:

1. *External oblique* Most superficial, lateral abdominal muscle. Its dorsal half is fleshy; its ventral half is an aponeurosis. Its fibers run dorsally and caudally. Make a very shallow longitudinal incision in the external oblique from the ribs to the pelvis. Transect and reflect. Find the aponeurosis of the external oblique.
2. *Internal oblique* Deep to the external oblique. Its fibers run ventrally and cranially, opposite those of the external oblique. Make a very shallow longitudinal incision in the internal oblique muscle until you can see transverse fibers. They belong to the transversus abdominis muscle. Reflect the internal oblique and find its aponeurosis.
3. *Transversus abdominis* This is the third and deepest layer of the abdominal wall and is deep to the internal oblique. Its fibers run transversely. Locate the aponeurosis of the transversus abdominis.
4. *Rectus abdominis* Longitudinal band of muscle just lateral to the linea alba. The muscle is visible when the external oblique aponeurosis is reflected.

The abdominal muscles are similar to those in humans.

d. MUSCLES OF THE NECK (FIGURE 10.24a AND b)

1. *Sternomastoid* Band of muscle extending from the manubrium to the mastoid portion of the skull. The muscle is deep to the external jugular vein. Dissect.
2. *Cleidomastoid* Narrow band of muscle that is between the sternomastoid and clavotrapezius. Most of the cleidomastoid is deep to the clavotrapezius. Dissect.
3. *Sternohyoid* Narrow, straight muscle on either side of the midline covering the trachea. Caudal end partially covered by the sternomastoid. Dissect.
4. *Sternothyroid* Narrow band of muscle lateral and dorsal to the sternohyoid. The sternohyoid must be lifted and pushed aside to expose the sternothyroid. The thyroid gland is deep to the sternothyroid.
5. *Thyrohyoid* Small muscle on the lateral side of the thyroid cartilage of the larynx. Its location is just cranial to the sternothyroid.
6. *Geniohyoid* Straplike muscle cranial to the sternohyoid. The mylohyoid must be transected and reflected to locate the geniohyoid. Transect and reflect the geniohyoid.
7. *Mylohyoid* Thin, sheetlike muscle seen after transection and reflection of the digastric muscle. Transect and reflect.
8. *Digastric* Superficial muscle attached to the ventral border of the mandible. Transect and reflect.
9. *Masseter* Large muscle mass at the angle of the jaw behind and below the ear.
10. *Temporalis* A muscle just above and caudal to the eye and medial to the ear (see Figure 10.23a).
11. *Hyoglossus* Short, obliquely directed muscle lateral to the geniohyoid. Dissect.

[3]Not illustrated.

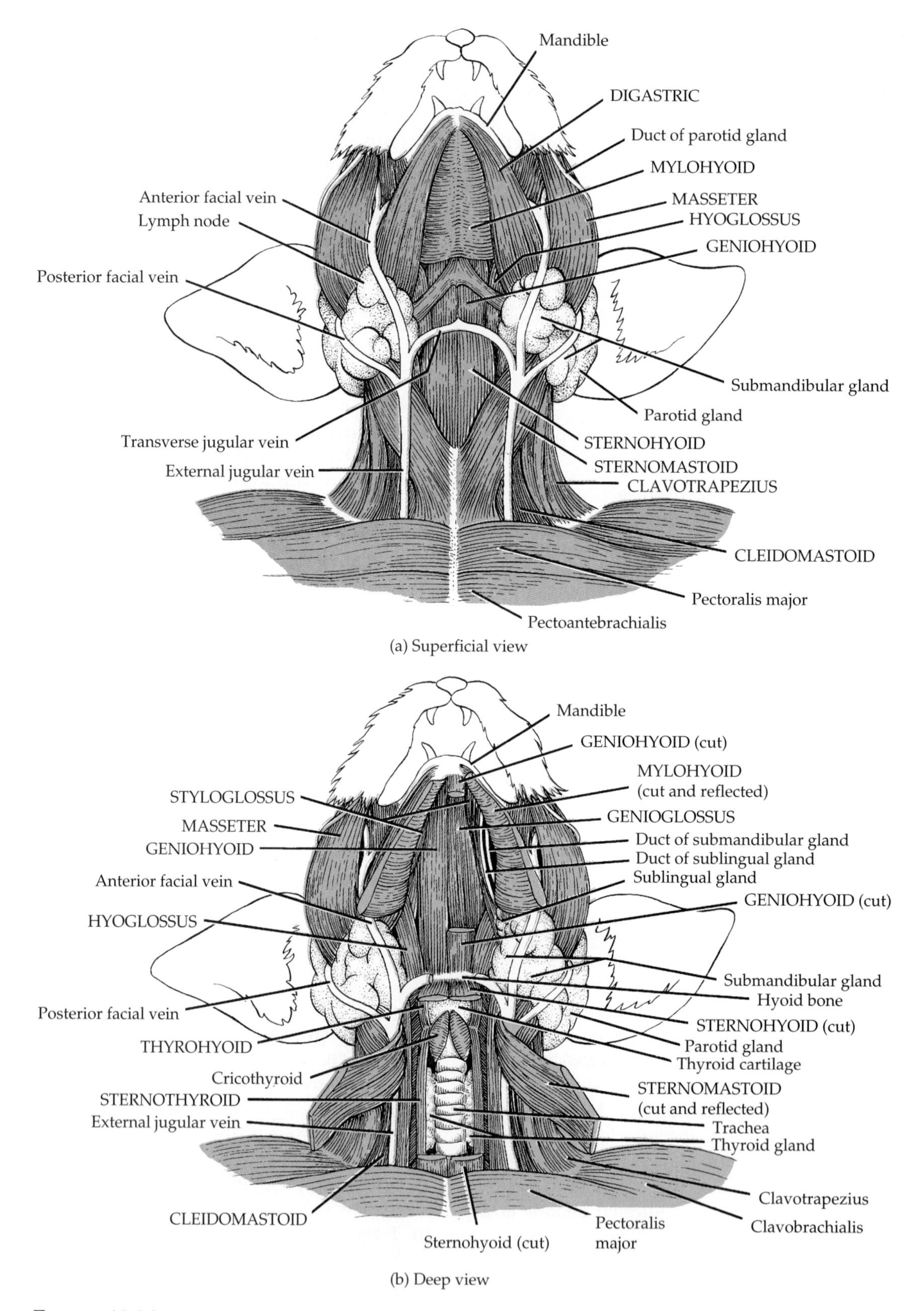

FIGURE 10.24 Ventral view of neck muscles.

12. *Styloglossus* Long muscle that runs parallel to the inside of the body and the mandible. Dissect.
13. *Genioglossus* A muscle lying deep to the geniohyoid.

e. SUPERFICIAL SHOULDER MUSCLES (FIGURE 10.25a)

1. *Clavotrapezius* Large, broad muscle on the back and side of the neck. Dissect, transect, and reflect.
2. *Levator scapulae ventralis* A muscle lying caudal to the clavotrapezius and passing underneath the clavotrapezius. Dissect.
3. *Clavobrachialis (clavodeltoid)* Triangular muscle extending onto the shoulder as a continuation of the clavotrapezius. Dissect, transect, and reflect.
4. *Acromiodeltoid* A muscle caudal to the clavobrachialis and lateral to the levator scapulae ventralis. Dissect, transect, and reflect.
5. *Spinodeltoid* Short, thick muscle caudal to the acromiodeltoid. Dissect, transect, and reflect.
6. *Acromiotrapezius* Thin, flat muscle extending from the middle of the back to the shoulder. The muscle is caudal to the clavotrapezius. Dissect, transect, and reflect.
7. *Spinotrapezius* A muscle lying caudal to the acromiotrapezius, partly covered by the acromiotrapezius. Dissect, transect, reflect.
8. *Latissimus dorsi* Large, triangular muscle caudal to the spinotrapezius. This muscle extends from the middle of the back to the humerus. Note its attachment to the lumbodorsal fascia. Dissect, transect, and reflect.

f. DEEP SHOULDER MUSCLES (FIGURE 10.25b)

1. *Rhomboideus capitis* Narrow muscle extending from the back of the skull to the scapula. Dissect.
2. *Rhomboideus* A muscle lying deep to medial portions of the acromiotrapezius and the spinotrapezius, between the scapula and the body wall. Dissect.
3. *Splenius* Broad muscle deep to the clavotrapezius and the rhomboideus capitis. Dissect.
4. *Supraspinatus* A muscle deep to the acromiotrapezius, located in the supraspinous fossa of the scapula. Dissect.
5. *Infraspinatus* A muscle caudal to the supraspinatus, occupying the infraspinous fossa of the scapula. Dissect.
6. *Teres major* Band of thick muscle caudal to the infraspinatus muscle and covering the axillary border of the scapula. Dissect.
7. *Teres minor* Tiny, round muscle caudal and adjacent to the insertion of the infraspinatus.[4] Dissect.
8. *Subscapularis* Large, flat muscle in the subscapular fossa. Dissect (see Figure 10.27b).

g. BACK MUSCLES (FIGURE 10.25a AND b)

1. *Serratus dorsalis cranialis* Four or five short muscle slips above and caudal to the scapula.
2. *Serratus dorsalis caudalis* Four or five short muscle slips caudal to the serratus dorsalis cranialis and medial to the upper part of the latissimus dorsi.
3. *Sacrospinalis* Complex muscle extending along the spine. Its three divisions are the lateral *iliocostalis,* which is deep to the serratus dorsalis cranialis and the serratus dorsalis caudalis; the intermediate *longissimus dorsi,* which fills the spaces between the spines and transverse processes of the thoracic and lumbar vertebrae; and the medial *spinalis dorsi,* next to the spines of the thoracic vertebrae. Dissect these components.
4. *Multifidus spinae* Narrow band of muscle adjacent to the vertebral spines extending the length of the vertebral column and best seen in the sacral region. Its cranial portion is called the *semispinalis cervicis*[5] muscle, which is under the splenius.

h. MUSCLES OF THE ARM: MEDIAL SURFACE (FIGURE 10.26b).

1. *Epitrochlearis* Flat band of muscle along the medial surface of the arm between the biceps brachii and the triceps brachii. Dissect, transect, and reflect.
2. *Biceps brachii* Located on the ventromedial surface of the arm. Dissect.

i. MUSCLES OF THE ARM: LATERAL SURFACE (FIGURE 10.27)

1. *Triceps brachii* Muscle with three main heads covering the caudal surface and much of the sides of the arm. The *long head,* located on the caudal surface of the arm, is the largest. The *lateral head* covers much of the lateral surface of the arm. Dissect, transect, and reflect the lateral head. The *medial head,* found on the medial surface of the arm, is located between the long and lateral heads.

[4]Not illustrated.
[5]Not illustrated.

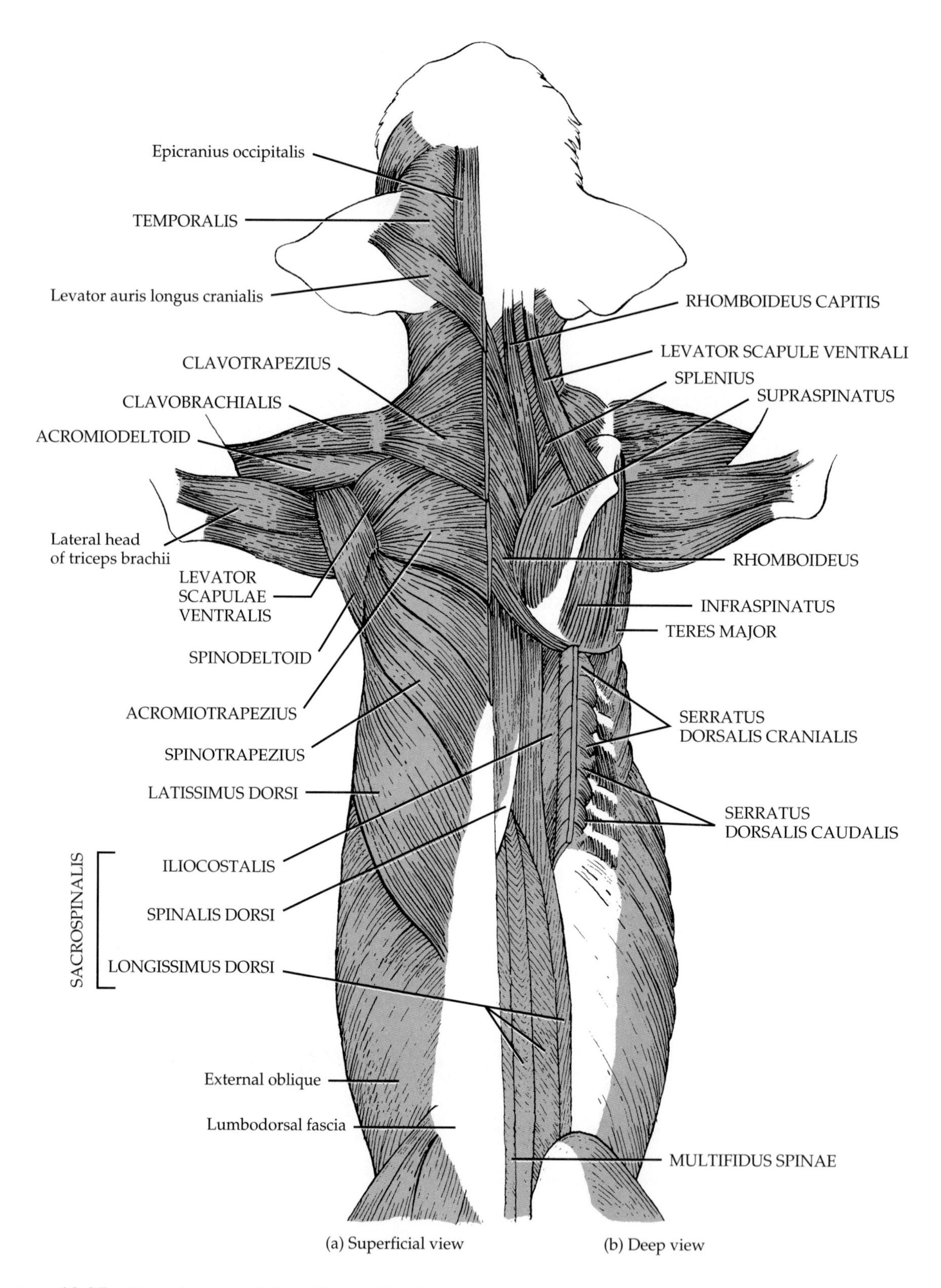

FIGURE 10.25 Dorsal view of shoulder and back muscles.

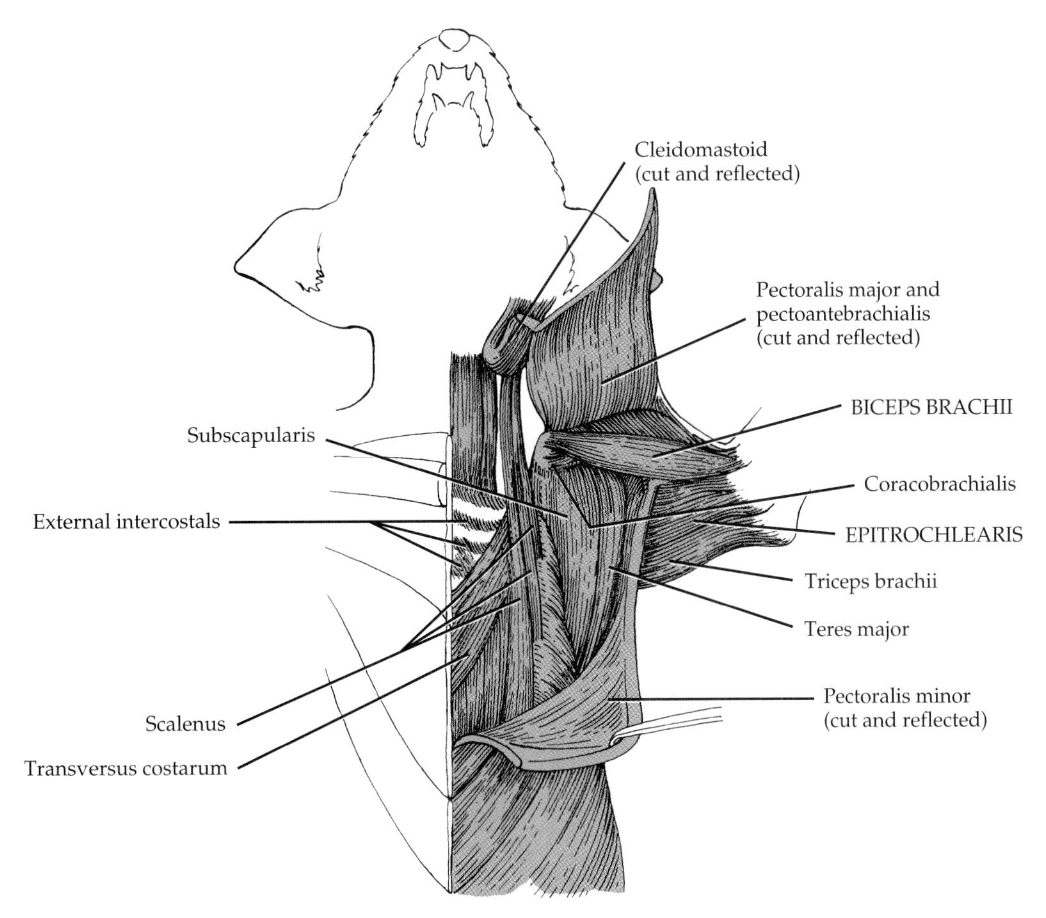

FIGURE 10.26 Muscles of medial surface of arm.

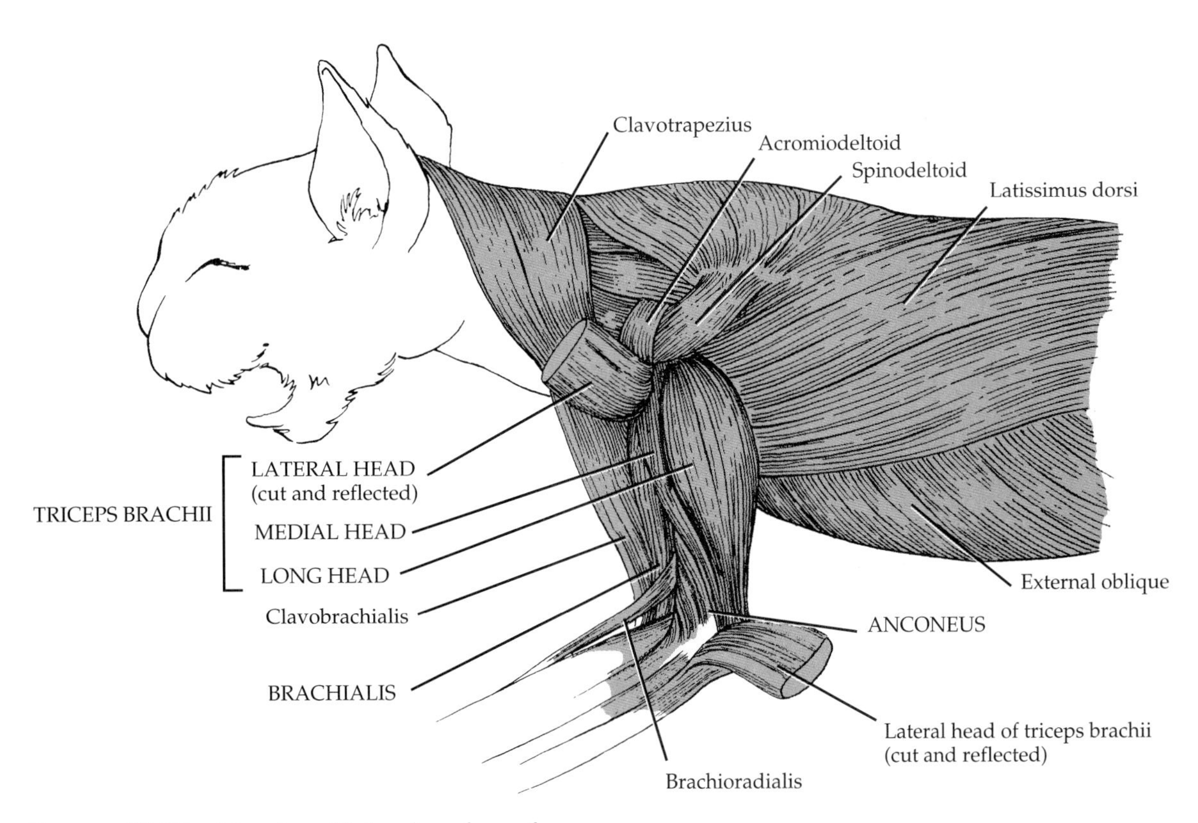

FIGURE 10.27 Muscles of lateral surface of arm.

2. ***Anconeus*** A tiny triangular muscle lying deep to the distal end of the lateral head of the triceps brachii. This is a fourth head of the triceps brachii. Dissect.
3. ***Brachialis*** A muscle located on the ventrolateral surface of the arm, cranial to the lateral head of the triceps brachii. Dissect.

j. MUSCLES OF THE FOREARM: VENTRAL SURFACE (FIGURE 10.28).

1. ***Pronator teres*** Small, triangular muscle that originates from the medial epicondyle of the humerus. Dissect.
2. ***Flexor carpi radialis*** Long, spindle-shaped muscle above the pronator teres. Dissect.
3. ***Palmaris longus*** Large, flat, broad muscle near the center of the medial surface. Dissect, transect, and reflect.
4. ***Flexor digitorum sublimis (superficialis)*** A muscle deep to the palmaris longus. Dissect.
5. ***Flexor carpi ulnaris*** Band of muscle above the palmaris longus consisting of two heads. Dissect.
6. ***Flexor digitorum profundus*** A muscle with five heads of origin, deep to the palmaris longus. Dissect.
7. ***Pronator quadratus*** Flat, quadrilateral muscle extending across the ventral surface of about the distal half of the ulna and radius. Dissect.

k. MUSCLES OF THE FOREARM: DORSAL SURFACE (FIGURE 10.29a AND b)

1. ***Brachioradialis*** Narrow, ribbonlike muscle on the lateral surface of the forearm. Dissect.
2. ***Extensor carpi radialis longus*** Larger muscle adjacent to the brachioradialis. Dissect, transect, and reflect.
3. ***Extensor carpi radialis brevis*** Muscle usually covered by the extensor carpi radialis longus at the proximal end. Dissect.
4. ***Extensor digitorum communis*** Large muscle running entire length of the lateral surface of forearm. Dissect, transect, and reflect.
5. ***Extensor digitorum lateralis*** Muscle with about the same structure as the extensor digitorum communis muscle. It lies lateral to the ulna. Dissect, transect, and reflect.
6. ***Extensor carpi ulnaris*** Muscle adjacent to the extensor digitorum lateralis, extending down the forearm laterally. Dissect, transect, and reflect.
7. ***Extensor indicis proprius*** Narrow muscle deep to the extensor carpi ulnaris.[6] Dissect.
8. ***Extensor pollicis brevis*** Muscle extending obliquely across the forearm deep to the extensor carpi ulnaris, extensor digitorum lateralis, and extensor digitorum communis. Dissect.
9. ***Supinator*** Muscle that passes obliquely across the radius deep to the extensor digitorum communis.

[6]Not illustrated.

l. MUSCLES OF THE THIGH: MEDIAL SURFACE (FIGURE 10.30a AND b)

1. ***Sartorius*** Straplike muscle on the cranial half of the medial side of the thigh. Dissect, transect, and reflect.
2. ***Gracilis*** Wide, flat muscle covering most of the caudal portion of the medial side of the thigh. Dissect, transect, and reflect.
3. ***Semimembranosus*** Thick muscle deep to the gracilis. Dissect.
4. ***Adductor femoris*** Triangular muscle deep to the gracilis and lateral to the semimembranosus. Dissect.
5. ***Adductor longus*** Small, triangular muscle cranial to the adductor femoris. Dissect.
6. ***Pectineus*** Very small triangular muscle cranial to the adductor femoris. Dissect.
7. ***Iliopsoas*** Long, cylindrical muscle that originates on the last two thoracic and lumbar vertebrae and passes along the medial surface of the thigh adjacent to the pectineus. Dissect.
8. ***Rectus femoris*** Narrow band of muscle deep to the sartorius, one of four components of the ***quadriceps femoris*** muscle. The others are the vastus medialis, vastus intermedius, and vastus lateralis. Dissect the rectus femoris.
9. ***Vastus medialis*** Large muscle on the medial surface of the thigh caudal to the rectus femoris and deep to the sartorius. Dissect.
10. ***Vastus intermedius*** Best seen by separating the rectus femoris and vastus lateralis.[7]

m. MUSCLES OF THE THIGH: LATERAL SURFACE (FIGURE 10.31a AND b)

1. ***Tensor fasciae latae*** Triangular mass of muscle cranial to the biceps femoris. Dissect, transect, and reflect. Reflect the tough, white fasciae latae.
2. ***Biceps femoris*** Muscle caudal to the tensor fasciae latae, covers most of the lateral surface of the the thigh. Dissect, transect, and reflect.
3. ***Semitendinosus*** Large band of muscle on the ventral border of the thigh between the biceps femoris and semimembranosus. Dissect.

[7]Not illustrated.

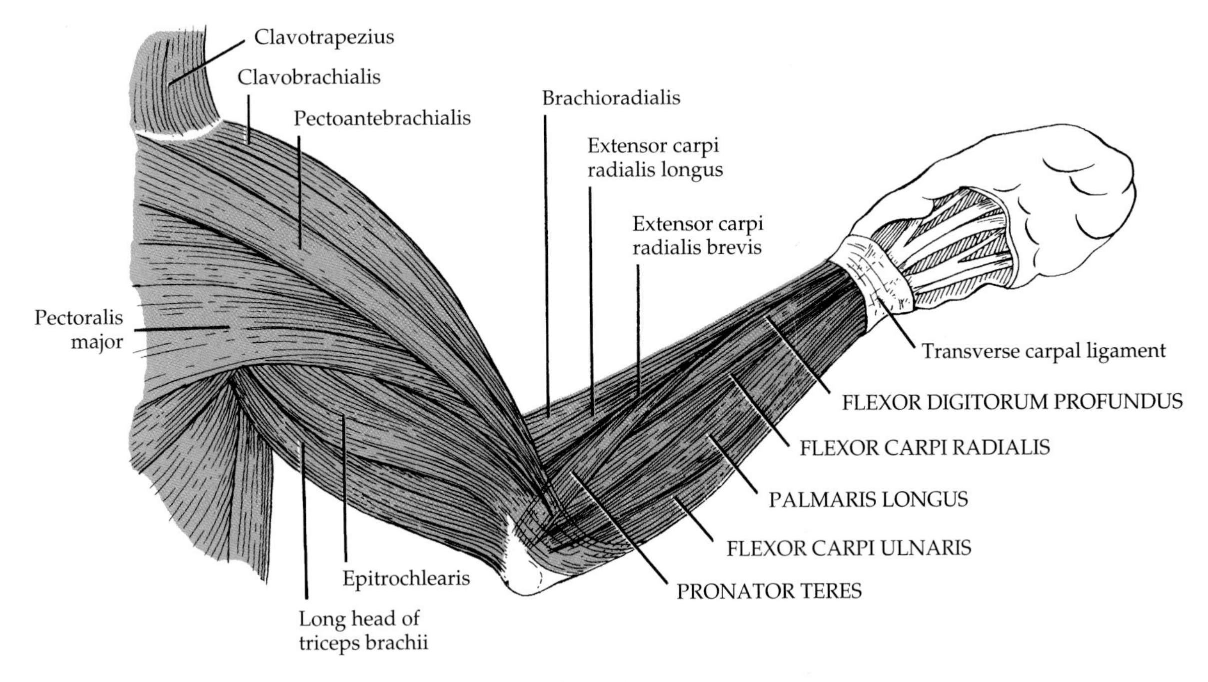

(a) Superficial view

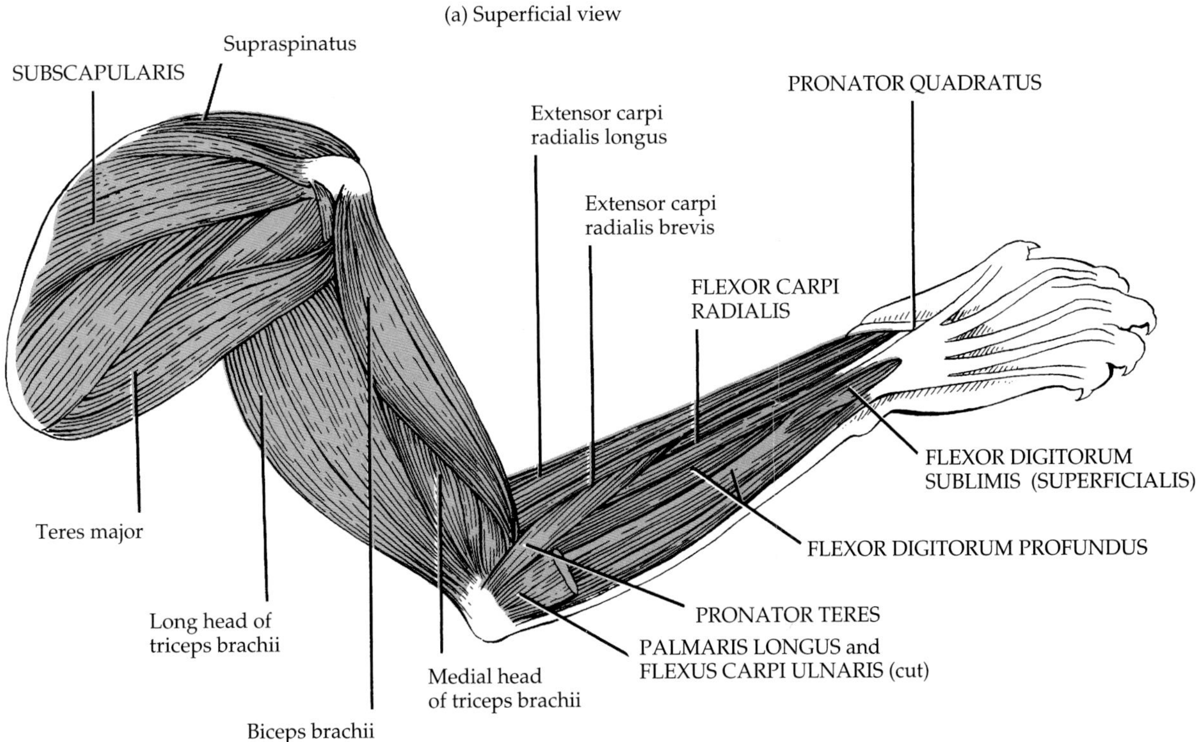

(b) Deep view

FIGURE 10.28 Ventral view of muscles of forearm.

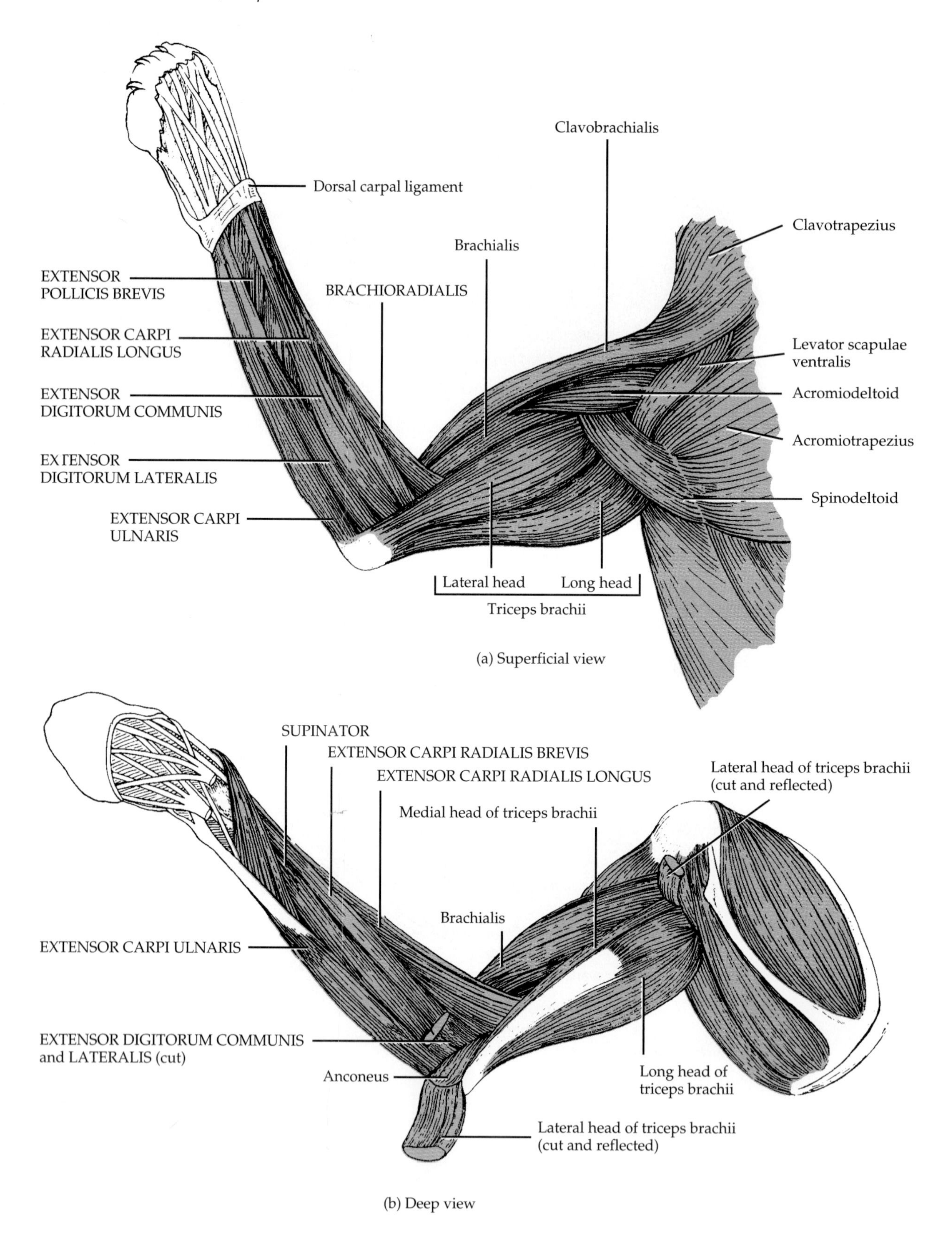

FIGURE 10.29 Dorsal view of muscles of forearm.

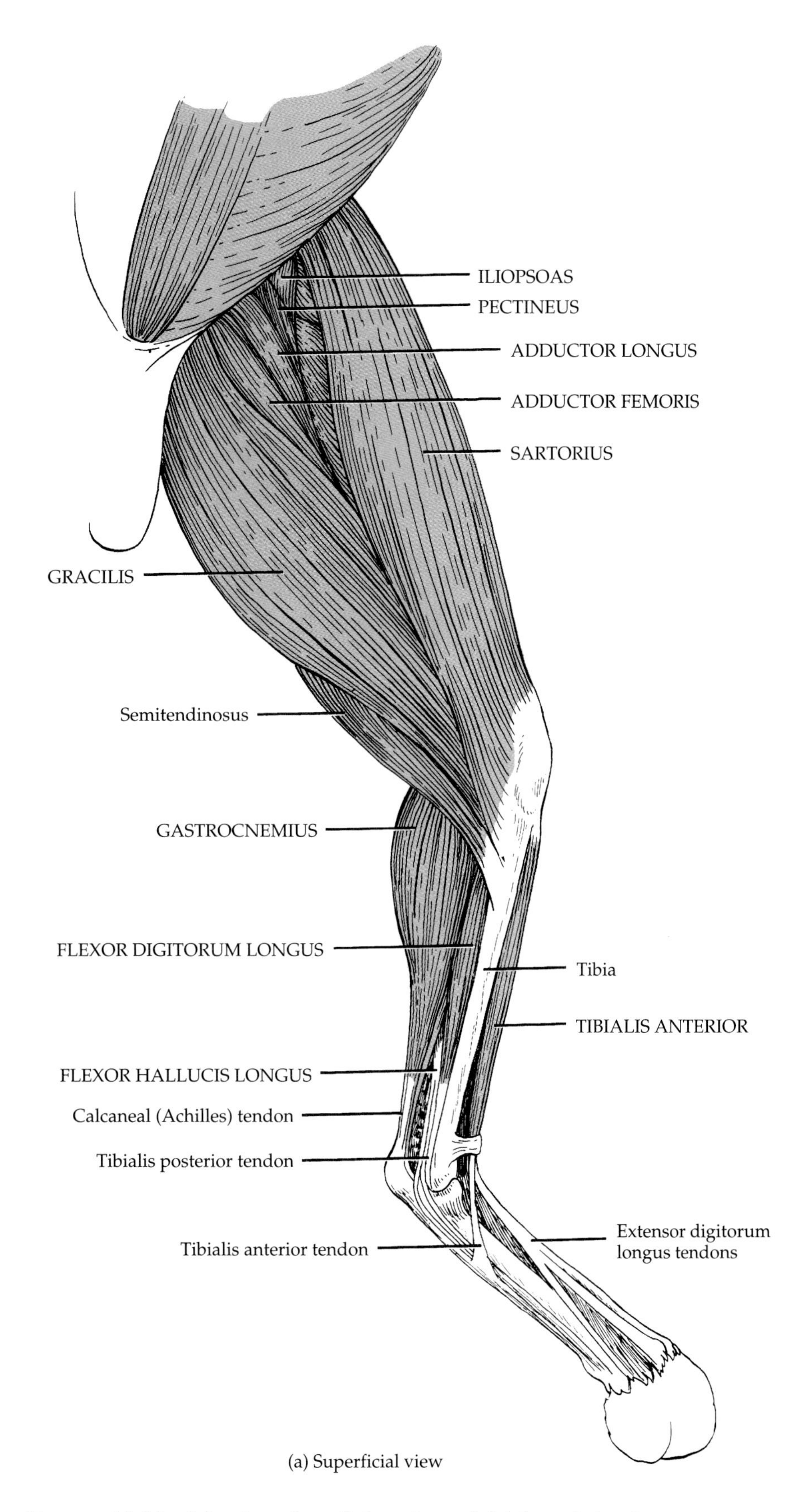

(a) Superficial view

FIGURE 10.30 Muscles of medial surface of thigh and shank.

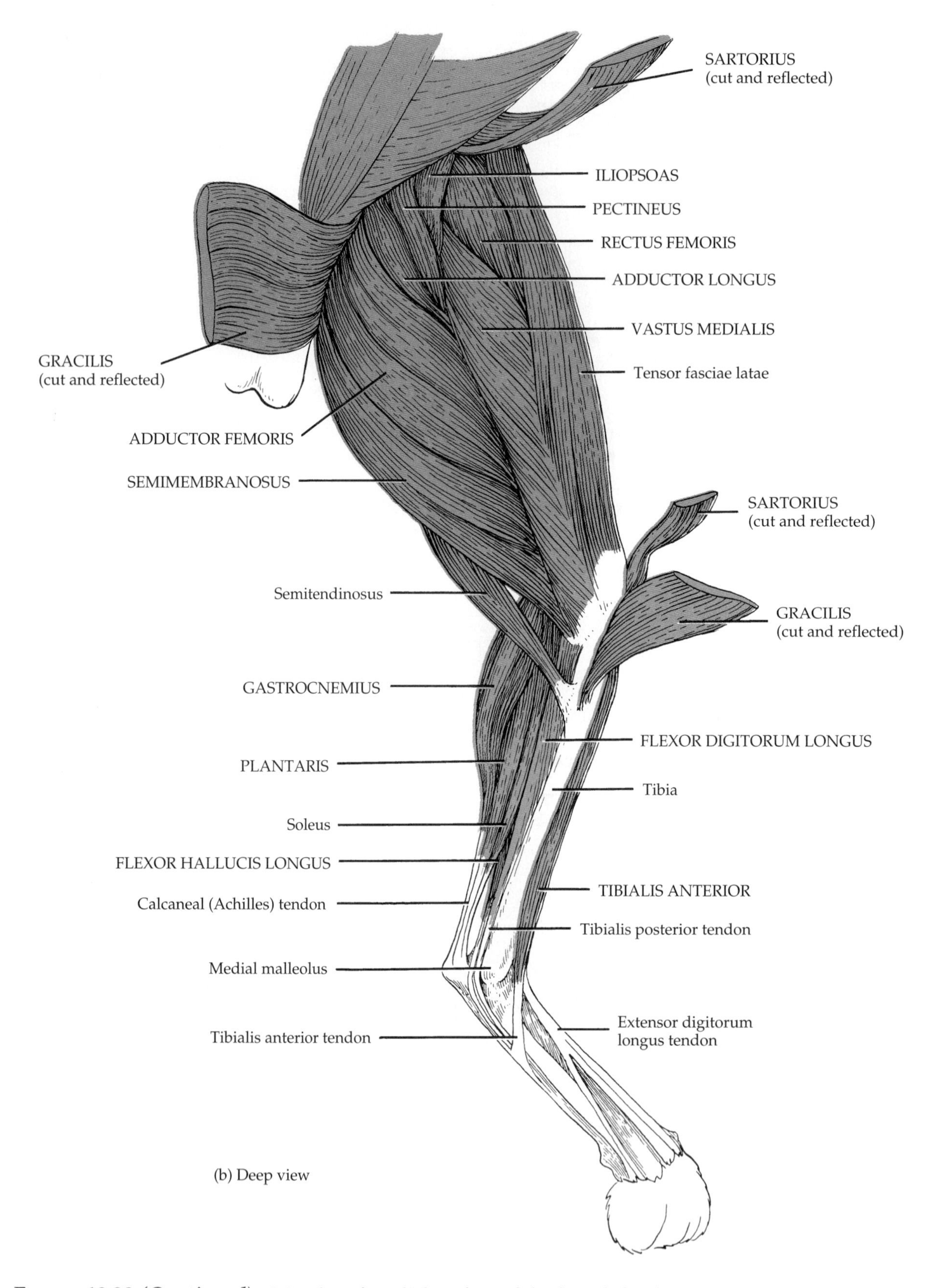

Figure 10.30 ***(Continued)*** Muscles of medial surface of thigh and shank.

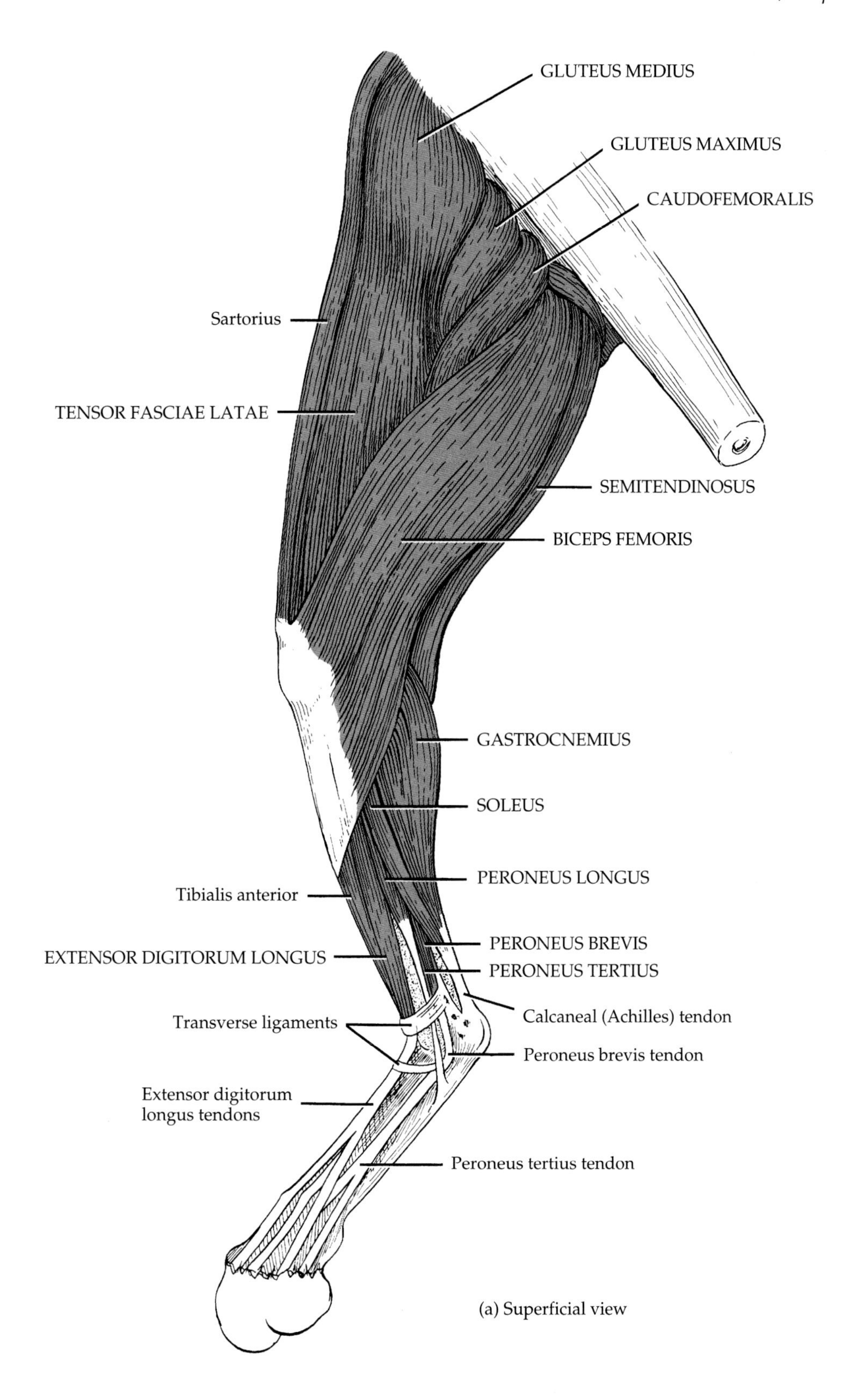

FIGURE 10.31 Muscles of lateral surface of thigh and shank.

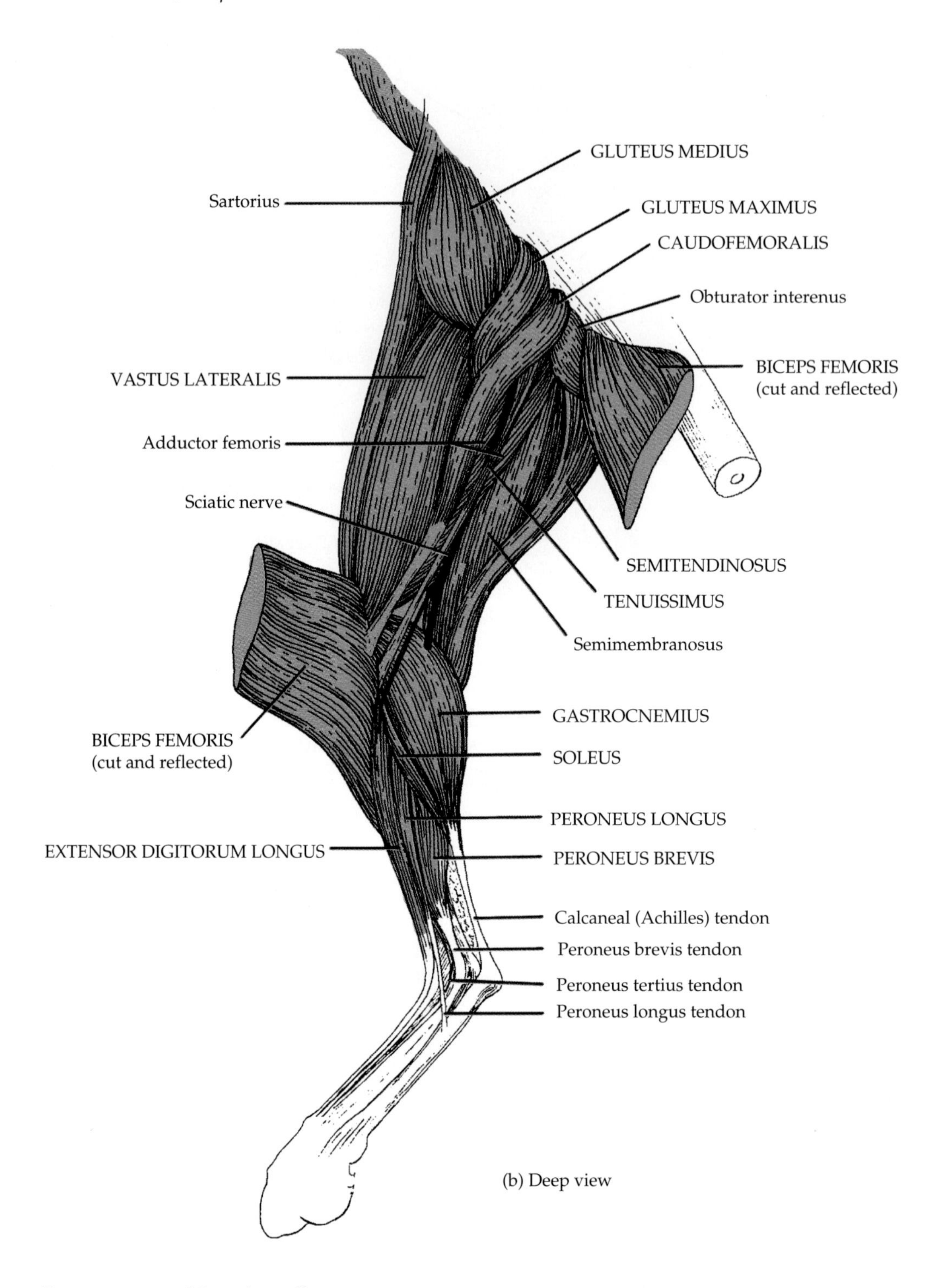

FIGURE 10.31 ***(Continued)*** Muscles of lateral surface of thigh and shank.

4. ***Caudofemoralis*** Band of muscle cranial and dorsal to the biceps femoris. Dissect.
5. ***Tenuissimus*** Very slender band of muscle deep to the biceps femoris. The sciatic nerve runs parallel to it. Dissect.
6. ***Gluteus maximus*** Triangular mass of muscle just cranial to the caudofemoralis. Dissect.
7. ***Gluteus medius*** Relatively large, somewhat triangular muscle cranial to the gluteus maximus. Dissect.
8. ***Vastus lateralis*** Large muscle deep to the tensor fasciae latae that occupies the craniolateral surface of the thigh. Dissect.

n. MUSCLES OF THE SHANK: MEDIAL SURFACE (FIGURE 10.30a AND b)

1. ***Gastrocnemius (medial head)*** Muscle originating on medial epicondyle of femur. The medial and lateral heads unite to form a large muscle mass that inserts into the calcaneus by the ***calcaneal (Achilles) tendon.*** Separate the medial head from the lateral head and reflect the medial head.
2. ***Flexor digitorum longus*** Muscle consisting of two heads. The first is partly covered by the soleus muscle. The second is between the first head and the medial head of the gastrocnemius. Dissect, transect, and reflect.
3. ***Flexor hallucis longus*** Long, heavy muscle deep to the flexor digitorum longus. Dissect.
4. ***Tibialis anterior*** Tapered band of muscle on the anterior aspect of the tibia. Dissect.
5. ***Plantaris*** Strong, round muscle between the two heads of the gastrocnemius. Also inserts into the calcaneal tendon. Dissect.

o. MUSCLES OF THE SHANK: LATERAL SURFACE (FIGURE 10.31a AND b)

1. ***Gastrocnemius (lateral head)*** Muscle originating on the lateral epicondyle of the femur. Dissect, transect, and reflect.
2. ***Soleus*** Small band of muscle next to the lateral head of the gastrocnemius and inserting into the calcaneal (Achilles) tendon. Dissect.
3. ***Extensor digitorum longus*** Muscle posterior to the tibialis anterior. Dissect
4. ***Peroneus longus*** Slender, superficial muscle caudal to the extensor digitorum longus. Dissect, transect, and reflect.
5. ***Peroneus brevis*** Short, thick muscle with the lateral surface covered by the peroneus longus. Dissect.
6. ***Peroneus tertius*** Slender muscle adjacent to the extensor digitorum longus. Dissect.

ANSWER THE LABORATORY REPORT QUESTIONS AT THE END OF THE EXERCISE.

LABORATORY REPORT QUESTIONS

Skeletal Muscles 10

Student ______________________ Date ______________________

Laboratory Section ______________________ Score/Grade ______________________

Part 1. Multiple Choice

_______ **1.** The connective tissue covering that encloses the entire skeletal muscle is the (a) perimysium (b) endomysium (c) epimysium (d) mesomysium

_______ **2.** A cord of connective tissue that attaches a skeletal muscle to the periosteum of bone is called a(n) (a) ligament (b) aponeurosis (c) perichondrium (d) tendon

_______ **3.** A skeletal muscle that decreases the angle at a joint is referred to as a(n) (a) flexor (b) abductor (c) pronator (d) evertor

_______ **4.** The name *abductor* means that a muscle (a) produces a downward movement (b) moves a part away from the midline (c) elevates a body part (d) increases the angle at a joint

_______ **5.** Which connective tissue layer directly encircles the fascicles of skeletal muscles? (a) epimysium (b) endomysium (c) perimysium (d) mesomysium

_______ **6.** Which muscle is *not* associated with a movement of the eyeball? (a) superior rectus (b) superior oblique (c) medial rectus (d) external oblique

_______ **7.** Of the following, which muscle is involved in compression of the abdomen? (a) external oblique (b) superior oblique (c) medial rectus (d) genioglossus

_______ **8.** A muscle directly concerned with breathing is the (a) masseter (b) mentalis (c) brachialis (d) external intercostal.

_______ **9.** Which muscle is *not* related to movement of the wrist? (a) extensor carpi ulnaris (b) flexor carpi radialis (c) supinator (d) flexor carpi ulnaris

_______ **10.** A muscle that helps move the thigh is the (a) piriformis (b) triceps brachii (c) hypoglossus (d) peroneus tertius

_______ **11.** Which muscle is *not* related to mastication? (a) temporalis (b) masseter (c) lateral rectus (d) medial pterygoid

_______ **12.** Which muscle elevates the tongue? (a) genioglossus (b) styloglossus (c) hyoglossus (d) omohyoid

_______ **13.** Which muscle does *not* belong with the others because of its relationship to the hyoid bone? (a) digastric (b) mylohyoid (c) geniohyoid (d) thyrohyoid

_______ **14.** Which muscle is *not* a component of the anterolateral abdominal wall? (a) external oblique (b) psoas major (c) rectus abdominis (d) internal oblique

_______ **15.** All are components of the pelvic diaphragm *except* the (a) anconeus (b) coccygeus (c) iliococcygeus (d) pubococcygeus

_______ **16.** Which is *not* a flexor of the forearm? (a) biceps brachii (b) brachialis (c) brachioradialis (d) triceps brachii

_______ **17.** The abductor pollicis brevis, opponeus pollicis, flexor pollicis brevis, and adductor pollicis are all components of the (a) thenar eminence (b) midpalmar muscles (c) hypothenar eminence (d) erector spinae

_______ **18.** Which muscle is *not* involved in moving the vertebral column? (a) splenius (b) longissimus (c) spinalis (d) sartorius

_______ **19.** Which muscle flexes the wrist? (a) palmaris longus (b) extensor carpi radialis longus (c) supinator (d) extensor indicis

_______ **20.** Which muscle is *not* involved in flexion of the thigh? (a) rectus femoris (b) sartorius (c) biceps femoris (d) vastus intermedius

_______ **21.** Of the muscles that move the foot and toes, the anterior compartment muscles are involved in (a) plantar flexion (b) dorsiflexion (c) abduction (d) adduction

_______ **22.** Which muscle is deepest? (a) plantar interosseous (b) adductor hallucis (c) quadratus plantae (d) abductor hallucis

PART 2. Matching

Identify the characteristic(s) used to name the following muscles:

_______ **23.** Supinator
_______ **24.** Deltoid
_______ **25.** Stylohyoid
_______ **26.** Flexor carpi radialis
_______ **27.** Gluteus maximus
_______ **28.** External oblique
_______ **29.** Triceps brachii
_______ **30.** Adductor longus
_______ **31.** Temporalis
_______ **32.** Trapezius

A. Location
B. Shape
C. Size
D. Direction of fibers
E. Action
F. Number of origins
G. Insertion and origin

PART 3. Completion

33. The principal muscle used in compression of the cheek is the ____________________.

34. The muscle that protracts the tongue is the ____________________.

35. The eye muscle that moves the eyeball inferiorly, medially, and counterclockwise is the ____________________.

36. The ____________________ muscle flexes the neck on the chest.

37. The abdominal muscle that flexes the vertebral column is the ____________________.

38. The muscle of the pectoral (shoulder) girdle that depresses the clavicle is the ____________________.

39. Flexion, adduction, and medial rotation of the arm are accomplished by the ____________________ muscle.

40. The ____________________ muscle is the most important extensor of the forearm.

41. The muscle that flexes and abducts the wrist is the ____________________.

42. The four muscles that extend the legs are the vastus lateralis, vastus medialis, vastus intermedius, and ____________________.

43. Muscles that lie inferior to the hyoid bone are referred to as ____________________ muscles.

44. The neck region is divided into two principal triangles by the ____________________ muscle.

45. Muscles that move the pectoral (shoulder) girdle originate on the axial skeleton and insert on the clavicle or ____________________.

46. The muscles that move the humerus (arm) and do not originate on the scapula are called ____________________ muscles.

47. Together, the subscapularis, supraspinatus, infraspinatus, and teres major muscles form the ____________________.

48. The posterior muscles involved in moving the wrist, hand, and fingers function in extension and ____________________.

49. The posterior muscles of the thigh are involved in ____________________ of the leg.

50. Together, the biceps femoris, semitendinosus, and semimembranosus are referred to as the ____________________ muscles.

51. ____________________ fascicles attach obliquely from many directions to several tendons.

52. Movement of the thumb medially across the palm is called ____________________.

Surface Anatomy

11

Now that you have studied the skeletal and muscular systems, you will be introduced to the study of ***surface anatomy***, in which you will study the form and markings of the surface of the body.[1] A knowledge of surface anatomy will help you identify certain superficial structures by visual inspection or palpation through the skin. ***Palpation*** means to feel with the hand. Knowledge of surface anatomy is important in health-related activities such as taking a pulse, listening to internal organs, drawing blood, and inserting needles and tubes.

A convenient way to study surface anatomy is first to divide the body into its principal regions: head, neck, trunk, and upper and lower limbs. These may be reviewed in Figure 2.2.

A. HEAD

The ***head*** (cephalic region or caput) is divisible into the cranium and face. Several surface features of the head are

1. ***Cranium (skull,*** or ***brain case)***

   a. ***Frontal region*** Front of skull (sinciput) that includes forehead.
   b. ***Parietal region*** Crown of skull (vertex).
   c. ***Temporal region*** Side of skull (tempora).
   d. ***Occipital region*** Base of skull (occiput).

2. ***Face (facies)***

   a. ***Orbital*** or ***ocular region*** Includes eyeballs (bulbi oculorum), eyebrows (supercilia), and eyelids (palpebrae).
   b. ***Nasal region*** Nose (nasus).
   c. ***Infraorbital region*** Inferior to orbit.
   d. ***Oral region*** Mouth.
   e. ***Mental region*** Anterior part of mandible.
   f. ***Buccal region*** Cheek.
   g. ***Parotid-masseteric region*** External to parotid gland and masseter muscle.
   h. ***Zygomatic region*** Inferolateral to orbit.
   i. ***Auricular region*** Ear.

Using your textbook as an aid, label Figure 11.1.

Using a mirror, examine the various features of the head just described. Working with a partner, be sure that you can identify the regions by both common *and* anatomical names.

The surface anatomy features of the eyeball and accessory structures are presented in Figure 14.9, of the ear in Figure 14.17, and of the nose in Figure 20.3.

B. NECK

The ***neck*** (collum) can be divided into an ***anterior cervical region,*** two ***lateral cervical regions,*** and a ***posterior (nuchal) region.*** Among the surface features of the neck are

1. ***Thyroid cartilage (Adam's apple)*** Triangular laryngeal cartilage in the midline of the anterior cervical region.
2. ***Hyoid bone*** First resistant structure palpated in the midline inferior to the chin, lying just superior to the thyroid cartilage opposite the superior border of C4.
3. ***Cricoid cartilage*** Inferior laryngeal cartilage that attaches larynx to trachea. This structure can be palpated by running your fingertip down from your chin over the thyroid cartilage. (After you pass the cricoid cartilage, your fingertip sinks in.) This cartilage is used as a landmark in locating the rings of cartilage in the trachea (windpipe) when performing a tracheostomy. The incision is made through the second, third, or fourth tracheal rings and a tube is inserted to assist breathing.

[1]At the discretion of your instructor, surface anatomy may be studied either before or in conjunction with your study of various body systems. This exercise can be used as an excellent review of many topics already studied.

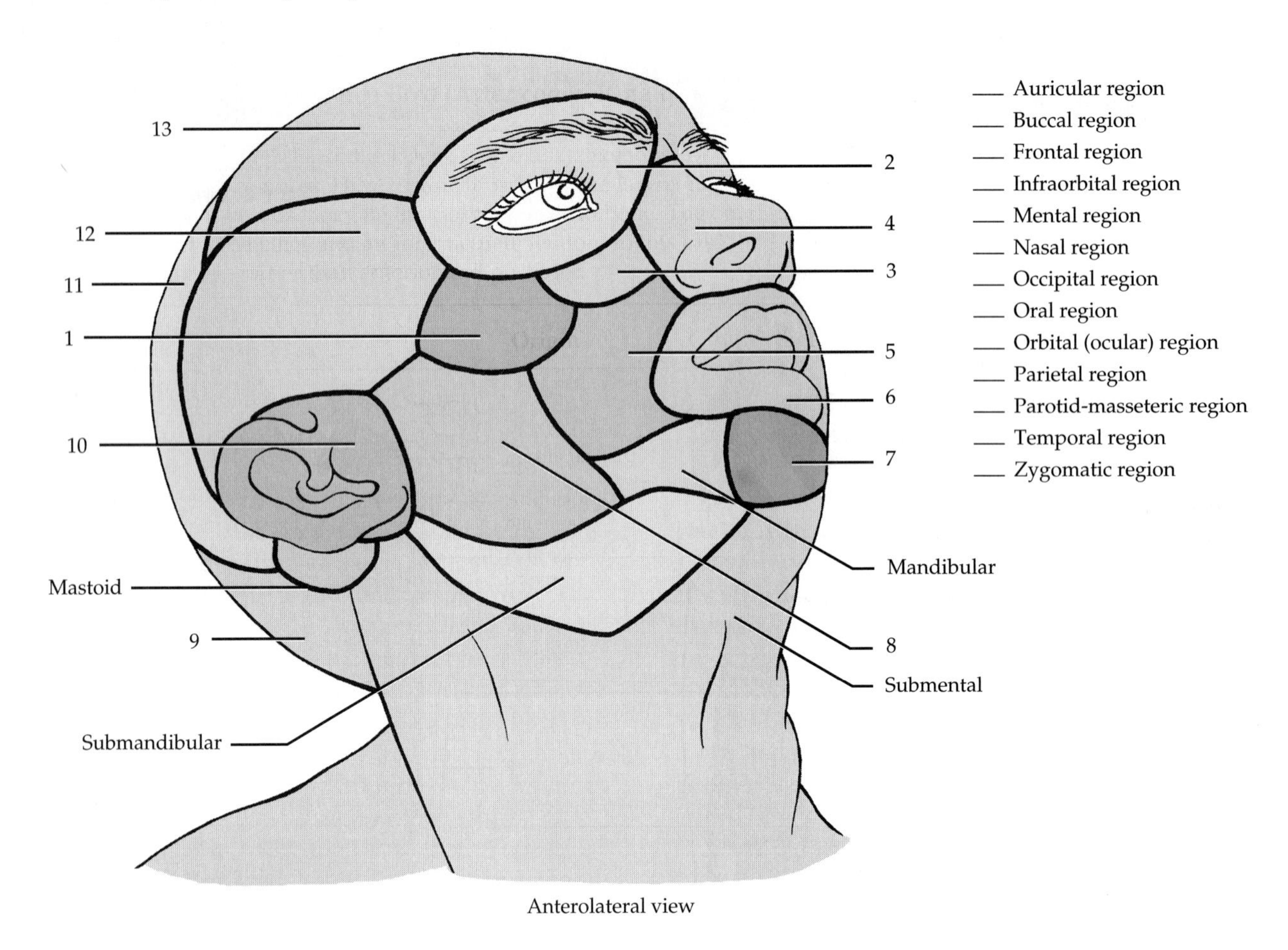

FIGURE 11.1 Regions of the cranium and face.

4. ***Sternocleidomastoid muscles*** Form major portion of lateral cervical regions, extending from mastoid process of temporal bone (felt as bump behind auricle of ear) to sternum and clavicle. Each muscle divides the neck into an anterior and posterior triangle. Carotid (neck) pulse is felt along the anterior border of the sternocleidomastoid muscle.
5. ***Trapezius muscles*** Form portion of lateral cervical region, extending inferiorly and laterally from base of skull. "Stiff neck" is frequently associated with inflammation of these muscles.
6. ***Anterior triangle of neck*** Bordered superiorly by mandible, inferiorly by sternum, medially by cervical midline, and laterally by anterior border of sternocleidomastoid muscle.
7. ***Posterior triangle of neck*** Bordered inferiorly by clavicle, anteriorly by posterior border of sternocleidomastoid muscle, and posteriorly by anterior border of trapezius muscle.
8. ***External jugular veins*** Prominent veins along lateral cervical regions, readily seen when a person is angry or a collar fits too tightly.

Using a mirror and working with a partner, use your textbook as an aid in identifying the surface features of the neck just described. Then label Figure 11.2.

C. TRUNK

The ***trunk*** is divided into the back, chest, abdomen, and pelvis. Its surface features include

1. ***Back (dorsum)***

   a. ***Spinous processes (spines)*** Posteriorly pointed projections of vertebrae. The spinous process of C2 is the first bony prominence encountered when the finger is drawn inferiorly along the midline; the spinous process of C7 ***(vertebra prominens)*** is the superior of the two prominences found at the base of the neck; the spinous process of T1 is the lower prominence at the base of the neck; the spinous process of T3 is at about the same level as the spinous process of the scapula; the spinous process of T7 is about opposite the infe-

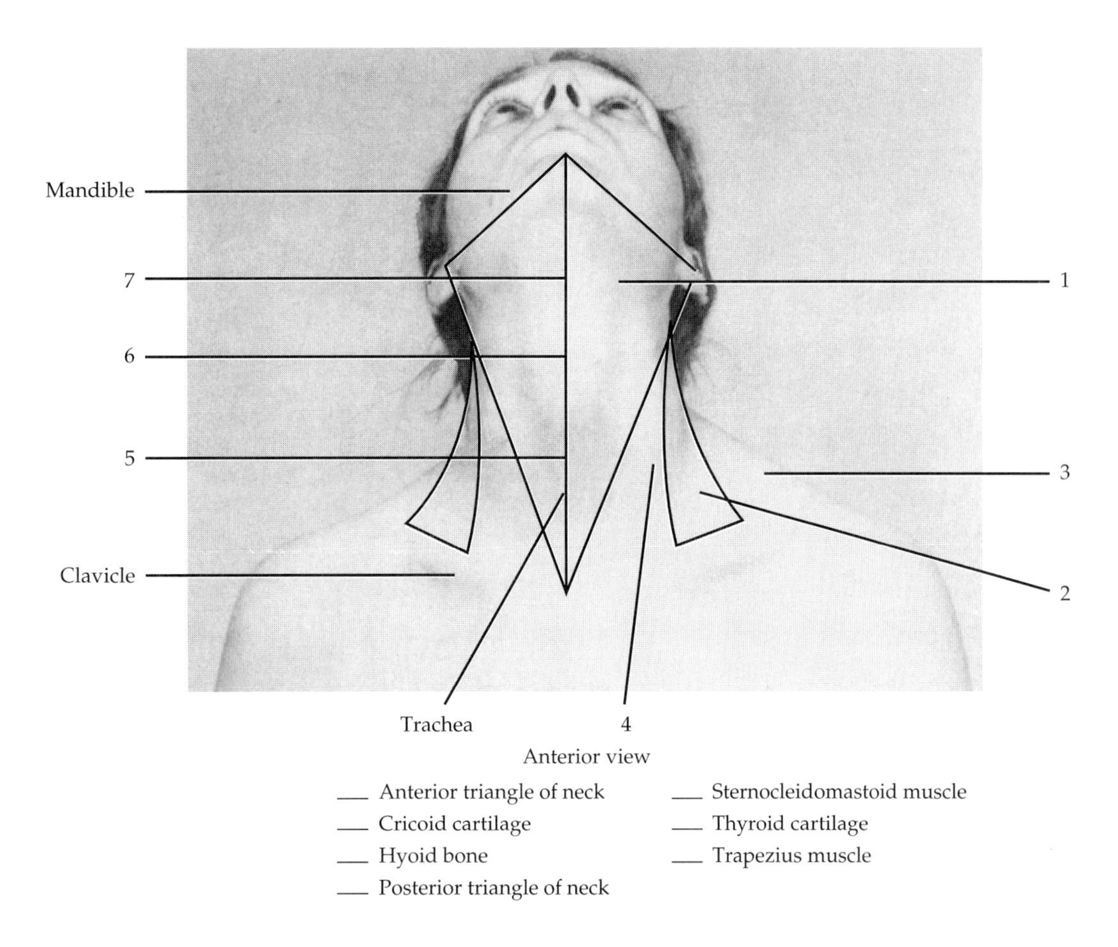

FIGURE 11.2 Surface anatomy of neck.

rior angle of the scapula; a line passing through the highest points of the iliac crests, called the supracristal line, passes through the spinous process of L4.

b. ***Scapula*** Shoulder blade. Several parts of the scapula, such as the medial (axillary) border, lateral (vertebral) border, inferior angle, spine, and acromion, may be observed or palpated.
c. ***Ribs*** These may be seen in thin individuals.
d. ***Muscles*** Among the visible superficial back muscles are the ***latissimus dorsi*** (covers lower half of back), ***erector spinae*** (on either side of vertebral column), ***infraspinatus*** (inferior to spine of scapula), ***trapezius***, and ***teres major*** (inferior to infraspinatus).
e. ***Posterior axillary fold*** Formed by the latissimus dorsi and teres major muscles; can be palpated between the finger and thumb.
f. ***Triangle of auscultation*** Triangle formed by latissimus dorsi muscle, trapezius muscle, and vertebral border of scapula. The space between the muscles in the region permits respiratory sounds to be heard clearly with a stethoscope.

Using your textbook as an aid, label Figure 11.3.

2. *Chest (thorax)*

a. ***Clavicles*** Collarbones. These lie in superior region of thorax and can be palpated along their entire length.
b. ***Sternum*** Breastbone. Lies in midline of chest. The following parts of the sternum are important surface features:
   i. ***Suprasternal notch*** Depression on superior surface of manubrium of sternum between medial ends of clavicles. The trachea can be palpated in the notch.
   ii. ***Manubrium of sternum*** Superior portion of sternum at the same levels as the bodies of the third and fourth thoracic vertebrae and anterior to the arch of the aorta.
   iii. ***Body of sternum*** Midportion of sternum anterior to heart and the vertebral bodies of T5–T8.
   iv. ***Sternal angle*** Formed by junction of manubrium and body of sternum, about 4 cm (1½ in.) inferior to suprasternal

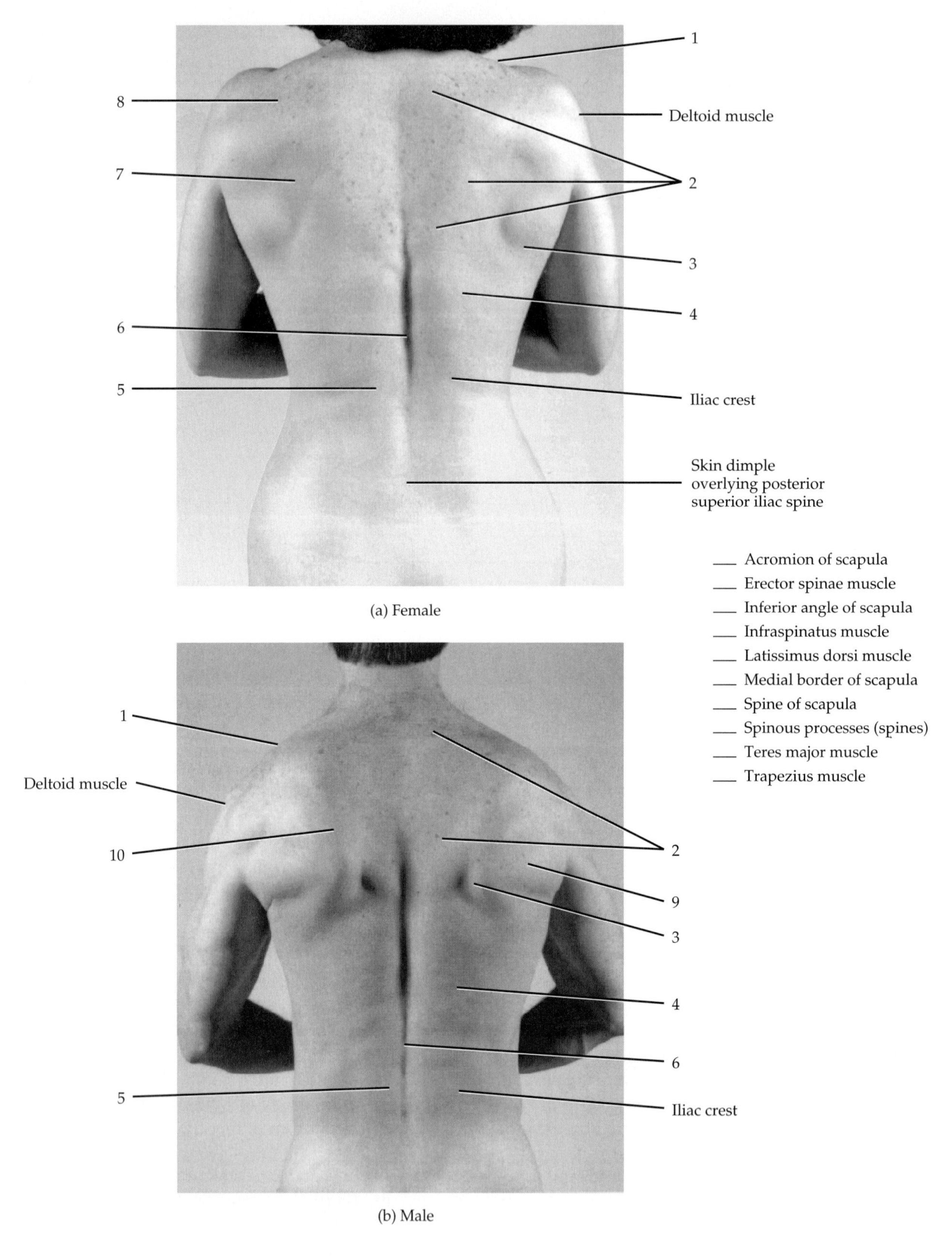

FIGURE 11.3 Surface anatomy of back.

notch. This is palpable under the skin, locates the costal cartilage of the second rib, and is the starting point from which the ribs are counted.

v. ***Xiphoid process of sternum*** Inferior portion of sternum medial to the seventh costal cartilages.

c. ***Ribs*** Form bony cage of thoracic cavity. The apex beat of the heart in adults is heard in the left fifth intercostal space, just medial to the left midclavicular line.

d. ***Costal margins*** Inferior edges of costal cartilages of ribs 7 through 10. The first costal cartilage lies inferior to the medial end of the clavicle; the seventh costal cartilage is the most inferior to articulate directly with the sternum; the tenth costal cartilage forms the most inferior part of the costal margin, when viewed anteriorly.

e. ***Muscles*** Among the superficial chest muscles that can be seen are the ***pectoralis major*** (principal upper chest muscle) and ***serratus anterior*** (inferior and lateral to pectoralis major).

f. ***Mammary glands*** Accessory organs of the female reproductive system located inside the breasts. They overlie the pectoralis major muscle (two-thirds) and serratus anterior muscle (one-third). After puberty, they enlarge to their hemispherical shape, and in young adult females, they extend from the second through sixth ribs and from the lateral margin of the sternum to the midaxillary line.

g. ***Nipples*** Superficial to fourth intercostal space or fifth rib about 10 cm (4 in.) from the midline in males and most females. The position of the nipples in females is variable, depending on the size and pendulousness of the breasts. The right dome of the diaphragm is just inferior to the right nipple, the left dome is about 2–3 cm (1 in.) inferior to the left nipple, and the central tendon is at the level of the junction of the body and xiphoid process of the sternum.

h. ***Anterior axillary fold*** Formed by the lateral border of the pectoralis major muscle; can be palpated between the fingers and thumb.

Using your textbook as an aid, label Figure 11.4.

3. *Abdomen* and *Pelvis*

a. ***Umbilicus*** Also called ***navel;*** previous site of attachment of umbilical cord to fetus. It is level with the intervertebral disc between the bodies of L3 and L4 and is the most obvious surface marking on the abdomen of most individuals. The abdominal aorta bifurcates into the right and left common iliac arteries anterior to the body of vertebra L4. The inferior vena cava lies to the right of the abdominal aorta and is wider; it rises anterior to the body of vertebra L5.

b. ***Linea alba*** Flat, tendonous raphe forming a furrow along midline between rectus abdominis muscles. The furrow extends from the xiphoid process to the pubic symphysis. It is particularly obvious in thin, muscular individuals. It is broad superior to the umbilicus and narrow inferior to it. The linea alba is a frequently selected site for abdominal surgery since an incision through it severs no muscles and only a few blood vessels and nerves.

c. ***Tendinous intersections*** Fibrous bands that run transversely across the rectus abdominis muscle. Three or more are visible in muscular individuals. One intersection is at the level of the umbilicus, one at the level of the xiphoid process, and one midway between.

d. ***Muscles*** Among the superficial abdominal muscles are the ***external oblique*** (inferior to serratus anterior) and ***rectus abdominis*** (just lateral to midline of abdomen).

e. ***Pubic symphysis*** Anterior joint of hipbones. This structure is palpated as a firm resistance in the midline at the inferior portion of the anterior abdominal wall.

Using your textbook as an aid, label Figure 11.5 on page 213.

D. UPPER LIMB (EXTREMITY)

The ***upper limb (extremity)*** consists of the armpit, shoulder, arm, elbow, forearm, wrist, and hand (palm and fingers).

1. Major surface features of the ***shoulder (acromial)*** region are

a. ***Acromioclavicular joint*** Slight elevation at lateral end of clavicle.

b. ***Acromion*** Expanded lateral end of spine of scapula. This is clearly visible in some individuals and can be palpated about 2.5 cm (1 in.) distal to acromioclavicular joint (see Figure 11.3).

c. ***Deltoid muscle*** Triangular muscle that forms rounded prominence of shoulder. This is a frequent site for intramuscular injections (see Figure 11.3).

2. Major surface features of the ***arm (brachium)*** and ***elbow (cubitus)*** are

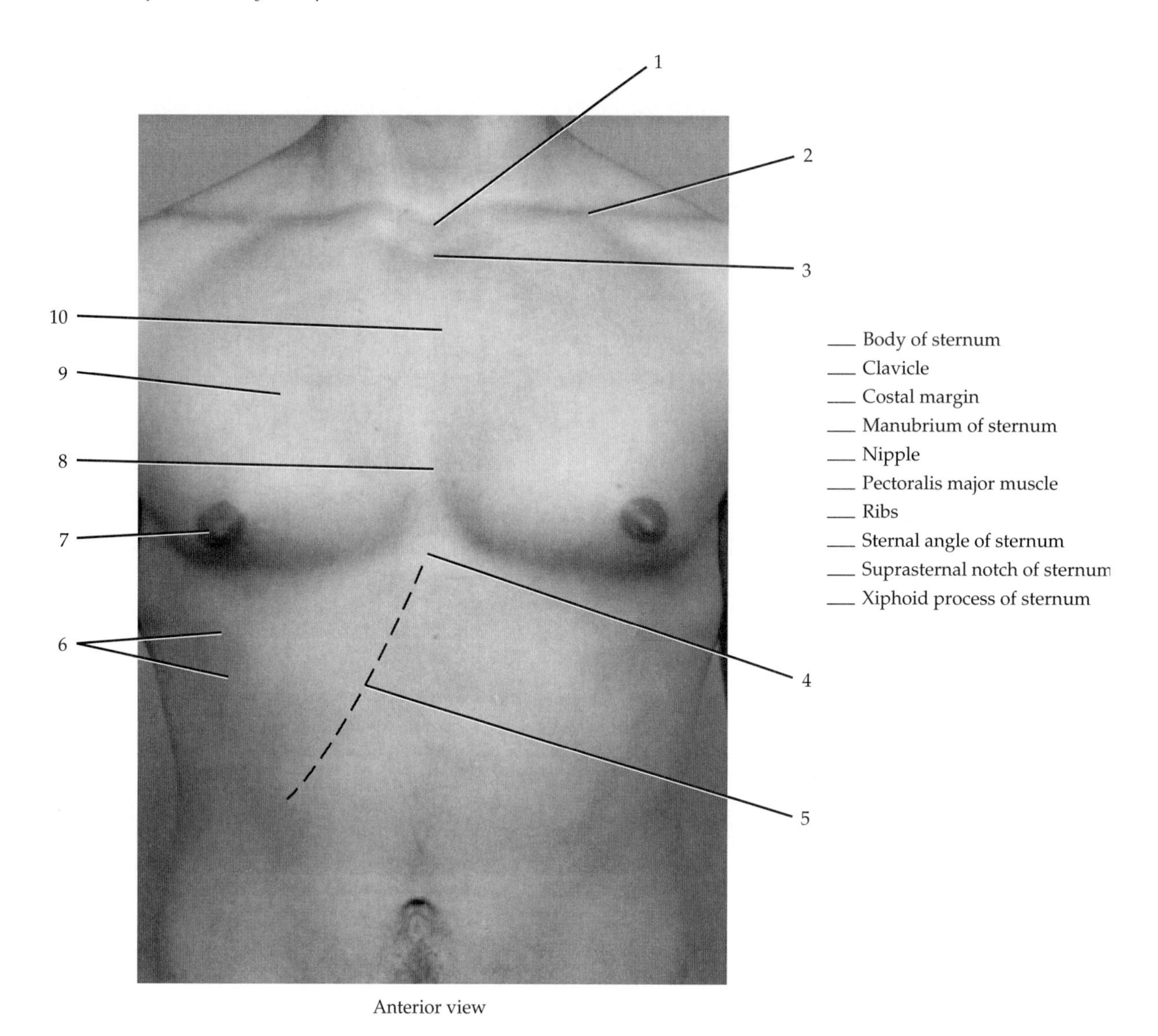

FIGURE 11.4 Surface anatomy of chest. (The serratus anterior muscle is shown in Figure 11.5.)

a. ***Biceps brachii muscle*** Forms bulk of anterior surface of arm.
b. ***Triceps brachii muscle*** Forms bulk of posterior surface of arm.
c. ***Medial epicondyle*** Medial projection at distal end of humerus.
d. ***Ulnar nerve*** Can be palpated as a rounded cord in a groove posterior to the medial epicondyle.
e. ***Lateral epicondyle*** Lateral projection at distal end of humerus.
f. ***Olecranon*** Projection of proximal end of ulna that lies between and slightly superior to epicondyles when forearm is extended; forms elbow.
g. ***Cubital fossa*** Triangular space in anterior region of elbow bounded proximally by an imaginary line between humeral epicondyles, laterally by the medial border of the brachioradialis muscle, and medially by the lateral border of the pronator teres muscle; contains tendon of biceps brachii muscle, brachial artery and its terminal branches (radial and ulnar arteries), and parts of median and radial nerves.
h. ***Median cubital vein*** Crosses cubital fossa obliquely. This vein is frequently selected for removal of blood.
i. ***Brachial artery*** Continuation of axillary artery that passes posterior to coracobrachialis

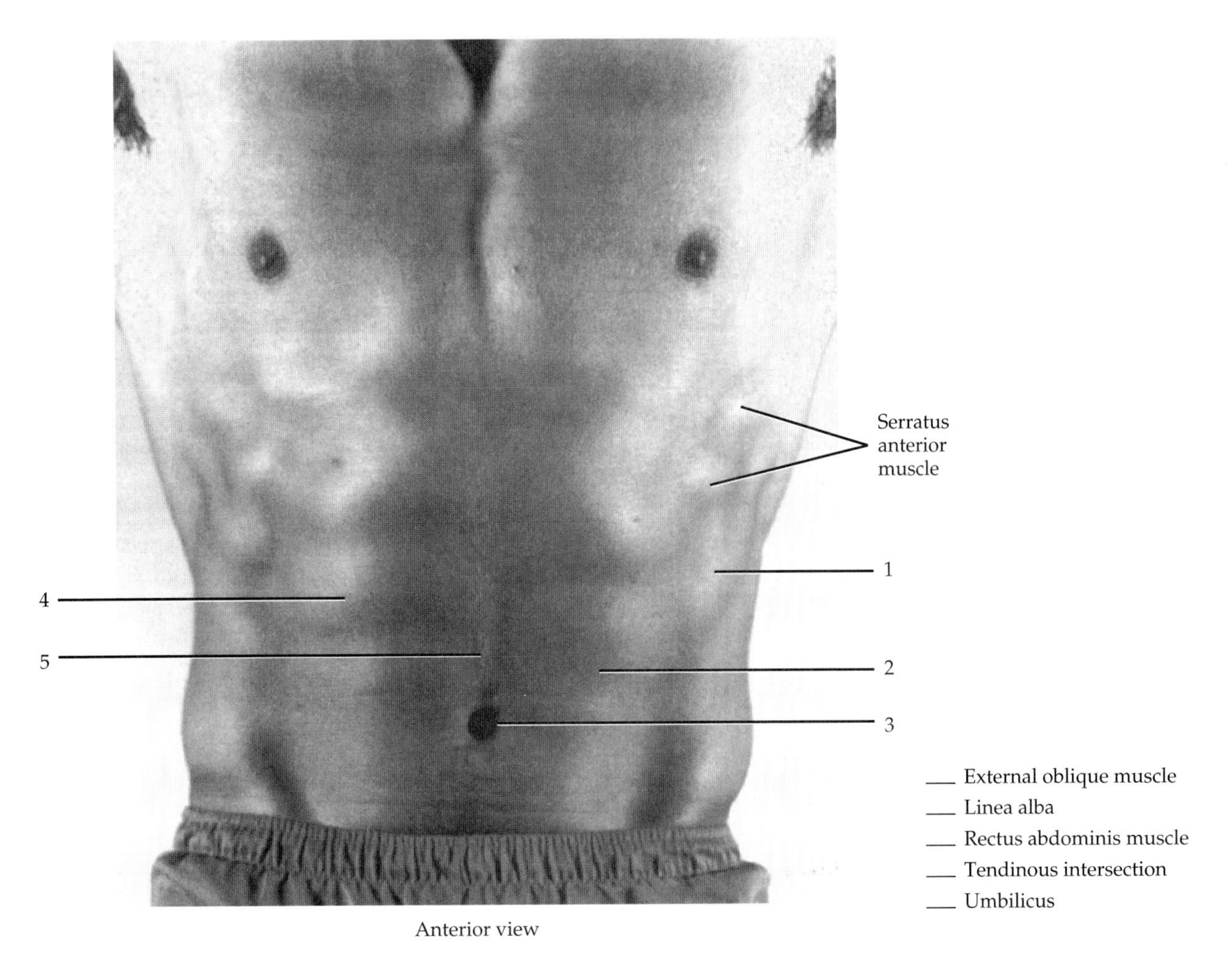

FIGURE 11.5 Surface anatomy of abdomen.

muscle and then medial to biceps brachii muscle. It enters the middle of the cubital fossa and passes inferior to the bicipital aponeurosis, which separates it from the median cubital vein. The artery is frequently used to take blood pressure. Pressure may be applied to it in cases of severe hemorrhage in the forearm and hand.

j. ***Bicipital aponeurosis*** An aponeurotic band that inserts the biceps brachii muscle into the deep fascia in the medial aspect of the forearm. It can be felt when the muscle contracts.

Using your textbook as an aid, label Figure 11.6.

3. Major surface features of the ***forearm (antebrachium)*** and ***wrist (carpus)*** are

a. ***Styloid process of ulna*** Projection of distal end of ulna at medial side of wrist.

b. ***Styloid process of radius*** Projection of distal end of radius at lateral side of wrist.

c. ***Brachioradialis muscle*** Located at superior and lateral aspect of forearm.

d. ***Flexor carpi radialis muscle*** Located along midportion of forearm.

e. ***Flexor carpi ulnaris muscle*** Located at medial aspect of forearm.

f. ***Tendon of palmaris longus muscle*** Located on anterior surface of wrist near ulna. When you make a fist, you can see this tendon. The palmaris longus muscle is absent in about 13% of the population.

g. ***Tendon of flexor carpi radialis muscle*** Tendon on anterior surface of wrist lateral to tendon of palmaris longus.

h. ***Radial artery*** Can be palpated just medial to styloid process of radius; this artery is frequently used to take pulse.

i. ***Pisiform bone*** Medial bone of proximal carpals. The bone is easily palpated as a

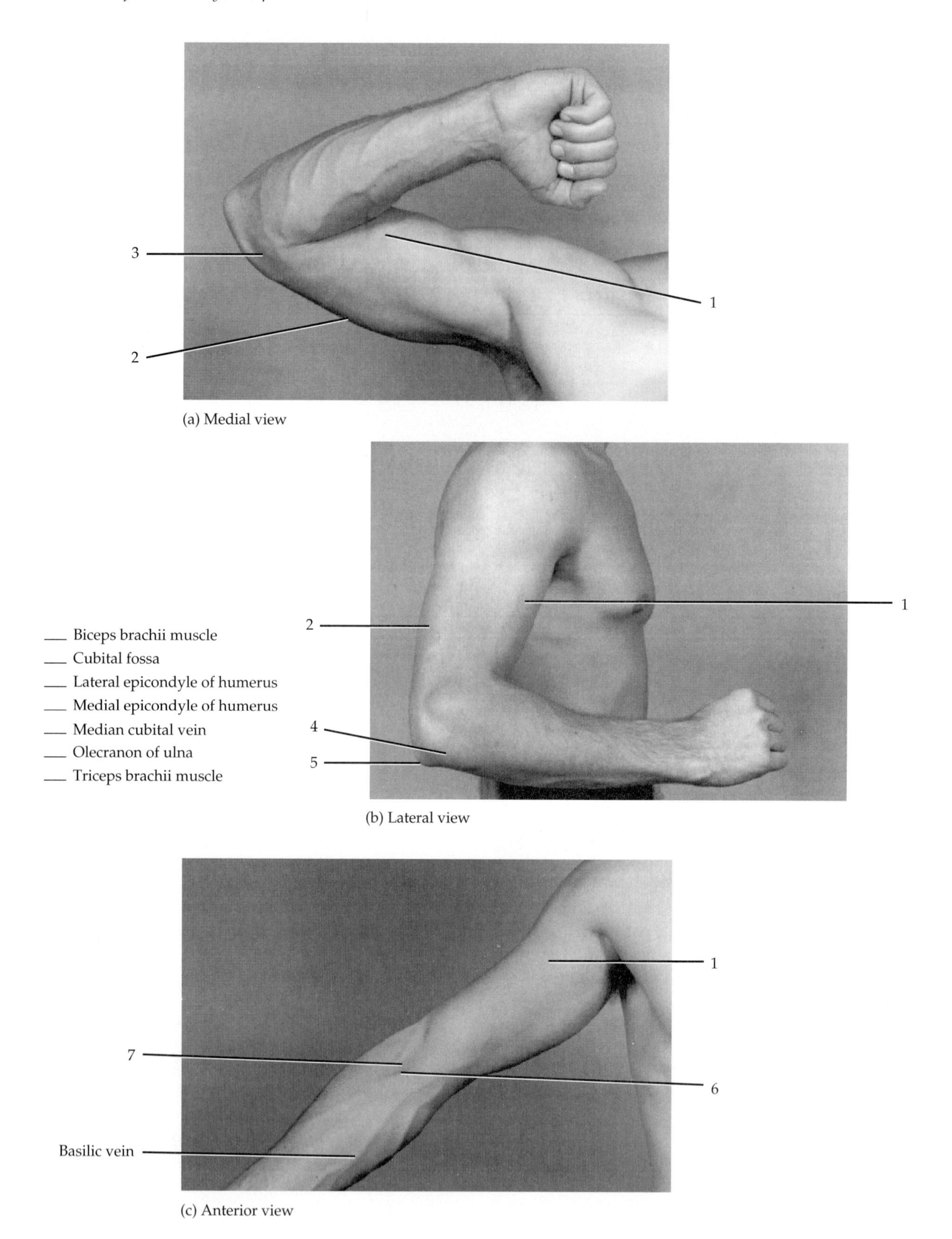

FIGURE 11.6 Surface anatomy of arm and elbow.

projection distal and anterior to styloid process of ulna.

j. ***Tendon of extensor pollicis brevis muscle*** Tendon close to styloid process of radius along posterior surface of wrist, best seen when thumb is bent backward.
k. ***Tendon of extensor pollicis longus muscle*** Tendon closer to styloid process of ulna along posterior surface of wrist, best seen when thumb is bent backward.
l. ***"Anatomical snuffbox"*** Depression between tendons of extensor pollicis brevis and extensor pollicis longus muscles. Styloid process of the radius, the base of the first metacarpal, trapezium, scaphoid, and radial artery can all be palpated in the depression.
m. ***Wrist creases*** Three more or less constant lines on anterior aspect of wrist where skin is firmly attached to underlying deep fascia.

Using your textbook as an aid, label Figure 11.7.

4. Major surface features of the ***hand (manus)*** are

a. ***"Knuckles"*** Commonly refers to dorsal aspects of distal ends of metacarpals II, III, IV, and V. Term also includes dorsal aspects of metacarpophalangeal and interphalangeal joints.
b. ***Thenar eminence*** Lateral rounded contour on anterior surface of hand formed by muscles of thumb.
c. ***Hypothenar eminence*** Medial rounded contour on anterior surface of hand formed by muscles of little finger.
d. ***Skin creases*** Several more or less constant lines on the anterior aspect of the palm ***(palmar flexion creases)*** and digits ***(digital flexion creases)*** where skin is firmly attached to underlying deep fascia.
e. ***Extensor tendons*** Besides the tendons of the extensor pollicis brevis and extensor pollicis longus muscles associated with the thumb, the following extensor tendons are also visible on the posterior aspect of the hand: ***extensor digiti minimi tendon*** in line with phalanx V (little finger) and ***extensor digitorum*** in line with phalanges II, III, and IV.
f. ***Dorsal venous arch*** Superficial veins on posterior surface of hand that form cephalic vein; displayed by compressing the blood vessels at the wrist for a few minutes as the hand is opened and closed.

With the aid of your textbook, label Figure 11.8 on page 217.

E. LOWER LIMB (EXTREMITY)

The ***lower limb (extremity)*** consists of the buttocks, thigh, knee, leg, ankle, and foot.

1. Major surface features of the ***buttocks (gluteal region)*** and ***thigh (femoral region)*** are

a. ***Iliac crest*** Superior margin of ilium of hipbone, forming outline of superior border of buttock. When you rest your hands on your hips, they rest on the iliac crests.
b. ***Posterior superior iliac spine*** Posterior termination of iliac crest; lies deep to a dimple (skin depression) about 4 cm (1.5 in.) lateral to midline; dimple forms because skin and underlying fascia are attached to bone. The spine marks the inferior limit of cerebrospinal fluid in the subarachnoid space around the spinal cord.
c. ***Gluteus maximus muscle*** Forms major portion of prominence of buttock.
d. ***Gluteus medius muscle*** Superolateral to gluteus maximus. This is a frequent site for intramuscular injections.
e. ***Gluteal (natal) cleft*** Depression along midline that separates the buttocks; it extends as high as the fourth or third sacral vertebra.
f. ***Gluteal fold*** Inferior limit of buttock formed by inferior margin of gluteus maximus muscle.
g. ***Ischial tuberosity*** Bony prominence of ischium of hipbone bears weight of body when seated.
h. ***Greater trochanter*** Projection of proximal end of femur on lateral surface of thigh felt and seen in front of hollow on side of hip. This can be palpated about 20 cm (8 in.) inferior to iliac crest.
i. ***Anterior thigh muscles*** Include ***sartorius*** (runs obliquely across thigh) and ***quadriceps femoris***, which consists of ***rectus femoris*** (midportion of thigh), ***vastus lateralis*** (anterolateral surface of thigh), ***vastus medialis*** (medial inferior portion of thigh), and ***vastus intermedius*** (deep to rectus femoris). Vastus lateralis is frequent injection site for diabetics.
j. ***Posterior thigh muscles*** Include the ***hamstrings (semitendinosus, semimembranosus,*** and ***biceps femoris).***

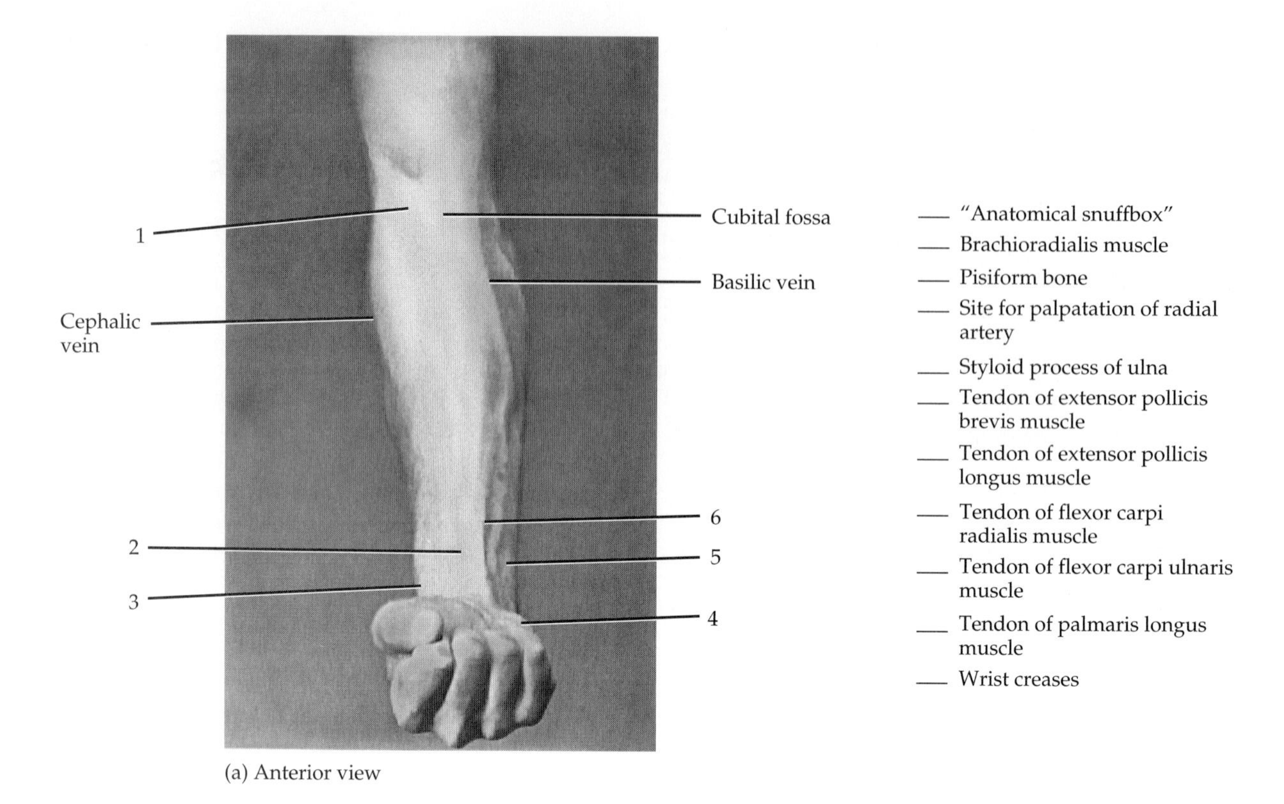

(a) Anterior view

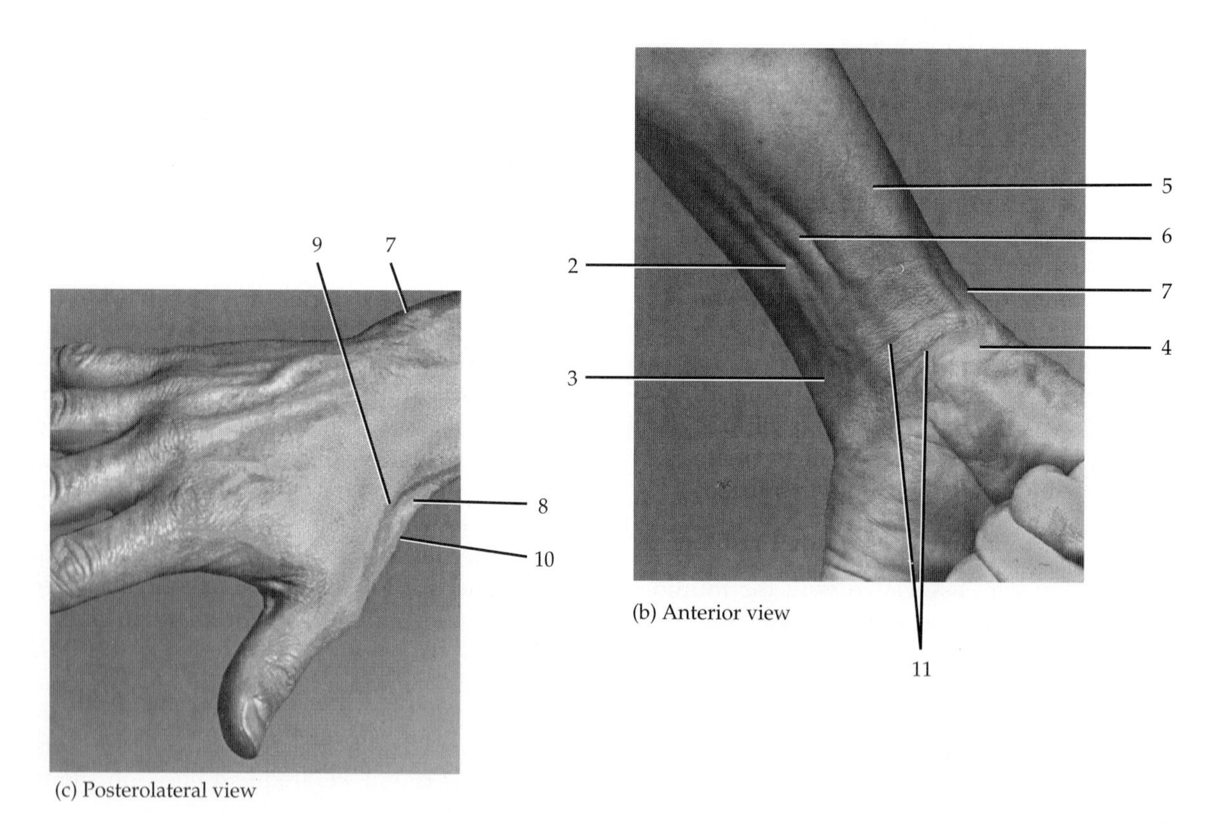

(b) Anterior view

(c) Posterolateral view

FIGURE 11.7 Surface anatomy of forearm and wrist.

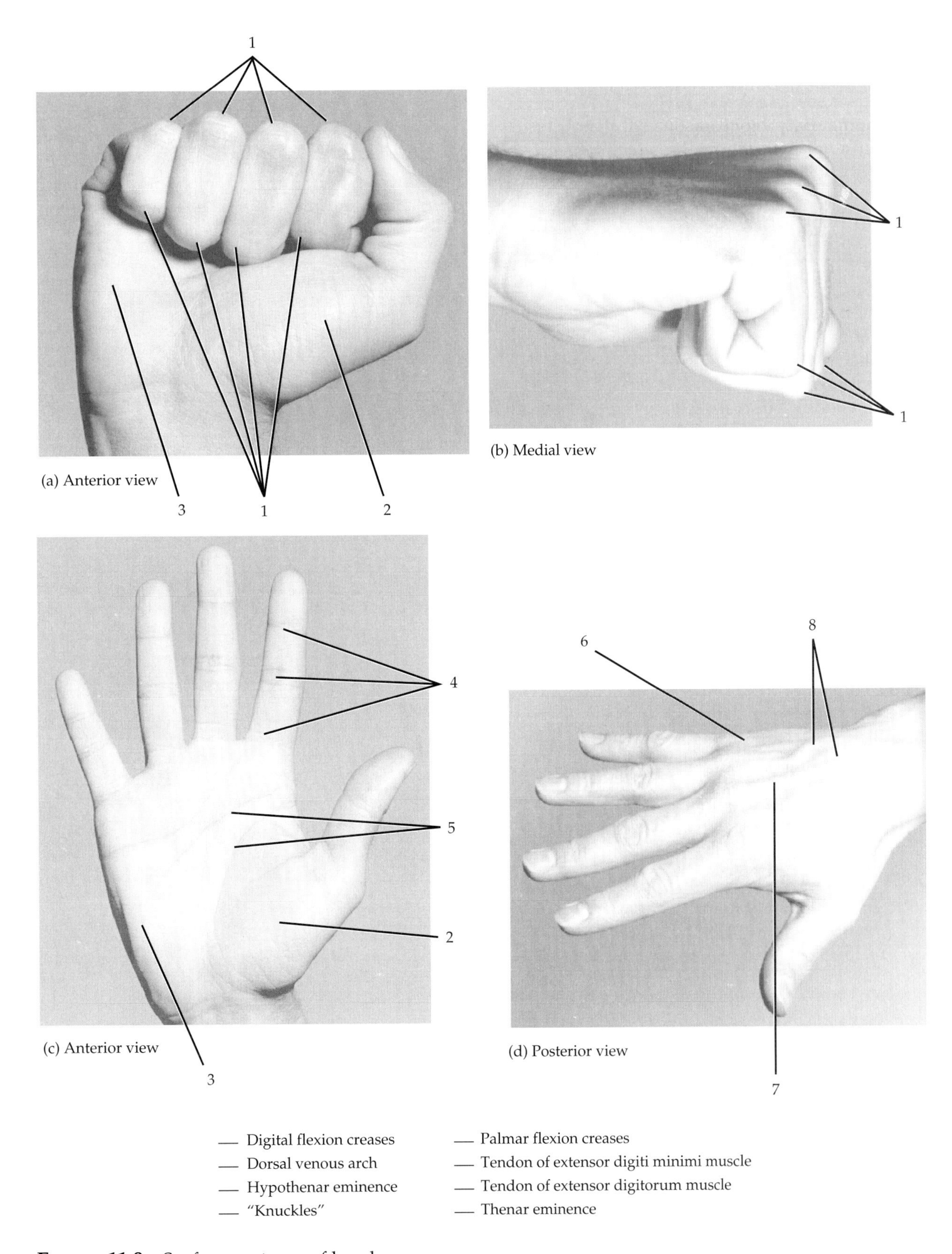

— Digital flexion creases
— Dorsal venous arch
— Hypothenar eminence
— "Knuckles"
— Palmar flexion creases
— Tendon of extensor digiti minimi muscle
— Tendon of extensor digitorum muscle
— Thenar eminence

FIGURE 11.8 Surface anatomy of hand.

k. ***Medial thigh muscles*** Include the ***adductor magnus, adductor brevis, adductor longus, gracilis, obturator externus,*** and ***pectineus.***

With the aid of your textbook, label Figure 11.9.

2. Major surface anatomy features of the ***knee (genu)*** are

a. ***Patella*** Kneecap, located within quadriceps femoris tendon on anterior surface of knee

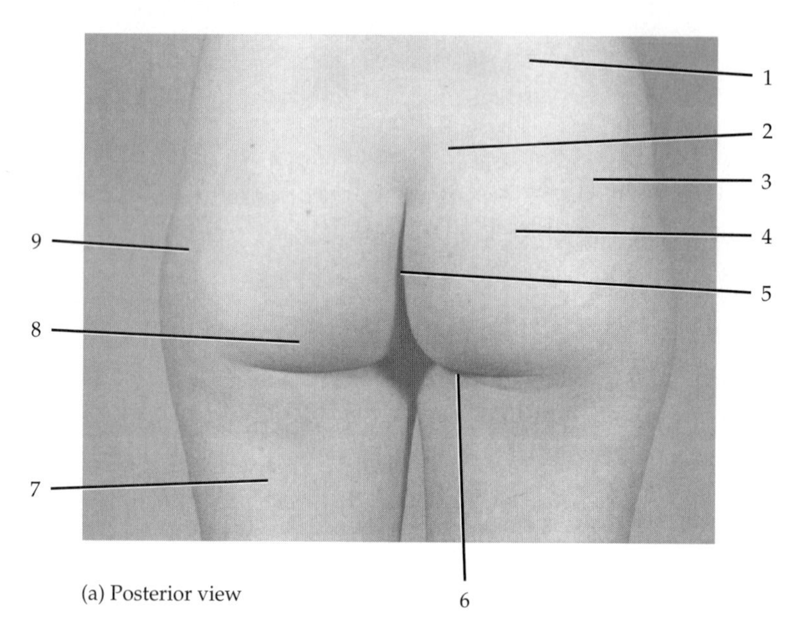

(a) Posterior view

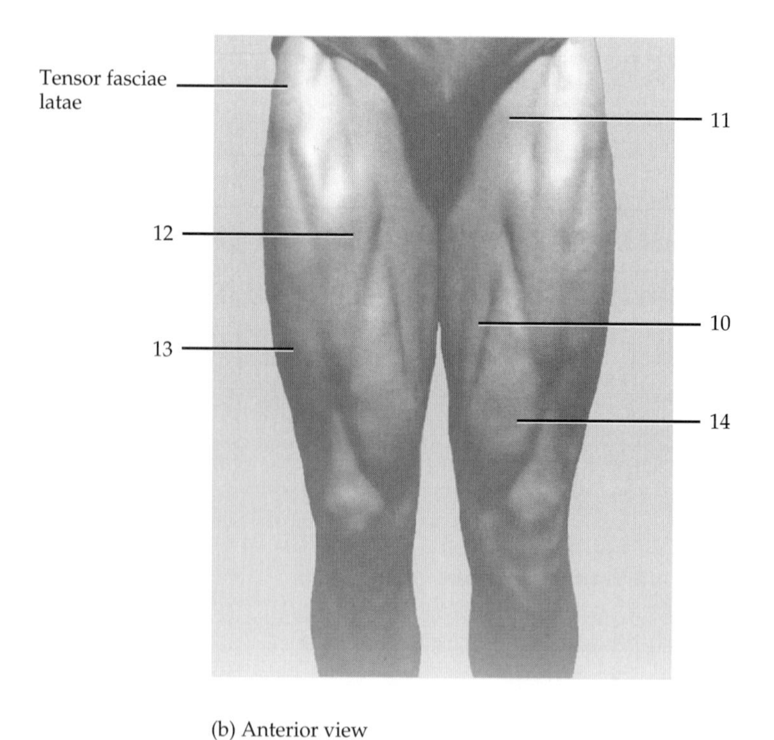

(b) Anterior view

___ Adductor magnus muscle
___ Gluteal (natal) cleft
___ Gluteal fold
___ Gluteus maximus muscle
___ Gluteus medius muscle
___ Greater trochanter of femur
___ Hamstrings
___ Iliac crest
___ Posterior superior iliac spine
___ Rectus femoris muscle
___ Sartorius muscle
___ Site of ischial tuberosity
___ Vastus lateralis muscle
___ Vastus medialis muscle

FIGURE 11.9 Surface anatomy of buttocks and thigh.

along midline. Margins of condyles (described shortly) can be felt on either side of it.

b. ***Patellar ligament*** Continuation of quadriceps femoris tendon inferior to patella.
c. ***Popliteal fossa*** Diamond-shaped space on posterior aspect of knee visible when knee is flexed. Fossa is bordered superolaterally by the biceps femoris muscle, superomedially by the semimembranosus and semitendinosus muscles, and inferolaterally and inferomedially by the lateral and medial heads of the gastrocnemius muscle, respectively.
d. ***Medial condyles of femur and tibia*** Medial projections just inferior to patella. Superior part of projection belongs to distal end of femur; inferior part of projection belongs to proximal end of tibia.
e. ***Lateral condyles of femur and tibia*** Lateral projections just inferior to patella. Superior part of projections belongs to distal end of femur. Inferior part of projections belongs to proximal end of tibia.

With the aid of your textbook, label Figure 11.10.

3. Major surface anatomy features of the ***leg (crus)*** and ***ankle (tarsus)*** are

a. ***Tibial tuberosity*** Bony prominence of tibia inferior to patella into which patellar ligament inserts (Figure 11.10a).
b. ***Medial malleolus of tibia*** Projection of distal end of tibia that forms medial prominence of ankle.
c. ***Lateral malleolus of fibula*** Projection of distal end of fibula that forms lateral prominence of ankle.
d. ***Calcaneal (Achilles) tendon*** Tendon of insertion into calcaneus for gastrocnemius and soleus muscles.
e. ***Tibialis anterior muscle*** Located on anterior surface of leg.
f. ***Gastrocnemius muscle*** Forms bulk of middle and superior portions of posterior surface of leg.
g. ***Soleus muscle*** Mostly deep to gastrocnemius, visible on either side of gastrocnemius below middle of leg.

With the aid of your textbook, label Figure 11.11 on page 221.

4. Major surface anatomy features of the ***foot (pes)*** are

a. ***Calcaneus*** Heel bone to which calcaneal (Achilles) tendon inserts.
b. ***Tendon of extensor hallucis longus muscle*** Visible in line with phalanx of great toe. Pulsations in the dorsalis pedis artery may be felt in most people just lateral to this tendon when the blood vessel passes over the navicular and cuneiform bones.
c. ***Tendons of extensor digitorum longus muscles*** Visible in line with phalanges II through V.
d. ***Dorsal venous arch*** Superficial veins on dorsum of foot that unite to form small and great saphenous veins.

With the aid of your textbook, label Figure 11.12 on page 221.

ANSWER THE LABORATORY REPORT QUESTIONS AT THE END OF THE EXERCISE.

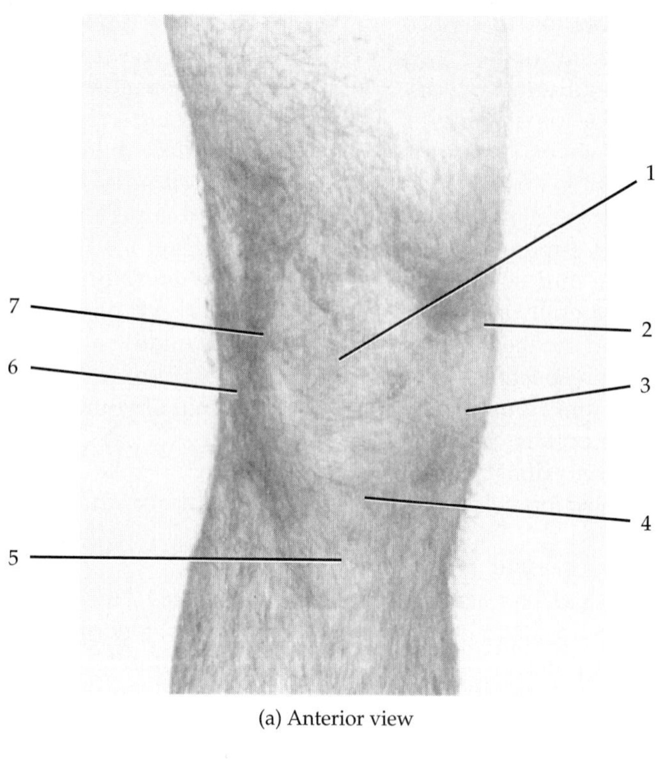

(a) Anterior view

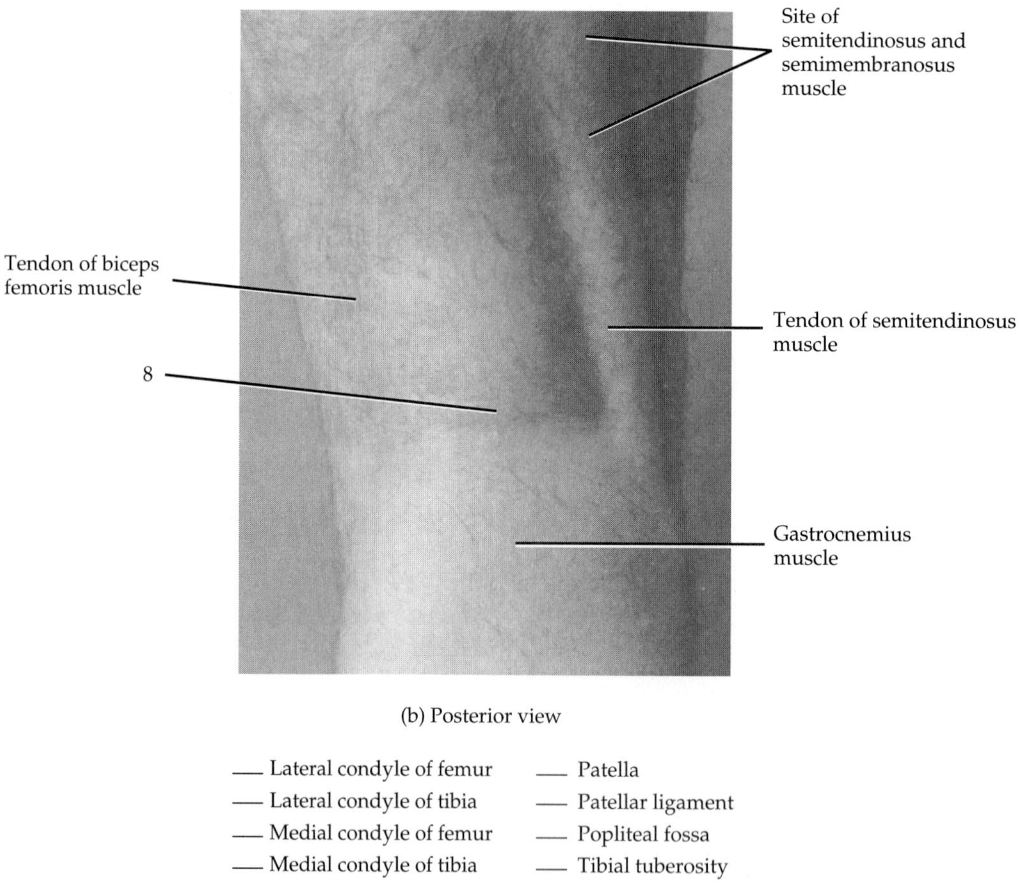

(b) Posterior view

___ Lateral condyle of femur	___ Patella
___ Lateral condyle of tibia	___ Patellar ligament
___ Medial condyle of femur	___ Popliteal fossa
___ Medial condyle of tibia	___ Tibial tuberosity

FIGURE 11.10 Surface anatomy of knee.

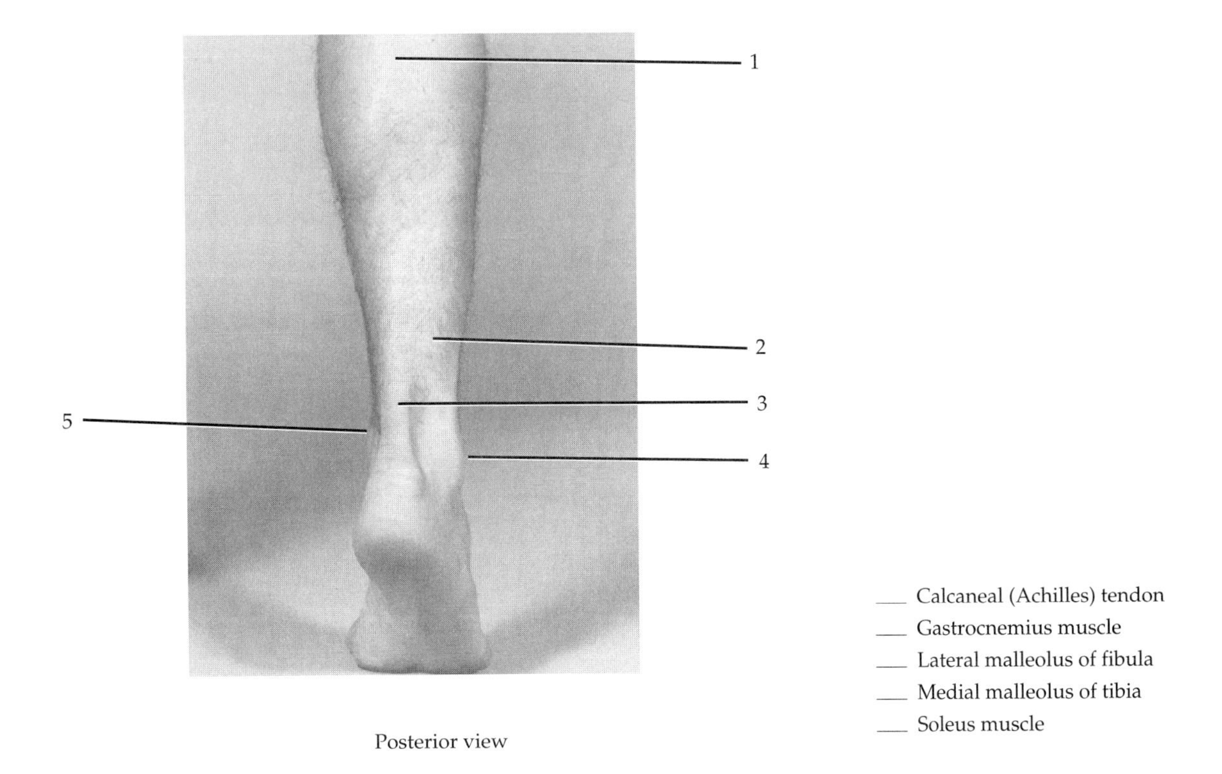

FIGURE 11.11 Surface anatomy of leg.

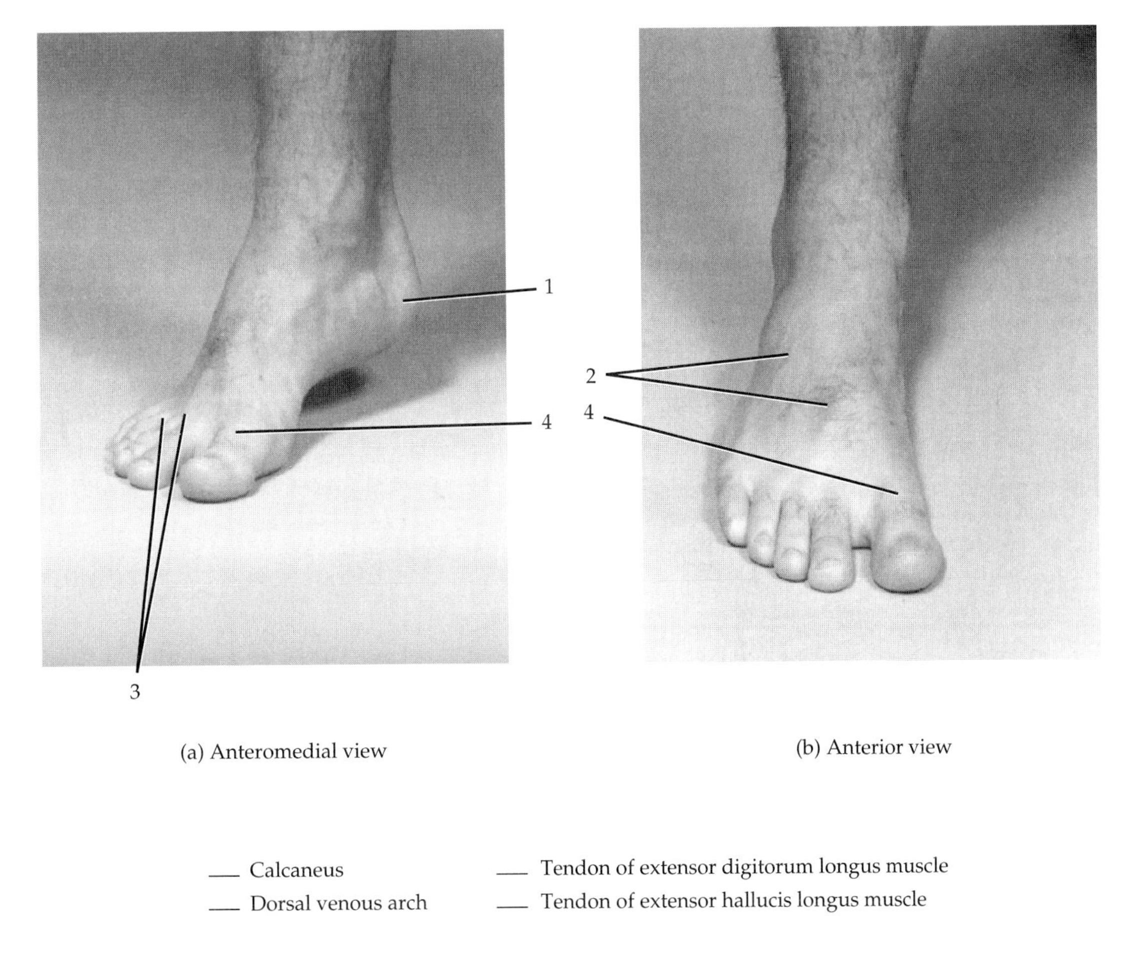

FIGURE 11.12 Surface anatomy of foot.

LABORATORY REPORT QUESTIONS

Surface Anatomy 11

Student ______________________ Date ______________________

Laboratory Section ______________________ Score/Grade ______________________

PART 1. Multiple Choice

_______ 1. The term used to refer to the crown of the skull is (a) occipital (b) mental (c) parietal (d) zygomatic

_______ 2. The laryngeal cartilage in the midline of the anterior cervical region known as the Adam's apple is the (a) cricoid (b) epiglottis (c) arytenoid (d) thyroid

_______ 3. Inflammation of which muscle is associated with "stiff neck"? (a) teres major (b) trapezius (c) deltoid (d) cervicalis

_______ 4. The skeletal muscle located directly on either side of the vertebral column is the (a) serratus anterior (b) infraspinatus (c) teres major (d) erector spinae

_______ 5. The suprasternal notch and xiphoid process are associated with the (a) sternum (b) scapula (c) clavicle (d) ribs

_______ 6. The expanded end of the spine of the scapula is the (a) acromion (b) linea alba (c) olecranon (d) superior angle

_______ 7. Which nerve can be palpated as a rounded cord in a groove posterior to the medial epicondyle? (a) median (b) radial (c) ulnar (d) brachial

_______ 8. Which carpal bone can be palpated as a projection distal to the styloid process of the ulna? (a) trapezoid (b) trapezium (c) hamate (d) pisiform

_______ 9. The lateral rounded contour on the anterior surface of the hand (at the base of the thumb) formed by the muscles of the thumb is the (a) "anatomical snuffbox" (b) thenar eminence (c) hypothenar eminence (d) dorsal venous arch

_______ 10. The superior margin of the hipbone is the (a) pubic symphysis (b) iliac spine (c) acetabulum (d) iliac crest

_______ 11. Which bony structure bears the weight of the body when a person is seated? (a) greater trochanter (b) iliac crest (c) ischial tuberosity (d) gluteal fold

_______ 12. Which muscle is *not* a component of the quadriceps femoris group? (a) biceps femoris (b) vastus lateralis (c) vastus medialis (d) rectus femoris

_______ 13. The diamond-shaped space on the posterior aspect of the knee is the (a) cubital fossa (b) posterior triangle (c) popliteal fossa (d) nuchal groove

_______ 14. The projection of the distal end of the tibia that forms the prominence on one side of the ankle is the (a) medial condyle (b) medial malleolus (c) lateral condyle (d) lateral malleolus

_______ 15. The tendon that can be seen in line with the great toe belongs to which muscle? (a) extensor digiti minimi (b) extensor digitorum longus (c) extensor hallucis longus (d) extensor carpi radialis

PART 2. Completion

16. The laryngeal cartilage that connects the larynx to the trachea is the ____________ cartilage.
17. The ____________ triangle is bordered by the mandible, sternum, cervical midline, and sternocleidomastoid muscle.
18. The depression on the superior surface of the sternum between the medial ends of the clavicles is the ____________.
19. The principal superficial chest muscle is the ____________.
20. Tendinous intersections are associated with the ____________ muscle.
21. The muscle that forms the rounded prominence of the shoulder is the ____________ muscle.
22. The triangular space in the anterior aspect of the elbow is the ____________.
23. The "anatomical snuffbox" is bordered by the tendons of the extensor pollicis brevis muscle and the ____________ muscle.
24. The dorsal aspects of the distal ends of metacarpals II through V are commonly referred to as ____________.
25. The tendon of the ____________ muscle is in line with phalanx V.
26. The dimple that forms about 4 cm lateral to the midline just above the buttocks lies superficial to the ____________.
27. The femoral projection that can be palpated about 20 cm (8 in.) inferior to the iliac crest is the ____________.
28. The continuation of the quadriceps femoris tendon inferior to the patella is the ____________.
29. The tendon of insertion for the gastrocnemius and soleus muscles is the ____________ tendon.
30. Superficial veins on the dorsum of the foot that unite to form the small and great saphenous veins belong to the ____________.
31. The prominent veins along the lateral cervical regions are the ____________ veins.
32. The pronounced vertebral spine of C7 is the ____________.
33. The most reliable surface anatomy feature of the chest is the ____________.
34. A slight groove extending from the xiphoid process to the pubic symphysis is the ____________.
35. The vein that crosses the cubital fossa and is frequently used to remove blood is the ____________ vein.

PART 3. MATCHING

______	**36.** Mental region	A.	Inferior edges of costal cartilages of ribs 7 through 10
______	**37.** Xiphoid process	B.	Forms elbow
______	**38.** Arm	C.	Inferior portion of sternum
______	**39.** Nucha	D.	Manus
______	**40.** Shoulder	E.	Crus
______	**41.** Gluteus maximus muscle	F.	Brachium
______	**42.** Costal margin	G.	Tarsus
______	**43.** Olecranon	H.	Anterior part of mandible
______	**44.** Wrist	I.	Brain case
______	**45.** Auricular region	J.	Component of hamstrings
______	**46.** Semitendinosus muscle	K.	Forms main part of prominence of buttock
______	**47.** Leg	L.	Carpus
______	**48.** Cranium	M.	Posterior neck region
______	**49.** Ankle	N.	Acromial
______	**50.** Hand	O.	Ear

Nervous Tissue

12

The ***nervous system*** has three basic functions: (1) sensory, (2) integrative, and (3) motor.

1. ***Sensory function*** It *senses* certain changes (stimuli), both within your body (the internal environment), such as stretching of your stomach or an increase in blood acidity, and outside your body (the external environment), such as a raindrop landing on your arm or the aroma of a rose.
2. ***Integrative function*** It *analyzes* the sensory information, *stores* some aspects, and *makes decisions* regarding appropriate behaviors.
3. ***Motor function*** It may *respond* to stimuli by initiating muscular contractions or glandular secretions.

The branch of medical science that deals with the normal functioning and disorders of the nervous system is called ***neurology*** (noo-ROL-ō-jē; *neuro* = nerve or nervous system; *logos* = study of).

A. NERVOUS SYSTEM DIVISIONS

The two principal divisions of the nervous system are the ***central nervous system (CNS)*** and the ***peripheral*** (pe-RIF-er-al) ***nervous system (PNS).*** The CNS consists of the ***brain*** and ***spinal cord.*** Within the CNS, various sorts of incoming sensory information are integrated and correlated, thoughts and emotions are generated, and memories are formed and stored. Most nerve impulses that stimulate muscles to contract and glands to secrete originate in the CNS.

The CNS is connected to sensory receptors, muscles, and glands in peripheral parts of the body by the PNS. The PNS consists of ***cranial nerves*** that arise from the brain and ***spinal nerves*** that emerge from the spinal cord. Portions of these nerves carry nerve impulses into the CNS while other portions carry impulses out of the CNS.

The input component of the PNS consists of nerve cells called ***sensory*** or ***afferent*** (AF-er-ent; *ad* = toward; *ferre* = to carry) ***neurons.*** They conduct nerve impulses from sensory receptors in various parts of the body to the CNS and end within the CNS. The output component consists of nerve cells called ***motor*** or ***efferent*** (EF-er-ent; *ex* = away from; *ferre* = to carry) ***neurons.*** They originate within the CNS and conduct nerve impulses from the CNS to muscles and glands.

The PNS may be subdivided further into a ***somatic*** (*soma* = body) ***nervous system (SNS)*** and an ***autonomic*** (*auto* = self; *nomos* = law) ***nervous system (ANS).*** The major difference between the two is the ***effector,*** the tissue that receives stimulation or inhibition from the nervous system (skeletal muscle, smooth muscle, cardiac muscle, or a gland). The SNS consists of sensory neurons that convey information from cutaneous and special sense receptors primarily in the head, body wall, and limbs to the CNS and motor neurons from the CNS that conduct impulses to *skeletal muscles* only. Because these motor responses can be consciously controlled, this portion of the SNS is *voluntary*.

The ANS consists of sensory neurons that convey information from receptors primarily in the viscera to the CNS and motor neurons from the CNS that conduct impulses to smooth muscle, cardiac muscle, and glands. Since its motor responses are not normally under conscious control, the ANS is *involuntary*.

The motor portion of the ANS consists of two branches, the ***sympathetic division*** and the ***parasympathetic division.*** With few exceptions, the viscera receive instructions from both. Usually, the two divisions have opposing actions. For example, sympathetic neurons speed the heartbeat while parasympathetic neurons slow it down. Processes promoted by sympathetic neurons often involve expenditure of energy while those promoted by parasympathetic neurons restore and conserve body energy.

In this exercise you will identify the parts of a neuron and the components of a reflex arc.

B. HISTOLOGY OF NERVOUS TISSUE

Despite its complexity, the nervous system consists of only two principal kinds of cells: neurons and neuroglia. ***Neurons*** (nerve cells) constitute the nervous tissue and are highly specialized for nerve impulse conduction. Mature neurons have only limited capacity for replacement or repair. ***Neuroglia*** (noo-ROG-lē-a; *neuro* = nerve; *glia* = glue) can divide and multiply and support, nurture, and protect neurons and maintain homeostasis of the fluid that bathes neurons. They do not transmit nerve impulses. Brain tumors are commonly derived from neuroglia. Such tumors, called ***gliomas***, are highly malignant and rapidly enlarging.

A neuron consists of the following parts:

1. ***Cell body*** Contains a nucleus, cytoplasm, lysosomes, mitochondria, Golgi apparatus, chromatophilic substance (Nissl bodies), and neurofibrils.
2. ***Dendrites*** (*dendro* = tree) Usually short, highly branched extensions of the cell body that conduct nerve impulses toward the cell body.
3. ***Axon*** (*axon* = axis) Single, usually relatively long process that conducts nerve impulses away from the cell body to another neuron, muscle fiber, or gland cell. An axon, in turn, consists of the following:
   a. ***Axon hillock*** (*hilloc* = small hill) The origin of an axon from the cell body represented as a small cone-shaped region.
   b. ***Initial segment*** First portion of a neuron. Except in sensory neurons, nerve impulses arise at the junction of the axon hillock and initial segment, a region called the ***trigger zone***.
   c. ***Axoplasm*** Cytoplasm of an axon.
   d. ***Axolemma*** (*lemma* = sheath or husk) Plasma membrane around the axoplasm.
   e. ***Axon collateral*** Side branch of an axon.
   f. ***Axon terminals*** Fine, branching filaments of an axon or axon collateral.
   g. ***Synaptic end bulbs*** Bulblike structures at distal end of axon terminals that contain storage sacs ***(synaptic vesicles)*** for neurotransmitters.
   h. ***Myelin sheath*** Multilayered, lipid and protein segmented covering of many axons, especially large peripheral ones; the myelin sheath is produced by peripheral nervous system neuroglia called ***neurolemmocytes (Schwann cells)*** and central nervous system neuroglia called ***oligodendrocytes*** (described shortly).
   i. ***Neurolemma (sheath of Schwann)*** Peripheral, nucleated cytoplasmic layer of the neurolemmocyte (Schwann cell) that encloses the myelin sheath. It is found only around axons in the peripheral nervous system.
   j. ***Neurofibral node (node of Ranvier)*** Unmyelinated gap between segments of the myelin sheath.

Using your textbook and models of neurons as a guide, label Figure 12.1.

Now obtain a prepared slide of an ox spinal cord (transverse section), human spinal cord (transverse and longitudinal sections), nerve endings in skeletal muscle, and a nerve trunk (transverse and longitudinal sections). Examine each under high power and identify as many parts of the neuron as you can.

Neurons may be classified on the basis of structure and function. Structural classification is based on the number of processes extending from the cell body. Structurally, neurons are ***multipolar*** (several dendrites and one axon), ***bipolar*** (one dendrite and one axon), and ***unipolar*** (a single process that branches into an axon and a dendrite). Functional classification is based on the type of information carried and the direction in which the information is carried. Functionally, neurons are classified as ***sensory (afferent)***, which carry nerve impulses toward the central nervous system; ***motor (efferent)***, which carry nerve impulses away from the central nervous system; and ***association (connecting*** or ***interneurons)***, neurons that are located in the central nervous system and carry nerve impulses between sensory and motor neurons. The neuron you have already labeled in Figure 12.1 is a motor neuron.

Using your textbook and models of neurons as a guide, label Figure 12.2 on page 230.

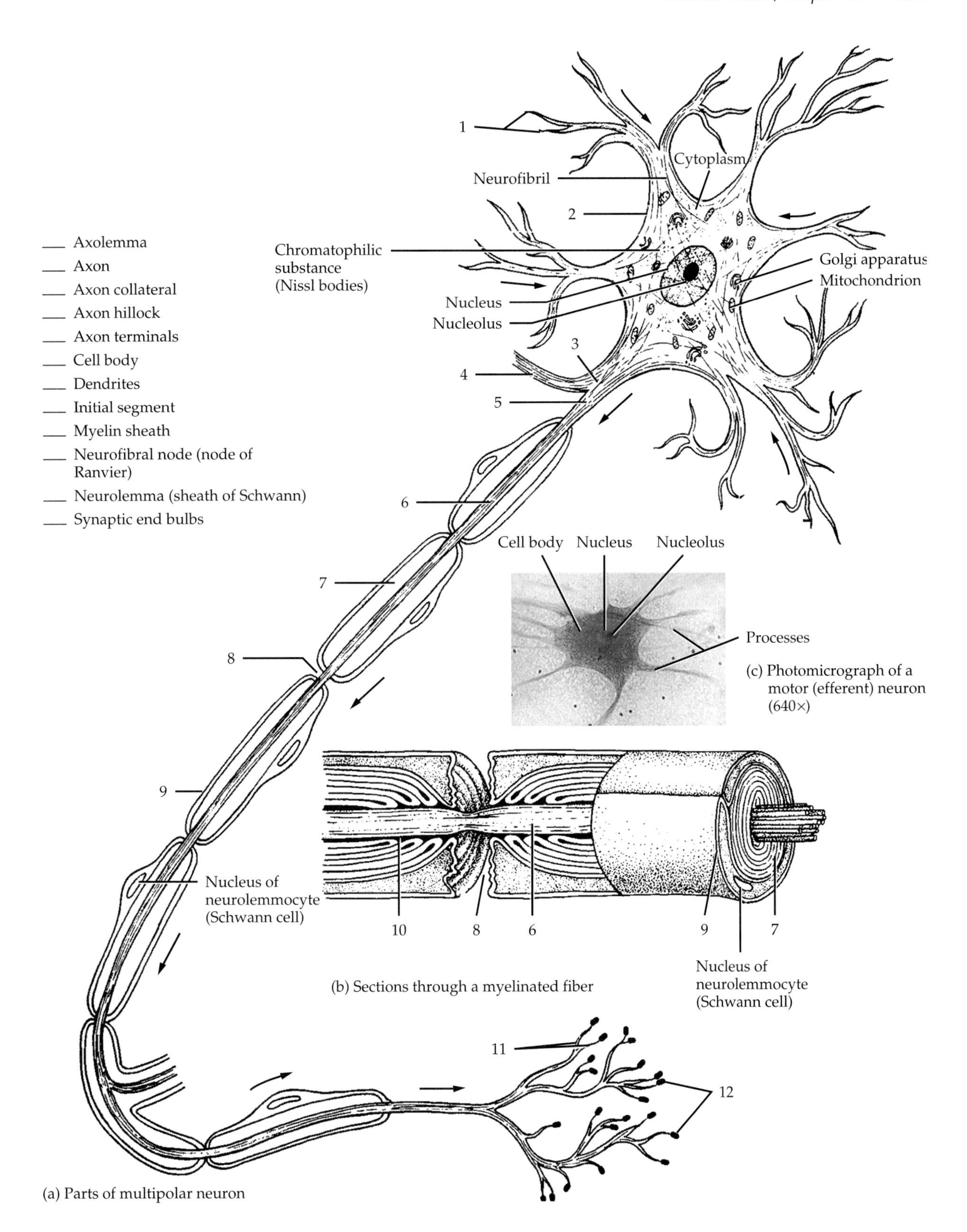

FIGURE 12.1 Structure of a neuron. In (a), arrows indicate the direction in which the nerve impulse travels.

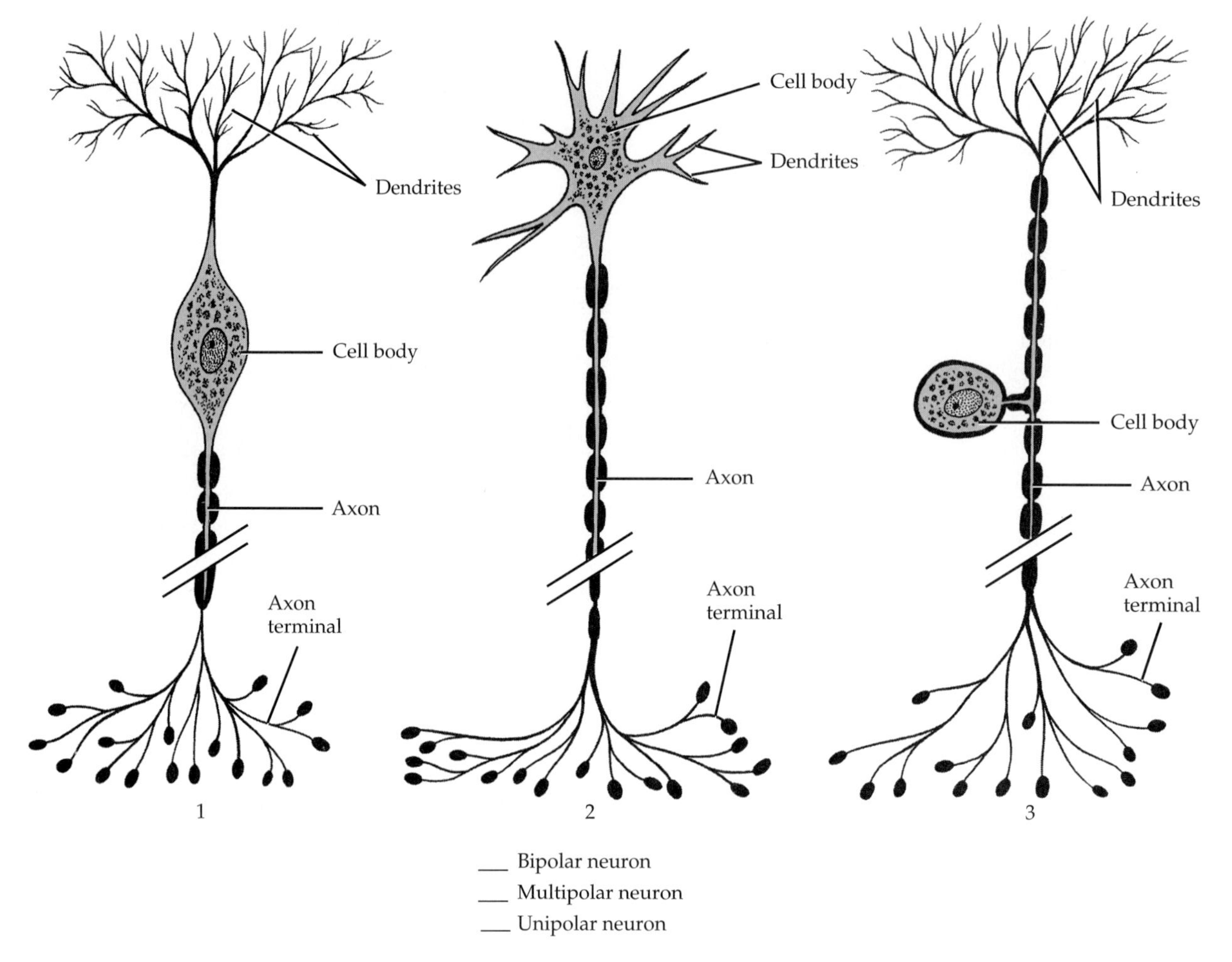

FIGURE 12.2 Structural classification of neurons.

C. HISTOLOGY OF NEUROGLIA

Among the types of neuroglial cells are the following:

1. ***Astrocytes*** (AS-trō-sīts; *astro* = star; *cyte* = cell) Participate in the metabolism of neurotransmitters (glutamate and γ-aminobutyric acid) and maintain the proper balance of potassium ions (K^+) for generation of nerve impulses by CNS neurons; participate in brain development by assisting migration of neurons; help form the blood-brain barrier, which regulates the passage of substances into the brain; twine around neurons to form a supporting network; and provide a link between neurons and blood vessels.

   a. ***Protoplasmic astrocytes*** are found in the gray matter of the central nervous system.
   b. ***Fibrous astrocytes*** are found in the white matter of the central nervous system.

2. ***Oligodendrocytes*** (ol'-i-gō-DEN-drō-sīts; *oligo* = few; *dendro* = tree) Resemble astrocytes but with fewer and shorter processes; they provide support in the CNS and produce a myelin sheath on axons of neurons of the CNS.
3. ***Microglia*** (mī-KROG-lē-a; *micro* = small) Small cells derived from monocytes with few processes; although normally stationary, they can migrate to damaged nervous tissue and there carry on phagocytosis; they are also called ***brain macrophages.***
4. ***Ependymal*** (e-PEN-di-mal) ***cells*** Epithelial cells arranged in a single layer that range from squamous to columnar in shape; many are ciliated; they form a continuous epithelial lining for the central canal of the spinal cord and for the ventricles of the brain, spaces that contain networks of capillaries that form cerebrospinal fluid; ependymal cells probably assist in circulating cerebrospinal fluid in these areas.
5. ***Neurolemmocytes (Schwann cells)*** Flattened cells arranged around axons. Produce a phos-

pholipid myelin sheath around axons and dendrites of neurons of PNS.

6. ***Satellite cells*** Flattened cells arranged around the cell bodies of ganglia (collections of neuron cell bodies outside the CNS). Support neurons in ganglia of PNS.

Obtain prepared slides of astrocytes (protoplasmic and fibrous), oligodendrocytes, microglia, ependymal cells, neurolemmocytes, and satellite cells. Using your textbook, Figure 12.3, and models of neuroglia as a guide, identify the various kinds of cells. In the spaces provided, draw each of the cells.

Protoplasmic astrocyte

Fibrous astrocyte

Oligodendrocyte

Microglial cell

Ependymal cell

Neurolemmocyte

Satellite cell

D. NEURONAL CIRCUITS

The CNS contains billions of neurons organized into complicated patterns called ***neuronal pools.*** Each pool differs from all others and has its own role in regulating homeostasis. A neuronal pool may contain thousands or even millions of neurons.

The functional contact between two neurons or between a neuron and an effector is called a ***synapse.*** At a synapse, the neuron sending the signal is called a ***presynaptic neuron,*** and the neuron receiving the message is called a ***postsynaptic neuron.***

Neuronal pools in the CNS are arranged in patterns called ***circuits*** over which the nerve impulses are conducted. In ***simple series circuits*** a presynaptic neuron stimulates only a single neuron in a pool. The single neuron then stimulates another, and so on. Most circuits, however, are more complex.

A single presynaptic neuron may synapse with several postsynaptic neurons. Such an arrangement, called ***divergence,*** permits one presynaptic neuron to influence several postsynaptic neurons or several muscle fibers or gland cells at the same time. In a ***diverging circuit,*** the nerve impulse from a single presynaptic neuron causes the stimulation of increasing numbers of cells along the circuit. For example, a small number of neurons in the brain that govern a particular body movement stimulate a much larger number of neurons in the spinal cord. Sensory signals also feed into diverging circuits and are often relayed to several regions of the brain.

In another arrangement, called ***convergence,*** several presynaptic neurons synapse with a single postsynaptic neuron. This arrangement permits more effective stimulation or inhibition of the postsynaptic neuron. In one type of ***converging circuit,*** the postsynaptic neuron receives nerve impulses from several different sources. For example, a single motor neuron that synapses with skeletal muscle fibers at neuromuscular junctions receives input from several pathways that originate in different brain regions.

Some circuits in your body are constructed so that once the presynaptic cell is stimulated, it will cause the postsynaptic cell to transmit a series of nerve impulses. One such circuit is called a ***reverberating (oscillatory) circuit.*** In this pattern, the incoming impulse stimulates the first neuron, which stimulates the second, which stimulates the third, and so on. Branches from later neurons synapse with earlier ones, however, sending the

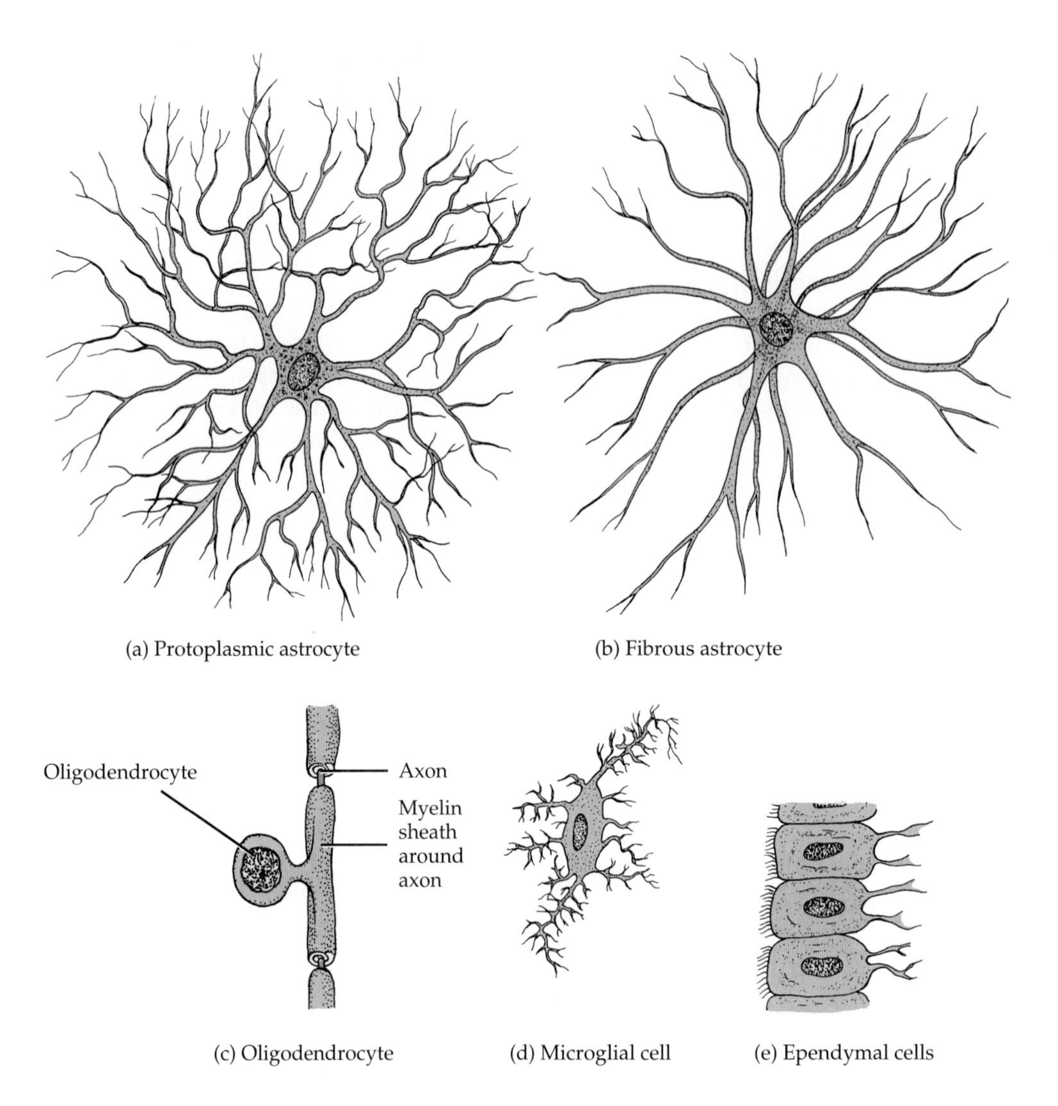

FIGURE 12.3 Neuroglia.

impulse back through the circuit again and again. The output signal may last from a few seconds to many hours, depending on the number of synapses and the arrangement of neurons in the circuit. Inhibitory neurons may turn off a reverberating circuit after a period of time. Among the body responses thought to be the result of output signals from reverberating circuits are breathing, coordinated muscular activities, waking up, sleeping (when reverberation stops), and short-term memory. One form of epilepsy (grand mal) is probably caused by abnormal reverberating circuits.

A fourth type of circuit is the ***parallel after-discharge circuit.*** In this circuit, a single presynaptic cell stimulates a group of neurons, each of which synapses with a common postsynaptic cell. If the input is excitatory, the postsynaptic neuron then can send out a stream of impulses in quick succession. It is thought that parallel after-discharge circuits may be employed for precise activities such as mathematical calculations.

Using your textbook as a guide, label Figure 12.4.

E. REFLEX ARC

A ***reflex*** is a fast, predictable, automatic response to a change in the environment (stimulus). The stimulus is applied to the periphery and conducted either to the brain or spinal cord. Reflexes serve to restore functions to homeostasis. For a reflex to occur, a stimulus must elicit a response in

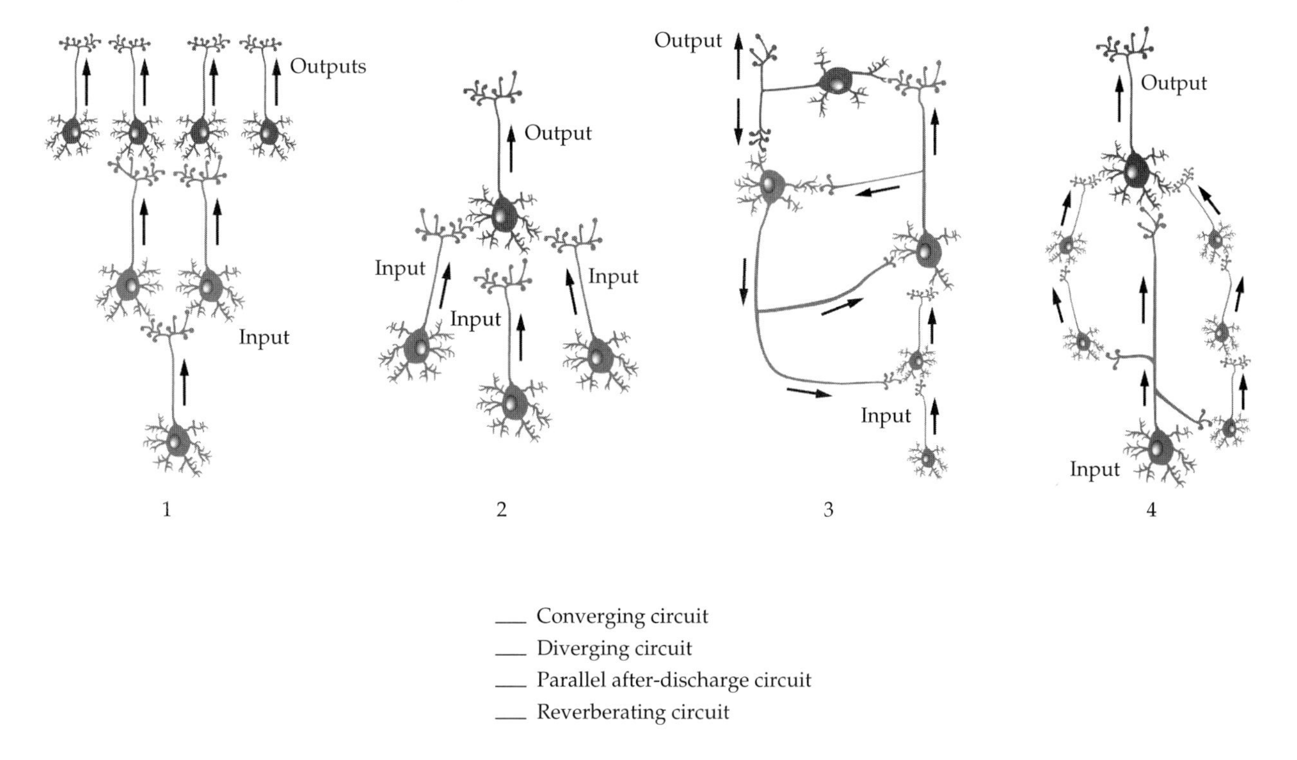

FIGURE 12.4 Neuronal circuits.

one or more structural units within the body termed ***reflex arcs.*** A reflex arc consists of the following components:

1. ***Receptor*** The distal end of a sensory neuron (dendrite) or an associated sensory structure that serves as a receptor. A receptor converts a stimulus from its particular form of energy (such as temperature, pressure, or stretch) into the electrical energy utilized by neurons. If this localized depolarization is of threshold value, an action potential (nerve impulse) will be initiated in a sensory neuron.
2. ***Sensory (afferent) neuron*** A nerve cell that carries the action potential from the receptor to the central nervous system (CNS).
3. ***Integrating center*** Region in the CNS where the sensory neuron makes a functional connection with one or more neurons. This CNS synapse may be with an association neuron or a motor neuron. It is here at the CNS synapse where the initial processing of the sensory information occurs: the more complex the reflex involved, the greater the number of CNS synapses involved in a reflex arc.
4. ***Motor (efferent) neuron*** A nerve cell that transmits the action potential generated by the sensory neuron or an association neuron away from the CNS to the effector.
5. ***Effector*** The part of the body, either a muscle or a gland, that responds to the motor neuron impulse and thus the stimulus.

Label the components of a reflex arc in Figure 12.5.

ANSWER THE LABORATORY REPORT QUESTIONS AT THE END OF THE EXERCISE.

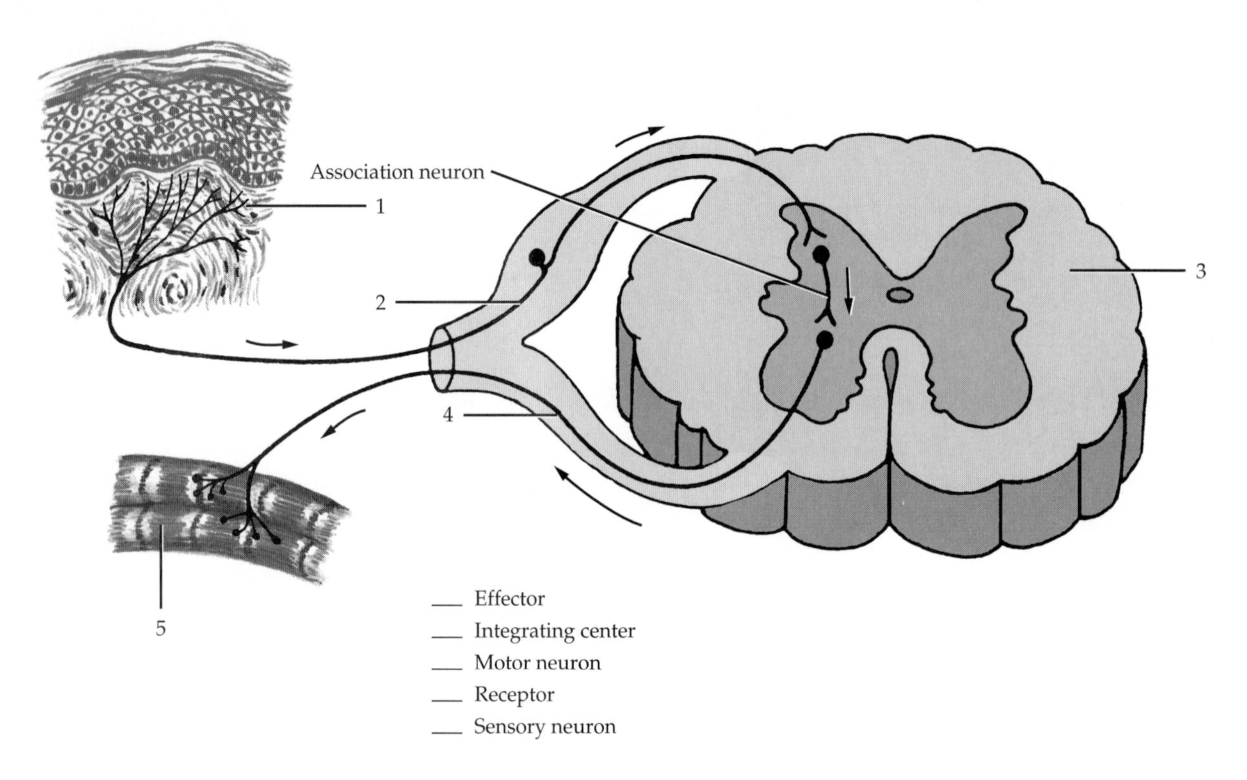

FIGURE 12.5 Components of a reflex arc.

LABORATORY REPORT QUESTIONS

Nervous Tissue 12

Student ______________________ **Date** ______________________

Laboratory Section ______________________ **Score/Grade** ______________________

PART 1. Multiple Choice

_______ **1.** The portion of a neuron that conducts nerve impulses away from the cell body is the (a) dendrite (b) axon (c) receptor (d) effector

_______ **2.** The fine branching filaments of an axon are called (a) myelin sheaths (b) axolemmas (c) axon terminals (d) axon hillocks

_______ **3.** The component of a reflex arc that responds to a motor impulse is the (a) integrating center (b) receptor (c) sensory neuron (d) effector

_______ **4.** Which type of neuron conducts nerve impulses toward the central nervous system? (a) sensory (b) association (c) connecting (d) motor

_______ **5.** In a reflex arc, the nerve impulse is transmitted directly to the effector by the (a) sensory neuron (b) motor neuron (c) integrating center (d) receptor

_______ **6.** Bulblike structures at the distal ends of axon terminals that contain storage sacs for neurotransmitters are called (a) dendrites (b) synaptic end bulbs (c) axon collaterals (d) neurofibrils

_______ **7.** Which neuroglial cell is phagocytic? (a) oligodendrocyte (b) protoplasmic astrocyte (c) microglial cell (d) fibrous astrocyte

PART 2. Completion

8. A neuron that contains several dendrites and one axon is classified as ______________________.

9. The portion of a neuron that contains the nucleus and cytoplasm is the ______________________.

10. The lipid and protein covering around many peripheral axons is called the ______________________.

11. The two types of cells that compose the nervous system are neurons and ______________________.

12. The peripheral, nucleated layer of the neurolemmocyte (Schwann cell) that encloses the myelin sheath is the ______________________.

13. The side branch of an axon is referred to as the ______________________.

14. The part of a neuron that conducts nerve impulses toward the cell body is the ______________________.

15. Neurons that carry nerve impulses between sensory neurons and motor neurons are called

_____________________ neurons.

16. Neurons with one dendrite and one axon are classified as _____________________.

17. Unmyelinated gaps between segments of the myelin sheath are known as _____________________.

18. The neuroglial cell that produces a myelin sheath around axons of neurons of the central nervous system is called a(n) _____________________.

19. In a reflex arc, the muscle or gland that responds to a motor impulse is called the

_____________________.

20. The functional contact between two neurons or between a neuron and an effector is called a(n)

_____________________.

21. The circuit that probably plays a role in breathing, waking up, sleeping, and short-term memory is a(n) _____________________ circuit.

Nervous System

13

In this exercise, you will examine the principal structural features of the spinal cord and spinal nerves, identify the principal structural features of the brain, trace the course of cerebrospinal fluid, identify the cranial nerves, and examine the structure and function of the autonomic nervous system. You will also dissect and study the cat nervous system and the sheep brain.

A. SPINAL CORD AND SPINAL NERVES

1. Meninges

The ***meninges*** (me-NIN-jēz) are connective tissue coverings that run continuously around the spinal cord and brain (***meninx,*** pronounced MĒ-ninks, is singular). They protect the central nervous system. The spinal meninges are:

a. ***Dura mater*** (DYOO-ra MĀ-ter; *dura* = tough; *mater* = mother) The most superficial meninx composed of dense irregular connective tissue. Between the wall of the vertebral canal and the dura mater is the ***epidural space,*** which is filled with fat, connective tissue, and blood vessels.

b. ***Arachnoid*** (a-RAK-noyd; *arachne* = spider) The middle meninx is an avascular covering composed of very delicate collagen and elastic fibers. Between the arachnoid and the dura mater is a space called the ***subdural space,*** which contains interstitial fluid.

c. ***Pia mater*** (PĒ-a MĀ-ter; *pia* = delicate) The deep meninx is a thin, transparent connective tissue layer that adheres to the surface of the brain and spinal cord. It consists of interlacing collagen and a few elastic fibers and contains blood vessels. Between the pia mater and the arachnoid is a space called the ***subarachnoid space*** where cerebrospinal fluid circulates. Extensions of the pia mater called ***denticulate*** (den-TIK-yoo-lāt; *denticulus* = a small tooth) ***ligaments*** are attached laterally to the dura mater along the length of the spinal cord and suspend the spinal cord and afford protection against shock and sudden displacement.

Label the meninges, subarachnoid space, and denticulate ligament in Figure 13.1.

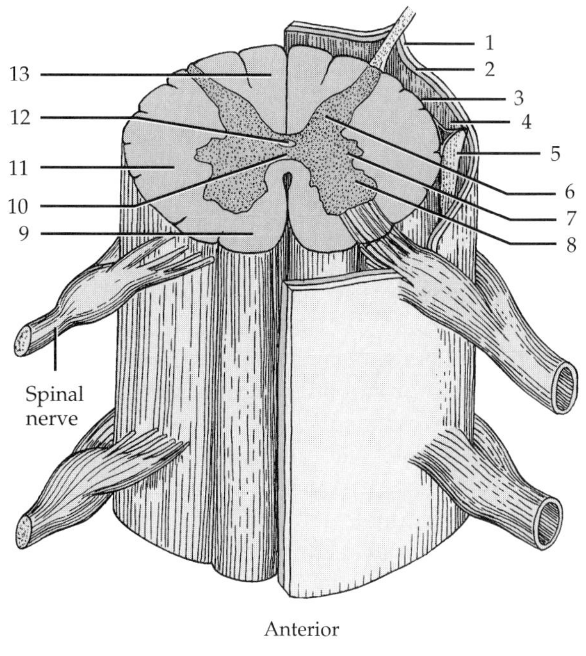

___ Anterior gray horn
___ Anterior white column
___ Arachnoid
___ Central canal
___ Denticulate ligament
___ Dura mater
___ Gray commissure
___ Lateral gray horn
___ Lateral white column
___ Pia mater
___ Posterior gray horn
___ Posterior white column
___ Subarachnoid space

FIGURE 13.1 Transverse (cross) section of spinal cord showing meninges on right side of figure.

2. General Features

Obtain a model or preserved specimen of the spinal cord and identify the following general features:

a. ***Cervical enlargement*** Between vertebrae C4 and T1; origin of nerves to upper limbs.
b. ***Lumbar enlargement*** Between vertebrae T9 and T12; origin of nerves to lower limbs.
c. ***Conus medullaris*** (KŌ-nus med-yoo-LAR-is; *konos* = cone). Tapered conical portion of spinal cord that ends at the intervertebral disc between L1 and L2.
d. ***Filum terminale*** (FĪ-lum ter-mi-NAL-ē; *filum* = filament; *terminale* = terminal) Nonnervous fibrous tissue arising from the conus medullaris and extending inferiorly to attach to the coccyx; consists mostly of pia mater.
e. ***Cauda equina*** (KAW-da ē-KWĪ-na) Spinal nerves that angle inferiorly in the vertebral canal giving the appearance of wisps of coarse hair.
f. ***Anterior median fissure*** Deep, wide groove on the anterior surface of the spinal cord.
g. ***Posterior median sulcus*** Shallow, narrow groove on the posterior surface of the spinal cord.

After you have located the parts on a model or preserved specimen of the spinal cord, label the spinal cord in Figure 13.2.

3. Transverse Section of Spinal Cord

In a freshly dissected section of the brain or spinal cord, some regions look white and glistening whereas others appear gray. ***White matter*** refers to aggregations of myelinated processes from many neurons. The whitish color of myelin gives white matter its name. The ***gray matter*** of the nervous system contains either nerve cell bodies, dendrites, and axon terminals or bundles of unmyelinated axons and neuroglia. They look grayish, rather than white, because there is no myelin in these areas.

Obtain a model or specimen of the spinal cord in transverse section and note the gray matter, shaped like a letter H or a butterfly. Identify the following parts:

a. ***Gray commissure*** (KOM-mi-shur) Cross bar of the letter H.
b. ***Central canal*** Small space in the center of the gray commissure that contains cerebrospinal fluid.
c. ***Anterior gray horn*** Anterior region of the upright portion of the H.
d. ***Posterior gray horn*** Posterior region of the upright portion of the H.
e. ***Lateral gray horn*** Intermediate region between the anterior and posterior gray horns present in the thoracic, upper lumbar, and sacral segments of the spinal cord.
f. ***Anterior white column*** Anterior region of white matter.
g. ***Posterior white column*** Posterior region of white matter.
h. ***Lateral white column*** Intermediate region of white matter between the anterior and posterior white columns.

Label these parts of the spinal cord in transverse section in Figure 13.1.

Examine a prepared slide of a spinal cord in transverse section and see how many structures you can identify.

4. Spinal Nerve Attachments

Spinal nerves are the paths of communication between the spinal cord tracts and most of the body. The 31 pairs of spinal nerves are named and numbered according to the region of the spinal cord from which they emerge. The first cervical pair emerges between the atlas and occipital bone; all other spinal nerves leave the vertebral column from intervertebral foramina between adjoining vertebrae. There are 8 pairs of cervical nerves, 12 pairs of thoracic nerves, 5 pairs of lumbar nerves, 5 pairs of sacral nerves, and 1 pair of coccygeal nerves. Label the spinal nerves in Figure 13.2.

Each pair of spinal nerves is connected to the spinal cord by two points of attachment called roots. The ***posterior (sensory) root*** contains sensory nerve fibers only and conducts nerve impulses from the periphery to the spinal cord. Each posterior root has a swelling, the ***posterior (sensory) root ganglion,*** which contains the cell bodies of the sensory neurons from the periphery. Fibers extend from the ganglion into the posterior gray horn. The other point of attachment, the ***anterior (motor) root,*** contains motor nerve fibers only and conducts nerve impulses from the spinal cord to the periphery. The cell bodies of the motor neurons are located in lateral or anterior gray horns.

Label the posterior root, posterior root ganglion, anterior root, spinal nerve, cell body of sensory neuron, axon of sensory neuron, cell body of motor neuron, and axon of motor neuron in Figure 13.3 on page 240.

5. Components and Coverings of Spinal Nerves

The posterior and anterior roots unite to form a spinal nerve at the intervertebral foramen. Because

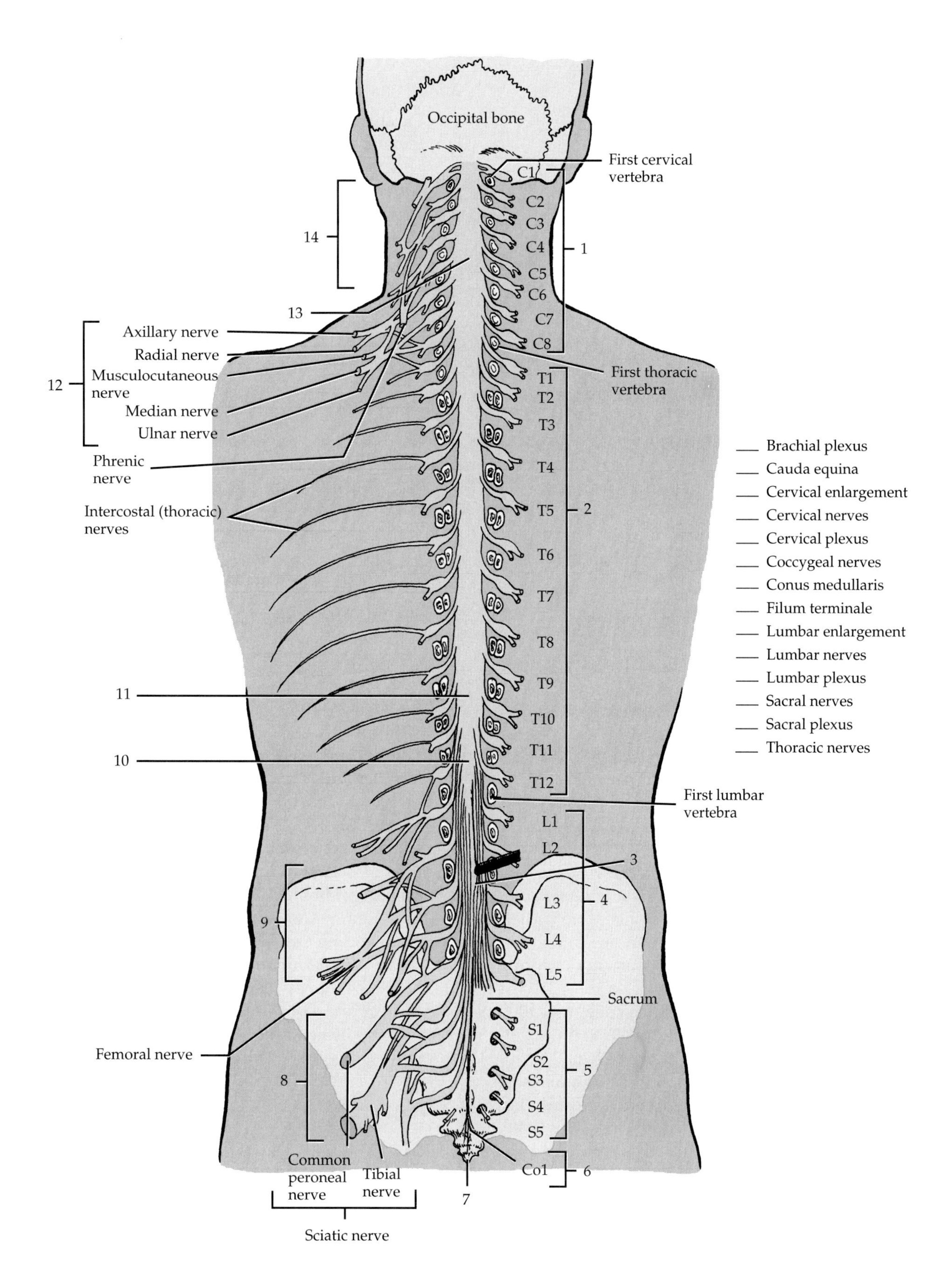

FIGURE 13.2 Spinal cord.

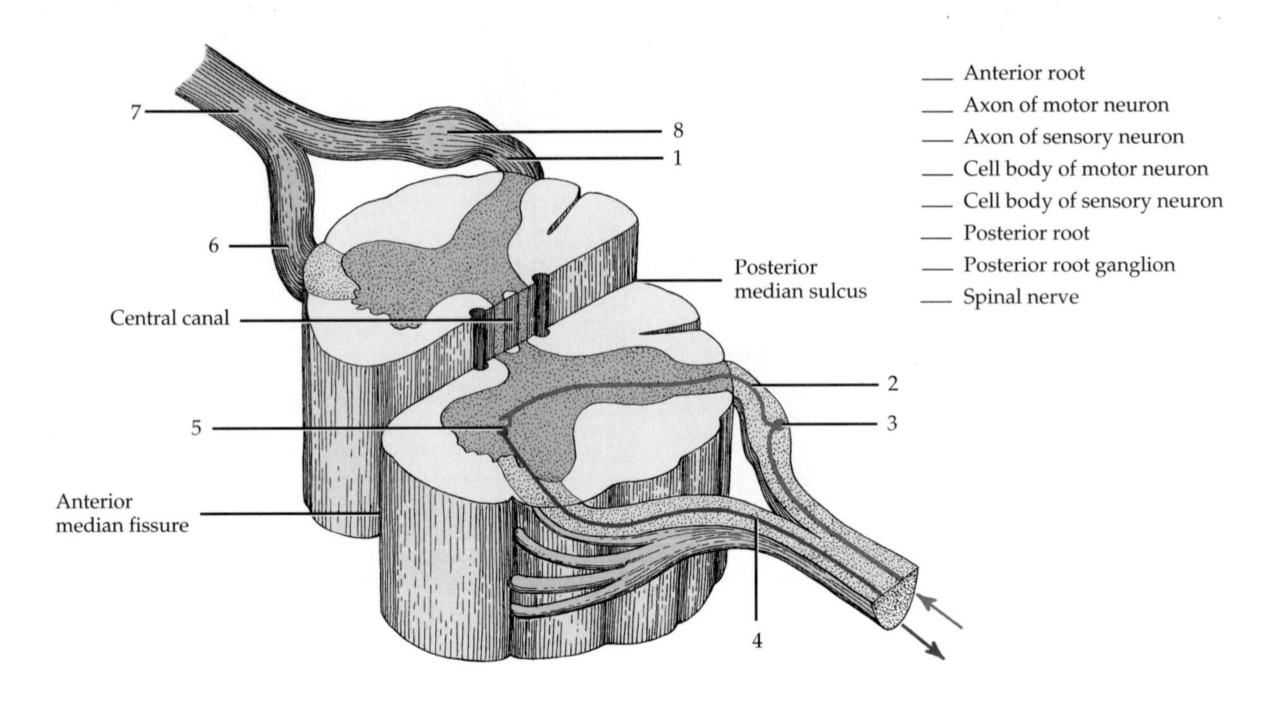

FIGURE 13.3 Spinal nerve attachments.

the posterior root contains sensory nerve fibers and the anterior root contains motor nerve fibers, all spinal nerves are ***mixed nerves.***

Spinal nerves are covered by several connective tissue layers. Individual nerve fibers within a nerve, whether myelinated or unmyelinated, are wrapped in a covering called the ***endoneurium*** (en'-dō-NOO-rē-um). Groups of fibers with their endoneurium are arranged in bundles called ***fascicles,*** and each bundle is wrapped in a covering called the ***perineurium*** (per'-i-NOO-rē-um). All the fascicles, in turn, are wrapped in a covering called the ***epineurium*** (ep'-i-NOO-rē-um). This is the outermost covering around the entire nerve.

Obtain a prepared slide of a nerve in transverse section and identify the fibers, endoneurium, perineurium, epineurium, and fascicles. Now label Figure 13.4.

6. Branches of Spinal Nerves

Shortly after leaving its intervertebral foramen, a spinal nerve divides into several branches called ***rami*** (singular is ***ramus*** [RĀ-mus]):

a. ***Dorsal ramus*** Innervates (supplies) deep muscles and skin of the dorsal surface of the back.

b. ***Ventral ramus*** Innervates superficial back muscles and all structures of the limbs and lateral and ventral trunk; except for thoracic nerves T2–T11, the ventral rami of the other spinal nerves form plexuses before innervating their structures.

c. ***Meningeal branch*** Innervates vertebrae, vertebral ligaments, blood vessels of the spinal cord, and meninges.

d. ***Rami communicantes*** (RĀ-mē ko-myoo-nē-KAN-tēz) Gray and white rami communicantes are components of the autonomic nervous system; they connect the ventral rami with sympathetic trunk ganglia.

7. Plexuses

The ventral rami of spinal nerves, except for T2–T11, do not go directly to body structures they supply. Instead, they join with adjacent nerves on either side of the body to form networks called ***plexuses*** (PLEK-sus-ēz; *plexus* = braid).

a. ***Cervical plexus*** (SER-vi-kul PLEK-sus) Formed by the ventral rami of the first four cervical nerves (C1–C4) with contributions from C5; one is located on each side of the neck alongside the

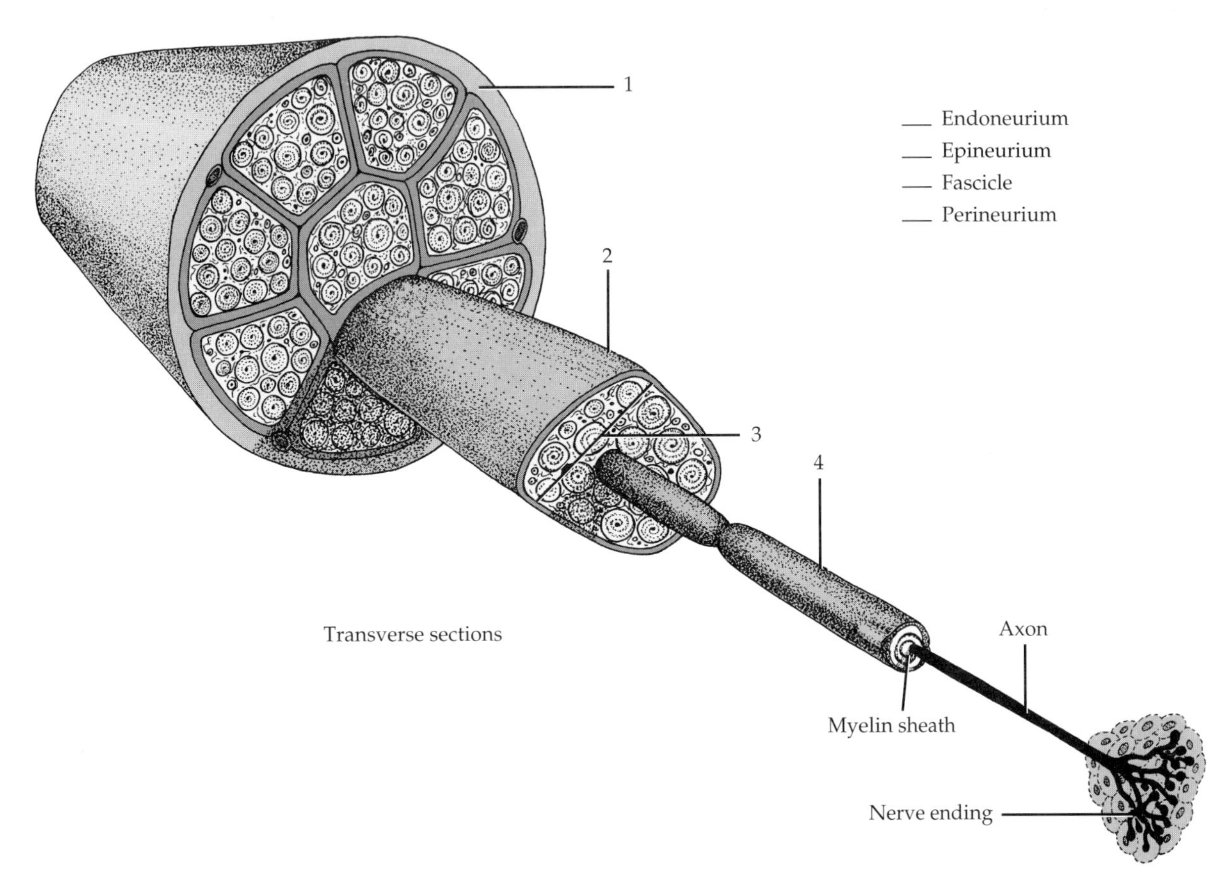

FIGURE 13.4 Coverings of a spinal nerve.

first four cervical vertebrae; the plexus supplies the skin and muscles of the head, neck, and upper part of shoulders.

Using your textbook as a guide, label the nerves of the cervical plexus in Figure 13.5.

b. ***Brachial*** (BRĀ-kē-al) ***plexus*** Formed by the ventral rami of spinal nerves C5–C8 and T1 with contributions from C4 and T2; each is located on either side of the last four cervical and first thoracic vertebrae and extends inferiorly and laterally, superior to the first rib posterior to the clavicle, and into the axilla; the plexus constitutes the entire nerve supply for the upper limbs and shoulder region.

Using your textbook as a guide, label the nerves of the brachial plexus in Figure 13.6.

c. ***Lumbar*** (LUM-bar) ***plexus*** Formed by the ventral rami of spinal nerves L1–L4; each is located on either side of the first four lumbar vertebrae posterior to the psoas major muscle and anterior to the quadratus lumborum muscle; the plexus supplies the anterolateral abdominal wall, external genitals, and part of the lower limbs.

Using your textbook as a guide, label the nerves of the lumbar plexus in Figure 13.7 on page 244.

d. ***Sacral*** (SĀ-kral) ***plexus*** Formed by the ventral rami of spinal nerves L4–L5 and S1–S4; each is located largely anterior to the sacrum; the plexus supplies the buttocks, perineum, and lower limbs.

Using your textbook as a guide, label the nerves of the sacral plexus in Figure 13.8 on page 244.

Label the cervical, brachial, lumbar, and sacral plexuses in Figure 13.2. Also note the names of some of the major peripheral nerves that arise from the plexuses.

For each nerve listed in Table 13.1 on page 246, indicate the plexus to which it belongs and the structure(s) it innervates.

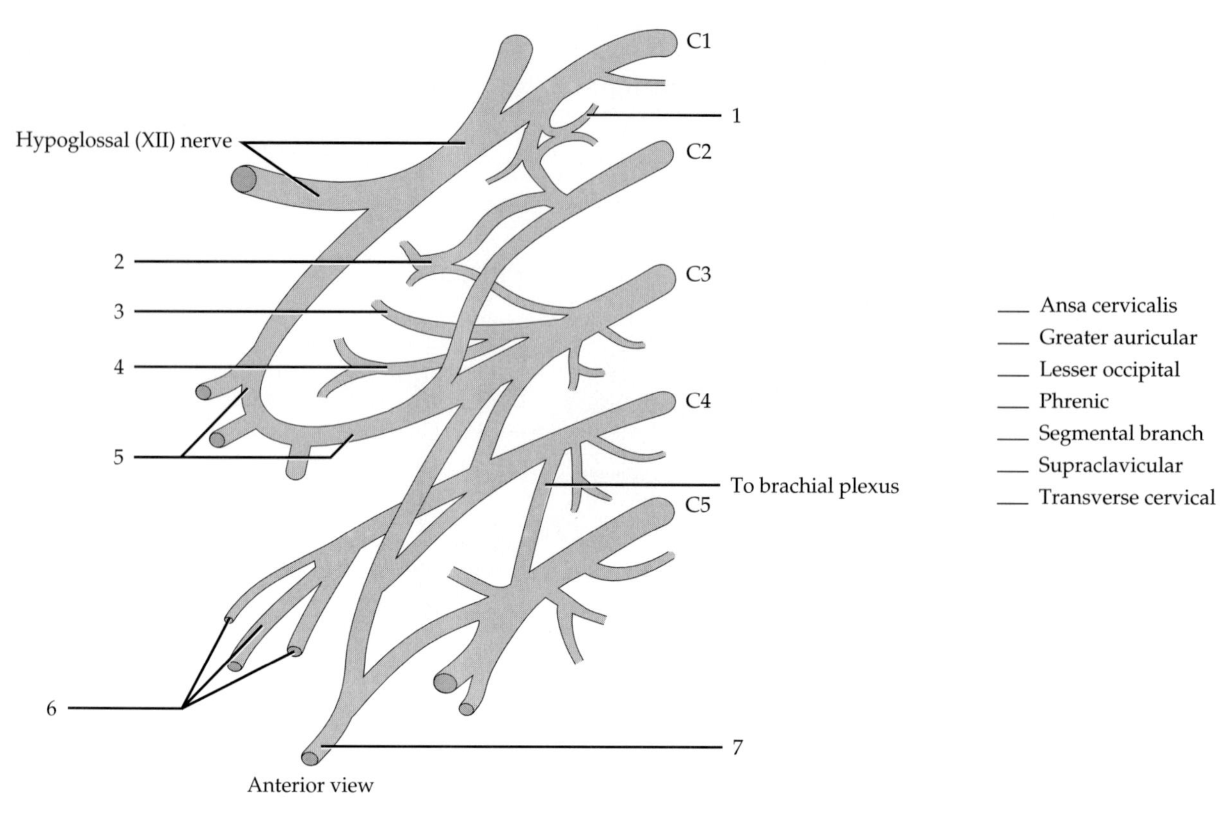

FIGURE 13.5 Cervical plexus.

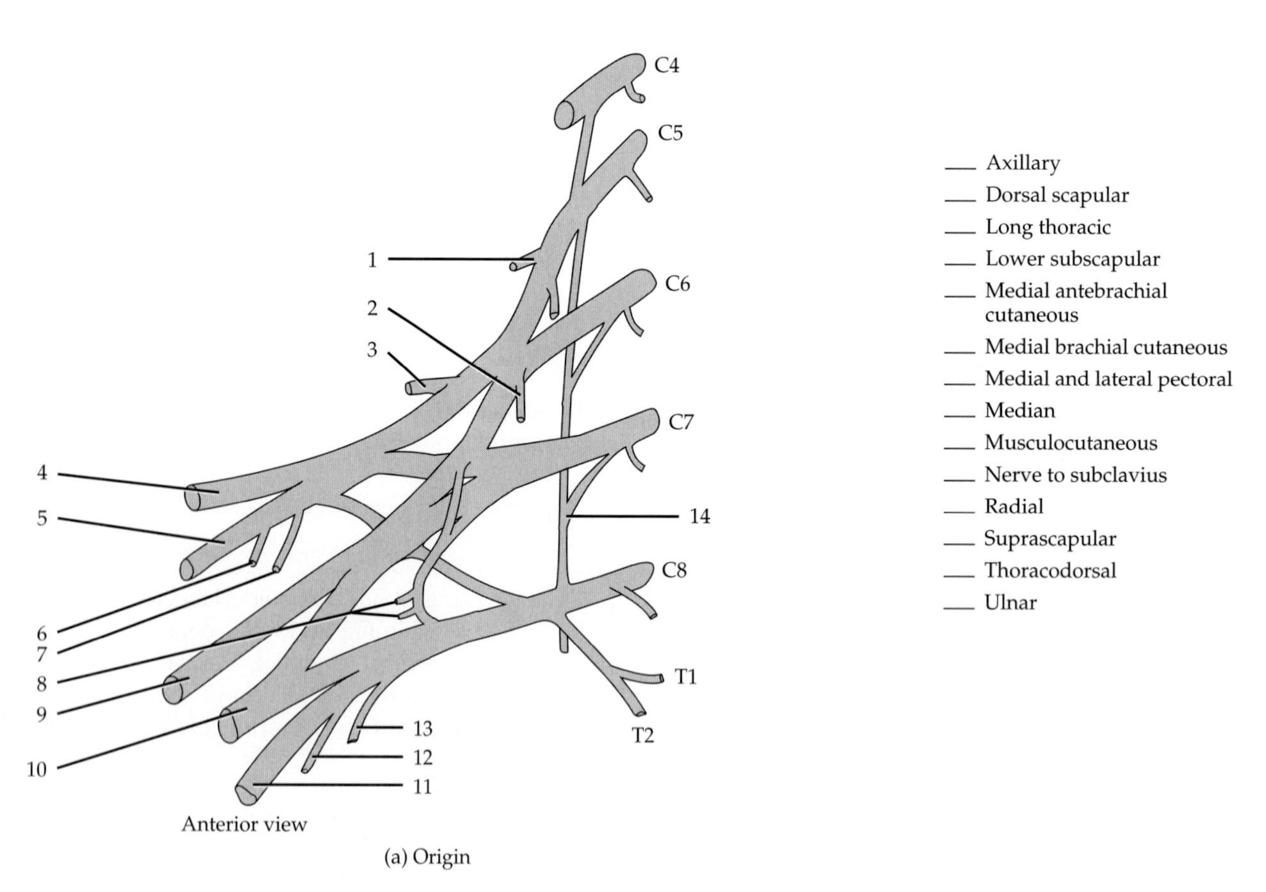

FIGURE 13.6 Brachial plexus.

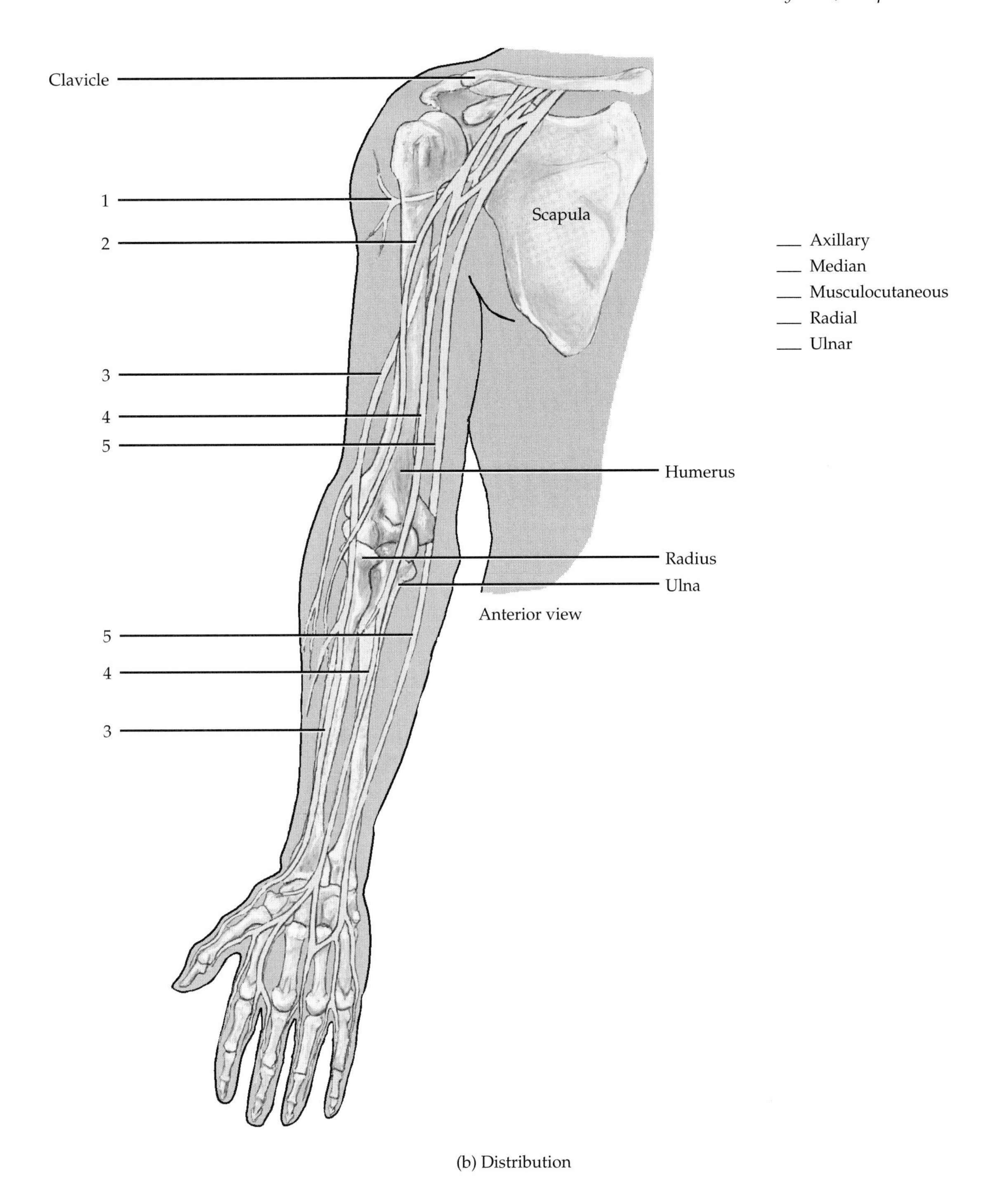

(b) Distribution

FIGURE 13.6 ***(Continued)*** Brachial plexus.

8. Spinal Cord Tracts

The vital function of conveying sensory and motor information to and from the brain is carried out by sensory (ascending) and motor (descending) tracts and pathways in the spinal cord. The names of the tracts and pathways indicate the white column in which the tract travels, where the cell bodies of the tract originate, and where the axons of the tract terminate. For example, the anterior spinothalamic tract is located in the ***anterior*** white column, it originates in the ***spinal cord,*** and it terminates in the ***thalamus*** of the brain. Since it conveys nerve impulses from the spinal cord upward to the brain, it is a sensory (ascending) tract.

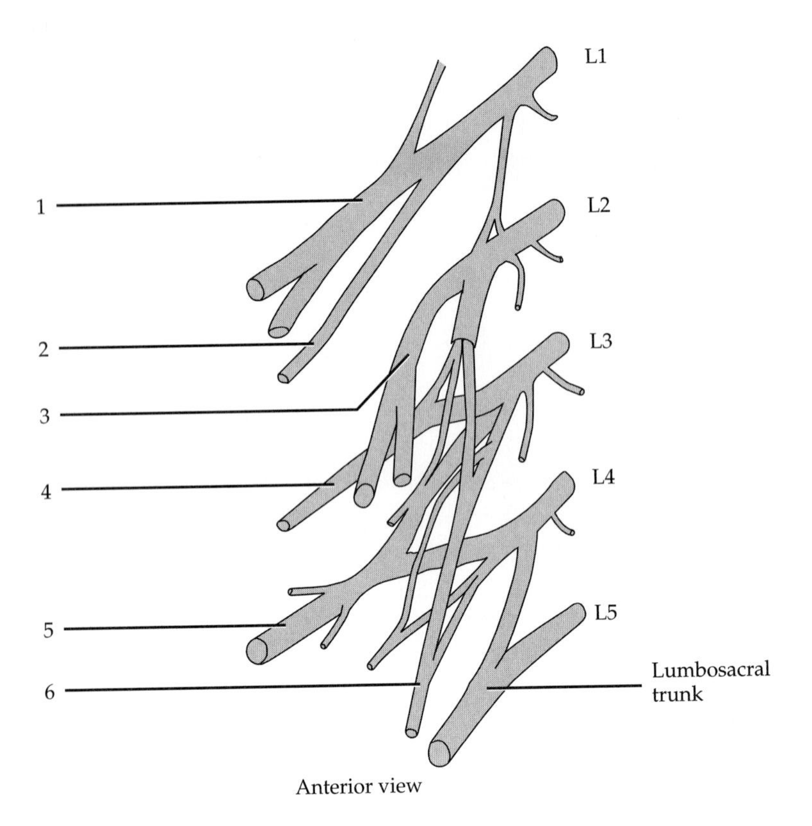

___ Femoral
___ Genitofemoral
___ Iliohypogastric
___ Ilioinguinal
___ Lateral femoral cutaneous
___ Obturator

FIGURE 13.7 Lumbar plexus.

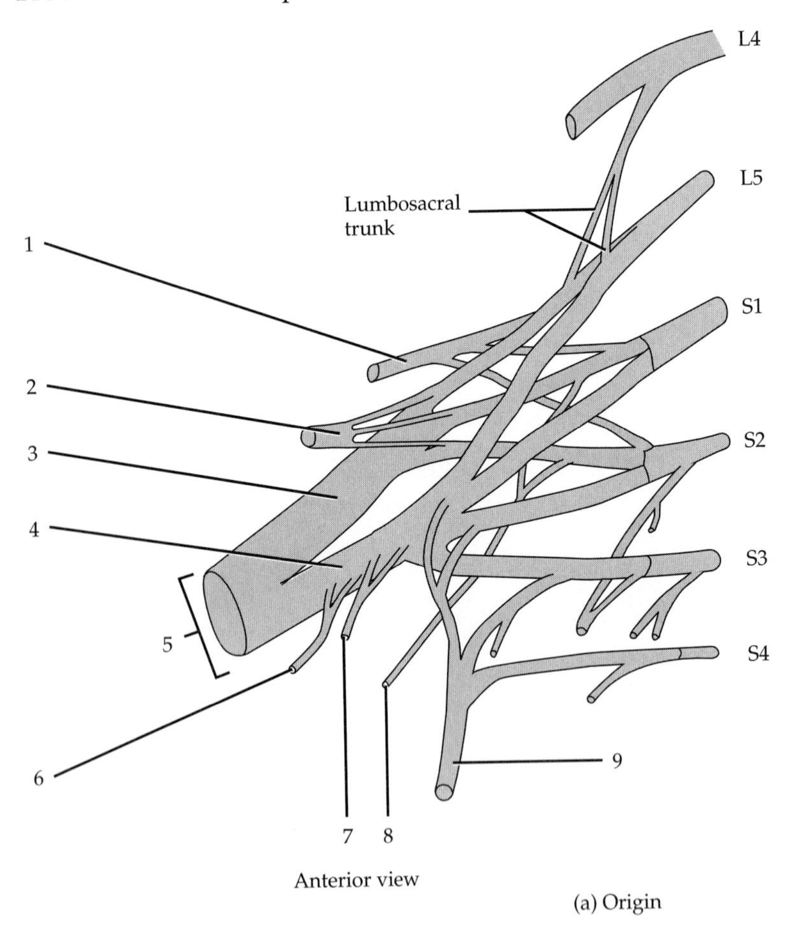

___ Common peroneal
___ Inferior gluteal
___ Nerve to obturator internus and superior gemellus
___ Nerve to quadratus femoris and inferior gemellus
___ Posterior femoral cutaneous
___ Pudendal
___ Sciatic
___ Superior gluteal
___ Tibial

FIGURE 13.8 Sacral plexus.

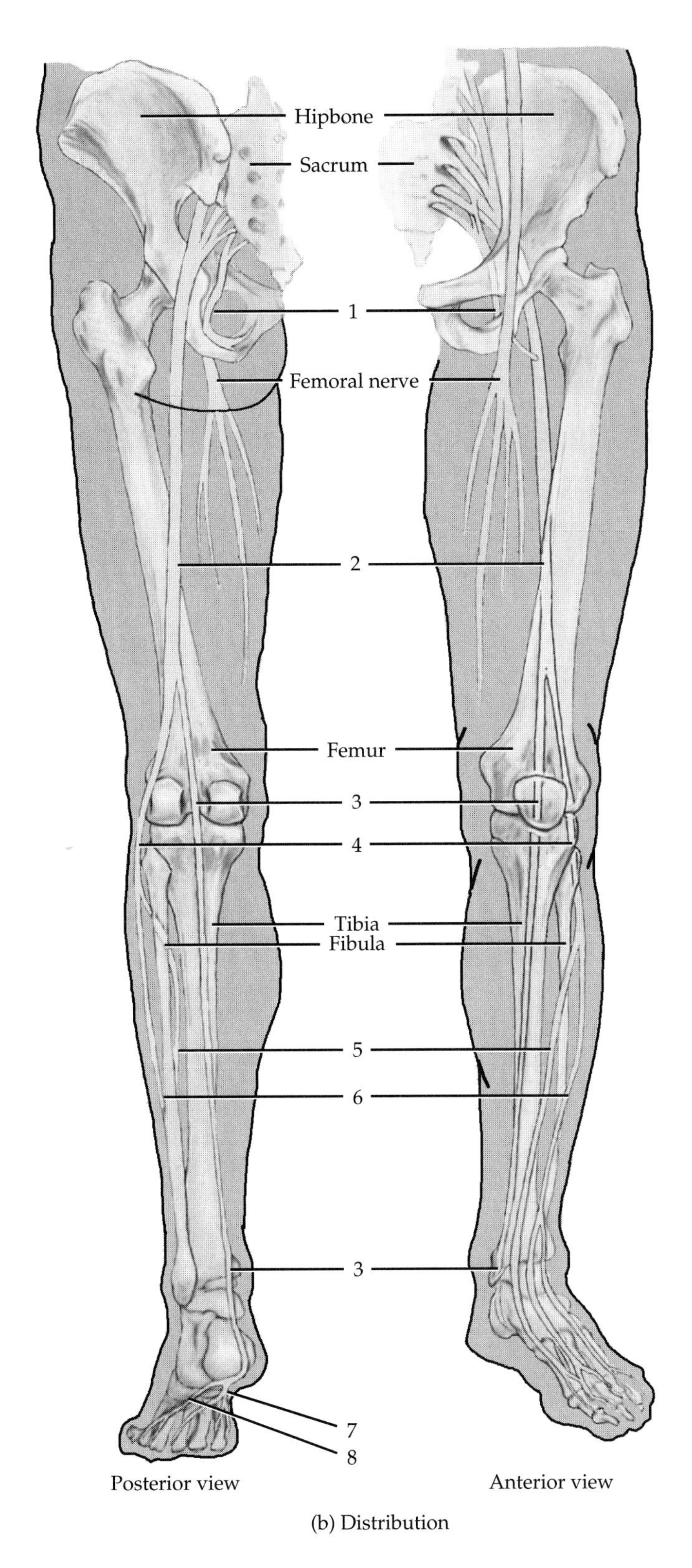

___ Common peroneal
___ Deep peroneal
___ Lateral plantar
___ Medial plantar
___ Pudendal
___ Sciatic
___ Superficial peroneal
___ Tibial

FIGURE 13.8 ***(Continued)*** Sacral plexus.

TABLE 13.1 Nerves of Plexuses and Innervations

Nerve	Plexus	Innervation
Musculocutaneous (mus'-kyoo-lō-kyoo-TĀN-ē-us)		
Femoral (FEM-or-al)		
Phrenic (FREN-ik)		
Pudendal (pyoo-DEN-dal)		
Axillary (AK-si-lar-ē)		
Transverse cervical (SER-vi-kul)		
Radial (RĀ-dē-al)		
Obturator (OB-too-rā-tor)		
Tibial (TIB-ē-al)		
Thoracodorsal (thō-RA-kō-dor-sal)		
Perforating cutaneous (PER-fō-rā-ting kyoo'-TĀ-nē-us)		
Ulnar (UL-nar)		
Long thoracic (thō-RAS-ik)		
Median (MĒ-dē-an)		
Iliohypogastric (il'-ē-ō-hī-pō-GAS-trik)		
Deep peroneal (per'-ō-NĒ-al)		
Ansa cervicalis (AN-sa ser-vi-KAL-is)		
Sciatic (sī-AT-ik)		

a. SOMATIC SENSORY PATHWAYS

Somatic sensory pathways from receptors to the cerebral cortex involve three-neuron sets. Axon collaterals (branches) of somatic sensory neurons simultaneously carry signals into the cerebellum and the reticular formation of the brain stem.

1. ***First-order neurons*** Carry signals from the somatic receptors into either the brain stem or spinal cord. From the face, mouth, teeth, and eyes, somatic sensory impulses propagate along *cranial nerves* into the brain stem. From the back of the head, neck, and body, somatic sensory impulses propagate along *spinal nerves* into the spinal cord.
2. ***Second-order neurons*** Carry signals from the spinal cord and brain stem to the thalamus. Axons of second-order neurons cross over (decussate) to the opposite side in the spinal cord or brain stem before ascending to the thalamus.
3. ***Third-order neurons*** Project from the thalamus to the primary somatosensory area of the cortex (postcentral gyrus; see Figure 13.14), where conscious perception of the sensations results.

There are two general pathways by which somatic sensory signals entering the spinal cord ascend to the cerebral cortex: the ***posterior column–medial lemniscus pathway*** and the ***anterolateral (spinothalamic) pathways.***

1. Posterior Column–Medial Lemniscus Pathway to the Cortex Nerve impulses for conscious proprioception and most tactile sensations ascend to the cortex along a common pathway formed by three-neuron sets. First-order neurons extend from sensory receptors into the spinal cord and up to the medulla oblongata on the same side of the body. The cell bodies of these first-order neurons are in the posterior (dorsal) root ganglia of spinal nerves. Their axons form the ***posterior column*** (***fasciculus gracilis;*** fa-SIK-yoo-lus gras-I-lus and ***fasciculus cuneatus;*** kyoo-nē-ĀT-us) in the spinal cord. The axon terminals form synapses with second-order neurons in the medulla. The cell body of a second-order neuron is located in the nucleus cuneatus (which receives input conducted along axons in the fasciculus cuneatus from the neck, upper limbs, and upper chest) or nucleus gracilis (which receives input conducted along axons in the fasciculus gracilis from the trunk and lower limbs). The axon of the second-order neuron crosses to the opposite side of the medulla and enters the ***medial lemniscus,*** a projection tract that extends from the medulla to the thalamus. In the thalamus, the axon terminals of second-order neurons synapse with third-order neurons, which project their axons to the somatosensory area of the cerebral cortex.

Impulses conducted along the posterior column-medial lemniscus pathway give rise to several highly evolved and refined sensations. These are:

Discriminative touch The ability to recognize the exact location of a light touch and to make two-point discriminations.
Stereognosis The ability to recognize by feel the size, shape, and texture of an object. Examples are identifying (with closed eyes) a paper clip put into your hand or reading braille.
Proprioception The awareness of the precise position of body parts, and ***kinesthesia,*** the awareness of directions of movement.
Weight discrimination The ability to assess the weight of an object.
Vibratory sensations The ability to sense rapidly fluctuating touch.

2. Anterolateral (Spinothalamic) Pathways to the Cortex The ***anterolateral*** (***spinothalamic;*** spī-nō-THAL-am-ik) ***pathways*** carry mainly pain and temperature impulses. In addition, they relay the sensations of tickle and itch and some tactile impulses, which give rise to a very crude, not well-localized touch or pressure sensation. Like the posterior column–medial lemniscus pathway, the anterolateral pathways are also composed of three-neuron sets. The first-order neuron connects a receptor of the neck, trunk, or limbs with the spinal cord. The cell body of the first-order neuron is in the posterior root ganglion. The axon of the first-order neuron synapses with the second-order neuron, which is located in the posterior gray horn of the spinal cord. The axon of the second-order neuron continues to the opposite side of the spinal cord and passes superiorly to the brain stem in either the ***lateral spinothalamic tract*** or the ***anterior spinothalamic tract.*** The axon from the second-order neuron ends in the thalamus. There, it synapses with the third-order neuron. The axon of the third-order neuron projects to the somatosensory area of the cerebral cortex. The lateral spinothalamic tract conveys sensory impulses for pain and temperature whereas the anterior spinothalamic tract conveys impulses for tickle, itch, crude touch, and pressure.

3. Somatic Sensory Pathways to the Cerebellum Two tracts in the spinal cord, the ***posterior spinocerebellar*** (spī'-nō-ser-e-BEL-ar) ***tract*** and the

anterior spinocerebellar tract, are major routes for the subconscious proprioceptive input to reach the cerebellum. Sensory input conveyed to the cerebellum along these two pathways is critical for posture, balance, and coordination of skilled movements.

b. SOMATIC MOTOR PATHWAYS

The most direct somatic motor pathways extend from the cerebral cortex to skeletal muscles. Other pathways are less direct and include synapses in the basal ganglia, thalamus, reticular formation, and cerebellum.

1. Direct Pathways Voluntary motor impulses are propagated from the motor cortex to voluntary motor neurons (somatic motor neurons) that innervate skeletal muscles via the ***direct*** or ***pyramidal*** (pi-RAM-i-dal) ***pathways.*** The simplest of these pathways consists of sets of two neurons, upper motor neurons and lower motor neurons. About one million pyramidal-shaped cell bodies of direct pathway ***upper motor neurons (UMNs)*** are in the cortex. Their axons descend through the internal capsule of the cerebrum. In the medulla oblongata, the axon bundles form the ventral bulges known as the ***pyramids.*** About 90% of these axons also cross (decussate) to the opposite side in the medulla oblongata. They terminate in nuclei of cranial nerves or in the anterior gray horn of the spinal cord. ***Lower motor neurons (LMNs)*** extend from the motor nuclei of nine cranial nerves to muscles of the face and head and from the anterior horn of each spinal cord segment to skeletal muscle fibers of the trunk and limbs. Close to their termination point, most upper motor neurons synapse with an association neuron, which, in turn, synapses with a lower motor neuron. A few upper motor neurons synapse directly with lower motor neurons.

The direct pathways convey impulses from the cortex that result in precise, voluntary movements. The main parts of the body governed by the direct pathways are the face, vocal cords (for speech), and hands and feet of the limbs. They channel nerve impulses into three tracts:

a. ***Lateral corticospinal*** (kor'-ti-kō-SPĪ-nal) ***tracts*** These pathways begin in the right and left motor cortex and descend through the ***internal capsule*** of the cerebrum and through the cerebral peduncle of the midbrain and the pons on the same side. About 90% of the axons of upper motor neurons cross over to the opposite side in the medulla oblongata. These axons form the lateral corticospinal tracts in the right and left lateral white columns of the spinal cord. Thus the motor cortex of the right side of the brain controls muscles on the left side of the body, and vice versa. The lower motor neurons then receive input from both upper motor neurons and association neurons. Axons of lower motor neurons (somatic motor neurons) exit all levels of the spinal cord via the anterior roots of spinal nerves and terminate in skeletal muscles. These motor neurons control skilled movements of the distal portions of the limbs.
b. ***Corticobulbar*** (kor'-ti-kō-BUL-bar) ***tracts*** The axons of upper motor neurons of these tracts accompany the corticospinal tracts from the motor cortex through the internal capsule to the brain stem. There some cross whereas others remain uncrossed. They terminate in the nuclei of nine pairs of cranial nerves in the pons and medulla: the oculomotor (III), trochlear (IV), trigeminal (V), abducens (VI), facial (VII), glossopharyngeal (IX), vagus (X), accessory (XI), and hypoglossal (XII). The lower motor neurons of cranial nerves convey impulses that control voluntary movements of the eyes, tongue, and neck; chewing; facial expression; and speech.
c. ***Anterior corticospinal tracts*** About 10% of the axons of upper motor neurons do not cross in the medulla oblongata. They pass through the medulla oblongata, descend on the same side, and form the anterior corticospinal tracts in the right and left anterior white columns. At several spinal cord levels, some of the axons of these upper motor neurons cross. After crossing to the opposite side, they synapse with association or lower motor neurons in the anterior gray horn of the spinal cord. Axons of these lower motor neurons exit the cervical and upper thoracic segments of the cord via the anterior roots of spinal nerves. They terminate in skeletal muscles that control movements of the neck and part of the trunk, thus coordinating movements of the axial skeleton.

2. Indirect Pathways The ***indirect (extrapyramidal) pathways*** include all descending (motor) tracts other than the corticospinal and corticobulbar tracts. Nerve impulses conducted along the indirect pathways follow complex, polysynaptic circuits that involve the motor cortex, basal ganglia, limbic system, thalamus, cerebellum, reticular formation, and nuclei in the brain stem. Axons of upper motor neurons that carry motor signals from the indirect pathways descend from various nuclei of the brain stem into five major tracts of the spinal cord and terminate on association neurons or lower motor neurons.

Lower motor neurons receive both excitatory and inhibitory input from many presynaptic neurons in both direct and indirect pathways, an example of convergence. For this reason, lower motor neurons are also called the ***final common pathway.*** Most nerve impulses from the brain are conveyed to association neurons before being received by lower motor neurons. The sum total of the input from upper motor neurons and association neurons determines the final response of the lower motor neuron. It is not just a simple matter of the brain sending an impulse and the muscle always contracting.

The five major tracts of the indirect pathways are the ***rubrospinal*** (ROO-brō-spī-nal), ***tectospinal*** (TEK-tō-spī-nal), ***vestibulospinal*** (ves-TIB-yoo-lō-spī-nal), ***lateral reticulospinal*** (re-TIK-yoo-lō-spī-nal), and ***medial reticulospinal.***

Label the sensory (ascending) and motor (descending) tracts shown in Figure 13.9.

Indicate the function of each sensory and motor tract listed in Table 13.2.

B. BRAIN

1. Parts

The brain may be divided into four principal parts: (1) ***brain stem,*** which consists of the medulla oblongata, pons, and midbrain; (2) ***diencephalon*** (dī-en-SEF-a-lon; *enkephalos* = brain), which consists primarily of the thalamus and hypothalamus; (3) ***cerebellum,*** (ser′-e-BEL-um) which is posterior to the brain stem; and (4) ***cerebrum,*** (se-RĒ-brum) which is superior to the brain stem and comprises about seven-eighths of the total weight of the brain.

Examine a model and preserved specimen of the brain and identify the parts just described. Then refer to Figure 13.10 on page 251 and label the parts of the brain.

2. Meninges

As in the spinal cord, the brain is protected by ***meninges.*** The cranial meninges are continuous with the spinal meninges. The cranial meninges are the superficial ***dura mater,*** the middle ***arachnoid,*** and the deep ***pia mater.*** The cranial dura mater consists of two layers called the *periosteal layer* (adheres to the cranial bones and serves as a periosteum) and the *meningeal layer* (thinner, inner layer that corresponds to the spinal dura mater).

Refer to Figure 13.11 on page 252 and label all of the meninges.

3. Cerebrospinal Fluid (CSF)

The central nervous system is nourished and protected by ***cerebrospinal fluid (CSF).*** The fluid circulates through the subarachnoid space around the brain and spinal cord and through the ***ventricles*** (VEN-tri-kuls; *ventriculus* = little belly or cavity) of the brain. The ventricles are cavities in the brain that communicate with each other, with the central canal of the spinal cord, and with the subarachnoid space.

Cerebrospinal fluid is formed primarily by filtration and secretion from networks of capillaries and ependymal cells in the ventricles, called ***choroid*** (KŌ-royd; *chorion* = membrane) ***plexuses*** (see Figure 13.11). Each of the two ***lateral ventricles*** is located within a hemisphere (side) of the cerebrum under the corpus callosum. The fluid formed in the choroid plexuses of the lateral ventricles circulates through an opening called the ***interventricular foramen*** into the third ventricle. The ***third ventricle*** is a slit between and inferior to the right and left halves of the thalamus and between the lateral ventricles. More fluid is added by the choroid plexus of the third ventricle. Then the fluid circulates through a canal-like structure called the ***cerebral aqueduct*** (AK-we-dukt) into the fourth ventricle. The ***fourth ventricle*** lies between the inferior brain stem and the cerebellum. More fluid is added by the choroid plexus of the fourth ventricle. The roof of the fourth ventricle has three openings: one ***median aperture*** (AP-er-chur) and two ***lateral apertures.*** The fluid circulates through the apertures into the subarachnoid space around the posterior portion of the brain and inferiorly through the central canal of the spinal cord to the subarachnoid space around the posterior surface of the spinal cord, up the anterior surface of the spinal cord, and around the anterior part of the brain. Most of the cerebrospinal fluid is absorbed into a vein called the superior sagittal sinus through its arachnoid villi. Normally, cerebrospinal fluid is absorbed as rapidly as it is formed.

Refer to Figure 13.11 on page 252 and label the choroid plexus of the lateral ventricle, lateral ventricle, interventricular foramen, choroid plexus of third ventricle, third ventricle, cerebral aqueduct, choroid plexus of fourth ventricle, fourth ventricle, median aperture, lateral aperture, subarachnoid space of spinal cord, superior sagittal sinus, and arachnoid villus.

Note the arrows in Figure 13.11, which indicate the path taken by cerebrospinal fluid. With the aid of your textbook, starting at the choroid plexus of the lateral ventricle and ending at the superior sagittal sinus, see if you can follow the remaining path of the fluid.

Now complete Figure 13.12 on page 253.

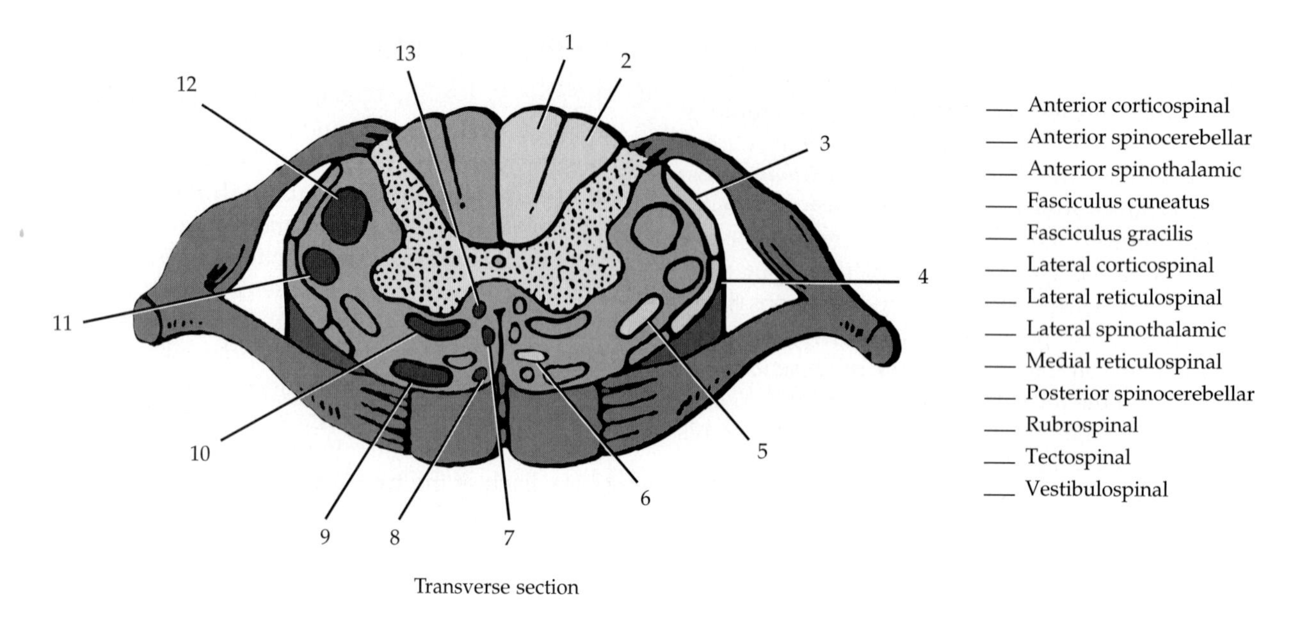

FIGURE 13.9 Selected sensory (ascending) and motor (descending) tracts of the spinal cord.

TABLE 13.2 Sensory and Motor Tracts and Their Functions

Sensory (ascending) tracts	Function
Posterior column	
Lateral spinothalamic	
Anterior spinothalamic	
Posterior spinocerebellar	
Anterior spinocerebellar	

Motor (descending) tracts	Function
Direct (pyramidal)	
Lateral corticospinal	
Anterior corticospinal	
Corticobulbar	
Indirect (extrapyramidal)	
Rubrospinal	
Tectospinal	
Vestibulospinal	
Lateral reticulospinal	
Medial reticulospinal	

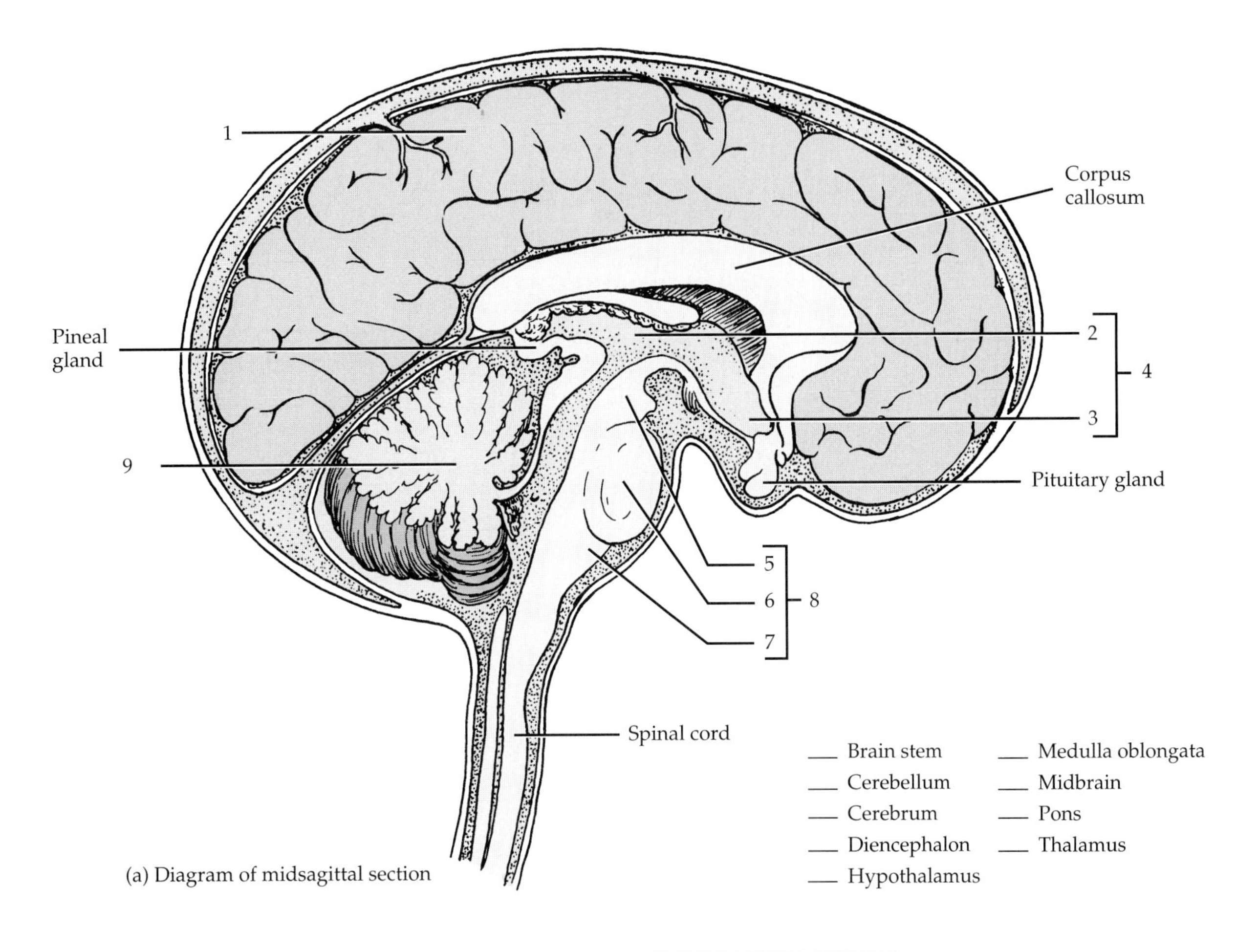

(a) Diagram of midsagittal section

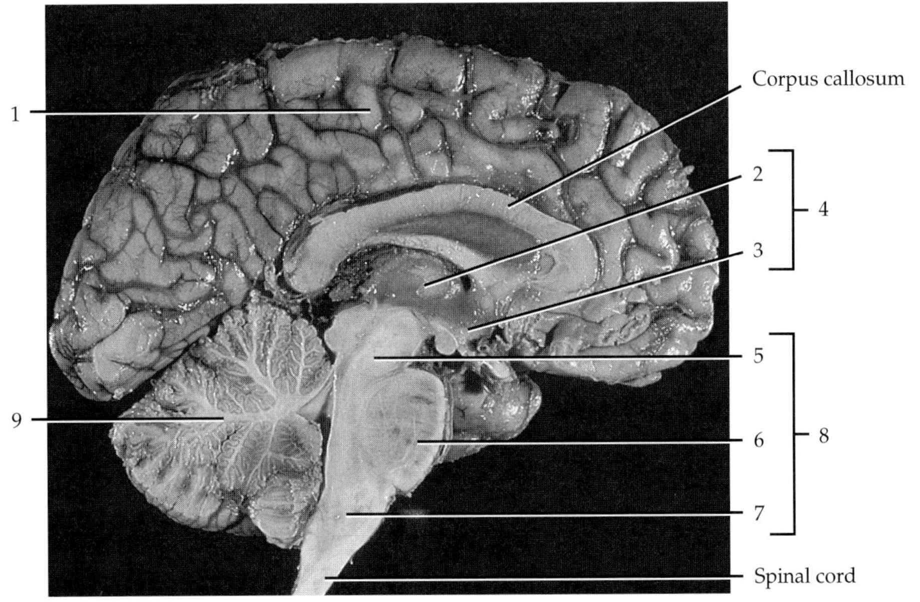

(b) Photograph of midsagittal section

FIGURE 13.10 Principal parts of the brain.

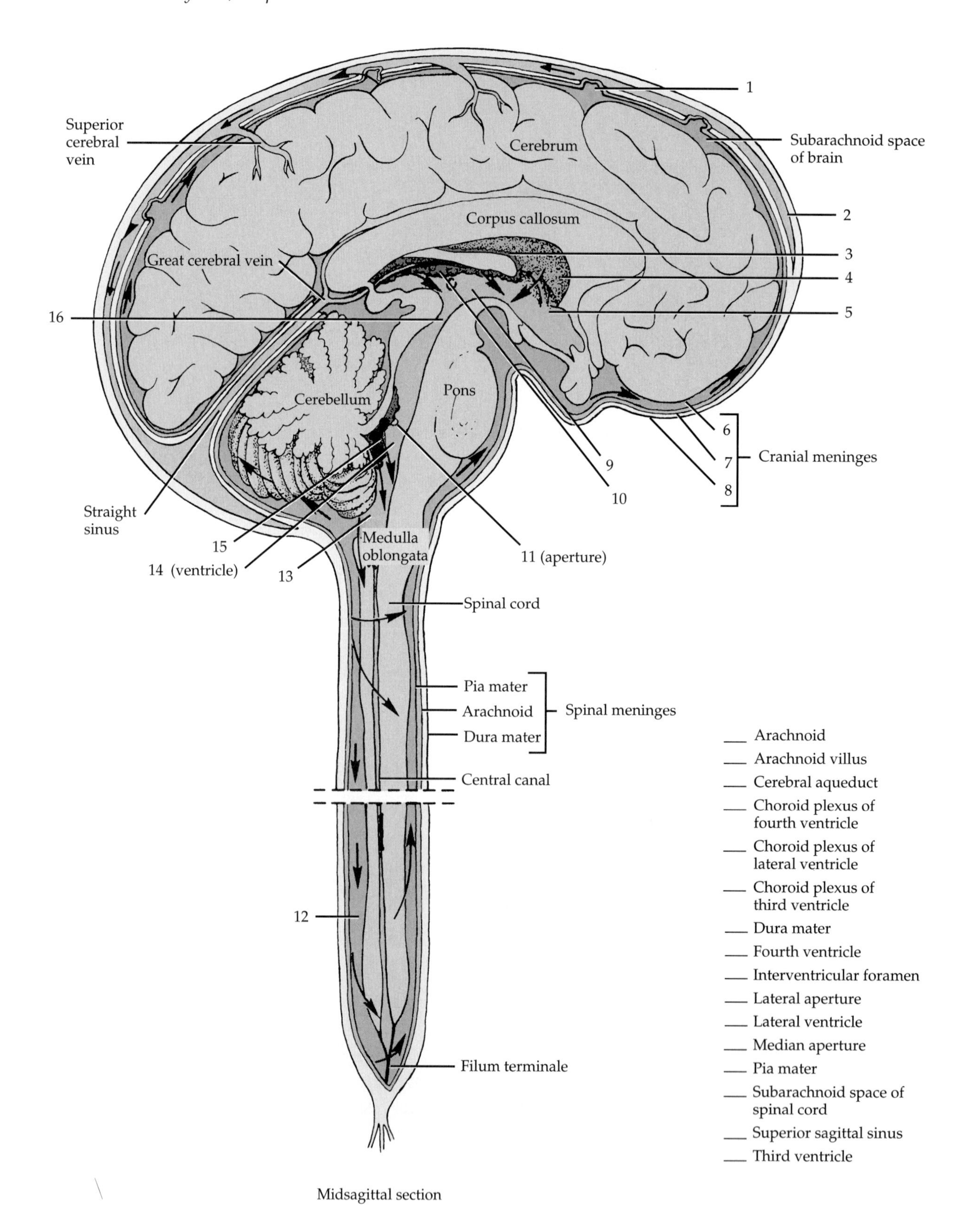

___ Arachnoid
___ Arachnoid villus
___ Cerebral aqueduct
___ Choroid plexus of fourth ventricle
___ Choroid plexus of lateral ventricle
___ Choroid plexus of third ventricle
___ Dura mater
___ Fourth ventricle
___ Interventricular foramen
___ Lateral aperture
___ Lateral ventricle
___ Median aperture
___ Pia mater
___ Subarachnoid space of spinal cord
___ Superior sagittal sinus
___ Third ventricle

FIGURE 13.11 Brain and spinal cord.

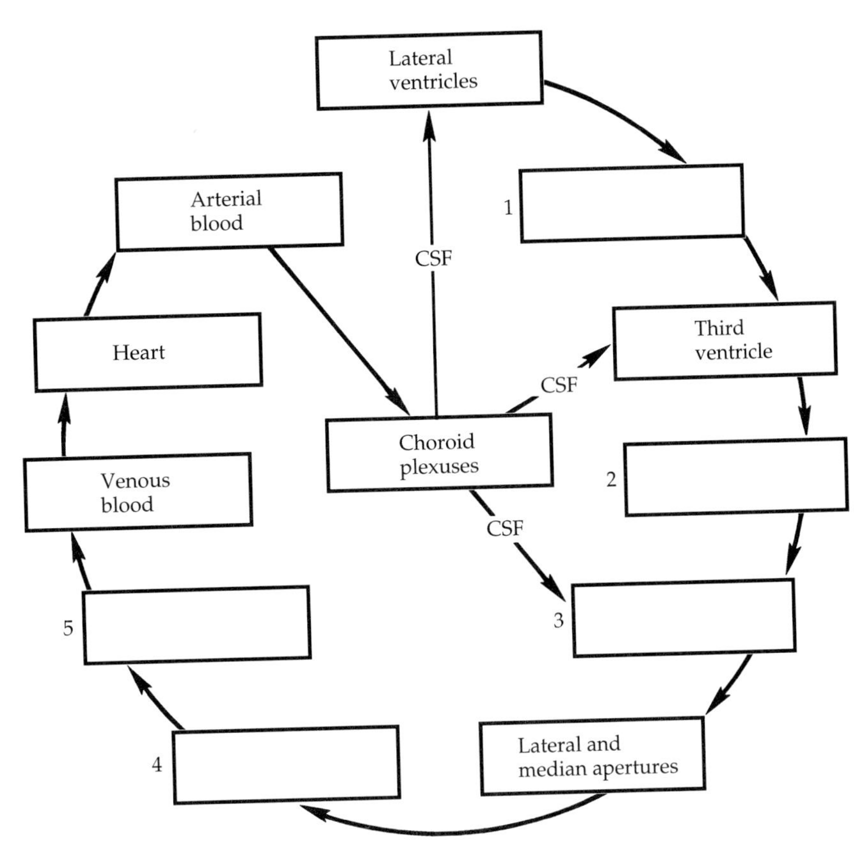

FIGURE 13.12 Formation, circulation, and absorption of cerebrospinal fluid (CSF).

4. Medulla Oblongata

The *medulla oblongata* (me-DULL-la ob'-long-GA-ta), or just simply *medulla,* is a continuation of the superior part of the spinal cord and forms the inferior part of the brain stem. The medulla contains all sensory and motor tracts that communicate between the spinal cord and various parts of the brain. On the ventral side of the medulla are two roughly triangular bulges called *pyramids.* They contain the largest motor tracts that run from the cerebral cortex to the spinal cord. Most fibers in the left pyramid cross to the right side of the spinal cord and most fibers in the right pyramid cross to the left side of the spinal cord. This crossing is called the *decussation* (dē'-ku-SĀ-shun) *of pyramids* and explains why motor areas of one side of the cerebral cortex control muscular movements on the opposite side of the body. The medulla contains three vital reflex centers called the *cardiac center, respiratory center,* and *vasomotor center.* The dorsal side of the medulla contains two pairs of prominent nuclei: *nucleus gracilis* and *nucleus cuneatus.* Most sensory impulses initiated on one side of the body cross to the thalamus on the opposite side either in the spinal cord or in these medullary nuclei. The medulla also contains the nuclei of origin of several cranial nerves. These are the cochlear and vestibular branches of the vestibulocochlear (VIII) nerves, glossopharyngeal (IX) nerves, vagus (X) nerves, spinal portions of the accessory (XI) nerves, and hypoglossal (XII) nerves. Examine a model or specimen of the brain and identify the parts of the medulla. Then refer to Figure 13.10 and locate the medulla.

5. Pons

The *pons* (*pons* = bridge) lies directly superior the medulla oblongata and anterior to the cerebellum. The pons contains fibers that connect parts of the cerebellum and medulla with the cerebrum. The pons contains the nuclei or origin of the following pairs of cranial nerves: trigeminal (V) nerves, abducens (VI) nerves, facial (VII) nerves, and vestibular branch of the vestibulocochlear (VIII) nerves. Other important nuclei in the pons are the *pneumotaxic* (noo-mō-TAK-sik) *area* and *apneustic* (ap-NOO-stik) *area* that help control respiration.

Identify the pons on a model or specimen of the brain. Locate the pons in Figure 13.10.

6. Midbrain

The *midbrain* extends from the pons to the inferior portion of the cerebrum. The ventral portion of the midbrain contains the paired *cerebral peduncles* (pe-DUNG-kulz; *pedunculus* = stemlike portion), which connect the superior parts of the brain to inferior parts of the brain and spinal cord. The dorsal part of the midbrain contains four rounded elevations called the *corpora quadrigemina* (KOR-por-ra kwad-ri-JEM-in-a; *corpus* = body; *quadrigeminus* = group of four). Two of the elevations, the *superior colliculi* (ko-LIK-yoo-lī) serve as reflex centers for movements of the eyeballs and the head in response to visual and other stimuli. *Colliculus,* which is singular, means small mound. The other two elevations, the *inferior colliculi,* serve as reflex centers for movements of the head and trunk in response to auditory stimuli. The midbrain contains the nuclei of origin for two pairs of cranial nerves: oculomotor (III) and trochlear (IV).

Identify the parts of the midbrain on a model or specimen of the brain. Locate the midbrain in Figure 13.10.

7. Thalamus

The *thalamus* (THAL-a-mus; *thalamos* = inner chamber) is a large oval structure that comprises about 80% of the diencephalon. It is located superior to the midbrain and consists of two masses of gray matter organized into nuclei and covered by a layer of white matter. The two masses are joined by a bridge of gray matter called the *intermediate mass.* The thalamus contains numerous nuclei that serve as relay stations for all sensory impulses. The most prominent are the *medial geniculate* (je-NIK-yoo-lāt) *nuclei* (hearing), *lateral geniculate nuclei* (vision), *ventral posterior nuclei* (general sensations and taste), *ventral lateral nuclei* (voluntary motor actions), and the *ventral anterior nuclei* (voluntary motor actions and arousal). The thalamus also allows crude appreciation of some sensations, such as pain, temperature, and pressure. Precise localization of such sensations depends on nerve impulses being relayed from the thalamus to the cerebral cortex.

Identify the thalamic nuclei on a model or specimen of the brain. Then refer to Figure 13.13 and label the nuclei.

8. Hypothalamus

The *hypothalamus* (*hypo* = under) is located inferior to the thalamus and forms the floor and part of the wall of the third ventricle. Among the functions served by the hypothalamus are the control and integration of the activities of the autonomic nervous system and parts of the endocrine system (pituitary gland) and the control of body temperature. The hypothalamus also assumes a role in feelings of rage and aggression, food intake, thirst, and the waking state and sleep patterns.

Identify the hypothalamus on a model or specimen of the brain. Locate the hypothalamus in Figure 13.10.

9. Cerebrum

The *cerebrum* is the largest portion of the brain and is supported on the brain stem. Its outer surface consists of gray matter and is called the *cerebral cortex* (*cortex* = rind or bark). Beneath the cerebral cortex is the cerebral white matter. The upfolds of the cerebral cortex are termed *gyri* (JĪ-rī) or *convolutions,* the deep downfolds are termed *fissures,* and the shallow downfolds are termed *sulci* (SUL-sī).

The *longitudinal fissure* separates the cerebrum into right and left halves called *hemispheres.* Each hemisphere is further divided into lobes by sulci or fissures. The *central sulcus* (SUL-kus) separates the *frontal lobe* from the *parietal lobe.* The *lateral cerebral sulcus* separates the frontal lobe from the *temporal lobe.* The *parietooccipital sulcus* separates the *parietal lobe* from the *occipital lobe.* Another prominent fissure, the *transverse fissure,* separates the cerebrum from the cerebellum. Another lobe of the cerebrum, the *insula,* lies deep within the lateral cerebral fissure under the parietal, frontal, and temporal lobes. It cannot be seen in external view. Two important gyri on either side of the *central sulcus* are the *precentral gyrus* and the *postcentral gyrus.* The olfactory (I) and optic (II) cranial nerves are associated with the cerebrum.

Examine a model of the brain and identify the parts of the cerebrum just described. Refer to Figure 13.14 on page 256 and label the lobes.

10. Functional Areas of Cerebral Cortex

The functions of the cerebrum are numerous and complex. In a general way, the cerebral cortex can be divided into sensory, motor, and association areas. The *sensory areas* interpret sensory impulses, the

motor areas control muscular movement, and the ***association areas*** are concerned with emotional and intellectual processes.

The principal sensory and motor areas of the cerebral cortex are indicated by numbers based on K. Brodmann's map of the cerebral cortex. His map, first published in 1909, attempts to correlate structure and function. Refer to Figure 13.15 on page 257 and match the name of the sensory or motor area next to the appropriate letter (A, B, C, etc.).

11. Cerebral White Matter

The white matter underlying the cerebral cortex consists of myelinated axons extending in three principal directions.

a. ***Association fibers*** connect and transmit nerve impulses between gyri in the same hemisphere.
b. ***Commissural fibers*** transmit impulses from the gyri in one cerebral hemisphere to the corresponding gyri in the opposite cerebral hemisphere. Three important groups of commissural fibers are the ***corpus callosum, anterior commissure,*** and ***posterior commissure.***
c. ***Projection fibers*** form motor and sensory tracts that transmit impulses from the cerebrum and other parts of the brain to the spinal cord or from the spinal cord to the brain.

12. Basal Ganglia

Basal ganglia (GANG-lē-a) or ***cerebral nuclei*** are groups of gray matter in the cerebral hemispheres. The largest of the basal ganglia is the ***corpus striatum*** (strī-Ā-tum; *corpus* = body; *striatum* = striped), which consists of the ***caudate*** (*cauda* = tail) ***nucleus*** and the ***lentiform*** (*lenticala* = shaped like a lentil or

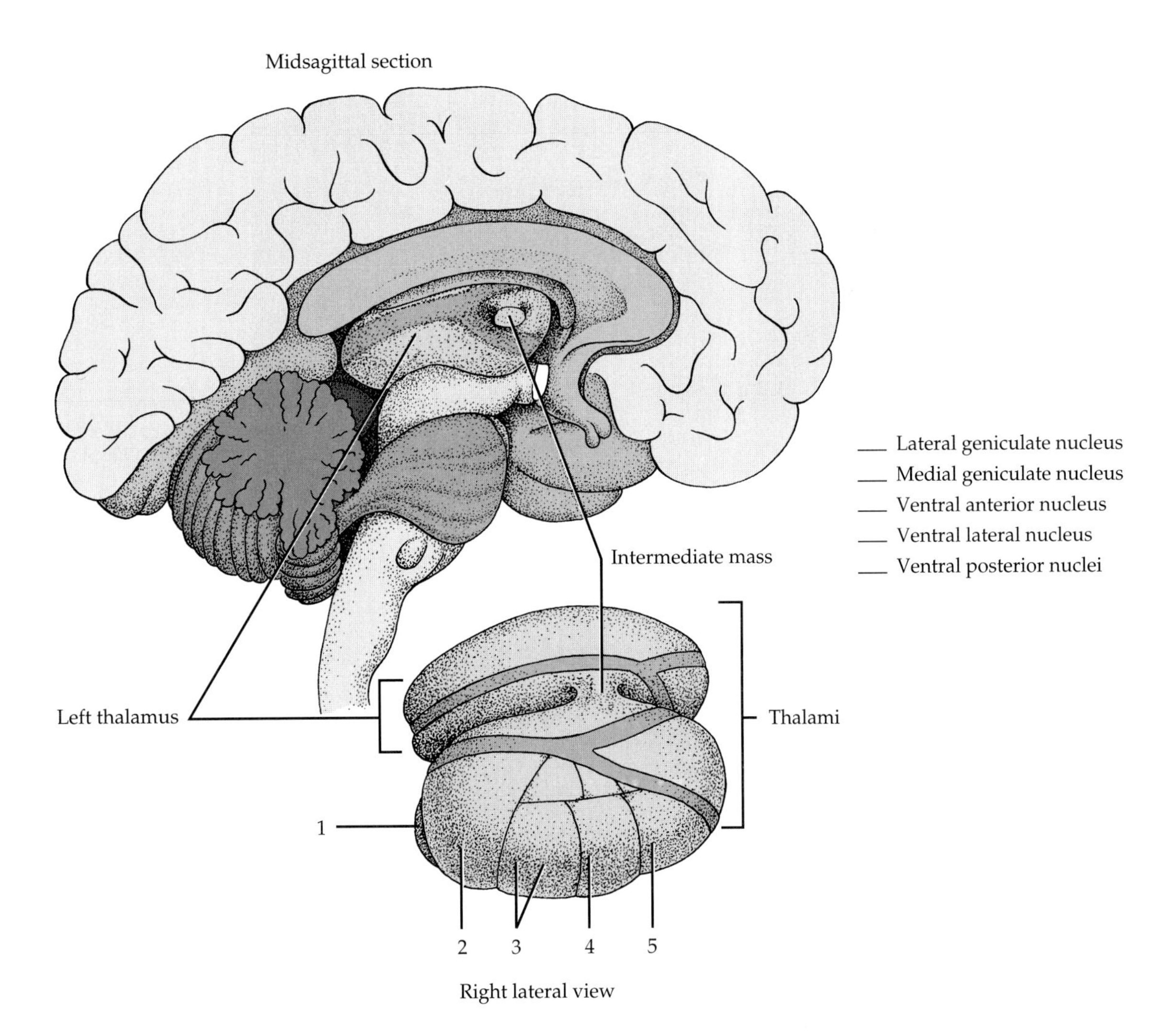

FIGURE 13.13 Thalamic nuclei.

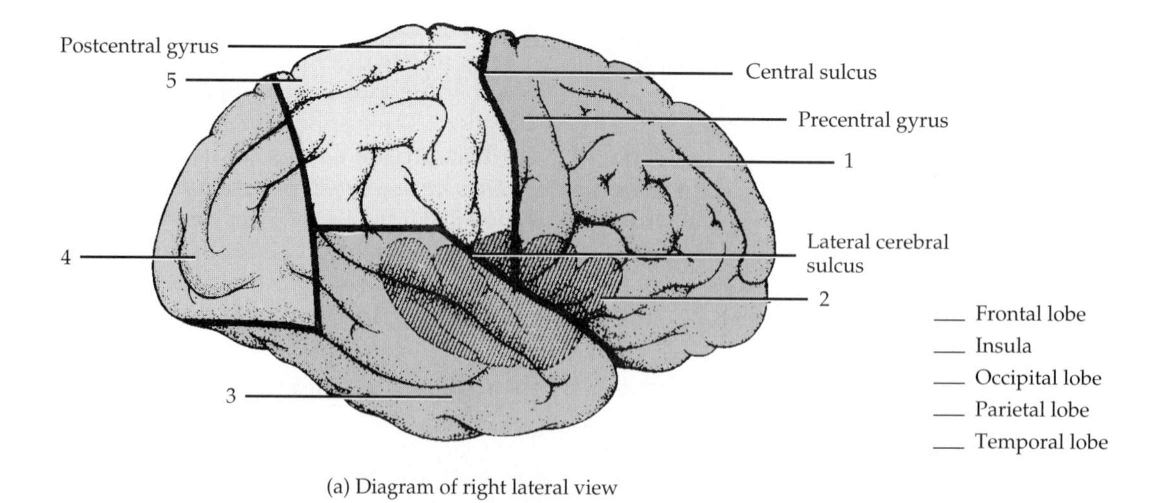

(a) Diagram of right lateral view

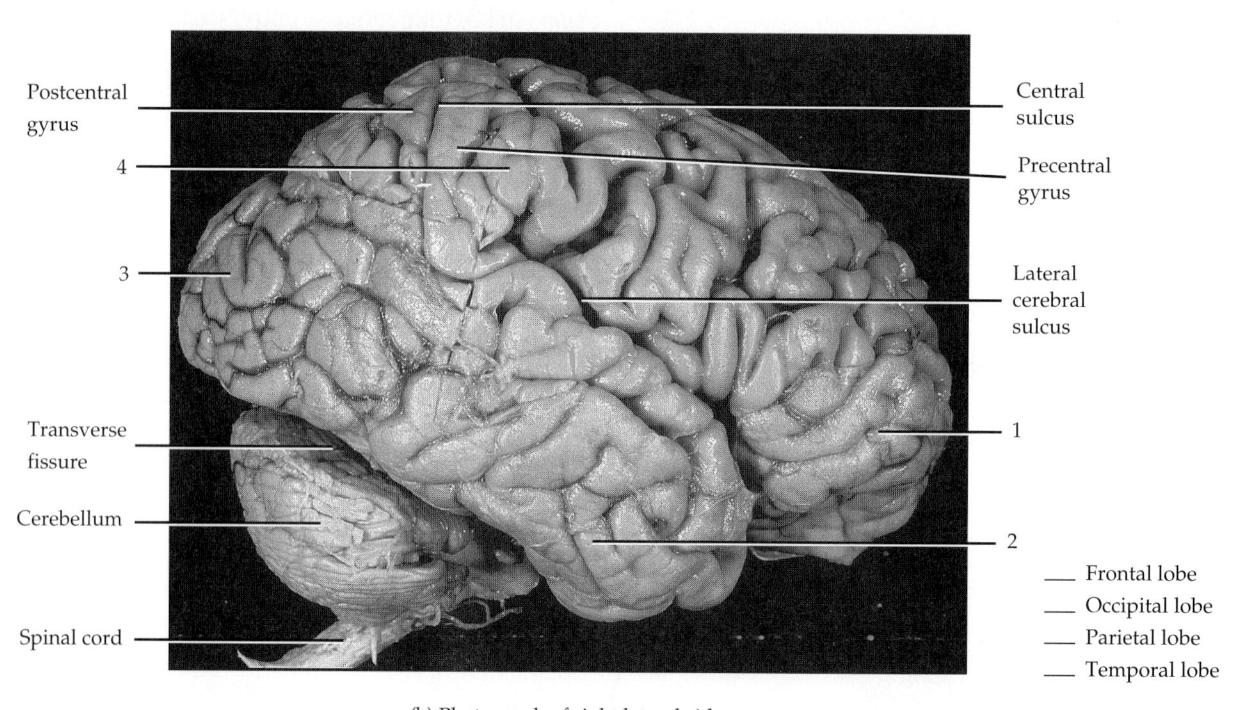

(b) Photograph of right lateral side

FIGURE 13.14 Lobes of cerebrum.

lens) or ***lenticular nucleus.*** The lentiform nucleus, in turn, is subdivided into a lateral ***putamen*** (pu-TĀ-men; *putamen* = shell) and a medial ***globus pallidus*** (*globus* = ball; *pallid* = pale).

The basal ganglia are interconnected by many nerve fibers. They also receive input from and provide output to the cerebral cortex, thalamus, and hypothalamus. The caudate nucleus and the putamen control large automatic movements of skeletal muscles, such as swinging the arms while walking. The globus pallidus is concerned with the regulation of muscle tone required for specific body movements.

Examine a model or specimen of the brain and identify the basal ganglia described. Now refer to Figure 13.16 on page 258 and label the basal ganglia shown.

13. Cerebellum

The ***cerebellum*** is inferior to the posterior portion of the cerebrum and separated from it by

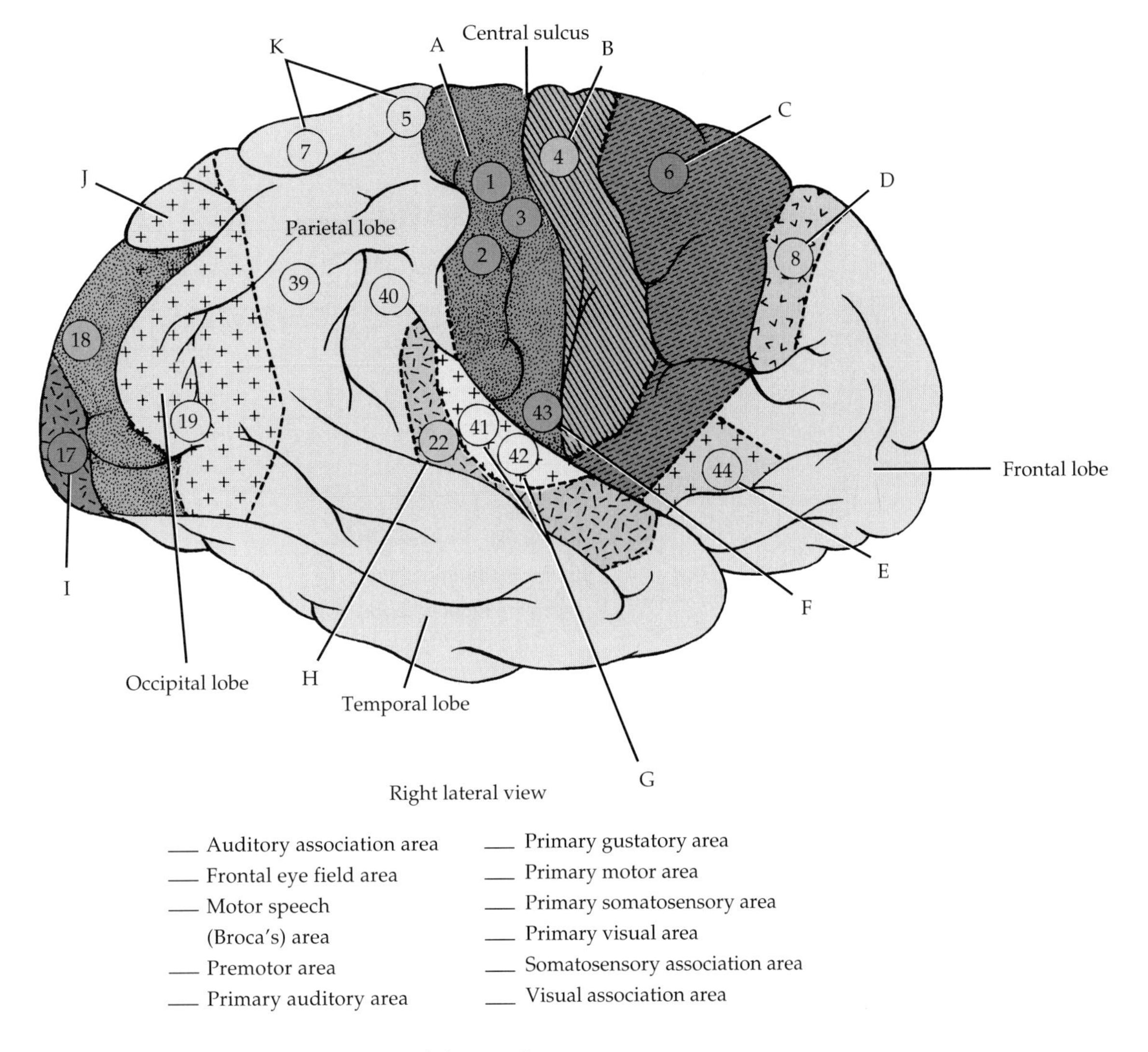

FIGURE 13.15 Functional areas of the cerebrum.

the transverse fissure. The central constricted area of the cerebellum is called the ***vermis*** (meaning worm-shaped) and the lateral portions are referred to as ***cerebellar hemispheres.*** The surface of the cerebellum, called the ***cerebellar cortex,*** consists of gray matter thrown into a series of slender parallel ridges called ***folia*** (*folia* = leaf). Deep to the gray matter are tracts of white matter called ***arbor vitae*** (meaning tree of life).

The cerebellum is attached to the brain stem by three paired bundles of fibers called ***cerebellar peduncles.*** The ***inferior cerebellar peduncles*** connect the cerebellum with the medulla oblongata at the base of the brain stem and with the spinal cord. The ***middle cerebellar peduncles*** connect the cerebellum with the pons. The ***superior cerebellar peduncles*** connect the cerebellum with the midbrain and thalamus.

The cerebellum functions to compare the intended movement determined by motor areas in the cerebrum with what is actually happening. The cerebellum constantly receives sensory input from proprioceptors in muscles, tendons, and joints, receptors for equilibrium, and visual receptors of the eyes. If the intent of the cerebral motor areas is not being attained by skeletal muscles, the cerebellum detects the variation and sends feedback signals to the motor areas to either stimulate or inhibit the activity of skeletal muscles. This interaction helps to smooth and coordinate complex sequences of skeletal muscle contractions. Besides coordinating skilled movements, the cerebellum is the main brain region

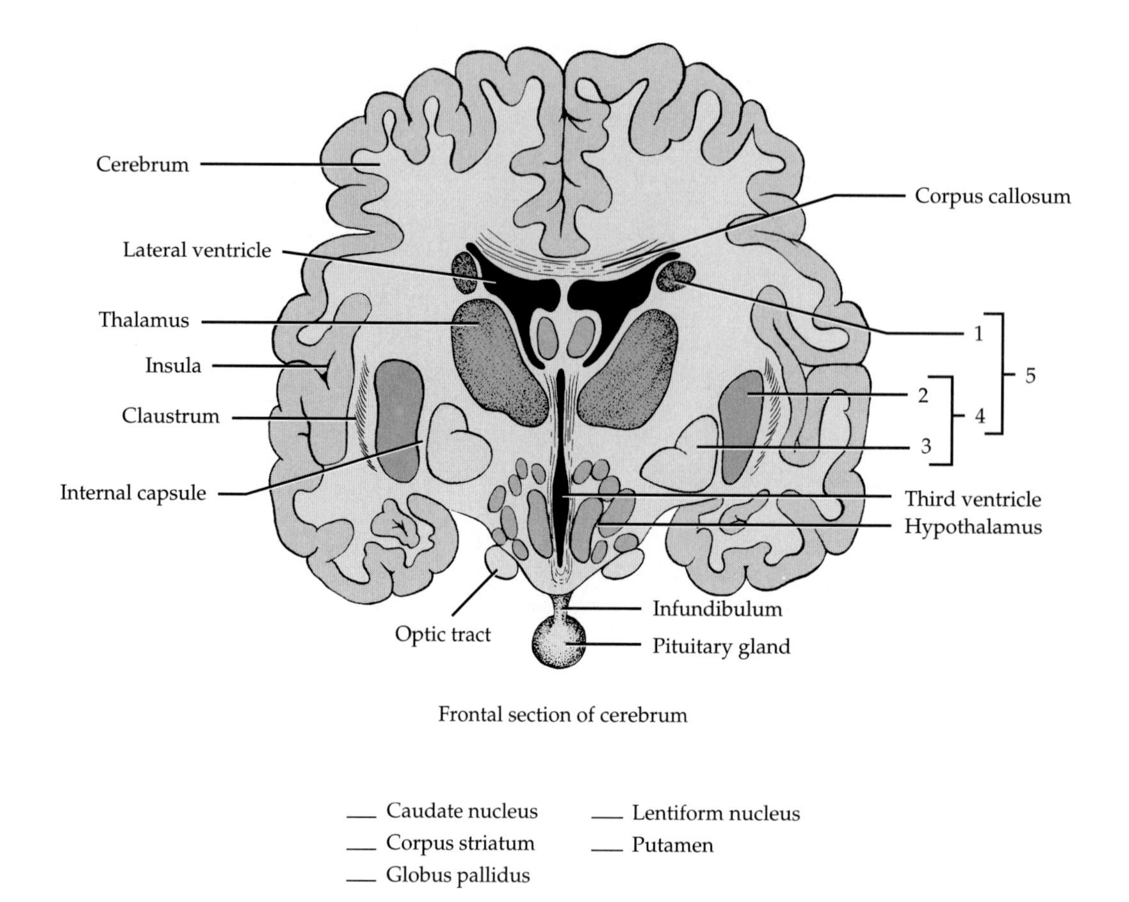

FIGURE 13.16 Basal ganglia.

that regulates posture and balance. These aspects of cerebellar function make possible all skilled motor activities, from catching a baseball to dancing.

Examine a model or specimen of the brain and locate the parts of the cerebellum. Identify the cerebellum in Figure 13.10.

C. CRANIAL NERVES

Of the 12 pairs of ***cranial nerves,*** 10 originate from the brain stem, but all pass through foramina in the base of the skull. The cranial nerves are designated by Roman numerals and names. The Roman numerals indicate the order in which the nerves arise from the brain, from anterior to posterior. The names indicate the distribution or function of the nerves.

Obtain a model of the brain and using your text and any other aids available identify the 12 pairs of cranial nerves. Now refer to Figure 13.17 and label the cranial nerves.

D. DISSECTION OF NERVOUS SYSTEM

CAUTION! *Please reread Section D, "Precautions Related to Dissection" at the beginning of the laboratory manual on page xiii before you begin your dissection.*

The following instructions for dissection of the mammalian brain can be used in cats, fetal pigs, and sheep, because their brains are structurally similar. One main difference is the surface (gyri and sulci) of the cerebrum. Another difference is the proportions of certain parts. For example, the human brain is basically the same as those of sheep, pig, and cat, but the cerebrum, the center of intelligence, is disproportionately larger.

Figures 13.18 through 13.21 are based on the sheep brain, but they can be used as references if you are to dissect the cat brain instead. Because the only major dissections needed of the brain are a transverse-sectional cut through the cerebrum

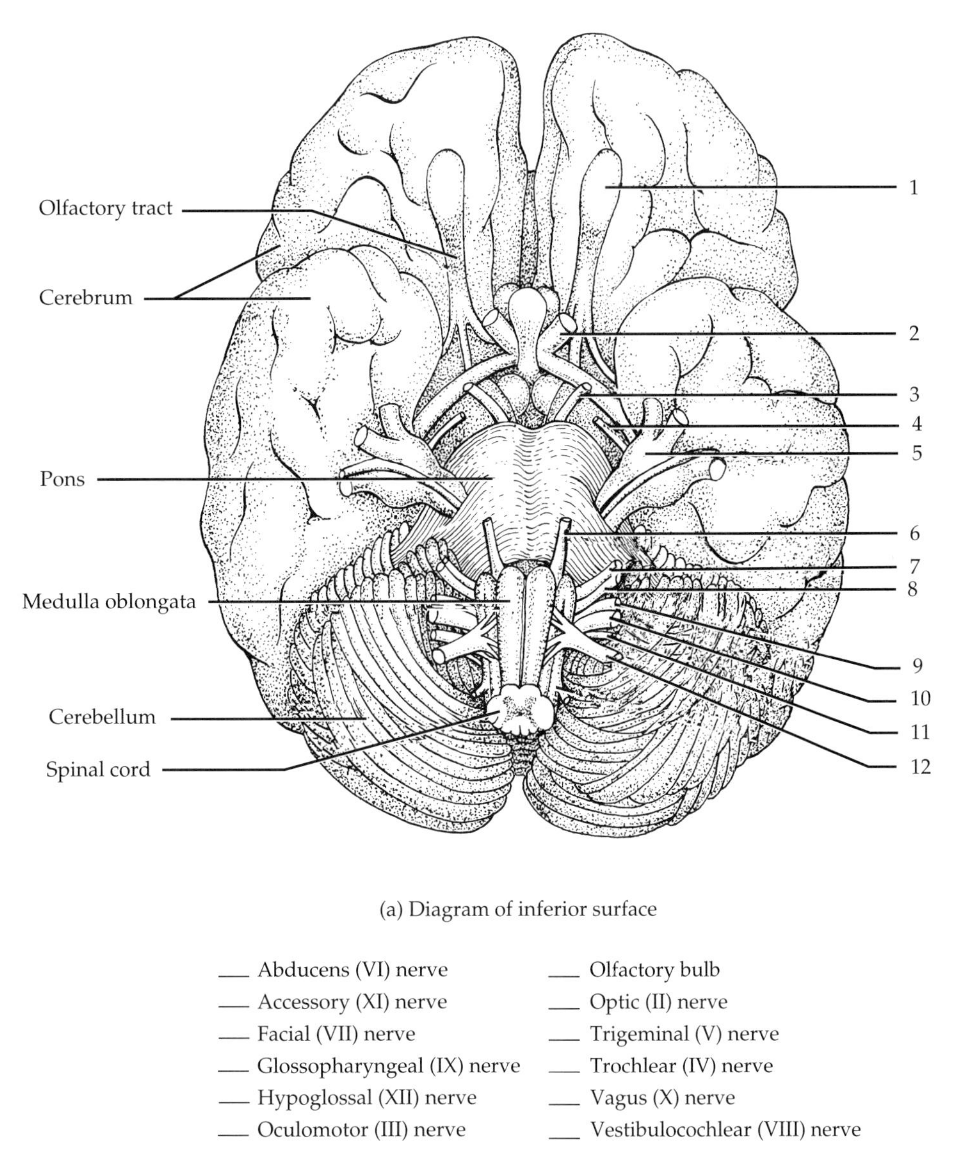

(a) Diagram of inferior surface

___ Abducens (VI) nerve	___ Olfactory bulb
___ Accessory (XI) nerve	___ Optic (II) nerve
___ Facial (VII) nerve	___ Trigeminal (V) nerve
___ Glossopharyngeal (IX) nerve	___ Trochlear (IV) nerve
___ Hypoglossal (XII) nerve	___ Vagus (X) nerve
___ Oculomotor (III) nerve	___ Vestibulocochlear (VIII) nerve

FIGURE 13.17 Cranial nerves of human brain.

and a midsagittal cut, brains can be returned to their preservative when you finish with them and used again. The instructor should provide brains that show cranial nerves, spinal cord, and meninges. Even if the cat brain is not to be dissected, the instructor might want to demonstrate a newly dissected one to the class.

PROCEDURE

1. If you are to dissect the brain from your cat, start by removing all of the muscles from the dorsal and lateral surfaces of the head.
2. *Very carefully*, using either bone shears or a bone saw, make an opening in the parietal bone. Do not damage the brain below.
3. Slowly chip away all the bone from the dorsal and lateral surfaces of the skull until the brain is exposed.
4. Between the cerebrum and the cerebellum is a transverse bony partition, the tentorium cerebelli, which should be removed carefully.
5. Cut the spinal cord transversely at the foramen magnum.
6. Gently lift the brain out of the floor of the cranium. As you do so, sever each cranial nerve as far from the brain as possible.

Study the following sections and dissect either the cat brain (Section D.2) or sheep brain (Section D.6) as your instructor directs.

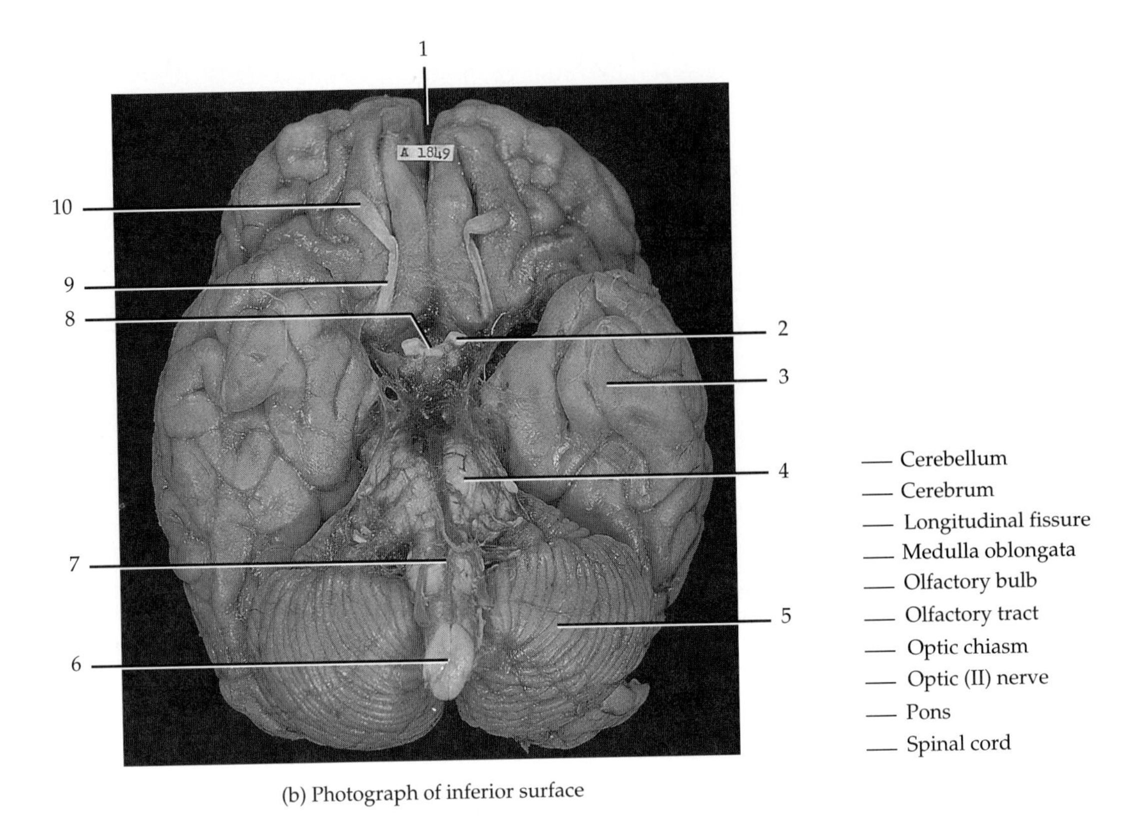

(b) Photograph of inferior surface

FIGURE 13.17 *(Continued)* Cranial nerves of human brain.

1. Meninges

On the brain that you are to use, locate the three layers of the meninges that cover the brain and the spinal cord: outer ***dura mater,*** middle ***arachnoid,*** and inner ***pia mater.*** Refer to pages 237 and 249 as you examine them. Note that the dura mater has two folds, the ***falx cerebri*** located between the cerebral hemispheres, and the ***tentorium cerebelli*** between the cerebral hemispheres and the cerebellum (the latter is ossified and fused to the parietal bones in the cat).

2. Dissection of Cat Brain

PROCEDURE

1. First identify on the dorsal surface of the brain the two large, cranial ***cerebral hemispheres*** (Figure 13.18). Together they constitute the ***cerebrum.*** The cerebral hemispheres are separated from each other by the deep ***longitudinal fissure.*** The surface of each hemisphere consists of upward folds called ***convolutions (gyri)*** and grooves called ***sulci.***
2. Spread the hemispheres and look into the longitudinal fissure. Note the ***corpus callosum,*** a band of transverse fibers that connects the two hemispheres internally.
3. Make a frontal section through a cerebral hemisphere. Note the ***gray matter*** near the surface of the cerebral cortex and the ***white matter*** beneath the gray matter.
4. Now locate the ***cerebellum*** caudal to the cerebral hemisphere and dorsal to the medulla oblongata. It is separated from the cerebral hemispheres by a deep ***transverse fissure.*** The cerebellum consists of a median ***vermis*** and two lateral ***cerebellar hemispheres.*** As does the cerebrum, the cerebellum has folia (convolutions), fissures, and sulci. The treelike arrangement of white matter tracts in the gray matter in the cerebellum is the ***arbor vitae*** ("tree of life").
5. Section a portion of the cerebellum longitudinally to examine the arbor vitae.
6. The cerebellum is connected to the brain stem by three prominent fiber tracts called ***peduncles.*** The ***superior cerebellar peduncle*** connects the cerebellum with the midbrain, the ***inferior cerebellar peduncle*** connects the cerebellum with the medulla, and the ***middle cerebellar peduncle*** connects the cerebellum with the pons.

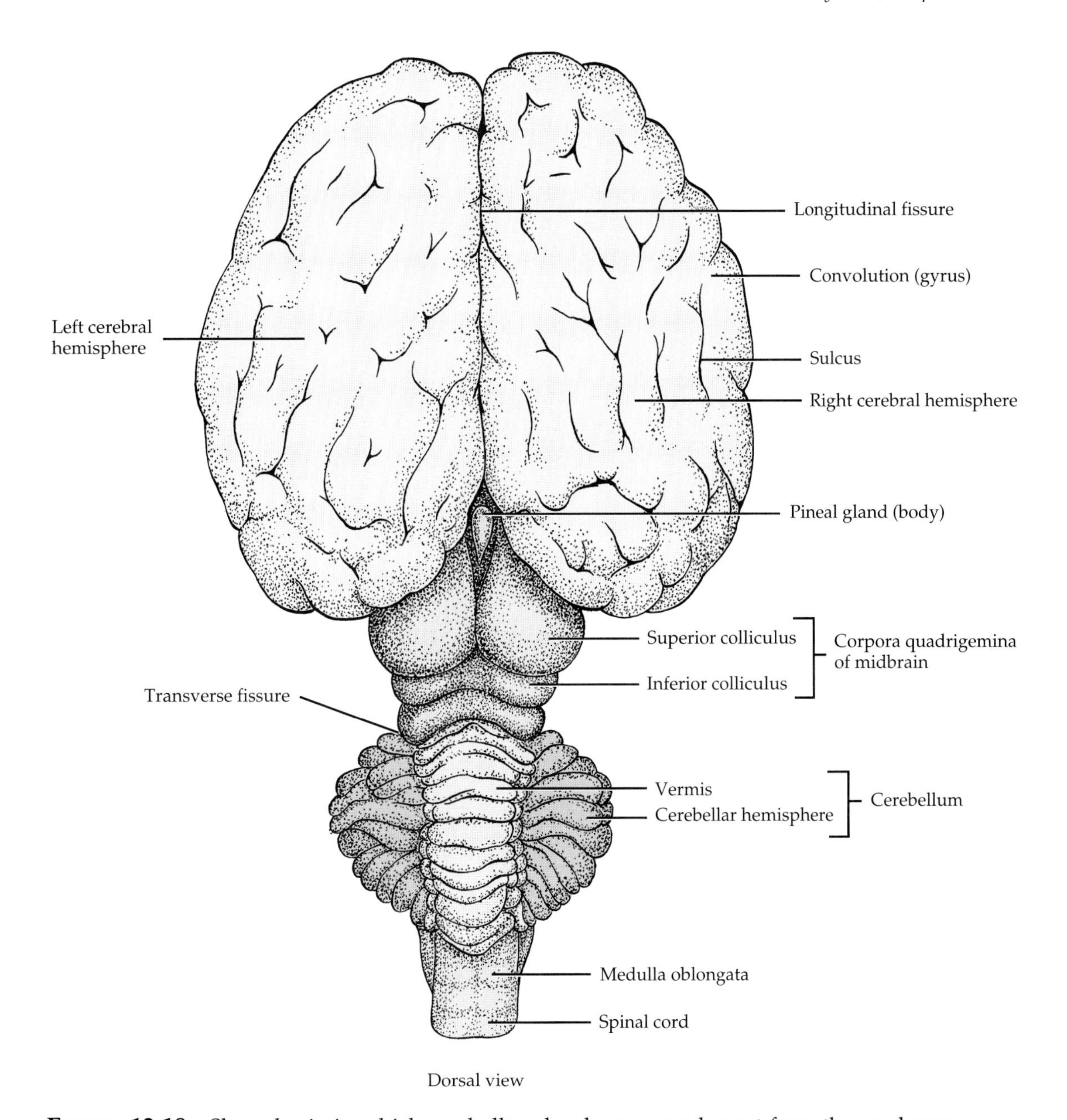

FIGURE 13.18 Sheep brain in which cerebellum has been spread apart from the cerebrum.

7. The dorsal portion (roof) of the ***midbrain*** can be examined if you spread the cerebral hemispheres apart from the cerebellum. The roof of the midbrain is formed by four rounded bulges, the ***corpora quadrigemina.*** The two larger cranial bulges are called the ***superior colliculi*** and the two smaller caudal bulges are called the ***inferior colliculi.*** If you look between the superior colliculi you should see the ***pineal gland (body).***
8. The ***medulla oblongata,*** or just simply medulla, is the most caudal portion of the brain and is continuous with the spinal cord.
9. To examine other parts of the brain, you will have to make your observations from its ventral surface (Figure 13.19). At the cranial end of the brain are the ***olfactory bulbs.*** The olfactory (I) nerves terminate on the bulbs after passing through the cribriform plate of the ethmoid bone. Bands of nerve fibers extending caudally and ventrally from the olfactory bulbs are the ***olfactory tracts.*** The tracts terminate in the cranial part of the cerebrum.
10. Caudal to the olfactory tracts is the ***hypothalamus.*** At the cranial border of the hypothalamus, the optic (II) nerves come together and partially

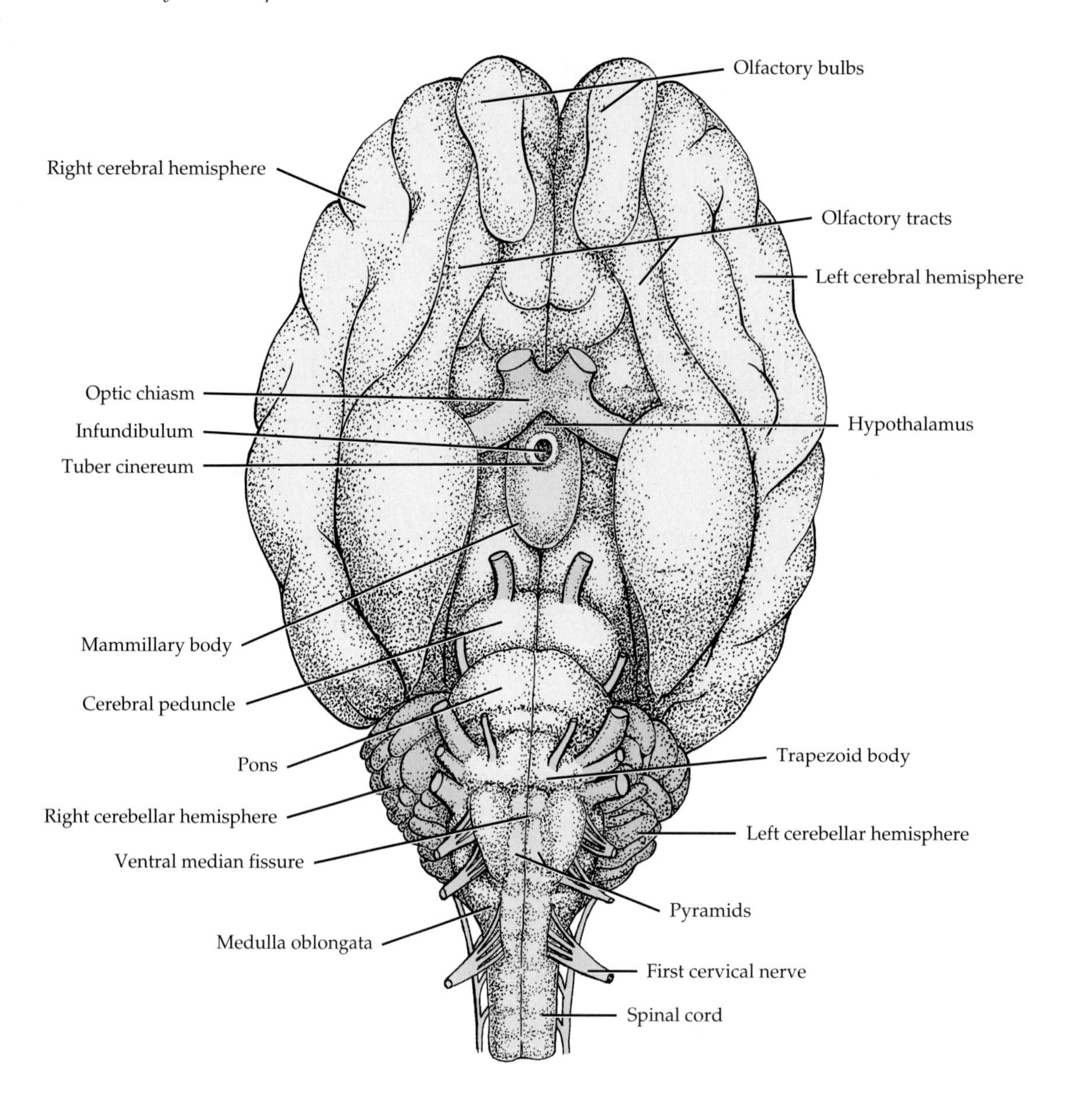

FIGURE 13.19 Sheep brain.

cross. This point of crossing is called the ***optic chiasm.*** The rest of the hypothalamus is caudal to the optic chiasm, covered mostly by the ***pituitary gland (hypophysis).*** The pituitary gland is attached to a small round elevation on the hypothalamus called the ***tuber cinereum*** by a narrow stalk called the ***infundibulum.***

11. Now examine the ventral portion (floor) of the midbrain called the ***cerebral peduncles.*** These bundles of white fibers connect the pons to the cerebrum. They run obliquely and cranially on either side of the hypophysis.
12. ***Caudal*** to the cerebral peduncles is the ***pons,*** which consists of transverse fibers that connect the cerebellar hemispheres.
13. Caudal to the pons is the ***medulla oblongata.*** Identify the ***ventral median fissure,*** a midline groove. The longitudinal bundles of fibers on either side of the ventral medial fissure are the ***pyramids.***
14. Lateral to the cranial end of the pyramids are the ***trapezoid bodies.*** These are narrow bands of transverse fibers that cross the cranial end of the medulla.

15. The remaining parts of the brain may be located by cutting the brain longitudinally into halves. Cut through the midsagittal plane. A long sharp knife may be used here.
16. Using Figure 13.20 as a guide, identify the ***corpus callosum,*** a transverse band of white fibers at the bottom of the cerebral hemispheres. The cranial portion of the corpus callosum is called the ***genu,*** the middle part is known as the ***body*** or ***trunk,*** and the caudal portion is called the ***splenium.*** The ventral deflection of the genu is known as the ***rostrum.*** Just beneath the splenium identify the ***pineal gland (body).***
17. Ventral to the corpus callosum is the ***fornix,*** a band of white fibers connecting the cerebral hemispheres. It is cranial to the third ventricle.
18. Between the fornix and corpus callosum is a thin vertical membrane, the ***septum pellucidum,*** which separates the two lateral ventricles of the brain. If you break the membrane, you can expose the ***lateral ventricle,*** a cavity in each cerebral hemisphere that contains cerebrospinal fluid in the living animal.
19. Identify the ***anterior commissure,*** a small group of fibers ventral to the fornix. Extending from the anterior commissure to the optic chiasm is the ***lamina terminalis.*** It forms the cranial wall of the third ventricle.
20. The ***thalamus*** forms the lateral walls of the ***third ventricle,*** another cavity containing cerebrospinal fluid. The thalamus is caudal to the fornix and extends ventrally to the infundibulum and caudally to the corpora quadrigemina. The third ventricle lies mostly in the thalamus and hypothalamus, and, in the living animal, is filled with cerebrospinal fluid. The lateral ventricles communicate with the third ventricle by an opening called the ***interventricular foramen.***
21. The ***intermediate mass*** is a circular area that extends across the third ventricle, connecting the two sides of the thalamus. The interventricular foramen lies in the depression cranial to the intermediate mass. The intermediate mass is absent in about 30% of human brains and, when present, is proportionately much smaller.
22. Identify the ***posterior commissure,*** a small round mass of fibers that crosses the midline. It is located just ventral to the pineal gland (body).
23. Now find the ***cerebral aqueduct,*** a narrow longitudinal channel that passes through the midbrain and connects the third and fourth ventricles. The roof of the cerebral aqueduct is called the ***lamina quadrigemina.***
24. The ***fourth ventricle*** is a cavity superior to the pons and medulla and inferior to the cerebellum. It is continuous with the cerebral aqueduct and contains cerebrospinal fluid in a living animal. A membrane that forms the roof of the fourth ventricle is called the ***velum.*** The fourth ventricle leads into the ***central canal*** of the spinal cord.

3. Cranial Nerves

Briefly review Section C, Cranial Nerves (page 258). Using Figure 13.21 on page 265 as a gude, identify the cranial nerves.

a. ***Olfactory (I) nerve*** Consists of fibers that pass from the olfactory bulb, through the foramina of the cribriform plate, to the mucous membrane of the nasal cavity. It functions in smell.
b. ***Optic (II) nerve*** Arises from the optic chiasm and passes through the optic foramen to the retina of the eye. It functions in vision.
c. ***Oculomotor (III) nerve*** Arises from the ventral surface of the cerebral peduncle, passes through the orbital fissure, and supplies the superior rectus, medial rectus, inferior rectus, inferior oblique, and levator palpebrae eye muscles.
d. ***Trochlear (IV) nerve*** Arises just caudal to the corpora quadrigemina, curves around the lateral aspect of the cerebral peduncle, passes through the orbital fissure, and supplies the superior oblique muscle of the eyeball.
e. ***Trigeminal (V) nerve*** Originates from the lateral surface of the pons by a large sensory root and a small motor root. It exits through the orbital fissure and foramen ovale. The three branches of the trigeminal nerve are the ***ophthalmic, maxillary,*** and ***mandibular.*** All three branches supply the skin of the head and face, the epithelium of the oral cavity, the anterior two-thirds of the tongue, and the teeth. The mandibular branch also supplies the muscles of mastication.
f. ***Abducens (VI) nerve*** Arises from between the trapezoid body and pyramid, passes through the orbital fissure, and supplies the lateral rectus muscle of the eyeball.
g. ***Facial (VII) nerve*** Arises caudal to the trigeminal nerve between the pons and trapezoid body, passes through the middle ear, and emerges at the stylomastoid foramen. It supplies many head and facial muscles and the muscosa of the tongue.
h. ***Vestibulocochlear (VIII) nerve*** Arises just caudal to the facial nerve, passes into the internal

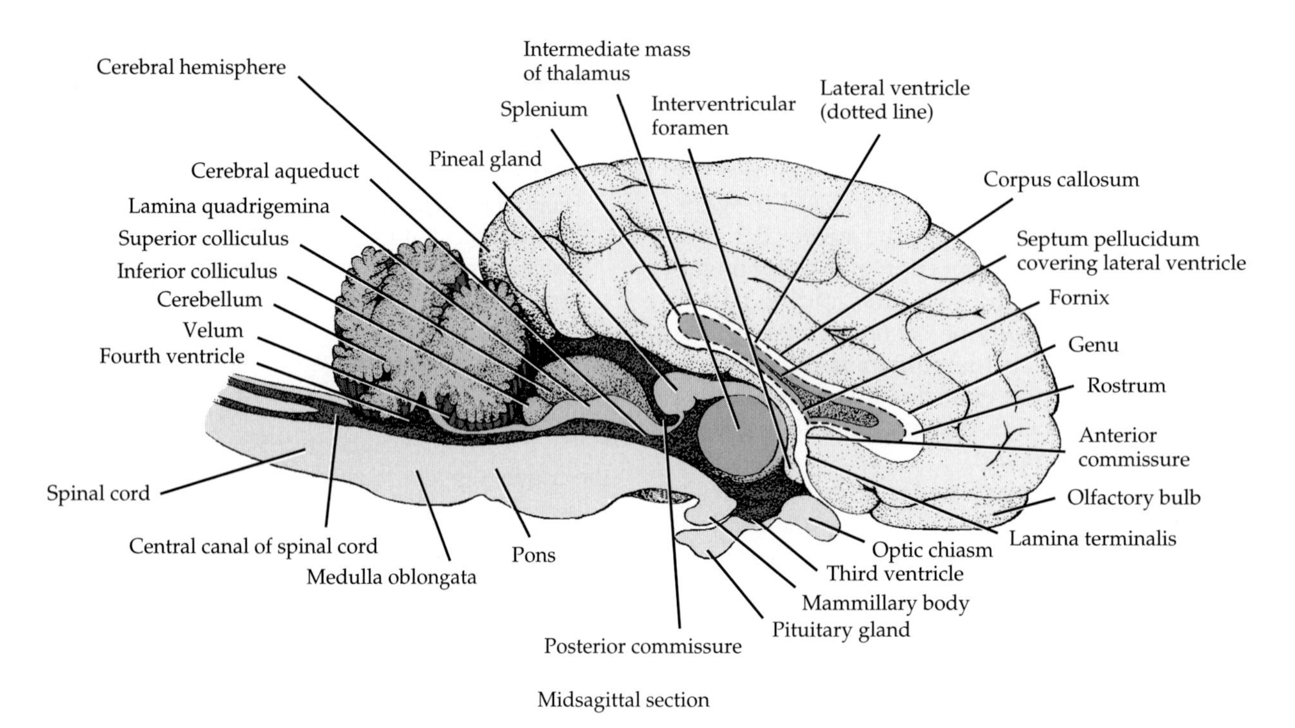

FIGURE 13.20 Sheep brain.

auditory meatus, and divides into two branches. One branch transmits auditory sensations from the cochlea. The other branch transmits sensations for equilibrium from the semicircular canals.

i. ***Glossopharyngeal (IX) nerve*** Arises just caudal to the vestibulocochlear nerve, emerges from the jugular foramen, and supplies the muscles of the pharynx and the taste buds.
j. ***Vagus (X) nerve*** Arises just caudal to the glossopharyngeal nerve and exits through the jugular foramen. It supplies most of the thoracic and abdominal viscera.
k. ***Accessory (XI) nerve*** Arises from the medulla and spinal cord, passes through the jugular foramen, and supplies the neck and shoulder muscles.
l. ***Hypoglossal (XII) nerve*** Arises from the ventral medulla, passes through the hypoglossal foramen, and supplies tongue muscles.

4. Dissection of Cat Spinal Cord

PROCEDURE

1. Remove the muscles dorsal to the vertebral column.
2. Using small bone clippers, cut through the laminae of the vertebrae and remove the neural arches of all the vertebrae starting at the foramen magnum.
3. Once the entire spinal cord is exposed, identify the following (Figure 13.22a on page 266):
   a. ***Cervical enlargement*** Between the fourth cervical and first thoracic vertebrae.
   b. ***Lumbar enlargement*** Between the third and seventh lumbar vertebrae.
   c. ***Conus medullaris*** Tapered conical tip of the spinal cord caudal to the lumbar enlargement.
   d. ***Filum terminale*** A slender strand continuous with the conus medullaris, which extends into the caudal region. The filum and caudal spinal nerves are referred to as the ***caudal equina.***
   e. ***Dorsal median sulcus*** Shallow groove on the dorsal surface of the spinal cord along the midline.
   f. ***Ventral median fissure*** Deeper groove on the ventral surface of the spinal cord along the midline (Figure 13.22b).
   g. ***Dorsolateral sulcus*** Groove parallel and lateral to the dorsal median sulcus. The dorsal roots of spinal nerves attach to this sulcus (see Figure 13.22b).
   h. ***Meninges*** These are similar in structure and location to those of the brain (see Figure 13.22b).
4. If you examine the cut end of the spinal cord at the foramen magnum with a hand lens,

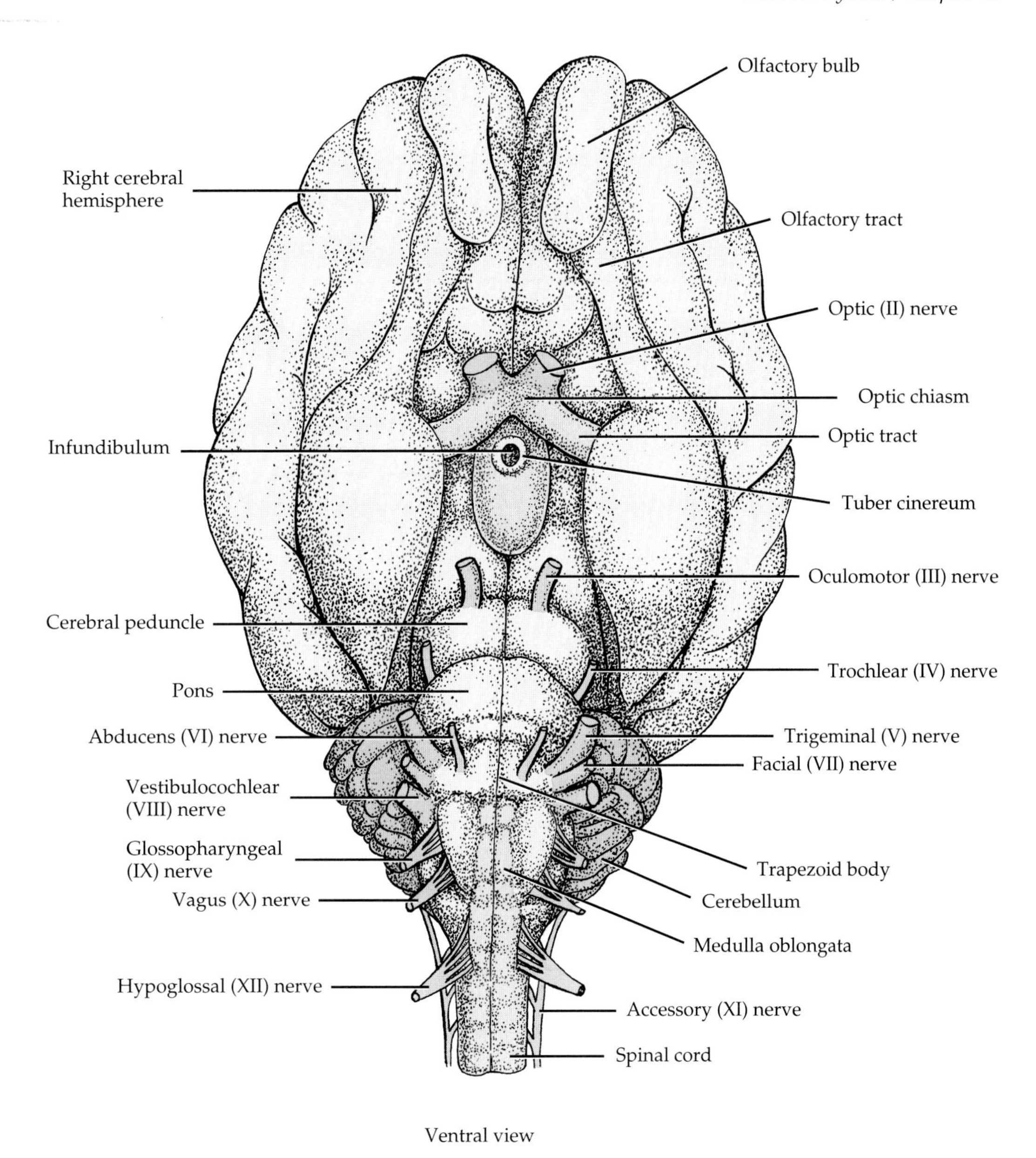

FIGURE 13.21 Cranial nerves of a sheep brain.

you can see the gray matter arranged like a letter H in the interior and the white matter external to the gray matter. Also identify the dorsal median sulcus and the ventral median fissure.

5. Spinal Nerves of Cat

The cat has 38 or 39 pairs of ***spinal nerves.*** These include 8 pairs of ***cervical,*** 13 pairs of ***thoracic,*** 7 pairs of ***lumbar,*** 3 pairs of ***sacral,*** and 7 or 8 pairs of ***caudal (coccygeal)*** (Figure 13.22a).

Each spinal nerve arises by a ***dorsal (sensory) root,*** which originates from the dorsolateral sulcus, and a ***ventral root*** (Figure 13.22b). The dorsal root has an oval swelling called the ***dorsal root ganglion.*** The first cervical nerve exits from the vertebral canal through the atlantal foramen. The second cervical nerve exits between the arches of the atlas and axis. All other spinal nerves exit through intervertebral foramina.

Just outside the intervertebral foramina, a spinal nerve divides into a ventral ramus and a dorsal ramus. The ***dorsal rami*** are small and are

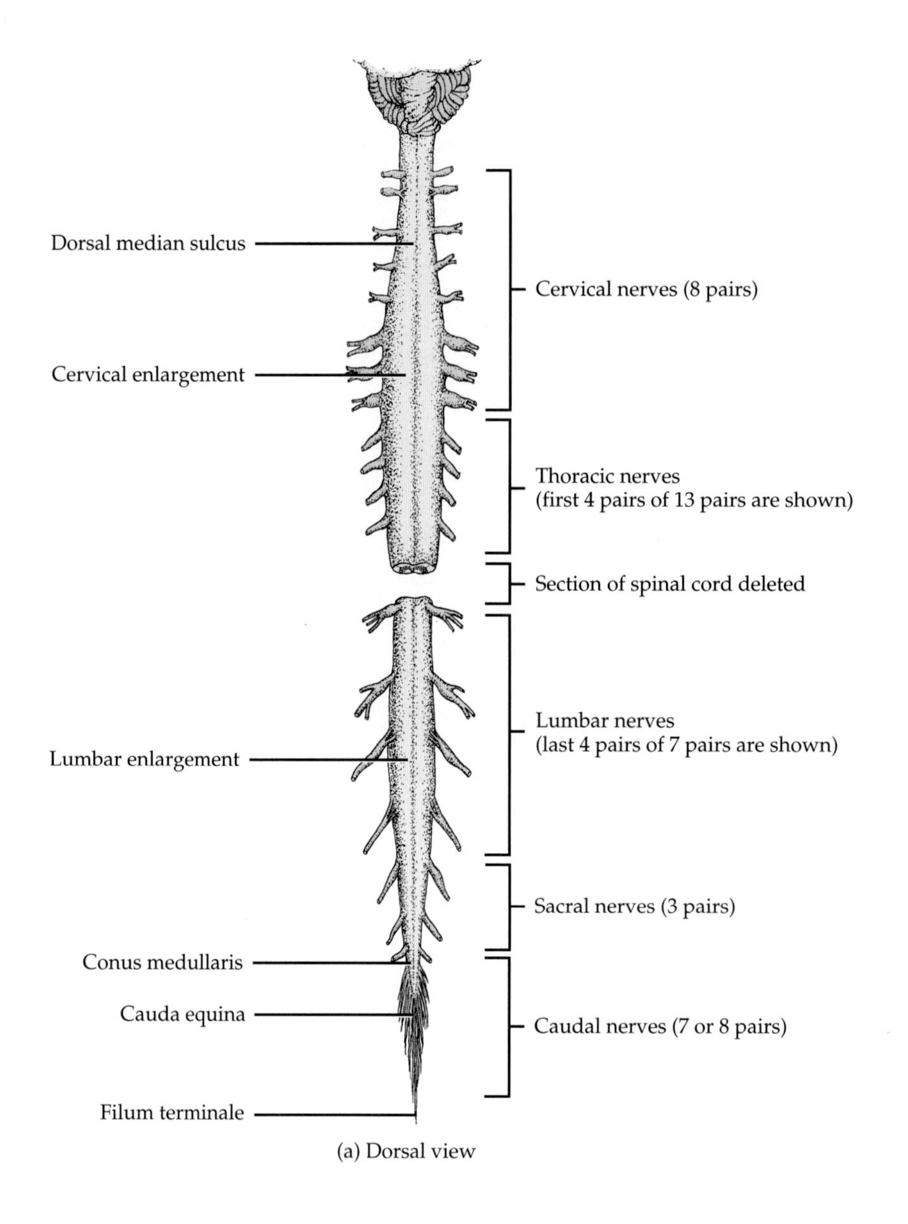

(a) Dorsal view

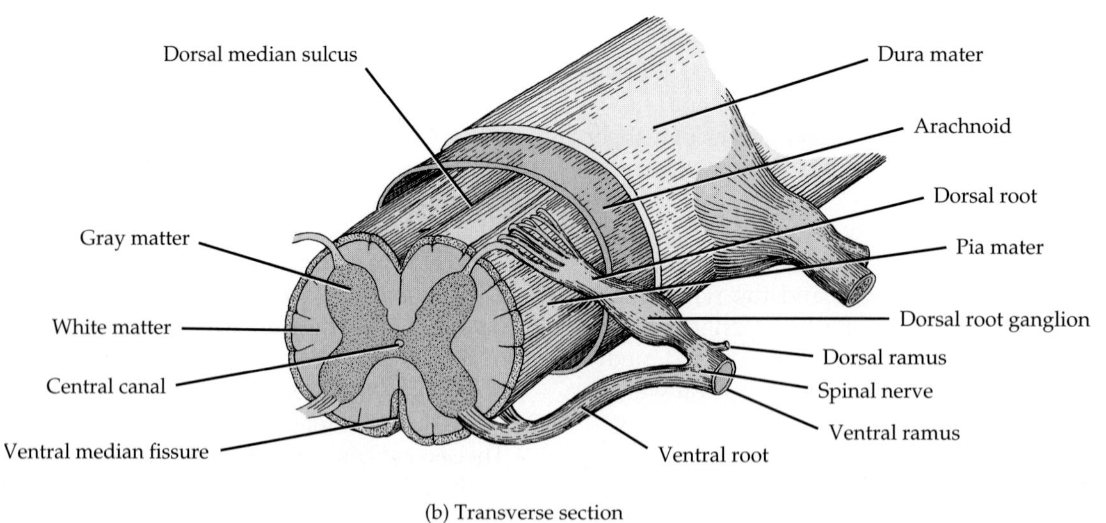

(b) Transverse section

FIGURE 13.22 The spinal cord.

distributed to the skin and muscles of the back. The ***ventral rami*** are large and supply the skin and muscles of the ventral part of the trunk.

Expose the roots of several spinal nerves and identify the parts just described.

6. Dissection of Sheep Brain

CAUTION! *Please reread Section D, "Precautions Related to Dissection" at the beginning of the laboratory manual on page xiii before you begin your dissection.*

The brains of the fetal pig, sheep, and human show many similarities. They possess the protective membranes called the ***meninges,*** which can easily be seen as you proceed with the dissections. The outermost layer is the ***dura mater.*** It is the toughest, protective one, and may be missing in the preserved sheep brain. The middle membrane is the ***arachnoid,*** and the inner one containing blood vessels and adhering closely to the surface of the brain itself is the ***pia mater.***

PROCEDURE

1. Use Figures 13.18 through 13.20 as references for this dissection.
2. The most prominent external parts of the brain are the pair of large ***cerebral hemispheres*** and the posterior ***cerebellum*** on the dorsal surface. These large hemispheres are separated from each other by the ***longitudinal fissure.*** The ***transverse fissure*** separates the hemispheres from the cerebellum. The surfaces of these hemispheres form many ***gyri,*** or raised ridges, that are separated by grooves, or ***sulci.***
3. If you spread the hemispheres gently apart you can see, deep in the longitudinal fissure, thick bundles of white transverse fibers. These bundles form the ***corpus callosum,*** which connects the hemispheres.
4. Most of the following structures can be identified by examining a midsagittal section of the sheep brain, or by cutting an intact brain along the longitudinal fissure completely through the corpus callosum.
5. If you break through the thin ventral wall, the ***septum pellucidum*** of the corpus callosum, you can see part of a large chamber, the ***lateral ventricle,*** inside the hemisphere.
6. Each hemisphere has one of these ventricles. Ventral to the septum pellucidum, locate a smaller band of white fibers called the ***fornix.*** Close by where the fornix disappears is a small, round bundle of fibers called the ***anterior commissure.***
7. The ***third ventricle*** and the ***thalamus*** are located ventral to the fornix. The third ventricle is outlined by its shiny epithelial lining, and the thalamus forms the lateral walls of this ventricle. This ventricle is crossed by a large circular mass of tissue, the ***intermediate mass,*** which connects the two sides of the thalamus. Each lateral ventricle communicates with the third ventricle through an opening, the ***interventricular foramen,*** which lies in a depression anterior to the intermediate mass and can be located with a dull probe.
8. Spreading the cerebral hemispheres and the cerebellum apart reveals the roof of the midbrain (mesencephalon), which is seen as two pairs of round swellings collectively called the ***corpora quadrigemina.*** The larger, more anterior pair are the ***superior colliculi.*** The smaller posterior pair are the ***inferior colliculi.*** The ***pineal gland (body)*** is seen directly between the superior colliculi. Just posterior to the inferior colliculi, appearing as a thin white strand, is the ***trochlear (IV) nerve.***
9. The cerebellum is connected to the brain stem by three prominent fiber tracts called ***peduncles.*** The ***superior cerebellar peduncle*** connects the cerebellum with the midbrain, the ***inferior cerebellar peduncle*** connects the cerebellum with the medulla, and the ***cerebellar peduncle*** connects the cerebellum with the pons.
10. Most of the following parts can be located on the ventral surface of the intact brain.
11. Just beneath the cerebral hemispheres are two ***olfactory bulbs,*** which continue posteriorly as two ***olfactory tracts.*** Posterior to these tracts, the ***optic (II) nerves*** undergo a crossing (decussation) known as the ***optic chiasm.***
12. Locate the ***pituitary gland (hypophysis)*** just posterior to the chiasm. This gland is connected to the ***hypothalamus*** portion of the diencephalon by a stalk called the ***infundibulum.*** The ***mammillary body*** appears immediately posterior to the infundibulum.
13. Just posterior to this body are the paired ***cerebral peduncles,*** from which arise the large ***oculomotor (III) nerves.*** They may be partially covered by the pituitary gland.
14. The ***pons*** is a posterior extension of the hypothalamus and the ***medulla oblongata*** is a posterior extension of the pons.
15. The ***cerebral aqueduct*** dorsal to the peduncles runs posteriorly and connects the third ventricle

with the ***fourth ventricle,*** which is located dorsal to the medulla and ventral to the cerebellum.

16. The medulla merges with the ***spinal cord,*** and is separated by the ***ventral median fissure.*** The ***pyramids*** are the longitudinal bands of tissue on either side of this fissure.
17. Identify the remaining ***cranial nerves*** on the ventral surface of the brain. They are the trigeminal (V), abducens (VI), facial (VII), vestibulocochlear (VIII), glossopharyngeal (IX), vagus (X), accessory (XI), and hypoglossal (XII). The previously identified cranial nerves are the olfactory (I), optic (II), oculomotor (III), and trochlear (IV), for a total of twelve.
18. A transverse section through a cerebral hemisphere reveals ***gray matter*** near the surface of the ***cerebral cortex*** and ***white matter*** beneath this layer.
19. A transverse section through the spinal cord reveals the ***central canal,*** which is connected to the fourth ventricle and contains ***cerebrospinal fluid.***
20. A midsagittal section through the cerebellum reveals a treelike arrangement of gray and white matter called the ***arbor vitae*** (tree of life).

E. AUTONOMIC NERVOUS SYSTEM

The ***autonomic nervous system (ANS)*** regulates the activities of smooth muscle, cardiac muscle, and certain glands, usually involuntarily. In the somatic nervous system (SNS), which is voluntary, the cell bodies of the motor neurons are in the CNS, and their axons extend all the way to skeletal muscles in spinal nerves. The ANS always has two motor neurons in the pathway. The first motor neuron, the ***preganglionic neuron,*** has its cell body in the CNS. Its axon leaves the CNS and synapses in an autonomic ganglion with the second neuron called the ***postganglionic neuron.*** The cell body of the postganglionic neuron is inside an autonomic ganglion, and its axon terminates in a ***visceral effector*** (muscle or gland).

Label the components of the autonomic pathway shown in Figure 13.23.

The ANS consists of two divisions: sympathetic and parasympathetic (Figure 13.24 on page 270). Most viscera are innervated by both divisions. In general, nerve impulses from one division stimulate a structure, whereas nerve impulses from the other division decrease its activity.

In the ***sympathetic division,*** the cell bodies of the preganglionic neurons are located in the lateral gray horns of the spinal cord in the thoracic and first two lumbar segments. The axons of preganglionic neurons are myelinated and leave the spinal cord through the ventral (anterior) root of a spinal nerve. Each axon travels briefly in a ventral ramus and then through a small branch called a ***white ramus communicans*** to enter a sympathetic trunk ganglion. These ganglia lie in a vertical row, on either side of the vertebral column, from the base of the skull to the coccyx. In the ganglion, the axon may synapse with a postganglionic neuron, travel upward or downward through the sympathetic trunk ganglia to synapse with postganglionic neurons at different levels, or pass through the ganglion without synapsing to form part of the splanchnic nerves. If the preganglionic axon synapses in a sympathetic trunk ganglion, it reenters the ventral or dorsal ramus of a spinal nerve via a small branch called a ***gray ramus communicans.*** If the preganglionic axon forms part of the splanchnic nerves, it passes through the sympathetic trunk ganglion but synapses with a postganglionic neuron in a prevertebral (collateral) ganglion. These ganglia are anterior to the vertebral column close to large abdominal arteries from which their names are derived (celiac, superior mesenteric, and inferior mesenteric).

In the ***parasympathetic division,*** the cell bodies of the preganglionic neurons are located in nuclei in the brain stem and lateral gray horn of the second through fourth sacral segments of the spinal cord. The axons emerge as part of cranial or spinal nerves. The preganglionic axons synapse with postganglionic neurons in terminal ganglia, near or within visceral effectors.

The sympathetic division is primarily concerned with processes that expend energy. During stress, the sympathetic division sets into operation a series of reactions collectively called the ***fight-or-flight response,*** designed to help the body counteract the stress and return to homeostasis. During the fight-or-flight response, the heart and breathing rates increase and the blood sugar level rises, among other things.

The parasympathetic division is primarily concerned with activities that restore and conserve energy. It is thus called the ***rest-repose system.*** Under normal conditions, the parasympathetic division dominates the sympathetic division in order to maintain homeostasis.

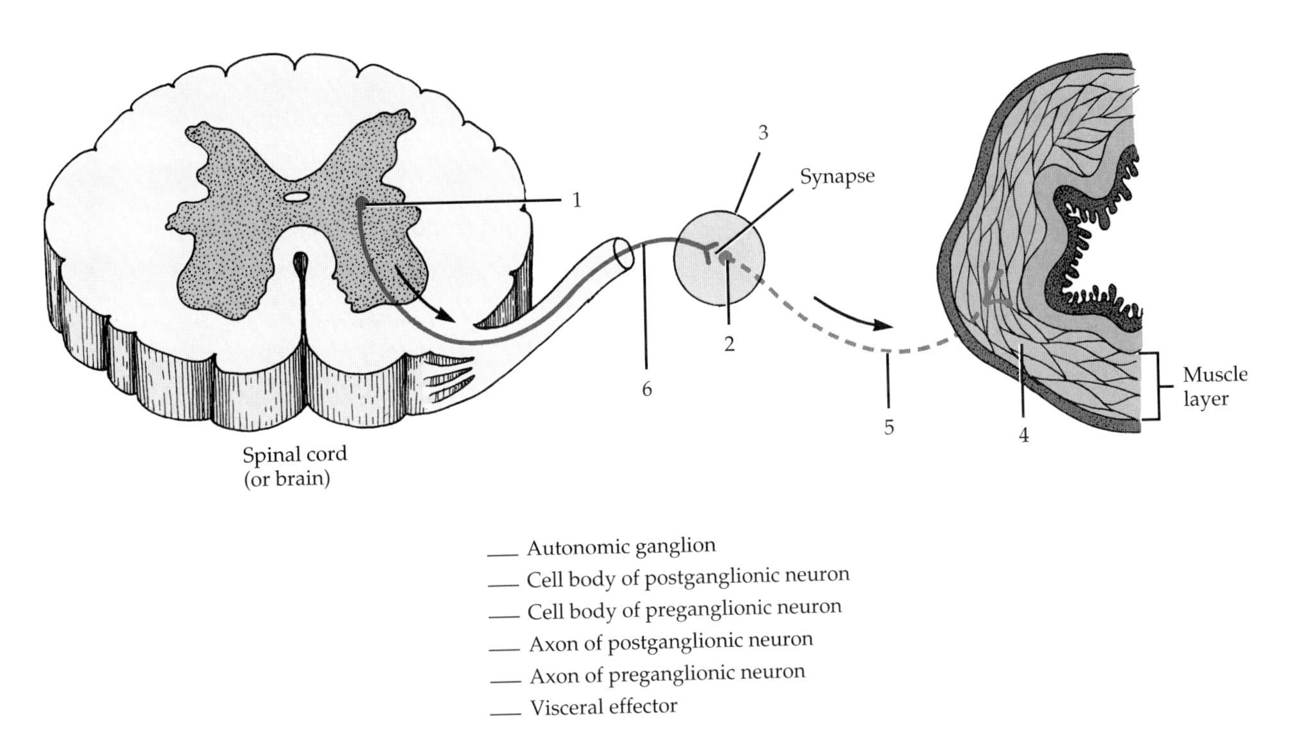

___ Autonomic ganglion
___ Cell body of postganglionic neuron
___ Cell body of preganglionic neuron
___ Axon of postganglionic neuron
___ Axon of preganglionic neuron
___ Visceral effector

FIGURE 13.23 Components of an autonomic pathway.

Using your textbook as a reference, write in the effects of sympathetic and parasympathetic stimulation for the visceral effectors listed in Table 13.3 on page 271.

ANSWER THE LABORATORY REPORT QUESTIONS AT THE END OF THE EXERCISE.

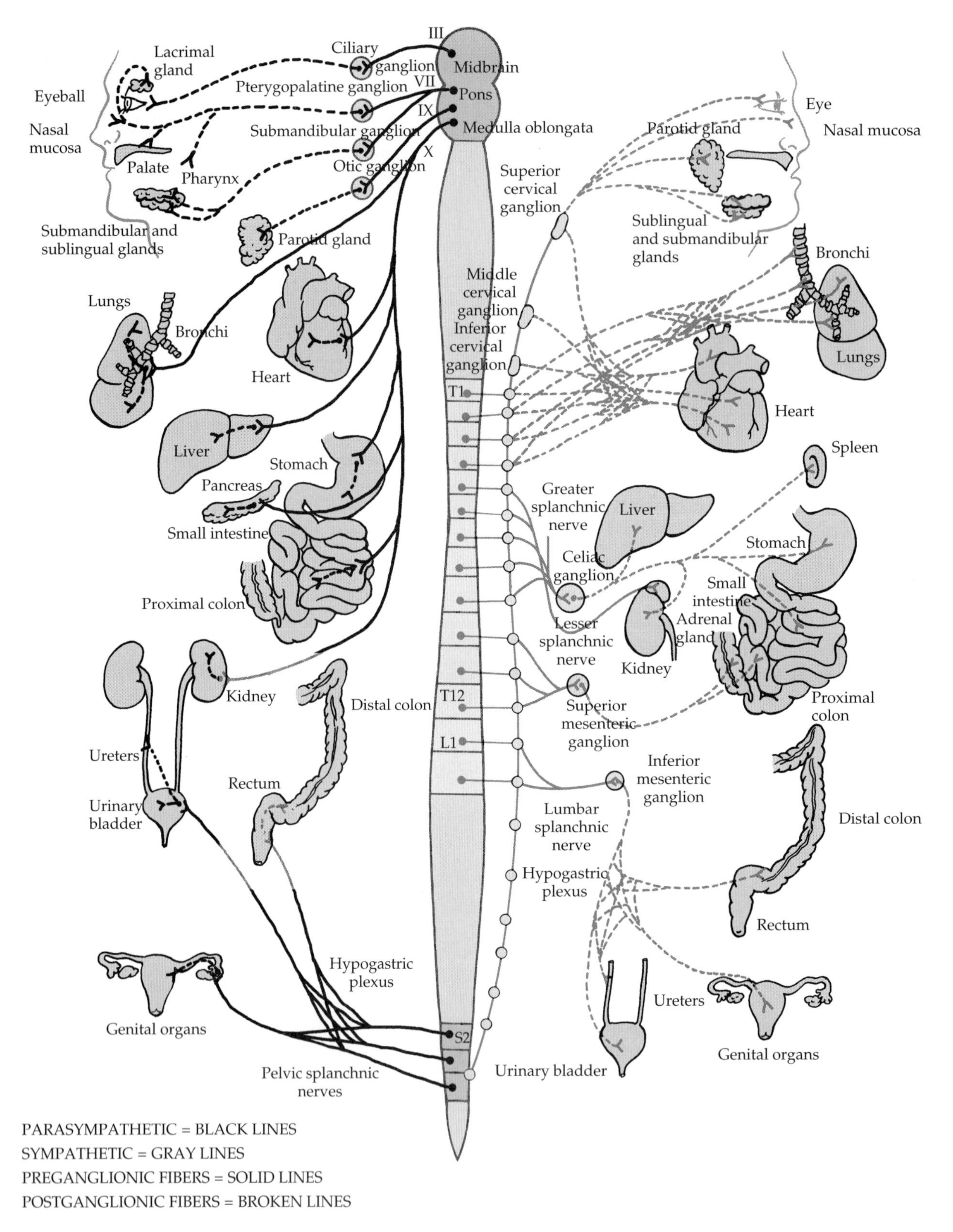

FIGURE 13.24 Structure of the autonomic nervous system.

TABLE 13.3
Activities of the Autonomic Nervous System

Visceral effector	Effect of sympathetic stimulation	Effect of parasympathetic stimulation
GLANDS		
Sweat		
Lacrimal (tear)		
Adrenal medulla		
Liver		
Kidney (juxtaglomerular cells)		
Pancreas		
SMOOTH MUSCLE		
Iris, radial muscle		
Iris, circular muscle		
Ciliary muscle of eye		
Salivary gland arterioles		
Gastric gland arterioles		
Intestinal gland arterioles		
Lungs, bronchial muscle		

TABLE 13.3 (*Continued*)

Visceral effector	Effect of sympathetic stimulation	Effect of parasympathetic stimulation
Heart arterioles		
Skin and mucosal arterioles		
Skeletal muscle arterioles		
Abdominal viscera arterioles		
Brain arterioles		
Systemic veins		
Gallbladder and ducts		
Stomach and intestines		
Kidney		
Ureter		
Spleen		
Urinary bladder		
Uterus		
Sex organs		
Hair follicles (arrector pili muscle)		
CARDIAC MUSCLE (HEART)		

LABORATORY REPORT QUESTIONS

Nervous System 13

Student ______________________ **Date** ______________________

Laboratory Section ______________________ **Score/Grade** ______________________

PART 1. Multiple Choice

_______ **1.** The tapered, conical portion of the spinal cord is the (a) filum terminale (b) conus medullaris (c) cauda equina (d) lumbar enlargement

_______ **2.** The superficial meninx composed of dense fibrous connective tissue is the (a) pia mater (b) arachnoid (c) dura mater (d) denticulate

_______ **3.** The portion of a spinal nerve that contains motor nerve fibers only is the (a) posterior root (b) posterior root ganglion (c) lateral root (d) anterior root

_______ **4.** The connective tissue covering around individual nerve fibers is the (a) endoneurium (b) epineurium (c) perineurium (d) ectoneurium

_______ **5.** On the basis of organization, which does *not* belong with the others? (a) pons (b) medulla oblongata (c) thalamus (d) midbrain

_______ **6.** The lateral ventricles are connected to the third ventricle by the (a) interventricular foramen (b) cerebral aqueduct (c) median aperture (d) lateral aperture

_______ **7.** The vital centers for heartbeat, respiration, and blood vessel diameter regulation are found in the (a) pons (b) cerebrum (c) cerebellum (d) medulla oblongata

_______ **8.** The reflex centers for movements of the head and trunk in response to auditory stimuli are located in the (a) inferior colliculi (b) medial geniculate nucleus (c) superior colliculi (d) ventral posterior nucleus

_______ **9.** Which thalamic nucleus controls general sensations and taste? (a) medial geniculate (b) ventral posterior (c) ventral lateral (d) ventral anterior

_______ **10.** Integration of the autonomic nervous system, control of body temperature, and the regulation of food intake and thirst are functions of the (a) pons (b) thalamus (c) cerebrum (d) hypothalamus

_______ **11.** The left and right cerebral hemispheres are separated from each other by the (a) central sulcus (b) transverse fissure (c) longitudinal fissure (d) insula

_______ **12.** Which structure does *not* belong with the others? (a) putamen (b) caudate nucleus (c) insula (d) globus pallidus

_______ **13.** Which peduncles connect the cerebellum with the midbrain? (a) superior (b) inferior (c) middle (d) lateral

_______ **14.** Which cranial nerve has the most anterior origin? (a) XI (b) IX (c) VII (d) IV

_______ **15.** Extensions of the pia mater that suspend the spinal cord and protect against shock are the (a) choroid plexuses (b) pyramids (c) denticulate ligaments (d) superior colliculi

_______ **16.** Which branch of a spinal nerve enters into formation of plexuses? (a) meningeal (b) dorsal (c) rami communicantes (d) ventral

_______ **17.** Which plexus innervates the upper limbs and shoulders? (a) sacral (b) brachial (c) lumbar (d) cervical

_______ **18.** How many pairs of thoracic spinal nerves are there? (a) 1 (b) 5 (c) 7 (d) 12

PART 2. Completion

19. The narrow, shallow groove on the posterior surface of the spinal cord is the _______________.

20. The space between the dura mater and wall of the vertebral canal is called the _______________.

21. In a spinal nerve, the cell bodies of sensory neurons are found in the _______________.

22. The superficial connective tissue covering around a spinal nerve is the _______________.

23. The middle meninx is referred to as the _______________.

24. The nuclei of origin for cranial nerves IX, X, XI, and XII are found in the _______________.

25. The portion of the brain containing the cerebral peduncles is the _______________.

26. Cranial nerves V, VI, VII, and VIII have their nuclei of origin in the _______________.

27. A shallow downfold of the cerebral cortex is called a(n) _______________.

28. The _______________ separates the frontal lobe of the cerebrum from the parietal lobe.

29. White matter tracts of the cerebellum are called _______________.

30. The space between the dura mater and the arachnoid is referred to as the _______________.

31. Together, the thalamus and hypothalamus constitute the _______________.

32. Cerebrospinal fluid passes from the third ventricle into the fourth ventricle through the

_______________.

33. The cerebrum is separated from the cerebellum by the _______________ fissure.

34. The branches of a spinal nerve that are components of the autonomic nervous system are known as

_______________.

35. The plexus that innervates the buttocks, perineum, and lower limbs is the _______________ plexus.

36. There are _______________ pairs of spinal nerves.

37. The part of the brain that coordinates subconscious movements in skeletal muscles is the

_______________.

38. The cell bodies of _______________ neurons of the ANS are found inside autonomic ganglia.

39. The portion of the ANS concerned with the fight-or-flight response is the _______________ division.

40. The autonomic ganglia that are anterior to the vertebral column and close to large abdominal

arteries are called _______________ ganglia.

41. The ______________________ nerve innervates the diaphragm.

42. The ______________________ tract conveys nerve impulses for touch and pressure.

43. The flexor muscles of the thigh and extensor muscles of the leg are innervated by the ______________________ nerve.

44. The ______________________ tract conveys nerve impulses related to muscle tone and posture.

45. The ______________________ nerve supplies the extensor muscles of the arm and forearm.

46. The ______________________ area of the cerebral cortex receives sensations from cutaneous, muscular, and visceral receptors in various parts of the body.

47. The portion of the cerebral cortex that translates thoughts into speech is the ______________________ area.

48. The term ______________________ refers to aggregations of myelinated processes from many neurons.

49. A(n) ______________________ spinal nerve contains preganglionic autonomic nervous system neurons.

50. ______________________-order neurons carry sensory information from the spinal cord and brain stem to the thalamus.

51. The part of the brain that allows crude appreciation of some sensations, such as pain, temperature, and pressure, is the ______________________.

52. During stress, the sympathetic division of the autonomic nervous system sets into operation a series of reactions collectively called the ______________________.

PART 3. Matching

Using B for brachial, C for cervical, L for lumbar, and S for sacral, indicate to which plexus the following nerves belong.

________ **53.** Sciatic
________ **54.** Femoral
________ **55.** Radial
________ **56.** Ansa cervicalis
________ **57.** Median
________ **58.** Obturator
________ **59.** Pudendal
________ **60.** Tibial
________ **61.** Ilioinguinal
________ **62.** Ulnar
________ **63.** Phrenic

PART 4. Matching

_______ **64.** Posterior column–medial meniscus pathway

_______ **65.** Spinocerebellar tracts

_______ **66.** Lateral corticospinal tracts

_______ **67.** Lateral spinothalamic tract

_______ **68.** Anterior corticospinal tracts

_______ **69.** Tectospinal tract

_______ **70.** Lateral reticulospinal tract

A. Conduct subconscious proprioceptive input to the cerebellum

B. Convey impulses that control movements of the neck and part of the trunk

C. Conducts impulses for conscious proprioception and most tactile sensations

D. Conveys impulses that facilitate flexon reflexes, inhibit extensor reflexes, and decrease muscle tone in muscles of the axial skeleton and proximal limbs

E. Control skilled movements of the distal limbs

F. Conduct mainly pain and temperature impulses

G. Conveys impulses that move the head and eyes in response to visual stimuli

General and Special Senses

14

Consider what would happen if you could not feel the pain of a hot pot handle or an inflamed appendix or if you could not see, hear, smell, taste, or maintain your balance. In short, if you could not "sense" your environment and make the necessary adjustments, you could not survive very well on your own.

Sensation refers to the awareness of external or internal conditions of the body. ***Perception*** refers to the conscious awareness and interpretation of sensations.

For a sensation to be detected (either consciously or subconsciously) four events must occur:

1. ***Stimulation*** A ***stimulus,*** or change in the environment, capable of activating certain sensory neurons must be present.
2. ***Transduction*** A ***sensory receptor*** or ***sense organ*** must receive the stimulus and ***transduce*** (convert) it to a receptor potential or generator potential. A sensory receptor or sense organ is a specialized structure or a specialized type of neuron that is selectively sensitive to particular stimuli.
3. ***Impulse generation and conduction*** When a receptor or generator potential reaches threshold, it elicits action potentials (nerve impulses) that are conducted along a sensory neural pathway to the central nervous system. The function of a receptor or generator potential is to convert a stimulus to an action potential.
4. ***Integration*** A region of the CNS must integrate the information carried via the action potentials into a sensation. Most conscious sensations or perceptions occur in the cerebral cortex of the brain after passing through the thalamus. In other words, you see, hear, and feel in the brain. You seem to see with your eyes, hear with your ears, and feel pain in an injured part of your body because sensory impulses from each part of the body arrive in a specific region in the cerebral cortex, which interprets the sensation as coming from the stimulated sensory receptors.

A. CHARACTERISTICS OF SENSATIONS

One characteristic of many sensations is ***adaptation,*** that is, a change in sensitivity, usually a decrease, even though a stimulus is still being applied. For example, when you first get into a tub of hot water, you might feel an intense burning sensation. However, after a brief period of time the sensation decreases to one of comfortable warmth, even though the stimulus (hot water) is still present and has not diminished in intensity.

A second characteristic is that of ***afterimages,*** that is, the persistence of a sensation after the stimulus has been removed. One common example of an afterimage occurs when you look at a bright light and then look away. You will still see the light for several seconds afterward.

A third characteristic of sensations is that of ***modality.*** Modality is the distinct quality by which one sensation may be distinguished from another. For example, temperature change, pain, touch, body position, hearing, vision, smell, and taste are all distinctive sensations because the brain perceives each differently. A given sensory neuron carries only one modality.

B. CLASSIFICATION OF RECEPTORS

One convenient method of classifying receptors is by their location:

1. ***Exteroceptors*** (EKS'-ter-ō-sep'-tors), located at or near the body surface, provide information about the *external* environment. They transmit sensations such as hearing, sight, smell, taste, touch, pressure, temperature, and pain.
2. ***Interoceptors*** or ***visceroceptors*** (VIS-er-ō-sep'-tors), located in blood vessels and viscera,

provide information about the *internal* environment. These produce sensations such as pain, taste, fatigue, hunger, thirst, and nausea.

3. ***Proprioceptors*** (PRŌ-prē-ō-sep'-tors; *proprio* = one's own) or stretch receptors, located in muscles, tendons, joints, and the internal ear, are stimulated by stretching or movement. They provide us with equilibrium and with an awareness of the location and movement of body parts, a phenomenon called ***proprioception.***

Another classification of receptors is based on the type of stimuli they detect:

1. ***Mechanoreceptors*** detect mechanical pressure or stretching. They provide sensations of touch, pressure, vibration, proprioception, hearing, equilibrium, and blood pressure.
2. ***Thermoreceptors*** detect changes in temperature.
3. ***Nociceptors*** (NŌ-sē-sep'-tors; *noci* = harmful) detect pain.
4. ***Photoreceptors*** detect light that strikes the retina of the eye.
5. ***Chemoreceptors*** detect chemicals in the mouth (taste), nose (smell), and body fluids.

Receptors vary considerably in complexity. The anatomically simplest receptors are called free nerve endings because they have no apparent structural specializations. Examples are receptors for pain and temperature. Other receptors, such as those for touch, pressure, vibration, and proprioception are somewhat more complex and have distinctive structure. Receptors for pain, temperature, touch, pressure, and vibration are associated with ***somatic (general) senses.*** There are still other receptors that are quite complex and have distinctive structure. These receptors are involved with smell, taste, vision, hearing, and equilibrium and are associated with ***special senses.***

Following is a classification of receptors associated with somatic (general) senses and special senses.

1. Receptors associated with somatic (general) senses
   *a. ***Tactile*** (TAK-tīl; *tact* = touch) ***receptors*** detect sensations related to touch, pressure, and vibration.
   *b. ***Thermoreceptors*** detect sensations related to warmth and coolness.
   *c. ***Nociceptors*** detect sensations related to pain.
   d. ***Proprioceptors*** (prō'-prē-ō-SEP-tors) detect sensations related to proprioception.

   * Taken together, tactile, thermoreceptive, and pain sensations are referred to as ***cutaneous*** (kyoo-TĀ-nē-us; *cutis* = skin) ***sensations.***

2. Receptors associated with special senses.
   a. ***Olfactory*** (ōl-FAK-tō-rē) ***receptors*** detect sensations related to smell.
   b. ***Gustatory*** (GUS-ta-tō-rē) ***receptors*** detect sensations related to taste.
   c. ***Visual receptors*** detect sensations related to sight.
   d. ***Auditory*** (AW-di-tō-rē) ***receptors*** detect sensations related to hearing.
   e. ***Equilibrium*** (ē'-kwi-LIB-rē-um) ***receptors*** detect sensations related to orientation of the body.

C. RECEPTORS FOR GENERAL SENSES

1. Tactile Receptors

Although touch, pressure, and vibration are classified as separate sensations, all are detected by mechanoreceptors.

Touch sensations generally result from stimulation of tactile receptors in the skin or in tissues immediately deep to the skin. *Crude touch* refers to the ability to perceive that something has touched the skin, although the size or texture cannot be determined. *Discriminative touch* refers to the ability to recognize exactly what point on the body is touched. Receptors for touch include corpuscles of touch, hair root plexuses, and type I and type II cutaneous mechanoreceptors. ***Corpuscles of touch*** or ***Meissner's*** (MĪS-ners) ***corpuscles*** are receptors for discriminative touch that are found in dermal papillae. They have already been discussed in Exercise 5. ***Hair root plexuses*** are dendrites arranged in networks around hair follicles that detect movement when hairs are disturbed. ***Type I cutaneous mechanoreceptors,*** also called ***tactile*** or ***Merkel*** (MER-kel) ***discs,*** are the flattened portions of dendrites of sensory neurons that make contact with epidermal cells of the stratum basale called Merkel cells. They are distributed in many of the same locations as corpuscles of touch and also function in discriminative touch. ***Type II cutaneous mechanoreceptors,*** or ***end organs of Ruffini,***

are embedded deeply in the dermis and in deeper tissues of the body. They detect heavy and continuous touch sensations.Touch receptors are most numerous in the fingertips, palms, and soles. They are also abundant in the eyelids, tip of the tongue, lips, nipples, clitoris, and tip of the penis.

Examine prepared slides of corpuscles of touch (Meissner's corpuscles), hair root plexuses, type I cutaneous mechanoreceptors (tactile or Merkel discs), and type II cutaneous mechanoreceptors (end organs of Ruffini). With the aid of your textbook, draw each of the receptors in the spaces that follow.

Corpuscles of touch (Meissner's corpuscles)

Hair root plexuses

Type I cutaneous mechanoreceptors (tactile or Merkel discs)

Type II cutaneous mechanoreceptors (end organs of Ruffini)

Pressure sensations generally result from stimulation of tactile receptors in deeper tissues. ***Pressure*** is a sustained sensation that is felt over a larger area than touch. Receptors for pressure sensations include lamellated corpuscles and type II cutaneous mechanoreceptors. ***Lamellated (Pacinian) corpuscles*** are found in the subcutaneous layer and have already been discussed in Exercise 5.

Pressure receptors are found in the subcutaneous tissue under the skin, in the deep subcutaneous tissues that lie under mucous membranes, around joints and tendons, in the perimysium of muscles, in the mammary glands, in the external genitals of both sexes, and in some viscera.

Examine a prepared slide of lamellated (Pacinian) corpuscles. With the aid of your textbook, draw the receptors in the space that follows.

Lamellated (Pacinian) corpuscles

Vibration sensations result from rapidly repetitive sensory signals from tactile receptors. Receptors for vibration include corpuscles of touch (Meissner's corpuscles) that detect low-frequency vibration and lamellated (Pacinian) corpuscles that detect high-frequency vibration.

2. Thermoreceptors

The cutaneous receptors for the sensation of warmth and coolness are free (naked) nerve endings that are widely distributed in the dermis and subcutaneous connective tissue. They are also located in the cornea of the eye, tip of the tongue, and external genitals.

3. Nociceptors

Receptors for ***pain,*** called ***nociceptors,*** are free (naked) nerve endings (see Figure 5.1). Pain receptors are found in practically every tissue of the body and adapt only slightly or not at all. They may be excited by any type of stimulus. Excessive stimulation of any sense organ causes pain. For example, when stimuli for other sensations such as touch, pressure, heat, and cold reach a certain threshold, they stimulate pain receptors as well. Pain receptors, because of their sensitivity to all stimuli, have a general protective function of informing us of changes that could be potentially dangerous to health or life. Adaptation to pain does not readily occur. This low level of adaptation is important, because pain indicates disorder or disease. If we became used to it and ignored it, irreparable damage could result.

4. Proprioceptors

An awareness of the activities of muscles, tendons, and joints and equilibrium is provided by the ***proprioceptive (kinesthetic) sense.*** It informs us of the degree to which tendons are tensed and muscles are contracted. The proprioceptive sense enables us to recognize the location and rate of movement of one part of the body in relation to other parts. It also allows us to estimate weight and to determine the muscular effort necessary to perform a task. With the proprioceptive sense, we can judge the position and movements of our limbs without using our eyes when we walk, type, play a musical instrument, or dress in the dark.

Proprioceptors are located in skeletal muscles and tendons, in and around joints, and in the internal ear. They adapt only slightly. This slight adaptation is beneficial because the brain must be apprised of the status of different parts of the body at all times so that adjustments can be made to ensure coordination.

Receptors for proprioception are as follows. The ***joint kinesthetic*** (kin'-es-THET-ik) ***receptors*** are located in the articular capsules of joints and ligaments about joints. These receptors provide feedback information on the degree and rate of angulation (change of position) of a joint. ***Muscle spindles*** are specialized muscle fibers (cells) that consist of endings of sensory neurons. They are located in nearly all skeletal muscles and are more numerous in the muscles of the limbs. Muscle spindles provide feedback information on the degree of muscle stretch. This information is relayed to the central nervous system to assist in the coordination and efficiency of muscle contraction. ***Tendon organs (Golgi tendon organs)*** are located at the junction of a skeletal muscle and tendon. They function by sensing the tension applied to a tendon and the force of contraction of associated muscles. The information is translated by the central nervous system.

Proprioceptors in the internal ear are the maculae and cristae that function in equilibrium. These are discussed at the end of the exercise.

Examine prepared slides of joint kinesthetic receptors, muscle spindles, and tendon organs (Golgi tendon organs). With the aid of your textbook, draw the receptors in the spaces that follow.

Joint kinesthetic receptors

Muscle spindles

Tendon organs (Golgi tendon organs)

D. SOMATIC SENSORY PATHWAYS

Most input from somatic receptors on one side of the body crosses over to the opposite side in the spinal cord or brain stem before ascending to the thalamus. It then projects from the thalamus to the somatosensory area of the cerebral cortex, where conscious sensations result. Axon collaterals (branches) of somatic sensory neurons also carry signals into the cerebellum and the reticular formation of the brain stem. Two general pathways lead from sensory receptors to the cerebral cortex: the posterior column–medial lemniscus pathway and the anterolateral (spinothalamic) pathway.

1. Posterior Column–Medial Lemniscus Pathway

Nerve impulses for conscious proprioception and most tactile sensations ascend to the cerebral cortex along a common pathway formed by three-neuron sets.

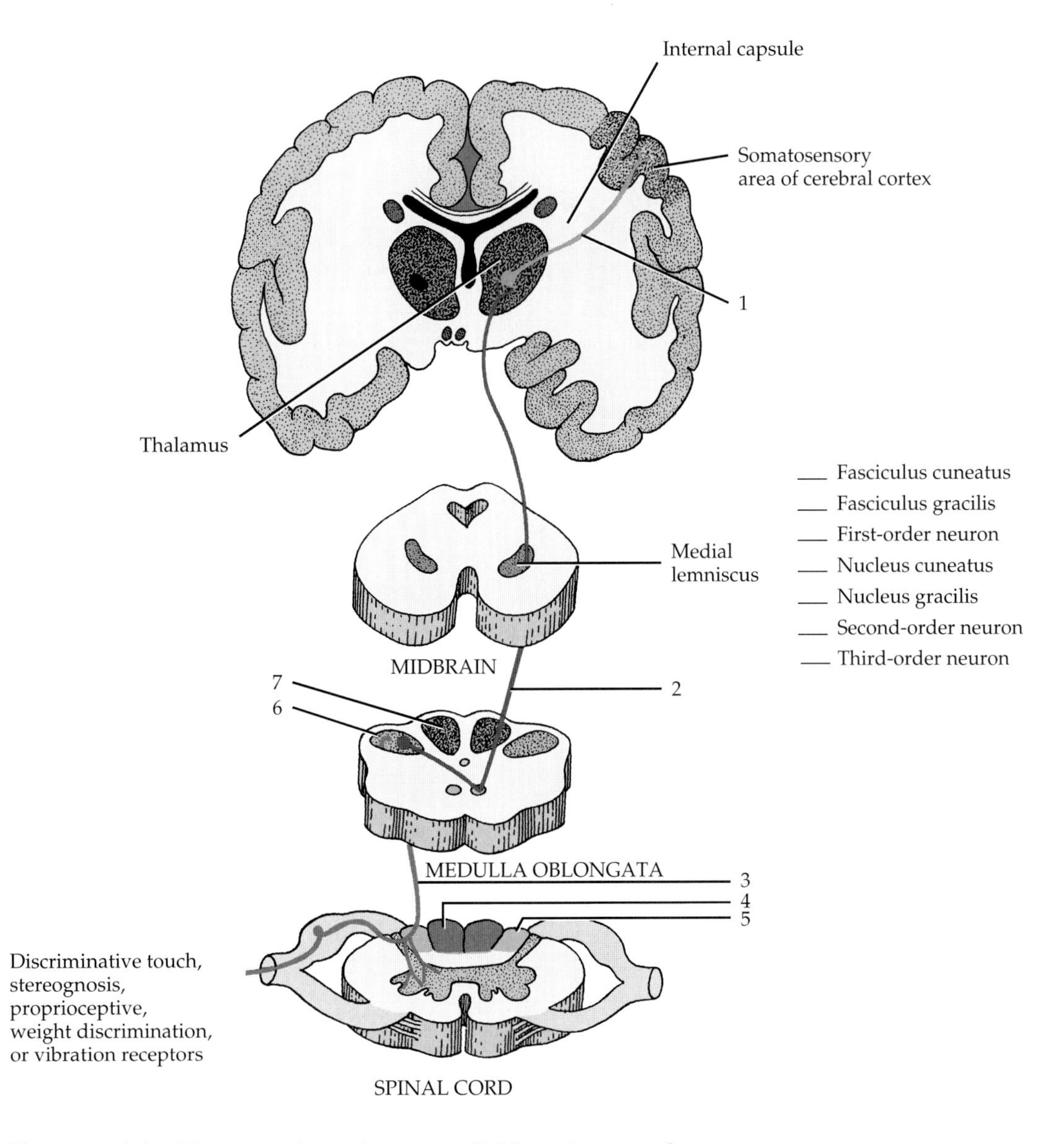

FIGURE 14.1 The posterior column–medial lemniscus pathway.

Based on the description provided on page 247, label the components of the posterior column–medial lemniscus pathway in Figure 14.1.

2. Anterolateral (Spinothalamic) Pathways

The *anterolateral (spinothalamic) pathways* carry mainly pain and temperature impulses. In addition, they relay tickle, itch, and some tactile impulses, which give rise to a very crude, not well-localized touch or pressure sensation. Like the posterior column–medial lemniscus pathway, the anterolateral pathways are also composed of three-neuron sets. The axon of the second-order neuron extends to the opposite side of the spinal cord and passes superiorly to the brain stem in either the *lateral spinothalamic tract* or *anterior spinothalamic tract*. The lateral spinothalamic tract conveys sensory impulses for pain and temperature, whereas the anterior spinothalamic tract conveys tickle, itch, and crude touch and pressure impulses.

Based on the descriptions provided on page 247, label the components of the lateral spinothalamic tract in Figure 14.2 on page 282 and the anterior spinothalamic tract in Figure 14.3 on page 283.

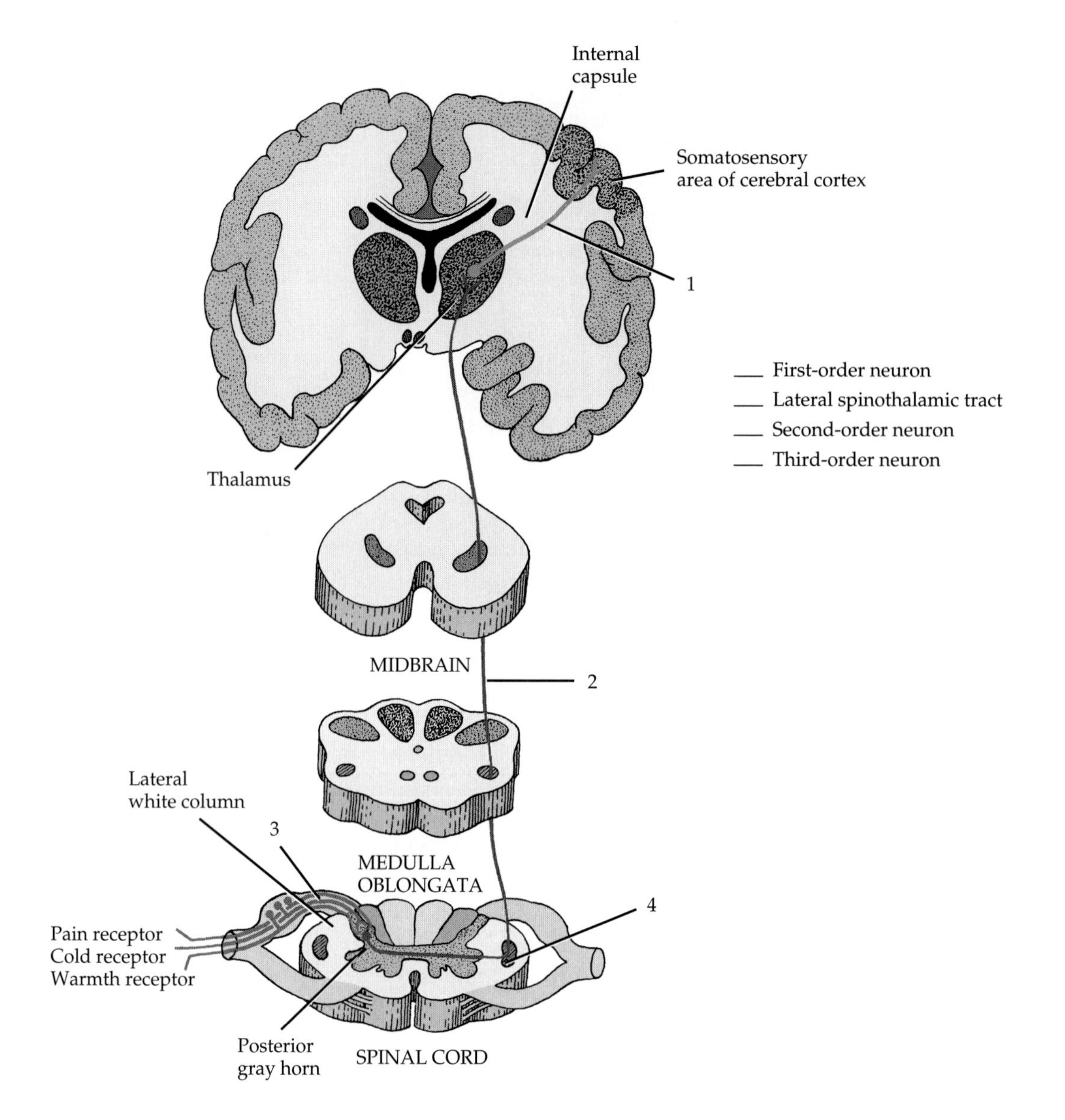

FIGURE 14.2 The lateral spinothalamic pathway.

E. OLFACTORY SENSATIONS

1. Olfactory Receptors

The receptors for the ***olfactory*** (ol-FAK-tō-rē; *olfactus* = smell) ***sense*** are found in the nasal epithelium in the superior portion of the nasal cavity on either side of the nasal septum. The nasal epithelium consists of three principal kinds of cells: olfactory receptors, supporting cells, and basal cells. The ***olfactory receptors (cells)*** are bipolar neurons. Their cell bodies lie between the supporting cells. The distal (free) end of each olfactory cell contains a knob-shaped dendrite from which six to eight cilia, called ***olfactory hairs,*** protrude. The ***supporting (sustentacular) cells*** are columnar epithelial cells of the mucous membrane that lines the nose. ***Basal cells*** lie between the bases of the supporting cells and produce new olfactory receptors. Within the connective tissue deep to the olfactory epithelium are ***olfactory (Bowman's) glands*** that secrete mucus.

The unmyelinated axons of the olfactory receptors unite to form the ***olfactory (I) nerves,*** which pass through foramina in the cribriform plate of the ethmoid bone. The olfactory nerves terminate in paired masses of gray matter, the ***olfactory bulbs,*** which lie inferior to the frontal

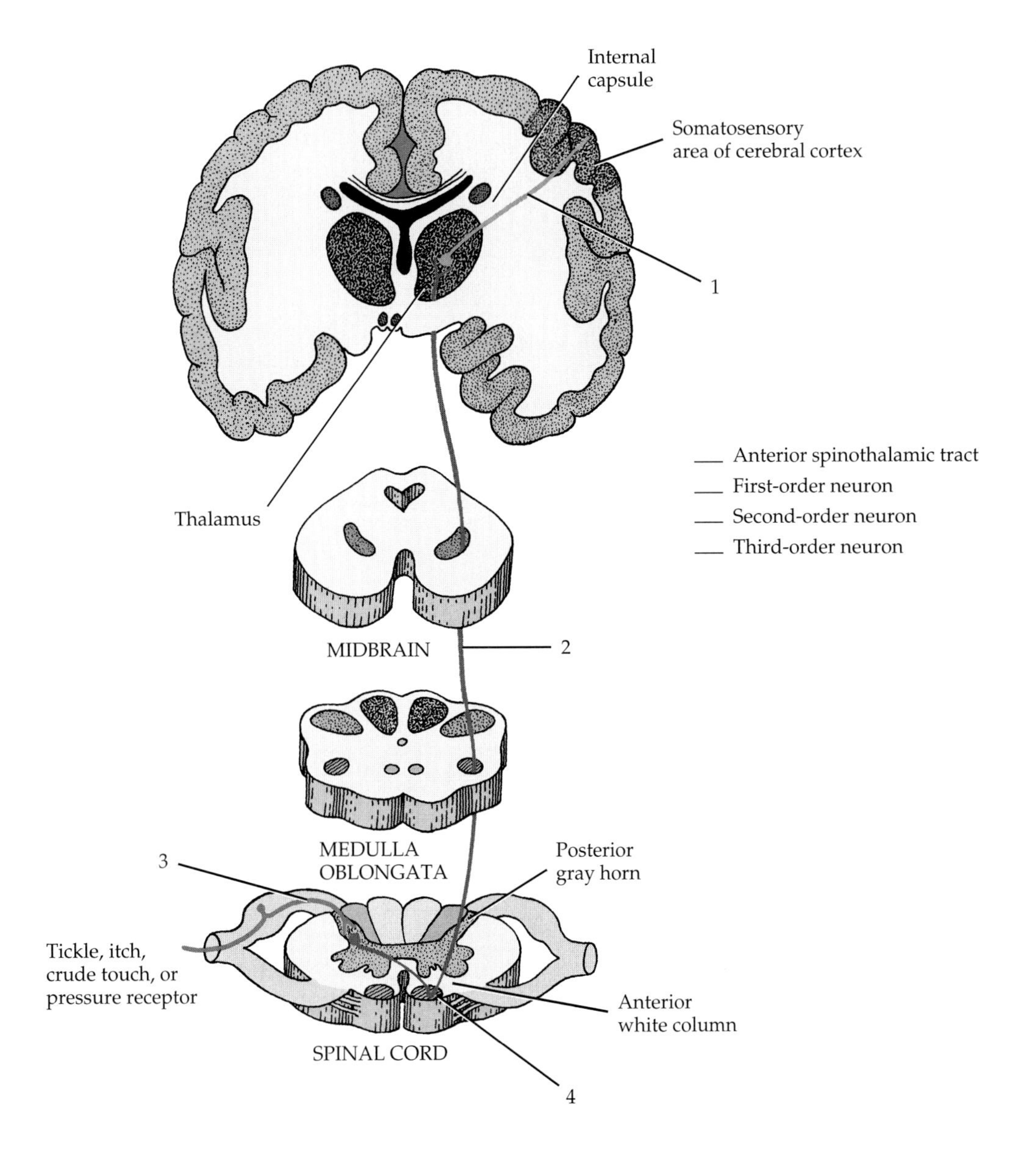

FIGURE 14.3 The anterior spinothalamic pathway.

lobes of the cerebrum on either side of the crista galli of the ethmoid bone. The first synapse of the olfactory neural pathway occurs in the olfactory bulbs between the axons of the olfactory (I) nerves and the dendrites of neurons inside the olfactory bulbs. Axons of these neurons run posteriorly to form the ***olfactory tract.*** From here, nerve impulses are conveyed to the primary olfactory area in the temporal lobe of the cerebral cortex. In the cortex, the nerve impulses are interpreted as odor and give rise to the sensation of smell.

Adaptation happens quickly, especially adaptation to odors. For this reason, we become accustomed to some odors and are also able to endure unpleasant ones. Rapid adaptation also accounts for the failure of a person to detect gas that accumulates slowly in a room.

Label the structures associated with olfaction in Figure 14.4 on page 284.

Now examine a slide of the olfactory epithelium under high power. Identify the olfactory receptors and supporting cells and label Figure 14.5 on page 285.

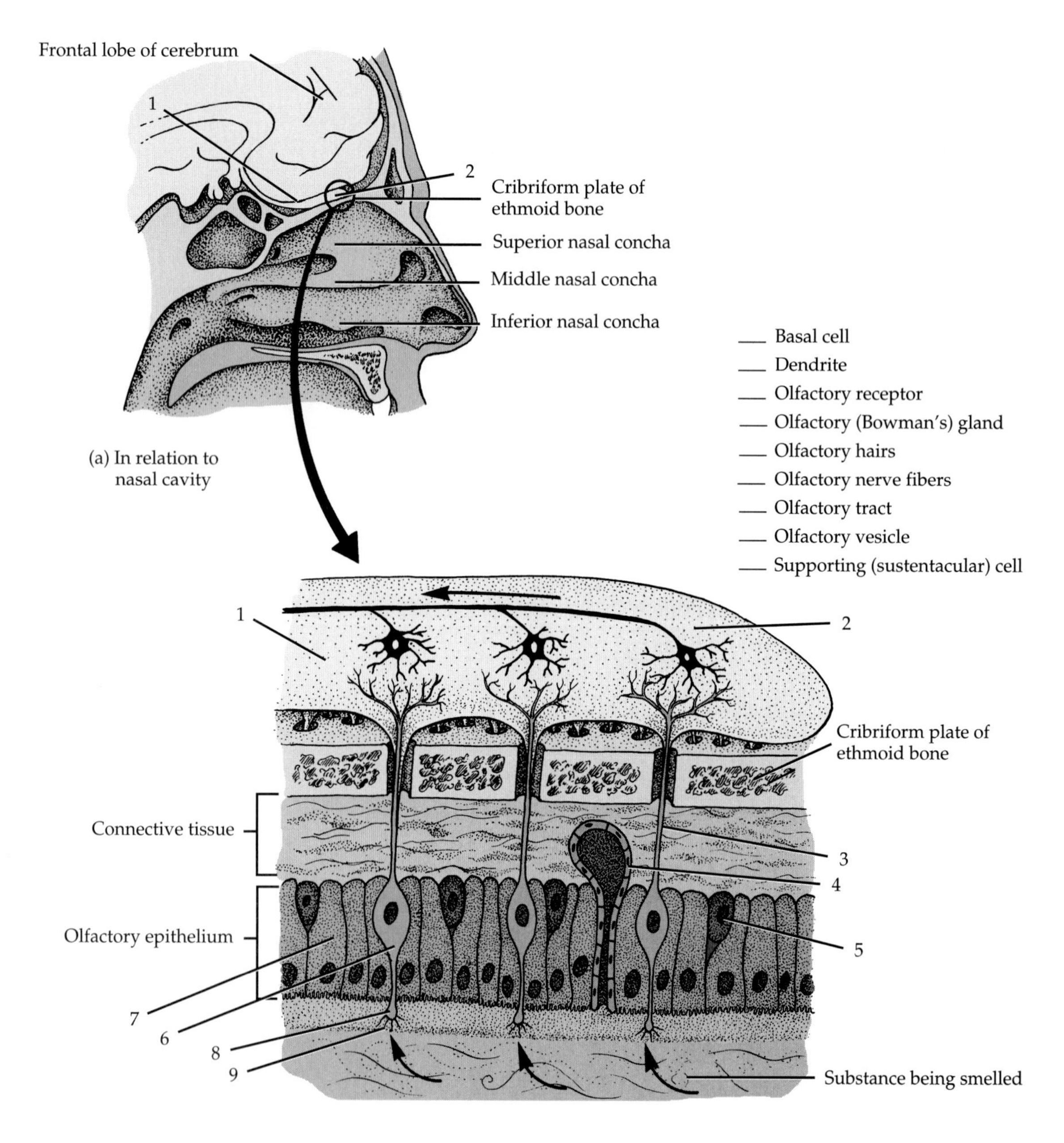

FIGURE 14.4 Olfactory receptors.

F. GUSTATORY SENSATIONS

1. Gustatory Receptors

The receptors for ***gustatory*** (GUS-ta-tō-rē; *gusto* = taste) ***sensations,*** or sensations of taste, are located in the taste buds. Although taste buds are most numerous on the tongue, they are also found on the soft palate, pharynx and larynx. ***Taste buds*** are oval bodies consisting of three kinds of cells: supporting cells, gustatory receptors, and basal cells. The ***supporting (sustentacular) cells*** are specialized epithelial cells that form a capsule. Inside each capsule are about 50 ***gustatory receptors (cells).***

Each gustatory receptor contains a hairlike process ***(gustatory hair)*** that projects to the surface through an opening in the taste bud called the ***taste pore.*** Gustatory cells make contact with taste stimuli through the taste pore. ***Basal cells*** are found at the periphery of the taste bud and produce new supporting cells, which then develop into gustatory receptors.

Examine a slide of taste buds and label the structures associated with gustation in Figure 14.6.

Taste buds are located in some connective tissue elevations on the tongue called **papillae** (pa-PILL-ē). They give the upper surface of the tongue its rough texture and appearance. ***Circumvallate*** (ser-kum-VAL-āt) ***papillae*** are circular and form an inverted V-shaped row at the posterior portion of the tongue. ***Fungiform*** (FUN-ji-form) ***papillae*** are knoblike elevations scattered over the entire surface of the tongue. All circumvallate and most fungiform papillae contain taste buds. ***Filiform*** (FIL-i-form) ***papillae*** are threadlike structures that are also distributed over the entire surface of the tongue.

Have your partner protrude his or her tongue and examine its surface with a hand lens to identify the shape and position of the papillae.

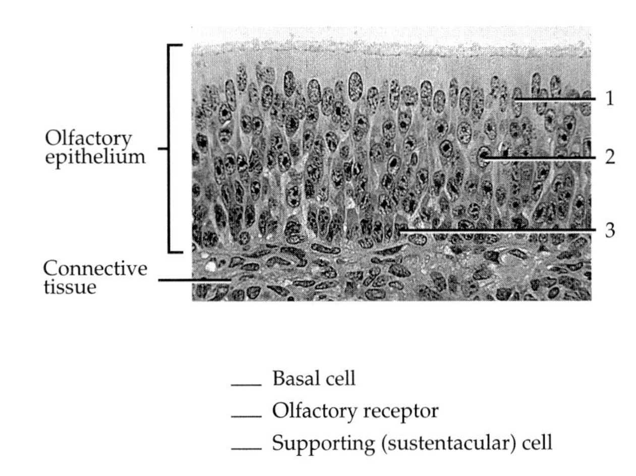

FIGURE 14.5 Photomicrograph of the olfactory epithelium.

G. VISUAL SENSATIONS

Structures related to ***vision*** are the eyeball (which is the receptor organ for visual sensations), optic (II) nerve, brain, and accessory structures. The extrinsic muscles of the eyeball may be reviewed in Figure 10.4.

1. Accessory Structures

Among the ***accessory structures*** are the eyebrows, eyelids, eyelashes, lacrimal (tearing) apparatus, and extrinsic eye muscles. ***Eyebrows*** protect the eyeball from falling objects, prevent perspiration from getting into the eye, and shade the eye from the direct rays of the sun. ***Eyelids,*** or ***palpebrae*** (PAL-pe-brē), consist primarily of skeletal muscle covered externally by skin. The underside of the muscle is lined by a mucous membrane called the ***palpebral conjunctiva*** (kon-junk-TĪ-va). The ***bulbar (ocular) conjunctiva*** covers the surface of the eyeball. Also within the eyelids are ***tarsal (Meibomian) glands,*** modified sebaceous glands whose oily secretion keeps the eyelids from adhering to each other. Infection of these glands produces a ***chalazion*** (cyst) in the eyelid. Eyelids shade the eyes during sleep, protect the eyes from light rays and foreign objects, and spread lubricating secretions over the surface of the eyeballs. Projecting from the border of each eyelid is a row of short, thick hairs, the ***eyelashes.*** Sebaceous glands at the base of the hair follicles of the eyelashes, called ***sebaceous ciliary glands (glands of Zeis),*** pour a lubricating fluid into the follicles. An infection of these glands is called a ***sty.***

The ***lacrimal*** (LAK-ri-mal; *lacrima* = tear) ***apparatus*** consists of a group of structures that manufacture and drain tears. Each ***lacrimal gland*** is located at the superior lateral portion of both orbits. Leading from the lacrimal glands are 6 to 12 ***excretory lacrimal ducts*** that empty tears onto the surface of the conjunctiva of the upper lid. From here, the tears pass medially and enter two small openings called ***lacrimal puncta*** that appear as two small pores, one in each papilla of the eyelid, at the medial commissure of the eye. The tears then pass into two ducts, the ***lacrimal canals,*** and are next conveyed into the lacrimal sac. The ***lacrimal sac*** is the superior expanded portion of the ***nasolacrimal duct,*** a canal that carries the tears into the nasal cavity. Tears clean, lubricate, and moisten the external surface of the eyeball.

Label the parts of the lacrimal apparatus in Figure 14.7 on page 287.

2. Structure of the Eyeball

The eyeball can be divided into three principal layers: (1) fibrous tunic, (2) vascular tunic, and (3) retina (nervous tunic). See Figure 14.8 on page 288.

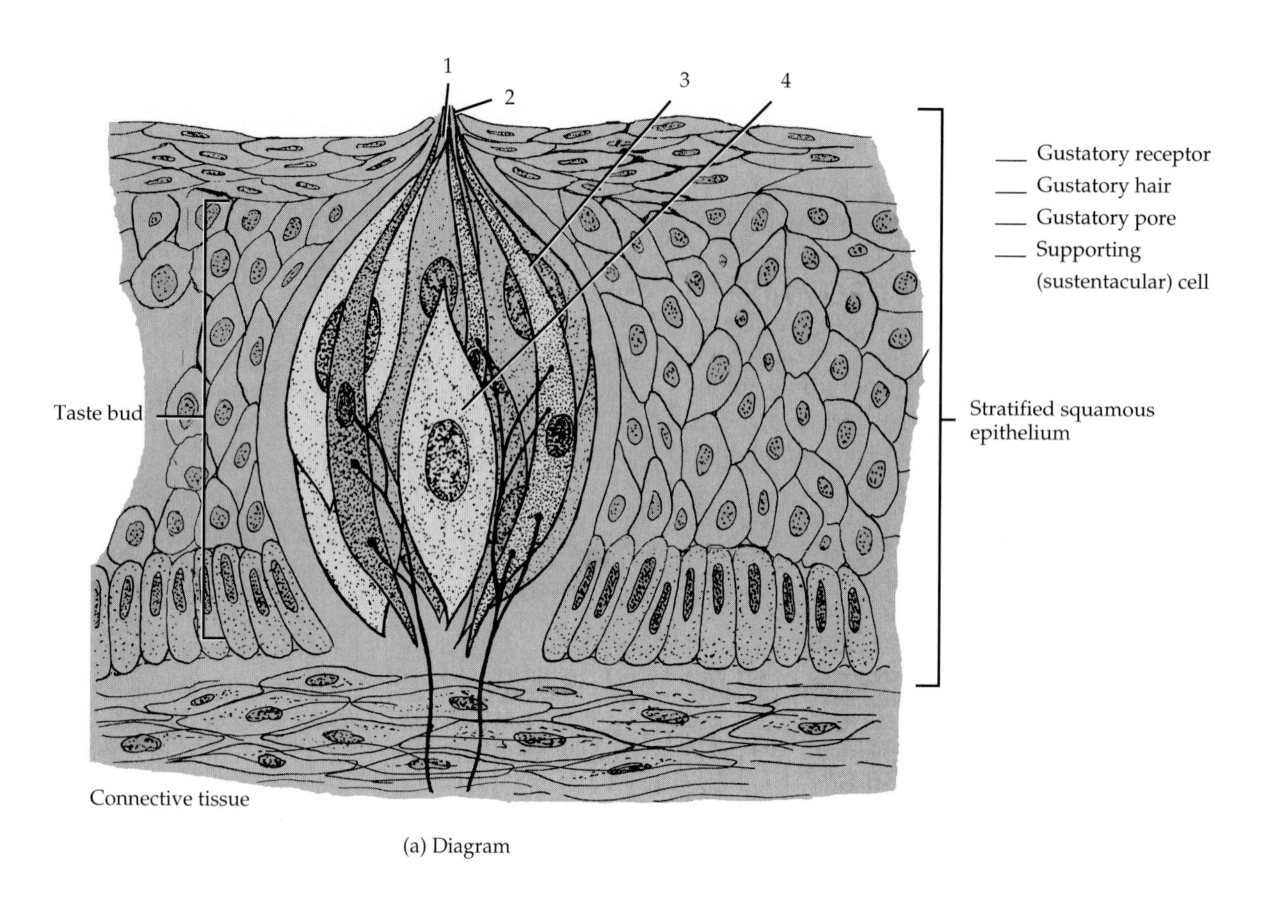

(a) Diagram

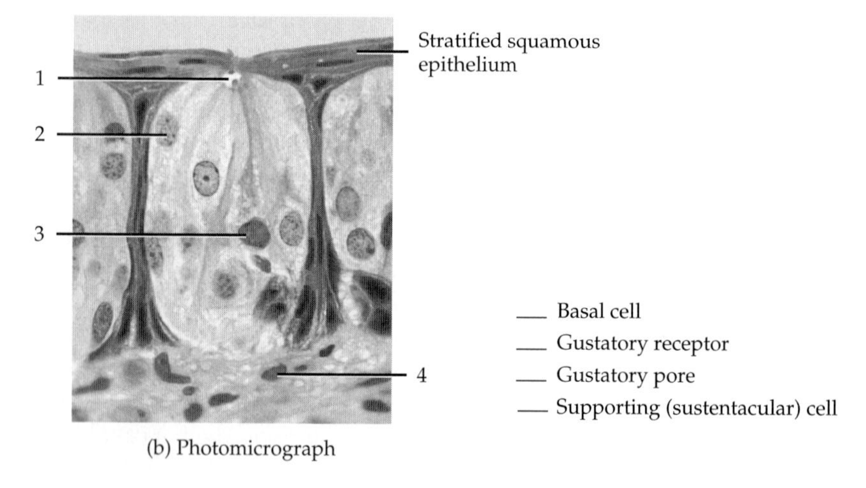

(b) Photomicrograph

FIGURE 14.6 Structure of a tastebud.

a. FIBROUS TUNIC

The ***fibrous tunic*** is the outer coat of the eyeball. It is divided into the posterior sclera and the anterior cornea. The ***sclera*** (SKLE-ra; *skleros* = hard), called the "white of the eye," is a coat of dense connective tissue that covers all the eyeball except the most anterior portion (cornea). The sclera gives shape to the eyeball and protects its inner parts. The anterior portion of the fibrous tunic is known as the ***cornea*** (KOR-nē-a). This nonvascular, transparent coat covers the colored iris. Because it is curved, the cornea helps focus light. The outer surface of the cornea contains epithelium that is continuous with the epithelium of the bulbar conjunctiva. At the junction of the sclera and cornea is the ***scleral venous sinus (canal of Schlemm).***

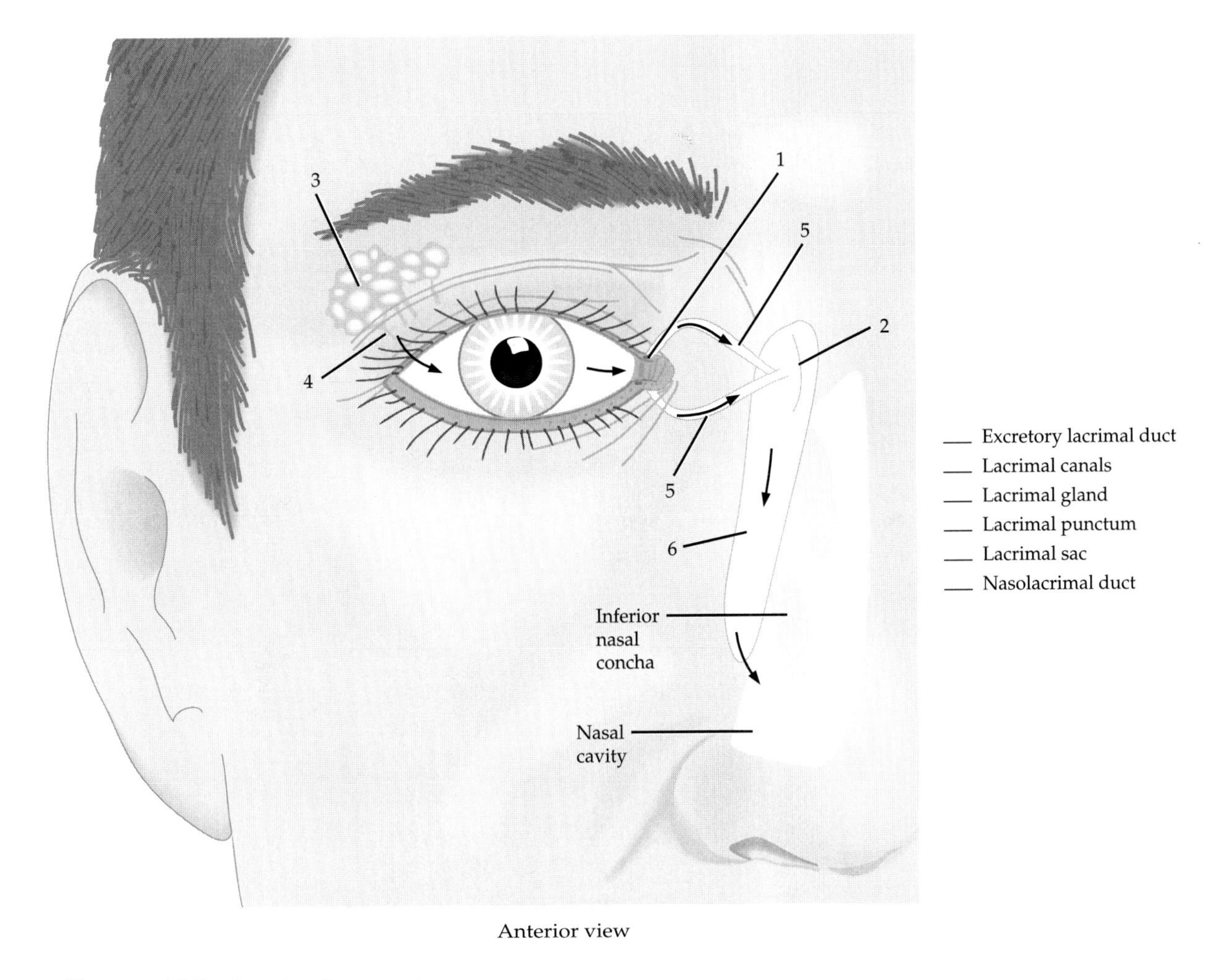

FIGURE 14.7 Lacrimal apparatus.

b. VASCULAR TUNIC

The ***vascular tunic*** is the middle layer of the eyeball and consists of three portions: choroid, ciliary body, and iris. The ***choroid*** (KŌ-royd) is the posterior portion of the vascular tunic. It is a thin, dark brown membrane that lines most of the internal surface of the sclera and contains blood vessels and melanin. The choroid absorbs light rays so they are not reflected back out of the eyeball and maintains the nutrition of the retina. The anterior portion of the choroid is the ***ciliary*** (SIL-ē-ar'-ē) ***body,*** the thickest portion of the vascular tunic. It extends from the ***ora serrata*** (Ō-ra ser-R-Ā-ta) of the retina (inner tunic) to a point just behind the sclerocorneal junction. The ora serrata is the jagged margin of the retina. The ciliary body consists of the ***ciliary processes*** (folds of the ciliary body that secrete aqueous humor) and the ***ciliary muscle*** (a circular band of smooth muscle that alters the shape of the lens for near or far vision). The iris (*irid* = colored circle), the third portion of the vascular tunic, is the colored portion of the eyeball and consists of circular and radial smooth-muscle fibers arranged to form a doughnut-shaped structure. The hole in the center of the iris is the ***pupil,*** through which light enters the eyeball. One function of the iris is to regulate the amount of light entering the eyeball.

c. RETINA

The third and inner coat of the eyeball, the ***retina (nervous tunic),*** is found only in the posterior portion of the eye. Its primary function is image formation. It consists of a pigment epithelium and a neural portion. The outer ***pigment epithelium*** (nonvisual portion) consists of melanin-containing epithelial cells in contact with the choroid. The inner ***neural portion*** (visual portion) is composed of three zones of neurons. Named in the order in which they conduct nerve impulses, these are the ***photoreceptor layer, bipolar cell layer,*** and ***ganglion***

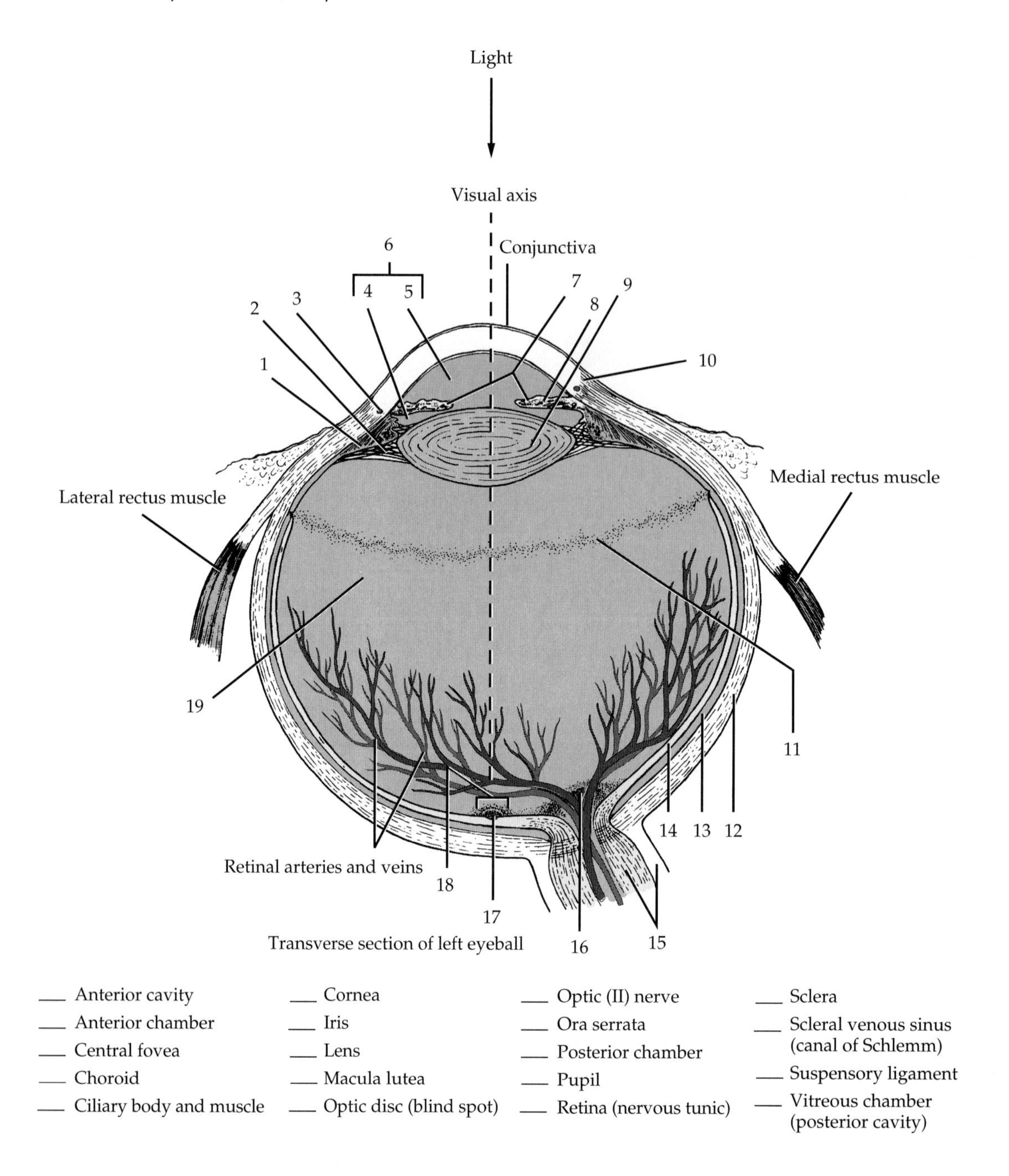

FIGURE 14.8 Parts of the eyeball.

cell layer. Structurally, the photoreceptor layer is just internal to the pigment epithelium, which lies adjacent to the choroid. The ganglion cell layer is the innermost zone of the neural portion.

The two types of photoreceptors are called rods and cones because of their respective shapes. ***Rods*** are specialized for vision in dim light. In addition, they allow discrimination between different shades of dark and light and permit discernment of shapes and movement. ***Cones*** are specialized for color vision and for sharpness of vision, that is, ***visual acuity.*** Cones are stimulated only by bright light and are most densely concentrated in the ***central fovea,*** a small depression in the center of the macula lutea. The ***macula lutea*** (MAK-yoo-la LOO-tē-a), or yellow spot, is situated in the exact center of the posterior portion of the retina and corresponds to the visual axis of the eye. The central fovea is the area of sharpest vision because of the high concentration of cones. Rods are absent from the fovea and macula but increase in density toward the periphery of the retina.

When light stimulates photoreceptors, impulses are conducted across synapses to the bipolar neurons in the intermediate zone of the neural portion of the retina. From there, the impulses pass to the ganglion cell layer. Axons of the ganglion neurons extend posteriorly to a small area of the retina called the ***optic disc (blind spot).*** This region contains openings through which fibers of the ganglion neurons exit as the ***optic (II) nerve.*** Because this area contains neither rods nor cones, and only nerve fibers, no image is formed on it. For this reason it is called the blind spot.

d. LENS

The eyeball itself also contains the lens, just behind the pupil and iris. The ***lens*** is constructed of numerous layers of protein fibers arranged like the layers of an onion. Normally, the lens is perfectly transparent and is enclosed by a clear capsule and held in position by the ***suspensory ligaments.*** A loss of transparency of the lens is called a ***cataract.***

e. INTERIOR

The interior of the eyeball contains a large cavity divided into two smaller cavities. These are called the anterior cavity and the vitreous chamber (posterior cavity) and are separated from each other by the lens. The ***anterior cavity,*** in turn, has two subdivisions known as the anterior chamber and the posterior chamber. The ***anterior chamber*** lies posterior to the cornea and anterior to the iris. The ***posterior chamber*** lies posterior to the iris and anterior to the suspensory ligaments and lens. The anterior cavity is filled with a clear, watery fluid known as the ***aqueous*** (*aqua* = water) ***humor,*** which is secreted by the ciliary processes posterior to the iris. From the posterior chamber, the fluid permeates the posterior cavity and then passes anteriorly between the iris and the lens, through the pupil into the anterior chamber. From the anterior chamber, the aqueous humor is drained off into the scleral venous sinus and passes into the blood. Pressure in the eye, called ***intraocular pressure (IOP),*** is produced mainly by the aqueous humor. Intraocular pressure keeps the retina smoothly applied to the choroid so that the retina may form clear images. Abnormal elevation of intraocular pressure, called ***glaucoma*** (glaw-KŌ-ma), results in degeneration of the retina and blindness.

The second, larger cavity of the eyeball is the ***vitreous chamber (posterior cavity).*** It is located between the lens and retina and contains a soft, jellylike substance called the ***vitreous body.*** This substance contributes to intraocular pressure, helps to prevent the eyeball from collapsing, and holds the retina flush against the internal portions of the eyeball.

Label the parts of the eyeball in Figure 14.8.

3. Surface Anatomy

Refer to Figure 14.9 for a summary of several surface anatomy features of the eyeball and accessory structures of the eye.

4. Dissection of Vertebrate Eye (Beef or Sheep)

CAUTION! *Please reread Section D, "Precautions Related to Dissection" at the beginning of the laboratory manual on page xiii before you begin your dissection.*

a. EXTERNAL EXAMINATION

PROCEDURE

1. Note any ***fat*** on the surface of the eyeball that protects the eyeball from shock in the orbit. Remove the fat.
2. Locate the ***sclera,*** the tough external white coat, and the ***conjunctiva,*** a delicate membrane that covers the anterior surface of the eyeball and is attached near the edge of the cornea. The ***cornea*** is the anterior, transparent portion of the sclera. It is probably opaque in your specimen due to the preservative.
3. Locate the ***optic (II) nerve,*** a solid, white cord of nerve fibers on the posterior surface of the eyeball.
4. If possible, identify six ***extrinsic eye muscles*** that appear as flat bands near the posterior part of the eyeball.

b. INTERNAL EXAMINATION

PROCEDURE

1. With a sharp scalpel, make an incision about 0.6 cm (¼ in.) lateral to the cornea (Figure 14.10 on page 291).
2. Insert scissors into the incision and carefully and slowly cut all the way around the corneal region. The eyeball contains fluid, so take care

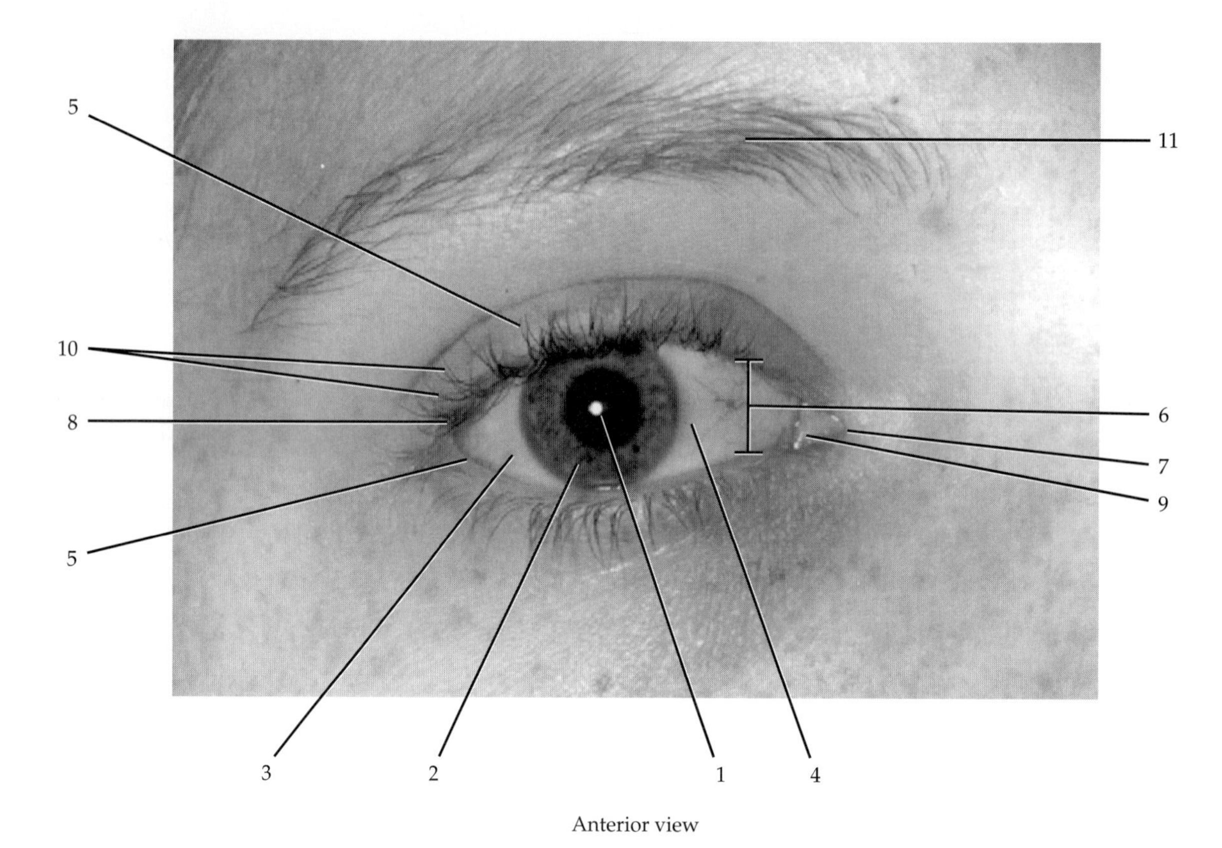

1. **Pupil.** Opening of center of iris of eyeball for light transmission.
2. **Iris.** Circular pigmented muscular membrane behind cornea.
3. **Sclera.** "White" of eye, a coat of fibrous tissue that covers entire eyeball except for cornea.
4. **Conjunctiva.** Membrane that covers exposed surface of eyeball and lines eyelids.
5. **Palpebrae (eyelids).** Folds of skin and muscle lined by conjunctiva.
6. **Palpebral fissure.** Space between eyelids when they are open.
7. **Medial commissure.** Site of union of upper and lower eyelids near nose.
8. **Lateral commissure.** Site of union of upper and lower eyelids away from nose.
9. **Lacrimal caruncle.** Fleshy, yellowish projection of medial commissure that contains modified sweat and sebaceous glands.
10. **Eyelashes.** Hairs on margins of eyelids, usually arranged in two or three rows.
11. **Eyebrows.** Several rows of hair superior to upper eyelids.

FIGURE 14.9 Surface anatomy of eyeball and accessory structures.

that it does not squirt out when you make your first incision. Examine the inside of the anterior part of the eyeball.

3. The ***lens*** is held in position by ***suspensory ligaments,*** which are delicate fibers. Around the outer margin of the lens, with a pleated appearance, is the black ***ciliary body,*** which also functions to hold the lens in place. Free the lens and notice how hard it is.
4. The ***iris*** can be seen just anterior to the lens and is also heavily pigmented or black.
5. The ***pupil*** is the circular opening in the center of the iris.
6. Examine the inside of the posterior part of the eyeball, identifying the thick ***vitreous body*** that fills the space between the lens and retina.
7. The ***retina*** is the white inner coat beneath the choroid coat and is easily separated from it.
8. The ***choroid coat*** is a dark, iridescent-colored tissue that gets its iridescence from a special structure called the ***tapetum lucidum.*** The tapetum lucidum, which is not present in the human eye, functions to reflect some light back onto the retina.
9. Finally, identify the ***blind spot,*** the point at which the retina is attached to the back of the eyeball.

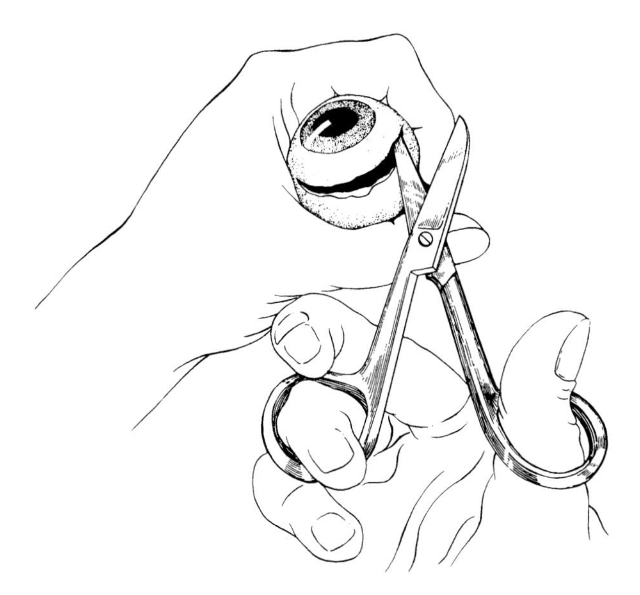

FIGURE 14.10 Procedure for dissecting a vertebrate eye.

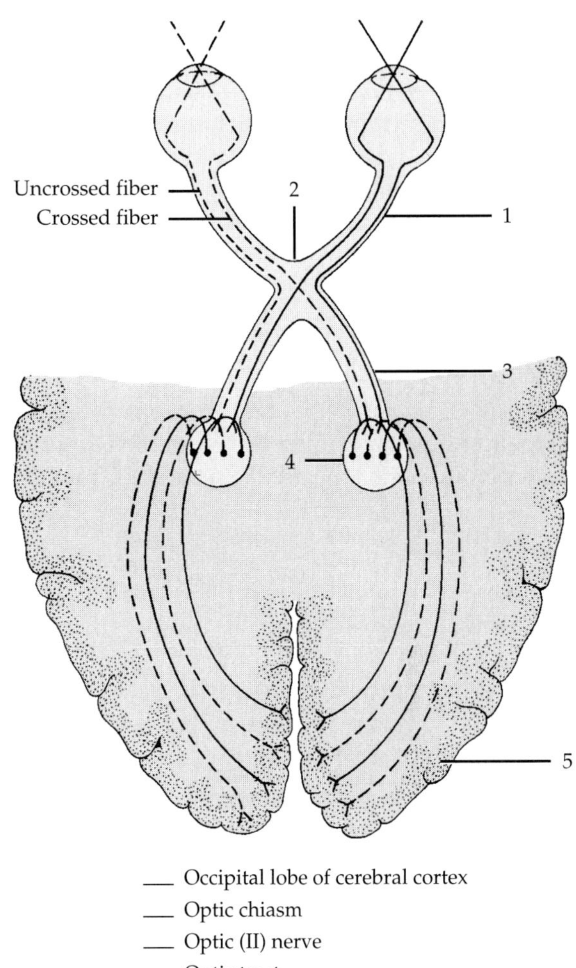

___ Occipital lobe of cerebral cortex
___ Optic chiasm
___ Optic (II) nerve
___ Optic tract
___ Thalamus

FIGURE 14.11 Visual pathway.

5. Visual Pathway

From the rods and cones, impulses are transmitted through bipolar cells to ganglion cells. The cell bodies of the ganglion cells lie in the retina and their axons leave the eye via the ***optic (II) nerve.*** The axons pass through the ***optic chiasm*** (kī-AZ-em), a crossing point of the optic nerves. Fibers from the medial retina cross to the opposite side. Fibers from the lateral retina remain uncrossed. Upon passing through the optic chiasm, the fibers, now part of the ***optic tract,*** enter the brain and terminate in the lateral geniculate nucleus of the thalamus. Here the fibers synapse with the neurons whose axons pass to the visual centers located in the occipital lobes of the cerebral cortex. Label the visual pathway in Figure 14.11.

H. AUDITORY SENSATIONS AND EQUILIBRIUM

In addition to containing receptors for sound waves, the ***ear*** also contains receptors for equilibrium. The ear is subdivided into three principal regions: (1) external (outer) ear, (2) middle ear, and (3) internal (inner) ear.

1. Structure of Ear

a. EXTERNAL (OUTER) EAR

The ***external (outer) ear*** collects sound waves and directs them inward. Its structure consists of the auricle, external auditory canal, and tympanic membrane. The ***auricle (pinna)*** is a trumpet-shaped flap of elastic cartilage covered by thick skin. The rim of the auricle is called the ***helix,*** and the inferior portion is referred to as the ***lobule.*** The auricle is attached to the head by ligaments and muscles. The ***external auditory canal (meatus)*** is a tube, about 2.5 cm (1 in.) in length that leads from the auricle to the eardrum. The walls of the canal consist of bone lined with cartilage that is continuous with the cartilage of the auricle. Near the exterior opening, the canal contains a few hairs and specialized sebaceous glands called ***ceruminous*** (se-ROO-me-nus) ***glands,*** which secrete ***cerumen*** (earwax). The combination of hairs and cerumen prevents foreign objects from entering the ear. The ***eardrum,*** or ***tympanic*** (tim-PAN-ik) ***membrane,*** is a thin, semitransparent partition of fibrous connective tissue located between the external auditory canal and middle ear.

Examine a model or charts and label the parts of the external ear in Figure 14.12.

b. MIDDLE EAR

The ***middle ear (tympanic cavity)*** is a small, epithelium-lined, air-filled cavity hollowed out of the temporal bone. The area is separated from the external ear by the eardrum and from the internal ear by a very thin bony partition that contains two small membrane-covered openings, called the oval window and the round window. The posterior wall of the middle ear communicates with the mastoid cells of the temporal bone through a chamber called the ***tympanic antrum.***

The anterior wall of the middle ear contains an opening that leads into the ***auditory (Eustachian) tube.*** The auditory tube connects the middle ear with the nose and nasopharynx. The function of the tube is to equalize air pressure on both sides of the eardrum. Any sudden pressure changes against the eardrum may be equalized by deliberately swallowing.

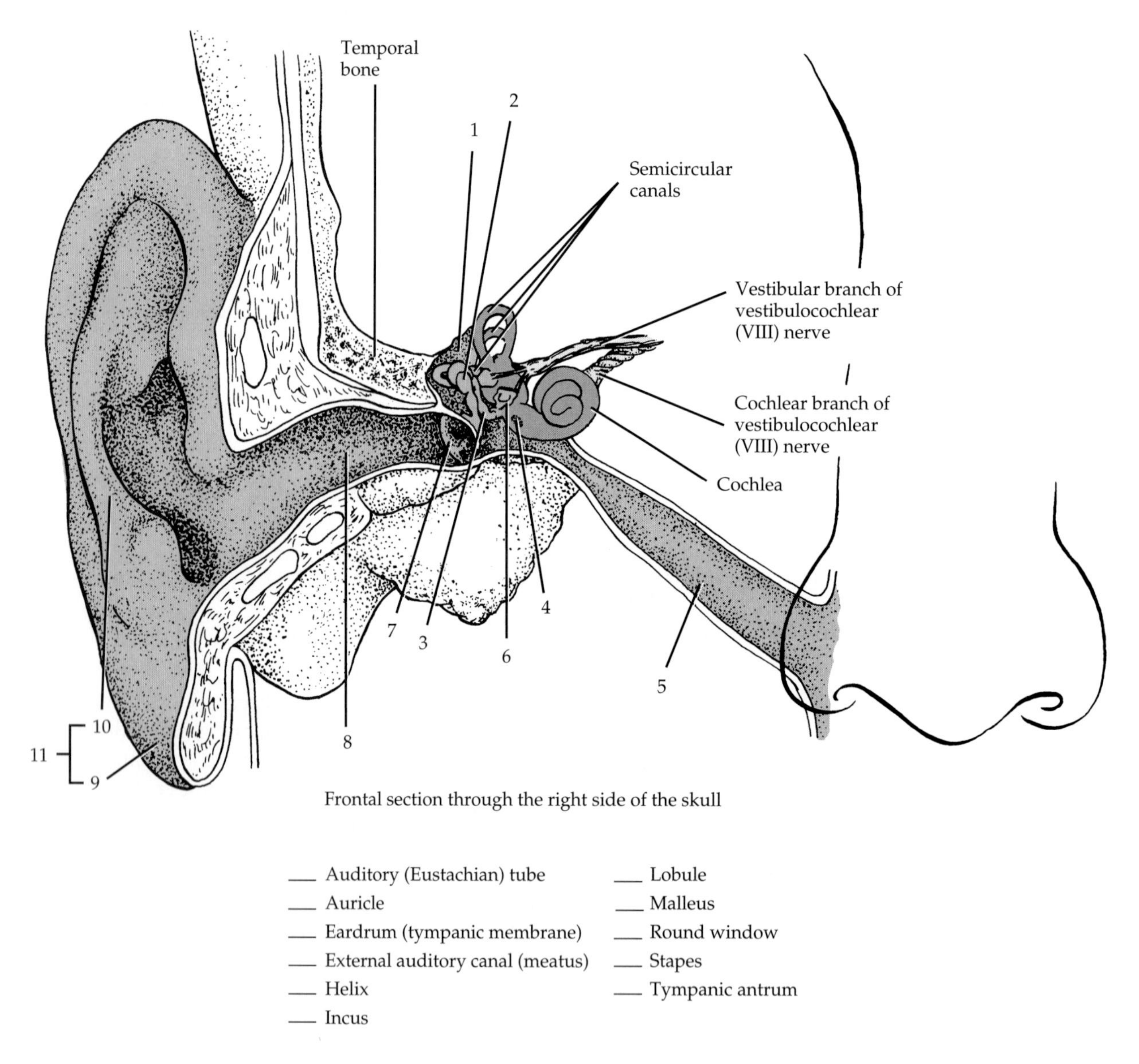

___ Auditory (Eustachian) tube
___ Auricle
___ Eardrum (tympanic membrane)
___ External auditory canal (meatus)
___ Helix
___ Incus
___ Lobule
___ Malleus
___ Round window
___ Stapes
___ Tympanic antrum

FIGURE **14.12** Principal subdivisions of ear.

Extending across the middle ear are three exceedingly small bones called ***auditory ossicles*** (OS-si-kuls). These are known as the malleus, incus, and stapes. Based on their shape, they are commonly named the hammer, anvil, and stirrup, respectively. The "handle" of the ***malleus*** is attached to the internal surface of the eardrum. Its head articulates with the base of the ***incus,*** the intermediate bone in the series, which articulates with the stapes. The base of the ***stapes*** fits into a small opening between the middle and inner ear called the ***oval window.*** Directly inferior to the oval window is another opening, the ***round window.*** This opening, which separates the middle and inner ears, is enclosed by a membrane called the ***secondary tympanic membrane.***

Examine a model or charts and label the parts of the middle ear in Figures 14.12 and 14.13.

C. INTERNAL (INNER) EAR

The ***internal (inner) ear*** is also known as the ***labyrinth*** (LAB-i-rinth). Structurally, it consists of two main divisions: (1) an outer bony labyrinth and (2) an inner membranous labyrinth that fits within the bony labyrinth. The ***bony labyrinth*** is a series of cavities within the petrous portion of the temporal bone that can be divided into three regions, named on the basis of shape: vestibule, cochlea, and semicircular canals. The bony labyrinth is lined with periosteum and contains a fluid called the ***perilymph.*** This fluid surrounds the ***membranous labyrinth,*** a series of sacs and tubes lying inside and having the same general form as the bony labyrinth. Epithelium lines the membranous labyrinth, which is filled with a fluid called the ***endolymph.***

The ***vestibule*** is the oval, central portion of the bony labyrinth (see Figure 14.13). The membranous labyrinth within the vestibule consists of two sacs called the ***utricle*** (YOO-tri-kul) and ***saccule*** (SAK-yool). These sacs are connected to each other by a small duct.

Projecting superiorly and posteriorly from the vestibule are the three bony ***semicircular canals*** (see Figure 14.13). Each is arranged at approximately right angles to the other two. They are called the anterior, posterior, and lateral canals. One end of each canal enlarges into a swelling called the ***ampulla*** (am-POOL-la; = little jar). Inside the bony

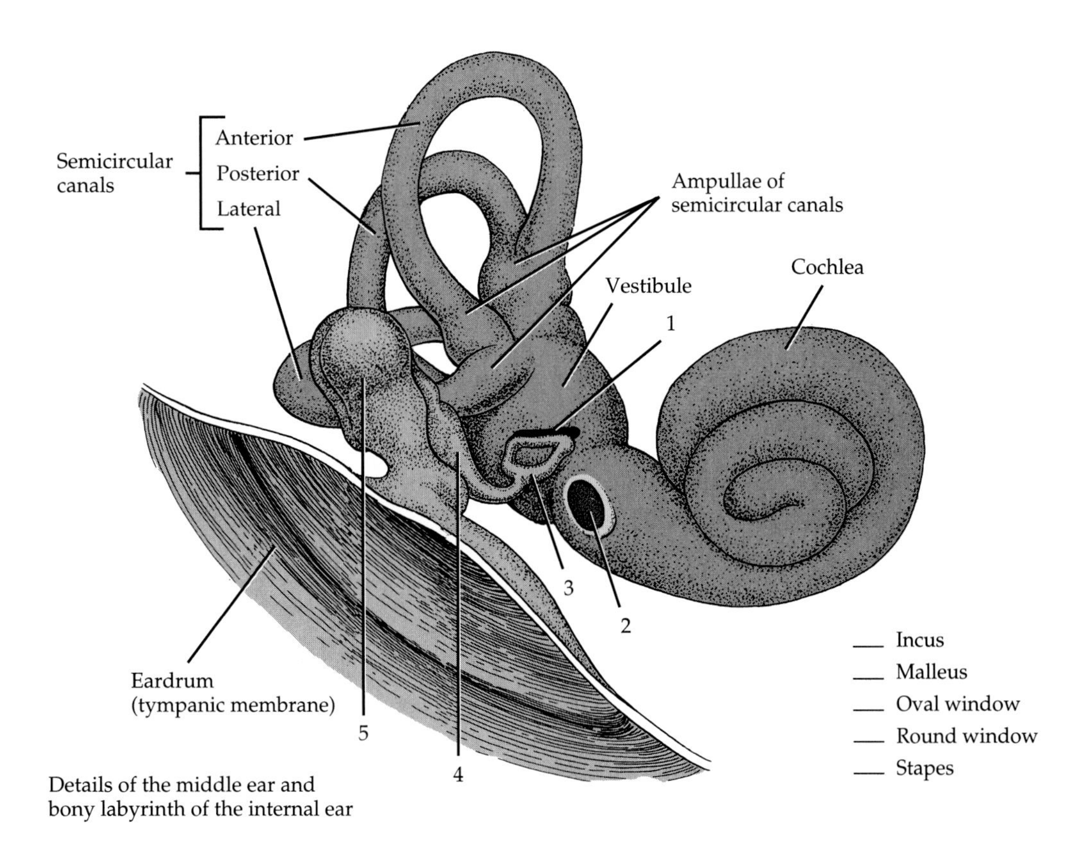

FIGURE 14.13 Ossicles of middle ear.

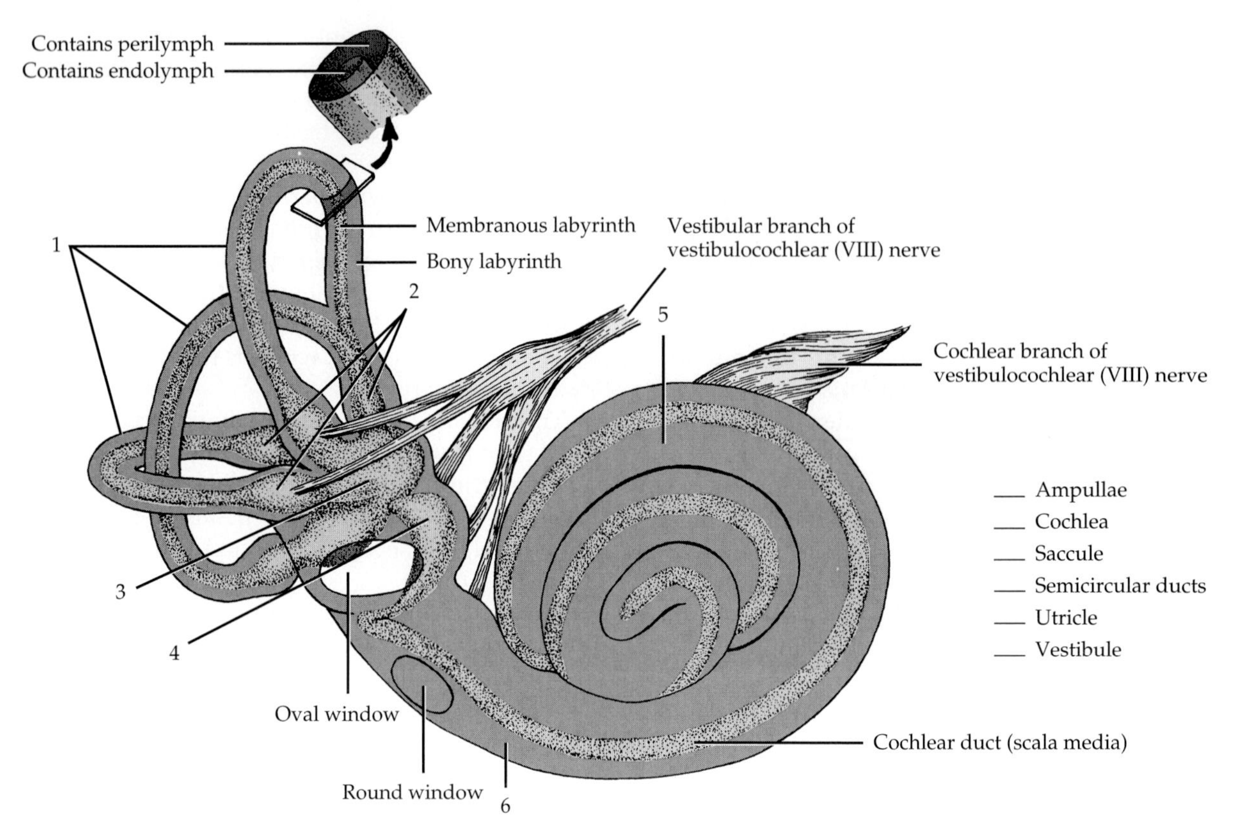

FIGURE 14.14 Details of inner ear.

semicircular canals lie portions of the membranous labyrinth, the ***semicircular ducts (membranous semicircular canals).*** These structures communicate with the utricle of the vestibule. Label these structures in Figure 14.14.

Anterior to the vestibule is the ***cochlea*** (KOK-lē-a; = snail's shell) (label it in Figure 14.14). The cochlea consists of a bony spiral canal that makes about three turns around a central bony core called the ***modiolus.*** A transverse section through the cochlea shows that the canal is divided by partitions into three separate channels resembling the letter Y lying on its side. The stem of the Y is a bony shelf that protrudes into the canal. The wings of the Y are composed of the vestibular and basilar membranes. The channel above the partition is called the ***scala vestibuli.*** The channel below is known as the ***scala tympani.*** The cochlea adjoins the wall of the vestibule, into which the scala vestibuli opens. The scala tympani terminates at the round window. The perilymph of the vestibule is continuous with that of the scala vestibuli. The third channel (between the wings of the Y) is the ***cochlear duct (scala media).*** This duct is separated from the scala vestibuli by the ***vestibular membrane*** and from the scala tympani by the ***basilar membrane.*** Resting on the basilar membrane is the ***spiral organ (organ of Corti),*** the organ of hearing. Label these structures in Figure 14.15.

The spiral organ (organ of Corti) is a coiled sheet of epithelial cells on the inner surface of the basilar membrane. This structure is composed of supporting cells and hair cells, which are receptors for auditory sensations. The hair cells have long hairlike processes at their free ends that extend into the endolymph of the cochlear duct. The basal ends of the hair cells are in contact with fibers of the cochlear branch of the vestibulocochlear (VIII) nerve. Projecting over and in contact with the hair cells of the spiral organ is the ***tectorial*** (*tectum* = cover) ***membrane,*** a very delicate and flexible gelatinous membrane. Label the tectorial membrane in Figure 14.15.

Obtain a prepared microscope slide of the spiral organ (organ of Corti) and examine under high power. Now label Figure 14.16 on page 296.

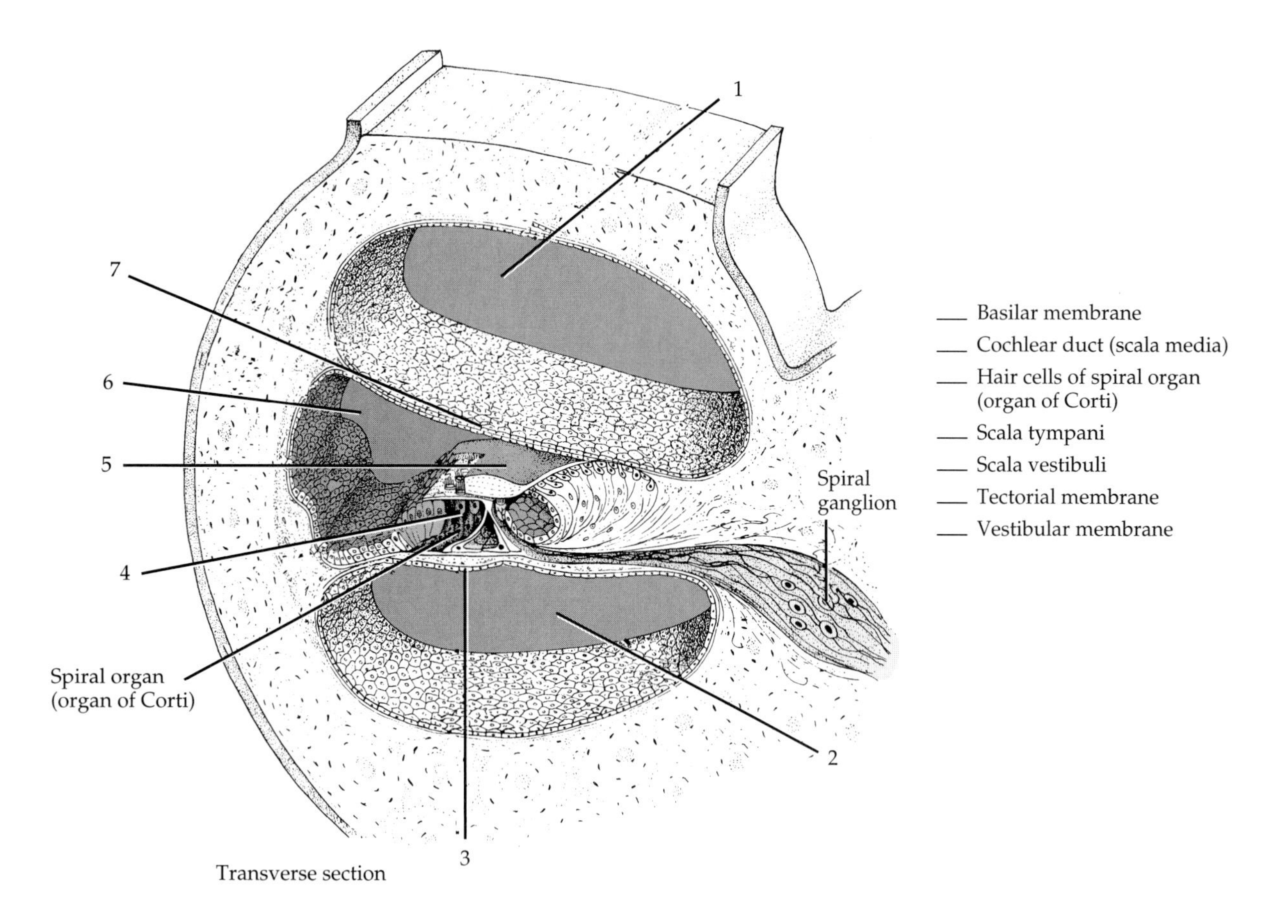

FIGURE 14.15 Cochlea.

2. Surface Anatomy

Refer to Figure 14.17 on page 297 for a summary of several surface anatomy features of the ear.

3. Equilibrium Apparatus

The term ***equilibrium*** (balance) has two meanings. One kind of equilibrium, called ***static equilibrium,*** refers to the position of the body (mainly the head) relative to the force of gravity. The second kind of equilibrium, called ***dynamic equilibrium,*** is the maintenance of the position of the body (mainly the head) in response to sudden movements (rotation, acceleration, and deceleration). Collectively, the receptor organs for equilibrium are called the ***vestibular apparatus,*** which includes the maculae in the saccule and utricle and the cristae in the semicircular ducts.

The ***maculae*** (MAK-yoo-lē) in the walls of the ***utricle*** and ***saccule*** are the receptors concerned mainly with static equilibrium. The maculae are small, flat regions that resemble the spiral organ (organ of Corti) microscopically. Maculae are located in planes perpendicular to each other and contain two kinds of cells: ***hair (receptor) cells*** and ***supporting cells.*** The hair cells project ***stereocilia*** (microvilli) and a ***kinocilium*** (conventional cilium). The columnar supporting cells are scattered among the hair cells. Floating over the hair cells is a thick, gelatinous glycoprotein layer, the ***otolithic membrane.*** A layer of calcium carbonate crystals, called ***otoliths*** (*oto* = ear; *lithos* = stone), extends over the entire surface of the otolithic membrane. When the head is tilted, the membrane slides over the hair cells in the direction determined by the tilt of the head. This sliding causes the membrane to pull on the stereocilia, thus initiating a nerve impulse that is conveyed via the vestibular branch of the vestibulocochlear (VIII) nerve to the cerebellum. The cerebellum sends continuous nerve impulses to the motor areas of the cerebral cortex in response to input from the maculae in the utricle and saccule, causing the

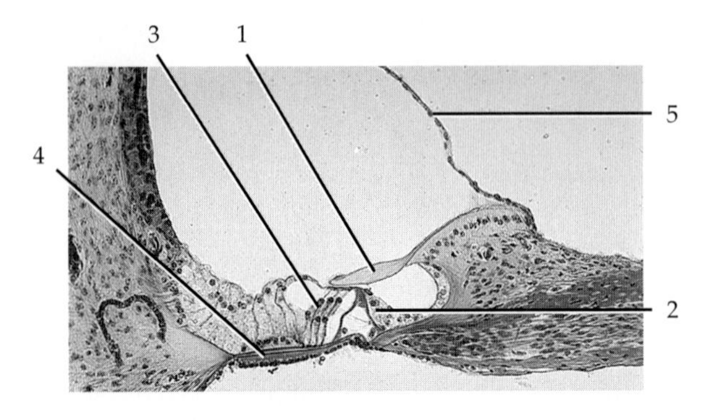

Section through the cochlea (120×)

___ Basilar membrane
___ Inner hair cell
___ Outer hair cell
___ Tectorial membrane
___ Vestibular membrane

FIGURE 14.16 Photomicrograph of the spiral organ (organ of Corti).

motor system to increase or decrease its nerve impulses to specific skeletal muscles to maintain static equilibrium.

Label the parts of the macula in Figure 14.18a on page 298.

Now consider the role of the cristae in the semicircular ducts in maintaining dynamic equilibrium. The three semicircular ducts are positioned at right angles to one another in three planes: the two vertical ones are called the ***anterior*** and ***posterior semicircular ducts*** and the horizontal one is called the ***lateral semicircular duct.*** This positioning permits correction of an imbalance in three planes. In the ***ampulla,*** the dilated portion of each duct, is a small elevation called the ***crista.*** Each crista is composed of a group of ***hair (receptor) cells*** and ***supporting cells*** covered by a mass of gelatinous material called the ***cupula.*** When the head moves, endolymph in the semicircular ducts flows over the hairs and bends them as water in a stream bends the plant life growing at its bottom. Movement of the hairs stimulates sensory neurons, and nerve impulses pass over the vestibular branch of the vestibulocochlear (VIII) nerve. The nerve impulses follow the same pathway as those involved in static equilibrium and are eventually sent to the muscles that contract to maintain body balance in the new position.

Label the parts of the crista in Figure 14.18b on page 298.

I. SENSORY-MOTOR INTEGRATION

Sensory systems provide the input that keeps the central nervous system informed of changes in the external and internal environment. Responses to this information are conveyed to motor systems, which enable us to move about, alter glandular secretions, and change our relationship to the world around us. As sensory information reaches the CNS, it becomes part of a large pool of sensory input. We do not actively respond to every bit of input the CNS receives. Rather, the incoming information is integrated with other information arriving from all other operating sensory receptors. The integration process occurs not just once, but at many stations along the pathways of the CNS and at both conscious and subconscious levels. It occurs within the spinal cord, brain stem, cerebellum, basal ganglia, and cerebral cortex. As a result, a motor response to make a muscle contract or a gland secrete can be modified at any of these levels. Motor portions of the cerebral motor cortex play the major role in initiating and controlling precise, discrete muscular movements. The basal ganglia largely integrate semivoluntary, automatic movements like walking, swimming, and laughing. The cerebellum assists the motor cortex and basal ganglia by making body movements smooth and coordinated and by contributing significantly to maintaining normal posture and balance.

J. SOMATIC MOTOR PATHWAYS

After receiving and interpreting sensory information, the CNS generates nerve impulses to direct responses to that sensory input. The nerve impulses are sent down the spinal cord in two major motor pathways: the direct (pyramidal) pathways and the indirect (extrapyramidal) pathways.

1. Direct Pathways

Voluntary motor impulses propagate from the motor cortex to voluntary motor neurons (somatic motor neurons) that innervate skeletal muscles via the ***direct*** or ***pyramidal*** (pi-RAM-i-dal) ***pathways.***

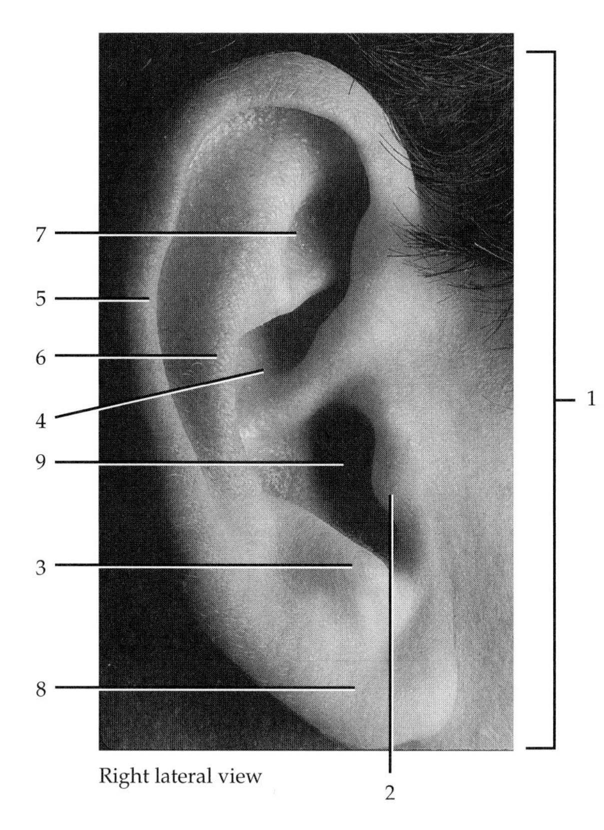

1. **Auricle.** Portion of external ear not contained in head, also called the pinna.
2. **Tragus.** Cartilaginous projecton.
3. **Antitragus.** Cartilaginous projection opposite tragus.
4. **Concha.** Hollow of auricle.
5. **Helix.** Superior and posterior free margin of auricle.
6. **Antihelix.** Semicircular ridge posterior and superior to concha.
7. **Triangular fossa.** Depression in superior portion of antihelix.
8. **Lobule.** Inferior portion of auricle devoid of cartilage.
9. **External auditory canal (meatus).** Canal extending from external ear to eardrum.

FIGURE 14.17 Surface anatomy of ear.

The direct pathways convey impulses from the cortex that result in precise, voluntary movements and include three tracts: lateral corticospinal (control muscles in the distal portions of the limbs), anterior corticospinal (control muscles of the neck and trunk), and corticobulbar (control muscles of the eyes, tongue, and neck; chewing, facial expression, and speech) (see page 248).

Based on the description provided on page 248, label the components of the lateral corticospinal tracts in Figure 14.19 on page 299.

2. Indirect Pathways

The *indirect* or *extrapyramidal pathways* include all motor tracts other than the corticospinal and corticobulbar tracts (see page 248). These are the rubrospinal, tectospinal, vestibulospinal, and lateral and medial veticulospinal.

ANSWER THE LABORATORY REPORT QUESTIONS AT THE END OF THE EXERCISE.

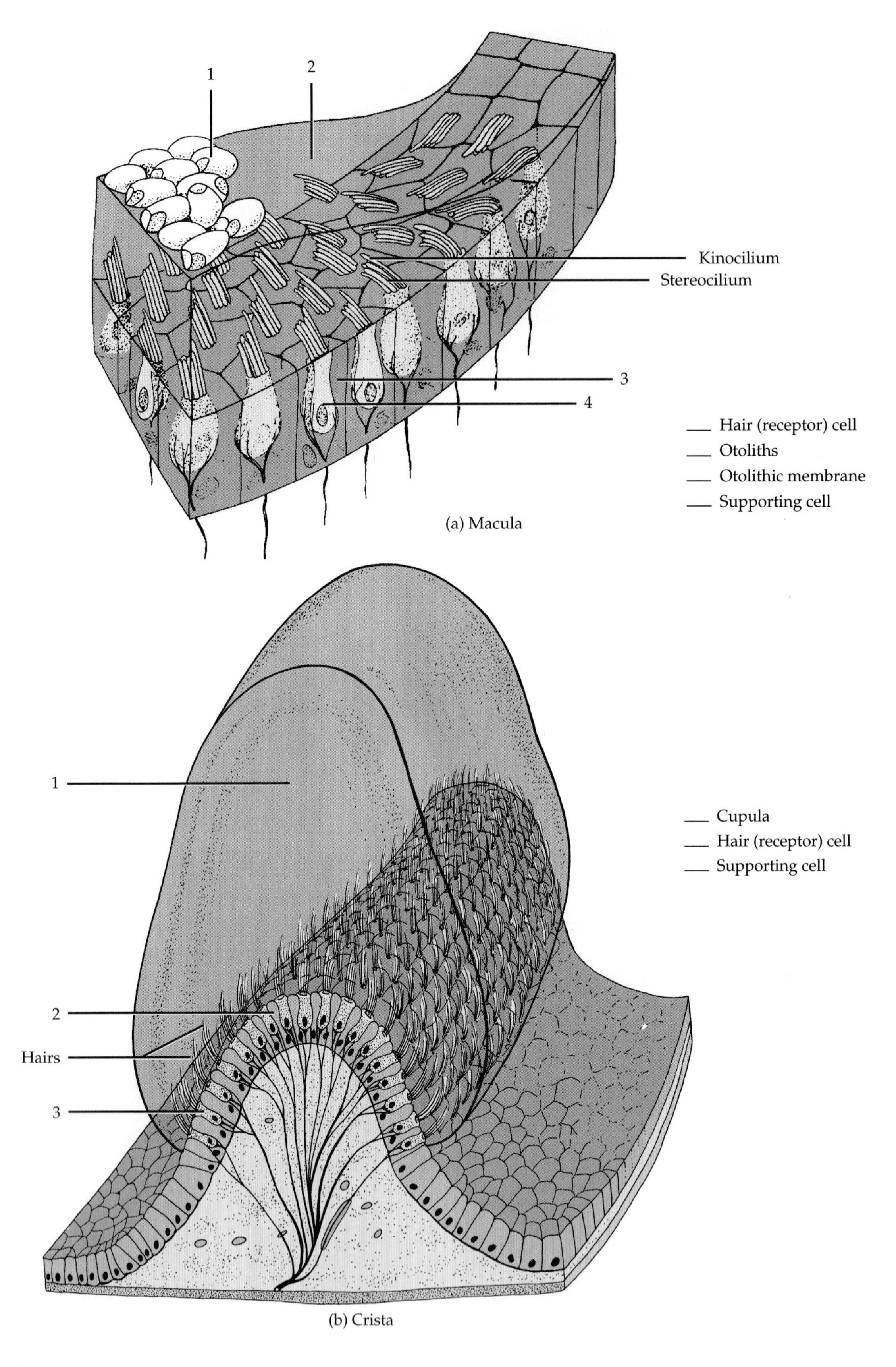

FIGURE 14.18 Equilibrium apparatus.

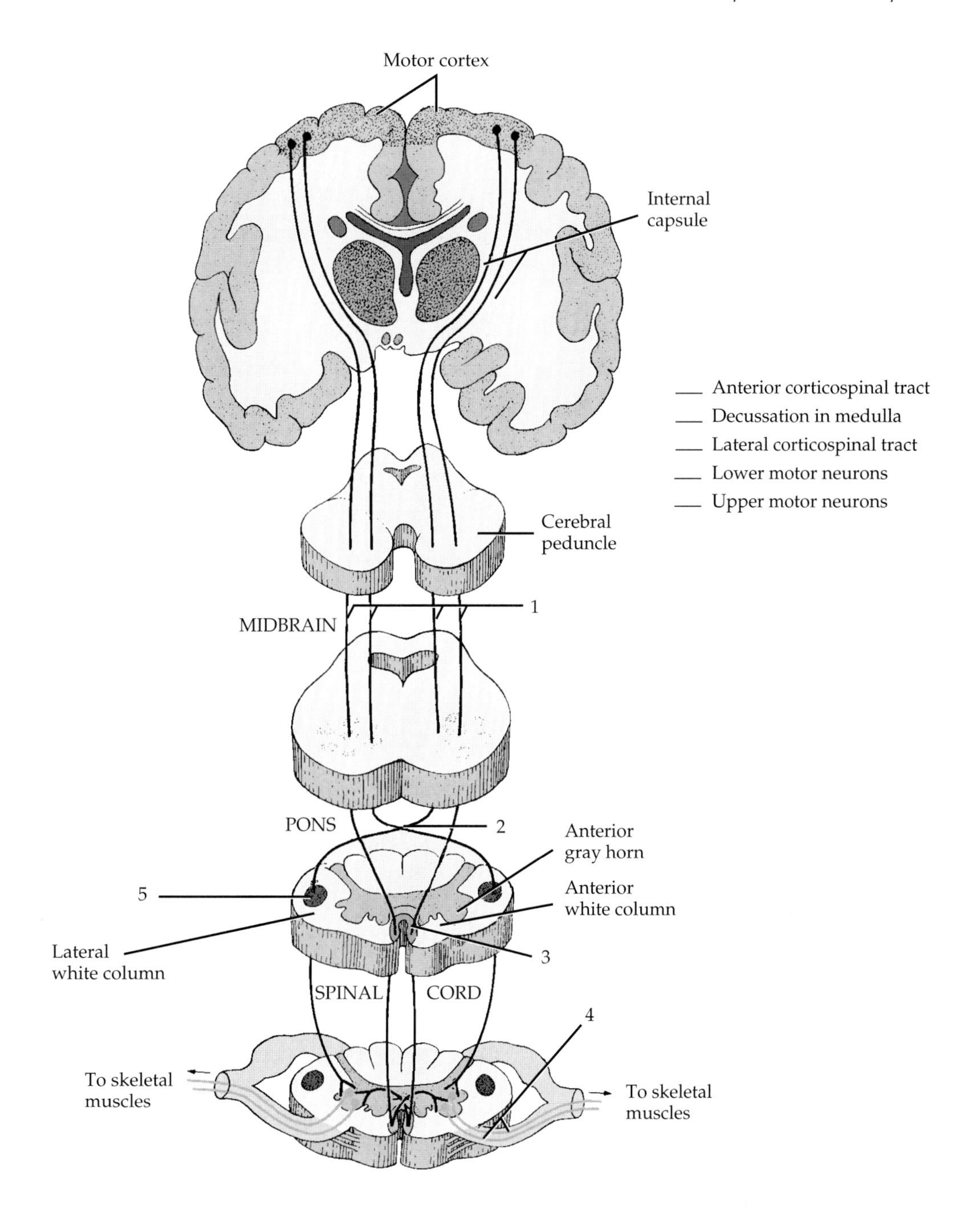

FIGURE 14.19 Direct (pyramidal) pathways: lateral and anterior corticospinal tracts.

LABORATORY REPORT QUESTIONS

General and Special Senses 14

Student ______________________ Date ______________________

Laboratory Section ______________________ Score/Grade ______________________

PART 1. Multiple Choice

_______ **1.** Conscious awareness and interpretation of sensations is called (a) integration (b) perception (c) transduction (d) adaptation

_______ **2.** An awareness of the activities of muscles, tendons, and joints is known as (a) referred pain (b) adaptation (c) refraction (d) proprioception

_______ **3.** The papillae located in an inverted V-shaped row at the posterior portion of the tongue are the (a) circumvallate (b) filiform (c) fungiform (d) gustatory

_______ **4.** The technical name for the "white of the eye" is the (a) cornea (b) conjunctiva (c) choroid (d) sclera

_______ **5.** Which is *not* a component of the vascular tunic? (a) choroid (b) macula lutea (c) iris (d) ciliary body

_______ **6.** The amount of light entering the eyeball is regulated by the (a) lens (b) iris (c) cornea (d) conjunctiva

_______ **7.** Which region of the eye is concerned primarily with image formation? (a) retina (b) choroid (c) lens (d) ciliary body

_______ **8.** The densest concentration of cones is found at the (a) blind spot (b) macula lutea (c) central fovea (d) optic disc

_______ **9.** Which region of the eyeball contains the vitreous body? (a) anterior chamber (b) vitreous chamber (c) posterior chamber (d) conjunctiva

_______ **10.** Among the structures found in the middle ear are the (a) vestibule (b) auditory ossicles (c) semicircular canals (d) external auditory canal

_______ **11.** The receptors for dynamic equilibrium are the (a) saccules (b) utricles (c) cristae in semicircular ducts (d) spiral organs (organs of Corti)

_______ **12.** The auditory ossicles are attached to the eardrum, to each other, and to the (a) semicircular ducts (b) semicircular canals (c) oval window (d) labyrinth

_______ **13.** Another name for the internal ear is the (a) labyrinth (b) fenestra (c) cochlea (d) vestibule

_______ **14.** Orientation of the body relative to the force of gravity is termed (a) the postural reflex (b) the tonal reflex (c) dynamic equilibrium (d) static equilibrium

_______ **15.** The inability to feel a sensation consciously even though a stimulus is still being applied is called (a) modality (b) projection (c) adaptation (d) afterimage formation

_______ **16.** Which sequence best describes the normal flow of tears from the eyes into the nose? (a) lacrimal canals, lacrimal sacs, nasolacrimal ducts (b) lacrimal sacs, lacrimal canals, nasolacrimal ducts (c) nasolacrimal ducts, lacrimal sacs, lacrimal canals (d) lacrimal sacs, nasolacrimal ducts, lacrimal canals

_______ **17.** A patient whose lens has lost transparency is suffering from (a) glaucoma (b) conjunctivitis (c) cataract (d) trachoma

_______ **18.** The portion of the eyeball that contains aqueous humor is the (a) anterior cavity (b) lens (c) posterior chamber (d) macula lutea

_______ **19.** The organ of hearing located within the inner ear is the (a) vestibule (b) oval window (c) modiolus (d) spiral organ (organ of Corti)

_______ **20.** The sense organs of static equilibrium are the (a) semicircular ducts (b) membranous labyrinths (c) maculae in the utricle and saccule (d) pinnae

_______ **21.** Which receptor does *not* belong with the others? (a) muscle spindle (b) tendon organ (Golgi tendon organ) (c) joint kinesthetic receptor (d) lamellated (Pacinian) corpuscle

_______ **22.** The membrane that is reflected from the eyelids onto the eyeball is the (a) retina (b) bulbar conjuctiva (c) sclera (d) choroid

_______ **23.** A characteristic of sensations by which one sensation may be distinguished from another is called (a) modality (b) perception (c) adaptation (d) afterimage formation

_______ **24.** Which are *not* cutaneous receptors? (a) hair root plexuses (b) muscle spindles (c) type I cutaneous mechanoreceptors (tactile or Merkel) discs (d) corpuscles of touch (Meissner's corpuscles)

_______ **25.** Which region of the tongue reacts strongest to bitter tastes? (a) tip (b) center (c) back (d) sides

_______ **26.** Technically, the eyelids are referred to as (a) palpebrae (b) papillae (c) vitreous chambers (d) pinnae

_______ **27.** The posterior portion of the middle ear communicates with the mastoid cells of the temporal bone through the (a) labyrinth (b) endolymph (c) perilymph (d) tympanic antrum

PART 2. Completion

28. Receptors found in blood vessels and viscera are classified as _______________.

29. Structures that collectively produce and drain tears are referred to as the _______________.

30. The three zones of the inner nervous layer of the retina (nervous tunic) are the photoreceptor layer, bipolar cell layer, and _______________.

31. The small area of the retina where no image is formed is referred to as the _______________.

32. Abnormal elevation of intraocular pressure (IOP) is called _______________.

33. In the visual pathway, nerve impulses pass from the optic chiasm to the _______________ before passing to the thalamus.

34. The openings between the middle and inner ears are the oval and _______________ windows.

35. The fluid within the bony labyrinth is called _______________.

36. Tactile sensations include touch, pressure, and _______________.

37. Receptors for pressure are type II cutaneous mechanoreceptors (end organs of Ruffini) and _______________ corpuscles.

38. Proprioceptive receptors that provide information about the degree and rate of angulations of joints are ______________________.

39. The neural pathway for olfaction includes olfactory receptors, olfactory bulbs, ______________________, and the cerebral cortex.

40. The posterior wall of the middle ear communicates with the mastoid air cells of the temporal bone through the ______________________.

41. A thin semitransparent partition of fibrous connective tissue that separates the external auditory meatus from the inner ear is the ______________________.

42. The fluid in the membranous labyrinth is called ______________________.

43. The cochlear duct is separated from the scala vestibuli by the ______________________.

44. The gelatinous glycoprotein layer over the hair cells in the maculae is called the ______________________.

45. The ______________________ separates the middle and inner ears.

46. ______________________ refers to the position of the body, mainly the head, relative to the force of gravity.

47. The ______________________ tract conveys sensory impulses for pain and temperature.

48. The ______________________ tract conveys motor impulses for precise contraction of muscles in the distal portions of the limbs.

49. Pain receptors are called ______________________.

50. ______________________ neurons extend from cranial nerve motor nuclei or spinal cord anterior horns to skeletal muscle fibers.

51. ______________________ tracts convey impulses that control voluntary movements of the head and neck.

Endocrine System

15

You have learned how the nervous system controls the body through nerve impulses that are delivered over neurons. Another system of the body, the ***endocrine system,*** is also involved in controlling bodily functions. The endocrine glands affect bodily activities by releasing chemical messengers, called ***hormones,*** into the blood stream (the term "hormone" means "to urge on"). The nervous and endocrine systems coordinate their regulatory activities via a complex series of interacting activities. Certain parts of the nervous system stimulate or inhibit the release of hormones. The hormones, in turn, are quite capable of stimulating or inhibiting the flow of particular nerve impulses.

The body contains two different kinds of glands: exocrine and endocrine. ***Exocrine glands*** secrete their products into ducts. The ducts then carry the secretions into body cavities, into the lumens of various organs, or to the external surface of the body. Examples are sudoriferous (sweat), sebaceous (oil), mucous, and digestive glands. ***Endocrine glands,*** by contrast, secrete their products (hormones) into the extracellular space around the secretory cells. The secretion then diffuses into blood vessels and the blood. Because they have no ducts, endocrine glands are also called ***ductless glands.***

A. ENDOCRINE GLANDS

The endocrine glands are the anterior pituitary gland, thyroid gland, parathyroid glands, adrenal cortex, adrenal medulla, pineal gland, and thymus gland. In addition, several organs of the body contain endocrine tissue, but are not endocrine glands exclusively. They include the pancreas, testes, ovaries, kidneys, hypothalamus, and placenta. The stomach, small intestine, skin, and heart also secrete hormone-like substances and are considered to have endocrine tissue with yet to be completely determined functions. The endocrine glands are organs that together form the ***endocrine system.***

Locate the endocrine glands on a torso, and, using your textbook or charts for reference, label Figure 15.1.

All hormones maintain homeostasis by changing the physiological activities of cells. A hormone may stimulate changes in the cells of one or more organs. The cells that respond to the effects of a hormone are called ***target cells.***

B. PITUITARY GLAND (HYPOPHYSIS)

The hormones of the ***pituitary gland,*** also called the ***hypophysis*** (hī-POF-i-sis), regulate so many body activities that the pituitary gland has been nicknamed the "master gland." This gland lies in the sella turcica of the sphenoid bone and is attached to the hypothalamus via a stalklike structure termed the ***infundibulum.*** Not only is the hypothalamus of the brain an important regulatory center in the nervous system; it is also a crucial endocrine gland. Cells in the hypothalamus synthesize at least nine different hormones, and the pituitary gland secretes seven more. Together, they play important roles in the regulation of virtually all aspects of growth, development, metabolism, and homeostasis.

The pituitary gland is divided structurally and functionally into an ***anterior pituitary gland,*** also called the ***adenohypophysis*** (ad′-i-nō-hī-POF-i-sis), and a ***posterior pituitary gland,*** also called the ***neurohypophysis*** (noo′-rō-hī-POF-i-sis). The anterior pituitary gland contains many glandular epitheloid (epithelial-like) cells and forms the glandular part of the pituitary gland. The hypothalamus is connected to the anterior pituitary gland by a series of blood vessels, the ***hypothalamic–hypophyseal portal system.*** The posterior pituitary gland contains the axon terminals of neurons whose cell bodies are in the

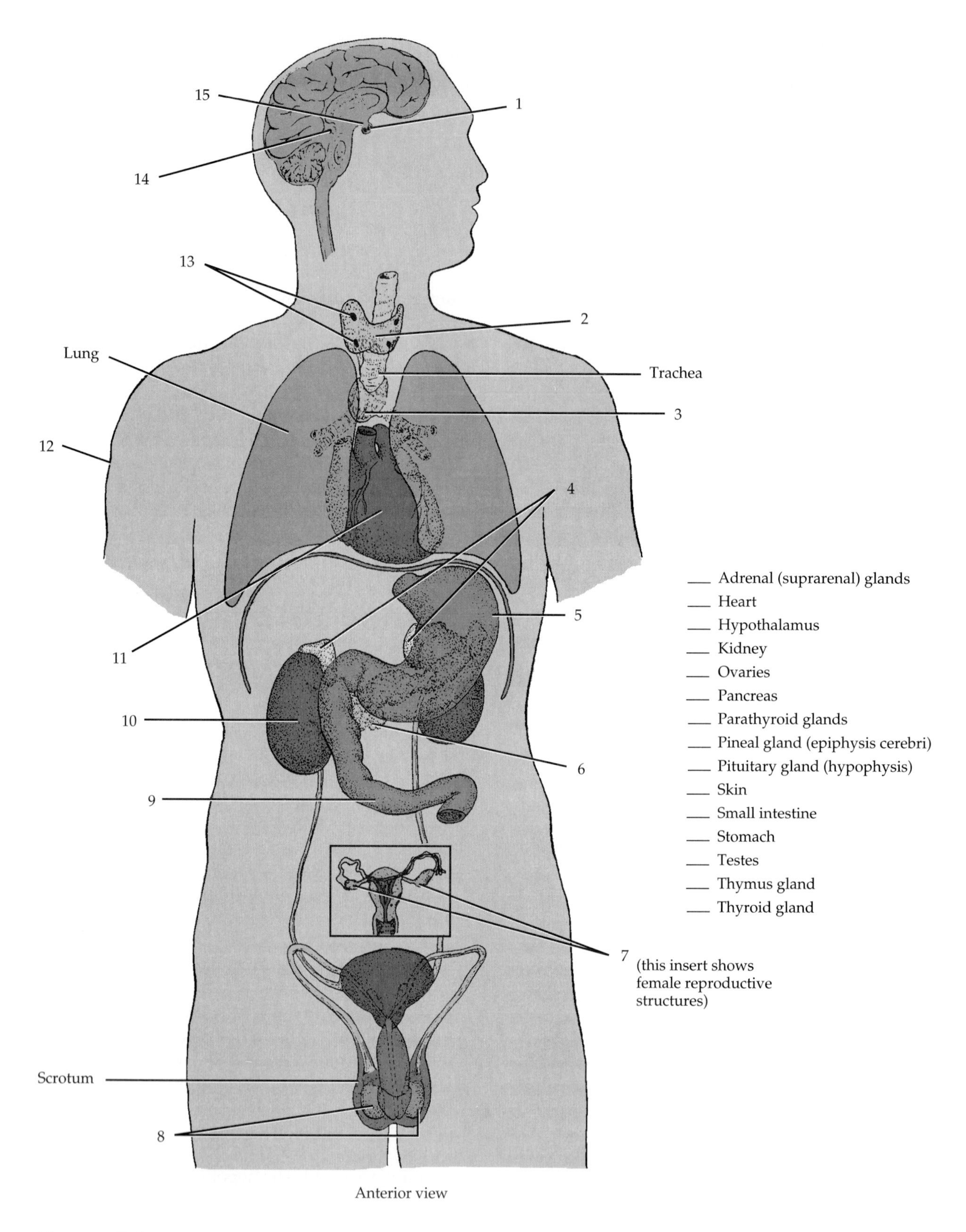

___ Adrenal (suprarenal) glands
___ Heart
___ Hypothalamus
___ Kidney
___ Ovaries
___ Pancreas
___ Parathyroid glands
___ Pineal gland (epiphysis cerebri)
___ Pituitary gland (hypophysis)
___ Skin
___ Small intestine
___ Stomach
___ Testes
___ Thymus gland
___ Thyroid gland

FIGURE 15.1 Location of endocrine glands, organs containing endocrine tissue, and associated structures.

hypothalamus. These terminals form the neural part of the pituitary gland. Between the anterior and posterior pituitary glands is a small, relatively avascular rudimentary zone, the ***pars intermedia***, whose role in humans is unknown.

1. Histology of the Pituitary Gland

The anterior pituitary gland releases hormones that regulate a whole range of body activities, from growth to reproduction. However, the release of these hormones is stimulated by ***releasing hormones*** or inhibited by ***inhibiting hormones*** that are produced by neurosecretory cells in the hypothalamus of the brain. This is one interaction between the nervous system and the endocrine system. The hypothalamic hormones reach the anterior pituitary gland through a network of blood vessels.

When the anterior pituitary gland receives proper stimulation from the hypothalamus via releasing hormones, its glandular cells secrete any one of seven hormones. Special staining techniques have established the division of glandular cells into five principal types:

1. ***Somatotrophs*** (*soma* = body; *tropos* = changing) produce ***human growth hormone (hGH)***, which stimulates general body growth and certain aspects of metabolism.
2. ***Lactotrophs*** (*lact* = milk) synthesize ***prolactin (PRL)***, which initiates milk production by suitably prepared mammary glands.
3. ***Corticolipotrophs*** (*cortex* = rind or bark) synthesize ***adrenocorticotropic*** (ad-rē-nō-kor'-ti-kō-TRŌ-pik) ***hormone (ACTH)***, which stimulates the adrenal cortex to secrete its hormones and ***melanocyte-stimulating hormone (MSH)***, which affects skin pigmentation.
4. ***Thyrotrophs*** (*thyreos* = shield) manufacture ***thyroid-stimulating hormone (TSH)***, which controls the thyroid gland.
5. ***Gonadotrophs*** (*gonas* = seed) produce ***follicle-stimulating hormone (FSH)*** and ***luteinizing hormone (LH)***. Together these hormones stimulate secretion of estrogens and progesterone by the ovaries and maturation of oocytes and secretion of testosterone and production of sperm in the testes.

Except for human growth hormone (hGH), melanocyte-stimulating hormone (MSH), and prolactin (PRL), all the secretions are referred to as ***tropins*** or ***tropic hormones***, which means that their target organs are other endocrine glands. Follicle-stimulating hormone (FSH) and luteinizing hormone (LH) are also called ***gonadotropic*** (gō-nad-ō-TRŌ-pik) ***hormones*** because they regulate the functions of the gonads (ovaries and testes). The gonads are the endocrine glands that produce sex hormones.

Examine a prepared slide of the anterior pituitary gland and identify as many types of cells as possible with the aid of your textbook or a histology textbook.

The posterior pituitary gland is not really an endocrine gland. Instead of synthesizing hormones, it stores and releases hormones synthesized by cells of the hypothalamus. The posterior pituitary gland consists of (1) cells called ***pituicytes*** (pi-TOO-i-sītz), which are similar in appearance to the neuroglia of the nervous system, and (2) axon terminations of secretory nerve cells of the hypothalamus. The cell bodies of the neurons, called ***neurosecretory cells***, originate in nuclei in the hypothalamus. The fibers project from the hypothalamus, form the ***supraopticohypophyseal*** (soo'-pra-op'ti-kō-hī'-pō-FIZ-ē-al) or ***hypothalamic–hypophyseal tract***, and terminate on blood capillaries in the posterior pituitary gland. The cell bodies of the neurosecretory cells produce the hormones ***oxytocin*** (ok'-sē-TŌ-sin), or ***OT***, and ***antidiuretic hormone (ADH)***. These hormones are transported in the neuron fibers into the posterior pituitary gland and are stored in the axon terminals resting on the capillaries. When properly stimulated, the hypothalamus sends impulses over the neurons. The impulses cause release of hormones from the axon terminals into the blood.

Examine a prepared slide of the posterior pituitary gland under high power. Identify the pituicytes and axon terminations of neurosecretory cells with the aid of your textbook.

2. Hormones of the Pituitary Gland

Using your textbook as a reference, give the major functions for the hormones listed below.

a. HORMONES SECRETED BY THE ANTERIOR PITUITARY GLAND

Human growth hormone (hGH) Also called ***somatotropin*** and ***somatotropic hormone (STH)***

____

____

____

Thyroid-stimulating hormone (TSH) Also called *thyrotropin* ______________________________

____

____

____

Adrenocorticotropic hormone (ACTH) ______________________________

____

____

____

Follicle-stimulating hormone (FSH)

In female: ______________________________

____

____

____

In male: ______________________________

____

____

____

Luteinizing hormone (LH)

In female: ______________________________

____

____

____

In male: ______________________________

____

____

____

Prolactin (PRL) Also called *lactogenic hormone*

____

____

Melanocyte-stimulating hormone (MSH) ______________________________

____

____

____

____

b. HORMONES STORED AND RELEASED BY THE POSTERIOR PITUITARY GLAND

Oxytocin (OT) ______________________________

____

____

____

Antidiuretic hormone (ADH) ______________________________

____

____

____

C. THYROID GLAND

The ***thyroid gland*** is the endocrine gland located just inferior to the larynx. The right and left ***lateral lobes*** lie lateral to the trachea and are connected by a mass of tissue called an ***isthmus*** (IS-mus) that lies anterior to the trachea just inferior to the cricoid cartilage. The ***pyramidal lobe,*** when present, extends superiorly from the isthmus.

1. Histology of the Thyroid Gland

Histologically, the thyroid gland consists of spherical sacs called ***thyroid follicles.*** The walls of each follicle consist of epithelial cells that reach the surface of the lumen of the follicle ***(follicular cells).*** In addition to the follicular cells the thyroid gland also contains ***parafollicular cells*** (also termed ***C cells***). These cells may be within a follicle, where they do not reach the surface of the lumen, or they may be found between follicles in groups of three to four cells. Follicular cells synthesize the hormones ***thyroxine*** (thī-ROX-sēn) (also termed ***T_4*** or ***tetraiodothyronine)*** and ***triiodothyronine*** (trī-ī-ōd-ō-THĪ-rō-nēn) ***(T_3).*** Together these hormones are referred to as the ***thyroid hormones.*** Approximately 90% of the hormone secreted by the follicular cells is thyroxine, while the remaining 10% is triiodothyronine. The functions of these two hormones are essentially the same, but they differ in rapidity and intensity of action. T_4 has a significantly longer life within the bloodstream, but is also significantly weaker than triiodothyronine. The parafollicular cells produce the hormone ***calcitonin*** (kal-si-TŌ-nin) ***(CT).*** Each thyroid follicle is filled with a glycoprotein called ***thyroglobulin (TBG),*** which is also called ***thyroid colloid.***

Examine a prepared slide of the thyroid gland under high power. Identify the thyroid follicles, epithelial cells forming the follicle, and thyroid colloid. Now label the photomicrograph in Figure 15.2.

2. Hormones of the Thyroid Gland

Using your textbook as a reference, give the major functions of the hormones listed below.

Thyroxine **(T_4)** and ***triiodothyronine*** **(T_3)**

____

____

____

Calcitonin (CT) ______________________________

____

____

____

D. PARATHYROID GLANDS

Attached to the posterior surfaces of the lateral lobes of the thyroid gland are small, round masses of tissue called the ***parathyroid*** (*para* = beside) ***glands.*** Typically two parathyroid glands, superior and inferior, are attached to each lateral thyroid lobe.

1. Histology of the Parathyroid Glands

Histologically, the parathyroid glands contain two kinds of epithelial cells. The first, a larger, more numerous cell called a ***principal (chief) cell,*** is believed to be the major synthesizer of ***parathyroid hormone (PTH).*** The second, a smaller cell called an ***oxyphil cell,*** may serve as a reserve cell for hormone synthesis.

Examine a prepared slide of the parathyroid glands under high power. Identify the principal and oxyphil cells. Now label the photomicrograph in Figure 15.3.

2. Hormone of the Parathyroid Glands

Using your textbook as a reference, give the major functions of the hormone listed below:

Parathyroid hormone (PTH) Also called ***parathormone*** ______________________________

____

____

____

E. ADRENAL (SUPRARENAL) GLANDS

The two ***adrenal (suprarenal) glands*** are superior to each kidney, and each is structurally and functionally differentiated into two separate endocrine

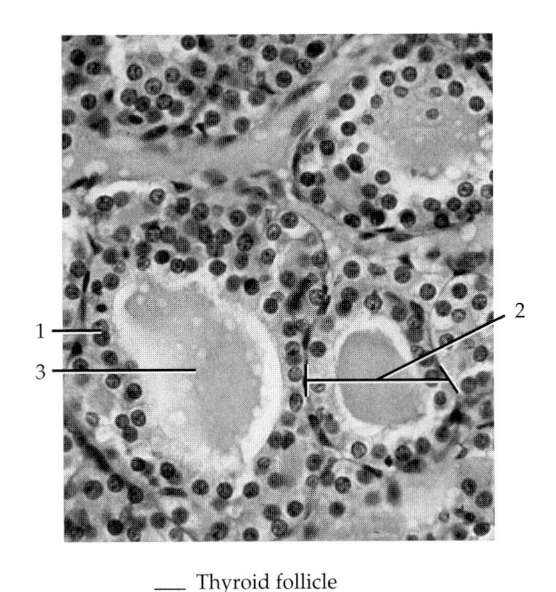

___ Thyroid follicle
___ Follicular cell
___ Thyroglobulin (TGB)

FIGURE 15.2 Histology of thyroid gland (400×).

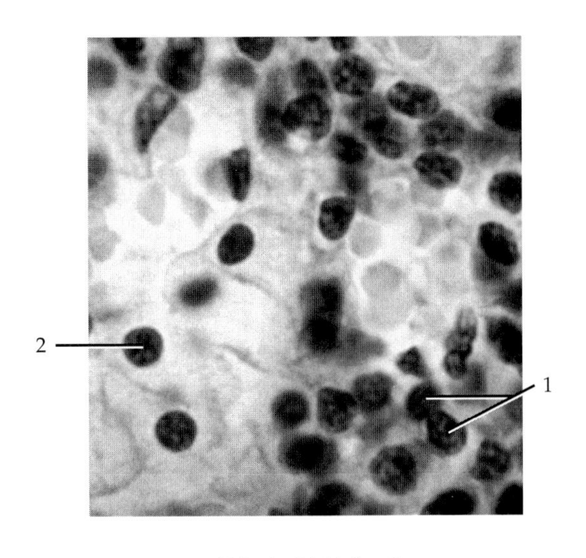

___ Principal (chief) cells
___ Oxyphil cell

FIGURE 15.3 Histology of parathyroid glands (320×).

glands: the superficial ***adrenal cortex***, which forms the bulk of the gland in humans, and the deeper ***adrenal medulla***. Covering the gland is a thick layer of fatty tissue and an outer thin fibrous connective tissue ***capsule***.

1. Histology of the Adrenal Cortex

Histologically, the adrenal cortex is subdivided into three zones, each of which has a different cellular arrangement and secretes different steroid hormones. The outer zone, immediately deep to the capsule, is called the ***zona glomerulosa*** (*glomerulus* = little ball). Its cells, arranged in arched loops or round balls, primarily secrete a group of hormones called ***mineralocorticoids*** (min'-er-al-ō-KOR-ti-koyds). The major mineralocorticoid produced by this region is ***aldosterone.***

The intermediate zone of the adrenal cortex is the ***zona fasciculata*** (*fasciculus* = little bundle). This zone, which is the widest of the three, consists of cells arranged in long, relatively straight cords. The zona fasciculata secretes mainly ***glucocorticoids*** (gloo'-ko-KOR-ti-koyds). The zona fasciculata secretes three glucocorticoids, 95% of which is ***cortisol*** (also known as ***hydrocortisone***). The remaining glucocorticoids synthesized include a small amount of ***corticosterone*** and a minute amount of ***cortisone.***

The deepest zone of the adrenal cortex is termed the ***zona reticularis*** (*reticular* = net). It is composed of frequently branching cords of cells. This zone synthesizes minute amounts of male hormones (***androgens***).

Examine a prepared slide of the adrenal cortex under high power. Identify the capsule, zona glomerulosa, zona fasciculata, and zona reticularis. Now label Figure 15.4.

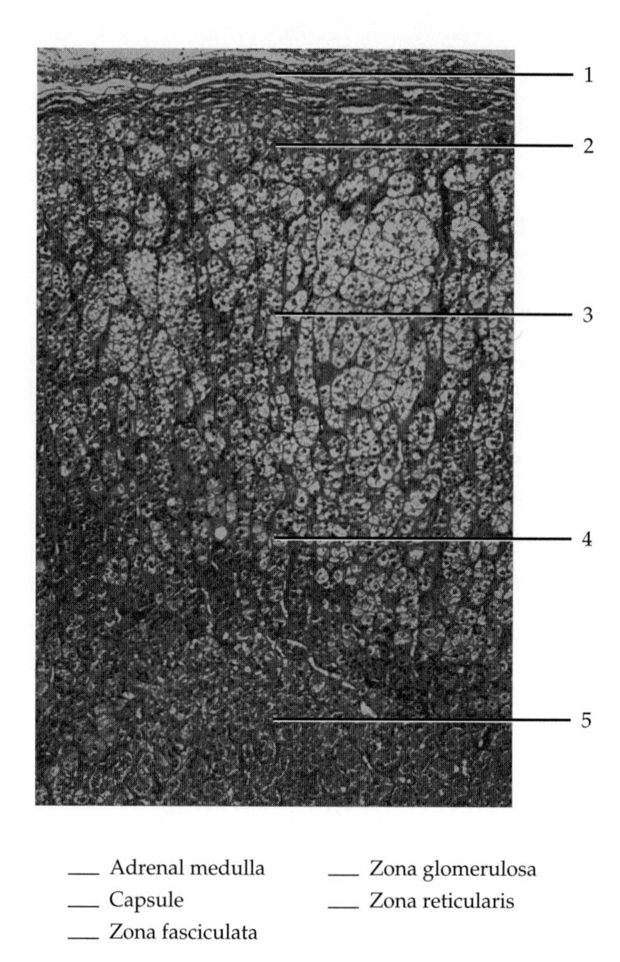

___ Adrenal medulla
___ Capsule
___ Zona fasciculata
___ Zona glomerulosa
___ Zona reticularis

FIGURE 15.4 Histology of adrenal (suprarenal) glands (10×).

2. Hormones of the Adrenal Cortex

Using your textbook as a reference, give the major functions of the hormones listed below.

Mineralocorticoids (mainly ***aldosterone***)

____

____

____

Glucocorticoids (mainly ***cortisol***) ______________

____

____

____

Androgens

In male: ______________________________

____

In female: ______________________________

____

3. Histology of the Adrenal Medulla

The adrenal medulla consists of hormone-producing cells called ***chromaffin*** (krō-MAF-in; *chroma* = color; *affinia* = affinity for) ***cells.*** These cells develop from the same embryonic tissue as the postganglionic cells of the sympathetic division of the nervous system. They are directly innervated by preganglionic cells of the sympathetic division of the autonomic nervous system and may be regarded as postganglionic cells that are specialized to secrete hormones. Secretion of hormones from the chromaffin cells is directly controlled by the sympathetic division of the autonomic nervous system, and innervation by the preganglionic fibers allows the gland to respond extremely rapidly to a stimulus. The adrenal medulla secretes the hormones ***epinephrine*** and ***norepinephrine (NE).***

Examine a prepared slide of the adrenal medulla under high power. Identify the chromaffin cells. Now locate the cells in Figure 15.4.

4. Hormones of the Adrenal Medulla

Using your textbook as a reference, give the major functions of the hormones listed below.

Epinephrine and ***norepinephrine (NE)*** ____________

____

____

____

F. PANCREAS

The ***pancreas*** is classified as both an endocrine and an exocrine gland. Thus, it is referred to as a ***heterocrine gland.*** We shall discuss only its endocrine functions now. The pancreas is a flattened organ located posterior and slightly inferior to the stomach. The adult pancreas consists of a head, body, and tail.

1. Histology of the Pancreas

The endocrine portion of the pancreas consists of clusters of cells called ***pancreatic islets (islets of Langerhans).*** They contain four kinds of cells: (1) ***alpha cells,*** which have more distinguishable plasma membranes, are usually peripheral in the islet, and secrete the hormone ***glucagon;*** (2) ***beta cells,*** which generally lie deeper within the islet and secrete the hormone ***insulin;*** (3) ***delta cells,*** which secrete ***somatostatin;*** and (4) ***F cells,*** which secrete pancreatic polypeptide. The pancreatic islets are surrounded by blood capillaries and by the cells called ***acini*** that form the exocrine part of the gland.

Examine a prepared slide of the pancreas under high power. Identify the alpha cells, beta cells, and acini (clusters of cells that secrete digestive enzymes) around the pancreatic islets. Now label Figure 15.5.

2. Hormones of the Pancreas

Using your textbook as a reference, give the major functions of the hormones listed below.

Glucagon ____________________________

____

____

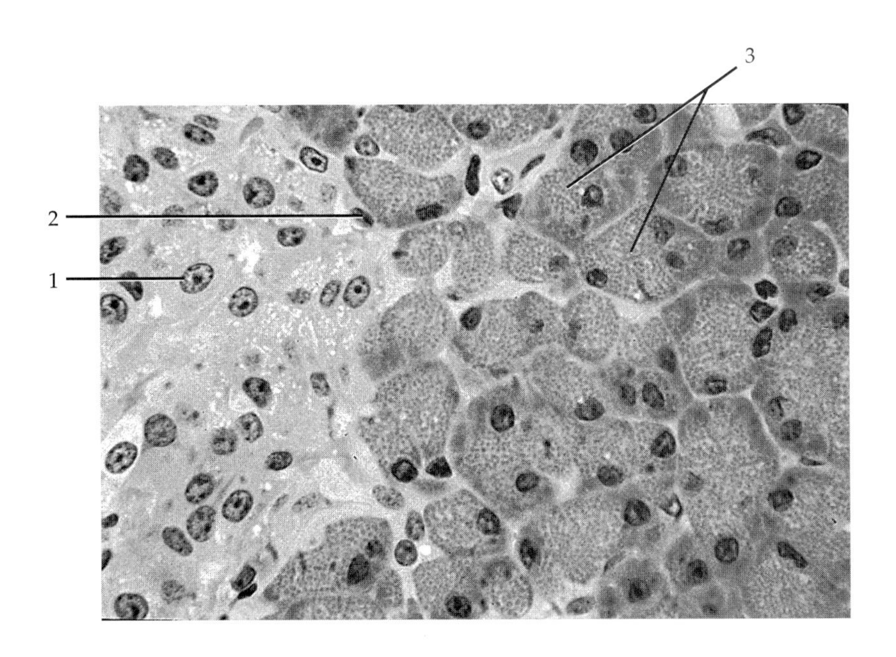

___ Acini
___ Alpha cell
___ Beta cell

FIGURE 15.5 Histology of pancreas (500×).

Insulin ______________________

____

____

Somatostatin ______________________

____

____

Pancreatic polypeptide ______________________

____

____

G. TESTES

The ***testes*** (male gonads) are paired oval glands enclosed by the scrotum. They are partially covered by a serous membrane called the ***tunica*** (*tunica* = sheath) ***vaginalis,*** which is derived from the peritoneum and forms during descent of the testes into the scrotum. Internal to the tunica vaginalis is a dense layer of white fibrous tissue, the ***tunica albuginea*** (al'-byoo-JIN-ē-a; *albus* = white), which extends inward and divides each testis into a series of internal compartments called ***lobules.*** Each of the 200–300 lobules contains one to three tightly coiled tubules, the convoluted ***seminiferous*** (*semen* = seed; *ferre* = to carry) ***tubules,*** which produce sperm by a process called ***spermatogenesis*** (sper'-ma-tō-JEN-e-sis).

1. Histology of the Testes

Spermatogenic cells are sperm-forming cells in various stages that undergo mitosis and differentiation to eventually produce sperm. Together with supporting cells, they line the seminiferous tubules. The most immature spermatogenic cells are called ***spermatogonia*** (sper'-ma-tō-GŌ-nē-a; *sperm* = seed; *gonium* = generation or offspring; the singular is ***spermatogonium***). They lie next to the basement membrane. Toward the lumen of the tubules are layers of progressively more mature cells. In order of advancing maturity, these are ***primary spermatocytes*** (SPER-ma-tō-sīts), ***secondary spermatocytes,*** and ***spermatids.*** By the time a ***sperm cell*** or ***spermatozoon*** (sper'-ma-tō-ZŌ-on; *zoon* = life; the plural is ***sperm*** or ***spermatozoa***), has nearly reached maturity, it is released into the lumen of the tubule and begins to move out of the rete testis.

Embedded among the spermatogenic cells in the tubules are large ***sustentacular*** (sus'-ten-TAK-yoo-lar; *sustentare* = to support), or ***Sertoli, cells,*** that extend from the basement membrane to the lumen of the tubule. Sustentacular cells support and protect developing spermatogenic cells; nourish spermatocytes, spermatids, and sperm; phagocytize excess spermatid cytoplasm as development proceeds; and mediate the effects of testosterone and follicle-stimulating hormone (FSH). Sustentacular cells also control movements of spermatogenic cells and the release of sperm into the lumen of the seminiferous tubule. They produce fluid for sperm transport and secrete the hormone inhibin, which helps regulate sperm production by inhibiting the secretion of FSH.

In the spaces between adjacent seminiferous tubules are clusters of cells called ***interstitial endocrinocytes,*** or ***Leydig cells.*** These cells secrete testosterone, the most important androgen (male sex hormone).

Examine a prepared slide of the testes under high power and identify all of the structures listed. Now label Figure 15.6.

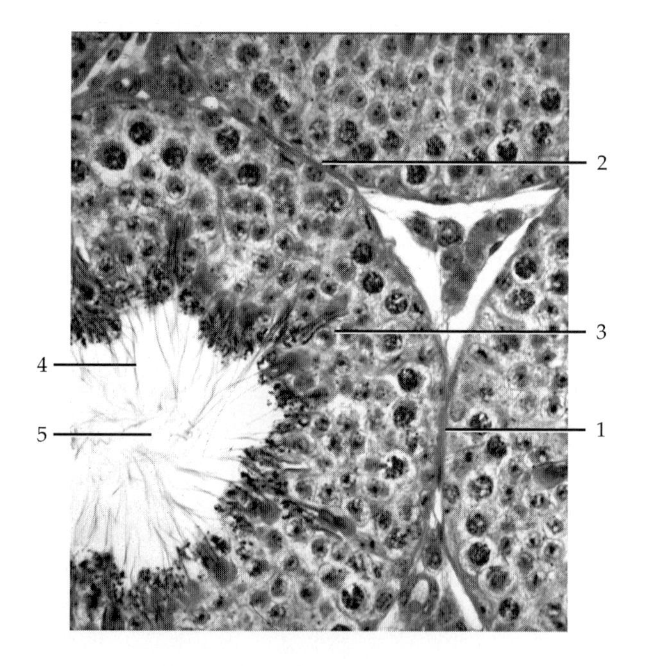

___ Basement membrane
___ Interstitial endocrinocyte (interstitial cell of Leydig)
___ Lumen of seminiferous tubule
___ Spermatid
___ Spermatogonium
___ Sperm cell

FIGURE 15.6 Histology of testes.

2. Hormones of the Testes

Using your textbook as a reference, give the major functions of the hormones listed below.

Testosterone ______________________________

____

____

Inhibin ______________________________

____

____

H. OVARIES

The ***ovaries*** (*ovarium* = egg receptacle), or female gonads, are paired glands resembling unshelled almonds in size and shape. They are positioned in the superior pelvic cavity, one on each side of the uterus, and are maintained in position by a series of ligaments. They are (1) attached to the ***broad ligament*** of the uterus, which is itself part of the parietal peritoneum, by a fold of peritoneum called the ***mesovarium;*** (2) anchored to the uterus by the ***ovarian ligament;*** and (3) attached to the pelvic wall by the ***suspensory ligament.*** Each ovary also contains a ***hilus,*** which is the point of entrance for blood vessels and nerves.

1. Histology of the Ovaries

Histologically, each ovary consists of the following parts:

1. ***Germinal epithelium*** A layer of simple epithelium (low cuboidal or squamous) that covers the free surface of the ovary and is continuous with the mesothelium that covers the mesovarium. The term *germinal epithelium* is a misnomer because it does not give rise to oocytes, although at one time it was believed that it did.
2. ***Tunica albuginea*** A whitish capsule of dense, irregular connective tissue immediately deep to the germinal epithelium.
3. ***Stroma*** A region of connective tissue deep to the tunica albuginea and composed of a superficial, dense layer called the ***cortex*** and a deep, loose layer known as the ***medulla.***
4. ***Ovarian follicles*** (*folliculus* = little bag) Lie in the cortex and consist of ***oocytes*** (immature ova) in various stages of development and their surrounding cells. When the surrounding cells form a single layer, they are called ***follicular cells.*** Later in development, when they form several layers, they are referred to as ***granulosa cells.*** The surrounding cells nourish the developing oocyte and begin to secrete estrogens as the follicle grows larger. Ovarian follicles undergo a series of changes prior to ovulation, progressing through several distinct stages. The most numerous and peripherally arranged follicles are termed ***primordial follicles.*** If a primordial follicle progresses to ovulation (release of a mature ova), it will sequentially transform into a ***primary (preantral) follicle,*** then a ***secondary (antral) follicle,*** and finally a ***mature (Graafian) follicle.***
5. ***Mature (Graafian) follicle*** A large, fluid-filled follicle that soon will rupture and expel a secondary oocyte, a process called ***ovulation.***
6. ***Corpus luteum*** (= yellow body) Contains the remnants of an ovulated mature follicle. The corpus luteum produces progesterone, estrogens, relaxin, and inhibin until it degenerates and turns into fibrous tissue called a ***corpus albicans*** (= white body).

Examine a prepared slide of the ovary under high power. (You may need to examine more than one slide to see all of the structures listed.) Label Figure 15.7.

2. Hormones of the Ovaries

Using your textbook as a reference, give the major functions of the hormones listed below.

Estrogens ______________________________

____

____

____

Progesterone ______________________________

____

____

____

Relaxin ______________________________

____

____

____

Inhibin ______________________________

____

____

____

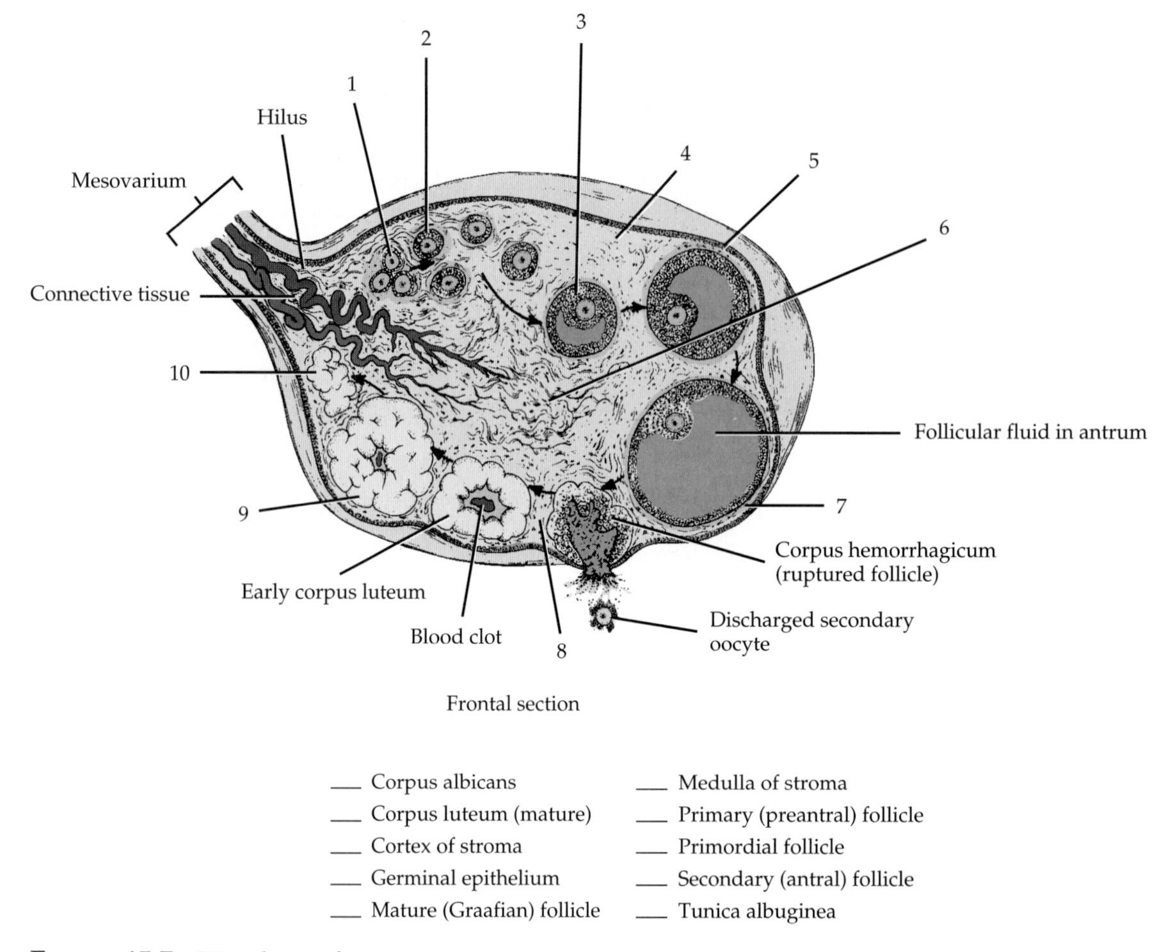

___ Corpus albicans
___ Corpus luteum (mature)
___ Cortex of stroma
___ Germinal epithelium
___ Mature (Graafian) follicle
___ Medulla of stroma
___ Primary (preantral) follicle
___ Primordial follicle
___ Secondary (antral) follicle
___ Tunica albuginea

FIGURE 15.7 Histology of ovary.

I. PINEAL GLAND (EPIPHYSIS CEREBRI)

The cone-shaped endocrine gland attached to the roof of the third ventricle is known as the ***pineal*** (PĪN-ē-al) ***gland (epiphysis cerebri).***

1. Histology of the Pineal Gland

The gland is covered by a capsule formed by the pia mater and consists of masses of neuroglial cells and secretory cells called ***pinealocytes*** (pin-ē-AL-ō-sīts). Around the cells are scattered preganglionic sympathetic fibers. The pineal gland starts to calcify at about the time of puberty. Such calcium deposits are called ***brain sand.*** Contrary to a once widely held belief, no evidence suggests that the pineal gland atrophies with age and that the presence of brain sand indicates atrophy. Rather, brain sand may even denote increased secretory activity.

The physiology of the pineal gland is still somewhat obscure. The gland secretes ***melatonin.***

2. Hormone of the Pineal Gland

Using your textbook as a reference, give the major functions of the hormone listed below.

Melatonin ______________________________

____

____

J. THYMUS GLAND

The ***thymus gland*** is a bilobed lymphatic gland located in the upper mediastinum posterior to the sternum and between the lungs. The gland is conspicuous in infants and, during puberty, reaches maximum size.

1. Histology of Thymus Gland

After puberty, thymic tissue, which consists primarily of ***lymphocytes,*** is replaced by fat. By the time a person reaches maturity, the gland has atrophied. Lymphoid tissue of the body consists primarily of lymphocytes that may be distinguished into two kinds: B cells and T cells. Both are derived originally in the embryo from lymphocytic stem cells in red bone marrow. Before migrating to their positions in lymphoid tissue, the descendants of the stem cells follow two distinct pathways. About half of them migrate to the thymus gland, where they are processed to become thymus-dependent lymphocytes, or ***T cells.*** The thymus gland confers on them the ability to destroy antigens (foreign microbes and substances). The remaining stem cells are processed in some as yet undetermined area of the body, possibly the fetal liver and spleen, and are known as ***B cells.*** Hormones produced by the thymus gland are ***thymosin, thymic humoral factor (THF), thymic factor (TF),*** and ***thymopoietin.***

2. Hormones of the Thymus Gland

Using your textbook as a reference, write the major functions of the hormones listed below:

Thymosin, thymic humoral factor (THF), thymic factor (TF), and thymopoietin ________________

____

____

K. OTHER ENDOCRINE TISSUES

Body tissues other than endocrine glands also secrete hormones. The gastrointestinal tract synthesizes several hormones that regulate digestion in the stomach and small intestine. Among these hormones are ***gastrin, secretin, cholecystokinin (CCK),*** and ***gastric inhibitory peptide (GIP).***

The placenta produces ***human chorionic gonadotropin (hCG), estrogens, progesterone (PROG), relaxin,*** and ***human chorionic somatomammotrophin (hCS),*** all of which are related to pregnancy.

When the kidneys (and liver, to a lesser extent) become hypoxic (subject to below normal levels of oxygen), they release an enzyme called ***renal erythropoietic factor.*** This is secreted into the blood where it acts on a plasma protein produced in the liver to form a hormone called ***erythropoietin*** (ē-rith'-rō-POY-ē-tin), or ***EPO,*** which stimulates red bone marrow to produce more red blood cells and hemoglobin. This ultimately reverses the original stimulus (hypoxia).

Vitamin D, produced by the skin, liver, and kidneys in the presence of sunlight, is converted to its active hormone, ***calcitriol,*** in the kidneys and liver.

The atria of the heart secrete a hormone called ***atrial natriuretic peptide (ANP),*** released in response to increased blood volume.

Using your textbook as a reference, write the major functions of the hormones listed below:

Gastrin ________________

____

Secretin ________________

____

Cholecystokinin (CCK) ________________

____

Gastric inhibitory peptide (GIP) ________________

____

Human chorionic gonadotropin (hCG) ________________

____

Human chorionic somatomammotropin (hCS) ____

____

Erythropoietin (EPO) ________________

____

Calcitriol ________________

____

Atrial natriuretic peptide (ANP) ________________

____

ANSWER THE LABORATORY REPORT QUESTIONS AT THE END OF THE EXERCISE.

LABORATORY REPORT QUESTIONS

Endocrine System 15

Student ______________________________ **Date** ______________________________

Laboratory Section ______________________ **Score/Grade** __________________________

PART 1. Multiple Choice

_______ **1.** Somatotrophs, gonadotrophs, and corticolipotrophs are associated with the (a) thyroid gland (b) anterior pituitary gland (c) parathyroid glands (d) adrenal glands

_______ **2.** The posterior pituitary gland is *not* an endocrine gland because it (a) has a rich blood supply (b) is not near the brain (c) does not make hormones (d) contains ducts

_______ **3.** Which hormone assumes a role in the development and discharge of a secondary oocyte? (a) hGH (b) TSH (c) LH (d) PRL

_______ **4.** The endocrine gland that is probably malfunctioning if a person has a high metabolic rate is the (a) thymus gland (b) posterior pituitary gland (c) anterior pituitary gland (d) thyroid gland

_______ **5.** The antagonistic hormones that regulate blood calcium level are (a) hGH-TSH (b) insulin-glucagon (c) aldosterone-cortisone (d) CT-PTH

_______ **6.** The endocrine gland that develops from the sympathetic nervous system is the (a) adrenal medulla (b) pancreas (c) thyroid gland (d) anterior pituitary gland

_______ **7.** The hormone that aids in sodium conservation and potassium excretion is (a) hydrocortisone (b) CT (c) ADH (d) aldosterone

_______ **8.** Which of the following hormones is sympathomimetic? (a) insulin (b) oxytocin (OT) (c) epinephrine (d) testosterone

_______ **9.** Which hormone lowers blood sugar level? (a) glucagon (b) melatonin (c) insulin (d) cortisone

_______ **10.** The endocrine gland that may assume a role in regulation of the menstrual cycle is the (a) pineal gland (b) thymus gland (c) thyroid gland (d) adrenal gland

PART 2. Completion

11. The pituitary gland is attached to the hypothalamus by a stalklike structure called the

______________________.

12. A hormone that acts on another endocrine gland and causes that gland to secrete its own hormones is called a ______________________.

13. ______________________ cells of the anterior pituitary gland synthesize ACTH.

14. The hormone that helps cause contraction of the smooth muscle of the pregnant uterus is

    ______________________.

15. Histologically, the spherical sacs that compose the thyroid gland are called thyroid

    ______________________.

16. The thyroid hormones associated with metabolism are triiodothyronine and ______________________.

17. Principal and oxyphil cells are associated with the ______________________ gland.

18. The zona glomerulosa of the adrenal cortex secretes a group of hormones called

    ______________________.

19. The hormones that promote normal metabolism, provide resistance to stress, and function as anti-inflammatories are ______________________.

20. The pancreatic hormone that raises blood sugar level is ______________________.

21. In spermatogenesis, the most immature cells near the basement membrane are called

    ______________________.

22. Cells within the testes that secrete testosterone are known as ______________________.

23. The ovaries are attached to the uterus by means of the ______________________ ligament.

24. The female hormones that help cause the development of secondary sex characteristics are called

    ______________________.

25. The endocrine gland that assumes a direct function in the proliferation and maturation of T cells is the ______________________.

26. Any hormone that regulates the functions of the gonads is classified as a(n) ______________________ hormone.

27. The hormone that is stored in the posterior pituitary gland that prevents excessive urine production is ______________________.

28. The region of the adrenal cortex that synthesizes androgens is the ______________________.

29. The hormone-producing cells of the adrenal medulla are called ______________________ cells.

30. Together, alpha cells, beta cells, delta cells, and F cells constitute the ______________________.

31. Regulating hormones are produced by the ______________________ and reach the anterior pituitary gland by a network of blood vessels.

32. FSH and LH are produced by ______________________ cells of the anterior pituitary gland.

33. The structure in an ovary that produces progesterone, estrogens, relaxin, and inhibin is the

    ______________________.

34. Calcium deposits in the pineal gland are referred to as ______________________.

35. A gland that is both an exocrine and endocrine gland is known as a ______________________ gland.

Blood

16

The blood, heart, and blood vessels constitute the ***cardiovascular system.*** In this exercise you will examine the characteristics of ***blood,*** a connective tissue also known as ***vascular tissue.***

A. COMPONENTS OF BLOOD

Blood is a liquid connective tissue that is composed of two portions: (1) ***plasma,*** a liquid that contains dissolved substances, and (2) ***formed elements,*** cells and cell fragments suspended in the plasma. In clinical practice, the most common classification of the formed elements of the blood is the following:

Erythrocytes (e-RITH-rō-sīts), or ***red blood cells***
Leukocytes (LOO-kō-sīts), or ***white blood cells***
Granular leukocytes (granulocytes)
Neutrophils
Eosinophils
Basophils
Agranular leukocytes (agranulocytes)
Lymphocytes (T cells, B cells, and natural killer cells)
Monocytes
Platelets (thrombocytes)

The origin and subsequent development of these formed elements can be seen in Figure 16.1. Blood cells are formed by a process called ***hemopoiesis*** (hē-mō-poy-Ē-sis; *hemo* = blood; *poiem* = to make), or ***hematopoiesis*** (hē-ma-tō-poy-Ē-sis), and the process by which erythrocytes are formed is called ***erythropoiesis*** (e-rith'-rō-poy-Ē-sis). The immature cells that are eventually capable of developing into mature blood cells are called ***hematopoietic stem cells,*** or ***hemocytoblasts*** (hē-mō-SĪ-tō-blasts) (see Figure 16.1). Mature blood cells are constantly being replaced as they die, so special cells called ***reticulo-endothelial cells*** (fixed macrophages that line liver sinusoids) have the responsibility of clearing away the dead, disintegrating cell bodies so that small blood vessels are not clogged.

The shapes of the nuclei, staining characteristics, and color of cytoplasmic granules are all useful in differentiation and identification of the various white blood cells. Red blood cells are biconcave discs without nuclei and can be identified easily.

B. PLASMA

When the formed elements are removed from blood, a straw-colored liquid called ***plasma*** is left. This liquid consists of about 91.5% water and about 8.5% solutes. Among the solutes are proteins (albumins, globulins, and fibrinogen), nonprotein nitrogen (NPN) substances (urea, uric acid, and creatine), foods (amino acids, glucose, fatty acids, glycerides, and glycerol), regulatory substances (enzymes and hormones), gases (oxygen and carbon dioxide), and electrolytes (Na^+, K^+, Ca^{2+}, Mg^{2+}, Cl^-, HCO_3^-, SO_4^{2-}, and HPO_4^{3-}).

C. ERYTHROCYTES

Erythrocytes (red blood cells, or ***RBCs)*** are biconcave in appearance, have no nucleus, and can neither reproduce nor carry on extensive metabolic activities (Figure 16.1). The interior of the cell contains a red pigment called ***hemoglobin,*** which is responsible for the red color of blood. The heme portion of hemoglobin combines with oxygen and, to a lesser extent, carbon dioxide and transports them through the blood vessels. An average red blood cell has a life span of about 120 days. A healthy male has about 5.4 million red blood cells per cubic millimeter (mm^3) or per deciliter (dl) of blood; a healthy female, about 4.8 million. Erythropoiesis and red blood cell destruction normally proceed at the same pace. A diagnostic test that informs the physician about the rate of erythropoiesis is the ***reticulocyte*** (re-TIK-yoo-lō-sīt) ***count. Reticulocytes*** (see Figure 16.1) are precursor cells of mature red blood cells. Normally, a reticulocyte count is 0.5 to 1.5%. The reticulocyte count is an important diagnostic tool because it is

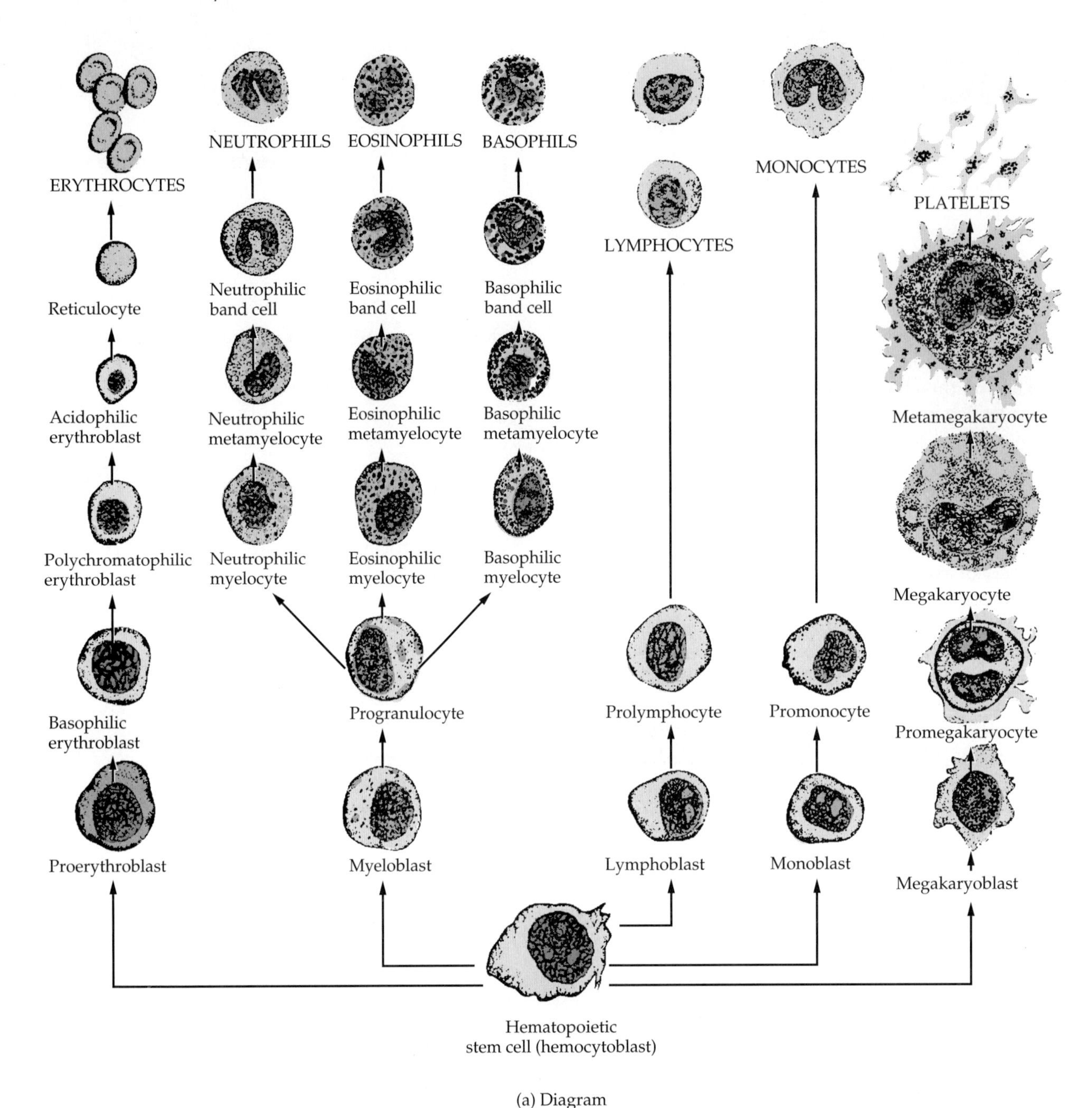

(a) Diagram

FIGURE 16.1 Origin, development, and structure of blood cells.

a relatively accurate predictor of the status of red blood cell production in red bone marrow. Normally, red bone marrow replaces about 1% of the adult red blood cells each day. A decreased reticulocyte count (reticulocytopenia) is seen in aplastic anemia and in conditions in which the red bone marrow is not producing red blood cells. An increase in reticulocytes (reticulocytosis) is found in acute and chronic blood loss and certain kinds of anemias such as iron-deficiency anemia.

D. LEUKOCYTES

Leukocytes (white blood cells, or ***WBCs)*** are different from red blood cells in that they have nuclei

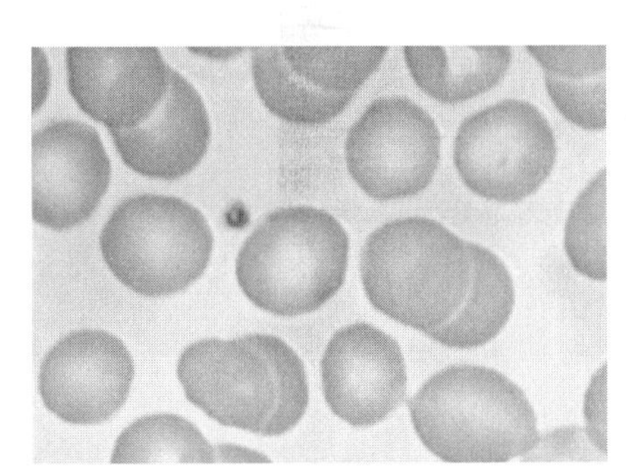

Erythrocytes and platelet

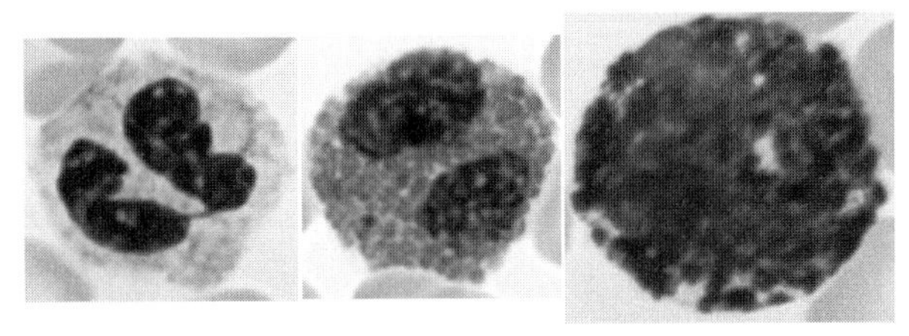

Neutrophil Eosinophil Basophil

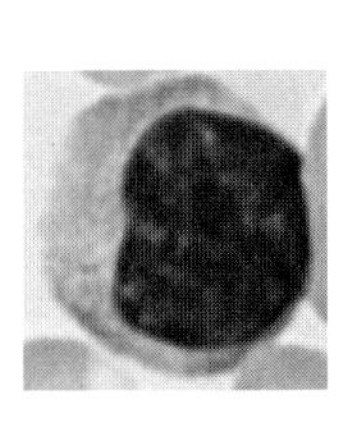

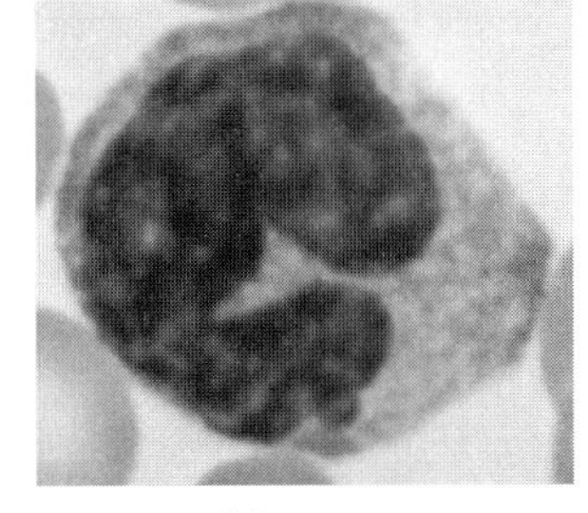

Lymphocyte Monocyte

(b) Photomicrographs

FIGURE 16.1 ***(Continued)***

and do not contain hemoglobin (see Figure 16.1). They are less numerous than red blood cells, ranging from 5000 to 10,000 cells per cubic millimeter (mm^3) or per deciliter (dl) of blood. The ratio, therefore, of red blood cells to white blood cells is about 700:1.

As Figure 16.1 shows, leukocytes can be differentiated by their appearance. They are divided into two major groups, granular leukocytes and agranular leukocytes. ***Granular leukocytes,*** which are formed from red bone marrow, have large granules in the cytoplasm that can be seen under a microscope and possess lobed nuclei. The three types of granular leukocytes are ***neutrophils, eosinophils,*** and ***basophils. Agranular leukocytes,*** which are also formed from red bone marrow, contain small granules that cannot be seen under a light microscope, and usually have spherical nuclei. The two types of agranular leukocytes are ***lymphocytes*** and ***monocytes.***

Leukocytes as a group function in phagocytosis, producing antibodies, and combating allergies. The life span of a leukocyte usually ranges from a few hours to a few months. Some lymphocytes, called T and B memory cells, can live throughout one's life once they are formed.

E. PLATELETS

Platelets (thrombocytes) are formed from fragments of the cytoplasm of megakaryocytes (see Figure 16.1). The fragments become enclosed in pieces of cell membrane from the megakaryocytes and develop into platelets. Platelets are very small, disc-shaped cell fragments without nuclei. Between 250,000 and 400,000 are found in each cubic millimeter or deciliter (dl) of blood. They function to prevent fluid loss by starting a chain of reactions that results in blood clotting. They have a short life span, probably only one week, because they are expended in clotting and are just too simple to carry on extensive metabolic activity.

F. DRAWINGS OF BLOOD CELLS

Examine a prepared slide of a stained smear of blood cells using the oil-immersion objective. With the aid of your textbook, identify red blood cells, neutrophils, basophils, eosinophils, lymphocytes, monocytes, and platelets. Using colored crayons or pencil crayons, draw and label all of the different blood cells on the blank page that follows. Be very accurate in drawing and coloring the granules and the nuclear shapes, because both of these are used to identify the various types of cells.

ANSWER THE LABORATORY REPORT QUESTIONS AT THE END OF THE EXERCISE.

LABORATORY REPORT QUESTIONS

Blood 16

Student ____________________ **Date** ____________________

Laboratory Section ____________________ **Score/Grade** ____________________

PART 1. Multiple Choice

_______ **1.** The process by which all blood cells are formed is called (a) hemocytoblastosis (b) erythropoiesis (c) hemopoiesis (d) leukocytosis

_______ **2.** An inability of body cells to receive adequate amounts of oxygen may indicate a malfunction of (a) neutrophils (b) leukocytes (c) lymphocytes (d) erythrocytes

_______ **3.** Special cells of the body that have the responsibility of clearing away dead, disintegrating bodies of red and white blood cells are called (a) agranular leukocytes (b) reticuloendothelial cells (c) erythrocytes (d) thrombocytes

_______ **4.** The name of the test procedure that informs the physician about the rate of erythropoiesis is called the (a) reticulocyte count (b) sedimentation rate (c) hemoglobin count (d) differential white blood cell count

_______ **5.** The normal red blood cell count per cubic millimeter or deciliter (dl) in a male is about (a) 5.4 million (b) 7 million (c) 4 million (d) more than 9 million

_______ **6.** The normal number of leukocytes per cubic millimeter or deciliter (dl) is (a) 5000 to 10,000 (b) 8000 to 12,000 (c) 2000 to 4000 (d) over 15,000

_______ **7.** Under the microscope, red blood cells appear as (a) circular discs with centrally located nuclei (b) circular discs with lobed nuclei (c) oval discs with many nuclei (d) biconcave discs without nuclei

_______ **8.** Platelets are formed from a special large cell that breaks up into small fragments. This cell is called a(n) (a) eosinophil (b) hemocytoblast (c) megakaryocyte (d) platelet

PART 2. Completion

9. Another name for red blood cells is ____________________.

10. Blood gets its red color from the presence of ____________________.

11. The life span of a red blood cell is approximately ____________________.

12. The normal ratio of red blood cells to white blood cells is about ____________________.

13. The granular leukocytes are formed from ____________________ tissue.

14. The number of platelets per cubic millimeter or deciliter (dl) found normally in blood is ____________________.

Heart

17

The ***heart*** is a hollow, muscular organ that pumps blood through miles and miles of blood vessels. This organ is located in the mediastinum, between the lungs, with two thirds of its mass lying to the left of the body's midline. Its pointed end, the ***apex,*** projects inferiorly to the left, and its broad end, the ***base,*** projects superiorly to the right. The main parts of the heart and associated structures to be discussed here are the pericardium, wall, chambers, great vessels, and valves.

A. PERICARDIUM

A loose-fitting serous membrane called the ***pericardium*** (*peri* = around; *cardio* = heart), or ***pericardial sac,*** encloses the heart (Figure 17.1). The membrane is composed of two layers, the fibrous pericardium and the serous pericardium. The ***fibrous pericardium*** forming the superficial layer is tough, inelastic, dense, irregular connective tissue that adheres to the parietal pleura and anchors the heart in the mediastinum. The deep layer, the ***serous pericardium,*** is a thinner, delicate membrane that is a double-layered structure. The outer ***parietal layer*** of the serous pericardium is directly beneath the fibrous pericardium. The inner ***visceral layer*** of the serous pericardium is also called the ***epicardium.*** Between the parietal and visceral layers of the serous pericardium is a potential space, the ***pericardial cavity,*** which contains pericardial fluid and functions to prevent friction between the layers as the heart beats. Identify these structures using a specimen, model, or chart of a heart and label Figure 17.1.

B. HEART WALL

Three layers of tissue compose the heart: the epicardium (external layer), the myocardium (middle layer), and the endocardium (inner layer). The ***epicardium*** (*epi* = above), which is also the visceral layer of the serous pericardium, is the thin, transparent superficial layer of the heart wall. The middle ***myocardium*** (*myo* = muscle), which is composed of cardiac muscle tissue, forms the bulk of the heart and is responsible for contraction. The ***endocardium*** (*endo* = within) is a thin, deep layer of endothelium and areolar connective tissue that lines the inside of

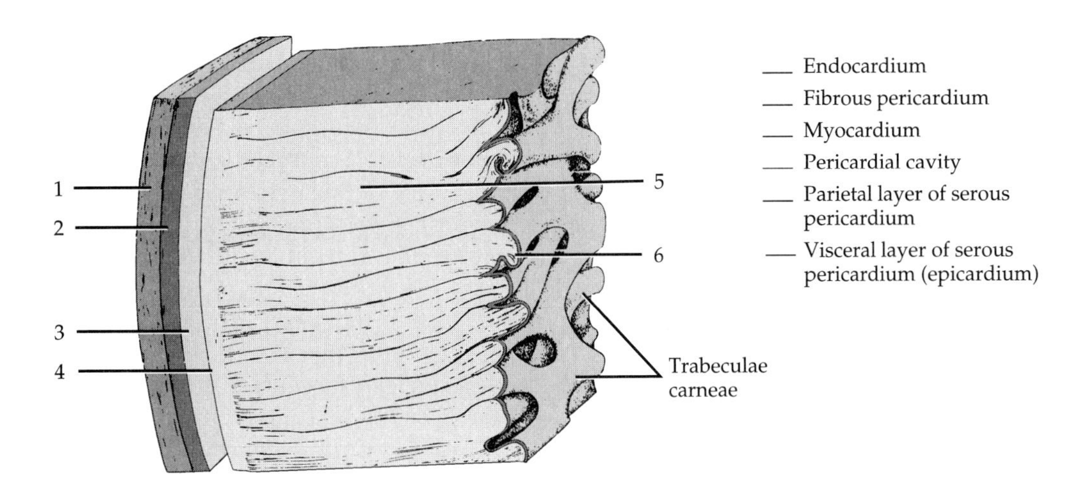

FIGURE 17.1 Structure of pericardium and heart wall.

the myocardium and covers the heart valves and the tendons that hold them open. Label the layers of the heart in Figure 17.1.

C. CHAMBERS OF HEART

The interior of the heart is divided into four cavities called ***chambers,*** which receive circulating blood. The two superior chambers are known as ***right*** and ***left atria*** (*atrium* = entry hall) and are separated internally by a partition called the ***interatrial septum*** (*septum* = partition). A prominent feature of this septum is an oval depression, the ***fossa ovalis,*** which corresponds to the site of the ***foramen ovale,*** an opening in the interatrial septum of the fetal heart that helps blood bypass the nonfunctioning lungs. Each atrium has an appendage called an ***auricle*** (OR-i-kul; *auris* = ear), so named because its shape resembles a dog's ear. The auricles increase the size of the atria so that they can hold greater volumes of blood. The lining of the atria is smooth, except for the anterior walls and linings of the auricles, which contain projecting muscle bundles called ***pectinate*** (PEK-ti-nāt) ***muscles.***

The two inferior and larger chambers, called the ***right*** and ***left ventricles*** (*ventricle* = little belly), are separated internally by a partition called the ***interventricular septum.*** The irregular surface of ridges and folds of the myocardium in the ventricles is known as the ***trabeculae carneae*** (tra-BEK-yoo-lē KAR-nē-ē; *trabecula* = little beam; *carneous* = fleshy). Externally, a groove known as the ***coronary sulcus*** (SUL-kus; plural is ***sulci;*** SUL-kē) separates the atria from the ventricles. The groove encircles the heart and houses the coronary sinus (a large cardiac vein) and the circumflex branch of the left coronary artery. The ***anterior interventricular sulcus*** and ***posterior interventricular sulcus*** separate the right and left ventricles externally. They also contain coronary blood vessels and a variable amount of fat.

Label the chambers and associated structures of the heart in Figures 17.2 on pages 327–329 and 17.3 on page 330.

D. GREAT VESSELS AND VALVES OF HEART

The right atrium receives *deoxygenated blood* (blood that gives up some of its oxygen to cells) through three veins: the ***superior vena cava (SVC)*** brings blood from most parts of the body superior to the heart, the ***inferior vena cava (IVC)*** brings blood from all parts of the body inferior to the diaphragm, and the ***coronary sinus*** brings blood from most of the vessels supplying the heart wall.

Deoxygenated blood is passed from the right atrium into the right ventricle through the atrioventricular valve called the ***tricuspid valve,*** which consists of three cusps (flaps). The right ventricle then pumps the blood through the ***pulmonary semilunar valve*** into the ***pulmonary trunk.*** The pulmonary trunk divides into a ***right*** and ***left pulmonary artery,*** each of which carries blood to the lungs where the blood releases its carbon dioxide and takes on oxygen. This blood, called *oxygenated blood,* returns to the heart via four ***pulmonary veins*** that empty the blood into the left atrium. The blood is then passed into the left ventricle through another atrioventricular valve called the ***bicuspid (mitral) valve,*** which consists of two cusps. The cusps of the tricuspid and bicuspid valves are connected to cords called ***chordae tendineae*** (KOR-dē TEN-din-ē-ē), which in turn attach to projections in the ventricular walls called ***papillary muscles.*** The left ventricle pumps oxygenated blood through the ***aortic semilunar valve*** into the ***ascending aorta.*** From this vessel, aortic blood is passed into the ***coronary arteries, arch of the aorta,*** and ***descending aorta.*** These blood vessels transport the blood to the body. The function of the heart valves is to permit the blood to flow in only one direction.

Label the great vessels and valves of the heart in Figures 17.2 and 17.3a and b.

E. BLOOD SUPPLY OF HEART

Because of its importance in myocardial infarction (heart attack), the blood supply of the heart will be described briefly at this point. The arterial supply of the heart is provided by the right and left coronary arteries. The ***right coronary artery*** originates as a branch of the ascending aorta, descends in the coronary sulcus, and gives off a ***marginal branch*** that supplies the right ventricle. The right coronary artery continues around the posterior surface of the heart in the posterior interventricular sulcus. This portion of the artery is known as the ***posterior interventricular branch*** and supplies the right and left ventricles and interventricular septum. In its course, the right coronary artery also supplies small branches (***atrial branches***) to the right atrium. The ***left coronary artery*** also originates as a branch of the ascending aorta. Between the pulmonary

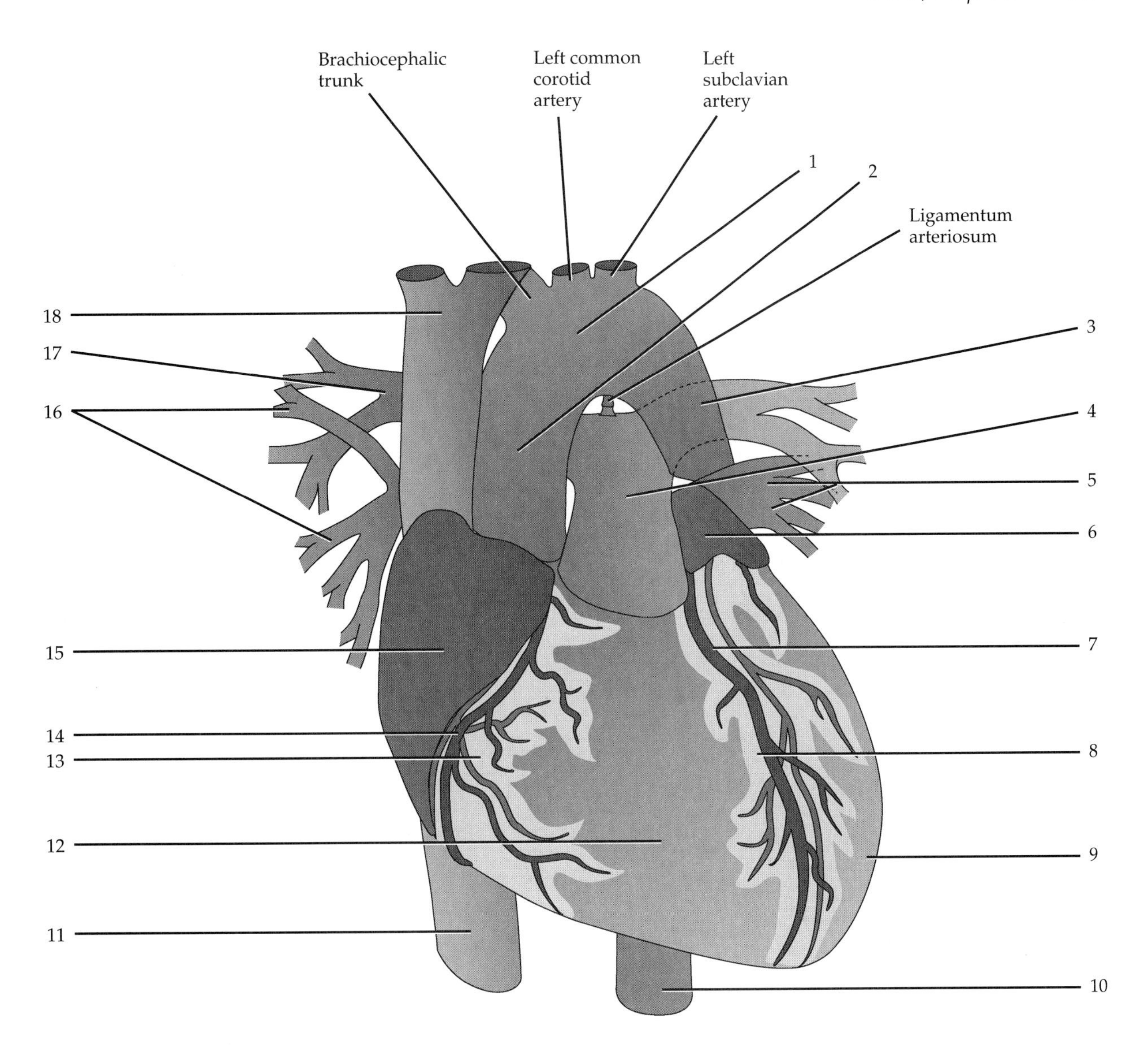

(a) Diagram of anterior external view

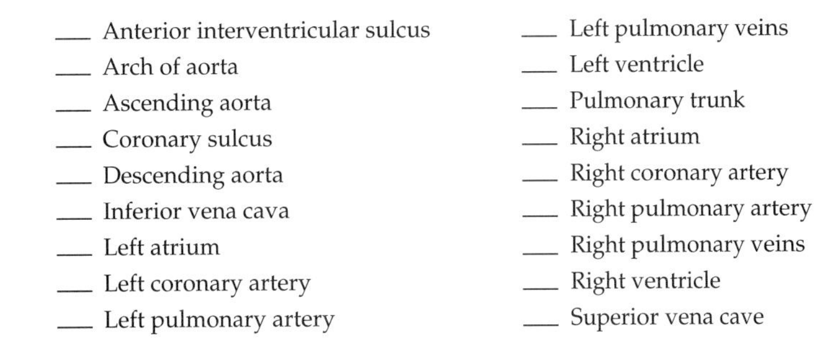

___ Anterior interventricular sulcus
___ Arch of aorta
___ Ascending aorta
___ Coronary sulcus
___ Descending aorta
___ Inferior vena cava
___ Left atrium
___ Left coronary artery
___ Left pulmonary artery
___ Left pulmonary veins
___ Left ventricle
___ Pulmonary trunk
___ Right atrium
___ Right coronary artery
___ Right pulmonary artery
___ Right pulmonary veins
___ Right ventricle
___ Superior vena cave

FIGURE 17.2 External surface of the human heart.

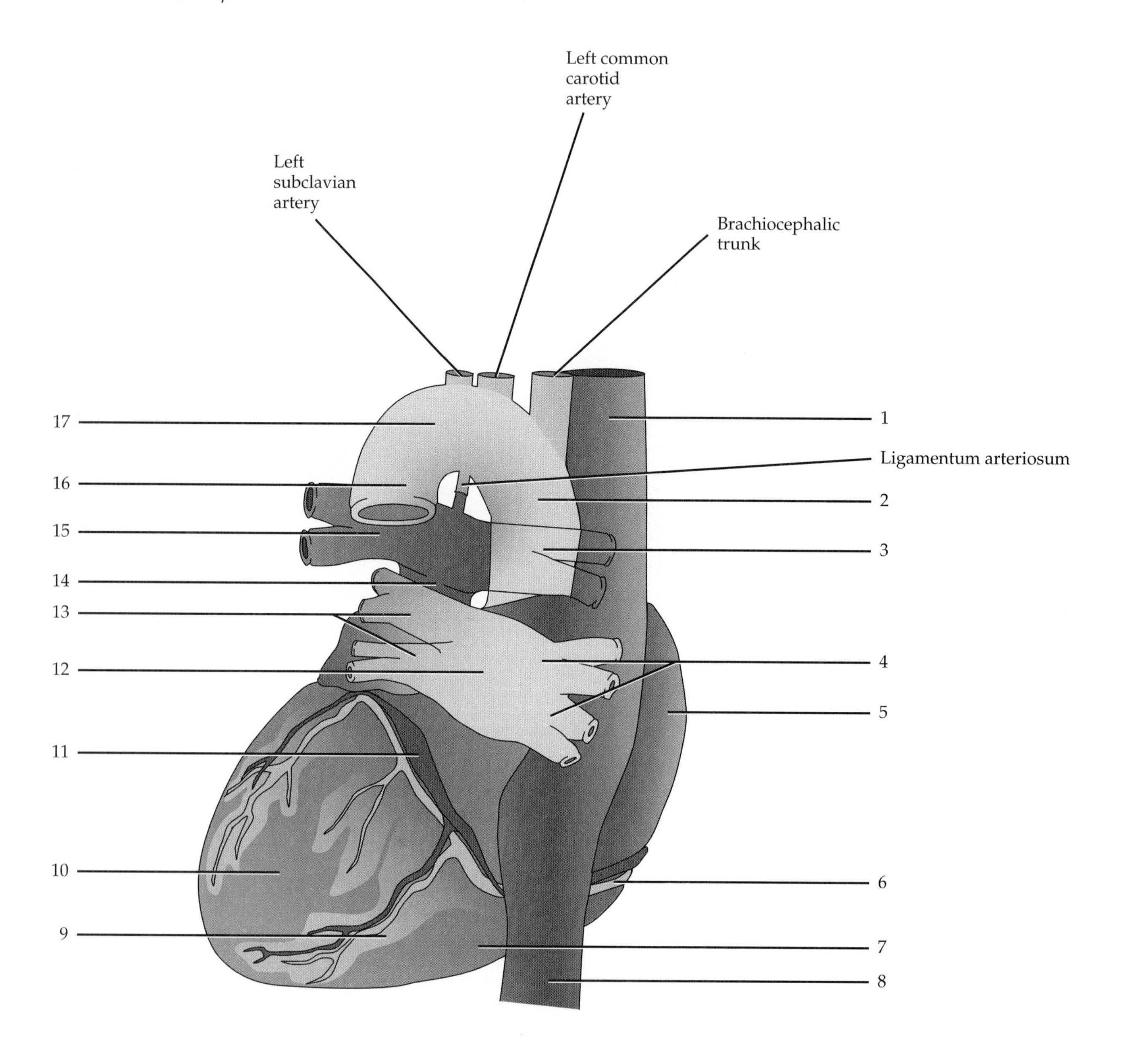

(b) Diagram of posterior external view

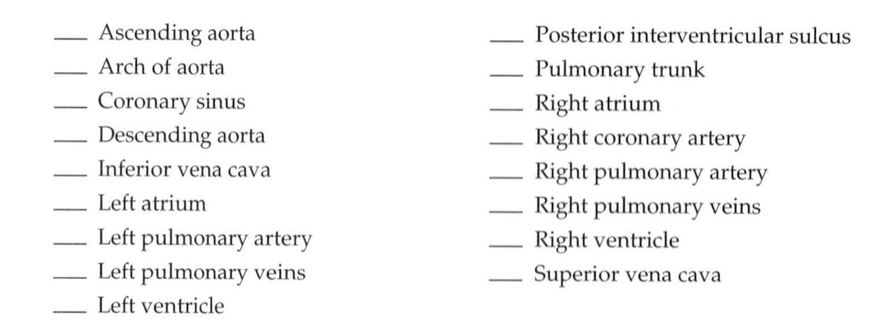

FIGURE 17.2 *(Continued)* External surface of the human heart.

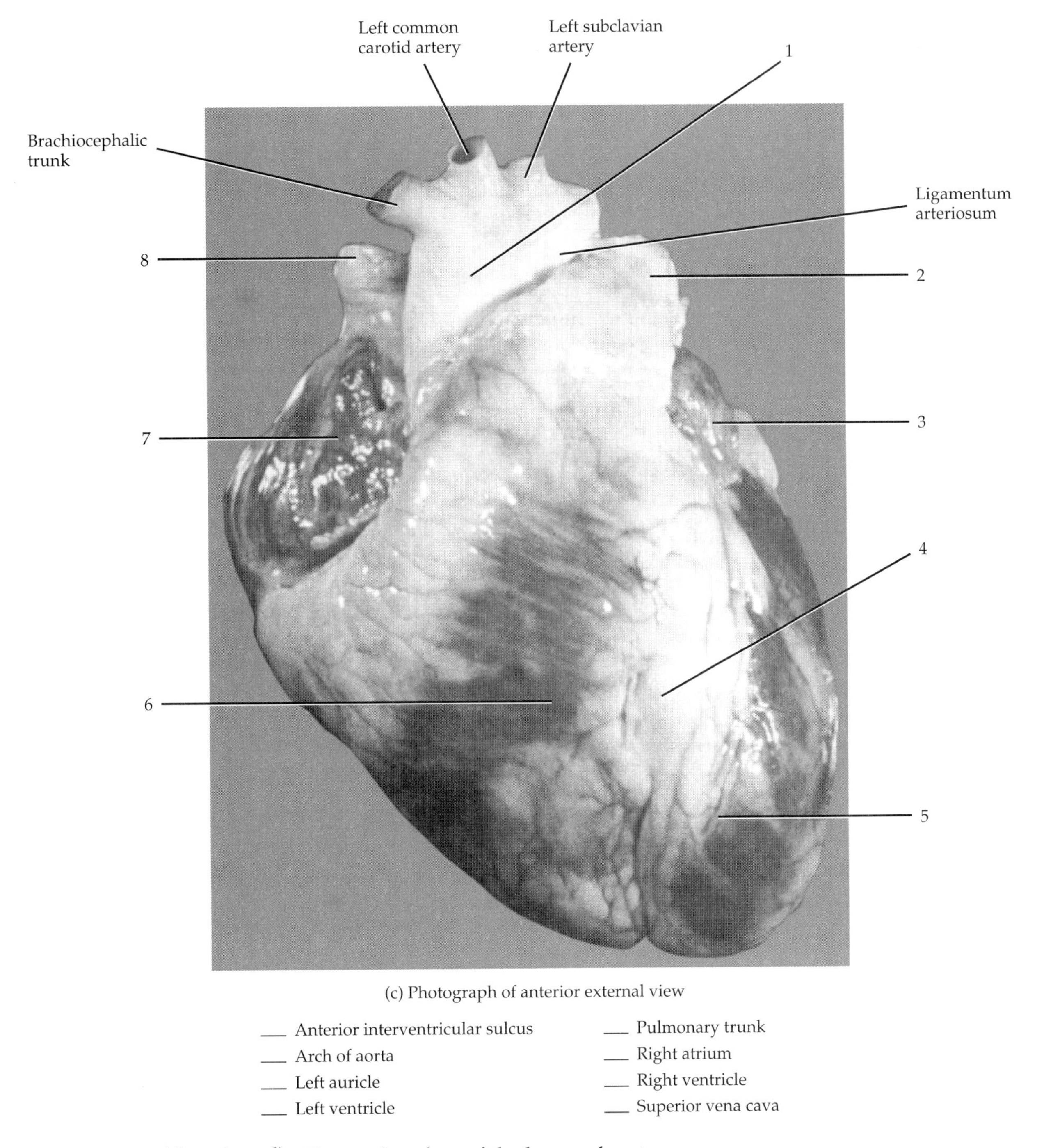

(c) Photograph of anterior external view

___ Anterior interventricular sulcus
___ Arch of aorta
___ Left auricle
___ Left ventricle
___ Pulmonary trunk
___ Right atrium
___ Right ventricle
___ Superior vena cava

FIGURE 17.2 *(Continued)* External surface of the human heart.

trunk and left auricle, the left coronary artery divides into two branches: anterior interventricular and circumflex. The ***anterior interventricular branch*** passes in the anterior interventricular sulcus and supplies the right and left ventricles and interventricular septum. The ***circumflex branch*** circles toward the posterior surface of the heart in the coronary sulcus and distributes blood to the left ventricle and left atrium.

Label the arteries in Figure 17.4a on page 331.

Most blood from the heart drains into the ***coronary sinus,*** a venous channel in the posterior portion of the coronary sulcus between the left atrium and left ventricle. The vein empties into the right atrium. The principal tributaries of the coronary sinus are the ***great cardiac vein,*** which drains the anterior aspect of the heart, and the ***middle cardiac vein,*** which drains the posterior aspect of the heart.

Label the veins in Figure 17.4b on page 331.

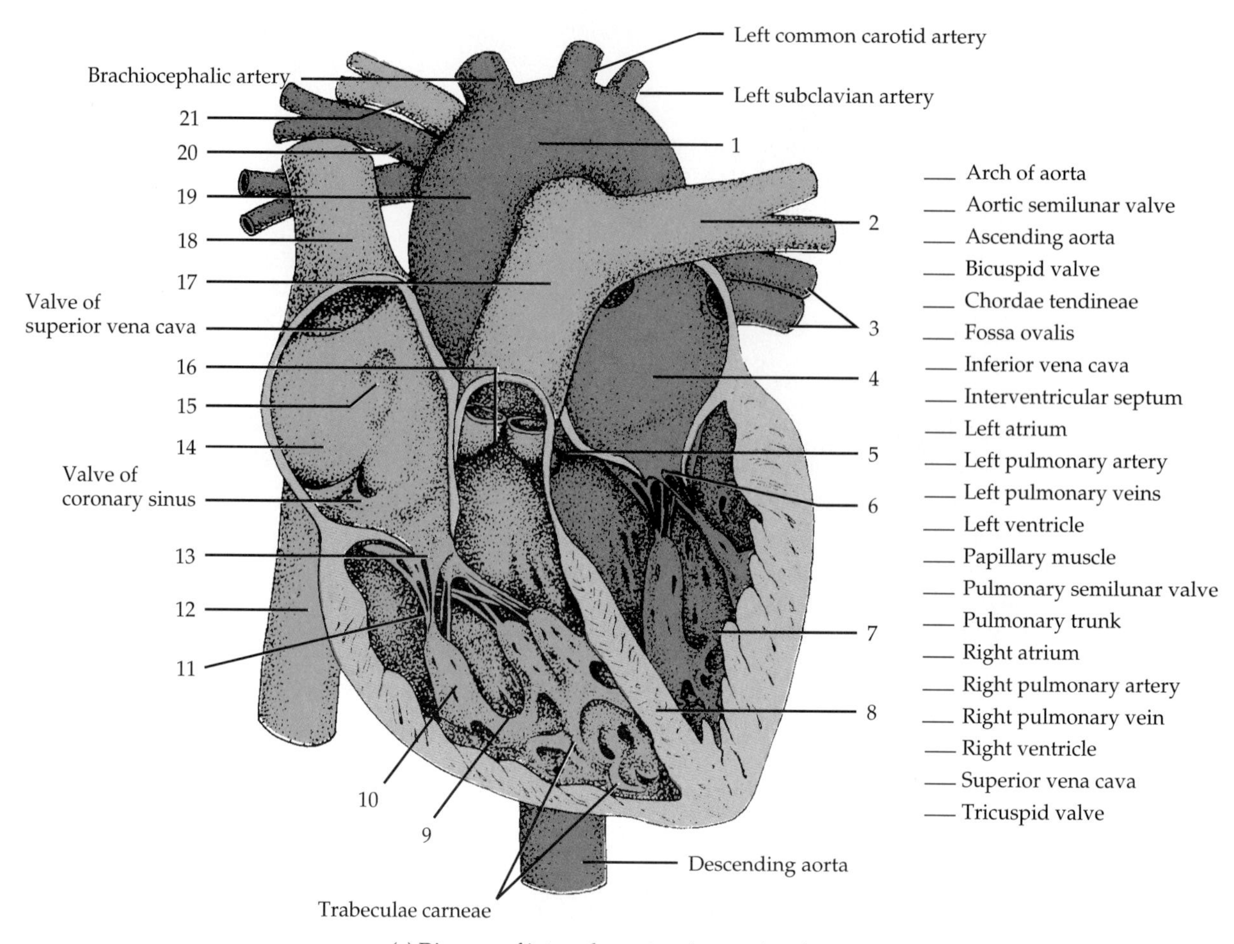

(a) Diagram of internal structure in anterior view

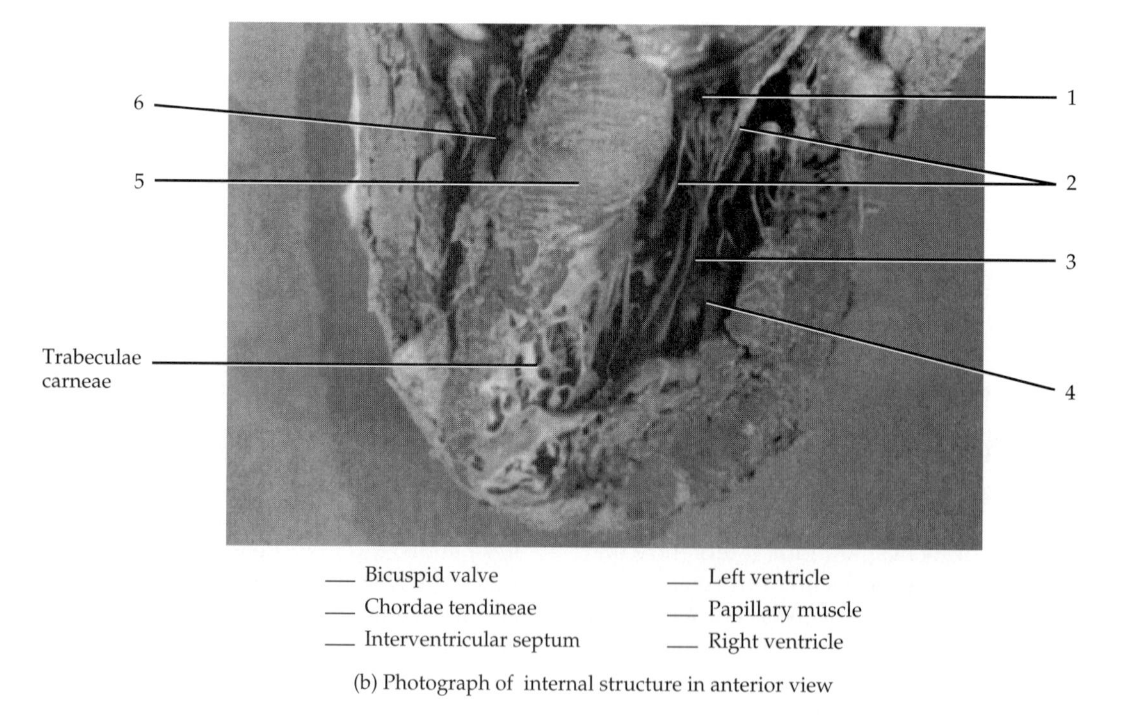

__ Bicuspid valve
__ Chordae tendineae
__ Interventricular septum
__ Left ventricle
__ Papillary muscle
__ Right ventricle

(b) Photograph of internal structure in anterior view

FIGURE 17.3 Structure of the human heart.

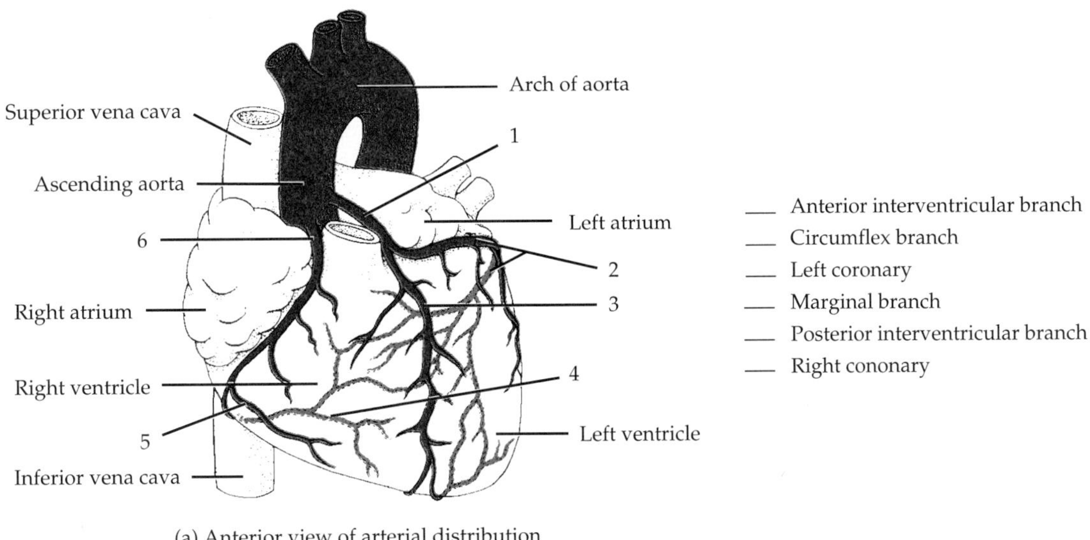

(a) Anterior view of arterial distribution

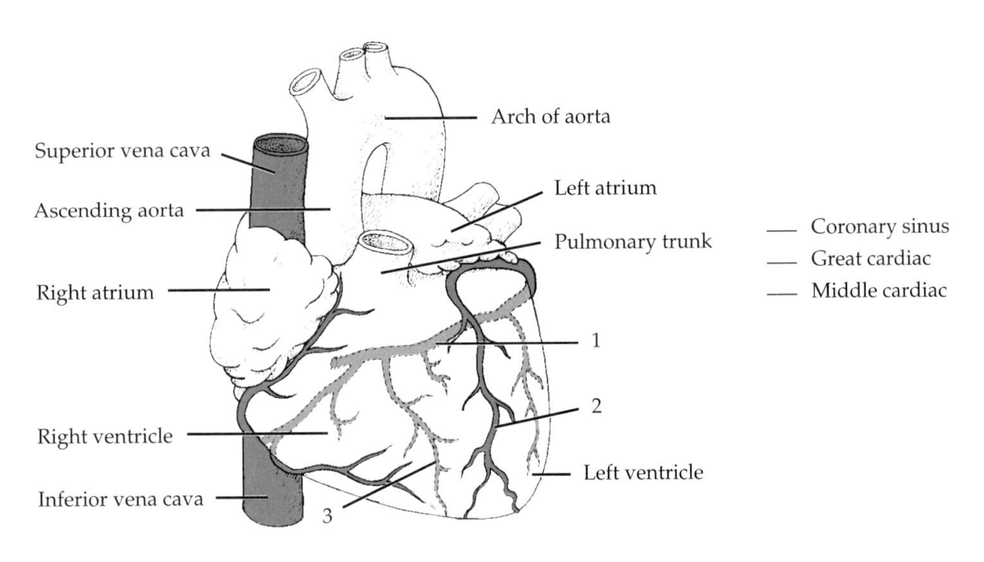

(b) Anterior view of venous drainage

FIGURE 17.4 Coronary (cardiac) circulation.

F. CARDIAC CONDUCTION SYSTEM

During embryonic development, about 1% of the cardiac muscle fibers become ***autorhythmic*** (self-excitable) ***cells:*** they repeatedly and rhythmically generate action potentials (nerve impulses). Autorhythmic fibers set the rhythm for the entire heart, and they form the ***conduction system,*** the route for conducting action potentials throughout the heart muscle. The conduction system ensures that cardiac chambers become excited to contract in a coordinated manner, which makes the heart an effective pump. Figure 17.5 shows the components of the conduction system.

Normally, cardiac excitation begins in the ***sinoatrial (SA) node,*** which is located in the right atrial wall just inferior to the opening of the superior vena cava. Each action potential from the SA node travels throughout both atria via gap junctions in the intercalated discs of atrial fibers. This causes the atria to contract. The action potential also spreads from the SA node down to the ***atrioventricular (AV) node,*** located in the septum between

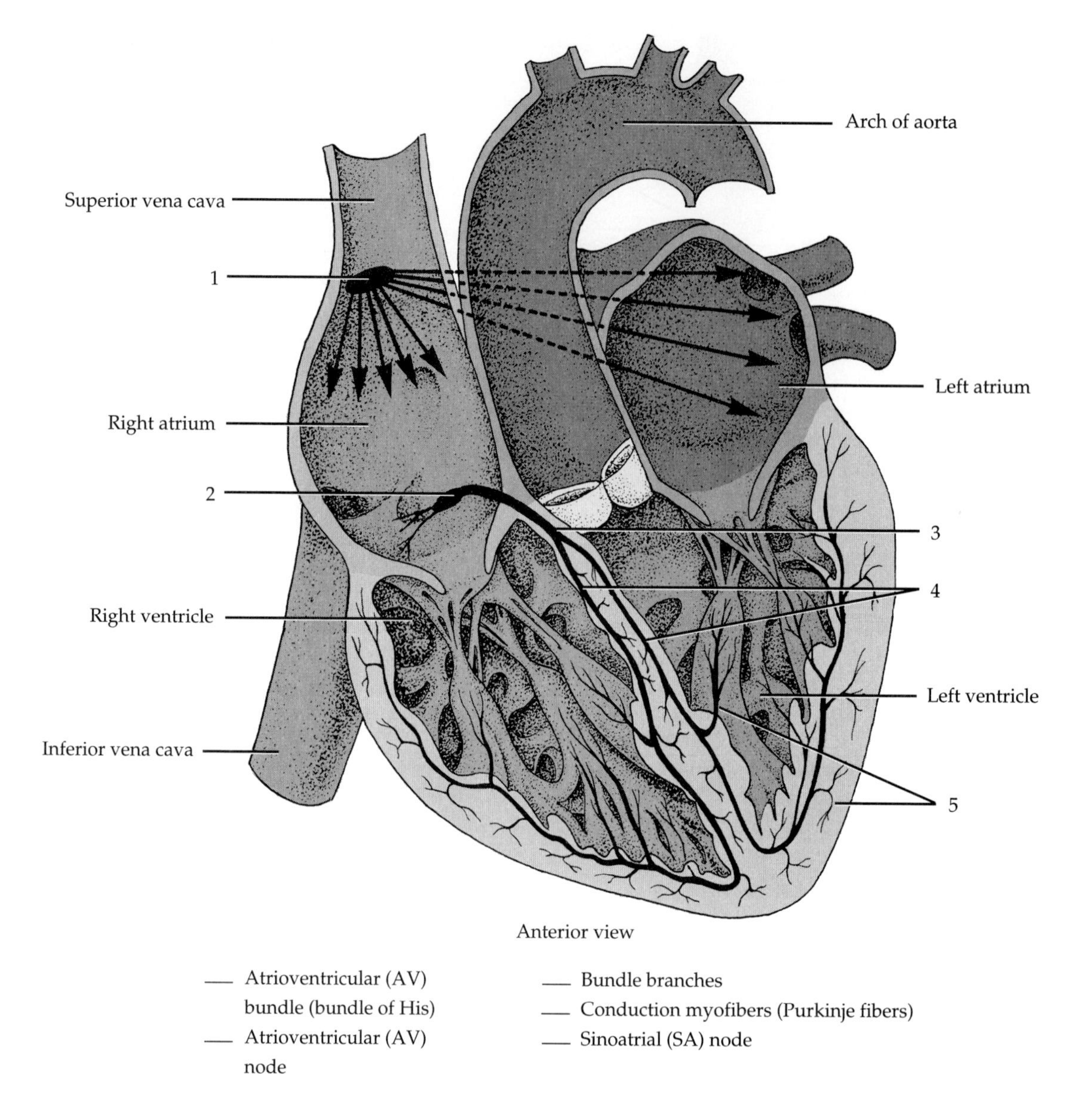

FIGURE 17.5 Conduction system of heart.

the two atria. From the AV node, the action potential enters the ***atrioventricular (AV) bundle (bundle of His).*** After conducting along the AV bundle, the action potential enters both the ***right*** and ***left bundle branches*** that course through the interventricular septum toward the apex of the heart. Finally, large-diameter ***conduction myofibers (Purkinje fibers)*** rapidly conduct the action potential first to the apex of the ventricular myocardium and then upward to the remainder of the ventricular myocardium. This causes the ventricles to contract.

Label the components of the cardiac conduction system shown in Figure 17.5.

Impulse conduction through the heart generates electrical currents that can be detected at the surface of the body. A recording of these electrical changes is called an ***electrocardiogram*** (e-lek'-trō-KAR-dē-ō-gram), abbreviated either ***ECG*** or ***EKG*** (from the German word *elektrokardiogram).* The ECG is a composite of action potentials produced by all the heart muscle fibers during each heartbeat. The instrument used to record the changes is an ***electrocardiograph.*** In clinical practice, the ECG is recorded by placing electrodes on the arms and legs (the limb leads) and at six positions on the chest. The electrocardiograph amplifies the heart's electrical activity and produces a series of up-and-down (vertical) tracings called ***deflection waves.*** By comparing these deflection waves with one another and with normal records, it is possible to determine (1) if the conduction pathway is abnormal, (2) if the heart is enlarged, and (3) if certain regions are damaged.

G. DISSECTION OF CAT HEART

CAUTION! *Please reread Section D, "Precautions Related to Dissection" at the beginning of the laboratory manual on page xi before you begin your dissection.*

PROCEDURE

1. Before examining the cat's heart and major blood vessels, you must open the thoracic cavity.
2. On the ventral surface of the animal, palpate the bottom edge of the rib cage. Using heavyduty scissors, penetrate the chest cavity immediately below the lowest rib and make a longitudinal incision about 1/2 in. to the right of the sternum.
3. Cut through the ribs, extending the incision anteriorly to the apex of the thorax. Be careful not to cut into any internal organs.
4. Spread apart the walls of the thorax and locate the diaphragm at the caudal end of the incision. *Be careful not to break any of the ribs,* because the sharp edge of a broken rib could interfere with your examination of the viscera.
5. Make two additional cuts laterally and dorsally toward the spine on both sides, keeping the incisions just anterior to the diaphragm.
6. Note the ***right*** and ***left atria*** and the ***right*** and ***left ventricles*** (Figure 17.6). Each atrium has an earlike flap on its ventral surface called an ***auricle.*** Identify the ***coronary sulcus*** between the right atrium and right ventricle and the ***anterior longitudinal sulcus*** between the right ventricle and left ventricle.
7. Carefully remove the thymus gland and any fat so that you can clearly see the ventral surface of the heart and its attached vessels (Figure 17.6). Be careful not to damage any nerves in this area.
8. The ***pericardium,*** the membrane surrounding the heart, consists of two layers. The outer ***parietal pericardium*** is a tough membrane that surrounds the heart and blood vessels that join the heart. This layer can be removed. The inner ***visceral pericardium*** invests the heart muscle (myocardium) closely and is very difficult to remove.
9. The ***apex*** of the heart is directed caudally and to the left and its ***base*** is directed cranially and to the right.
10. The cat's heart can be examined without removal. Leaving the heart in place will help you examine and identify the major blood vessels later on.
11. If a detailed dissection of the cat heart is to be done, follow the directions of the sheep heart dissection and identify the following structures of the cat heart, which correlate with the structures of the sheep heart. Because the cat heart is small, it may be difficult to identify all of them.
    a. Opening of the superior vena cava
    b. Opening of the inferior vena cava
    c. Fossa ovalis
    d. Opening of the coronary sinus
    e. Pectinate muscle
    f. Interatrial septum
    g. Right ventricle
        i. tricuspid valve
        ii. chordae tendineae
        iii. papillary muscles
        iv. interventricular septum
        v. pulmonary semilunar valve
    h. Left atrium: openings of pulmonary veins
    i. Left ventricle
        i. bicuspid (mitral) valve
        ii. chordae tendineae
        iii. papillary muscles
        iv. aortic semilunar valve

H. DISSECTION OF SHEEP HEART

The anatomy of the sheep heart closely resembles that of the human heart. Use Figures 17.6 and 17.7 as references for this dissection. In addition, models of human hearts can also be used as references.

CAUTION! *Please reread Section D, "Precautions Related to Dissection" at the beginning of the laboratory manual on page xi before you begin your dissection.*

First examine the ***pericardium,*** a fibroserous membrane that encloses the heart, which may have already been removed in preparing the sheep heart for preservation. The ***myocardium*** is the middle layer and constitutes the main muscle portion of the heart. The ***endocardium*** (the third layer) is the inner lining of the heart. Use the figures to

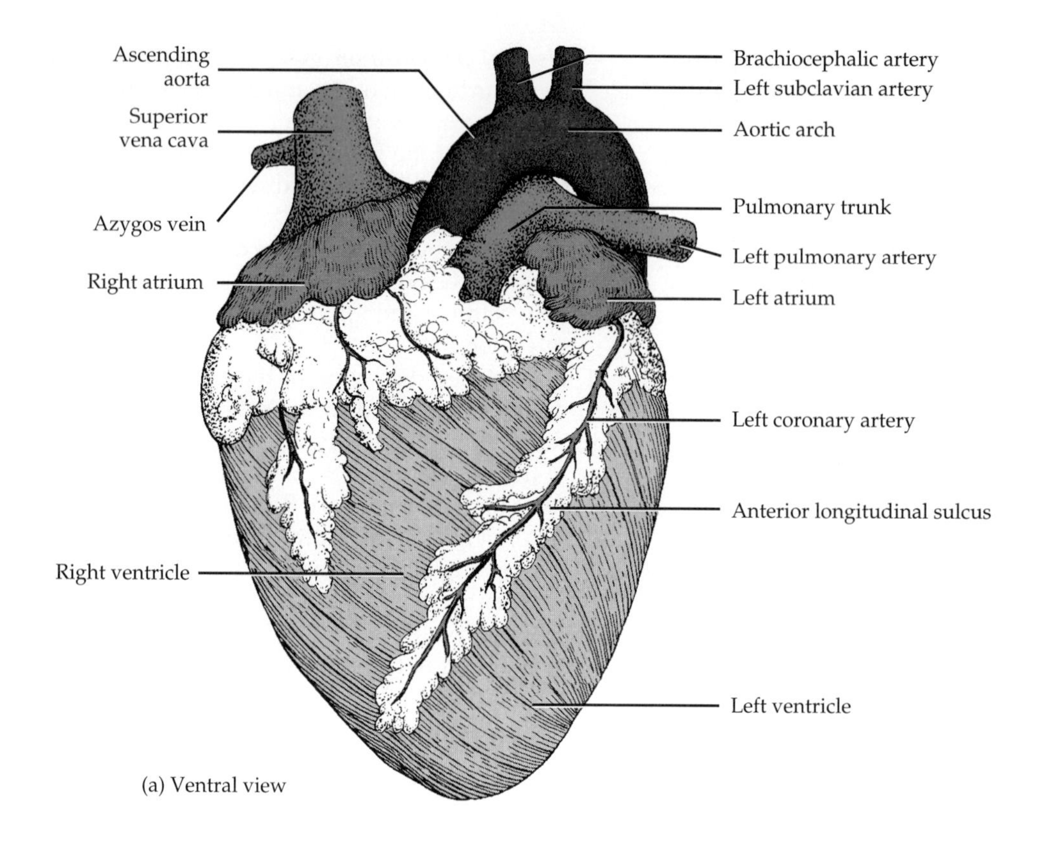

(a) Ventral view

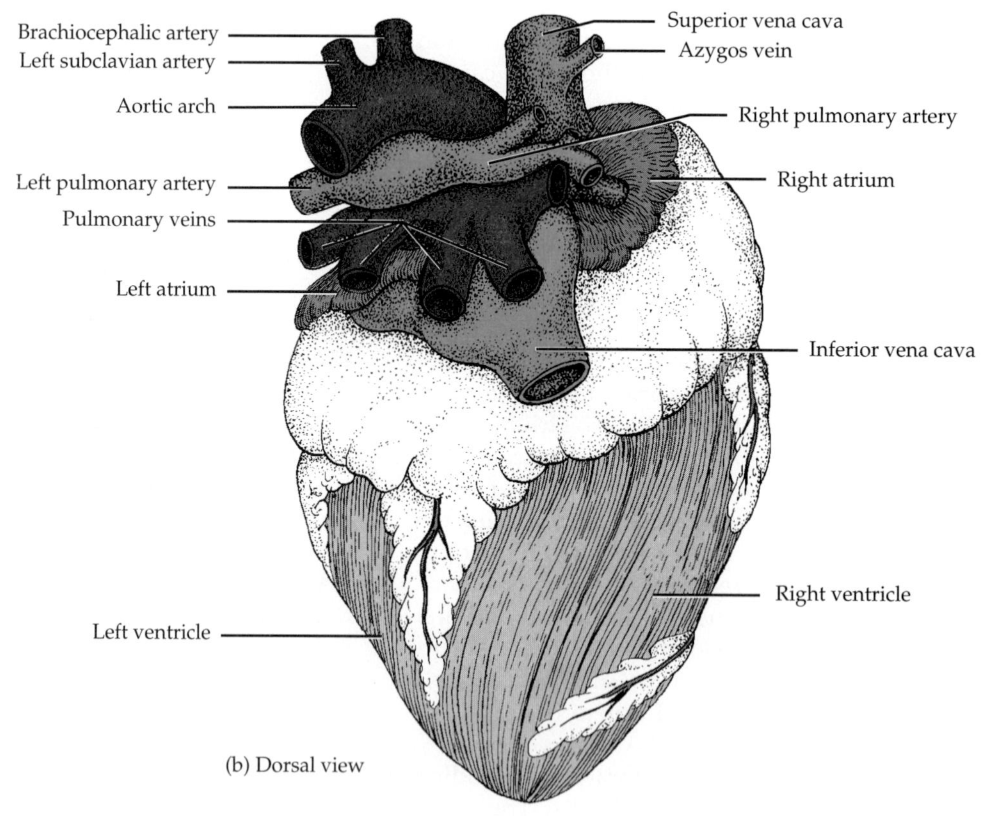

(b) Dorsal view

FIGURE 17.6 External structure of a cat or sheep heart.

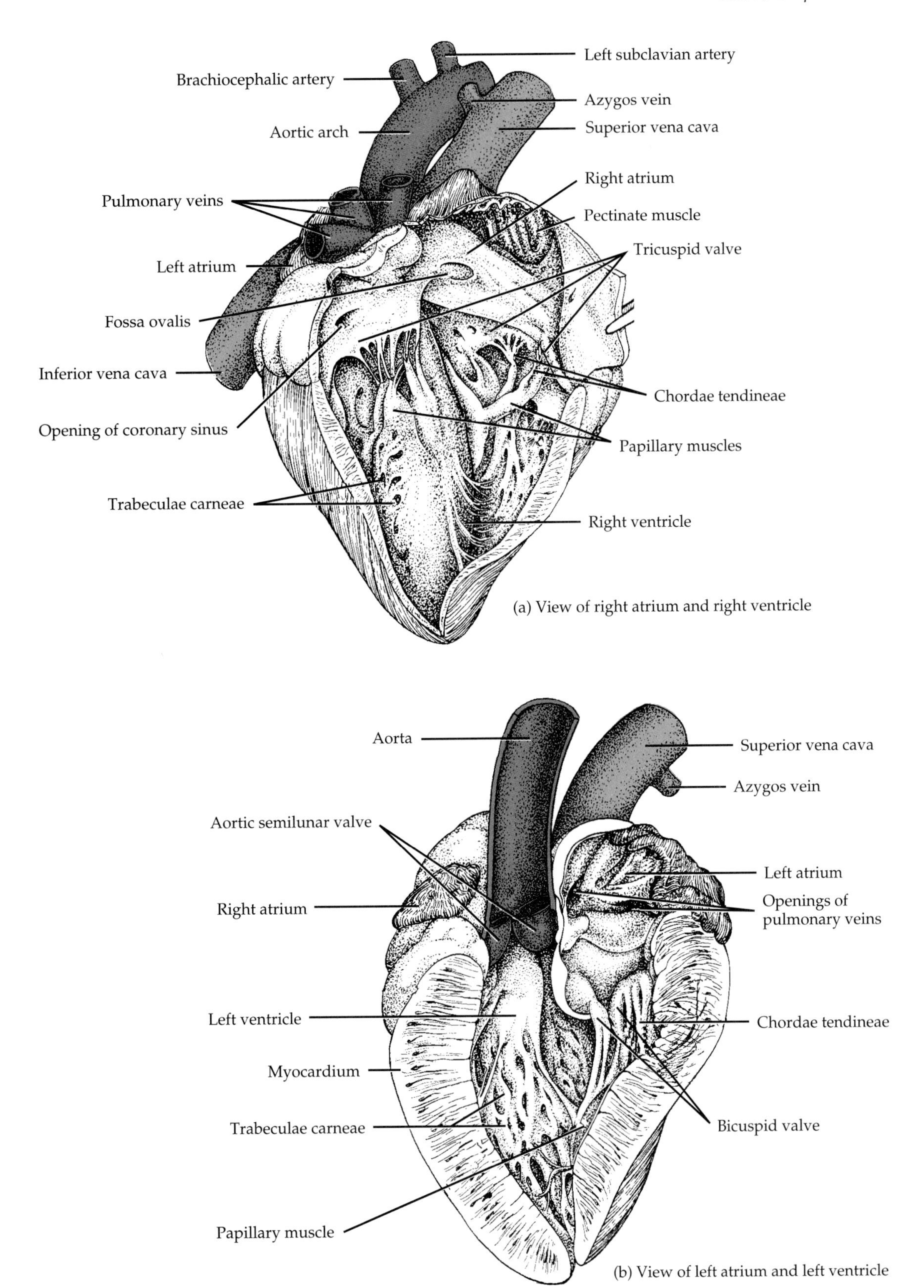

FIGURE 17.7 Internal structure of a cat or sheep heart.

determine which is the ventral surface of the heart and then identify the ***pulmonary trunk*** emerging from the anterior ventral surface, near the midline, and medial to the ***left auricle***. A longitudinal depression on the ventral surface, called the ***anterior longitudinal sulcus***, separates the right ventricle from the left ventricle. Locate the ***coronary blood vessels*** lying in this sulcus.

PROCEDURE

1. Remove any fat or pulmonary tissue that is present.
2. In cutting the sheep heart open to examine the chambers, valves, and vessels, the anterior longitudinal sulcus is used as a guide.
3. Carefully make a shallow incision through the ventral wall of the pulmonary trunk and the right ventricle, trying not to cut the dorsal surface of either structure.
4. The incision is best made *less than an inch to the right of, and parallel to,* the previously mentioned anterior longitudinal sulcus.
5. If necessary, the incision can be continued to where the pulmonary trunk branches into a ***right pulmonary artery***, which goes to the right lung, and a ***left pulmonary artery***, which goes to the left lung. The ***pulmonary semilunar valve*** of the pulmonary artery can be clearly seen upon opening it. In any of these internal dissections of the heart, any coagulated blood or latex should be immediately removed so that all important structures can be located and identified.
6. Keeping the cut still parallel to the sulcus, extend the incision around and through the dorsal ventricular wall until you reach the ***interventricular septum.***
7. Now examine the dorsal surface of the heart and locate the thin-walled ***superior vena cava*** directly above the ***right auricle***. This vein proceeds posteriorly straight into the right atrium.
8. Make a second longitudinal cut, this time through the superior vena cava (dorsal wall).
9. Extend the cut posteriorly through the right atrium on the left of the right auricle. Proceed posteriorly to the dorsal right ventricular wall and join your first incision.
10. The entire internal right side of the heart should now be clearly seen when carefully spread apart. The interior of the superior vena cava, right atrium, and right ventricle will now be examined. Start with the right auricle and locate the ***pectinate muscle***, the large opening of the ***inferior vena cava*** on the left side of the right atrium, and the opening of the ***coronary sinus*** just below the opening of the inferior vena cava. By using a dull probe and gentle pressure, most of the vessels can be traced to the dorsal surface of the heart.
11. Now find the wall that separates the two atria, the ***interatrial septum***. Also find the ***fossa ovalis***, an oval-shaped depression ventral to the entrance of the inferior vena cava.
12. The ***tricuspid valve*** between the right atrium and the right ventricle should be examined to locate the three cusps, as its name indicates. From the cusps of the valve itself, and tracing posteriorly, the ***chordae tendineae***, which hold the valve in place, should be identified. Still tracing posteriorly, the chordae are seen to originate from the ***papillary muscles***, which themselves originate from the wall of the right ventricle itself.
13. Look carefully again at the dorsal surface of the left atrium and locate as many ***pulmonary veins*** (normally, four) as possible.
14. Make your third longitudinal cut through the most lateral of the pulmonary veins that you have located.
15. Continue posteriorly through the left atrial wall and the left ventricle to the ***apex*** of the heart.
16. Compare the difference in the thickness of the wall between the right and left ventricles. Explain your answer.
17. Examine the ***bicuspid (mitral) valve***, again counting the cusps. Determine if the left side of the heart has basically the same structures as studied on the right side.
18. Probe from the left ventricle to the ***aorta*** as it emerges from the heart, examining the ***aortic semilunar valve***. Find the openings of the right and left main coronary arteries.
19. Locate now the ***brachiocephalic artery***, which is one of the first branches from the arch of the aorta. This artery continues branching and terminates by supplying the arms and head as its name indicates.
20. Connecting the aorta with the pulmonary artery is the remnant of the ***ductus arteriosus***, called the ***ligamentum arteriosum***. It may not be present in your sheep heart.

ANSWER THE LABORATORY REPORT QUESTIONS AT THE END OF THE EXERCISE.

LABORATORY REPORT QUESTIONS

Heart 17

Student ______________________ **Date** ______________________

Laboratory Section ______________________ **Score/Grade** ______________________

PART 1. Multiple Choice

_______ **1.** Which of the following veins drain the blood from most of the vessels supplying the heart wall? (a) vasa vasorum (b) superior vena cava (c) coronary sinus (d) inferior vena cava

_______ **2.** The atrioventricular valve on the same side of the heart as the origin of the aorta is the (a) aortic semilunar (b) tricuspid (c) bicuspid (d) pulmonary semilunar

_______ **3.** Which valve does the blood go through just before entering the pulmonary trunk on the way to the lungs? (a) tricuspid (b) pulmonary semilunar (c) aortic semilunar (d) bicuspid

_______ **4.** The pointed end of the heart that projects inferiorly and to the left is the (a) costal surface (b) base (c) apex (d) coronary sulcus

_______ **5.** Which of these structures is more internal? (a) fibrous pericardium (b) visceral layer of serous pericardium (c) parietal layer of serous pericardium (d) myocardium

_______ **6.** The musculature of the heart is referred to as the (a) endocardium (b) myocardium (c) epicardium (d) pericardium

_______ **7.** The depression in the interatrial septum corresponding to the foramen ovale of fetal circulation is the (a) interventricular sulcus (b) pectinate muscle (c) chordae tendineae (d) fossa ovalis

PART 2. Completion

8. Malfunction of the ______________________ valve would interfere with the flow of blood from the right atrium to the right ventricle.

9. Deoxygenated blood is sent to the lungs through the ______________________.

10. The loose-fitting serous membrane that encloses the heart is called the ______________________.

11. The two inferior chambers of the heart are separated by the ______________________.

12. The earlike flap of tissue on each atrium is called a(n) ______________________.

13. The large vein that drains blood from most parts of the body superior to the heart and empties into the right atrium is the ______________________.

14. The cusps of atrioventricular valves are prevented from inverting by the presence of cords called ______________________, which are attached to papillary muscle.

15. A groove on the surface of the heart that houses blood vessels and a variable amount of fat is called a(n) ______________________.

16. The branch of the left coronary artery that distributes blood to the left atrium and left ventricle is the ______________________.

17. Action potentials that stimulate cardiac muscle fibers to contract are distributed by the cardiac______________________ system.

18. A record of electrical events associated with a heartbeat is called a(n) ______________________.

PART 3. Special Exercise

Draw a model of the heart and carefully label the four chambers, the four valves in their proper places, and the major blood vessels entering and exiting from the heart.

Blood Vessels

18

Blood vessels are networks of tubes that carry blood throughout the body. They are called arteries, arterioles, capillaries, venules, or veins. In this exercise, you will study the histology of blood vessels and identify the principal arteries and veins of the cardiovascular system.

A. ARTERIES AND ARTERIOLES

Arteries (AR-ter-ēs; *aer* = air; *tereo* = to carry) are blood vessels that carry blood *away* from the heart to body tissues. Arteries are constructed of three coats of tissue called ***tunics*** and a hollow core, called a ***lumen,*** through which blood flows (Figure 18.1). The deep coat is called the ***tunica interna*** and consists of a lining of endothelium in contact with the blood and a layer of elastic tissue called the *internal elastic lamina.* The middle coat, or ***tunica media,*** is usually the thickest layer and consists of elastic fibers and smooth muscle fibers. This tunic is responsible for two major properties of arteries: ***elasticity*** and ***contractility.*** The superficial coat, or ***tunica externa,*** is composed principally of elastic and collagen fibers. An *external elastic lamina* may separate the tunica externa from the tunica media.

Obtain a prepared slide of a transverse section of an artery and identify the tunics using Figure 18.1 as a guide.

As arteries approach various tissues of the body, they become smaller and are known as ***arterioles*** (ar-TER-rē-ōls; *arteriola* = small artery). Arterioles play a key role in regulating blood flow from arteries into capillaries. When arterioles enter a tissue, they branch into countless microscopic blood vessels called capillaries.

B. CAPILLARIES

Capillaries (KAP-i-lar'-ēs; *capillaris* = hairlike) are microscopic blood vessels that connect arterioles and venules. Their function is to permit the exchange of nutrients and wastes between blood and body tissues. This function is related to the fact that capillaries consist of only a single layer of endothelium.

C. VENULES AND VEINS

When several capillaries unite, they form small veins called ***venules*** (VEN-yools; *venula* = little vein). They collect blood from capillaries and drain it into veins.

Veins (VĀNS) are composed of the same three tunics as arteries, but there are variations in their relative thicknesses. The tunica interna of veins is thinner than that of their accompanying arteries. In addition, the tunica media of veins is much thinner than that of accompanying arteries with relatively little smooth muscle or elastic fibers. The tunica externa is the thickest layer consisting of collagen and elastic fibers (Figure 18.1). Functionally, veins return blood from tissues *to* the heart.

Obtain a prepared slide of a transverse section of an artery and its accompanying vein and compare them, using Figure 18.1 as a guide.

D. CIRCULATORY ROUTES

The two basic postnatal (after birth) circulatory routes are systemic and pulmonary circulation (Figure 18.2 on page 341). Some other circulatory routes, which are all subdivisions of systemic circulation, include hepatic portal circulation, coronary (cardiac) circulation, fetal circulation, and the cerebral arterial circle (circle of Willis). The latter is found at the base of the brain (see Table 18.2 on page 344).

1. Systemic Circulation

The largest route is the ***systemic circulation*** (see Figures 18.3 through 18.11 and Tables 18.1 through

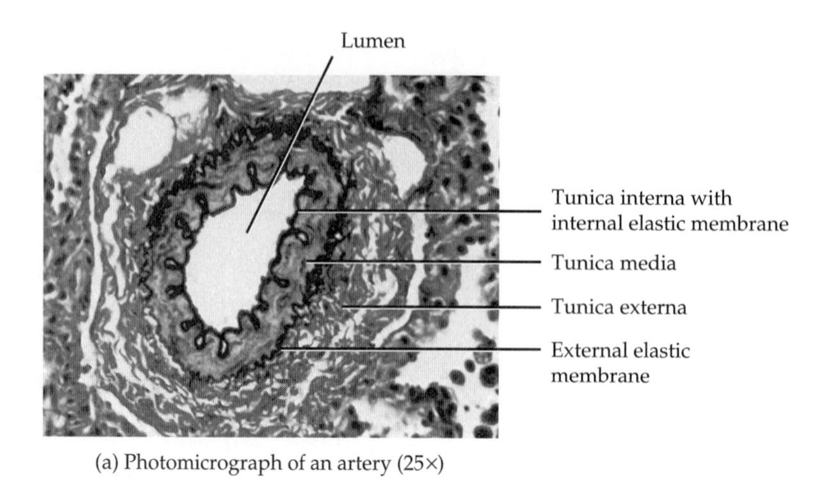

(a) Photomicrograph of an artery (25×)

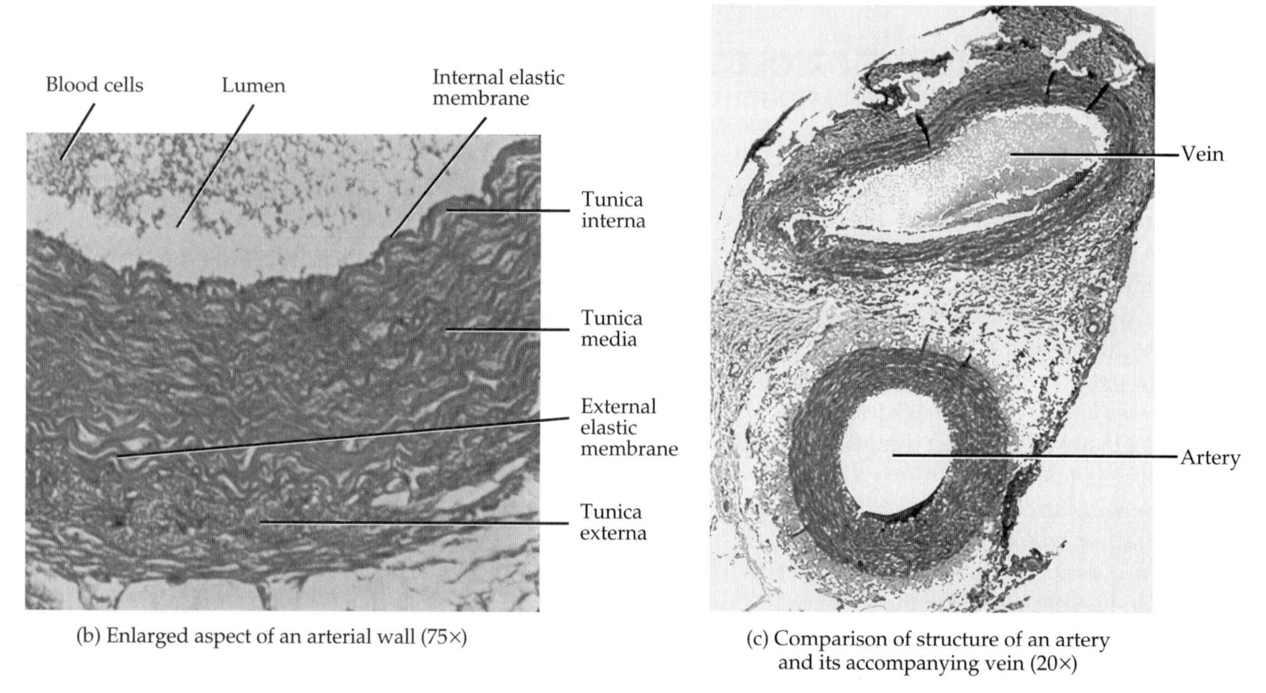

(b) Enlarged aspect of an arterial wall (75×)

(c) Comparison of structure of an artery and its accompanying vein (20×)

FIGURE 18.1 Histology of blood vessels.

18.11). This route includes the flow of blood from the left ventricle to all parts of the body. The function of the systemic circulation is to carry oxygen and nutrients to body tissues and to remove carbon dioxide and other wastes from them. All systemic arteries branch from the ***aorta***. As the aorta emerges from the left ventricle, it passes superiorly and posteriorly to the pulmonary trunk. At this point, it is called the ***ascending aorta***. The ascending aorta gives off two coronary branches (right and left coronary arteries) to the heart muscle. Then it turns to the left, forming the ***arch of the aorta*** before descending to the level of the intervertebral disc between the fourth and the fifth thoracic vertebrae as the ***descending aorta***. The descending aorta lies close to the vertebral bodies, passes through the diaphragm, and divides at the level of the fourth lumbar vertebra into two ***common iliac arteries***, which carry blood to the lower limbs. The section of the descending aorta between the arch of the aorta and the diaphragm is referred to as the ***thoracic aorta***. The section

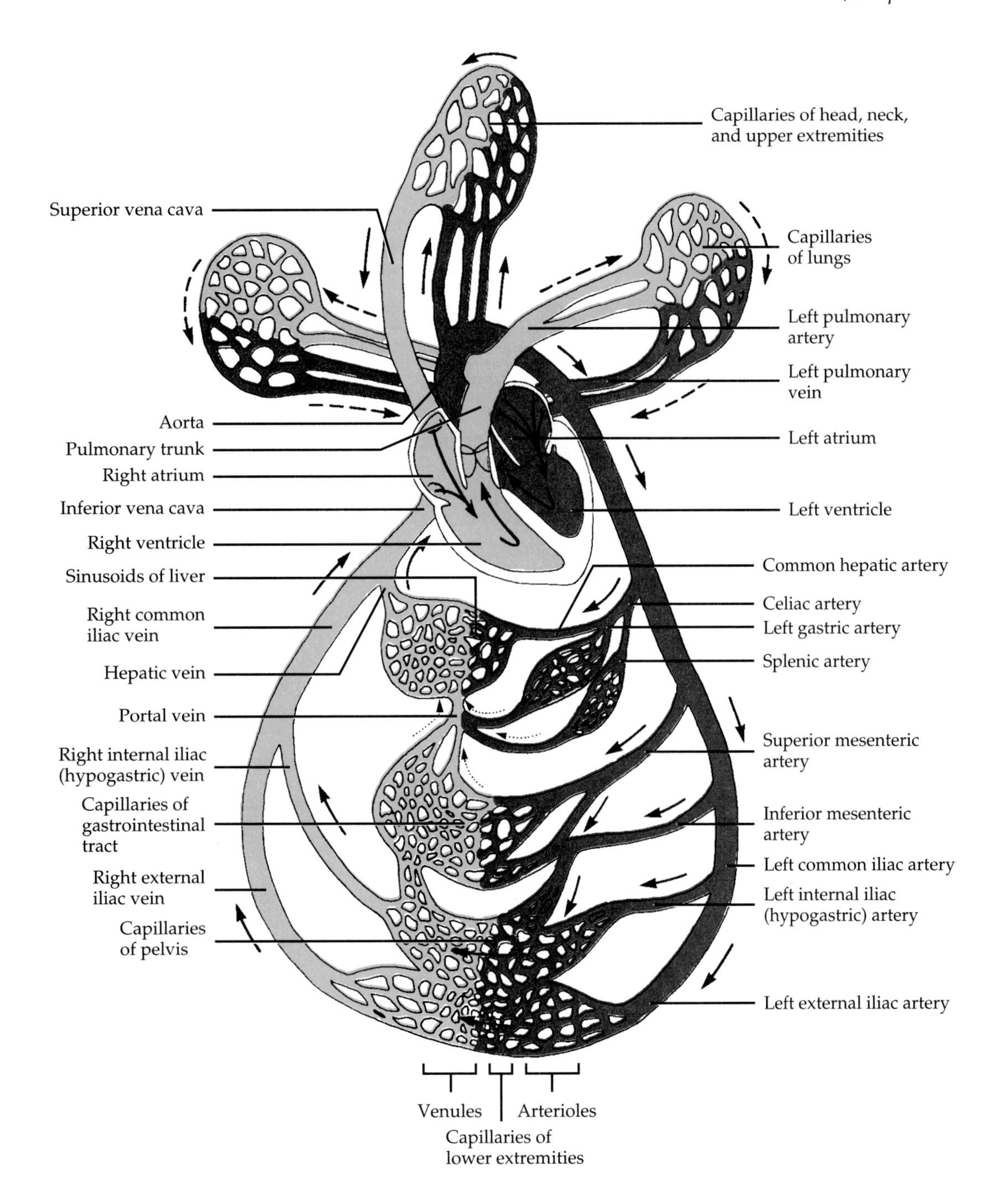

FIGURE 18.2 Circulatory routes. Systemic circulation is indicated by solid arrows; pulmonary circulation by broken arrows; and hepatic portal circulation by dotted arrows.

between the diaphragm and the common iliac arteries is termed the ***abdominal aorta.*** Each section of the aorta gives off arteries that continue to branch into distributing arteries leading to organs and finally into the arterioles and capillaries that service the systemic tissues (except the air sacs of the lungs).

Deoxygenated blood is returned to the heart through the systemic veins. All the veins of the systemic circulation flow into either the ***superior vena cava, inferior vena cava,*** or ***coronary sinus.*** They in turn empty into the right atrium.

Refer to Tables 18.1 through 18.11 and Figures 18.3 through 18.12.

TABLE 18.1
Aorta and Its Branches (Figure 18.3)

OVERVIEW: The ***aorta*** (ā-OR-ta) is the largest artery of the body, about 2 to 3 cm (0.8 to 1.2 in.) in diameter. It begins at the left ventricle and contains a valve at its origin, called the aortic semilunar valve (see Figure 17.3a), which prevents backflow of blood into the left ventricle during its diastole (relaxation). The principal divisions of the aorta are the ascending aorta, arch of the aorta, thoracic aorta, and abdominal aorta.

Division of aorta	Arterial branch		Region supplied
Ascending aorta (ā-OR-ta)	Right and left coronary		Heart
Arch of aorta	Brachiocephalic (brā'-kē-ō-se-FAL-ik) trunk	→ Right common carotid (ka-ROT-id)	Right side of head and neck
		→ Right subclavian (sub-KLĀ-vē-an)	Right upper limb
	Left common carotid		Left side of head and neck
	Left subclavian		Left upper limb
Thoracic (thō-RAS-ik) ***aorta***	Intercostals (in'-ter-KOS-tals)		Intercostal and chest muscles, pleurae
	Superior phrenics (FREN-iks)		Posterior and superior surfaces of diaphragm
	Bronchials (BRONG-kē-als)		Bronchi of lungs
	Esophageals (e-sof'-a-JĒ-als)		Esophagus
Abdominal (ab-DOM-i-nal) ***aorta***	Inferior phrenics (FREN-iks)		Inferior surface of diaphragm
	Celiac	→ Common hepatic (he-PAT-ik)	Liver
		→ Left gastric (GAS-trik)	Stomach and esophagus
		→ Splenic (SPLĒN-ik)	Spleen, pancreas, stomach
	Superior mesenteric (MES-en-ter'-ik)		Small intestine, cecum, ascending and transverse colons, and pancreas
	Suprarenals (soo'-pra-RĒ-nals)		Adrenal (suprarenal) glands
	Renals (RĒ-nals)		Kidneys
	Gonadals (gō-NAD-als)	→ Testiculars (tes-TIK-yoo-lars)	Testes
		or → Ovarians (ō-VA-rē-ans)	Ovaries
	Inferior mesenteric (MES-en-ter'-ik)		Transverse, descending, sigmoid colons and rectum
	Common iliacs (IL-ē-aks)	→ External iliacs	Lower limbs
		→ Internal iliacs (hypogastrics)	Uterus, prostate gland, muscles of buttocks, and urinary bladder

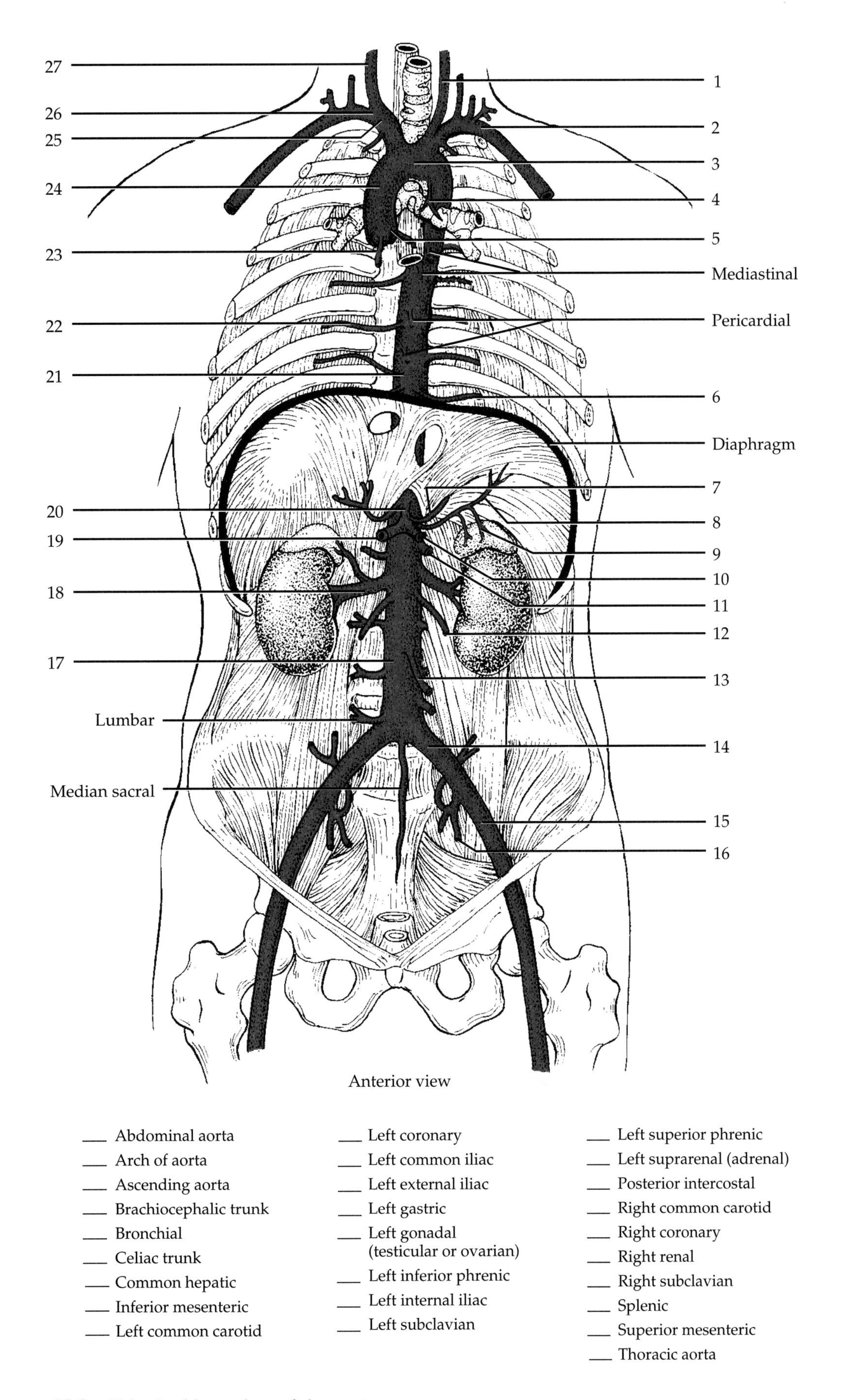

FIGURE 18.3 Principal branches of the aorta.

TABLE 18.2
Arch of Aorta (Figures 18.4, 18.5, and 18.6)

OVERVIEW: The ***arch of the aorta*** is about 4.5 cm (1.8 in.) in length and is the continuation of the ascending aorta that emerges from the pericardium behind the sternum at the level of the sternal angle. Initially, the arch is directed superiorly, posteriorly and to the left, and then inferiorly on the left side of the body of the fourth thoracic vertebra. Actually, the arch is directed not only from right to left, but from anterior to posterior as well. The arch of the aorta terminates at the level of the intervertebral disc between the fourth and fifth thoracic vertebrae, where it becomes the thoracic aorta. The thymus gland lies anterior to the arch of the aorta, while the trachea lies posterior to it.

Three major arteries branch from the arch of the aorta. In order of their origination, they are the brachiocephalic trunk, left common carotid artery, and left subclavian artery.

Branch	Description and region
Brachiocephalic (brā'-kē-ō-se-FAL-ik)	The ***brachiocephalic trunk,*** which is found only on the right side, is the first and largest branch off the arch of the aorta. It bifurcates (divides) to form the right subclavian artery and right common carotid artery. The ***right subclavian*** (sub-KLĀ-vē-an) ***artery*** extends from the brachiocephalic to the first rib and then passes into the armpit (axilla) and supplies the arm, forearm, and hand. Continuation of the right subclavian into the axilla is called the ***axillary*** (AK-si-ler'-ē) ***artery.*** From here, it continues into the arm as the ***brachial*** (BRĀ-kē-al) ***artery.*** At the bend of the elbow, the brachial artery divides into the medial ***ulnar*** (UL-nar) and lateral ***radial*** (RĀ-dē-al) ***arteries.*** These vessels pass inferiorly to the palm, one on each side of the forearm. In the palm, branches of the two arteries anastomose to form two palmar arches—the ***superficial palmar*** (PAL-mar) ***arch*** and the ***deep palmar arch.*** From these arches arise the ***digital*** (DIJ-i-tal) ***arteries,*** which supply the fingers and thumb (Figure 18.4). Before passing into the axilla, the right subclavian gives off a major branch to the brain called the ***right vertebral*** (VER-te-bral) ***artery.*** The right vertebral artery passes through the foramina of the transverse processes of the cervical vertebrae and enters the skull through the foramen magnum to reach the inferior surface of the brain. Here it unites with the left vertebral artery to form the ***basilar*** (BAS-i-lar) ***artery*** (Figures 18.5 and 18.6). The ***right common carotid artery*** passes superiorly in the neck. At the upper level of the larynx, it divides into the ***right external*** and ***right internal carotid*** (ka-ROT-id) ***arteries.*** The external carotid supplies the right side of the thyroid gland, tongue, throat, face, ear, scalp, and dura mater. The internal carotid supplies the brain, right eye, and right sides of the forehead and nose (Figure 18.5). Inside the cranium, anastomoses of the left and right internal carotids along with the basilar artery form an arrangement of blood vessels at the base of the brain near the sella turcica called the ***cerebral*** (se-RĒ-bral) ***arterial circle (circle of Willis).*** From this circle arise arteries supplying most of the brain. Essentially the cerebral arterial circle is formed by the union of the ***anterior cerebral arteries*** (branches of internal carotids) and ***posterior cerebral arteries*** (branches of basilar artery). Posterior cerebral arteries are connected with internal carotids by the ***posterior communicating*** (ko-MYOO-ni-kā'-ting) ***arteries.*** The anterior cerebral arteries are connected by the ***anterior communicating arteries.*** The ***internal carotid*** (ka-ROT-id) ***arteries*** are also considered part of the cerebral arterial circle. The function of the cerebral arterial circle is to equalize blood pressure to the brain and provide alternate routes for blood to the brain, should the arteries become damaged.
Left common carotid (ka-ROT-id)	The ***left common carotid*** is the second branch off the arch of the aorta. Corresponding to the right common carotid, it divides into basically the same branches with the same names, except that the arteries are now labeled "left" instead of "right."
Left subclavian (sub-KLĀ-vē-an)	The ***left subclavian artery*** is the third branch off the arch of the aorta. It distributes blood to the left vertebral artery and vessels of the left upper limb. Arteries branching from the left subclavian are named like those of the right subclavian.

Write the names of the missing arteries in the following scheme of circulation. Be sure to indicate left or right where applicable.

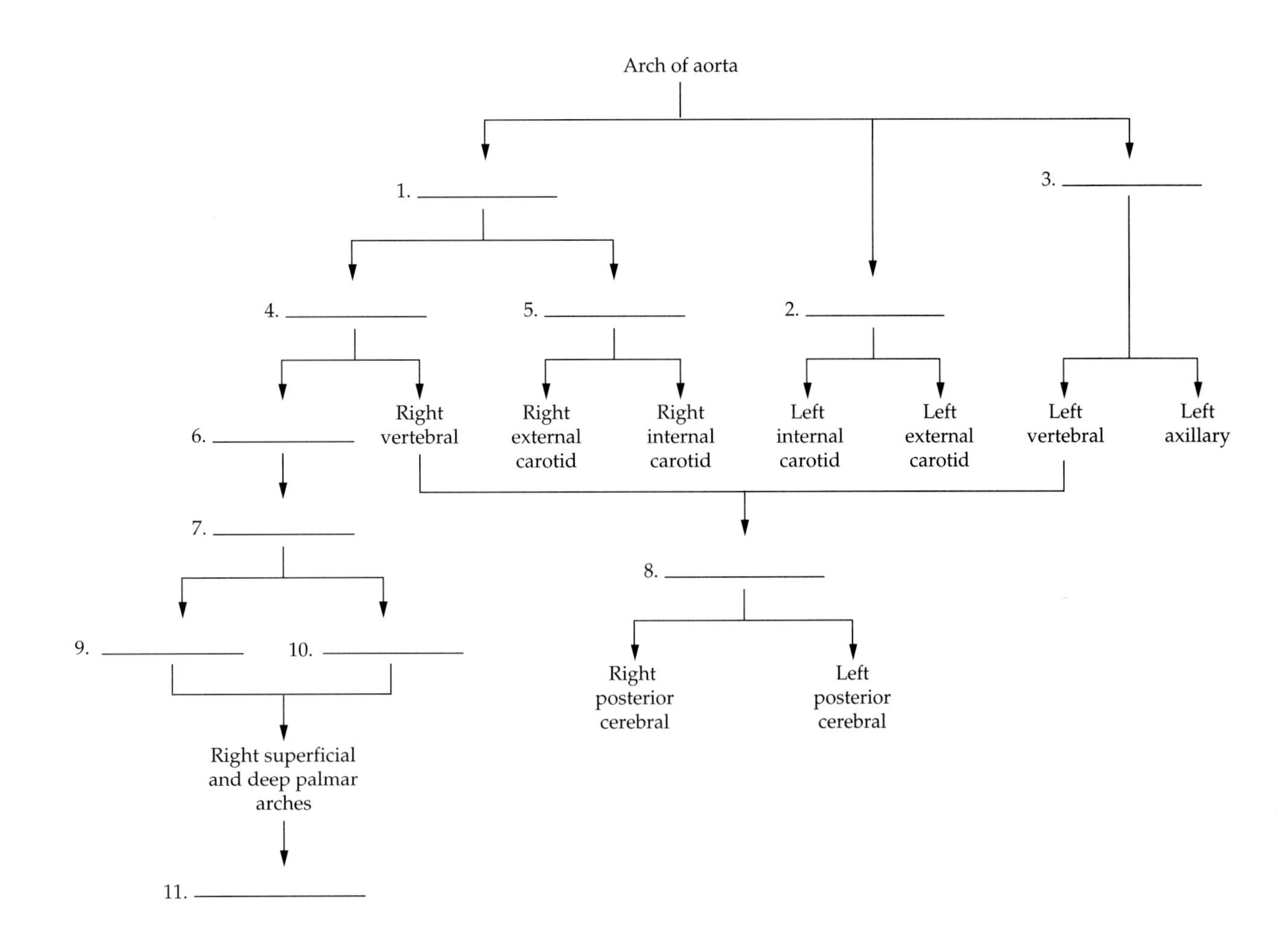

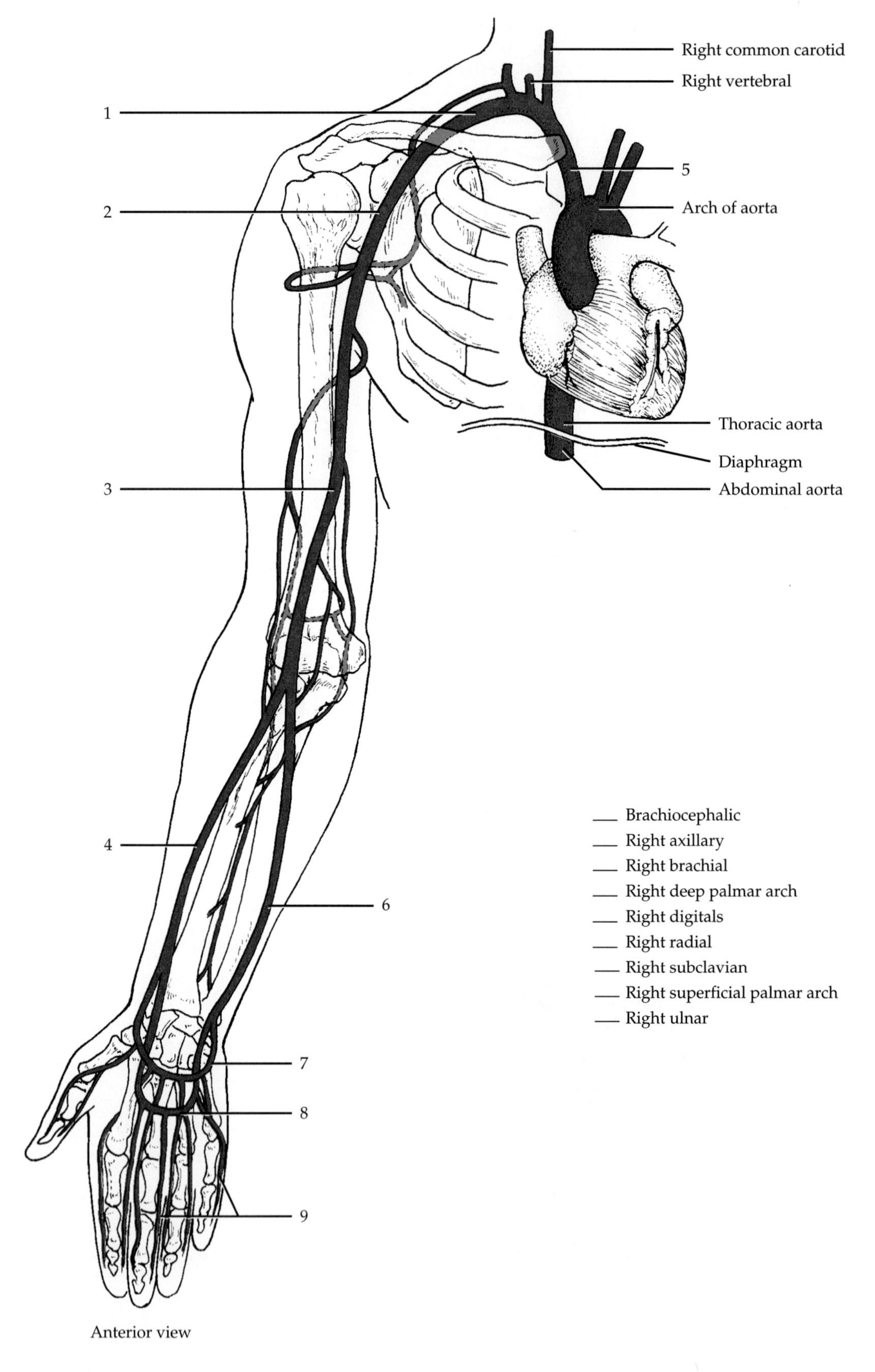

___ Brachiocephalic
___ Right axillary
___ Right brachial
___ Right deep palmar arch
___ Right digitals
___ Right radial
___ Right subclavian
___ Right superficial palmar arch
___ Right ulnar

FIGURE 18.4 Arteries of right upper limb.

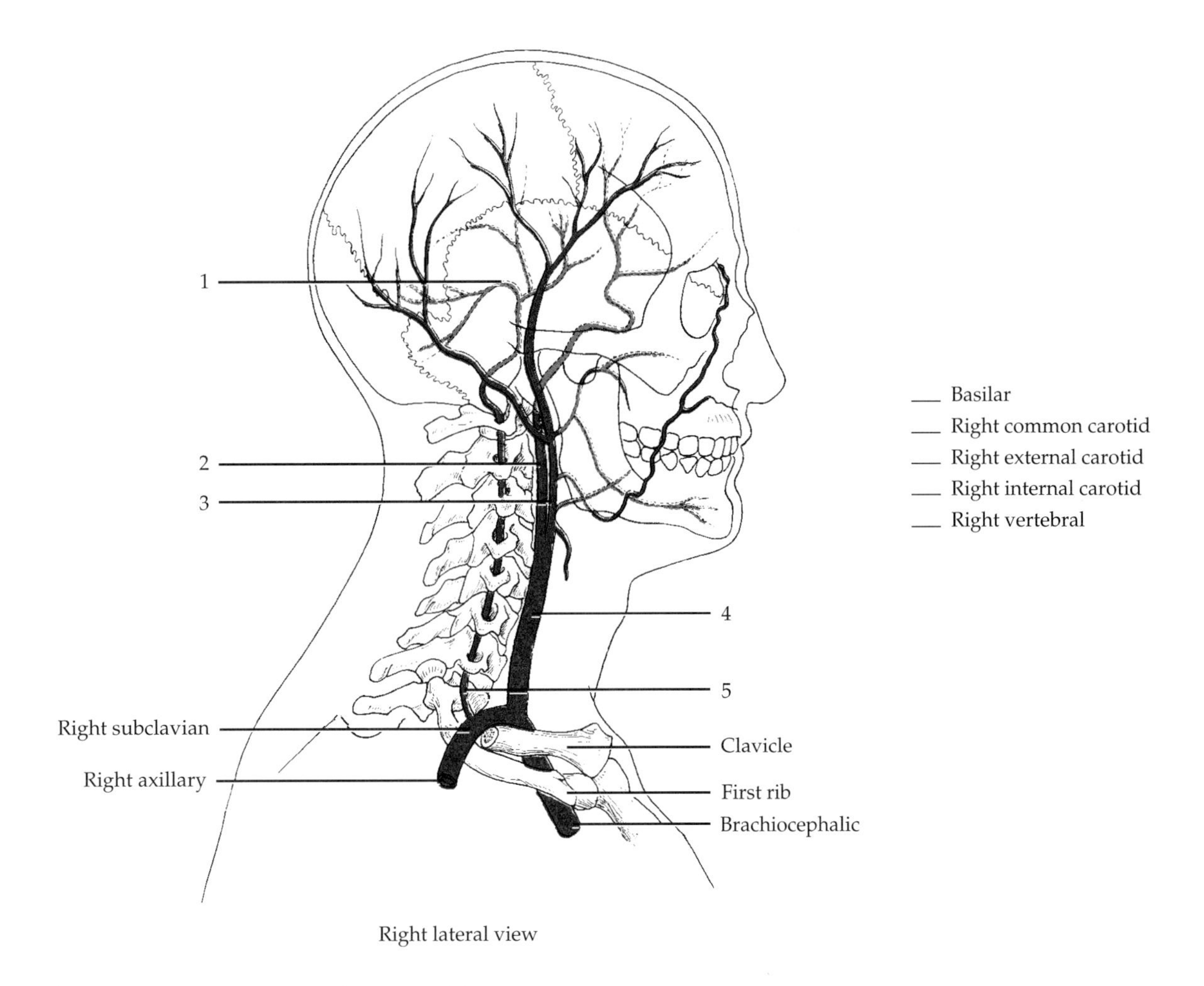

FIGURE 18.5 Arteries of neck and head.

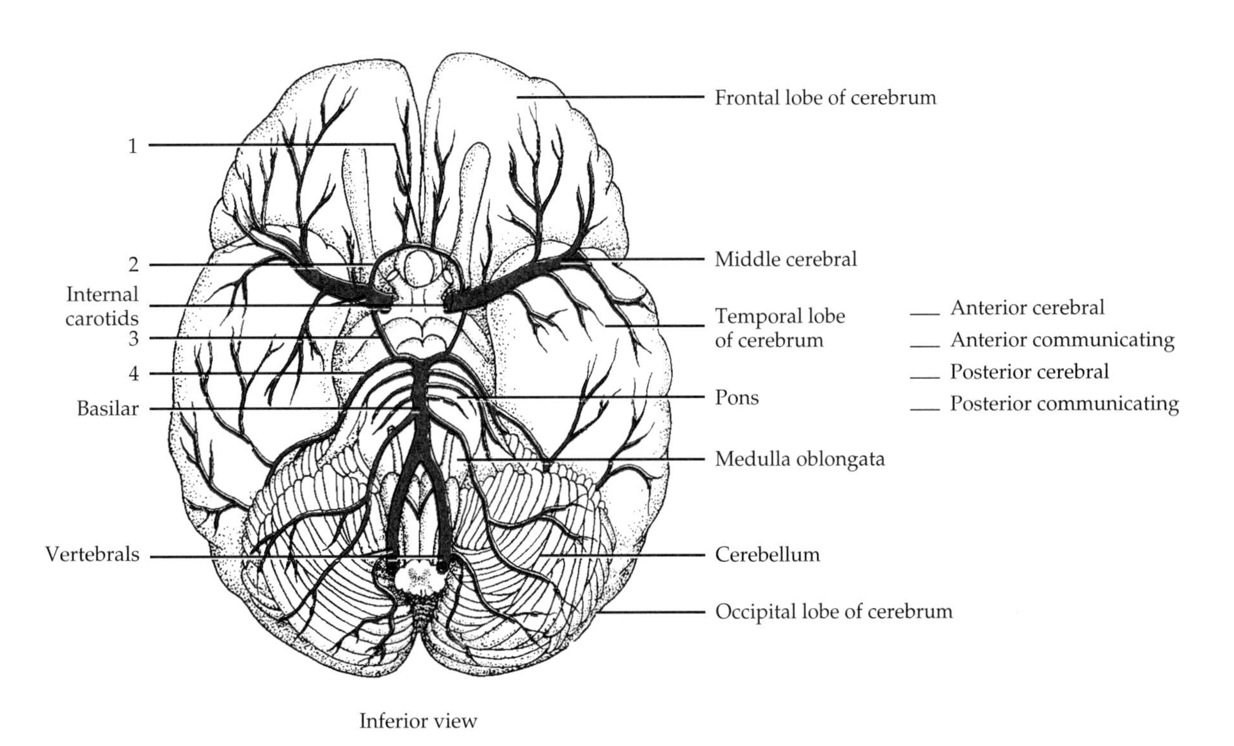

FIGURE 18.6 Arteries of base of brain.

TABLE 18.3
Thoracic Aorta (Figure 18.3)

OVERVIEW: The ***thoracic aorta*** is about 20 cm (8 in.) long and is a continuation of the arch of the aorta. It begins at the level of the intervertebral disc between the fourth and fifth thoracic vertebrae, where it lies to the left of the vertebral column. As it descends, it moves closer to the midline and terminates at an opening in the diaphragm (aortic hiatus) anterior to the vertebral column at the level of the intervertebral disc between the twelfth thoracic and first lumbar vertebrae.

Along its course, the thoracic aorta sends off numerous small arteries to the viscera ***(visceral branches)*** and body wall structures ***(parietal branches).***

Branch	**Description and region supplied**
VISCERAL	
Pericardial (per'-i-KAR-dē-al)	Several minute ***pericardial arteries*** supply blood to the posterior aspect of the pericardium.
Bronchial (BRONG-kē-al)	One right and two left ***bronchial arteries*** supply the bronchial tubes, visceral pleurae, bronchial lymph nodes, and esophagus. (Whereas the right bronchial artery arises from the third posterior intercostal artery, the two left bronchial arteries arise from the thoracic aorta.)
Esophageal (e-sof'-a-JĒ-al)	Four or five ***esophageal arteries*** supply the esophagus.
Mediastinal (mē'-dē-as-TĪ-nal)	Numerous small ***mediastinal arteries*** supply blood to structures in the posterior mediastinum.
PARIETAL	
Posterior intercostal (in'-ter-KOS-tal)	Nine pairs of ***posterior intercostal arteries*** supply the intercostal, pectoral, and abdominal muscles; overlying subcutaneous tissue and skin; mammary glands; and vertebral canal and its contents.
Subcostal (SUB-kos-tal)	The left and right ***subcostal arteries*** have a distribution similar to that of the posterior intercostals.
Superior phrenic (FREN-ik)	Small ***superior phrenic arteries*** supply the posterior and superior surfaces of the diaphragm.

TABLE 18.4
Abdominal Aorta (Figure 18.7)

OVERVIEW: The ***abdominal aorta*** is the continuation of the thoracic aorta. It begins at the aortic hiatus in the diaphragm and ends at about the level of the fourth lumbar vertebra, where it divides into right and left common iliac arteries. The abdominal aorta lies anterior to the vertebral column.

As with the thoracic aorta, the abdominal aorta gives off ***visceral*** and ***parietal branches.*** The unpaired visceral branches arise from the anterior surface of the aorta and include the celiac, superior mesenteric, and inferior mesenteric arteries. The paired visceral branches arise from the lateral surfaces of the aorta and include the adrenal (suprarenal), renal, and gonadal arteries. The paired parietal branches arise from the posterolateral surfaces of the aorta and include the inferior phrenic and lumbar arteries. The unpaired parietal artery is the median sacral.

Branch	Description and region supplied
VISCERAL	
Celiac (SĒ-lē-ak)	The ***celiac artery (trunk)*** is the first visceral aortic branch inferior to the diaphragm. It has three branches: (1) ***common hepatic*** (he-PAT-ik) ***artery,*** (2) ***left gastric*** (GAS-trik) ***artery,*** and (3) ***splenic*** (SPLĒN-ik) ***artery.*** The common hepatic artery has three main branches: (1) ***hepatic artery proper,*** a continuation of the common hepatic artery, which supplies the liver and gallbladder; (2) ***right gastric artery,*** which supplies the stomach and duodenum; and (3) ***gastroduodenal*** (gas'-trō-doo'-ō-DE-nal) ***artery,*** which supplies the stomach, duodenum, and pancreas. The left gastric artery supplies the stomach and its ***esophageal*** (e-sof'-a-JĒ-al) ***branch*** supplies the esophagus. The splenic artery supplies the spleen and has three main branches: (1) ***pancreatic*** (pan'-krē-AT-ik) ***arteries,*** which supply the pancreas; (2) ***left gastroepiploic*** (gas'-trō-ep'-i-PLŌ-ik) ***artery,*** which supplies the stomach and greater omentum; and (3) ***short gastric*** (GAS-trik) ***arteries,*** which supply the stomach.
Superior mesenteric (MES-en-ter'-ik)	The ***superior mesenteric artery*** anastomoses extensively and has several principal branches: (1) ***inferior pancreaticoduodenal*** (pan'-krē-at'-i-kō-doo'-ō-DĒ-nal) ***artery,*** which supplies the pancreas and duodenum; (2) ***jejunal*** (je-JOO-nal) and ***ileal*** (IL-ē-al) ***arteries,*** which supply the jejunum and ileum, respectively; (3) ***ileocolic*** (il'-ē-ō-KŌL-ik) ***artery,*** which supplies the ileum and ascending colon; (4) ***right colic*** (KŌL-ik) ***artery,*** which supplies the ascending colon; and (5) ***middle colic artery,*** which supplies the transverse colon.
Suprarenals (soo'-pra-RĒ-nals)	Right and left ***suprarenal arteries*** supply blood to the adrenal (suprarenal) glands. The glands are also supplied by branches of the renal and inferior phrenic arteries.
Renals (RĒ-nals)	Right and left ***renal arteries*** carry blood to the kidneys and adrenal (suprarenal) glands.
Gonadals (gō-NAD-als) [***testiculars*** (tes-TIK-yoo-lars) or ***ovarians*** (ō-VA-rē-ans)]	Right and left ***testicular arteries*** extend into the scrotum and terminate in the testes; right and left ***ovarian arteries*** are distributed to the ovaries.
Inferior mesenteric (MES-en-ter'-ik)	The principal branches of the ***inferior mesenteric artery,*** which also anastomoses, are the (1) ***left colic*** (KŌL-ik) ***artery,*** which supplies the transverse and descending colons; (2) ***sigmoid*** (SIG-moyd) ***arteries,*** which supply the descending and sigmoid colons; and (3) ***superior rectal*** (REK-tal) ***artery,*** which supplies the rectum.
PARIETAL	
Inferior phrenics (FREN-iks)	The ***inferior phrenic arteries*** are distributed to the inferior surface of the diaphragm and adrenal (suprarenal) glands.
Lumbars (LUM-bars)	The ***lumbar arteries*** supply the spinal cord and its meninges and the muscles and skin of the lumbar region of the back.
Median sacral (SĀ-kral)	The ***median sacral artery*** supplies the sacrum, coccyx, and rectum.

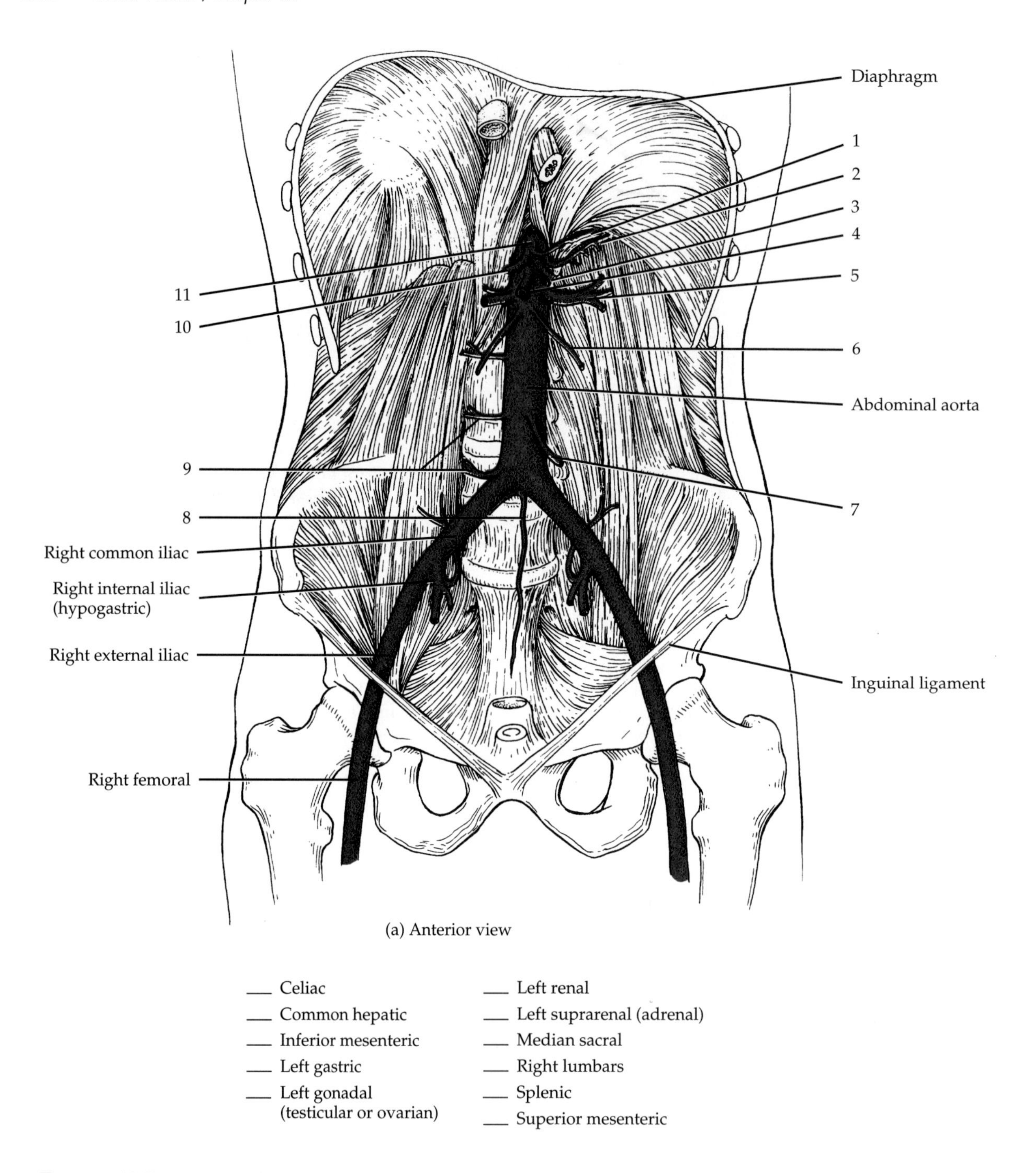

FIGURE 18.7 Abdominal arteries. (a) Abdominal aorta and its principal branches.

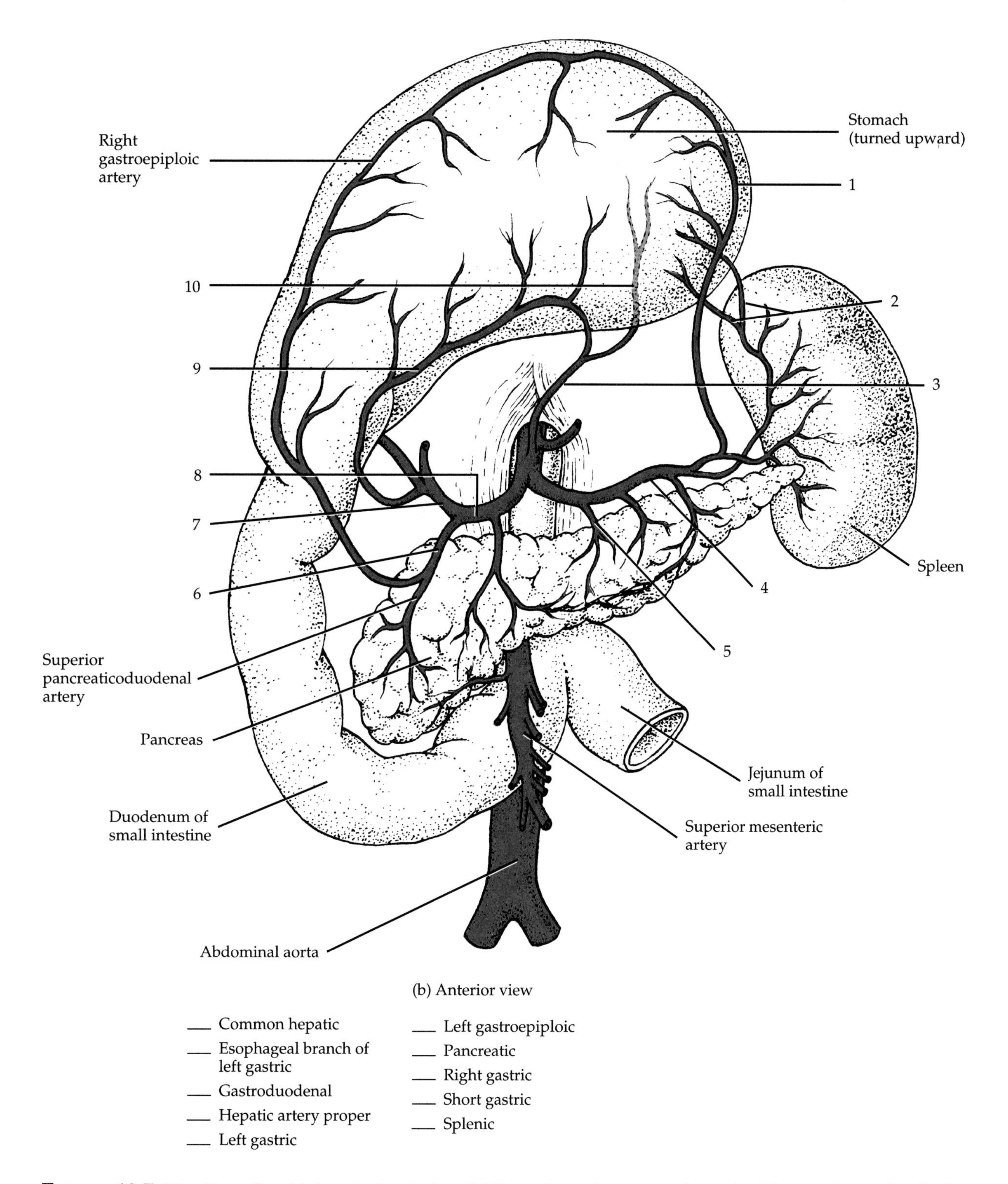

(b) Anterior view

___ Common hepatic
___ Esophageal branch of left gastric
___ Gastroduodenal
___ Hepatic artery proper
___ Left gastric
___ Left gastroepiploic
___ Pancreatic
___ Right gastric
___ Short gastric
___ Splenic

FIGURE 18.7 *(Continued)* Abdominal arteries. (b) Branches of common hepatic, left gastric, and splenic arteries.

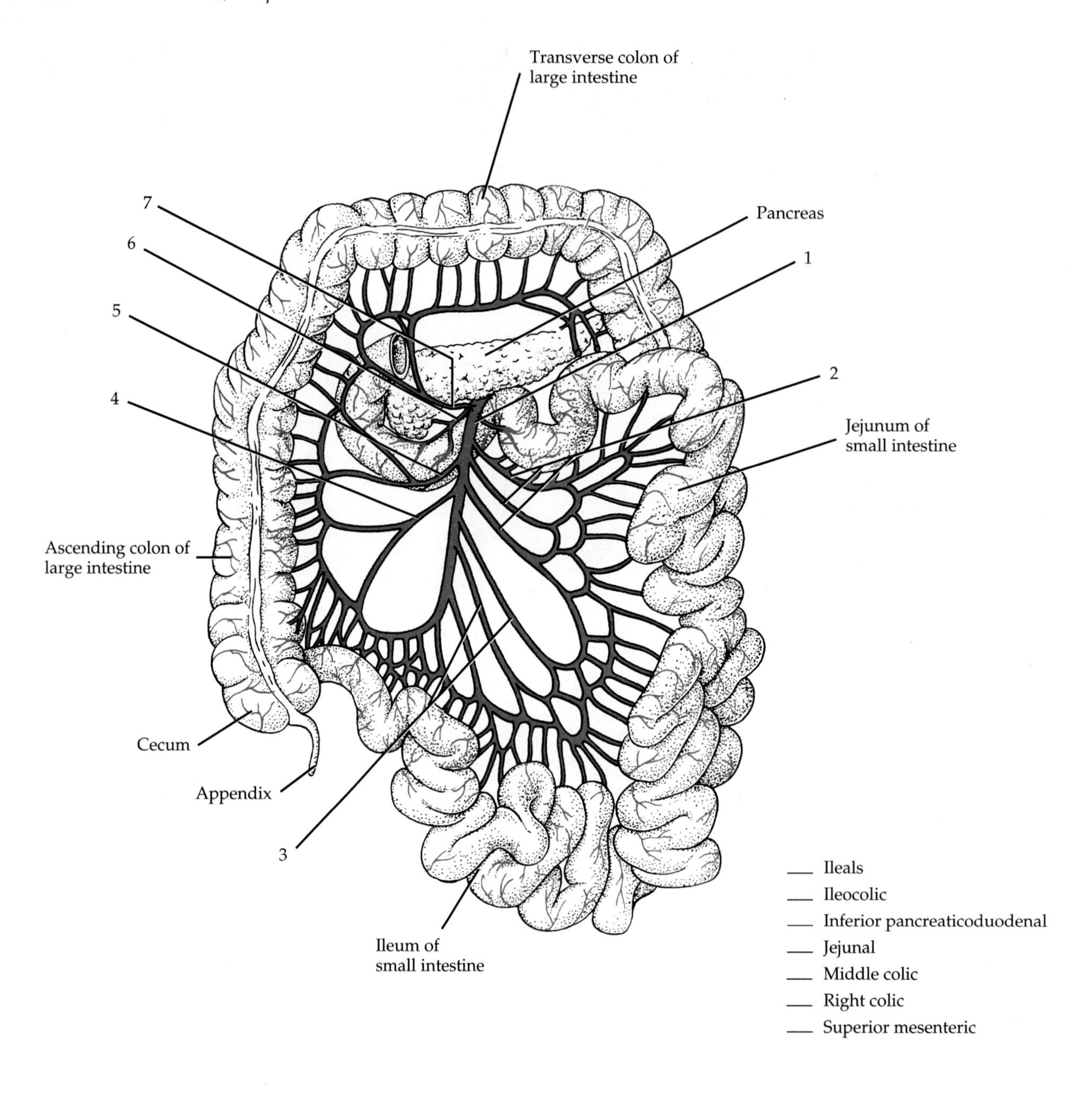

(c) Anterior view

FIGURE 18.7 ***(Continued)*** Abdominal arteries. (c) Branches of superior mesenteric artery.

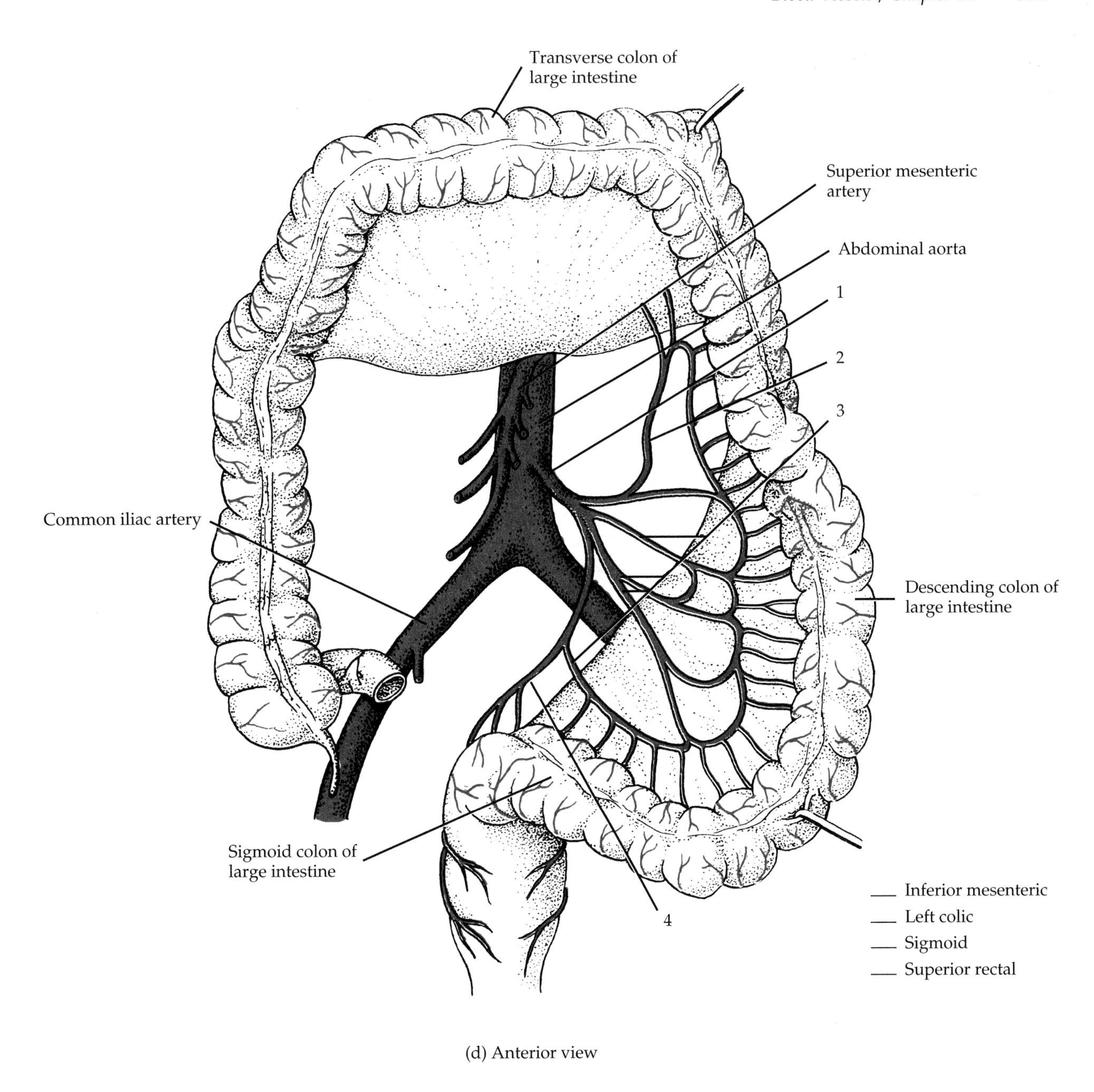

(d) Anterior view

FIGURE 18.7 ***(Continued)*** Abdominal arteries. (d) Branches of inferior mesenteric artery.

TABLE 18.5
Arteries of Pelvis and Lower Limbs (Extremities) (Figure 18.8)

OVERVIEW: The ***internal iliac arteries*** enter the pelvic cavity anterior to the sacroiliac joint and supply most of the blood to the pelvic viscera and wall. The ***external iliacs*** travel along the brim of the lesser (true) pelvis. Posterior to the midportion of the inguinal ligament, each external iliac artery enters the thigh where its name changes to the femoral artery.

Branch	Description and region supplied
Common iliacs (IL-ē-aks)	At about the level of the fourth lumbar vertebra, the abdominal aorta divides into the right and left ***common iliac arteries.*** Each passes inferiorly about 5 cm (2 in.) and gives rise to two branches: internal iliac and external iliac.
Internal iliacs	The ***internal iliac (hypogastric) arteries*** form branches that supply the psoas minor, gluteal muscles, quadratus lumborum, medial side of each thigh, urinary bladder, rectum, prostate gland, ductus (vas) deferens, uterus, and vagina.
External iliacs	The ***external iliac arteries*** diverge through the greater (false) pelvis and enter the thighs to become the right and left ***femoral*** (FEM-o-ral) ***arteries.*** Both femorals send branches back up to the genitals and the wall of the abdomen. Other branches run to the muscles of the thigh. The femoral continues down the medial and posterior side of the thigh posterior to the knee joint, where it becomes the ***popliteal*** (pop'-li-TĒ-al) ***artery.*** Between the knee and ankle, the popliteal runs down on the posterior aspect of the leg and is called the ***posterior tibial*** (TIB-ē-al) ***artery.*** Inferior to the knee, the ***peroneal*** (per'-ō-NĒ-al) ***artery*** branches off the posterior tibial to supply structures on the medial side of the fibula and calcaneus. In the calf, the ***anterior tibial artery*** branches off the popliteal and runs along the anterior surface of the leg. At the ankle, it becomes the ***dorsalis pedis*** (PED-is) ***artery.*** At the ankle, the posterior tibial divides into the ***medial*** and ***lateral plantar*** (PLAN-tar) ***arteries.*** The lateral plantar artery and the dorsalis pedis artery unite to form the ***plantar arch.*** From this arch, ***digital arteries*** supply the toes.

Write the names of the missing arteries in the following scheme of circulation. Be sure to indicate left or right where applicable.

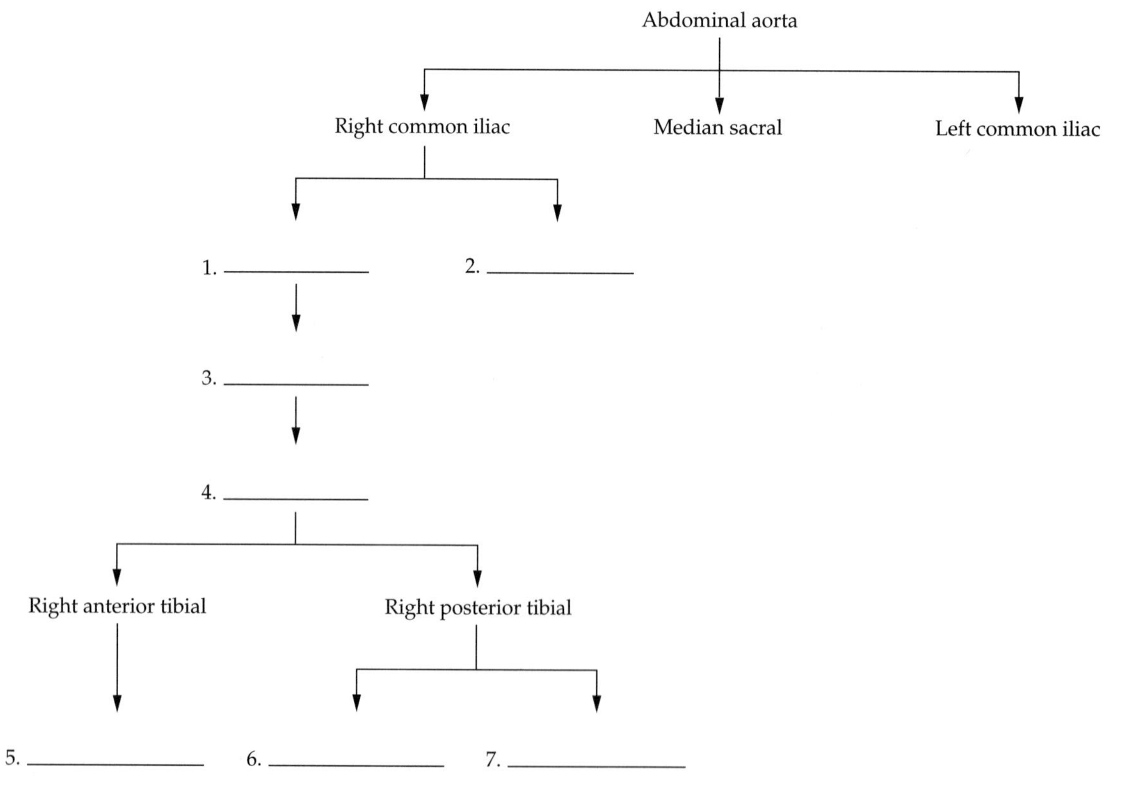

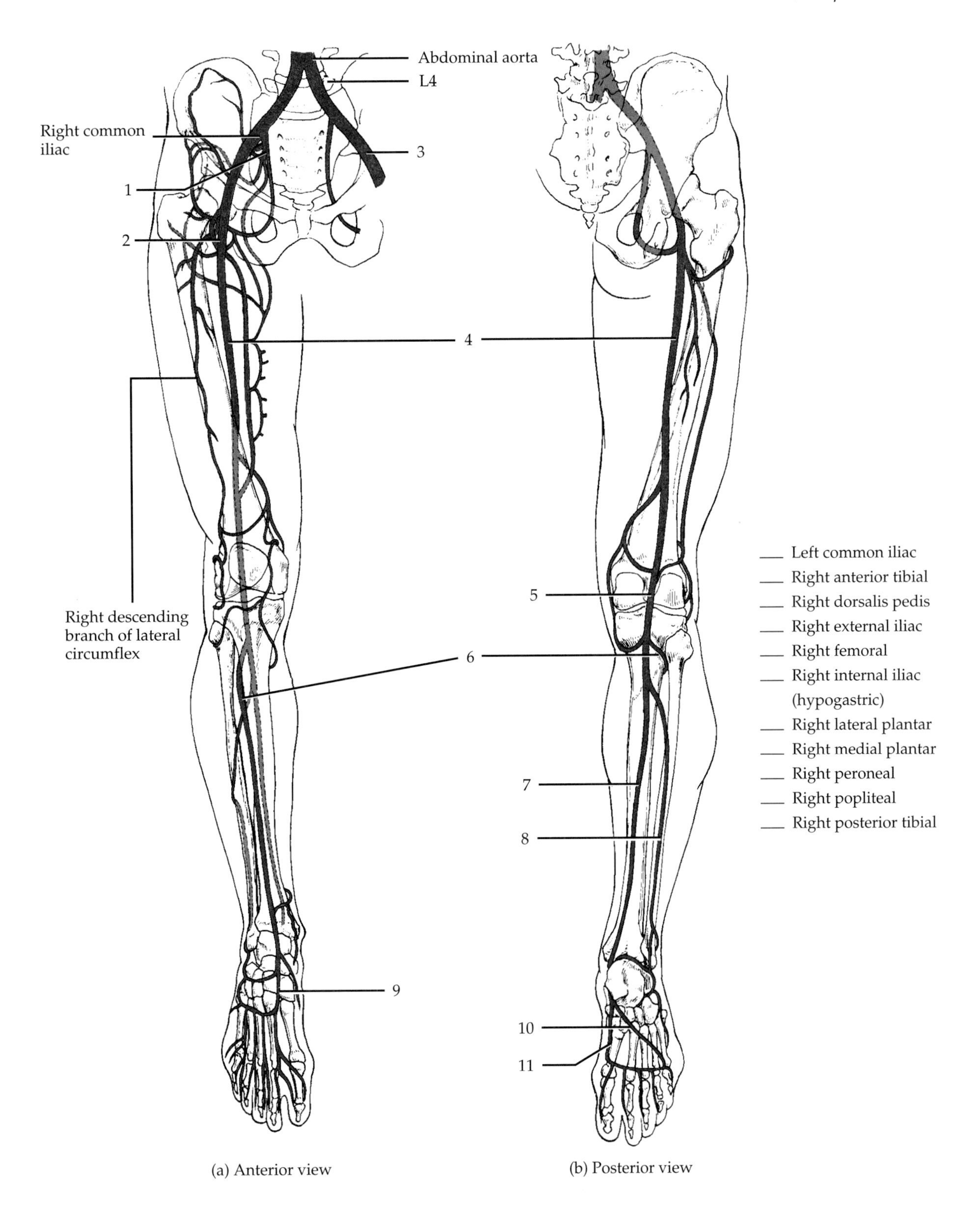

FIGURE 18.8 Arteries of pelvis and right lower limb.

TABLE 18.6
Veins of Systemic Circulation (see Figure 18.10)

OVERVIEW: Deoxygenated blood returns to the right atrium from three veins: the ***coronary sinus, superior vena cava,*** and ***inferior vena cava.*** The coronary sinus receives blood from the cardiac veins; the superior vena cava receives blood from veins superior to the diaphragm, except the air sacs of the lungs. This includes the head, neck, upper limbs, and thoracic wall. The inferior vena cava receives blood from veins inferior to the diaphragm. This includes the lower limbs, most of the abdominal walls, and abdominal viscera.

Vein	Description and region drained
Coronary (KOR-o-nar-ē) *sinus*	The ***coronary sinus*** receives almost all venous blood from the myocardium. It is located in the coronary sulcus (see Fig. 17.4b) and opens into the right atrium between the orifice of the inferior vena cava and the tricuspid valve.
Superior vena cava (VĒ-na CĀ-va) ***SVC***	The ***SVC*** is about 7.5 cm (3 in.) long and empties its blood into the upper part of the right atrium. It begins posterior to the right first costal cartilage by the union of the right and left brachiocephalic veins and ends at the level of the right third costal cartilage where it enters the right atrium.
Inferior vena cava (IVC)	The ***IVC*** is the largest vein in the body, about 3½ cm (1.4 in.) in diameter. It begins anterior to the fifth lumbar vertebra by the union of the common iliac veins, ascends behind the peritoneum to the right of the midline, pierces the costal tendon of the diaphragm at the level of the eighth thoracic vertebra, and enters the inferior part of the right atrium. The IVC is commonly compressed during the later stages of pregnancy owing to the enlargement of the uterus. This produces edema of the ankles and feet and temporary varicose veins.

TABLE 18.7
Veins of Head and Neck (Figure 18.9)

OVERVIEW: The majority of blood draining from the head passes into three pairs of veins: ***internal jugular, external jugular,*** and ***vertebral.*** Within the cranium, all veins lead to the external jugular veins.

Vein	Description and region drained
Internal jugulars (JUG-yoo-lars)	Right and left ***internal jugular veins*** receive blood from the face and neck. They arise as a continuation of the ***sigmoid*** (SIG-moyd) ***sinuses*** at the base of the skull. Intracranial vascular sinuses are located between layers of the dura mater and receive blood from the brain. Other sinuses that drain into the internal jugulars include the ***superior sagittal*** (SAJ-i-tal) ***sinus, inferior sagittal sinus, straight sinus,*** and ***transverse (lateral) sinuses.*** Internal jugulars descend on either side of the neck and pass posterior to the clavicles, where they join with the right and left subclavian veins. Unions of the internal jugulars and subclavians form the right and left ***brachiocephalic*** (brā'-kē-ō-se-FAL-ik) ***veins.*** From here blood flows into the superior vena cava.
External jugulars	Right and left ***external jugular veins*** run inferiorly in the neck along the outside of the internal jugulars. They drain blood from the parotid (salivary) glands, facial muscles, scalp, and other superficial structures into the subclavian veins. In cases of heart failure, the venous pressure in the right atrium may rise. In such patients the pressure in the column of blood in the external jugular vein rises so that, even with the patient at rest and sitting in a chair, the external jugular vein will be visibly distended. Temporary distention of the vein is often seen in healthy adults when the intrathoracic pressure is raised in coughing and physical exertion.
Vertebrals (VER-te-brals)	Right and left ***vertebral veins*** descend through the transverse foramina of the cervical vertebrae and enter the subclavian veins. They drain deep structures of the neck such as vertebrae and muscles.

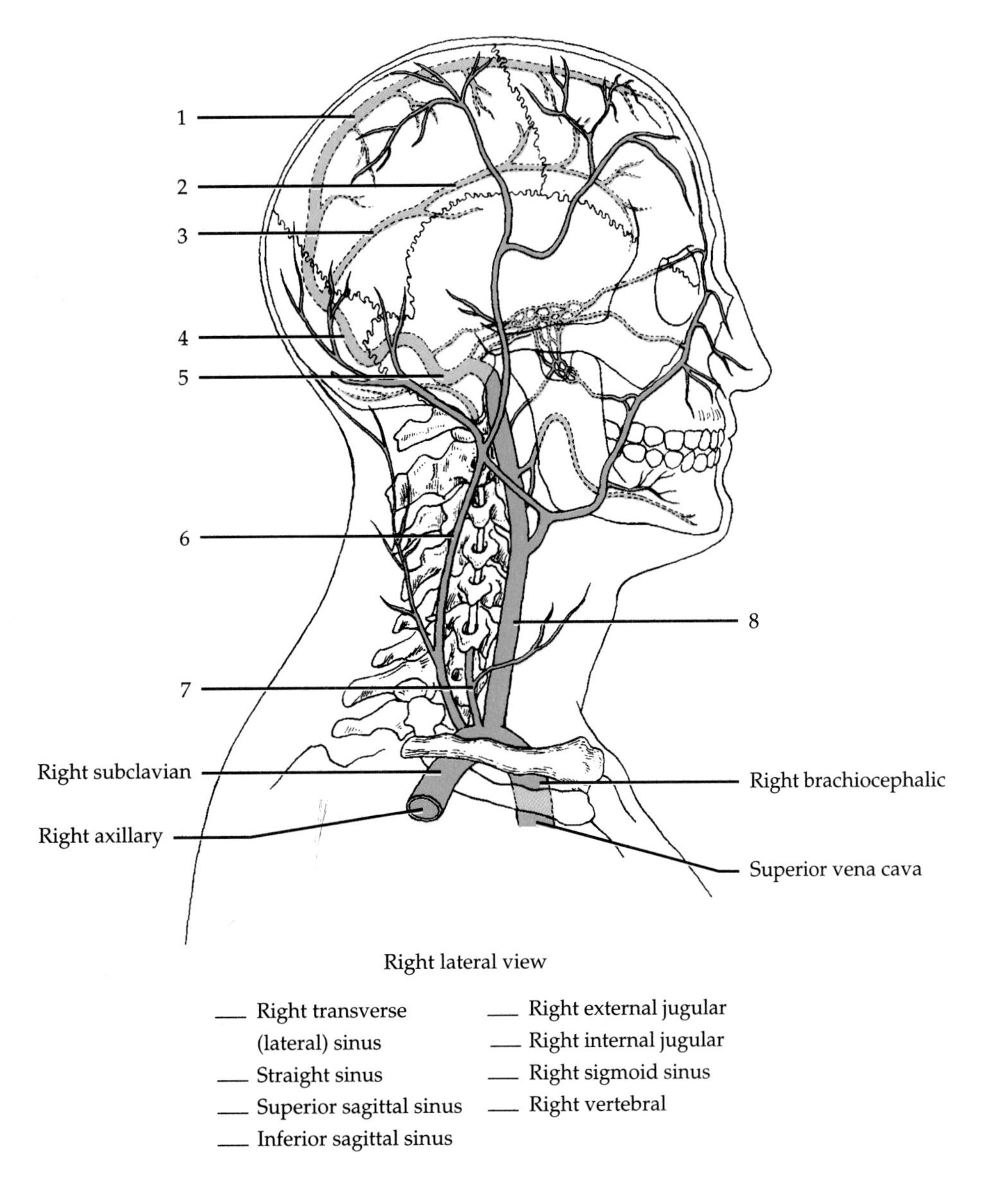

FIGURE 18.9 Veins of head and neck.

TABLE 18.8
Veins of Upper Limbs (Extremities) (Figure 18.10)

OVERVIEW: Blood from each upper limb is returned to the heart by superficial and deep veins. Both sets of veins contain valves. ***Superficial veins*** are located just deep to the skin and are often visible. They anastomose extensively with each other and with deep veins. ***Deep veins*** are located deep in the body. They usually accompany arteries, and many have the same names as corresponding arteries. Most deep veins are paired vessels.

Vein	Description and region drained
SUPERFICIAL	
Cephalics (se-FAL-iks)	The ***cephalic vein*** of each upper limb begins in the medial part of the ***dorsal venous*** (VĒ-nus) ***arch*** and winds superiorly around the radial border of the forearm. Anterior to the elbow, it is connected to the basilic vein by the ***median cubital*** (KYOO-bi-tal) ***vein.*** Just inferior to the elbow, the cephalic vein unites with the ***accessory cephalic vein*** to form the cephalic vein of the upper limb. Ultimately, the cephalic vein empties into the axillary vein.
Basilics (ba-SIL-iks)	The ***basilic vein*** of each upper limb originates in the ulnar part of the ***dorsal venous arch.*** It extends along the posterior surface of the ulna to a point near the elbow where it receives the ***median cubital vein.*** If a vein must be punctured for an injection, transfusion, or removal of a blood sample, the median cubitals are preferred. After receiving the median cubital vein, the basilic continues ascending on the medial side until it reaches the middle of the arm. There it penetrates the tissues deeply and runs alongside the brachial artery until it joins the brachial vein. As the basilic and brachial veins merge in the axillary area, they form the axillary vein.
Median antebrachials (an'-tē-BRĀ-kē-als)	The ***median antebrachial veins*** drain the ***palmar venous arch,*** ascend on the ulnar side of the anterior forearm, and end in the median cubital veins.
DEEP	
Radials (RĀ-dē-als)	***Radial veins*** receive the ***dorsal metacarpal*** (met'-a-KAR-pal) ***veins.***
Ulnars (UL-nars)	***Ulnar veins*** receive tributaries from the ***palmar venous arch.*** Radial and ulnar veins unite in the bend of the elbow to form the brachial veins.
Brachials (BRĀ-kē-als)	Located on either side of the brachial arteries, the ***brachial veins*** join into the axillary veins.
Axillaries (AK-si-ler'-ēs)	***Axillary veins*** are a continuation of brachials and basilics. Axillaries end at the first rib, where they become subclavians.
Subclavians (sub-KLĀ-vē-ans)	Right and left ***subclavian veins*** unite with the internal jugulars to form brachiocephalic veins. The thoracic duct of the lymphatic system delivers lymph into the left subclavian vein at the junction with the internal jugular. The right lymphatic duct delivers lymph into the right subclavian vein at the corresponding junction.

Write the names of the missing veins in the following scheme of circulation. Be sure to indicate left or right where applicable.

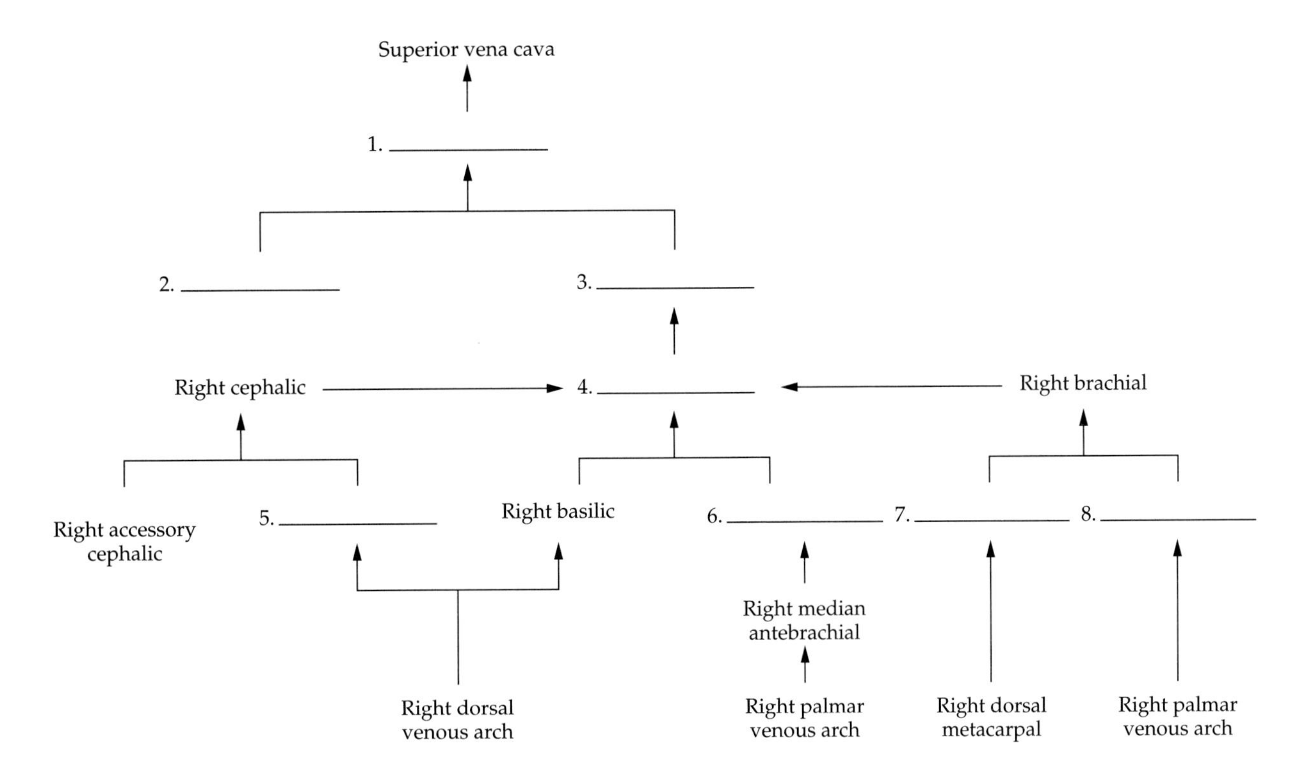

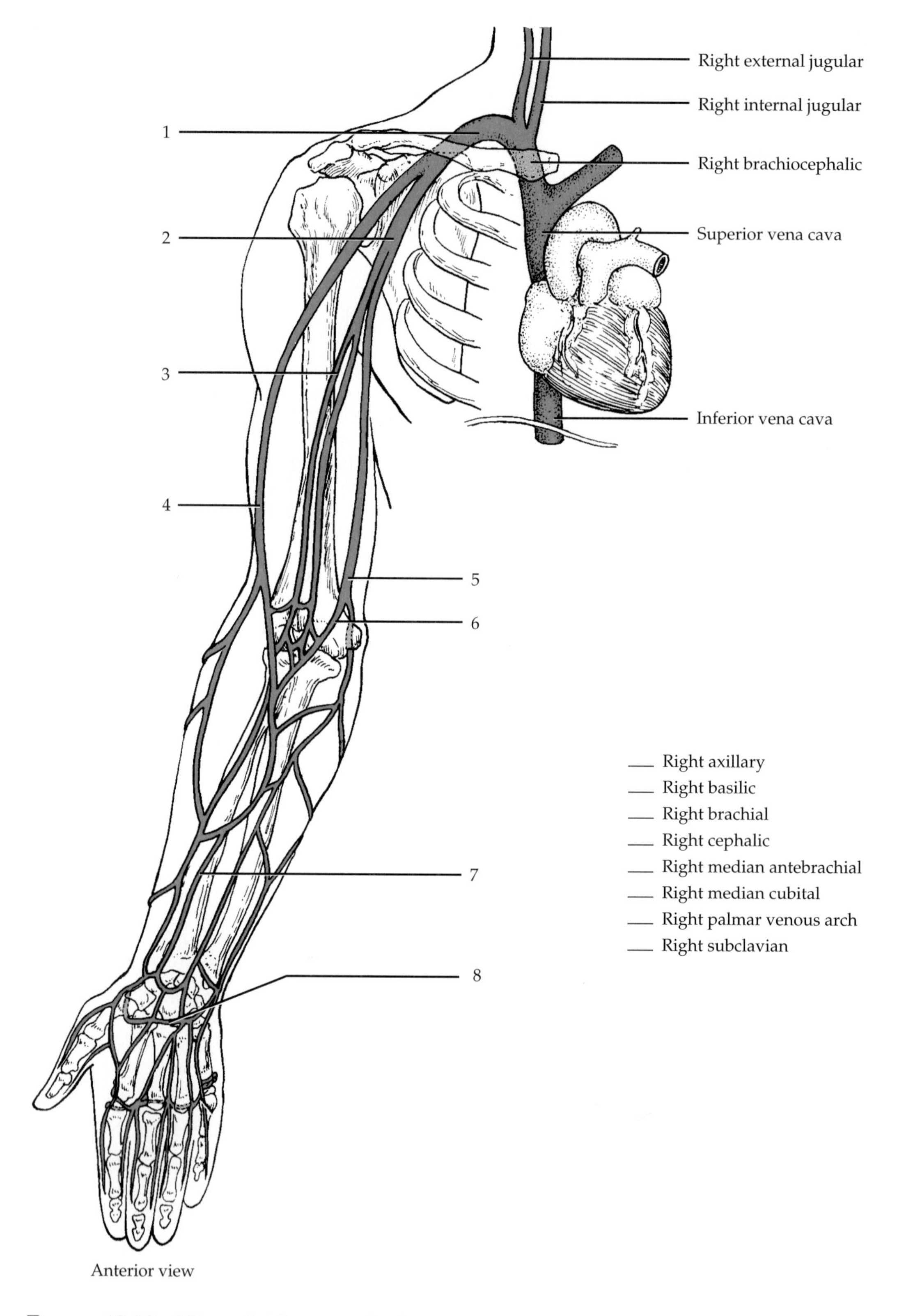

___ Right axillary
___ Right basilic
___ Right brachial
___ Right cephalic
___ Right median antebrachial
___ Right median cubital
___ Right palmar venous arch
___ Right subclavian

FIGURE 18.10 Veins of right upper limb.

Table 18.9
Veins of Thorax (Figure 18.11)

OVERVIEW: Although the brachiocephalic veins drain some portions of the thorax, most thoracic structures are drained by a network of veins called the ***azygos system***. This is a network of veins on each side of the vertebral column: azygos, hemiazygos, and accessory hemiazygos. They show considerable variation in origin, course, tributaries, anastomoses, and termination. Ultimately, they empty into the superior vena cava.

Vein	Description and region drained
Brachiocephalic (brā-kē-ō-se-FAL-ik)	Right and left ***brachiocephalic veins,*** formed by the union of the subclavians and internal jugulars, drain blood from the head, neck, upper limbs, mammary glands, and upper thorax. Brachiocephalics unite to form the superior vena cava.
Azygos (az-Ī-gos) ***system***	The ***azygos system,*** besides collecting blood from the thorax, may serve as a bypass for the inferior vena cava that drains blood from the lower body. Several small veins directly link the azygos system with the inferior vena cava. Large veins that drain the lower limbs and abdomen pass blood into the azygos system. If the inferior vena cava or hepatic portal vein becomes obstructed, the azygos system can return blood from the lower body to the superior vena cava.
Azygos	The ***azygos vein*** lies anterior to the vertebral column and slightly to the right of the midline. It begins as a continuation of the right ascending lumbar vein. It connects with the inferior vena cava, right common iliac, and lumbar veins. The azygos receives blood from the ***left intercostal*** (in'-ter-KOS-tal) ***veins*** that drain the chest muscles; from the hemiazygos and accessory hemiazygos veins; from several ***esophageal*** (e-sof'-a-JĒ-al), ***mediastinal*** (mē-dē-a-STĪ-nal), and ***pericardial*** (per'-i-KAR-dē-al) ***veins;*** and from the right ***bronchial*** (BRONG-kē-al) ***vein.*** The vein ascends to the fourth thoracic vertebra, arches over the right lung, and empties into the superior vena cava.
Hemiazygos (HEM-ē-az-ī-gos)	The ***hemiazygos vein*** is anterior to the vertebral column and slightly to the left of the midline. It begins as a continuation of the left ascending lumbar vein. It receives blood from the lower four or five left ***intercostal*** (in'-ter-KOS-tal) ***veins*** and some ***esophageal*** (e-sof'-a-JĒ-al) and ***mediastinal*** (mē-dē-a-STI-nal) ***veins.*** At the level of the ninth thoracic vertebra, it joins the azygos vein.
Accessory hemiazygos	The ***accessory hemiazygos vein*** is also anterior to the left of the vertebral column. It receives blood from three or four of the superior left ***intercostal*** (in'-ter-KOS-tal) ***veins*** and left ***bronchial*** (BRONG-kē-al) ***vein.*** It joins the azygos at the level of the eighth thoracic vertebra.

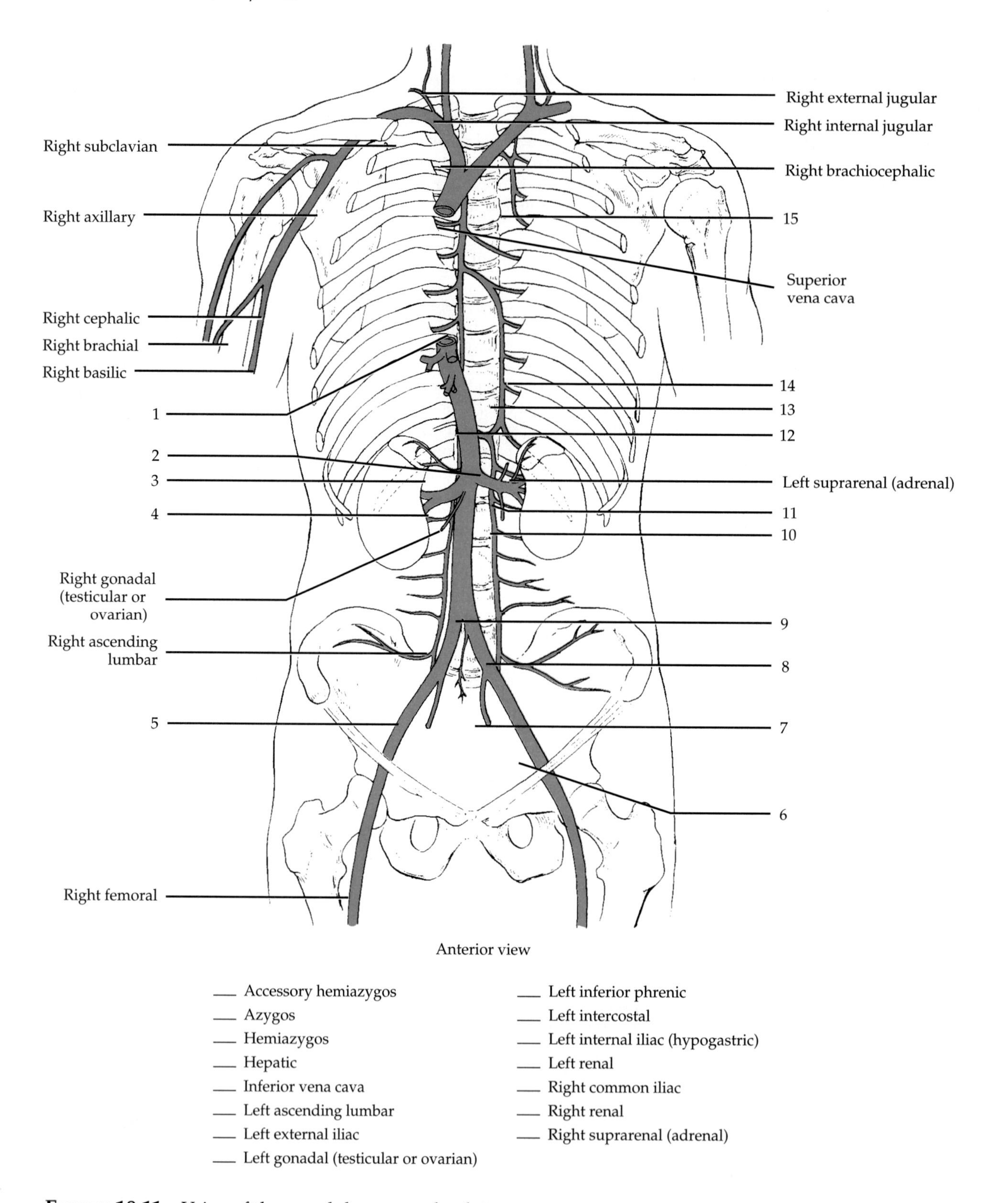

___ Accessory hemiazygos
___ Azygos
___ Hemiazygos
___ Hepatic
___ Inferior vena cava
___ Left ascending lumbar
___ Left external iliac
___ Left gonadal (testicular or ovarian)
___ Left inferior phrenic
___ Left intercostal
___ Left internal iliac (hypogastric)
___ Left renal
___ Right common iliac
___ Right renal
___ Right suprarenal (adrenal)

FIGURE 18.11 Veins of thorax, abdomen, and pelvis.

TABLE 18.10
Veins of Abdomen and Pelvis (Figure 18.11)

OVERVIEW: Blood from the abdominopelvic viscera and abdominal wall returns to the heart via the ***inferior vena cava***. Many small veins enter the inferior vena cava. Most carry return flow from parietal branches of the abdominal aorta and their names correspond to the names of the arteries. The inferior vena cava does not receive veins from the gastrointestinal tract, spleen, pancreas, and gallbladder. These organs pass their blood into a common vein, the hepatic portal vein, which delivers the blood to the liver. From here, the blood drains into the hepatic veins, which enter the inferior vena cava. This special flow of venous blood is called ***hepatic portal circulation***, which is described shortly.

Vein	Description and region drained
Inferior vena cava (VĒ-na CĀ-va)	The ***inferior vena cava*** is formed by the union of two common iliac veins that drain the lower limbs and abdomen. The inferior vena cava extends superiorly through the abdomen and thorax to the right atrium.
Common iliacs (IL-ē-aks)	The ***common iliac veins*** are formed by the union of the internal (hypogastric) and external iliac veins and represent the distal continuation of the inferior vena cava at its bifurcation (branching).
Internal iliacs	Tributaries of the ***internal iliac (hypogastric) veins*** basically correspond to branches of the internal iliac arteries. The internal iliacs drain the gluteal muscles, medial side of the thigh, urinary bladder, rectum, prostate gland, ductus (vas) deferens, uterus, and vagina.
External iliacs	The ***external iliac veins*** are a continuation of the femoral veins and receive blood from the lower limbs and inferior part of the anterior abdominal wall.
Renals (RĒ-nals)	The ***renal veins*** drain the kidneys.
Gonadals (gō-NAD-als) [*testicular* (tes-TIK-yoo-lor) or *ovarian* (ō-VA-rē-an)]	The ***testicular veins*** drain the testes (left testicular vein empties into the left renal vein and right testicular drains into the inferior vena cava); the ***ovarian veins*** drain the ovaries (left ovarian vein empties into the left renal vein and right ovarian drains into the inferior vena cava).
Suprarenals (soo'-pra-RĒ-nals)	The ***suprarenal veins*** drain the adrenal (suprarenal) glands (left suprarenal vein empties into the left renal vein and right suprarenal vein empties into the superior vena cava).
Inferior phrenics (FREN-iks)	The ***inferior phrenic veins*** drain the diaphragm (left interior phrenic vein sends a tributary to the left renal vein and right inferior phrenic vein empties into the superior vena cava).
Hepatics (he-PAT-iks)	The ***hepatic veins*** drain the liver.
Lumbars (LUM-bars)	A series of parallel ***lumbar veins*** drain blood from both sides of the posterior abdominal wall. The lumbars connect at right angles with the right and left ***ascending lumbar veins***, which form the origin of the corresponding azygos or hemiazygos vein. The lumbars drain blood into the ascending lumbars and then run to the inferior vena cava, where they release the remainder of the flow.

TABLE 18.11
Veins of Lower Limbs (Extremities) (Figure 18.12)

OVERVIEW: Blood from each lower limb is drained by ***superficial*** and ***deep veins***. The superficial veins often anastomose with each other and with deep veins along their length. Deep veins, for the most part, have the same names as their accompanying arteries. Both superficial and deep veins have valves.

Vein	Description and region drained
SUPERFICIAL VEINS	
Great saphenous (sa-FĒ-nus)	The ***great saphenous vein,*** the longest vein in the body, begins at the medial end of the ***dorsal venous arch*** of the foot. It passes anterior to the medial malleolus and then superiorly along the medial aspect of the leg and thigh just deep to the skin. It receives tributaries from superficial tissues and connects with the deep veins as well. It empties into the femoral vein in the groin. The great saphenous vein is frequently used for prolonged administration of intravenous fluids. This is particularly important in very young babies and in patients of any age who are in shock and whose veins are collapsed. It and the small saphenous vein are subject to varicosity.
Small saphenous	The ***small saphenous vein*** begins at the lateral end of the dorsal venous arch of the foot. It passes behind the lateral malleolus and ascends under the skin of the back of the leg. It receives blood from the foot and posterior portion of the leg. It empties into the popliteal vein posterior to the knee.
DEEP VEINS	
Posterior tibial (TIB-ē-al)	The ***posterior tibial vein*** is formed by the union of the ***medial*** and ***lateral plantar*** (PLAN-tar) ***veins*** posterior to the medial malleolus. It ascends deep in the muscle along the posterior aspect of the leg, receives blood from the ***peroneal*** (per'-ō-NĒ-al) ***vein,*** and unites with the anterior tibial vein just inferior to the knee.
Anterior tibial	The ***anterior tibial vein*** is the superior continuation of the ***dorsalis pedis*** (PED-is') ***veins*** in the foot. It runs between the tibia and fibula and unites with the posterior tibial to form the popliteal vein.
Popliteal (pop'-li-TĒ-al)	The ***popliteal vein,*** just posterior to the knee, receives blood from the anterior and posterior tibials and the small saphenous vein.
Femoral (FEM-o-ral)	The ***femoral vein*** is the superior continuation of the popliteal just superior to the knee. The femorals run up the posterior surface of the thighs and drain the deep structures of the thighs. After receiving the great saphenous veins in the groin, they continue as the right and left external iliac veins.

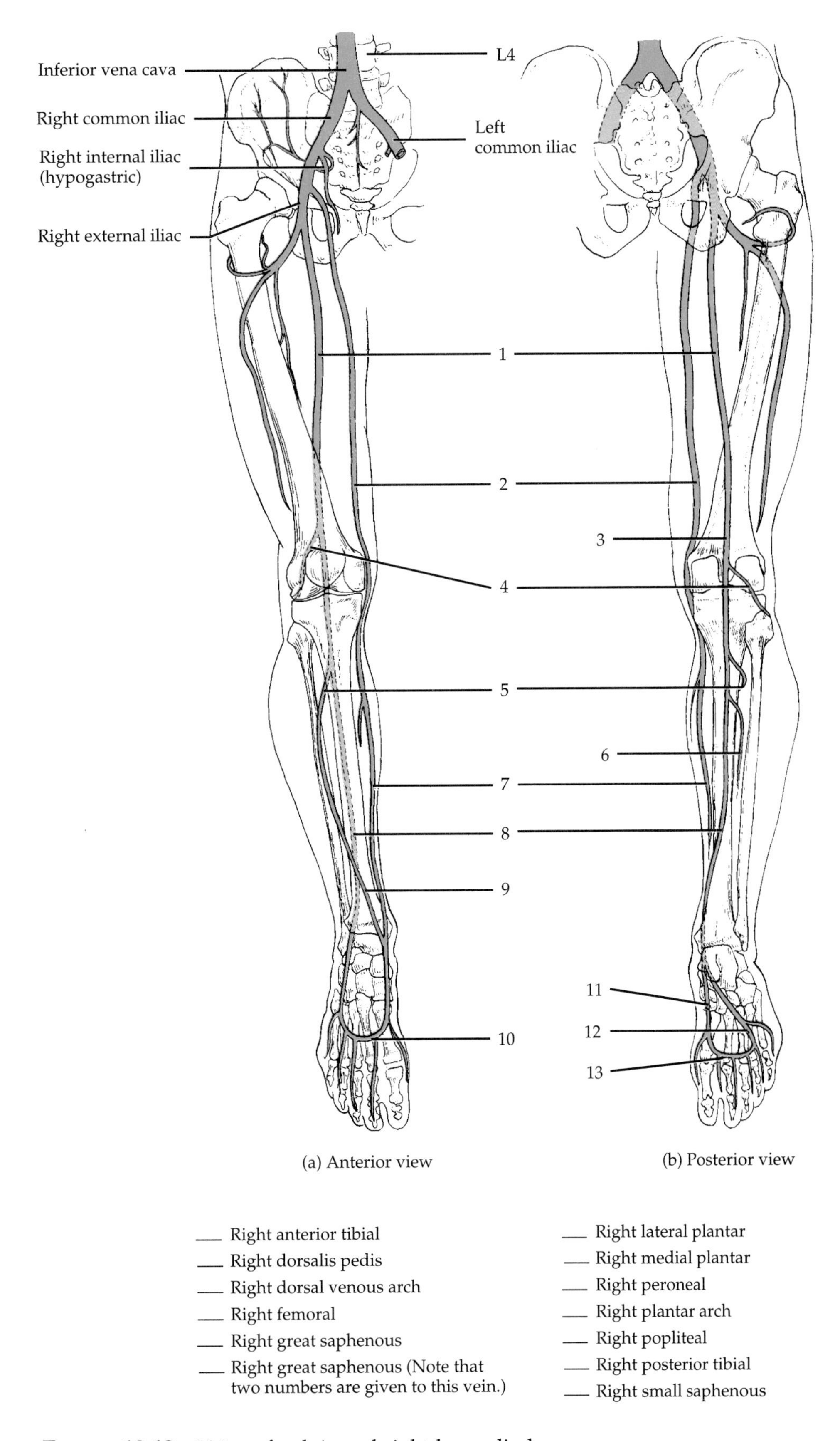

___ Right anterior tibial
___ Right dorsalis pedis
___ Right dorsal venous arch
___ Right femoral
___ Right great saphenous
___ Right great saphenous (Note that two numbers are given to this vein.)
___ Right lateral plantar
___ Right medial plantar
___ Right peroneal
___ Right plantar arch
___ Right popliteal
___ Right posterior tibial
___ Right small saphenous

FIGURE 18.12 Veins of pelvis and right lower limb.

Write the names of the missing veins in the following scheme of circulation. Be sure to indicate left or right where applicable.

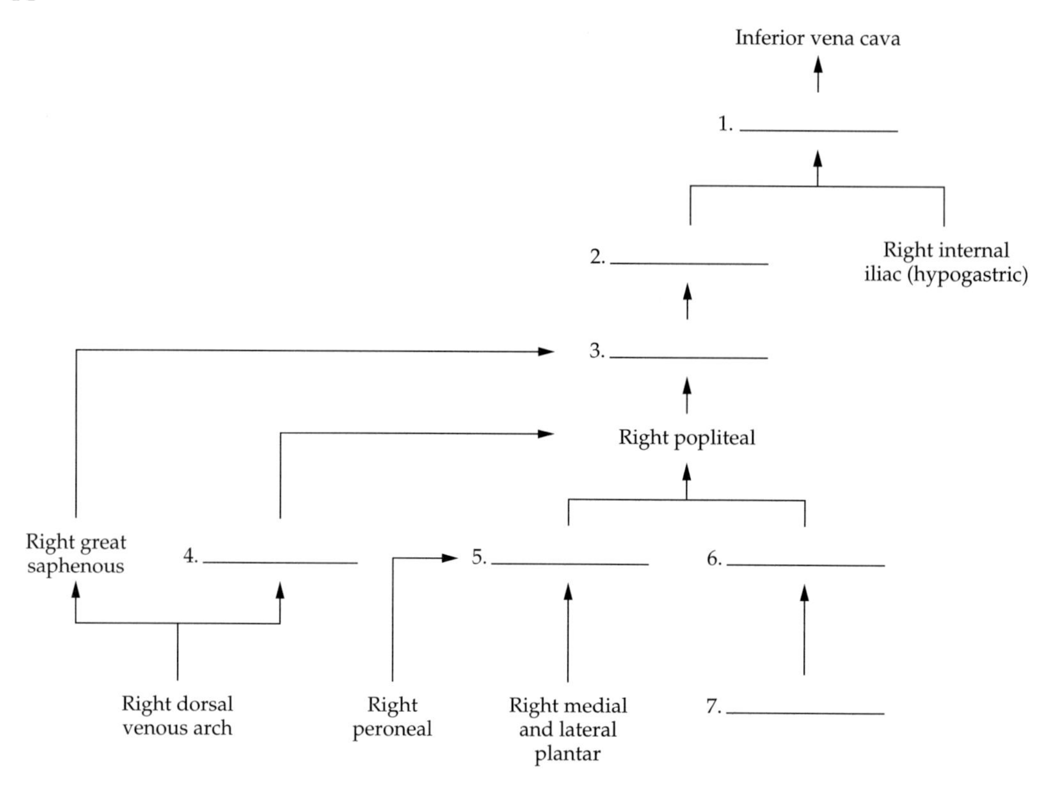

2. Hepatic Portal Circulation

The ***hepatic*** (*hepato* = liver) ***portal circulation*** detours venous blood from the gastrointestinal organs and spleen through the liver before it returns to the heart (Figure 18.13). A *portal system* carries blood between two capillary networks, from one location in the body to another without passing through the heart, in this case from capillaries of the gastrointestinal tract to sinusoids of the liver. After a meal, hepatic portal blood is rich with absorbed substances. The liver stores some and modifies others before they pass into the general circulation. For example, the liver converts glucose into glycogen for storage. It also modifies other digested substances so they may be used by cells, detoxifies harmful substances that have been absorbed by the gastrointestinal tract, and destroys bacteria by phagocytosis.

The ***hepatic portal vein*** is formed by the union of the (1) superior mesenteric and (2) splenic veins. The ***superior mesenteric vein*** drains blood from the small intestine, portions of the large intestine, stomach, and pancreas through the *jejunal, ileal, ileocolic, right colic, middle colic, pancreaticoduodenal,* and *right gastroepiploic veins.* The ***splenic vein*** drains blood from the stomach, pancreas, and portions of the large intestine through the *superior rectal, sigmoidal,* and *left colic veins.* The right and left gastric veins, which open directly into the hepatic portal vein, drain the stomach. The *cystic vein,* which also opens into the hepatic portal vein, drains the gallbladder.

At the same time the liver receives deoxygenated blood via the hepatic portal system, it also receives oxygenated blood from the systemic circulation via the hepatic artery. Ultimately, all blood leaves the liver through the ***hepatic veins,*** which drain into the inferior vena cava.

Using your textbook, charts, or models for reference, label Figure 18.13.

3. Pulmonary Circulation

The ***pulmonary*** (*pulmo* = lung) ***circulation*** carries deoxygenated blood from the right ventricle to the air sacs within the lungs and returns oxygenated blood from the air sacs within the lungs to the left atrium (Figure 18.14 on page 368). The ***pulmonary trunk*** emerges from the right ventricle and passes superiorly, posteriorly, and to the left. It then divides into two branches: the ***right pulmonary artery*** extends to the right lung; the ***left pulmonary artery*** goes to the left lung. The pulmonary arteries are the only postna-

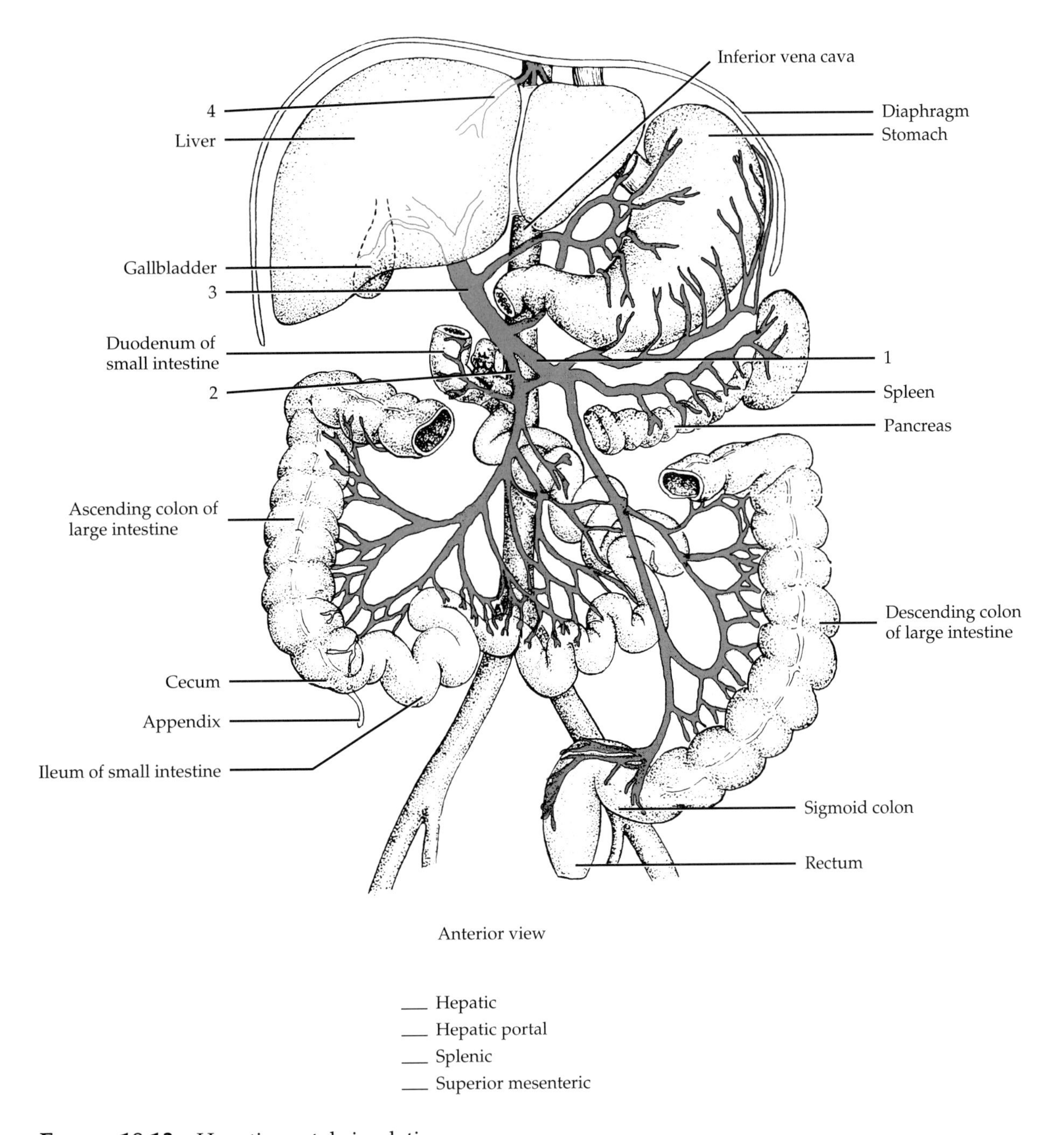

FIGURE 18.13 Hepatic portal circulation.

tal (after birth) arteries that carry deoxygenated blood. On entering the lungs, the branches divide and subdivide until finally they form capillaries around the air sacs within the lungs. CO_2 passes from the blood into these air sacs and is exhaled. Inhaled O_2 passes from the air sacs into the blood. The pulmonary capillaries unite, form venules and veins, and eventually two ***pulmonary veins*** exit from each lung and transport the oxygenated blood to the left atrium. The pulmonary veins are the only postnatal veins that carry oxygenated blood. Contractions of the left ventricle then send the blood into the systemic circulation.

Study a chart or model of the pulmonary circulation, trace the path of blood through it, and label Figure 18.14.

4. Fetal Circulation

The circulatory system of a fetus, called ***fetal circulation,*** differs from the postnatal circulation because the lungs, kidneys, and gastrointestinal tract begin to

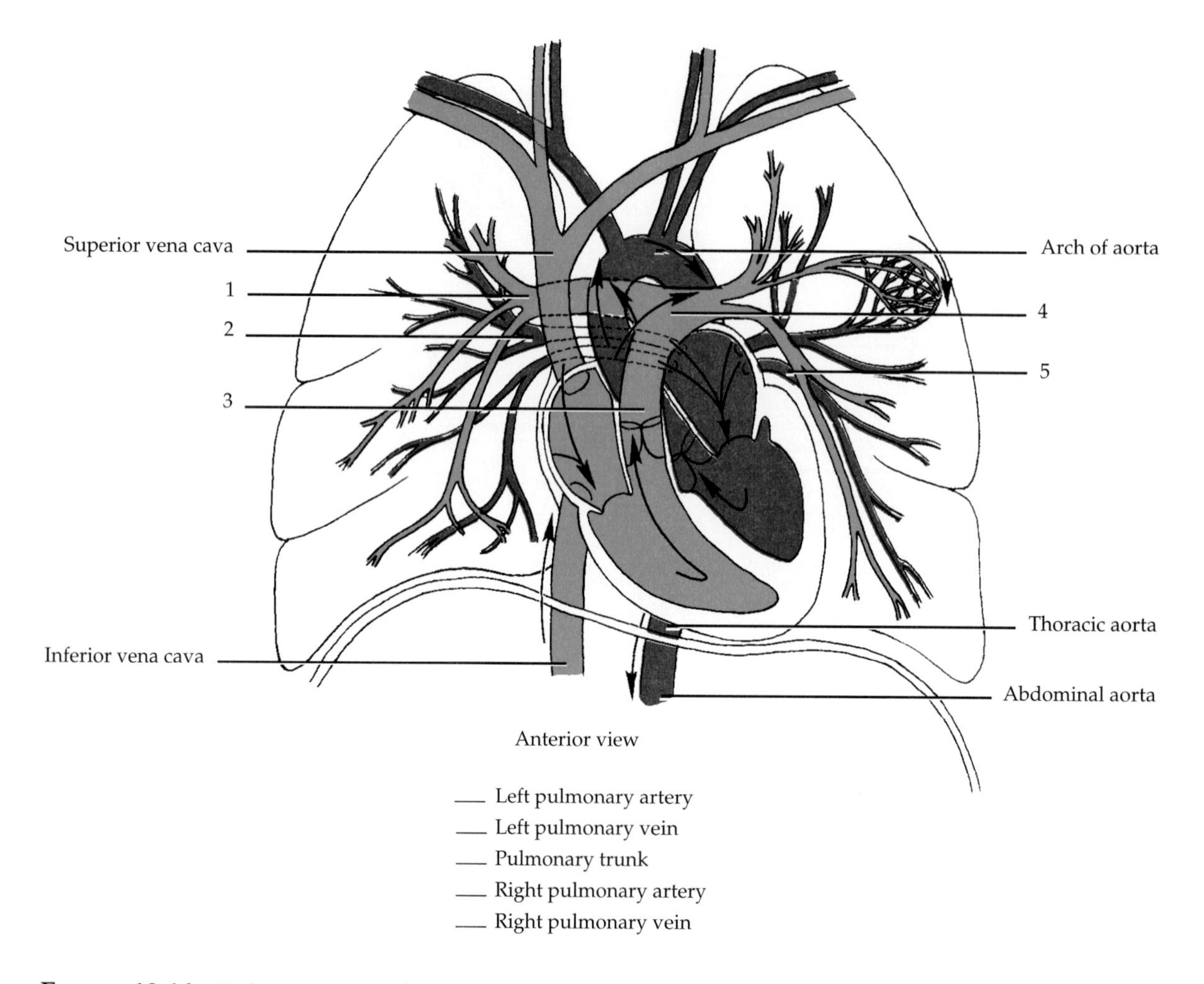

FIGURE 18.14 Pulmonary circulation.

function at birth. The fetus obtains its O_2 and nutrients by diffusion from maternal blood and eliminates its CO_2 and wastes by diffusion into maternal blood (Figure 18.15).

The exchange of materials between fetal and maternal circulation occurs through a structure called the ***placenta*** (pla-SEN-ta). It is attached to the umbilicus (navel) of the fetus by the umbilical (um-BIL-i-kal) cord, and it communicates with the mother through countless small blood vessels that emerge from the uterine wall. The umbilical cord contains blood vessels that branch into capillaries in the placenta. Wastes from fetal blood diffuse out of the capillaries, into spaces containing maternal blood (intervillous spaces) in the placenta, and finally into the mother's uterine blood vessels. Nutrients travel the opposite route—from the maternal blood vessels to the intervillous spaces to the fetal capillaries. Normally, there is no direct mixing of maternal and fetal blood since all exchanges occur by diffusion through capillary walls.

Blood passes from the fetus to the placenta via two ***umbilical arteries.*** These branches of the internal iliac (hypogastric) arteries are within the umbilical cord. At the placenta, fetal blood picks up O_2 and nutrients and eliminates CO_2 and wastes. The oxygenated blood returns from the placenta via a single ***umbilical vein.*** This vein ascends to the liver of the fetus, where it divides into two branches. Some blood flows through the branch that joins the hepatic portal vein and enters the liver. Most of the blood flows into the second branch, the ***ductus venosus*** (DUK-tus ve-NŌ-sus), which drains into the inferior vena cava. Thus a good portion of the O_2 and nutrients in the blood bypasses the fetal liver and is delivered to the developing fetal brain.

Circulation through other portions of the fetus is similar to postnatal circulation. Deoxygenated blood returning from the lower regions mingles with oxygenated blood from the ductus venosus in the inferior vena cava. This mixed blood then enters the right atrium. Deoxygenated blood returning

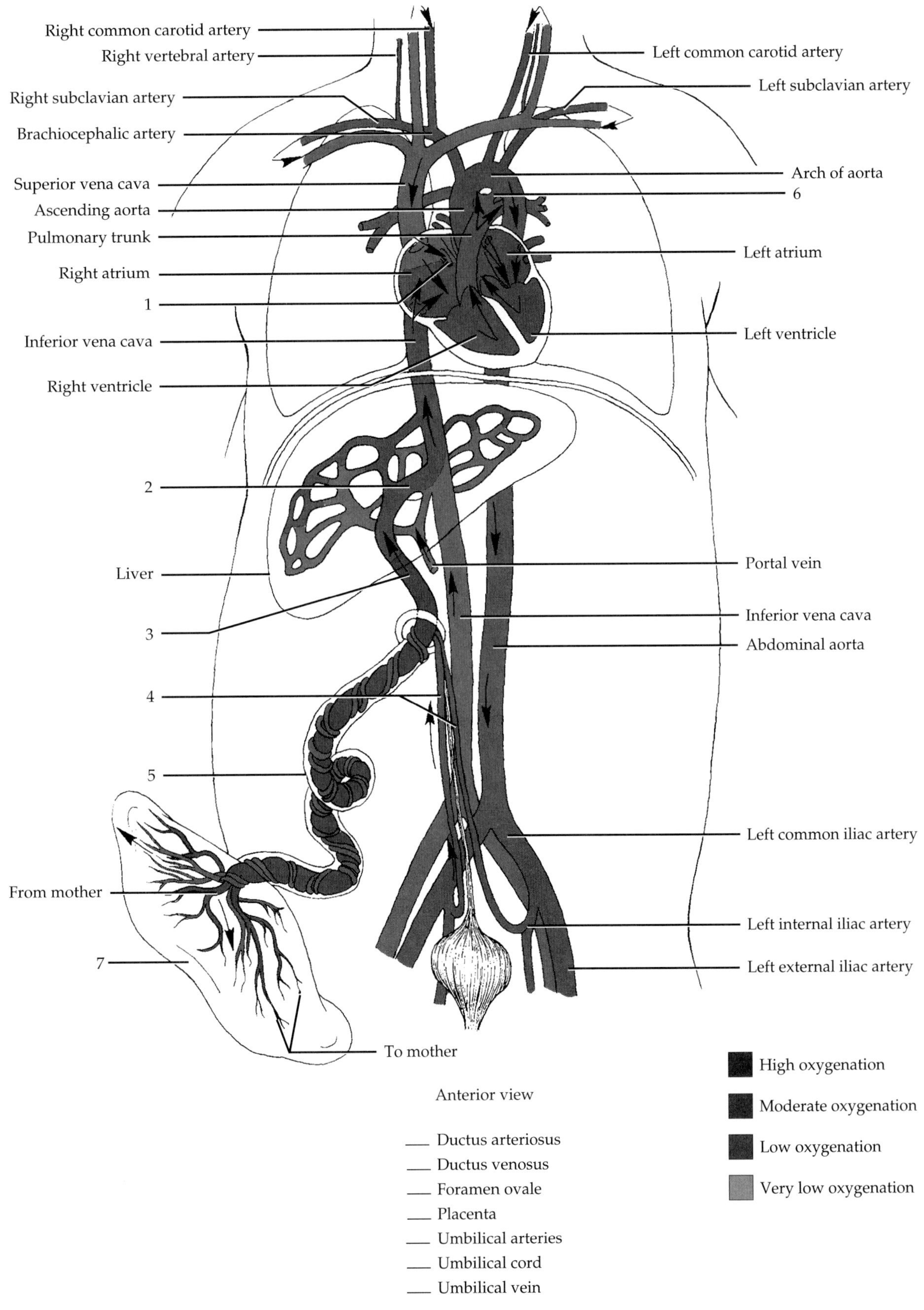

FIGURE 18.15 Fetal circulation.

from the upper regions of the fetus enters the superior vena cava and passes into the right atrium.

Most of the fetal blood does not pass from the right ventricle to the lungs, as it does in postnatal circulation, since the fetal lungs do not operate. In the fetus, an opening called the ***foramen ovale*** (fō-RĀ-men ō-VAL-ē) exists in the septum between the right and left atria. About one-third of the blood passes through the foramen ovale directly into the systemic circulation. The blood that does pass into the right ventricle is pumped into the pulmonary trunk, but little of this blood reaches the lungs. Most is sent through the ***ductus arteriosus*** (ar-tē-rē-Ō-sus). This vessel connects the pulmonary trunk with the aorta and allows most blood to bypass the fetal lungs. The blood in the aorta is carried to all parts of the fetus through the systemic circulation. When the common iliac arteries branch into the external and internal iliacs, part of the blood flows into the internal iliacs. It then goes to the umbilical arteries and back to the placenta for another exchange of materials. The only fetal vessel that carries fully oxygenated blood is the umbilical vein.

Label Figure 18.15.

E. BLOOD VESSEL EXERCISE

For each vessel listed, indicate the region supplied (if an artery) or the region drained (if a vein):

1. ***Coronary artery*** ______________________
2. ***Internal iliac veins*** ______________________
3. ***Lumbar arteries*** ______________________
4. ***Renal artery*** ______________________
5. ***Left gastric artery*** ______________________
6. ***External jugular vein*** ______________________
7. ***Left subclavian artery*** ______________________
8. ***Axillary vein*** ______________________
9. ***Brachiocephalic veins*** ______________________
10. ***Transverse sinuses*** ______________________
11. ***Hepatic artery*** ______________________
12. ***Inferior mesenteric artery*** ______________________
13. ***Suprarenal artery*** ______________________
14. ***Inferior phrenic artery*** ______________________
15. ***Great saphenous vein*** ______________________
16. ***Popliteal vein*** ______________________
17. ***Azygos vein*** ______________________
18. ***Internal iliac (hypogastric) artery*** ______________________
19. ***Internal carotid artery*** ______________________
20. ***Cephalic vein*** ______________________

F. DISSECTION OF CAT CARDIOVASCULAR SYSTEM

CAUTION! *Please reread Section D, "Precautions Related to Dissection" at the beginning of the laboratory manual on page xi before you begin dissection.*

In this section, you will dissect some of the principal blood vessels in the cat. Blood vessels should be studied in specimens that are at least doubly injected, meaning that the arteries have been injected with red latex and the veins have been injected with blue latex. Triply injected specimens are necessary to observe all of the hepatic portal system. This system is usually injected with yellow latex.

As you dissect blood vessels, bear in mind that they vary somewhat in position from one specimen to the next. Thus, the pattern of blood vessels in your specimen may not be exactly like those shown in the illustrations. Also remember that when you dissect blood vessels, you must free them from surrounding tissues so that they are clearly visible. When doing so, use forceps or a blunt probe, not a scalpel. Finally, you obtain best results if you dissect blood vessels on the side of the body opposite the side on which you dissected muscles.

1. Arteries

PROCEDURE

1. To examine the major blood vessels of the cat, extend the original thoracic longitudinal incisions posteriorly through the abdominal wall.
2. Using scissors, cut the wall from the diaphragm to the pubic region. Be careful not to damage the viscera below.
3. On both sides of the abdomen make two additional cuts, laterally and dorsally, toward the spine about 4 in. below the diaphragm. Reflect the flaps of the abdominal wall.
4. Locate the ***pulmonary trunk*** exiting from the right ventricle (see Figure 17.5). The trunk branches into a right and a left pulmonary artery. The ***left pulmonary artery*** passes ventral to the aorta to reach the left lung whereas the ***right pulmonary artery*** passes between the arch of the aorta and the heart to reach the right lung.
5. Dorsal to the pulmonary trunk is the ***ascending aorta,*** which arises from the left ventricle. At the base of the ascending aorta, try to locate the origins of the coronary arteries. The ***right coronary artery*** passes in the ***coronary sulcus,*** a groove that separates the right atrium from the right ventricle. The ***left coronary artery*** passes in the ***anterior longitudinal sulcus,*** a groove that separates the right ventricle from the left ventricle.
6. The ascending aorta continues cranially and then turns to the left. This part of the aorta is called the ***arch of the aorta.*** Originating from the arch of the aorta are two large arteries—the brachiocephalic and left subclavian.
7. The ***brachiocephalic artery*** divides at about the level of the second rib into the ***left common carotid artery*** and then into the ***right common carotid artery*** and the ***right subclavian artery*** (Figure 18.16). The left common carotid artery ascends in the neck between the internal jugular vein and the trachea. Among the branches of the left common carotid artery are (1) the ***superior thyroid artery,*** which originates at the level of the thyroid cartilage and supplies the thyroid gland and muscles of the neck; (2) the ***laryngeal artery,*** which supplies the larynx; and (3) the ***occipital artery,*** which supplies the back of the neck. The right common carotid artery supplies the same structures on the right side of the body. The right and left subclavian arteries will be discussed shortly.
8. At the level of the cranial border of the larynx, the common carotid arteries divide into two arteries. The large ***external carotid artery*** primarily supplies the head structures outside the cranial cavity. The very small ***internal carotid artery,*** which may be absent, primarily supplies the structures within the cranial cavity. At the bifurcation of the common carotid into external and internal carotids, the dilated area is called the ***carotid sinus.*** This area contains stretch receptors that inform the brain of fluctuations in blood pressure so that necessary adjustments can be made. The ***aortic sinus*** in the aorta has a similar function.
9. The right and left subclavian arteries supply the head, neck, thoracic wall, and forelimbs. Among their important branches are (1) the ***vertebral arteries,*** which supply the brain; (2) the ***internal mammary arteries,*** which primarily supply the ventral thoracic wall; (3) the ***costocervical trunk,*** which supplies the neck, back, and intercostal muscles; and (4) the ***thyrocervical trunk,*** which supplies some neck and back muscles.
10. The continuation of the subclavian artery out of the thoracic cavity across the axillary space is called the ***axillary artery*** (Figure 18.17 on page 373). Branches of the axillary artery include (1) the ***ventral thoracic artery,*** which supplies the pectoral muscles; (2) the ***long thoracic artery,*** which supplies the latissimus dorsi muscle; and (3) the ***subscapular artery,*** which supplies the shoulder muscles, dorsal muscles of the arm, and muscles of the scapula.
11. The continuation of the axillary artery into the arm is called the ***brachial artery.*** A branch of the brachial artery, the ***anterior humeral***

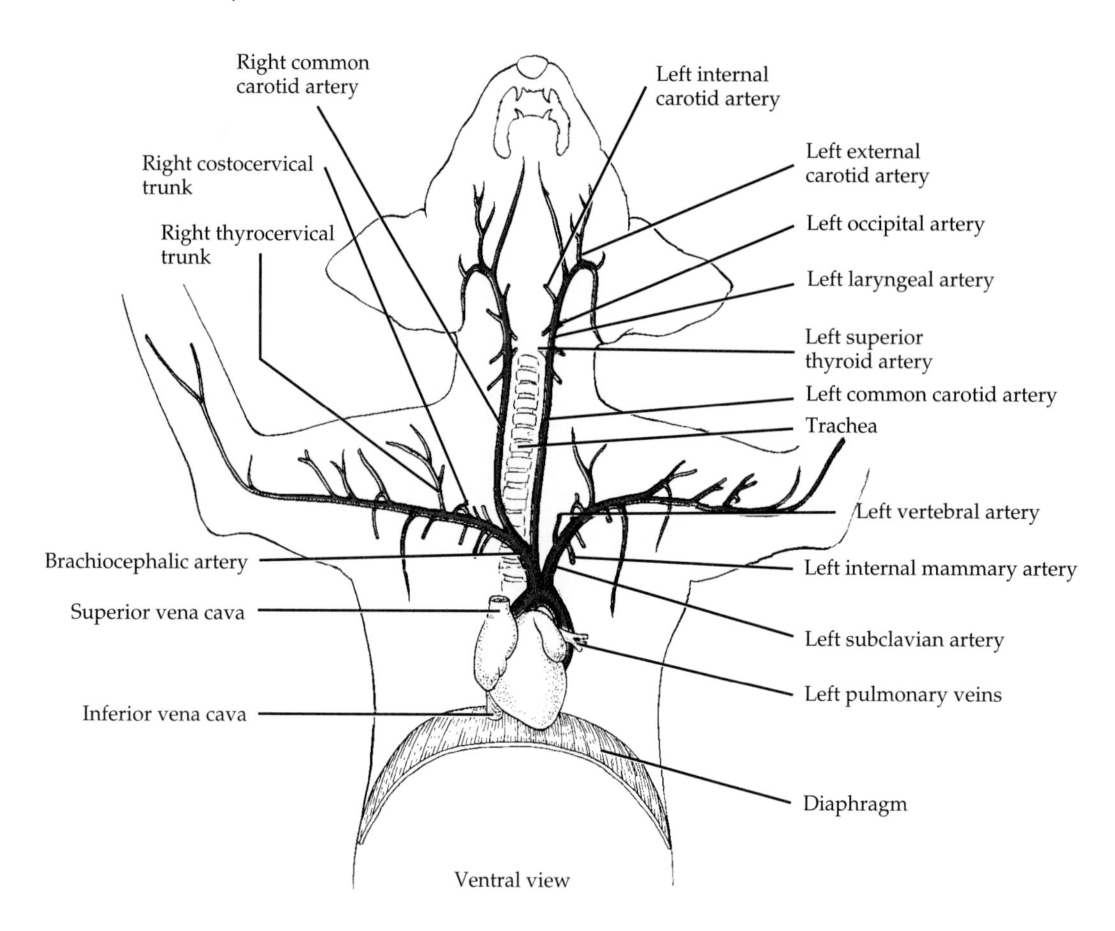

FIGURE 18.16 Arteries of forelimbs, neck, and head.

circumflex artery, arises just distal to the subscapular artery and supplies the biceps brachii muscle. Another branch of the brachial artery, the ***deep brachial artery,*** is distal to the anterior humeral circumflex. It supplies the triceps brachii, epitrochlearis, and latissimus dorsi muscles. Distal to the elbow, the brachial artery continues as the ***radial artery.*** In the forearm, the radial artery gives off a small branch, the ***ulnar artery.*** The radial and ulnar arteries supply the musculature of the forearm.

12. Continue your dissection of arteries by returning to the arch of the aorta. Follow it caudally as it passes through the thoracic cavity. At this point the aorta is known as the ***thoracic aorta.*** Arising from the thoracic aorta are (1) the ***intercostal arteries*** (10 pairs), which supply the intercostal muscles; (2) the ***bronchial arteries,*** which supply the bronchi and lungs; and (3) the ***esophageal arteries,*** which supply the esophagus.
13. The aorta passes through the diaphragm at the level of the second lumbar vertebra. The portion of the aorta between the diaphragm and pelvis is called the ***abdominal aorta.***
14. The first branch of the abdominal aorta immediately caudal to the diaphragm is the ***celiac trunk*** (Figure 18.18 on page 374). It consists of three branches. The ***hepatic artery*** supplies the liver, the ***left gastric artery*** along the lesser curvature of the stomach supplies the stomach, and the ***splenic artery*** supplies the spleen. See if you can also locate the ***cystic artery,*** a branch of the hepatic artery that supplies the gallbladder.
15. Caudal to the celiac trunk is the ***superior mesenteric artery.*** Its branches are the ***inferior pancreatoduodenal artery*** to the pancreas and duodenum, the ***ileocolic artery*** to the ileum and cecum, the ***right colic artery*** to the ascending colon, the ***middle colic artery*** to the transverse colon, and the ***intestinal arteries*** to the jejunum and ileum.
16. Just caudal to the superior mesenteric artery, the abdominal aorta gives off a pair of ***adrenolumbar arteries.*** Each gives off an ***inferior phrenic***

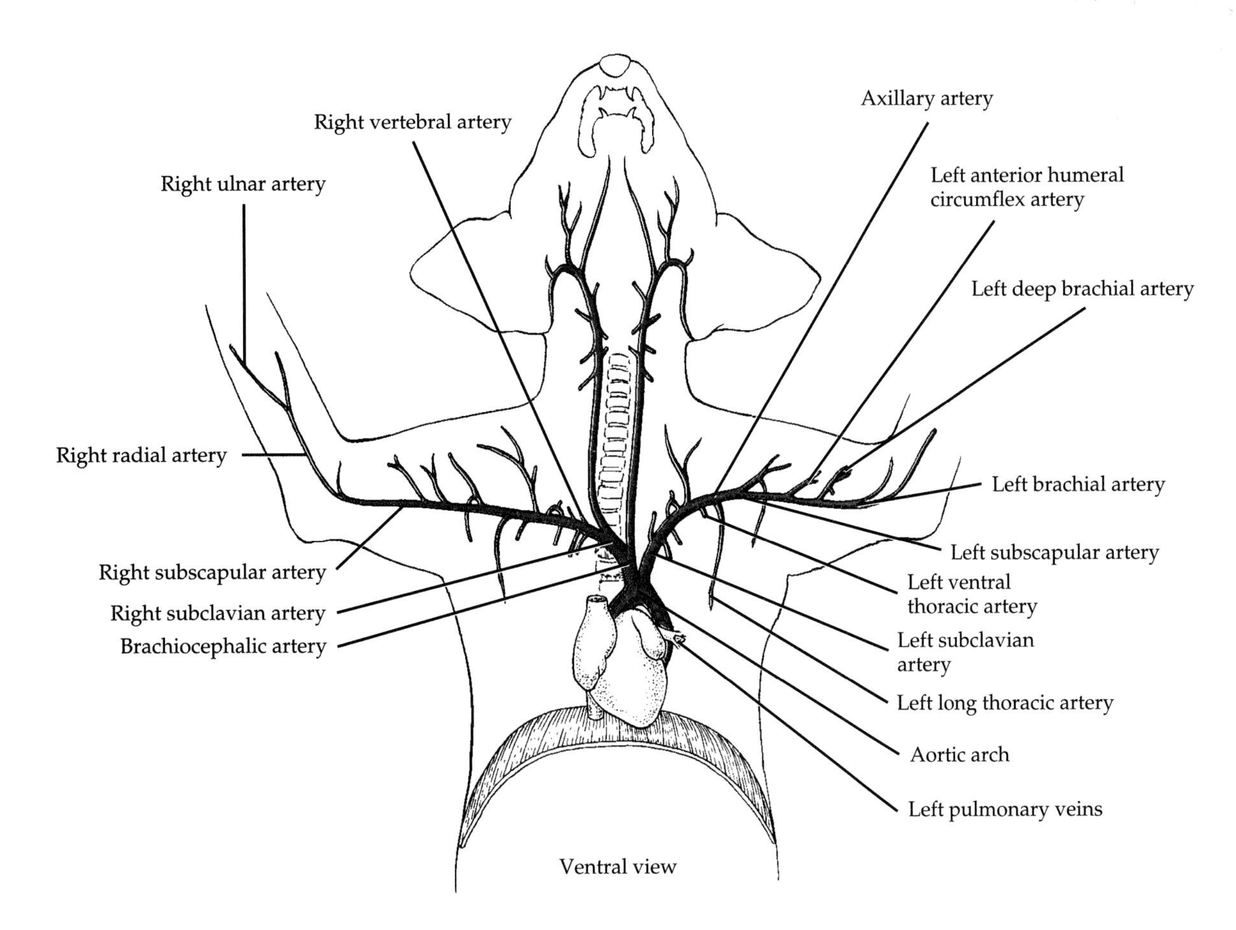

FIGURE 18.17 Arteries of forelimbs.

artery to the diaphragm, an ***adrenal (suprarenal) artery*** to the adrenal gland, and a ***parietal branch*** to the muscles of the back.

17. The paired ***renal arteries,*** caudal to the adrenolumbar arteries, supply the kidneys.
18. In the male, the paired ***internal spermatic arteries,*** or ***testicular arteries,*** arise from the abdominal aorta caudal to the renal arteries. The internal spermatic arteries pass caudally to the internal inguinal ring and through the inguinal canal where they accompany the ductus deferens of each testis.
19. In the female, the paired ***ovarian arteries*** arise from the abdominal aorta caudal to the renal arteries. They pass laterally in the broad ligament and supply the ovaries.
20. The unpaired ***inferior mesenteric artery*** arises caudal to the internal spermatic or ovarian arteries. Its branches include a ***left colic artery*** to the descending colon and a ***superior hemorrhoidal artery*** to the rectum.
21. Seven pairs of ***lumbar arteries*** arise from the dorsal surface of the abdominal aorta and supply the muscles of the dorsal abdominal wall. A pair of ***iliolumbar arteries*** arise near the inferior mesenteric artery and also supply the muscles of the dorsal abdominal wall.
22. Near the sacrum, the abdominal aorta divides into right and left ***external iliac arteries*** (Figure 18.19 on page 375). Also from this point of division, a short common artery extends caudally and gives rise to the paired ***internal iliac (hypogastric) arteries*** and unpaired ***median sacral artery*** that passes along the ventral aspect of the tail. The first branch off the internal iliac artery, the ***umbilical artery,*** supplies the urinary bladder. Another branch of the internal iliac artery, the ***superior gluteal artery,*** supplies the gluteal muscles and other muscles of the thigh. Still another branch of the internal iliac artery, the ***middle hemorrhoidal artery,*** supplies the rectum, anus, urethra, and

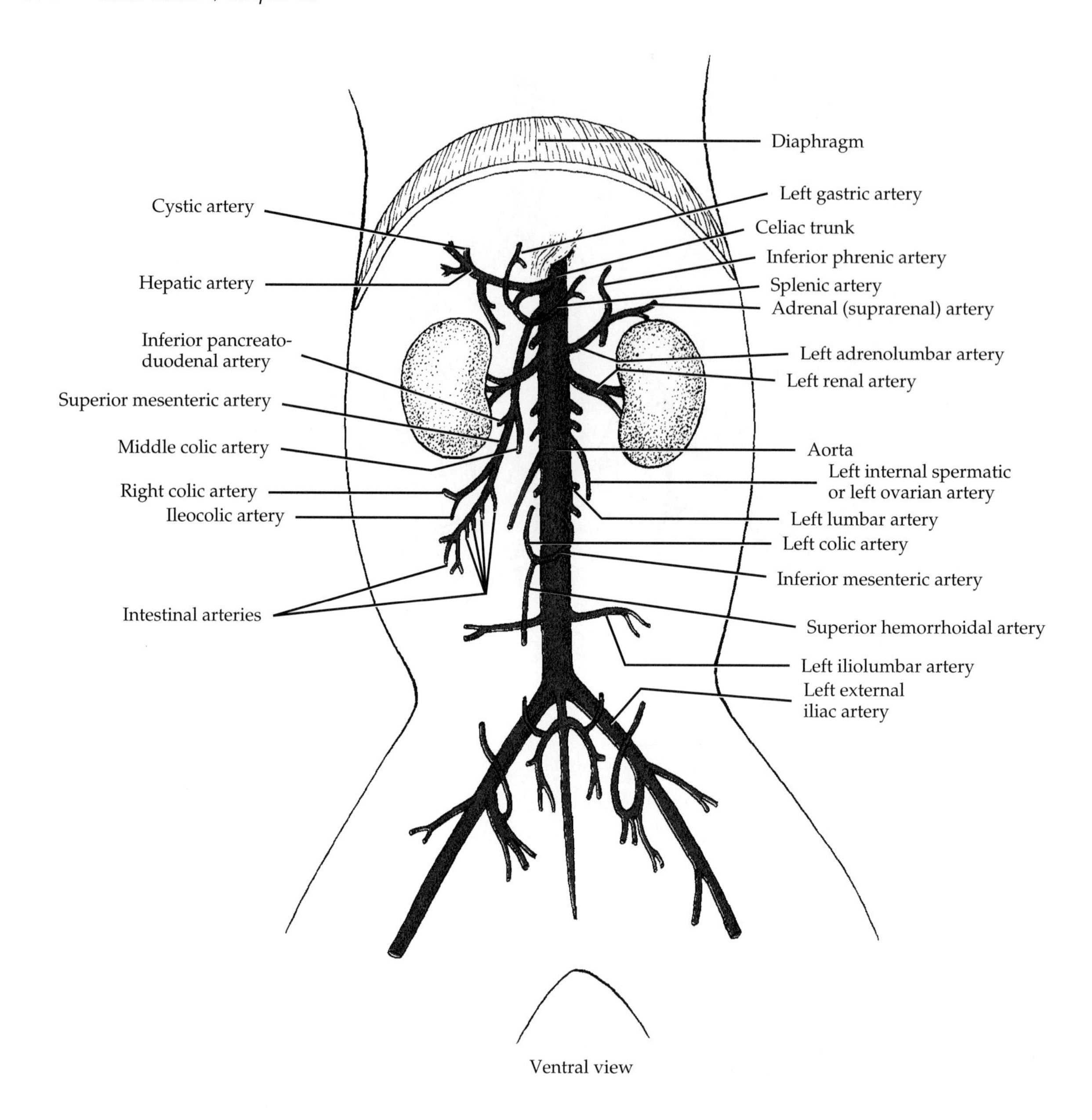

FIGURE 18.18 Arteries of abdomen.

adjacent structures. The terminal portion of the internal iliac artery, the ***inferior gluteal artery,*** supplies thigh muscles.

23. Now we can return to the external iliac arteries. Just before leaving the abdominal cavity, each external iliac artery gives off a branch, the ***deep femoral artery,*** which supplies the muscles of the thigh. Branches of the deep femoral artery supply the rectus abdominis muscle, the urinary bladder, and muscles on the medial aspect of the thigh.
24. The external iliac artery continues outside the abdominal cavity and into the thigh as the ***femoral artery.*** Passing through the thigh, it gives off one branch as the ***lateral femoral circumflex artery*** to the rectus femoris and vastus medialis. Just above the knee, the femoral artery gives off a branch as the ***superior articular artery*** and another as the ***saphenous artery.*** Both supply thigh muscles.
25. In the region of the knee, the femoral artery is known as the ***popliteal artery.*** It divides into an ***anterior tibial artery*** and a ***posterior tibial artery,*** which supply the leg and foot.

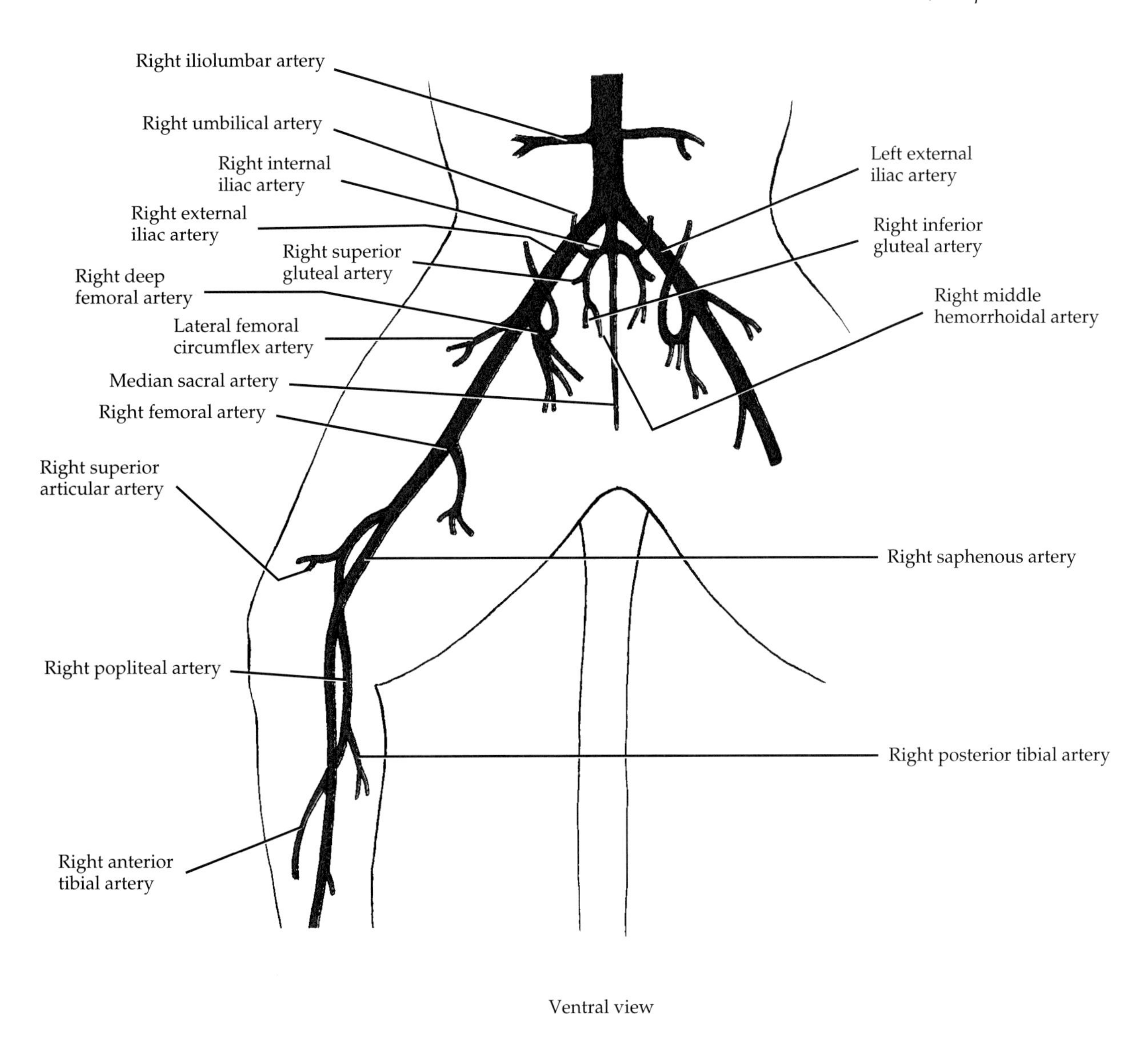

FIGURE 18.19 Arteries of right hindlimb.

2. Veins

PROCEDURE

1. In learning the arteries of the cat, we began our study with the heart and then worked outward from the heart to various regions of the body. In learning the veins of the cat, we will begin our study with the various regions of the body and work toward the heart. Whereas arteries supply various parts of the body with blood, veins drain various parts of the body. That is, veins drain blood from capillary beds and take it back to the heart.
2. The veins of the pelvis and thigh generally correspond to the arteries in those regions. Whereas the artery supplies the region with blood, the corresponding vein drains the same region. An important difference between the arteries and veins of the pelvis is that the internal iliac vein joins the external iliac vein directly to form the common iliac vein. The cat has no common iliac artery.
3. The deep veins of the leg are the ***anterior tibial vein*** and the ***posterior tibial vein*** (Figure 18.20). At the knee the two unite to form the ***popliteal vein,*** another deep vein. In the thigh,

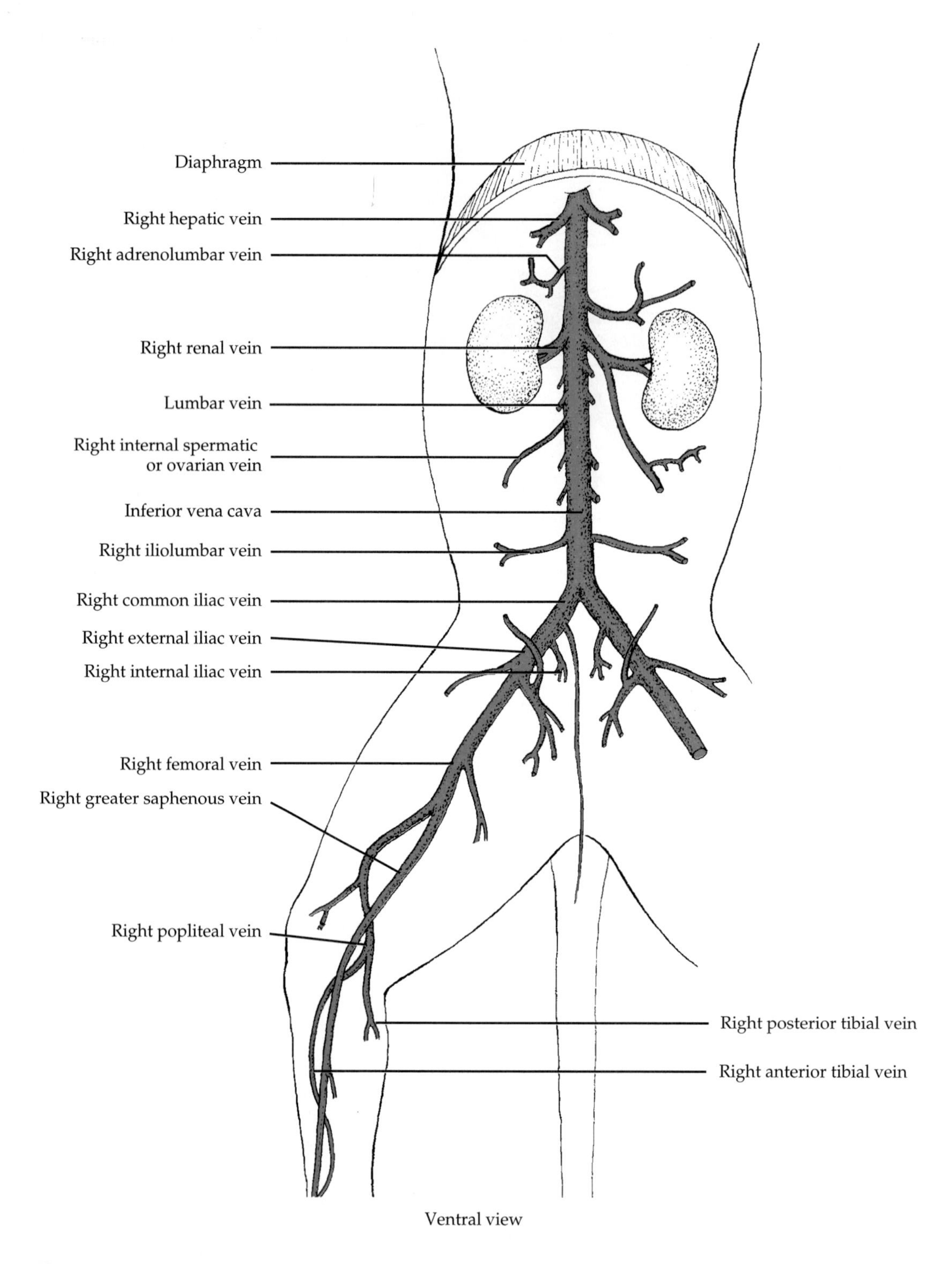

FIGURE 18.20 Veins of right hindlimb and abdomen.

the popliteal vein is known as the ***femoral vein.*** The superficial veins of the lower extremity are the ***greater saphenous vein,*** which runs along the medial surface and joins the femoral vein, and the ***lesser saphenous vein,*** which runs along the dorsal surface and joins a tributary of the internal iliac vein.

4. The ***external iliac vein*** is a continuation of the femoral vein. It has tributaries corresponding to the arteries and returns blood distributed by these arteries. The ***internal iliac (hypogastric) vein,*** with its tributaries, returns blood distributed by the internal iliac artery and its branches. The ***common iliac vein*** is formed by the union of the external and internal iliac veins.
5. The ***inferior vena cava (postcava),*** formed by the union of the common iliac veins, enters the caudal portion of the right atrium of the heart. The tributaries of the inferior vena cava, exclusive of the hepatic portal system, include the ***iliolumbar, lumbar, renal, adrenolumbar,*** and ***internal spermatic*** or ***ovarian veins.*** These correspond to the paired branches of the abdominal aorta. Cranial to the diaphragm, the inferior vena cava has no tributaries. After piercing the diaphragm, the inferior vena cava ascends and enters the right atrium.
6. The ***hepatic portal system*** consists of a series of veins that drain blood from the small and large intestines, stomach, spleen, and pancreas and convey the blood to the liver. The blood is delivered to the liver by the hepatic portal vein. Within the liver, the hepatic portal vein gives branches to each lobe. These branches terminate in capillaries called ***liver sinusoids.*** The sinusoids are drained by the hepatic veins, which leave the liver and join the inferior vena cava. Blood delivered to the liver by the hepatic artery also passes through the liver sinusoids.
7. Unless your specimen is triply injected, locating all the veins of the hepatic portal system will be very difficult. However, using Figure 18.21 as a guide, identify as many of the vessels as you can. Notice that the hepatic portal vein receives blood from the spleen and from all digestive organs except the liver. Blood is delivered to the hepatic portal vein by veins corresponding to the arteries that distribute blood to these viscera. In the cat, the ***hepatic portal vein*** is formed by the union of the ***gastrosplenic vein*** and the ***superior mesenteric vein.*** The superior mesenteric vein is formed by the ***inferior mesenteric vein, intestinal veins,*** and ***ileocolic vein,*** all of which drain the small intestine, and by the ***anterior pancreatoduodenal*** vein from the pancreas and small intestine (Figure 18.21).
8. In the forearm identify the ulnar and radial veins. The ***ulnar vein*** joins the ***radial vein*** (Figure 18.22). Also find the ***cephalic vein,*** which joins the ***transverse scapular vein,*** and the ***median cubital vein,*** a communicating vein between the cephalic and brachial veins.
9. In the arm, identify the ***brachial vein*** and its tributaries. In general, this vein parallels the brachial artery and its branches.
10. The ***axillary vein*** is a continuation of the brachial vein in the axilla. The tributaries to the axillary vein generally correspond to the branches of the axillary artery.
11. The axillary vein is joined by the ***subscapular vein*** in the axillary region. In the shoulder, the axillary vein is known as the ***subclavian vein.*** Follow the subclavian medially and locate the superior vena cava.
12. The ***superior vena cava (precava)*** enters the anterior part of the right atrium and returns blood to the heart from all areas cranial to the diaphragm, except the heart itself. The superior vena cava is formed by the union of the two ***brachiocephalic veins.***
13. Entering the right side of the superior vena cava immediately cranial to the heart is the single ***azygos vein.*** This vein collects blood from ***intercostal veins*** from the body wall, from the diaphragm, from the ***esophageal veins,*** and from the ***bronchial veins.*** Identify the azygos vein (Figure 18.22 on page 379).
14. The superior vena cava also receives blood from the chest wall from the ***internal mammary veins.*** These veins unite as the ***sternal vein*** before entering the superior vena cava. The ***right vertebral vein*** may join the superior vena cava or the brachiocephalic vein.
15. From the superior vena cava, trace the ***left brachiocephalic (innominate) vein*** cranially (Figure 18.23 on page 380). Identify the ***left vertebral vein.*** Note that the left brachiocephalic vein is formed by the union of the ***left external jugular vein,*** which drains the head outside the cranial cavity, and the ***left subclavian vein,*** which drains the arm. The ***right brachiocephalic vein*** is formed by the union of the right external jugular and right subclavian veins.

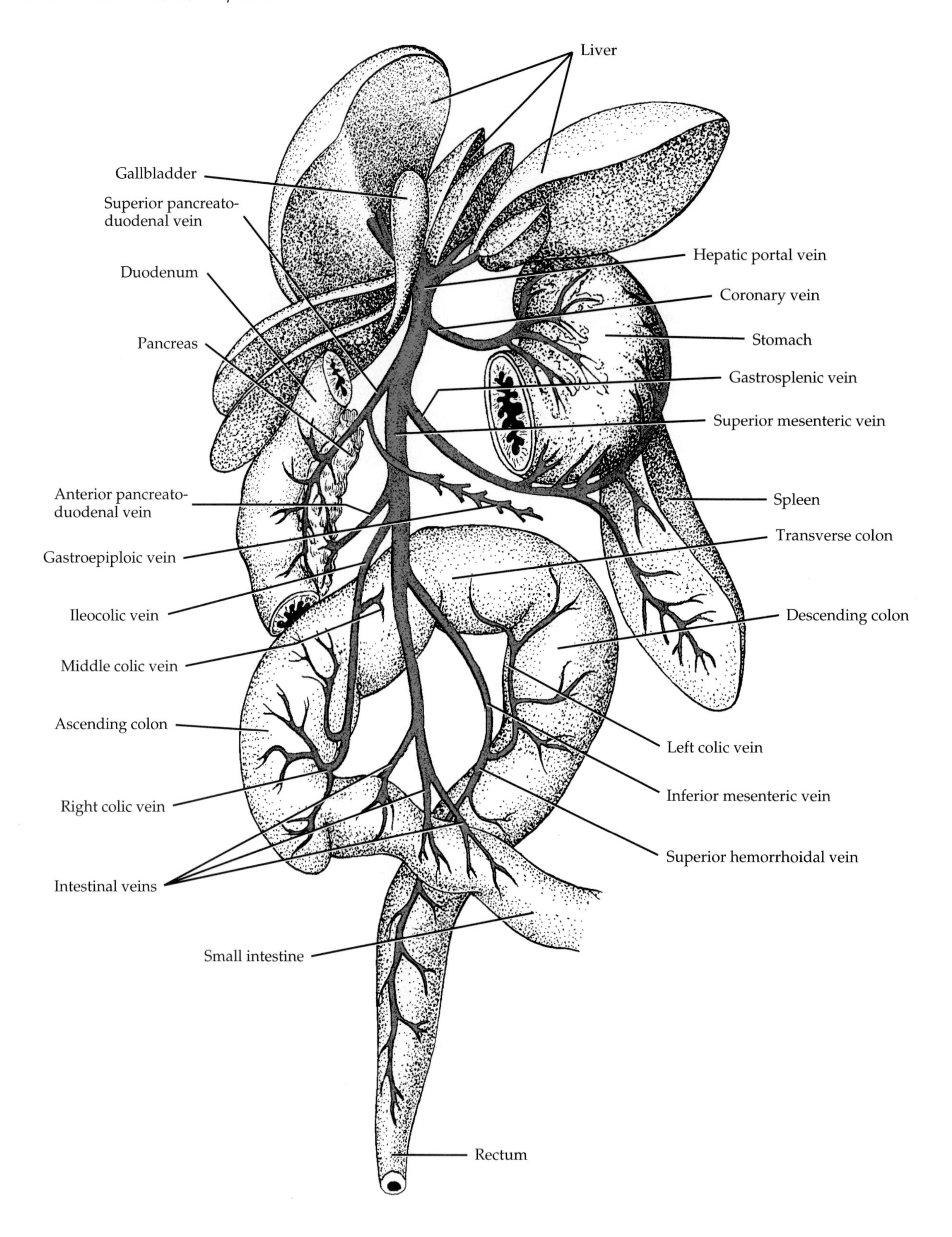

FIGURE 18.21 Hepatic portal system.

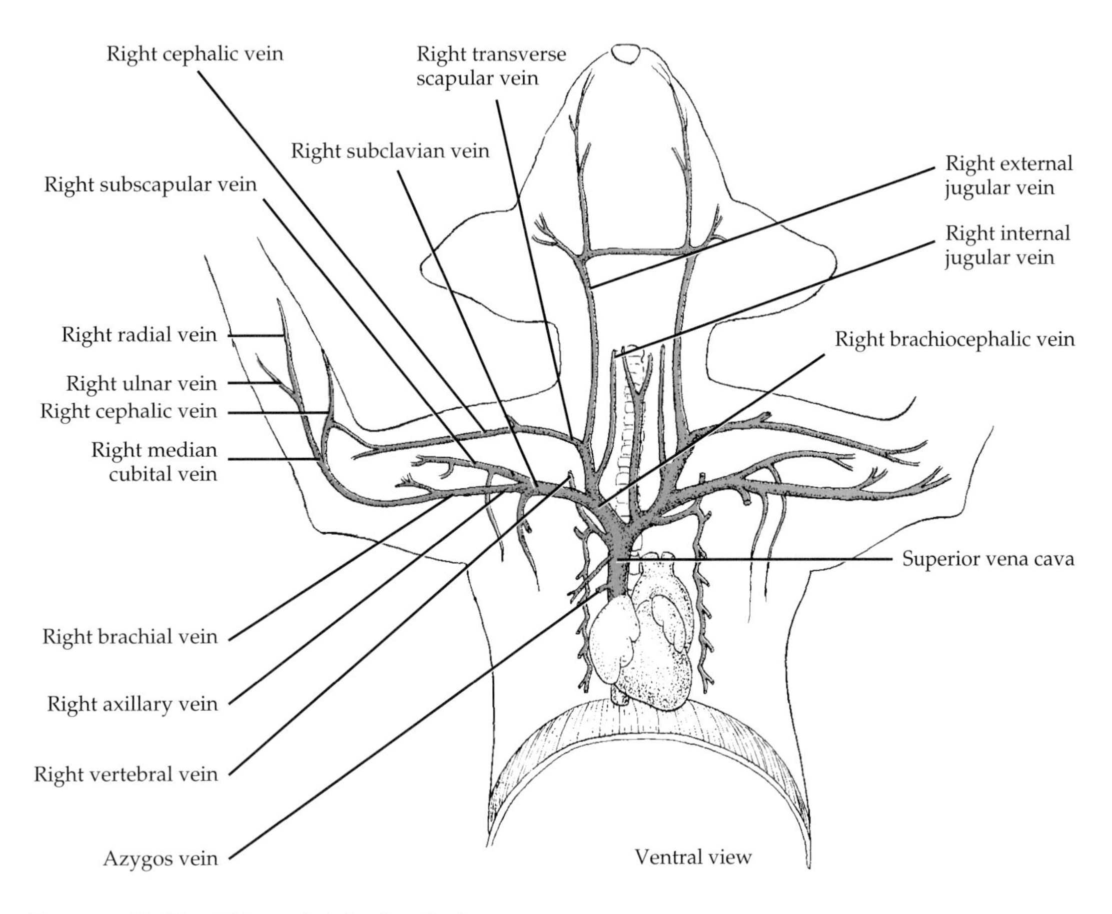

FIGURE 18.22 Veins of right forelimb.

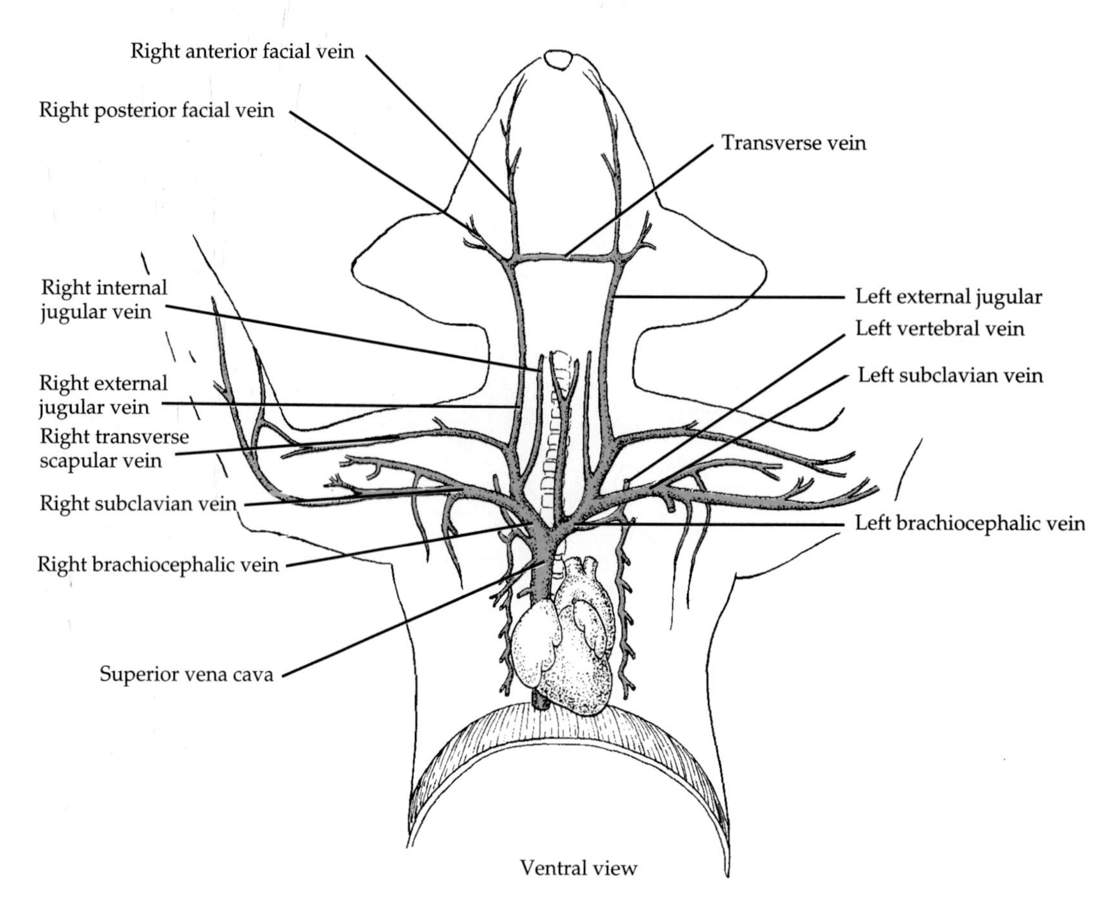

FIGURE 18.23 Veins of neck and head.

16. If you trace the external jugular vein cranially, you will note that it receives the smaller ***internal jugular vein*** above the point of union of the external jugular and subclavian veins. The internal jugular vein drains the brain. Cranial to the internal jugular vein is the ***transverse scapular vein,*** which joins the external jugular.
17. The ***external jugular vein*** is formed by the union of the ***anterior*** and ***posterior facial veins.*** These veins have tributaries that return blood distributed by the branches of the external carotid artery. The external jugular veins communicate near their points of formation by a large ***transverse vein*** that passes ventral to the neck.

ANSWER THE LABORATORY REPORT QUESTIONS AT THE END OF THE EXERCISE.

LABORATORY REPORT QUESTIONS

Blood Vessels 18

Student ________________ **Date** ________________

Laboratory Section ________________ **Score/Grade** ________________

PART 1. Multiple Choice

_______ **1.** The largest of the circulatory routes is (a) systemic (b) pulmonary (c) coronary (d) hepatic portal

_______ **2.** All arteries of systemic circulation branch from the (a) superior vena cava (b) aorta (c) pulmonary artery (d) coronary artery

_______ **3.** The arterial system that supplies the brain with blood is the (a) hepatic portal system (b) pulmonary system (c) cerebral arterial circle (circle of Willis) (d) carotid system

_______ **4.** An obstruction in the inferior vena cava would hamper the return of blood from the (a) head and neck (b) upper limbs (c) thorax (d) abdomen and pelvis

_______ **5.** Which statement best describes arteries? (a) all carry oxygenated blood to the heart (b) all contain valves to prevent the backflow of blood (c) all carry blood away from the heart (d) only large arteries are lined with endothelium

_______ **6.** Which statement is *not* true of veins? (a) they have less elastic tissue and smooth muscle than arteries (b) their tunica externa is the thickest coat (c) most veins in the limbs have valves (d) they always carry deoxygenated blood

_______ **7.** A thrombus in the first branch of the arch of the aorta would affect the flow of blood to the (a) left side of the head and neck (b) myocardium of the heart (c) right side of the head and neck and right upper limb (d) left upper limb

_______ **8.** If a vein must be punctured for an injection, transfusion, or removal of a blood sample, the likely site would be the (a) median cubital (b) subclavian (c) hemiazygos (d) anterior tibial

_______ **9.** In hepatic portal circulation, blood is eventually returned to the inferior vena cava through the (a) superior mesenteric vein (b) hepatic portal vein (c) hepatic artery (d) hepatic veins

_______ **10.** Which of the following are involved in pulmonary circulation? (a) superior vena cava, right atrium, and left ventricle (b) inferior vena cava, right atrium, and left ventricle (c) right ventricle, pulmonary artery, and left atrium (d) left ventricle, aorta, and inferior vena cava

_______ **11.** If a thrombus in the left common iliac vein dislodged, into which arteriole system would it first find its way? (a) brain (b) kidneys (c) lungs (d) left arm

_______ **12.** In fetal circulation, the blood containing the highest amount of oxygen is found in the (a) umbilical arteries (b) ductus venosus (c) aorta (d) umbilical vein

_______ **13.** The greatest amount of elastic tissue found in the arteries is located in which coat? (a) tunica interna (b) tunica media (c) tunica externa (d) tunica adventitia

_______ **14.** Which coat of an artery contains endothelium? (a) tunica interna (b) tunica media (c) tunica externa (d) tunica adventitia

_______ **15.** Permitting the exchange of nutrients and gases between the blood and tissue cells is the primary function of (a) capillaries (b) arteries (c) veins (d) arterioles

_______ **16.** The circulatory route that runs from the gastrointestinal tract to the liver is called (a) coronary circulation (b) pulmonary circulation (c) hepatic portal circulation (d) cerebral circulation

_______ **17.** Which of the following statements about systemic circulation is *not* correct? (a) its purpose is to carry oxygen and nutrients to body tissues and to remove carbon dioxide (b) all systemic arteries branch from the aorta (c) it involves the flow of blood from the left ventricle to all parts of the body except the lungs (d) it involves the flow of blood from the body to the left atrium

_______ **18.** The opening in the septum between the right and left atria of a fetus is called the (a) foramen ovale (b) ductus venosus (c) foramen rotundum (d) foramen spinosum

_______ **19.** The branch of the umbilical vein in the fetus that connects with the inferior vena cava, bypassing the liver, is the (a) foramen ovale (b) ductus venosus (c) ductus arteriosus (d) patent ductus

_______ **20.** Which of the vessels does *not* belong with the others? (a) brachiocephalic artery (b) left common carotid artery (c) celiac artery (d) left subclavian artery

PART 2. Matching

_______ **21.** Aortic branch that supplies the head and associated structures

_______ **22.** Artery that distributes blood to the small intestine and part of the large intestine

_______ **23.** Vessel into which veins of the head and neck, upper limbs, and thorax enter

_______ **24.** Vessel into which veins of the abdomen, pelvis, and lower limbs enter

_______ **25.** Vein that drains the head and associated structures

_______ **26.** Longest vein in the body

_______ **27.** Vein just behind the knee

_______ **28.** Artery that supplies a major part of the large intestine and rectum

_______ **29.** First branch off of the arch of the aorta

_______ **30.** Arteries supplying the heart

A. Inferior vena cava
B. Superior vena cava
C. Superior mesenteric
D. Common carotid
E. Jugular
F. Brachiocephalic
G. Coronary
H. Inferior mesenteric
I. Popliteal
J. Great saphenous

Lymphatic and Immune System

19

The *lymphatic* (lim-FAT-ik) *system* is composed of a pale yellow fluid called lymph, vessels that transport lymph called lymphatic vessels (lymphatics), a number of structures and organs that contain lymphatic tissue, and red bone marrow, which is the site of lymphocyte production. Lymphatic tissue is a specialized form of reticular connective tissue that contains large numbers of lymphocytes.

A. LYMPHATIC VESSELS

Lymphatic vessels originate as microscopic ***lymphatic capillaries*** in spaces between cells. They are found in most parts of the body; they are absent in avascular tissue, the central nervous system, splenic pulp, and bone marrow. They are slightly larger than blood capillaries and have a unique structure that permits interstitial fluid to flow into them but not out. Lymphatic capillaries also differ from blood capillaries in that they end blindly; blood capillaries have an arterial and a venous end. In addition, lymphatic capillaries are structurally adapted to ensure the return of proteins to the cardiovascular system when they leak out of blood capillaries.

Just as blood capillaries converge to form venules and veins, lymphatic capillaries unite to form larger and larger lymph vessels called ***lymphatic vessels*** (Figure 19.1). Lymphatic vessels

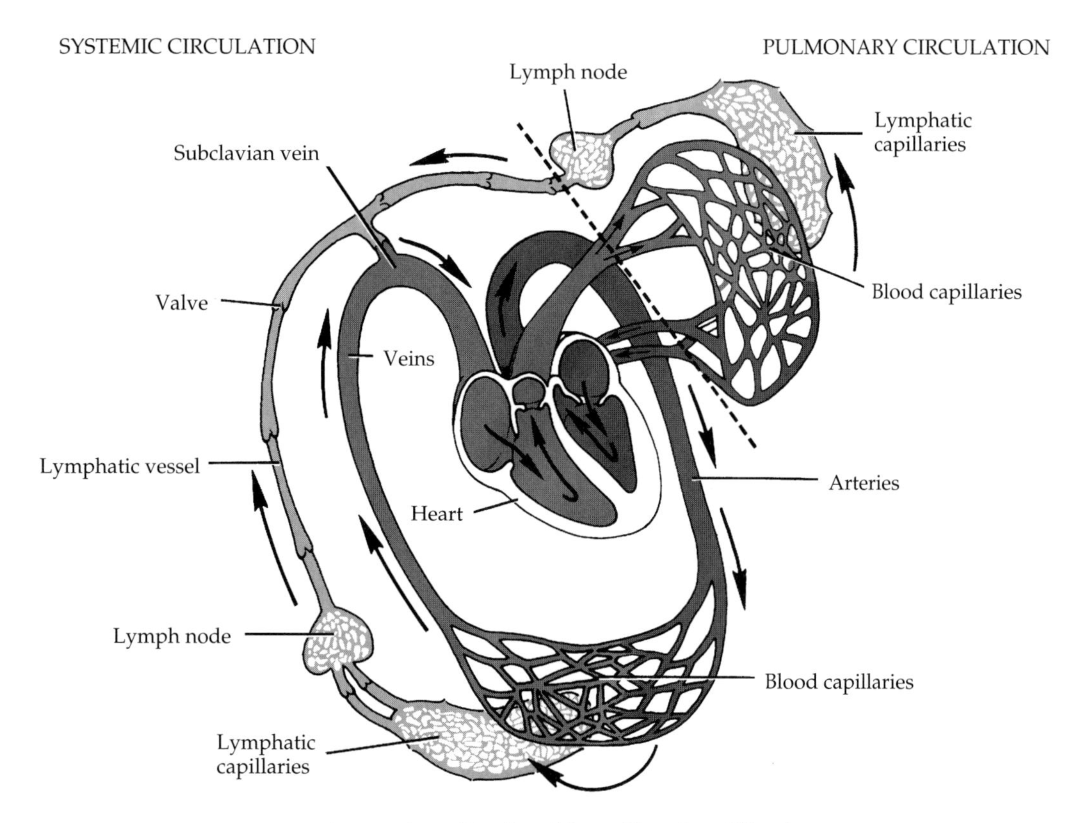

FIGURE 19.1 Relationship of lymphatic system to cardiovascular system.

resemble veins in structure, but have thinner walls and more valves, and contain lymph nodes at various intervals along their length (Figure 19.1). Ultimately, lymphatic vessels deliver lymph into two main channels—the thoracic duct and the right lymphatic duct. These will be described shortly.

Lymphangiography (lim-fan'-jē-OG-ra-fē) is the x-ray examination of lymphatic vessels and lymph organs after they are filled with a radiopaque substance. Such an x-ray is called a ***lymphangiogram*** (lim-FAN-jē-ō-gram). Lymphangiograms are useful in detecting edema and carcinomas, and in locating lymph nodes for surgical and radiotherapeutic treatment.

B. LYMPHATIC TISSUE

1. Lymph Nodes

The oval or bean-shaped structures located along the length of lymphatic vessels are called ***lymph nodes.*** A lymph node contains a slight depression on one side called a ***hilus*** (HĪ-lus), where blood vessels and efferent lymphatic vessels leave the node. Each node is covered by a ***capsule*** of dense connective tissue that extends into the node. The capsular extensions are called ***trabeculae*** (tra-BEK-yoo;-lē; *trabecula* = little beam). Internal to the capsule is a supporting network of reticular fibers and fibroblasts. The capsule, trabeculae, and reticular fibers and fibroblasts constitute the stroma (framework) of a lymph node. The parenchyma (functioning part) of a lymph node is specialized into two regions: cortex and medulla. The outer ***cortex*** contains densely packed lymphocytes arranged in masses called ***lymphatic follicles.*** The outer rim of each follicle contains T cells. The central area of each follicle, called the ***germinal center,*** is the site where B cells proliferate into antibody-secreting plasma cells. The inner region of a lymph node is called the ***medulla.*** In the medulla, the lymphocytes are arranged in strands called ***medullary cords.*** These cords also contain macrophages and plasma cells.

Lymph flows through a node in one direction. It enters through ***afferent*** (*ad* = to; *ferre* = to carry) ***lymphatic vessels*** that enter the convex surface of the node at several points. They contain valves that open toward the node so that the lymph is directed *inward.* Once inside the node, the lymph enters the sinuses, which are a series of irregular channels. Lymph from the afferent lymphatic vessels enters the ***cortical sinuses*** just inside the capsule. From here it circulates to the ***medullary sinuses*** between the medullary cords. From these sinuses the lymph usually circulates into one or two ***efferent*** (*ex* = away) ***lymphatic vessels,*** located at the hilus of the lymph node. Efferent lymphatic vessels are wider than afferent vessels and contain valves that open away from the node to convey lymph *out* of the node.

Lymph passing from tissue spaces through lymphatic vessels on its way back to the cardiovascular system is filtered through lymph nodes. As lymph passes through the nodes it is filtered of foreign substances. These substances are trapped by the reticular fibers within the node. Then, macrophages destroy some foreign substances by phagocytosis and lymphocytes bring about destruction of others by immune responses. Plasma cells and T cells that proliferate within lymph nodes can circulate to other parts of the body.

Label the parts of a lymph node in Figure 19.2 and the various groups of lymph nodes in Figure 19.3 on page 386.

2. Tonsils

Tonsils are multiple aggregations of large lymphatic nodules embedded in a mucous membrane. The tonsils are arranged in a ring at the junction of the oral cavity and pharynx. The tonsils are situated strategically to protect against invasion of foreign substances (Figure 19.3a). The ***pharyngeal*** (fa-RIN-jē-al) ***tonsil*** is embedded in the posterior wall of the nasopharynx. The paired ***palatine*** (PAL-a-tīn) ***tonsils*** are situated in the space between the pharyngopalatine and palatoglossal arches. They are the ones commonly removed by a tonsillectomy. The ***lingual*** (LIN-gwal) ***tonsil*** is located at the base of the tongue and may also have to be removed by a tonsillectomy. Functionally, the tonsils participate in immune responses by producing lymphocytes and antibodies.

3. Spleen

The oval ***spleen*** is the largest mass of lymphatic tissue in the body (Figure 19.3a). It is situated in the left hypochondriac region between the fundus of the stomach and diaphragm.

The splenic artery and vein and the efferent lymphatic vessels pass through the hilus. Since the spleen has no afferent lymphatic vessels or lymph sinuses, it does not filter lymph. One key

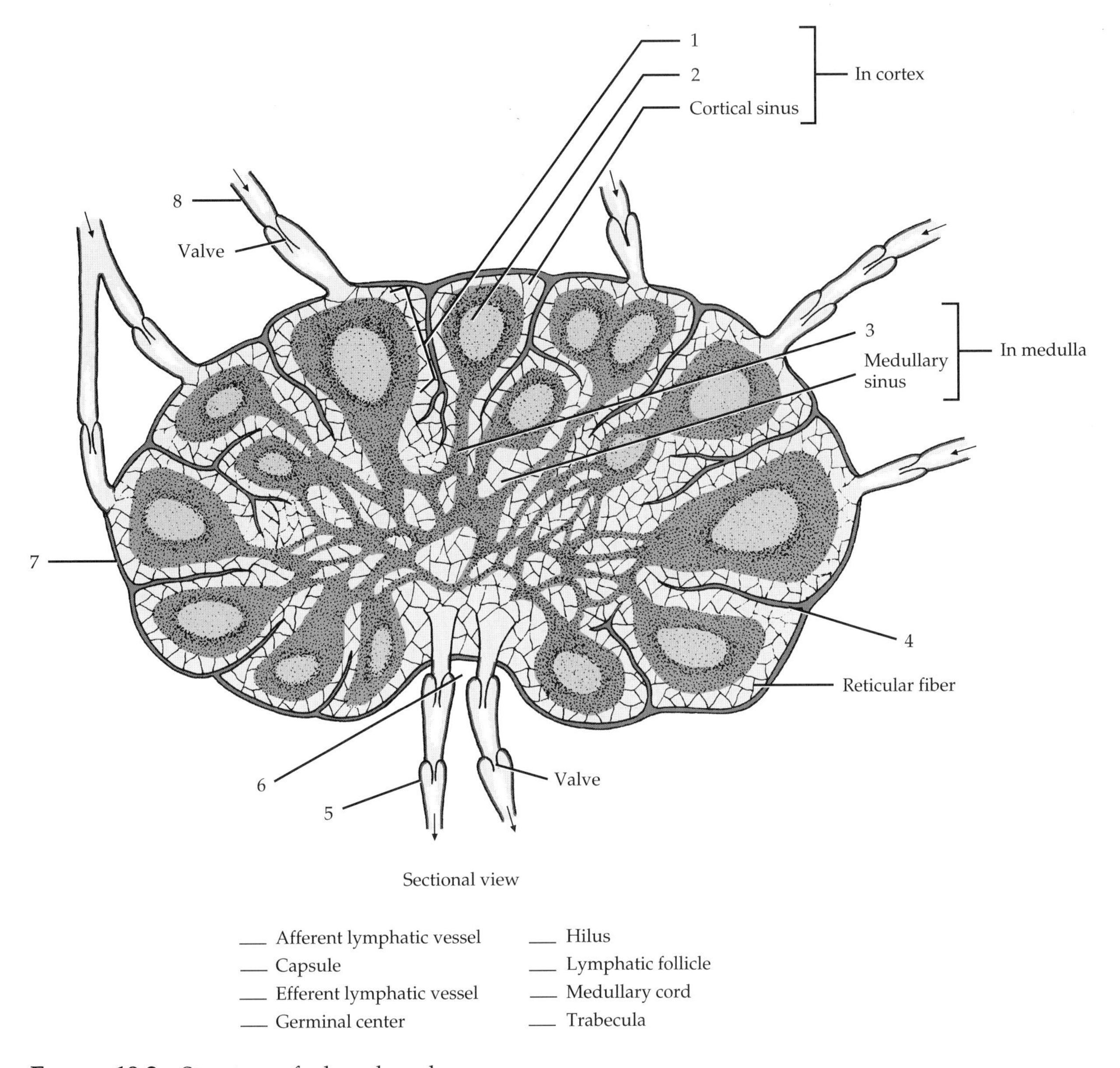

FIGURE 19.2 Structure of a lymph node.

splenic function related to immunity is the production of B cells, which develop into antibody-producing plasma cells. The spleen also phagocytizes bacteria and worn-out and damaged red blood cells and platelets. During early fetal development, the spleen participates in blood cell formation.

4. Thymus Gland

Usually a bilobed lymphatic organ, the ***thymus gland*** is located in the mediastinum, posterior to the sternum and between the lungs (Figure 19.3a). Its role in immunity is to synthesize hormones that help produce T cells that destroy invading microbes, including the AIDS (acquired immune deficiency syndrome) virus, directly or indirectly by producing various substances.

C. LYMPH CIRCULATION

When plasma is filtered by blood capillaries, it passes into the interstitial spaces; it is then known as interstitial fluid. When this fluid passes from interstitial spaces into lymph capillaries, it is called ***lymph*** (*lympha* = clear water). Lymph from lymph capillaries flows into lymphatic vessels that run toward lymph nodes. At the nodes, afferent vessels penetrate the capsules at numerous points,

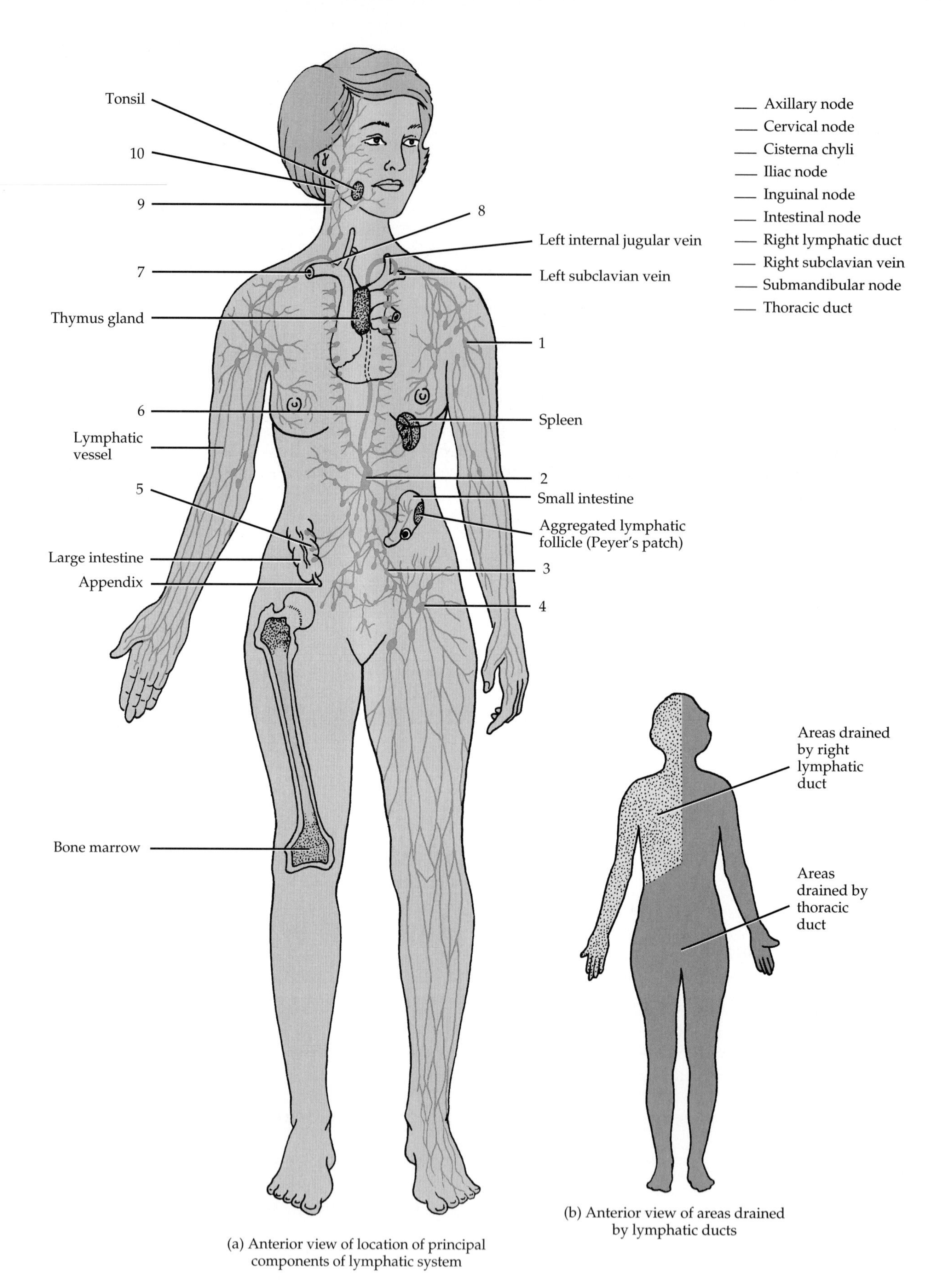

(a) Anterior view of location of principal components of lymphatic system

(b) Anterior view of areas drained by lymphatic ducts

FIGURE 19.3 Lymphatic system.

and the lymph passes through the sinuses of the nodes. Efferent vessels from the nodes unite to form ***lymph trunks.***

The principal trunks pass their lymph into two main channels, the thoracic duct and the right lymphatic duct. The ***thoracic (left lymphatic) duct*** begins as a dilation in front of the second lumbar vertebra called the ***cisterna chyli*** (sis-TER-na KĪ-lē). The thoracic duct is the main collecting duct of the lymphatic system and receives lymph from the left side of the head, neck, and chest, the left upper limb, and the entire body inferior to the ribs (Figure 19.3b).

The ***right lymphatic duct*** drains lymph from the upper right side of the body (Figure 19.3b). Ultimately, the thoracic duct empties all of its lymph into the junction of the left internal jugular vein and left subclavian vein, and the right lymphatic duct empties all of its lymph into the junction of the right internal jugular vein and right subclavian vein. Thus, lymph is drained back into the blood and the cycle repeats itself continuously.

Edema, an excessive accumulation of interstitial fluid in tissue spaces, may be caused by an obstruction, such as an infected node or a blockage of vessels, in the pathway between the lymphatic capillaries and the subclavian veins. Another cause is excessive lymph formation and increased permeability of blood capillary walls. A rise in capillary blood pressure, in which interstitial fluid is formed faster than it is passed into lymphatic vessels, also may result in edema.

D. DISSECTION OF CAT LYMPHATIC SYSTEM

CAUTION! *Please reread Section D, "Precautions Related to Dissection" at the beginning of the laboratory manual on page xi before you begin your dissection.*

Even if your cat has not been injected for examination of the lymphatic system, certain parts can be observed.

1. ***Lymphatic vessels*** The major lymphatic vessel is the ***thoracic duct,*** a brownish vessel in the left pleural cavity dorsal to the aorta. It passes deep to most of the blood vessels at the base of the neck and enters the left external jugular vein near the entrance of the internal jugular vein (Figure 19.4). Using the diagram as a guide, see if you can find the ***right lymphatic duct.***
2. ***Lymph nodes*** These are small ovoid or round masses of lymphatic tissue connected by lymph vessels. You may be able to identify lymph nodes in the axillary and inguinal regions, at the base of the neck, at the angle of the jaws, and in the mesentery of the small intestine (Figure 19.4).
3. ***Lymphatic organs*** Lymphatic organs include the ***thymus gland, tonsils,*** and ***spleen.***

ANSWER THE LABORATORY REPORT QUESTIONS AT THE END OF THE EXERCISE.

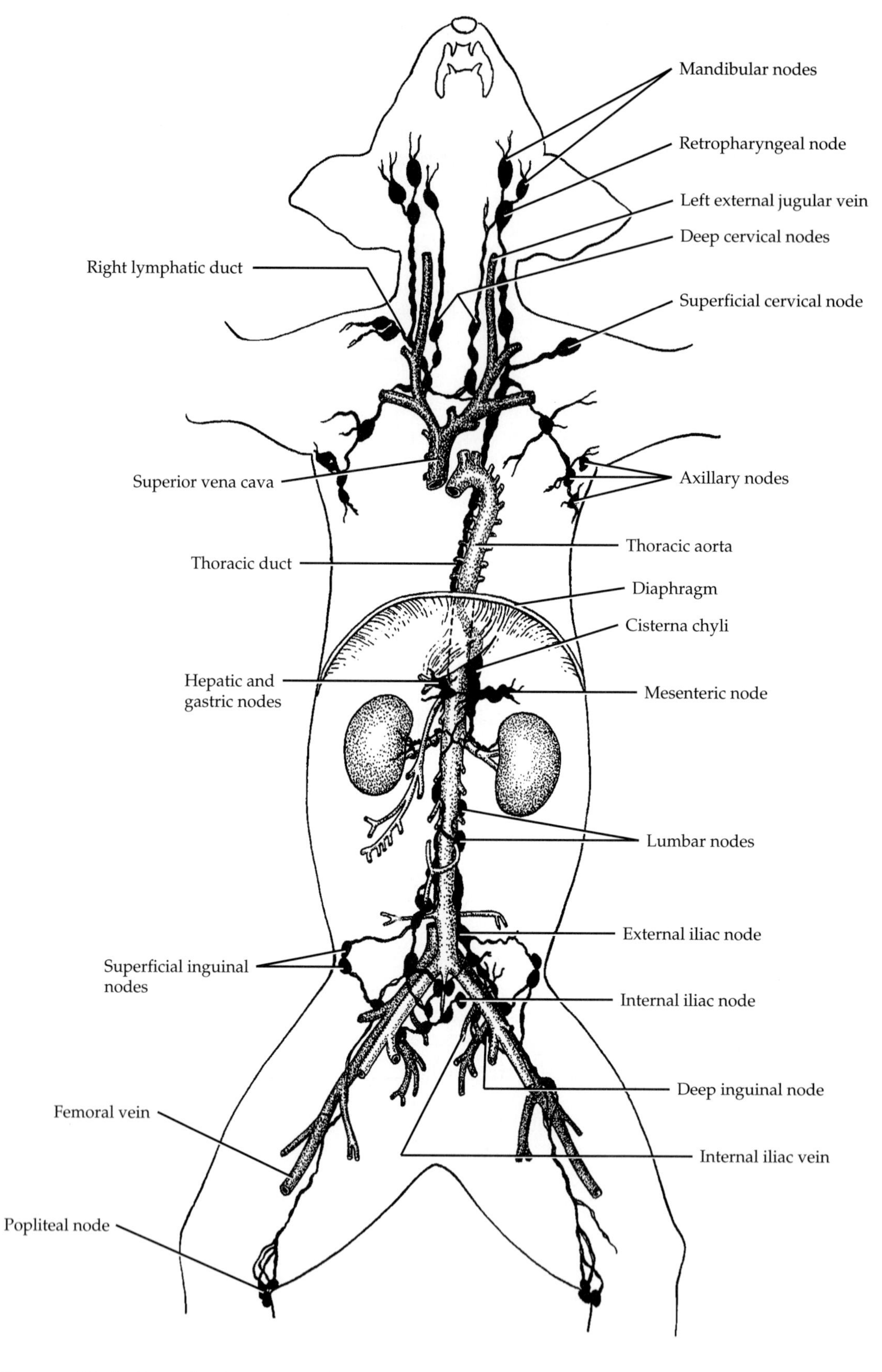

FIGURE 19.4 Lymphatic system of cat.

LABORATORY REPORT QUESTIONS

Lymphatic and Immune System 19

Student ______________________ **Date** ______________________

Laboratory Section ______________________ **Score/Grade** ______________________

PART 1. Completion

1. Small masses of lymphatic tissue located along the length of lymphatic vessels are called ______________________.
2. Lymphatic vessels have thinner walls than veins, but resemble veins in that they also have ______________________.
3. All lymphatic vessels converge, get larger, and eventually merge into two main channels, the thoracic duct and the ______________________.
4. Cells in the lymph nodes that carry on phagocytosis are ______________________.
5. Lymph is conveyed out of lymph nodes in ______________________ vessels.
6. B cells in lymph nodes produce certain cells that are responsible for the production of antibodies. These cells are called ______________________.
7. The x-ray examination of lymphatic vessels and lymph organs after they are filled with a radiopaque substance is called ______________________.
8. This x-ray examination is useful in detecting edema and ______________________.
9. The largest mass of lymphatic tissue is the ______________________.
10. The main collecting duct of the lymphatic system is the ______________________ duct.
11. ______________________ are the areas within a lymph node that contain T and B cells.
12. The ______________________ tonsils are commonly removed by a tonsillectomy.

Respiratory System 20

Cells continually use oxygen (O_2) for the metabolic reactions that release energy from nutrient molecules and produce ATP. At the same time, these reactions release carbon dioxide (CO_2). Since an excessive amount of CO_2 produces acidity that is toxic to cells, the excess CO_2 must be eliminated quickly and efficiently. The two systems that cooperate to supply O_2 and eliminate CO_2 are the cardiovascular system and the respiratory system. The respiratory system provides for gas exchange, intake of O_2, and elimination of CO_2, whereas the cardiovascular system transports the gases in the blood between the lungs and the cells. Failure of either system has the same effect on the body: disruption of homeostasis and rapid death of cells from oxygen starvation and buildup of waste products. In addition to functioning in gas exchange, the respiratory system also contains receptors for the sense of smell, filters inspired air, produces sounds, and helps eliminate wastes.

The *respiratory system* consists of the nose, pharynx (throat), larynx (voice box), trachea (windpipe), bronchi, and lungs (Figure 20.1). Structurally, the respiratory system consists of two portions. (1) The term *upper respiratory system* refers to the nose, pharynx, and associated structures. (2) The *lower respiratory system* refers to the larynx, trachea, bronchi, and lungs. Functionally, the respiratory system also consists of two portions. (1) The *conducting portion* consists of a series of interconnecting cavities and tubes—nose, pharynx, larynx, trachea, bronchi, and terminal bronchioles—that conduct air into the lungs. (2) The *respiratory portion* consists of those portions of the respiratory system where the exchange of gases occurs—respiratory bronchioles, alveolar ducts, alveolar sacs, and alveoli.

Respiration is the exchange of gases between the atmosphere, blood, and cells. It takes place in three basic steps:

1. *Pulmonary ventilation* The first process, *pulmonary* (*pulmo* = lung) *ventilation,* or breathing, is the inspiration (inflow) and expiration (outflow) of air between the atmosphere and the lungs.
2. *External (pulmonary) respiration* This is the exchange of gases between the air spaces of the lungs and blood in pulmonary capillaries. The blood gains O_2 and loses CO_2.
3. *Internal (tissue) respiration* The exchange of gases between blood in systemic capillaries and tissue cells is known as internal (tissue) respiration. The blood loses O_2 and gains CO_2. Within cells, the metabolic reactions that consume O_2 and produce CO_2 during production of ATP are termed *cellular respiration.*

Using your textbook, charts, or models for reference, label Figure 20.1.

A. ORGANS OF THE RESPIRATORY SYSTEM

1. Nose

The *nose* has an external portion and an internal portion inside the skull. The external portion consists of a supporting framework of bone and hyaline cartilage covered with muscle and skin and lined by mucous membrane. The bridge of the nose is formed by the nasal bones, which hold it in a fixed position. Because it has a framework of pliable hyaline cartilage, the rest of the external nose is somewhat flexible. On the undersurface of the external nose are two openings called the *external nares* (NA-rēz; singular is *naris*), or *nostrils.* The interior structures of the nose are specialized for three functions: (1) incoming air is warmed, moistened, and filtered; (2) olfactory stimuli are received; and (3) large, hollow resonating chambers modify speech sounds.

The internal portion of the nose is a large cavity in the skull that lies inferior to the anterior cranium and superior to the mouth. Anteriorly, the

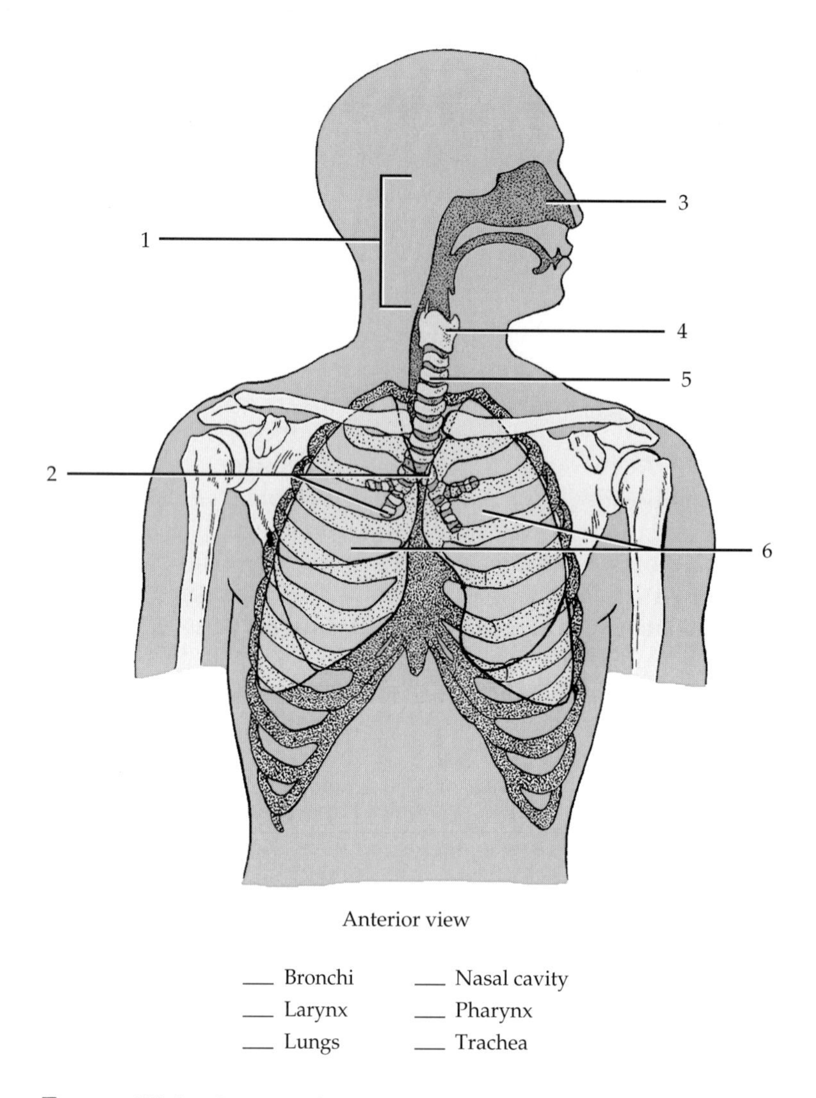

FIGURE 20.1 Organs of respiratory system.

internal nose merges with the external nose, and posteriorly it communicates with the pharynx through two openings called the ***internal nares (choanae).*** Ducts from the paranasal sinuses (frontal, sphenoidal, maxillary, and ethmoidal) and the nasolacrimal ducts also open into the internal nose. The lateral walls of the internal nose are formed by the ethmoid, maxillae, lacrimal, palatine, and inferior nasal conchae bones. The ethmoid also forms the roof of the internal nose. The floor of the internal nose is formed mostly by the palatine bones and palatine processes of the maxillae, which together comprise the hard palate.

The inside of both the external and internal nose is called the ***nasal cavity.*** It is divided into right and left sides by a vertical partition called the ***nasal septum.*** The anterior portion of the septum consists primarily of hyaline cartilage. The remainder is formed by the vomer, perpendicular plate of the ethmoid, maxillae, and palatine bones (see Figure 20.2). The anterior portion of the nasal cavity, just inside the nostrils, is called the ***vestibule*** and is surrounded by cartilage. The superior nasal cavity is surrounded by bone.

When air enters the nostrils, it passes first through the vestibule. The vestibule is lined by skin containing coarse hairs that filter out large dust particles. The air then passes into the superior nasal cavity. Three shelves formed by projections of the superior, middle, and inferior nasal conchae extend out of each lateral wall of the cavity. The conchae, almost reaching the septum, subdivide each side of the nasal cavity into a series of groovelike passageways—the ***superior, middle,*** and ***inferior meatuses*** (mē-Ā-tes-ez; *meatus* = passage; singular is ***meatus***). Mucous membrane lines the cavity and its shelves.

The olfactory receptors lie in the membrane lining the superior nasal conchae and adjacent septum.

This region is called the ***olfactory epithelium.*** Inferior to the olfactory epithelium, the mucous membrane contains capillaries and pseudostratified ciliated columnar epithelium with many goblet cells. As the air whirls around the conchae and meatuses, it is warmed by blood in the capillaries. Mucus secreted by the goblet cells moistens the air and traps dust particles. Drainage from the nasolacrimal ducts and perhaps secretions from the paranasal sinuses also help moisten the air. The

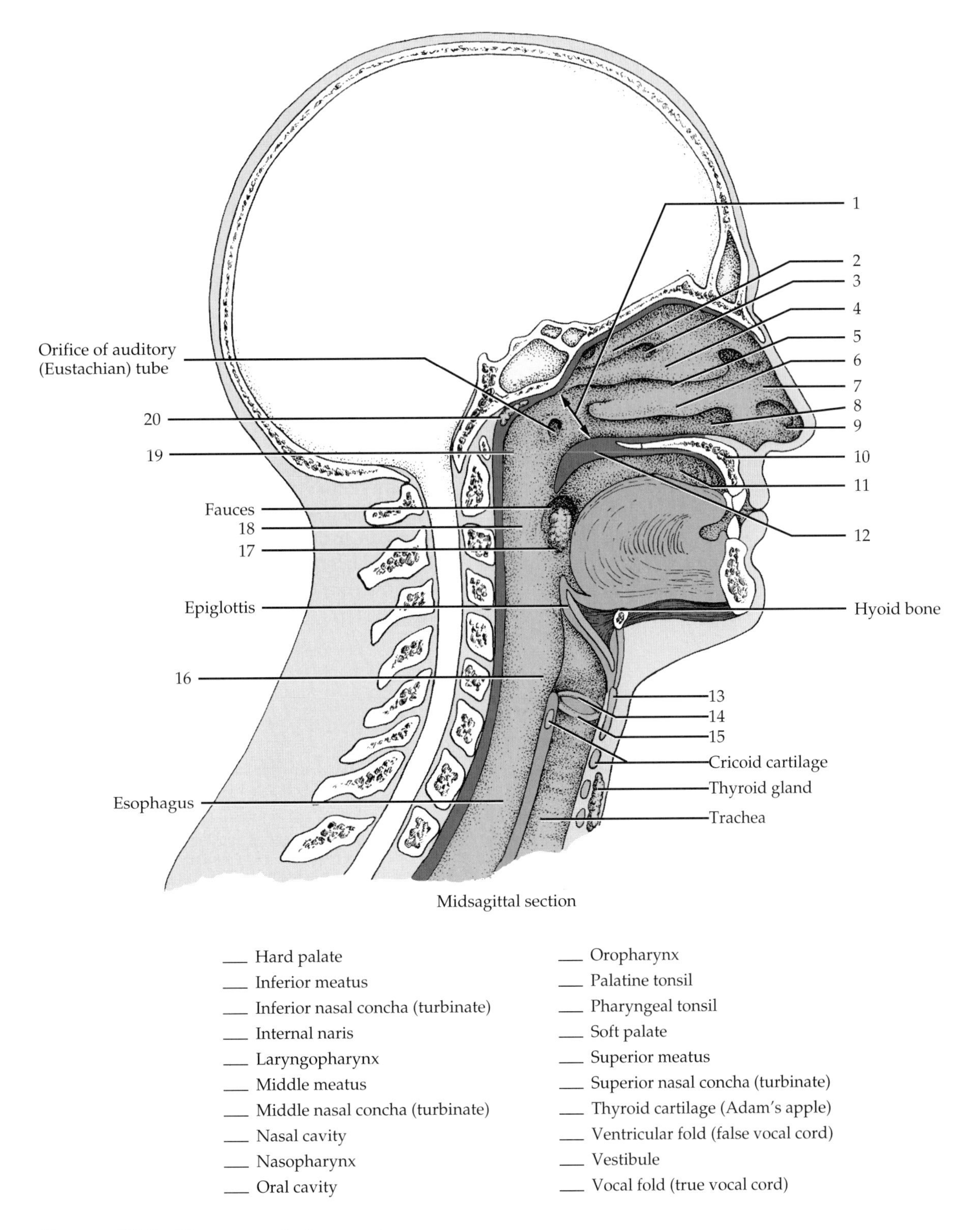

FIGURE 20.2 Upper respiratory system.

cilia move the mucus–dust particles toward the pharynx so that they can be eliminated from the respiratory tract by swallowing or expectoration (spitting). Substances in cigarette smoke inhibit movement of cilia. When this happens, only coughing can remove mucus–dust particles from the airways. This is one reason that smokers cough often.

Label the hard palate, inferior meatus, inferior nasal concha, internal naris, middle meatus, middle nasal concha, nasal cavity, oral cavity, soft palate, superior meatus, superior nasal concha, and vestibule in Figure 20.2.

The surface anatomy of the nose is shown in Figure 20.3.

2. Pharynx

The ***pharynx*** (FAR-inks) (throat) is a somewhat funnel-shaped tube about 13 cm (5 in.) long that starts at the internal nares and extends to the level of the cricoid cartilage, the most inferior cartilage of the larynx (voice box). Lying posterior to the nasal and oral cavities and just anterior to the cervical vertebrae, the pharynx is a passageway for air and food, a resonating chamber for speech sounds, and a housing for tonsils.

The pharynx is composed of a superior portion, called the ***nasopharynx,*** an intermediate portion, the ***oropharynx,*** and an inferior portion, the ***laryngopharynx*** (la-rin'-gō-FAR-inks) or ***hypopharynx.*** The nasopharynx consists of ***pseudostratified ciliated columnar epithelium*** and has four openings in its wall: two ***internal nares*** plus two openings into the ***auditory (Eustachian) tubes.*** The nasopharynx also contains the ***pharyngeal tonsil (adenoid).*** The oropharynx is lined by ***nonkeratinized stratified squamous epithelium*** and receives one opening: the ***fauces*** (FAW-sēz). The oropharynx contains the ***palatine*** and ***lingual tonsils.*** The laryngopharynx is also lined by ***nonkeratinized stratified squamous epithelium*** and becomes continuous with the esophagus posteriorly and the larynx anteriorly.

Label the laryngopharynx, nasopharynx, oropharynx, palatine tonsil, and pharyngeal tonsil in Figure 20.2.

3. Larynx

The ***larynx*** (LAIR-inks), or voice box, is a short passageway connecting the laryngopharynx with the trachea. Its wall is composed of nine pieces of cartilage.

a. ***Thyroid cartilage (Adam's apple)*** Large anterior piece that gives larynx its triangular shape.

b. ***Epiglottis*** (*epi* = above; *glotta* = tongue) Leaf-shaped cartilage on top of larynx that closes off the larynx so that foods and liquids are routed into the esophagus and kept out of the respiratory system.

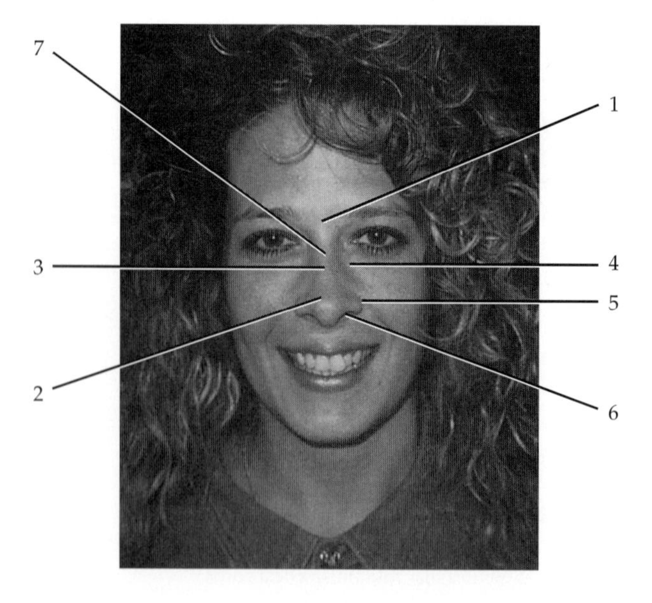

1. **Root.** Superior attachment of nose at forehead located between eyes.
2. **Apex.** Tip of nose.
3. **Dorsum nasi.** Rounded anterior border connecting root and apex; in profile, may be straight, convex, concave, or wavy.
4. **Nasofacial angle.** Point at which side of nose blends with tissues of face.
5. **Ala.** Convex flared portion of inferior lateral surface.
6. **External nares.** External openings into nose (nostrils).
7. **Bridge.** Superior portion of dorsum nasi, superficial to nasal bones.

FIGURE 20.3 Surface anatomy of nose.

c. ***Cricoid*** (KRĪ-koyd; *krikos* = ring) ***cartilage*** Ring of cartilage forming the inferior portion of the larynx that is attached to the first ring of tracheal cartilage.

d. ***Arytenoid*** (ar-i-TĒ-noyd; *arytaina* = ladle) ***cartilages*** Paired, pyramid-shaped cartilages at superior border of cricoid cartilage that attach vocal folds to the intrinsic pharyngeal muscles.

e. ***Corniculate*** (kor-NIK-yoo-lāt; *corniculate* = shaped like a small horn) ***cartilages*** Paired, horn-shaped cartilages at apex of arytenoid cartilages.

f. ***Cuneiform*** (kyoo-NĒ-i-form; *cuneus* = wedge) ***cartilages*** Paired, club-shaped cartilages anterior to the corniculate cartilages.

With the aid of your textbook, label the laryngeal cartilages shown in Figure 20.4. Also label the thyroid cartilage in Figure 20.2.

The mucous membrane of the larynx is arranged into two pairs of folds, a superior pair called the ***ventricular folds (false vocal cords)*** and an inferior pair called the ***vocal folds (true vocal cords)***. The space between the vocal folds when they are apart is called the ***rima glottidis***. Together, the vocal folds and rima glottidis are referred to as the ***glottis***. Movement of the vocal folds produces sounds; variations in pitch result from (1) varying degrees of tension and (2) varying lengths in males and females.

With the aid of your textbook, label the ventricular folds, vocal folds, and rima glottidis in Figure 20.5. Also label the ventricular and vocal folds in Figure 20.2.

4. Trachea

The ***trachea*** (TRĀ-kē-a) (windpipe) is a tubular air passageway about 12 cm (4½ in.) in length and 2.5 cm (1 in.) in diameter. It lies anterior to the esophagus and, at its inferior end (fifth thoracic vertebra), divides into right and left primary bronchi

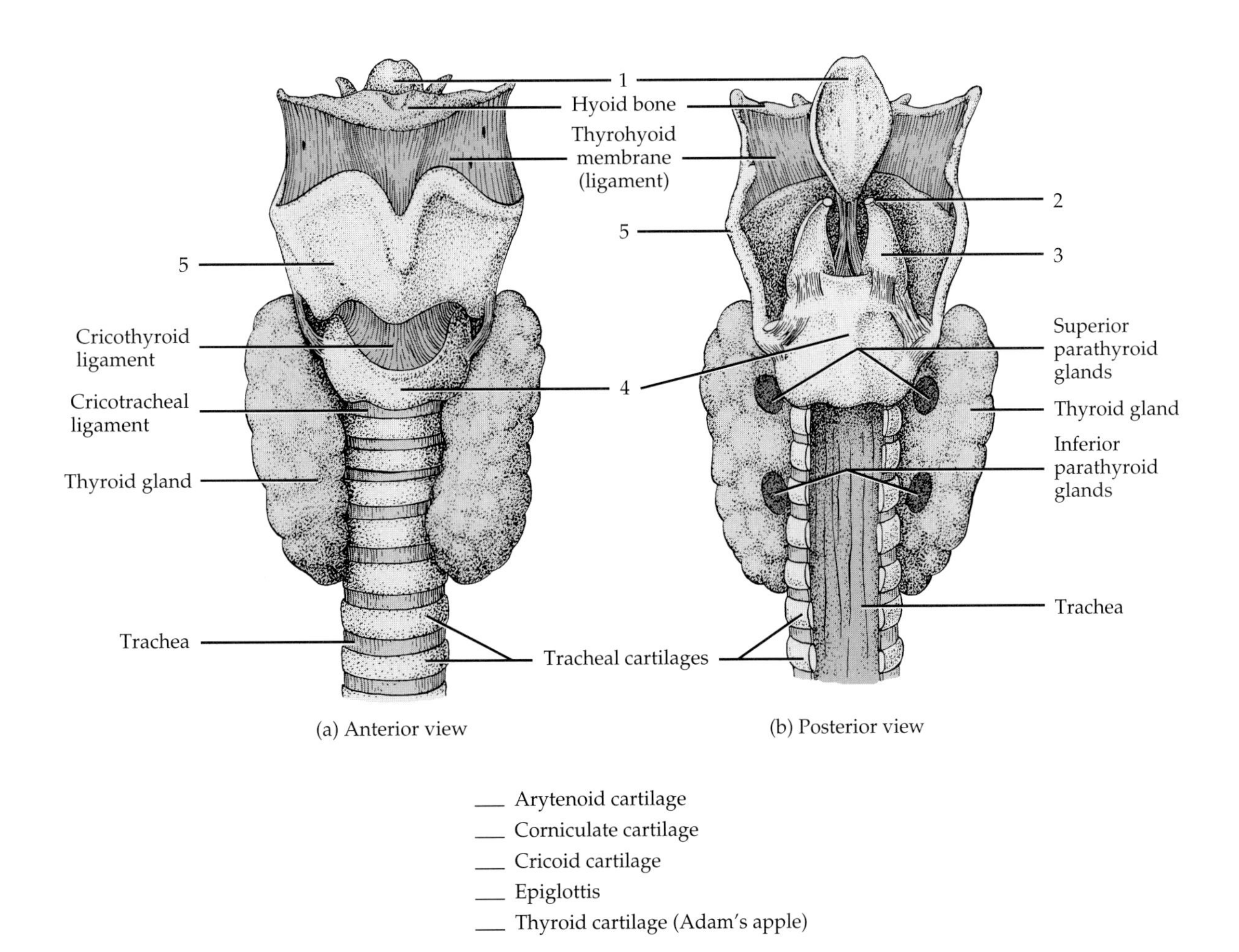

FIGURE 20.4 Larynx.

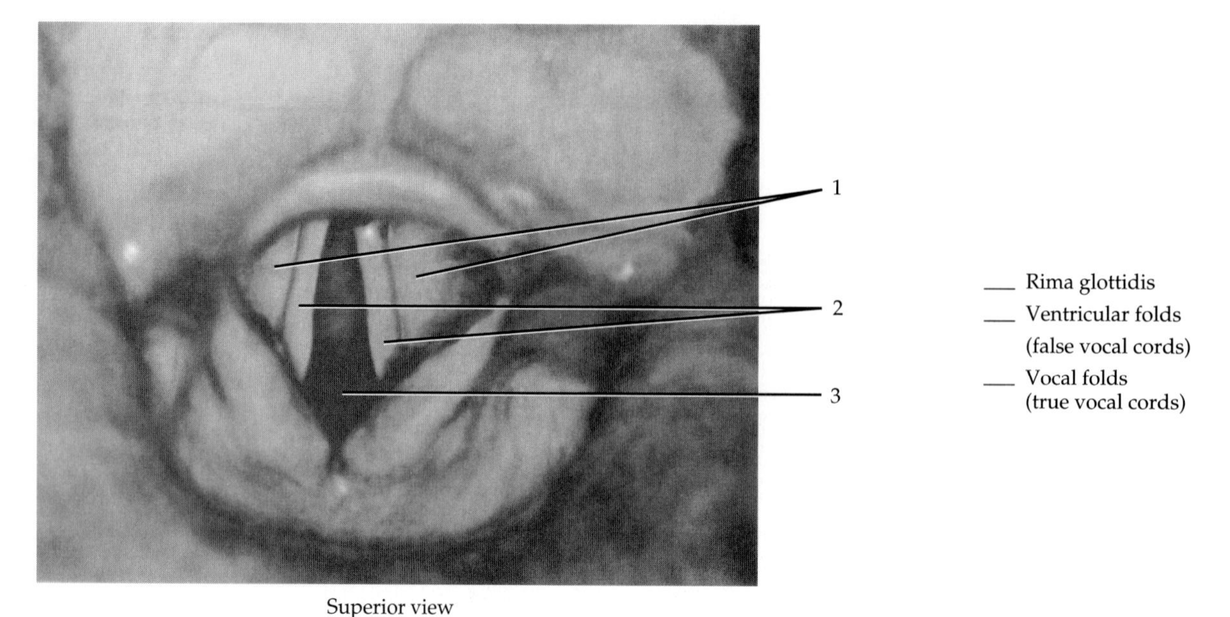

FIGURE 20.5 Photograph of larynx.

(Figure 20.6). The epithelium of the trachea consists of ***pseudostratified ciliated columnar epithelium.*** This epithelium contains ciliated columnar cells, goblet cells, and basal cells. The epithelium offers the same protection against dust as the membrane lining the nasal cavity and larynx.

Obtain a prepared slide of pseudostratified ciliated columnar epithelium from the trachea, and, with the aid of your textbook, label the ciliated columnar cells, cilia, goblet cells, and basal cells in Figure 20.7 on page 398.

The trachea consists of smooth muscle, elastic connective tissue, and incomplete ***rings of cartilage*** (hyaline) shaped like a series of letter Cs. The open ends of the Cs are held together by the ***trachealis muscle.*** The cartilage provides a rigid support so that the tracheal wall does not collapse inward and obstruct the air passageway, and, because the open parts of the Cs face the esophagus, the latter can expand into the trachea during swallowing. If the trachea should become obstructed, a ***tracheostomy*** (trā-kē-OS-tō-mē) may be performed. Another method of opening the air passageway is called ***intubation,*** in which a tube is passed into the mouth and down through the larynx and the trachea.

5. Bronchi

The trachea terminates by dividing into a ***right primary bronchus*** (BRONG-kus), going to the right lung, and a ***left primary bronchus,*** going to the left lung. They continue dividing in the lungs into smaller bronchi, the ***secondary (lobar) bronchi*** (BRONG-kē), one for each lobe of the lung. These bronchi, in turn, continue dividing into still smaller bronchi called ***tertiary (segmental) bronchi,*** which divide into ***bronchioles.*** The next division is into even smaller tubes called ***terminal bronchioles.*** This entire branching structure of the trachea down to the level of terminal bronchioles is commonly referred to as the ***bronchial tree.***

Label Figure 20.6.

Bronchography (brong-KOG-ra-fē) is a technique for examining the bronchial tree. With this procedure, an intratracheal catheter is passed transorally or transnasally through the rima glottidis into the trachea. Then an opaque contrast medium is introduced into the trachea and distributed through the bronchial branches. Radiographs of the chest in various positions are taken and the developed film, called ***bronchogram*** (BRONG-kō-gram), provides a picture of the bronchial tree.

6. Lungs

The ***lungs*** (*lunge* = light, since the lungs float) are paired, cone-shaped organs lying in the thoracic cavity (see Figure 20.1). The ***pleural*** (*pleura* = side) ***membrane*** encloses and protects each lung. Whereas the *superficial* ***parietal pleura*** lines the wall of the thoracic cavity, the *deep* ***visceral pleura*** covers the lungs; the potential space between parietal and visceral pleurae, the ***pleural cavity,*** contains a lubricating fluid to reduce friction as the lungs expand and recoil.

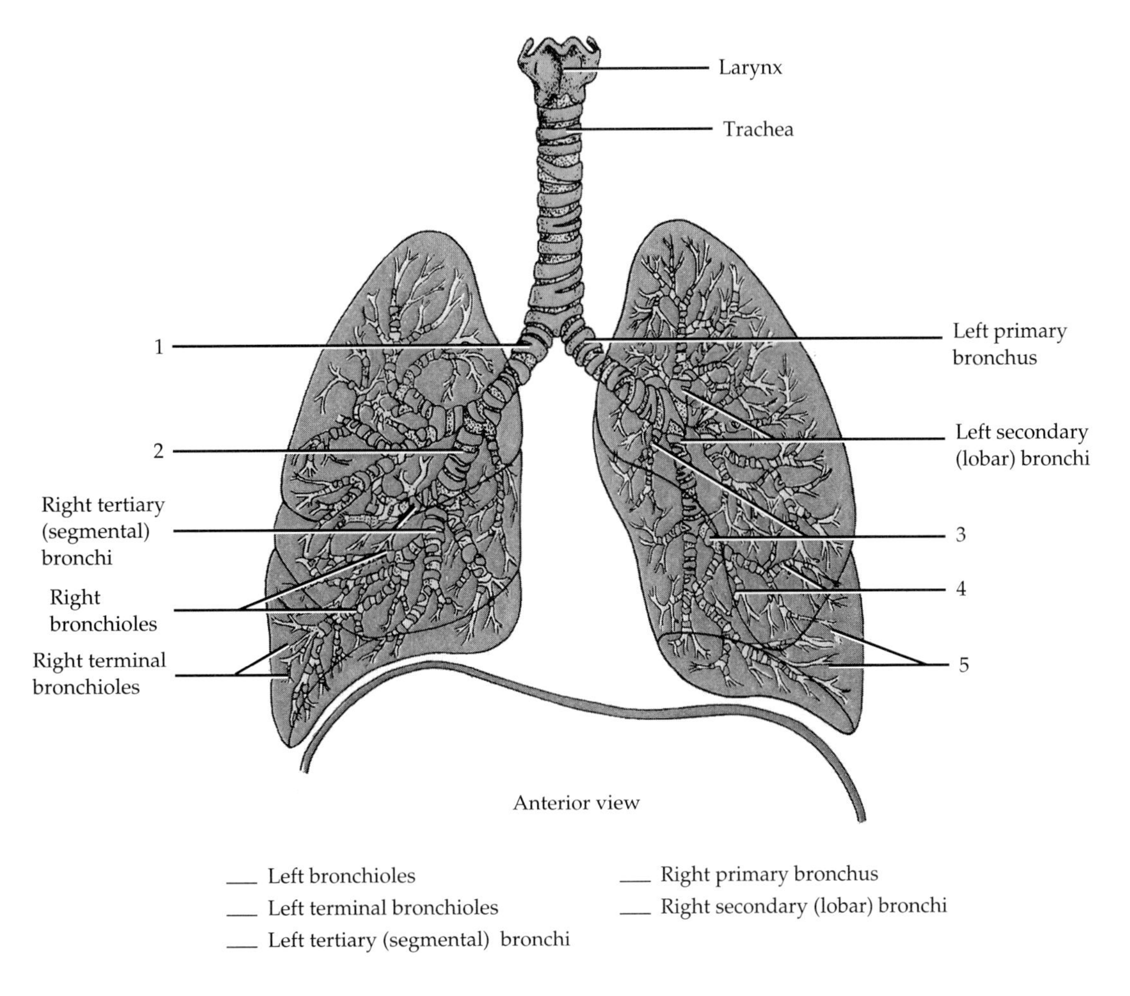

FIGURE 20.6 Air passageways to the lungs. Shown is the bronchial tree in relationship to lungs.

Major surface features of the lungs include

a. ***Base*** Broad inferior portion resting on diaphragm.
b. ***Apex*** Narrow superior portion just superior to clavicles.
c. ***Costal surface*** Surface lying against ribs.
d. ***Mediastinal surface*** Medial surface.
e. ***Hilus*** Region in mediastinal surface through which bronchial tubes, blood vessels, lymphatic vessels, and nerves enter and exit the lung.
f. ***Cardiac notch*** Medial concavity in left lung in which heart lies.

Each lung is divided into ***lobes*** by one or more ***fissures***. The right lung has three lobes, ***superior***, ***middle***, and ***inferior***; the left lung has two lobes, ***superior*** and ***inferior***. The ***horizontal fissure*** separates the superior lobe from the middle lobe in the right lung; an ***oblique fissure*** separates the middle lobe from the inferior lobe in the right lung and the superior lobe from the inferior lobe in the left lung. Each lobe receives its own secondary bronchus.

Using your textbook as a reference, label Figure 20.8a on page 399.

Each lobe of a lung is divided into regions called ***bronchopulmonary segments***, each supplied by a tertiary bronchus. Each bronchopulmonary segment is composed of many smaller compartments called ***lobules***. Each lobule is wrapped in elastic connective tissue and contains a lymphatic vessel, arteriole, venule, and branch from a terminal bronchiole. Terminal bronchioles divide into ***respiratory bronchioles***, which, in turn, divide into several ***alveolar*** (al-VĒ-ō-lar) ***ducts***. Around the circumference of alveolar ducts are numerous alveoli and alveolar sacs. ***Alveoli*** (al-VĒ-ō-lī) are cup-shaped outpouchings lined by epithelium and supported by a thin elastic membrane. The singular is ***alveolus***. ***Alveolar sacs*** are two or more alveoli that share a common opening. Over

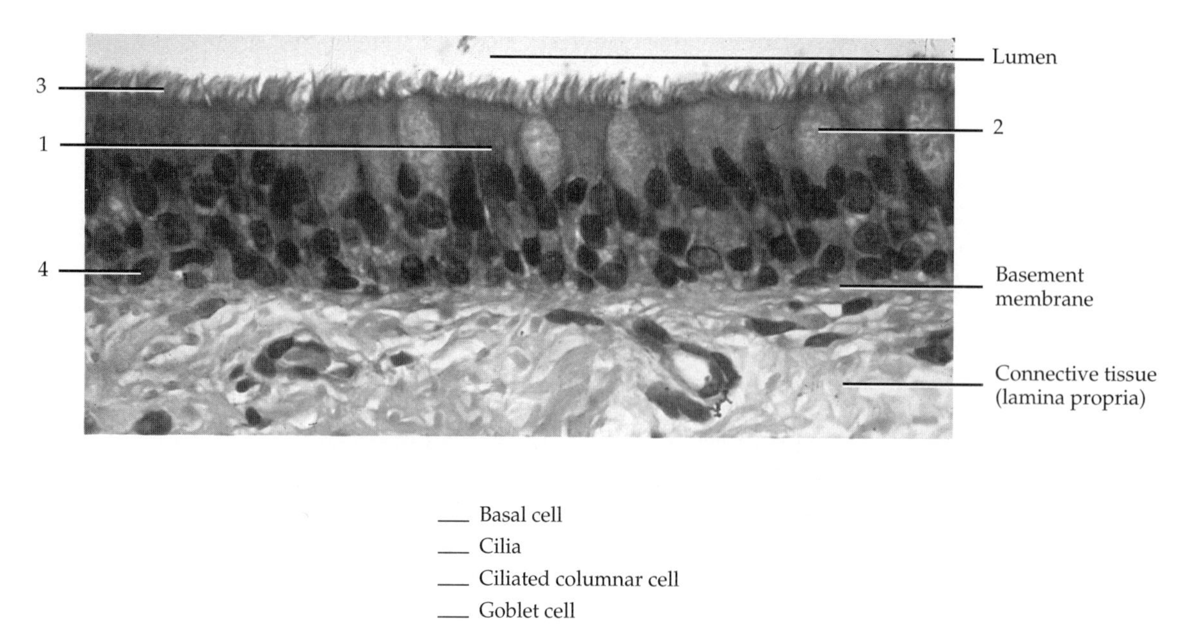

FIGURE 20.7 Histology of trachea.

the alveoli, an arteriole and venule disperse into a network of capillaries. Gas is exchanged between the lungs and blood by diffusion across the alveolar and the capillary walls.

Using your textbook as a reference, label Figure 20.8b.

Each alveolus consists of

a. ***Type I alveolar (squamous pulmonary epithelial) cells*** Cells that form a continuous lining of the alveolar wall, except for occasional type II alveolar (septal) cells.
b. ***Type II alveolar (septal) cells*** Cuboidal cells dispersed among type I alveolar cells that secrete a phospholipid and lipoprotein mixture called ***surfactant*** (sur-FAK-tant), a surface tension-lowering agent.
c. ***Alveolar macrophages (dust cells)*** Phagocytic cells that remove fine dust particles and other debris from the alveolar spaces.

Obtain a slide of normal lung tissue and examine it under high power. Using your textbook as a reference, see if you can identify a terminal bronchiole, respiratory bronchiole, alveolar duct, alveolar sac, and alveoli.

If available, examine several pathological slides of lung tissue, such as slides that show emphysema and lung cancer. Compare your observations to the normal lung tissue.

The exchange of respiratory gases between the lungs and blood takes place by diffusion across the alveolar and capillary walls. This membrane, through which the respiratory gases move, is collectively known as the ***alveolar-capillary (respiratory) membrane*** (Figure 20.9 on page 400). It consists of

1. A layer of type I alveolar (squamous pulmonary epithelial) cells with type II alveolar (septal) cells and alveolar macrophages (dust cells) that constitute the alveolar (epithelial) wall.
2. An epithelial basement membrane underneath the alveolar wall.
3. A capillary basement membrane that is often fused to the epithelial basement membrane.
4. The endothelial cells of the capillary.

Label the components of the alveolar-capillary (respiratory) membrane in Figure 20.9.

Examine the scanning electron micrograph of the cells of the alveolar wall in Figure 20.10 on page 400.

B. DISSECTION OF CAT RESPIRATORY SYSTEM

CAUTION! *Please reread Section D, "Precautions Related to Dissection" at the beginning of the laboratory manual on page xi before you begin your dissection.*

Air enters the respiratory system by one of two routes, the nose or the mouth. Air entering the external nares (nostrils) passes into the nasal cavity,

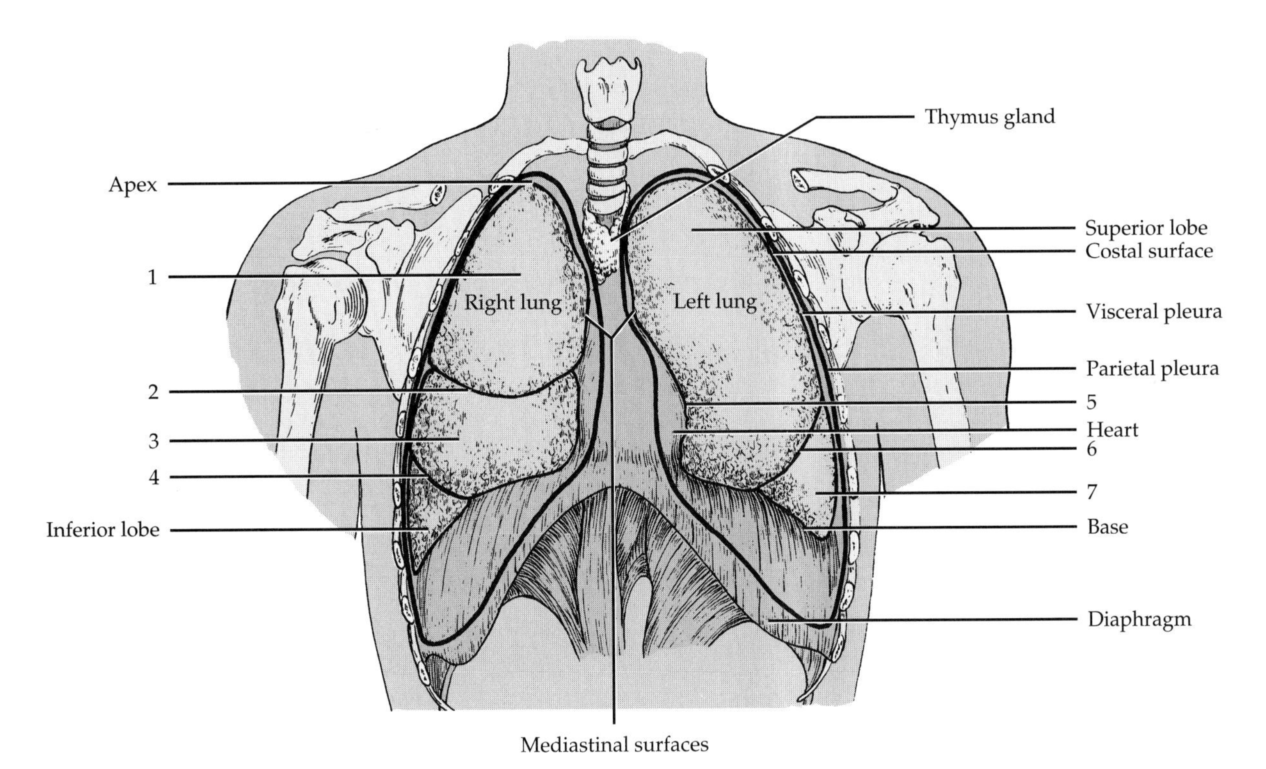

(a) Coverings and external anatomy in anterior view

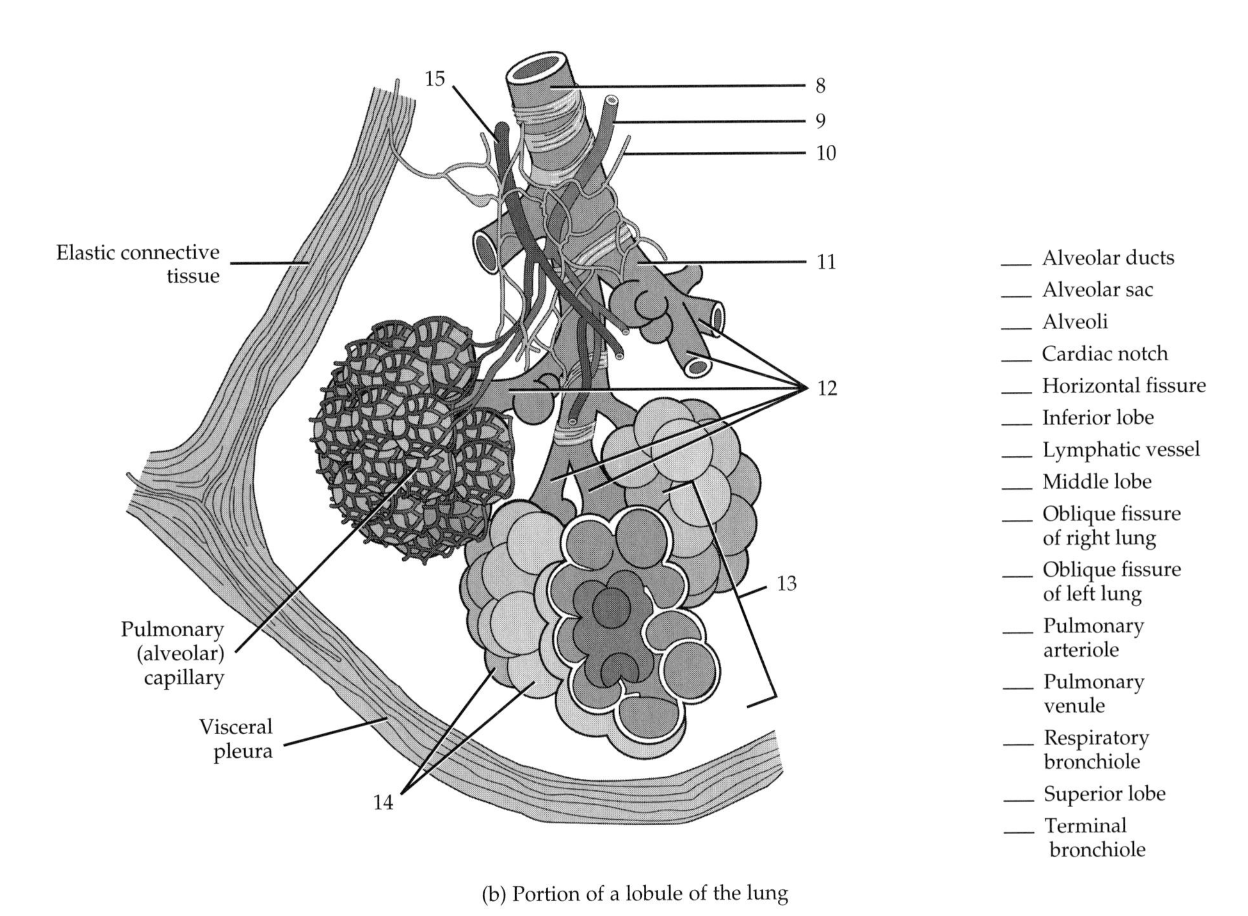

(b) Portion of a lobule of the lung

FIGURE 20.8 Lungs.

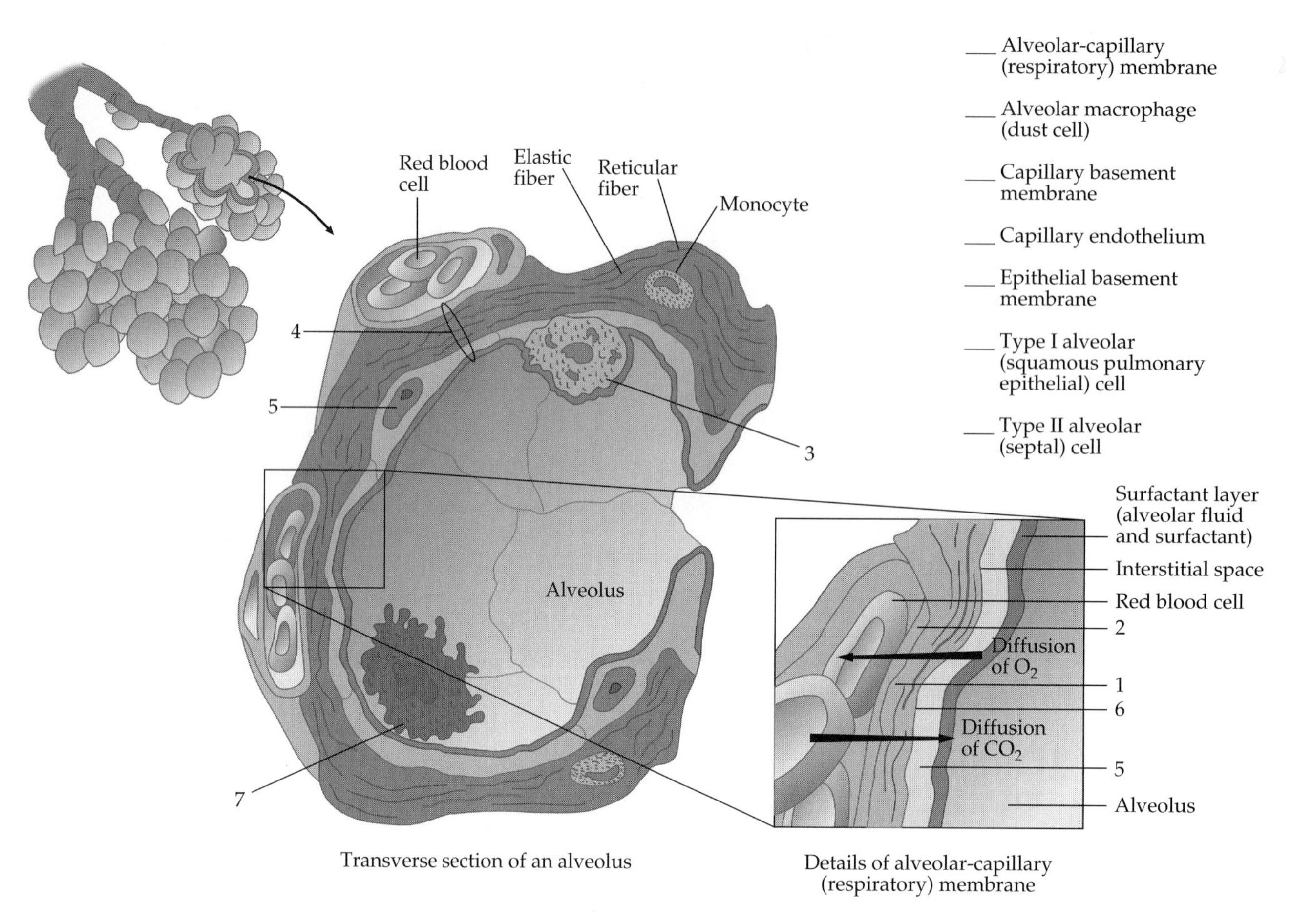

FIGURE 20.9 Alveolar-capillary (respiratory) membrane.

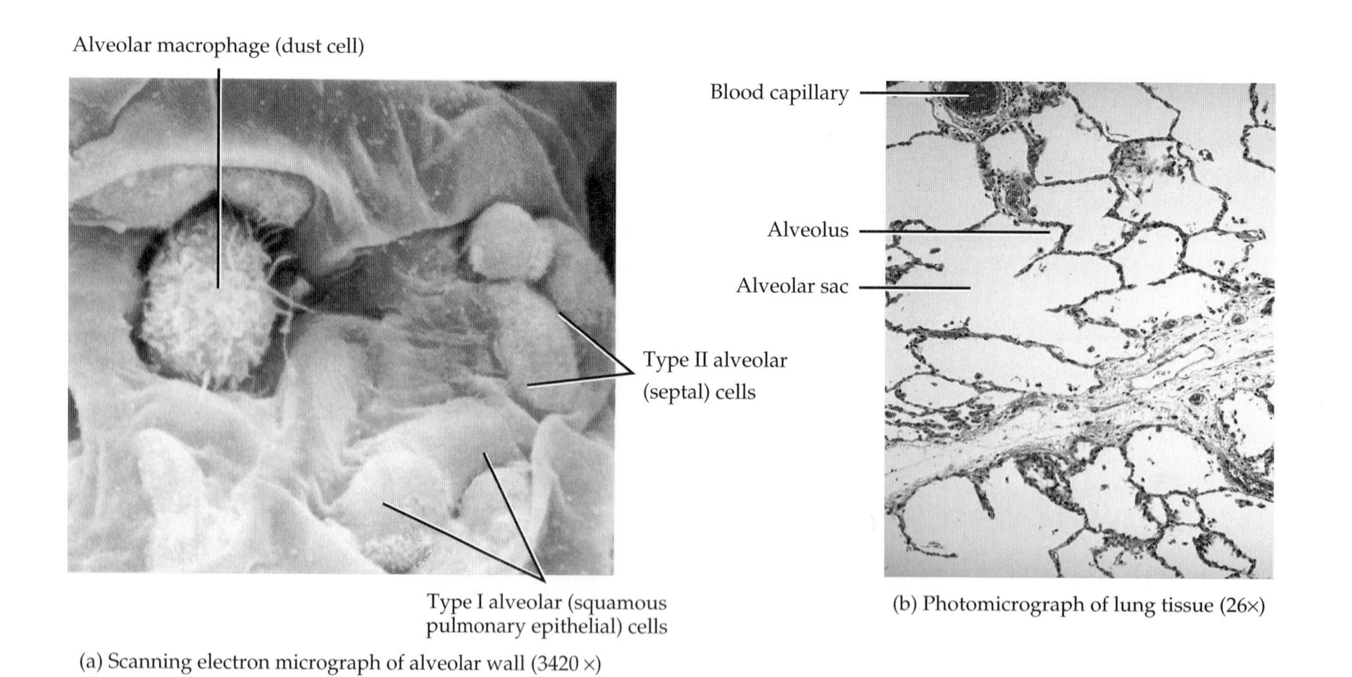

FIGURE 20.10 Histology of lung tissue.

through the internal nares (choanae), and into the nasopharynx, oropharynx, and laryngopharynx. From here it enters the larynx. Air entering the mouth passes through the vestibule, oral cavity, fauces, subdivisions of the pharynx, and into the larynx. The mouth and pharynx of the cat will be studied in detail in Exercise 21, Digestive System.

PROCEDURE

1. Larynx

1. The ***larynx*** (voice box) consists of five cartilages (Figure 20.11).
2. Expose the laryngeal cartilages by stripping off the surrounding tissues on the dorsal, ventral, and left lateral surfaces.
3. Identify the following parts of the larynx:
   a. ***Thyroid cartilage*** Forms most of the ventral and lateral walls.
   b. ***Cricoid cartilage*** Caudal to the thyroid cartilage; its expanded dorsal portion forms most of the dorsal wall of the larynx.
   c. ***Arytenoid cartilages*** Paired, small, triangular cartilages cranial to dorsal part of cricoid cartilage. These are best seen if the larynx is separated from the esophagus.
   d. ***Epiglottis*** Projects cranially from the thyroid cartilage to which it is attached; guards the opening into the larynx. To examine the cranial portion of the larynx, make a midventral incision through the larynx and cranial part of the trachea. Be careful not to damage the blood vessels on either side of the trachea. Spread the cut edges back and identify the following structures.
   e. ***False vocal cords*** Cranial pair of mucous membranes extending across the larynx from the arytenoid cartilages to the base of the epiglottis.
   f. ***True vocal cords*** Caudal, larger pair of mucous membranes extending across the larynx from the arytenoid cartilages to the thyroid cartilage; important in sound production.
   g. ***Rima glottidis*** Space between the free margins of the true vocal cords.
   h. ***Tracheal cartilages*** C-shaped, dorsally incomplete rings of cartilage that support the trachea and prevent it from collapsing.

2. Trachea

1. The ***trachea*** (windpipe) is a tube that extends from the larynx and bifurcates into the bronchial tubes (Figure 20.12a). Note the tracheal cartilages.
2. On either side of the trachea near the larynx you can locate the dark lobes of the ***thyroid gland.*** This gland belongs to the endocrine system. Its lobes are connected by a ventral strip of thyroid tissue called the ***isthmus.***
3. Also on either side of the trachea you can identify the large ***common carotid arteries,*** the smaller ***internal jugular veins,*** and the white,

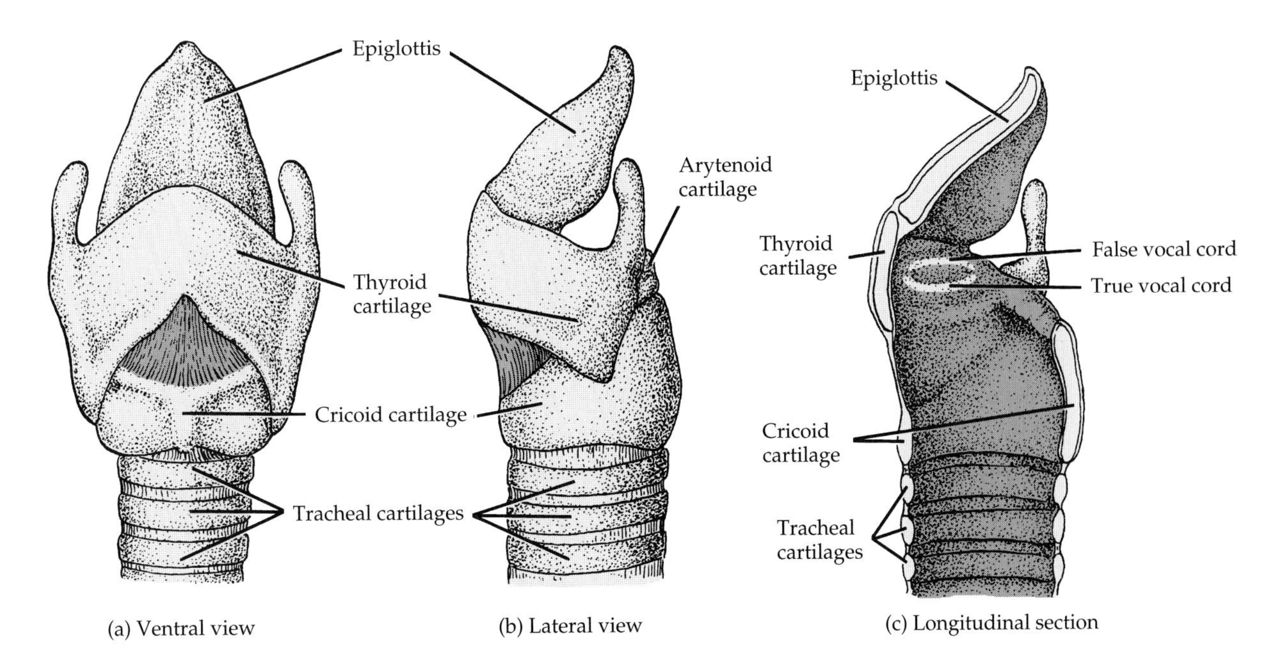

FIGURE 20.11 Larynx of cat.

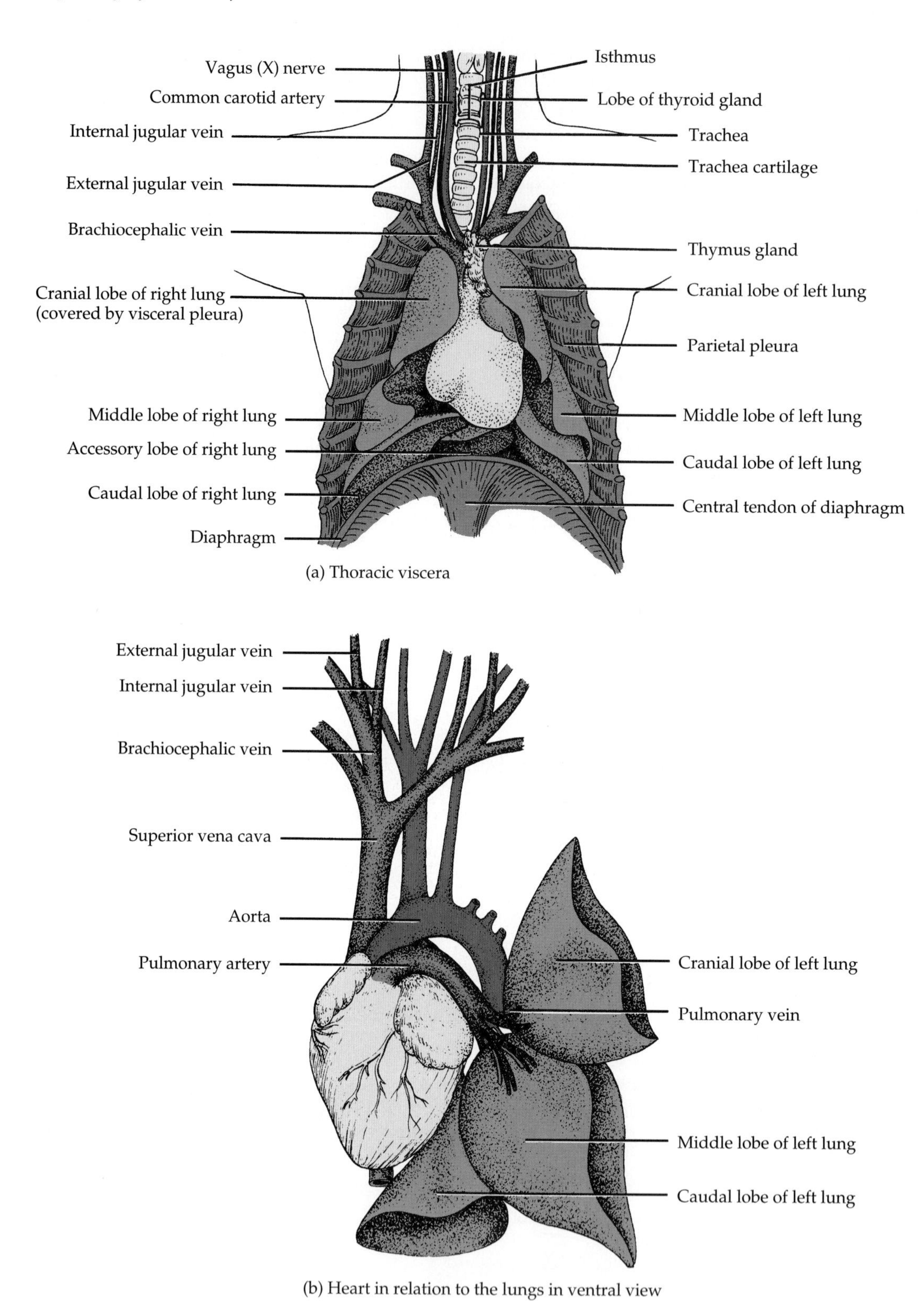

FIGURE 20.12 Respiratory system of cat.

threadlike ***vagus nerve*** adjacent to the common carotid artery.

3. Bronchi

1. The ***bronchi*** are two short tubes formed by the bifurcation of the trachea at the level of the sixth rib. One bronchus enters each lung. The bronchi, like the trachea, consist of C-shaped cartilages.
2. Within the lungs, the bronchi branch repeatedly into ***secondary bronchi.*** These branch into ***tertiary bronchi,*** which, in turn, branch into ***bronchioles.***
3. Ultimately, the bronchioles terminate in aveoli, or air sacs, where gases are exchanged.
4. After you examine the parts of the lungs, you can trace the bronchial tree into the substance of the lungs. Use Figure 20.6 for reference.

4. Lungs

1. The ***lungs*** are paired, lobed, spongy organs in which respiratory gases are exchanged (Figure 20.12).
2. Each lung is covered by ***visceral pleura,*** whereas ***parietal pleura*** lines the thoracic wall.
3. The potential space between the visceral pleurae of the lungs is called the ***mediastinum.*** This contains the thymus gland of the endocrine system, heart, trachea, esophagus, nerves, and blood vessels. In the midventral thorax, visceral and parietal pleurae meet to form a ventral partition called the ***mediastinal septum.*** This structure is caudal to the heart.
4. Note that the right lung has four lobes and the left lung has three lobes. Try to identify the ***pulmonary arteries*** entering the lungs. They are usually injected with blue latex.
5. Now try to find the ***pulmonary veins*** leaving the lungs. They are usually injected with red latex.
6. If you cut through a lobe of the lung, you should be able to identify branches of the pulmonary arteries, pulmonary veins, and bronchi. The lobes of the lung are labeled in Figure 20.12a.

C. DISSECTION OF SHEEP PLUCK

CAUTION! *Please reread Section D, "Precautions Related to Dissection" at the beginning of the laboratory manual on page xi before you begin your dissection.*

PROCEDURE

1. Preserved sheep pluck may or may not be available for dissection.
2. Pluck consists mainly of a sheep ***trachea, bronchi, lungs, heart,*** and ***great vessels,*** and a small portion of the ***diaphragm.*** It is a good demonstration because it is large and shows the close anatomical correlation between these structures and the systems to which they belong, namely, the respiratory and the cardiovascular systems.
3. The heart and its great blood vessels have been described in detail in Exercise 17.
4. Pluck can also be used to examine in great detail the trachea and its relationship to the development of the bronchi until they branch into each lung. In addition, this specimen is sufficiently large that the bronchial tree can be exposed by careful dissection.
5. This dissection is done by starting at the primary and secondary bronchi and slowly and carefully removing lung tissue as the trees form even smaller branches into lungs.
6. These specimens should be used primarily by the instructor for demonstration, but you can help expose the bronchial tree.

ANSWER THE LABORATORY REPORT QUESTIONS AT THE END OF THE EXERCISE.

LABORATORY REPORT QUESTIONS

Respiratory System 20

Student ______________________ Date ______________________

Laboratory Section ______________________ Score/Grade ______________________

PART 1. Multiple Choice

_______ 1. The overall exchange of gases between the atmosphere, blood, and cells is called (a) inspiration (b) respiration (c) expiration (d) none of these

_______ 2. The lateral walls of the internal nose are formed by the ethmoid bone, maxillae, lacrimal, inferior conchae, and the (a) hyoid bone (b) nasal bone (c) palatine bone (d) occipital bone

_______ 3. The portion of the pharynx that contains the pharyngeal tonsils is the (a) oropharynx (b) laryngopharynx (c) nasopharynx (d) pharyngeal orifice

_______ 4. The Adam's apple is a common term for the (a) thyroid cartilage (b) cricoid cartilage (c) epiglottis (d) none of these

_______ 5. The C-shaped rings of cartilage of the trachea not only prevent the trachea from collapsing but also aid in the process of (a) lubrication (b) removing foreign particles (c) gas exchange (d) swallowing

_______ 6. Of the following structures, the smallest in diameter is the (a) left primary bronchus (b) bronchioles (c) secondary bronchi (d) alveolar ducts

_______ 7. The structures of the lung that actually contain the alveoli are the (a) respiratory bronchioles (b) fissures (c) lobules (d) terminal bronchioles

_______ 8. From superficial to deep, the structure(s) that you would encounter first among the following is (are) the (a) bronchi (b) parietal pleura (c) pleural cavity (d) secondary bronchi

PART 2. Completion

9. An advantage of nasal breathing is that the air is warmed, moistened, and ______________________.

10. Improper fusion of the palatine and maxillary bones results in a condition called ______________________.

11. The protective lid of cartilage that prevents food from entering the trachea is the ______________________.

12. After removal of the ______________________ an individual would be unable to speak.

13. The upper respiratory tract is able to trap and remove dust because of its lining of ______________________.

14. The passage of a tube into the mouth and down through the larynx and trachea to bypass an obstruction is called ____________________.

15. A radiograph of the bronchial tree after administration of an iodinated medium is called a ____________________.

16. The sequence of respiratory tubes from largest to smallest is trachea, primary bronchi, secondary bronchi, bronchioles, ____________________, respiratory bronchioles, and alveolar ducts.

17. Both the external and internal nose are divided internally by a vertical partition called the ____________________.

18. The undersurface of the external nose contains two openings called the nostrils or ____________________.

19. The functions of the pharynx are to serve as a passageway for air and food and to provide a resonating chamber for ____________________.

20. An inflammation of the membrane that encloses and protects the lungs is called ____________________.

21. The anterior portion of the nasal cavity just inside the nostrils is called the ____________________.

22. Groovelike passageways in the nasal cavity formed by the conchae are called ____________________.

23. The portion of the pharynx that contains the palatine and lingual tonsils is the ____________________.

24. Each bronchopulmonary segment of a lung is subdivided into many compartments called ____________________.

25. A(n) ____________________ is an outpouching lined by epithelium and supported by a thin elastic membrane.

26. The ____________________ cartilage attaches the larynx to the trachea.

27. The portion of a lung that rests on the diaphragm is the ____________________.

28. Phagocytic cells in the alveolar wall are called ____________________.

29. The surface of a lung lying against the ribs is called the ____________________ surface.

30. The ____________________ is a structure in the medial surface of a lung through which bronchi, blood vessels, lymphatic vessels, and nerves pass.

31. The nose, pharynx, and associated structures comprise the ____________________ respiratory system.

32. The ____________________ is the rounded, anterior border of the nose that connects the root and apex.

33. The ____________________ cartilages of the larynx attach the vocal folds to the intrinsic pharyngeal muscles.

34. Each ____________________ of a lung is supplied by a tertiary bronchus.

Digestive System

21

Digestion occurs basically as two events—mechanical digestion and chemical digestion. ***Mechanical digestion*** consists of various movements of the gastrointestinal tract that help chemical digestion. These movements include the physical breakdown of food by the teeth and complete churning and mixing of this food with enzymes by the smooth muscles of the stomach and small intestine. ***Chemical digestion*** consists of a series of catabolic (hydrolysis) reactions that break down the large nutrient molecules that we eat, such as carbohydrates, lipids, and proteins, into much smaller molecules that can be absorbed and used by body cells.

A. GENERAL ORGANIZATION OF DIGESTIVE SYSTEM

Digestive organs are usually divided into two main groups. The first is the ***gastrointestinal (GI) tract,*** or ***alimentary*** (*alimentum* = nourishment) ***canal,*** a continuous tube running from the mouth to the anus, and measuring about 9 m (30 ft) in length in a cadaver. This tract is composed of the mouth, pharynx, esophagus, stomach, small intestine, and large intestine. The small intestine has three regions: duodenum, jejunum, and ileum. The large intestine has four regions: cecum, colon, rectum, and anal canal. The colon is divided into ascending colon, transverse colon, descending colon, and sigmoid colon.

The second group of organs composing the digestive system consists of the ***accessory structures*** such as the teeth, tongue, salivary glands, liver, gallbladder, and pancreas (see Figure 21.1).

Using your textbook, charts, or models for reference, label Figure 21.1.

The wall of the gastrointestinal tract, especially from the stomach to the anal canal, has the same basic arrangement of tissues. The four layers (tunics) of the tract, from deep to superficial, are the ***mucosa, submucosa, muscularis,*** and ***serosa*** (see Figure 21.9 on page 416).

Inferior to the diaphragm, the serosa is also called the ***peritoneum*** (per′-i-tō-NĒ-um; *peri* = around; *tonos* = tension). The peritoneum is composed of a layer of simple squamous epithelium (called mesothelium) and an underlying layer of connective tissue. The ***parietal peritoneum*** lines the wall of the abdominopelvic cavity, and the ***visceral peritoneum*** covers some of the organs in the cavity. The potential space between the parietal and visceral portions of the peritoneum is called the ***peritoneal cavity.*** Unlike the two other serous membranes of the body, the pericardium and the pleurae, which smoothly cover the heart and lungs, the peritoneum contains large folds that weave in between the viscera. The important extensions of the peritoneum are the ***mesentery*** (MEZ-en-ter′-ē; *meso* = middle; *enteron* = intestine) ***mesocolon, falciform*** (FAL-si-form) ***ligament, lesser omentum*** (ō-MENT-um), and ***greater omentum.***

Inflammation of the peritoneum, called ***peritonitis,*** is a serious condition because the peritoneal membranes are continuous with one another, enabling the infection to spread to all the organs in the cavity.

B. ORGANS OF DIGESTIVE SYSTEM

1. Mouth (Oral Cavity)

The ***mouth,*** also called the ***oral,*** or ***buccal*** (BUK-al; *bucca* = cheeks), ***cavity,*** is formed by the cheeks, hard and soft palates, and tongue. The ***hard palate*** forms the anterior portion of the roof of the mouth and the ***soft palate*** forms the posterior portion. The ***tongue*** forms the floor of the oral cavity and is composed of skeletal muscle covered by mucous membrane. Partial digestion of carbohydrates and triglycerides occurs in the mouth.

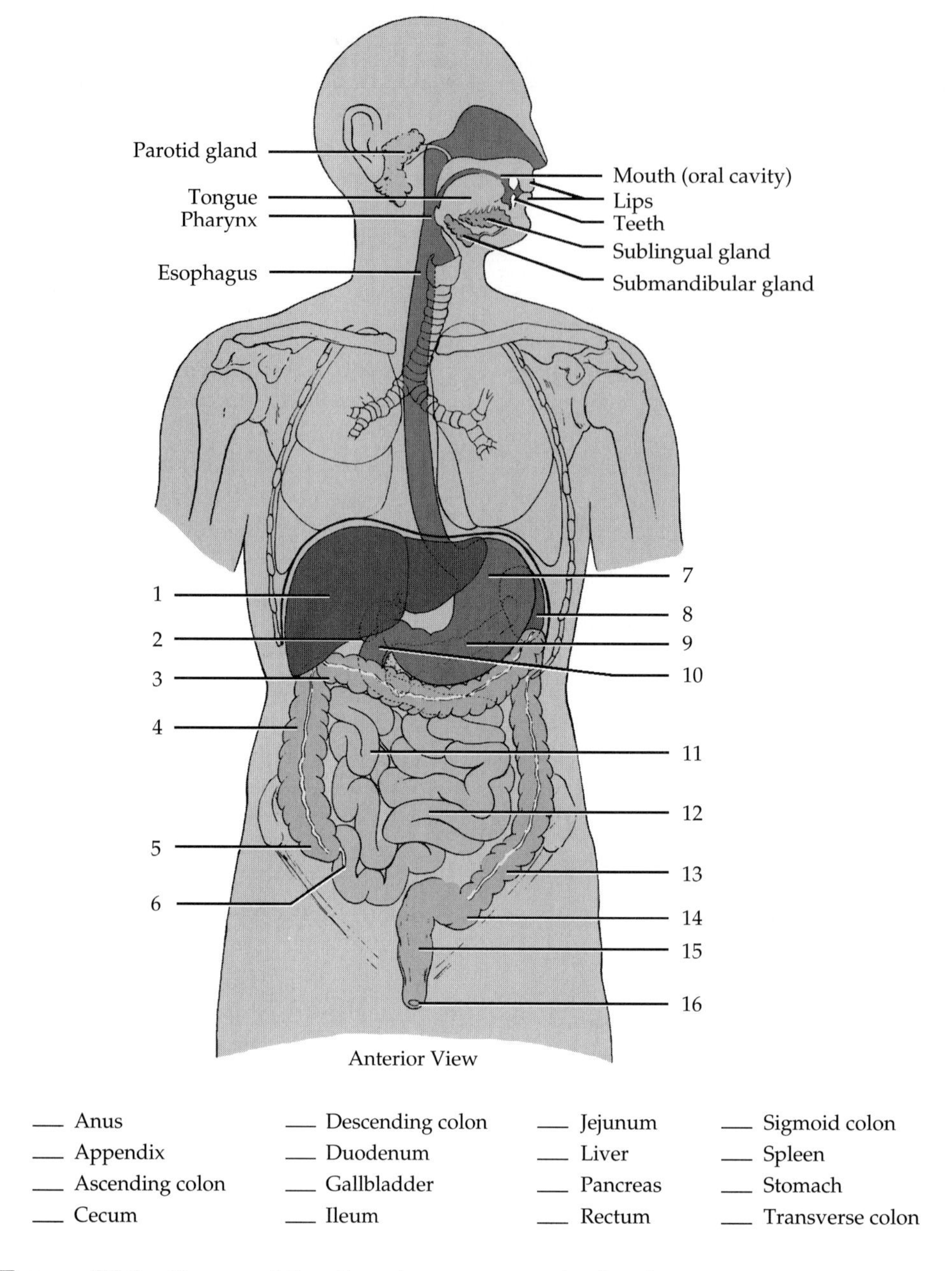

___ Anus
___ Appendix
___ Ascending colon
___ Cecum
___ Descending colon
___ Duodenum
___ Gallbladder
___ Ileum
___ Jejunum
___ Liver
___ Pancreas
___ Rectum
___ Sigmoid colon
___ Spleen
___ Stomach
___ Transverse colon

FIGURE 21.1 Organs of the digestive system and related structures.

a. ***Cheeks*** Lateral walls of oral cavity. Muscular structures covered by skin and lined by nonkeratinized stratified squamous epithelium; anterior portions terminate in the ***superior*** and ***inferior labia*** (lips).

b. ***Vermilion*** (ver-MIL-yon) Transition zone of lips where outer skin and inner mucous membranes meet.

c. ***Labial frenulum*** (LĀ-bē-al FREN-yoo-lum; *labium* = fleshy border; *frenulum* = small bridle) Midline fold of mucous membrane that attaches the inner surface of each lip to its corresponding gum.

d. ***Vestibule*** (= entrance to a canal) Space bounded externally by cheeks and lips and internally by gums and teeth.

e. ***Oral cavity proper*** Space extending from the gums and teeth to the ***fauces*** (FAW-sēz; *fauces* = passages), opening of oral cavity proper into pharynx. Area is enclosed by the dental arches.

f. ***Hard palate*** Formed by maxillae and palatine bones and covered by mucous membrane.

g. ***Soft palate*** Arch-shaped muscular partition between oropharynx and nasopharynx lined by mucous membrane. Hanging from free border of soft palate is a muscular projection, the ***uvula*** (YOU-vyoo-la = little grape).
h. ***Palatoglossal arch (anterior pillar)*** Muscular fold that extends inferiorly, laterally, and anteriorly to the side of the base of tongue.
i. ***Palatopharyngeal*** (PAL-a-tō-fa-rin′-jē-al) ***arch (posterior pillar)*** Muscular fold that extends inferiorly, laterally, and posteriorly to the side of pharynx. ***Palatine tonsils*** are between arches and ***lingual tonsil*** is at base of tongue.
j. ***Tongue*** Movable, muscular organ on floor of oral cavity. ***Extrinsic muscles*** originate outside tongue (to bones in the area), insert into connective tissues, and move tongue from side to side and in and out to maneuver food for chewing and swallowing; ***intrinsic muscles*** originate and insert into connective tissues within the tongue and alter shape and size of tongue for speech and swallowing.
k. ***Lingual*** (*lingua* = tongue) ***frenulum*** Midline fold of mucous membrane on undersurface of tongue that helps restrict its movement posteriorly.
l. ***Papillae*** (pa-PIL-ē = nipple-shaped projections) Projections of lamina propria on surface of tongue covered with epithelium; ***filiform*** (= threadlike) ***papillae*** are conical projections in parallel rows over anterior two-thirds of tongue; ***fungiform*** (= shaped like a mushroom) ***papillae*** are mushroomlike elevations distributed among filiform papillae and more numerous near tip of tongue (appear as red dots and most contain taste buds); ***circumvallate*** (*circum* = around; *vallare* = to wall) ***papillae*** are arranged in the form of an inverted V on the posterior surface of tongue (all contain taste buds).

Using a mirror, examine your mouth and locate as many of the structures (a through l) as you can. Label Figure 21.2.

2. Salivary Glands

Most saliva is secreted by the ***salivary glands,*** which lie outside the mouth and pour their contents into ducts that empty into the oral cavity. The carbohydrate-digesting enzyme in saliva is salivary amylase. The three pairs of salivary glands are the ***parotid*** (*para* = near; *otia* = ear) ***glands*** (anterior and inferior to the ears), which secrete into the oral cavity vestibule through ***parotid (Stensen's) ducts; submandibular glands*** (deep to the base of the tongue in the posterior part of the floor of the mouth), which secrete on either side of the lingual frenulum in the floor of the oral cavity through ***submandibular (Wharton's) ducts;*** and ***sublingual glands*** (superior to the submandibular glands), which secrete into the floor of the oral cavity through ***lesser sublingual (Rivinus') ducts.***

Label Figure 21.3 on page 411.

The parotid glands are compound tubuloacinar glands, whereas the submandibulars and sublinguals are compound acinar glands (see Figure 21.4).

Examine prepared slides of the three different types of salivary glands and compare your observations to Figure 21.4 on page 411.

3. Teeth

Teeth (dentes) are located in the sockets of the alveolar processes of the mandible and maxillae. The alveolar processes are covered by ***gingivae*** (jin-JĪ-vē) or gums, which extend slightly into each socket. The sockets are lined by a dense fibrous connective tissue called a ***periodontal*** (*peri* = around; *odous* = tooth) ***ligament,*** which anchors the teeth in position and acts as a shock absorber during chewing.

Following are the parts of a tooth:

a. ***Crown*** Exposed portion above level of gums.
b. ***Root*** One to three projections embedded in socket.
c. ***Neck*** Constricted junction line of the crown and root near the gum line.
d. ***Dentin*** Calcified connective tissue that gives teeth their basic shape and rigidity.
e. ***Pulp cavity*** Enlarged part of cavity in crown within dentin.
f. ***Pulp*** Connective tissue containing blood vessels, lymphatic vessels, and nerves.
g. ***Root canal*** Narrow extension of pulp cavity in root.
h. ***Apical foramen*** Opening in base of root canal through which blood vessels, lymphatic vessels, and nerves enter tooth.
i. ***Enamel*** Covering of crown that consists primarily of calcium phosphate and calcium carbonate.
j. ***Cementum*** Bonelike substance that covers and attaches root to periodontal ligament.

With the aid of your textbook, label the parts of a tooth shown in Figure 21.5 on page 412.

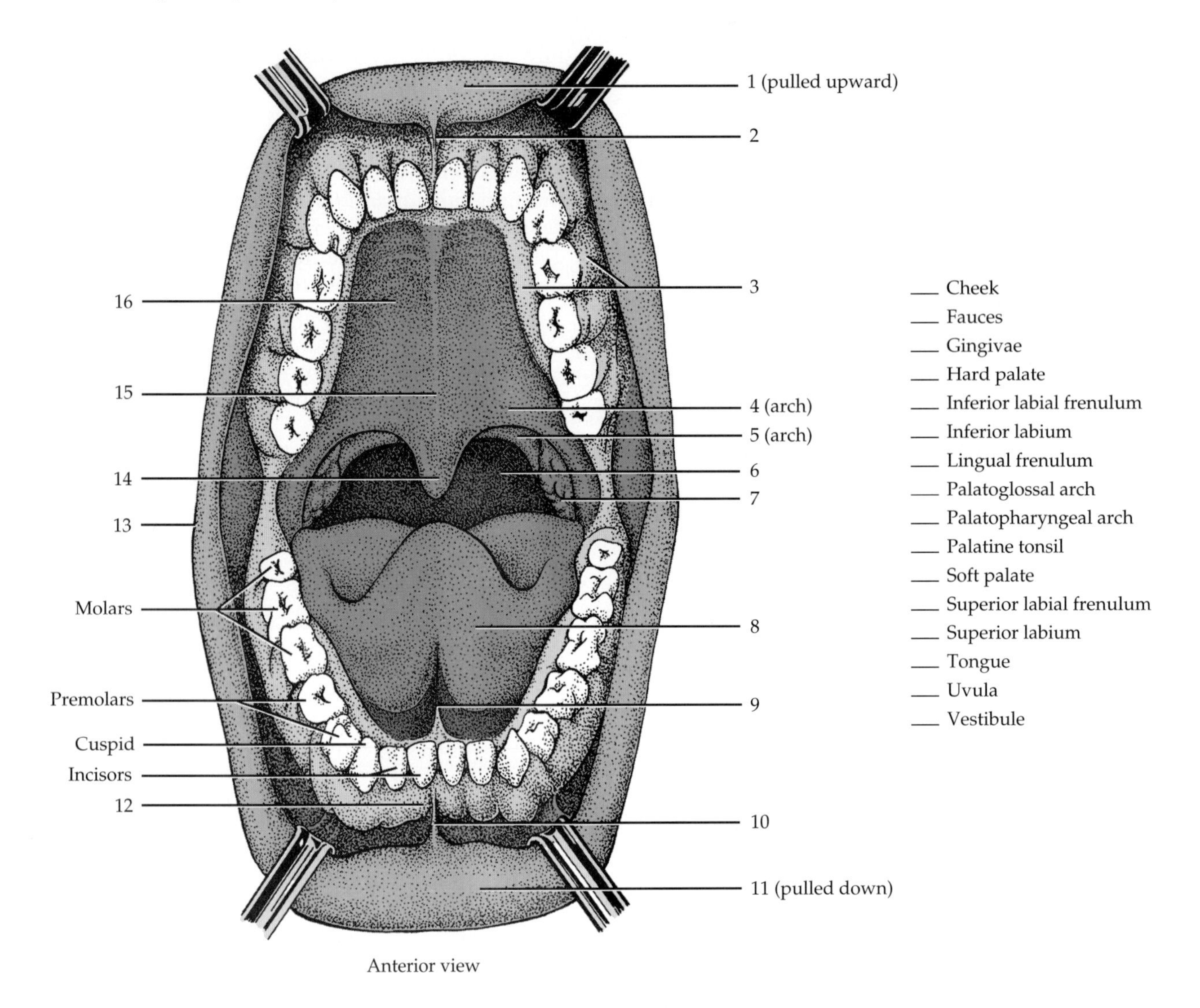

FIGURE 21.2 Mouth (oral cavity).

4. Dentitions

Dentitions (sets of teeth) are of two types: ***deciduous*** (baby) and ***permanent***. Deciduous teeth begin to erupt at about 6 months of age, and one pair appears at about each month thereafter until all 20 are present. The deciduous teeth are as follows:

- **a.** ***Incisors*** Central incisors closest to midline, with lateral incisors on either side. Incisors are chisel-shaped, adapted for cutting into food, have only one root.
- **b.** ***Cuspids (canines)*** Posterior to incisors. Cuspids have pointed surfaces (cusps) for tearing and shredding food, have only one root.
- **c.** ***Molars*** First and second molars posterior to canines. Molars crush and grind food. Upper (maxillary) molars have four cusps and three roots, lower (mandibular) molars have four cusps and two roots.

All deciduous teeth are usually lost between 6 and 12 years of age and replaced by permanent dentition consisting of 32 teeth that erupt between age 6 and adulthood. The permanent teeth are:

- **a.** ***Incisors*** Central incisors and lateral incisors replace those of deciduous dentition.
- **b.** ***Cuspids (canines)*** These replace those of deciduous dentition.
- **c.** ***Premolars (bicuspids)*** First and second premolars replace deciduous molars. Premolars crush and grind food, have two cusps and one root (upper first premolars have two roots).
- **d.** ***Molars*** These erupt behind premolars as jaw grows to accommodate them and do not

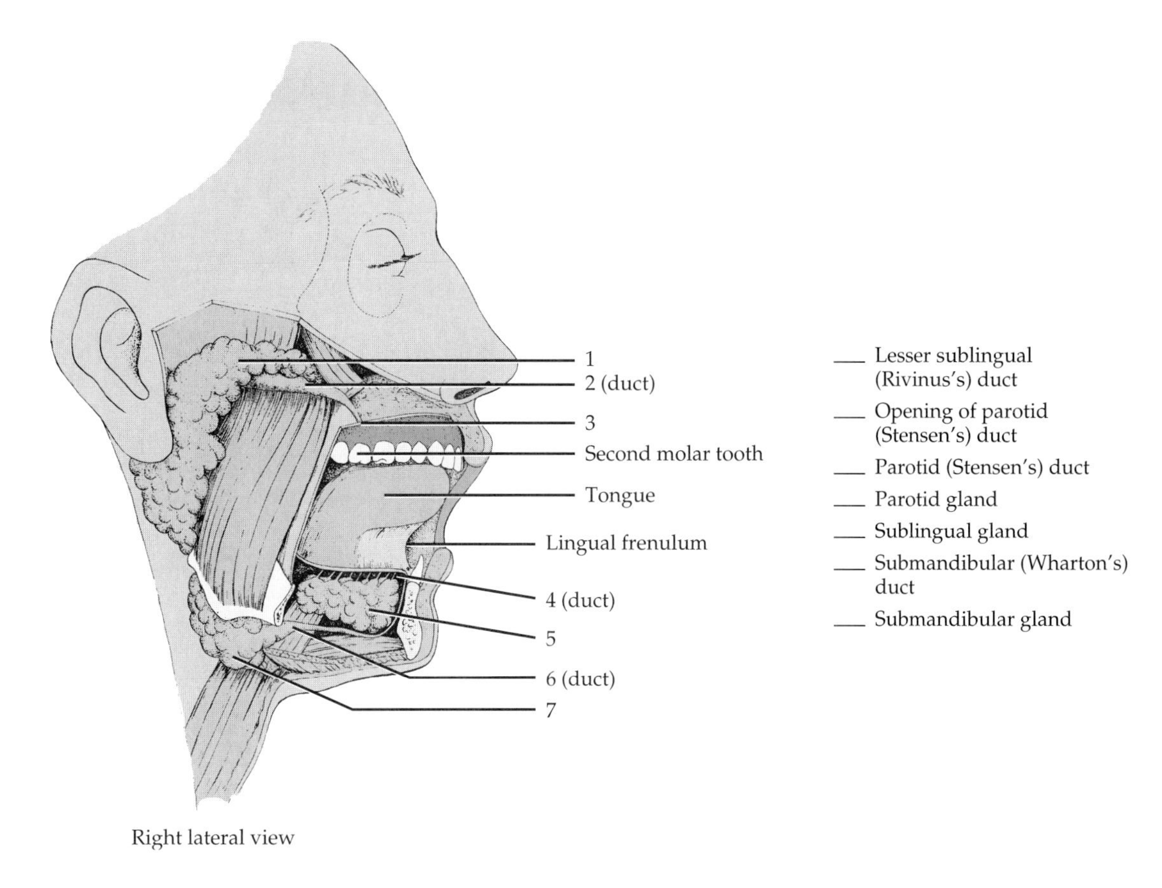

FIGURE 21.3 Location of salivary glands.

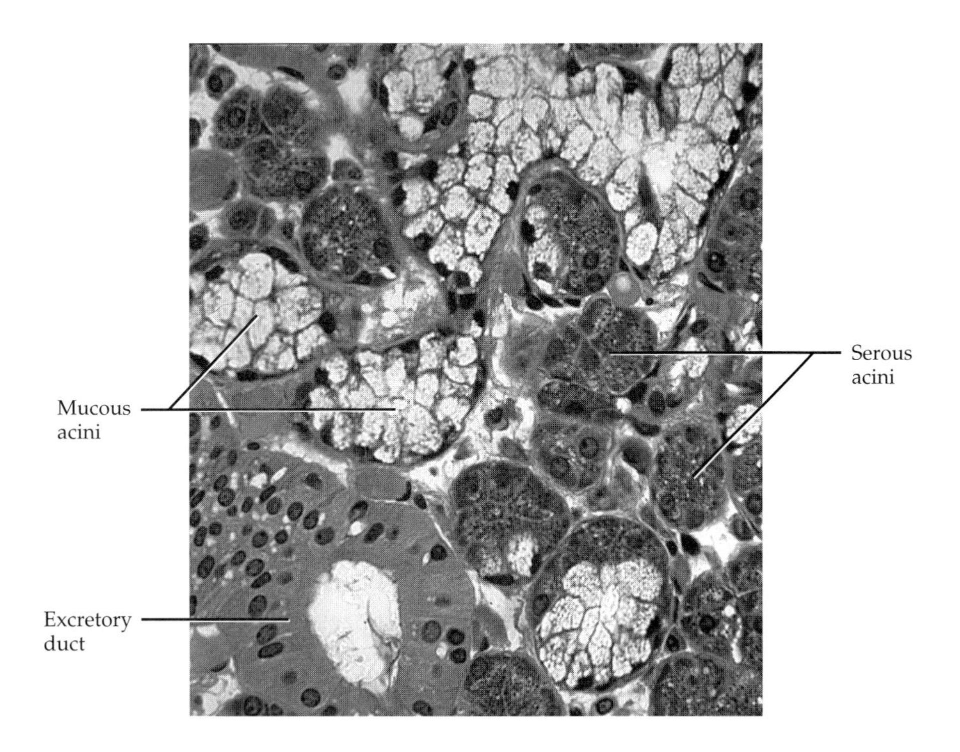

FIGURE 21.4 Histology of salivary glands (100×).

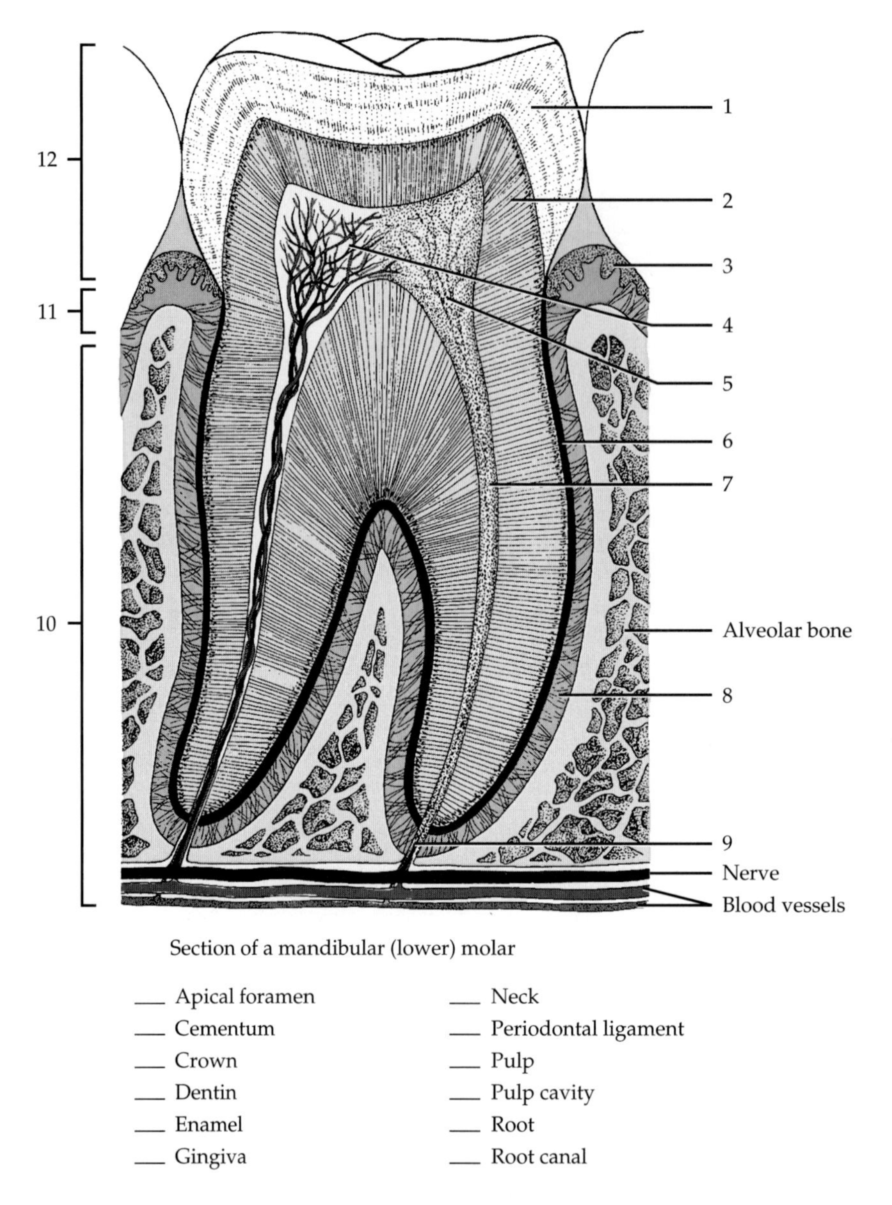

___ Apical foramen
___ Cementum
___ Crown
___ Dentin
___ Enamel
___ Gingiva
___ Neck
___ Periodontal ligament
___ Pulp
___ Pulp cavity
___ Root
___ Root canal

FIGURE 21.5 Parts of a tooth.

replace any deciduous teeth. First molars erupt at age 6, second at age 12, and third (wisdom teeth) after age 18.

Using a mirror, examine your mouth and locate as many teeth of the permanent dentition as you can.

With the aid of your textbook, label the deciduous and permanent dentitions in Figure 21.6.

5. Esophagus

The ***esophagus*** (e-SOF-a-gus; *oisein* = to carry; *phagema* = food) is a muscular, collapsible tube posterior to the trachea. The structure is 23 to 25 cm (10 in.) long and extends from the laryngopharynx through the mediastinum and esoph-ageal hiatus in the diaphragm and terminates in the superior portion of the stomach. The esophagus conveys food from the pharynx to the stomach by peristalsis.

Histologically, the esophagus consists of a ***mucosa*** (nonkeratinized stratified squamous epithelium, lamina propria, muscularis mucosae), ***submucosa*** (areolar connective tissue, blood vessels, mucous glands), ***muscularis*** (superior third striated, middle third striated and smooth, inferior third smooth), and ***adventitia*** (ad-ven-TISH-ya). The esophagus is not covered by a serosa.

Examine a prepared slide of a transverse section of the esophagus that shows its various coats. With the aid of your textbook, label Figure 21.7.

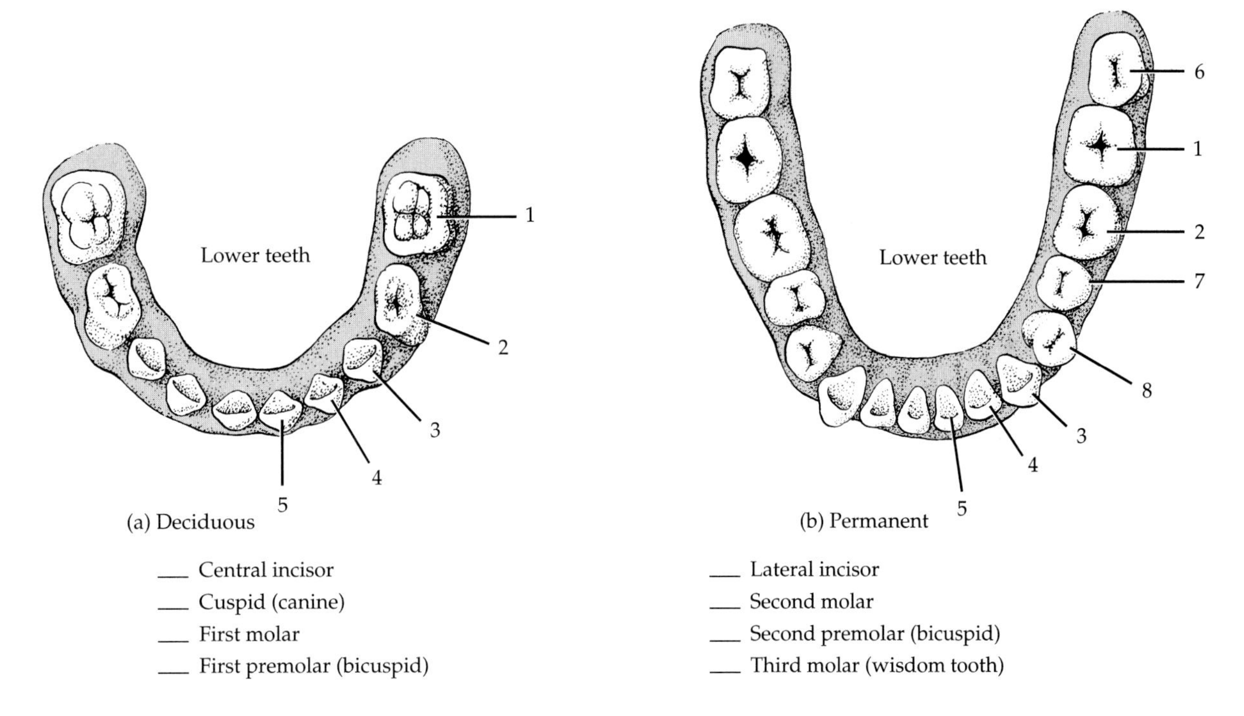

FIGURE 21.6 Dentitions.

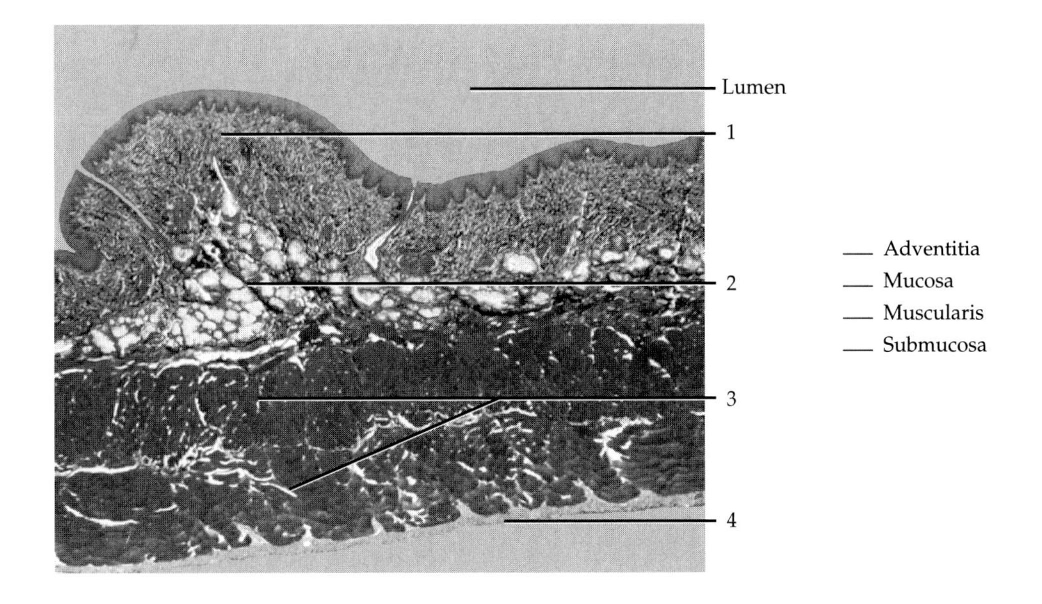

FIGURE 21.7 Histology of esophagus.

6. Stomach

The ***stomach*** is a J-shaped enlargement of the gastrointestinal tract inferior to the diaphragm (see Figure 21.1). It is in the epigastric, umbilical, and left hypochondriac regions of the abdomen. The superior part is connected to the esophagus; the inferior part empties into the duodenum, the first portion of the small intestine. The stomach is divided into four main areas: cardia, fundus, body, and pylorus. The ***cardia*** (CAR-dē-a) surrounds the lower esophageal sphincter, a physiological sphincter in the esophagus just superior to the diaphragm. The rounded portion superior to and to the left of the cardia is the ***fundus*** (FUN-dus). Inferior to the fundus, the large central portion of the stomach is called the ***body***. The narrow, inferior region is the ***pylorus*** (pī-LOR-us; *pyle* = gate; *ouros* = guard). The pylorus consists of a ***pyloric antrum*** (AN-trum = cave), which is closer to the body of the stomach, and a ***pyloric canal***, which is closer to the duodenum. The concave medial border of the stomach is called the ***lesser curvature***, and the convex lateral border is the ***greater curvature***. The pylorus communicates with the duodenum of the small intestine via a sphincter called the ***pyloric sphincter (valve)***. The main chemical activity of the stomach is to begin the digestion of proteins.

Label Figure 21.8.

The ***mucosa*** of the stomach consists of simple columnar epithelium, lamina propria, and muscularis mucosae. The mucosa is arranged in large folds called ***rugae*** (ROO-jē = wrinkles). The columnar epithelium of the mucosa contains many narrow channels that extend down into the lamina propria and are referred to as ***gastric glands (pits)***. The glands are lined with four kinds of cells: (1) ***chief (zymogenic) cells*** that secrete inactive pepsinogen, which is converted to active pepsin, a protein-digesting enzyme; (2) ***parietal (oxyntic) cells*** that secrete hydrochloric acid and intrinsic factor; (3) ***mucous neck cells*** that secrete mucus; and (4) ***enteroendocrine cells*** called ***G cells*** that secrete the hormone gastrin. The ***submucosa*** consists of areolar connective tissue. The ***muscularis*** has three layers of smooth muscle—superficial longitudinal, middle circular, and deep oblique. The oblique layer is limited chiefly to the body of the stomach. The ***serosa*** is part of the visceral peritoneum.

Examine a prepared slide of a section of the stomach that shows its various layers. With the aid of your textbook, label Figure 21.9 on page 416.

7. Pancreas

The ***pancreas*** (*pan* = all; *kreas* = flesh) is a retroperitoneal gland posterior to the greater curvature of the stomach (see Figure 21.1). The gland consists of a ***head*** (expanded portion near duodenum), ***body*** (central portion), and ***tail*** (terminal tapering portion).

Histologically, the pancreas consists of ***pancreatic islets (islets of Langerhans)*** that contain (1) glucagon-producing ***alpha cells***, (2) insulin-producing ***beta cells***, (3) somatostatin-producing ***delta cells***, and pancreatic polypeptide-producing ***F cells*** (see Figure 15.5). The pancreas also consists of ***acini*** that produce pancreatic juice (see Figure 15.5). Pancreatic juice contains enzymes that assist in the chemical breakdown of carbohydrates, proteins, triglycerides, and nucleic acids.

Pancreatic juice is delivered from the pancreas to the duodenum by a large main tube, the ***pancreatic duct (duct of Wirsung)***. This duct unites with the common bile duct from the liver and pancreas and enters the duodenum in a common duct called the ***hepatopancreatic ampulla (ampulla of Vater)***. The ampulla opens on an elevation of the duodenal mucosa, the ***duodenal papilla***. An ***accessory pancreatic duct (duct of Santorini)*** may also lead from the pancreas and empty into the duodenum about 2.5 cm (1 in.) superior to the hepatopancreatic ampulla. With the aid of your textbook, label the structures associated with the pancreas in Figure 21.10 on page 417.

8. Liver

The ***liver*** is located inferior to the diaphragm (see Figure 21.1). It occupies most of the right hypochondriac and part of the epigastric regions of the abdomen. The gland is divided into two principal lobes, the ***right lobe*** and ***left lobe***, separated by the ***falciform ligament***. The falciform ligament attaches the liver to the anterior abdominal wall and diaphragm. The right lobe consists of an inferior ***quadrate lobe*** and a posterior ***caudate lobe***.

Each lobe is composed of microscopic functional units called ***lobules***. Among the structures in a lobule are cords of ***hepatocytes (liver cells)*** arranged in a radial pattern around a ***central vein; sinusoids***, endothelial lined spaces between hepatocytes through which blood flows; and ***stellate reticuloendothelial (Kupffer) cells*** that destroy bacteria and worn-out blood cells by phagocytosis.

Examine a prepared slide of several liver lobules. Compare your observations with Figure 21.11 on page 417.

Bile is manufactured by hepatocytes and functions in the emulsification of triglycerides in the small intestine. The liquid is passed to the small intestine as follows: Hepatocytes secrete bile into ***bile canaliculi*** (kan'-a-LIK-yoo-lī = small canals) that empty into small ducts. The small ducts merge into larger ***right*** and ***left hepatic ducts***, one in each principal lobe of the liver. The right and left hepatic ducts unite outside the liver to form a single ***common hepatic duct***. This duct joins the ***cystic*** (*kystis* = bladder) ***duct*** from the gallbladder to become the ***common bile duct***, which empties

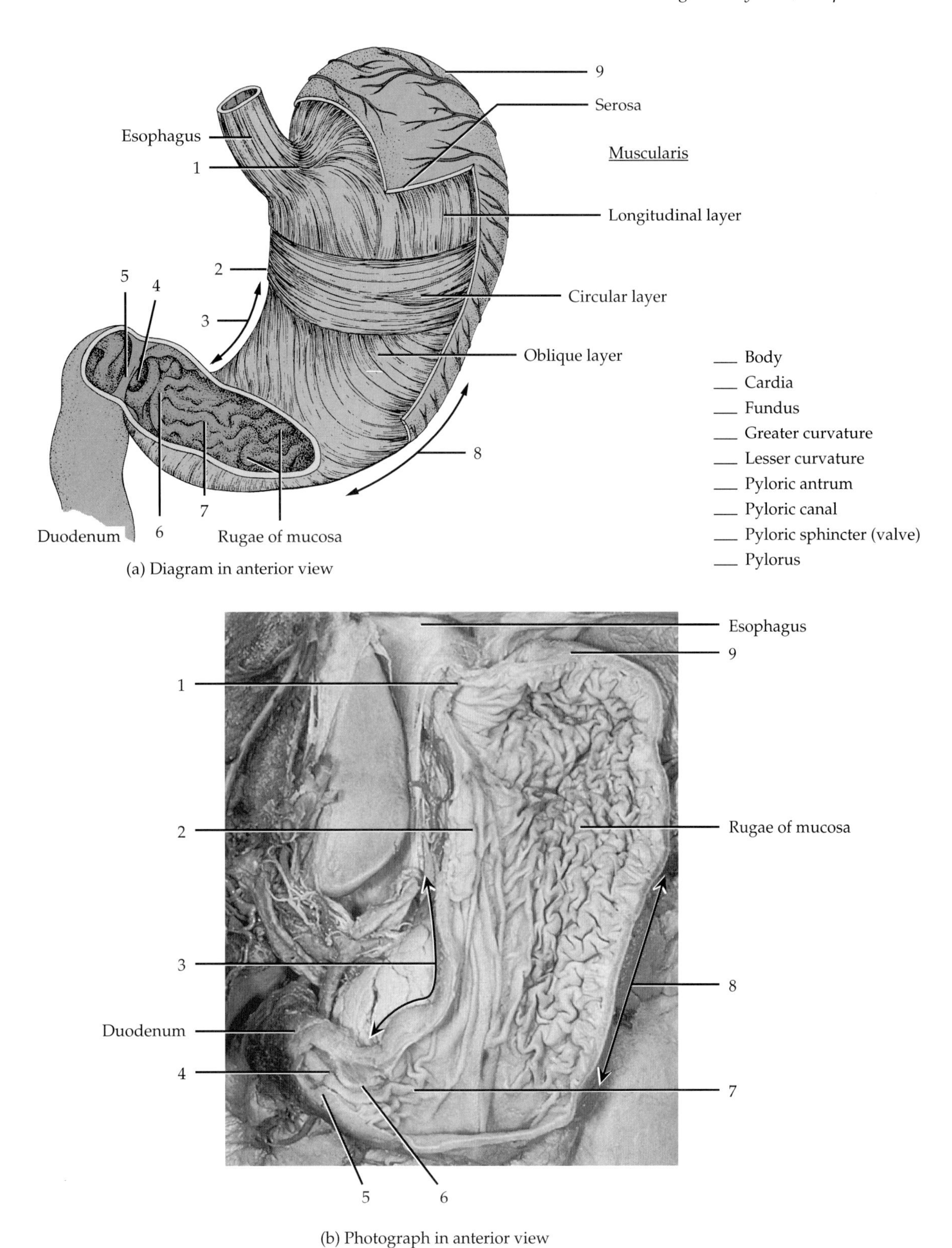

FIGURE 21.8 Stomach. External and internal anatomy.

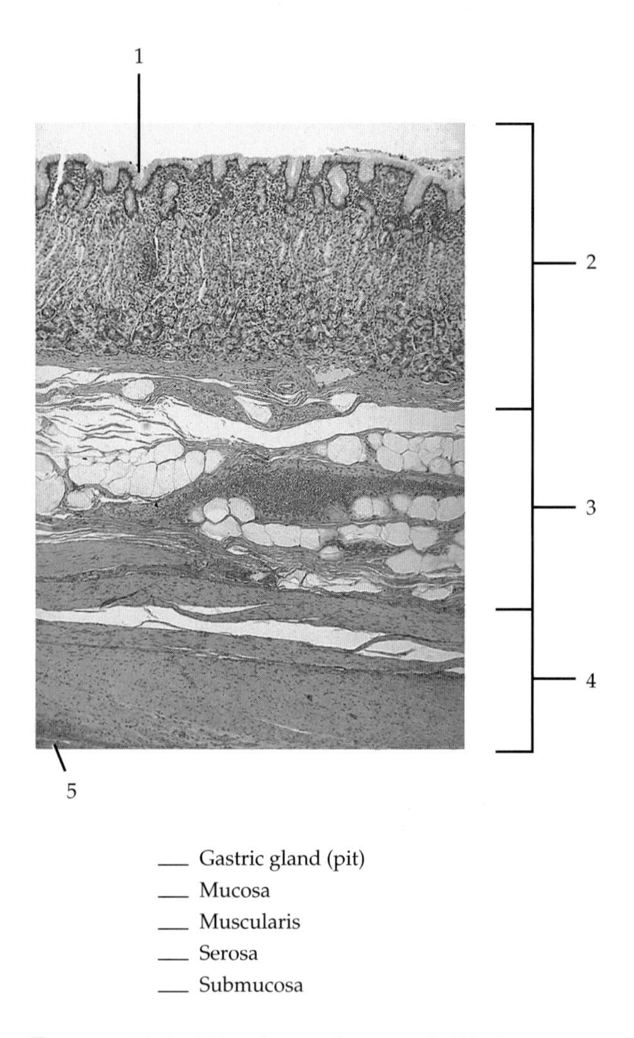

FIGURE 21.9 Histology of stomach (17×).

into the duodenum at the hepatopancreatic ampulla (ampulla of Vater). When triglycerides are not being digested, a valve around the hepatopancreatic ampulla, the ***sphincter of the hepatopancreatic ampulla (sphincter of Oddi),*** closes, and bile backs up into the gallbladder via the cystic duct. In the gallbladder, bile is stored and concentrated.

With the aid of your textbook, label the structures associated with the liver in Figure 21.10.

9. Gallbladder

The ***gallbladder*** (*galla* = bile) is a pear-shaped sac in a fossa along the posterior surface of the liver (see Figure 21.1). The gallbladder stores and concentrates bile. The cystic duct of the gallbladder and common hepatic duct of the liver merge to form the common bile duct. The ***mucosa*** of the gallbladder consists of simple columnar epithelium that contains rugae. The ***muscularis*** consists of smooth muscle and the outer coat consists of visceral peritoneum.

With the aid of your textbook, label the structures associated with the gallbladder in Figure 21.10.

10. Small Intestine

The bulk of digestion and absorption occurs in the ***small intestine,*** which begins at the pyloric sphincter (valve) of the stomach, coils through the central and inferior part of the abdomen, and joins the large intestine at the ileocecal sphincter (see Figure 21.1). The mesentery attaches the small intestine to the posterior abdominal wall. The small intestine is about 6.35 m (21 ft) long and is divided into three segments: ***duodenum*** (doo'-ō-DĒ-num), which begins at the stomach; ***jejunum*** (jē-JOO-num), the middle segment; and ***ileum*** (IL-ē-um), which terminates at the large intestine (Figure 21.12 on page 418).

Histologically, the ***mucosa*** contains many pits lined with glandular epithelium called ***intestinal glands (crypts of Lieberkühn);*** they secrete enzymes that digest carbohydrates, proteins, and nucleic acids (see Figure 21.13). Some of the simple columnar cells of the mucosa are ***goblet cells*** that secrete mucus (see Figure 21.13); others contain ***microvilli*** to increase the surface area for absorption (see Figure 3.1). The mucosa contains a series of fingerlike projections, the ***villi*** (see Figure 21.13). Each villus contains a blood capillary and a lymphatic vessel called a ***lacteal;*** these absorb digested nutrients. The 4 million to 5 million villi in the small intestine greatly increase the surface area for absorption. The mucosa and ***submucosa*** also contain deep permanent folds, the ***circular folds,*** or ***plicae circulares*** (PLĪ-kē SER-kyoo-lar-es), which also help to increase the surface area for absorption. In the submucosa of the duodenum are ***duodenal (Brunner's) glands,*** which secrete an alkaline mucus to protect the mucosa from excess acid and the action of digestive enzymes. The ***muscularis*** of the small intestine consists of an outer longitudinal layer and an inner circular layer of smooth muscle. Except for a major portion of the duodenum, the ***serosa*** (visceral peritoneum) completely covers the small intestine.

Obtain prepared slides of the small intestine (section through its tunics and villi) and identify as many structures as you can, using Figure 21.13 on page 419 and your textbook as references.

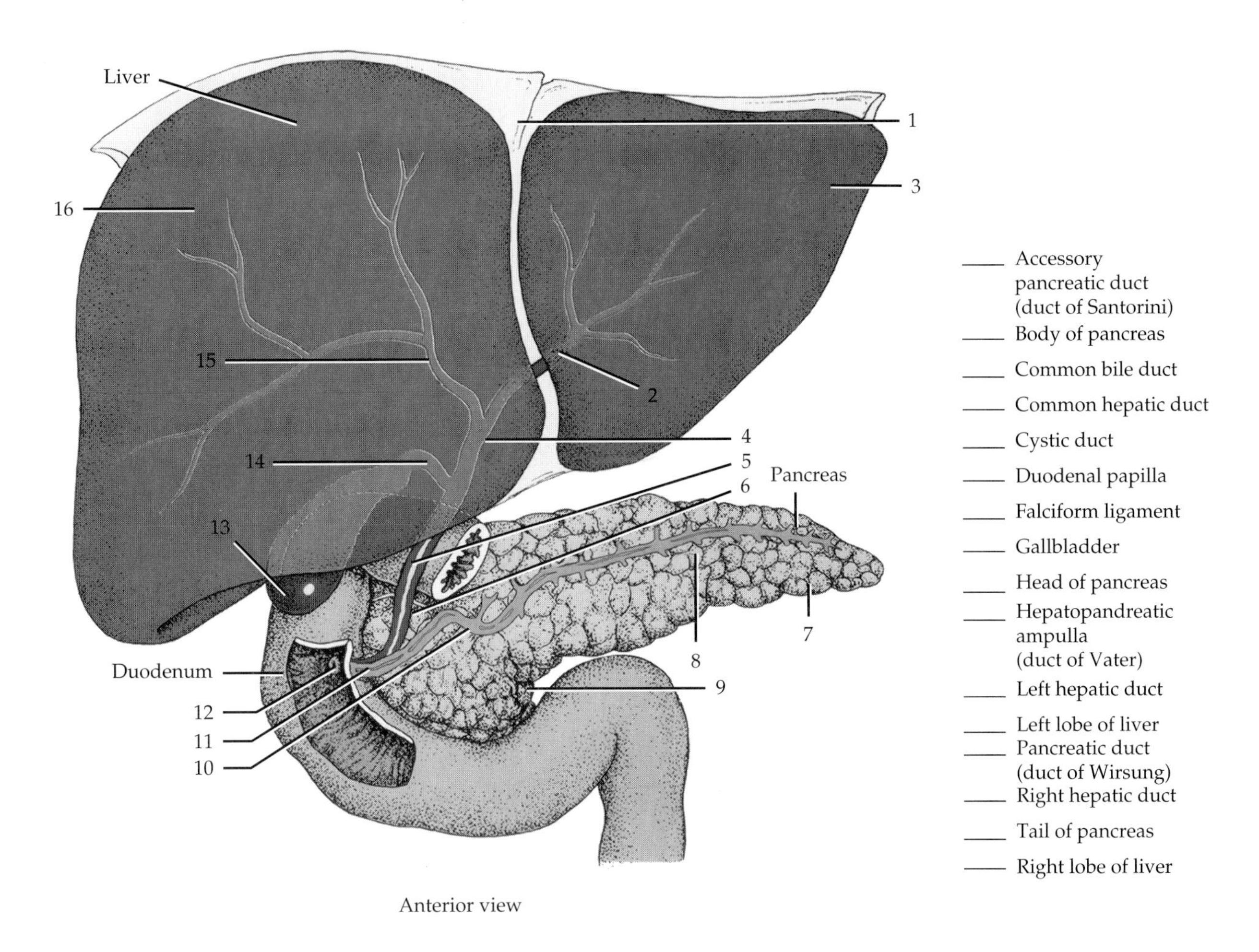

FIGURE 21.10 Relations of the liver, gallbladder, duodenum, and pancreas.

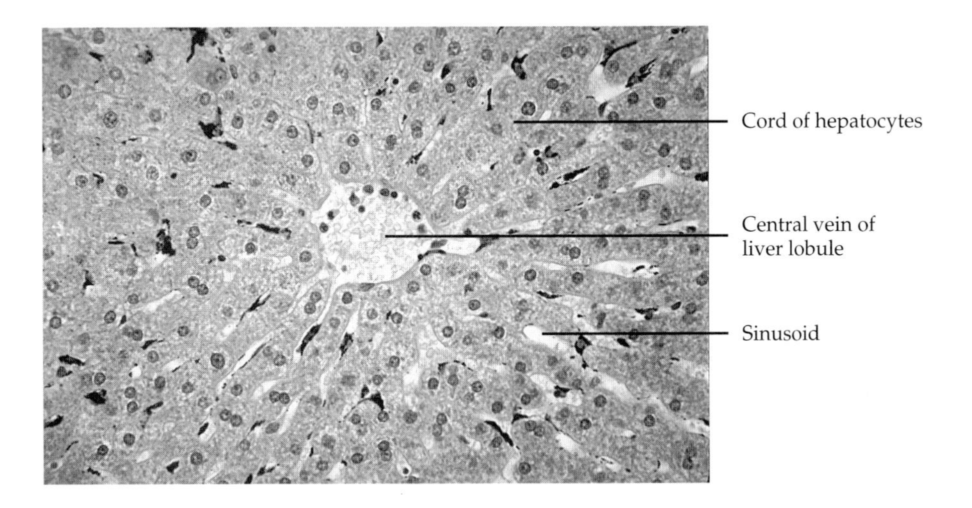

FIGURE 21.11 Photomicrograph of liver lobule (100×).

11. Large Intestine

The ***large intestine*** functions in the completion of absorption of water that leads to the formation of feces, and in the expulsion of feces from the body. Bacteria residing in the large intestine manufacture certain vitamins (some B vitamins and vitamin K). The large intestine is about 1½ m (5 ft) long and extends from the ileum to the anus (see Figure 21.12). It is attached to the posterior abdominal wall by an extension of visceral peritoneum called ***mesocolon***. The large intestine is divided into four principal regions: cecum, colon, rectum, and anal canal.

The opening from the ileum into the large intestine is guarded by a fold of mucous membrane, the ***ileocecal sphincter (valve).*** Hanging inferior to the valve is a blind pouch, the ***cecum,*** to which is attached the ***vermiform appendix*** (*vermis* = worm; *appendix* = appendage) by an extension of visceral peritoneum called the ***mesoappendix***. Inflammation of the vermiform appendix is called ***appendicitis***. The open end of the cecum merges with the ***colon***

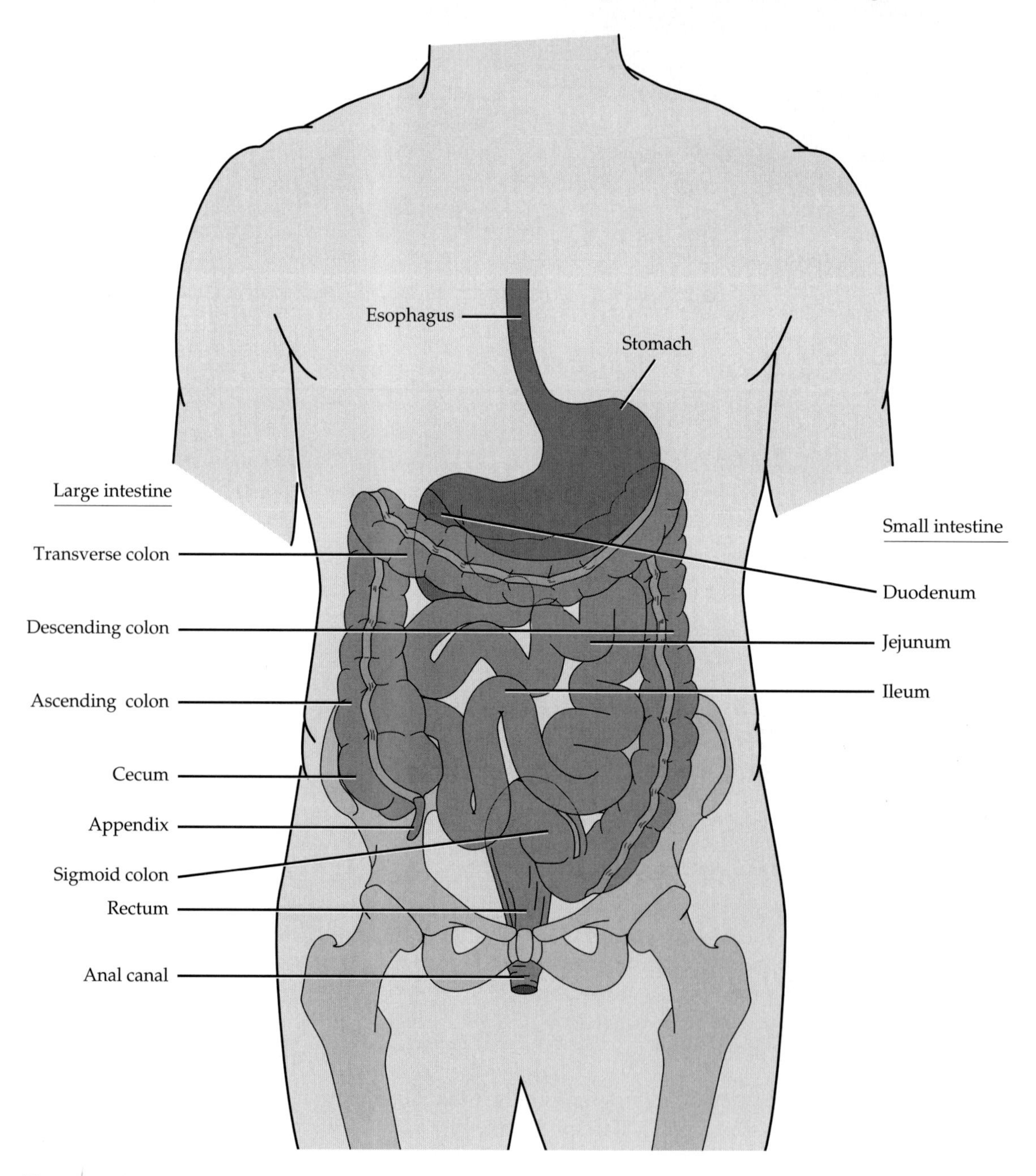

FIGURE 21.12 Intestines. Note the parts of the small intestine on the right side of the illustration and the parts of the large intestine on the left side.

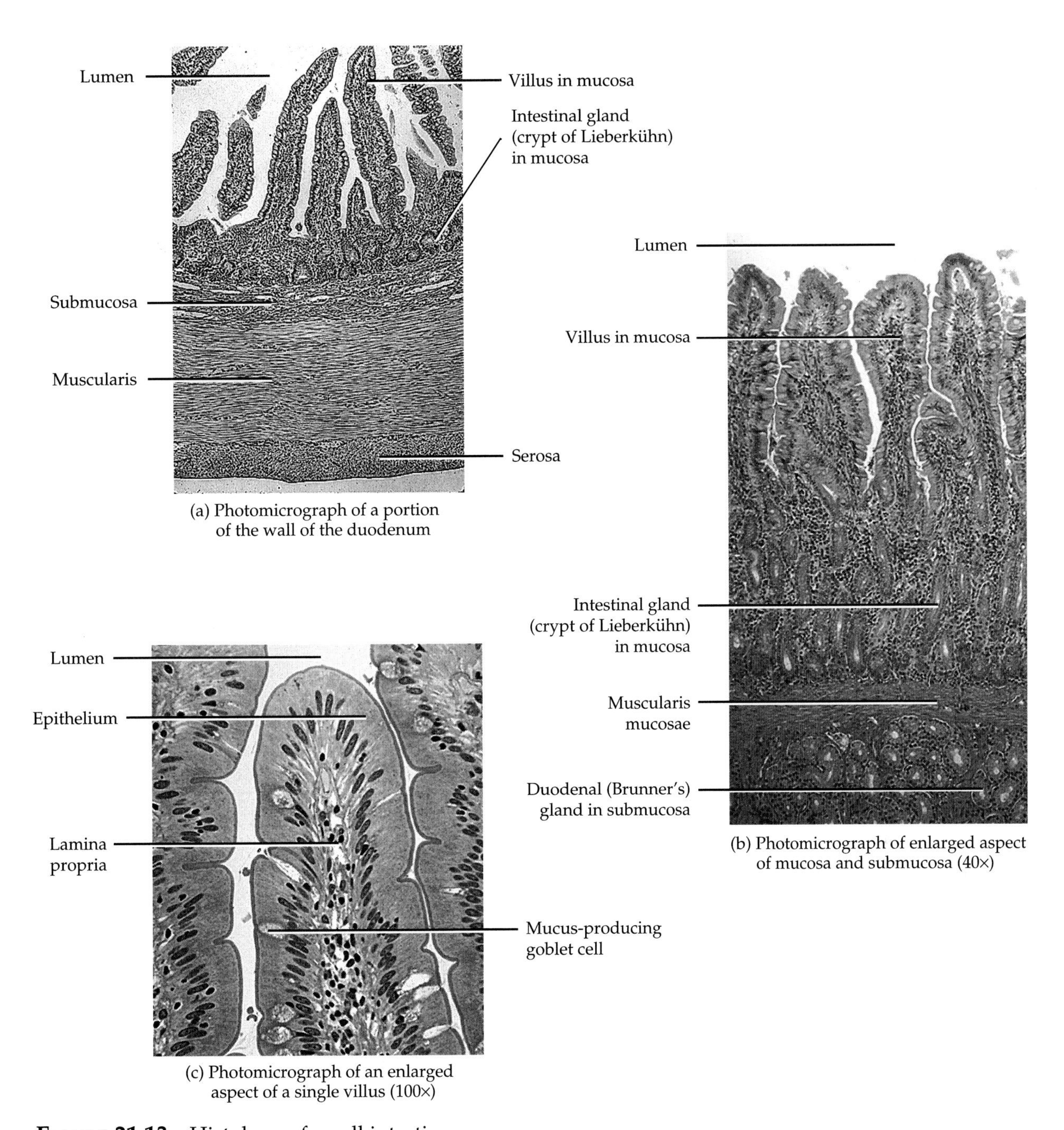

(a) Photomicrograph of a portion of the wall of the duodenum

(b) Photomicrograph of enlarged aspect of mucosa and submucosa (40×)

(c) Photomicrograph of an enlarged aspect of a single villus (100×)

FIGURE 21.13 Histology of small intestine.

(*kolon* = food passage). The first division of the colon is the ***ascending colon,*** which ascends on the right side of the abdomen and turns abruptly to the left at the inferior surface of the liver ***(right colic [hepatic] flexure).*** The ***transverse colon*** continues across the abdomen, curves at the inferior surface of the spleen ***(left colic [splenic] flexure),*** and passes down the left side of the abdomen as the ***descending colon.*** The ***sigmoid colon*** begins near the iliac crest, projects medially toward the midline, and terminates at the rectum at the level of the third sacral vertebra. The ***rectum*** is the last 20 cm (7 to 8 in.) of the gastrointestinal tract. Its terminal 2 to 3 cm (1 in.) is known as the ***anal canal.*** The opening of the anal canal to the exterior is the ***anus.***

With the aid of your textbook, label the parts of the large intestine in Figure 21.14.

The ***mucosa*** of the large intestine consists of simple columnar epithelium with numerous ***goblet cells*** (Figure 21.15 on page 421). The ***submucosa*** is similar to that in the rest of the gastrointestinal tract. The ***muscularis*** consists of an outer longitudinal

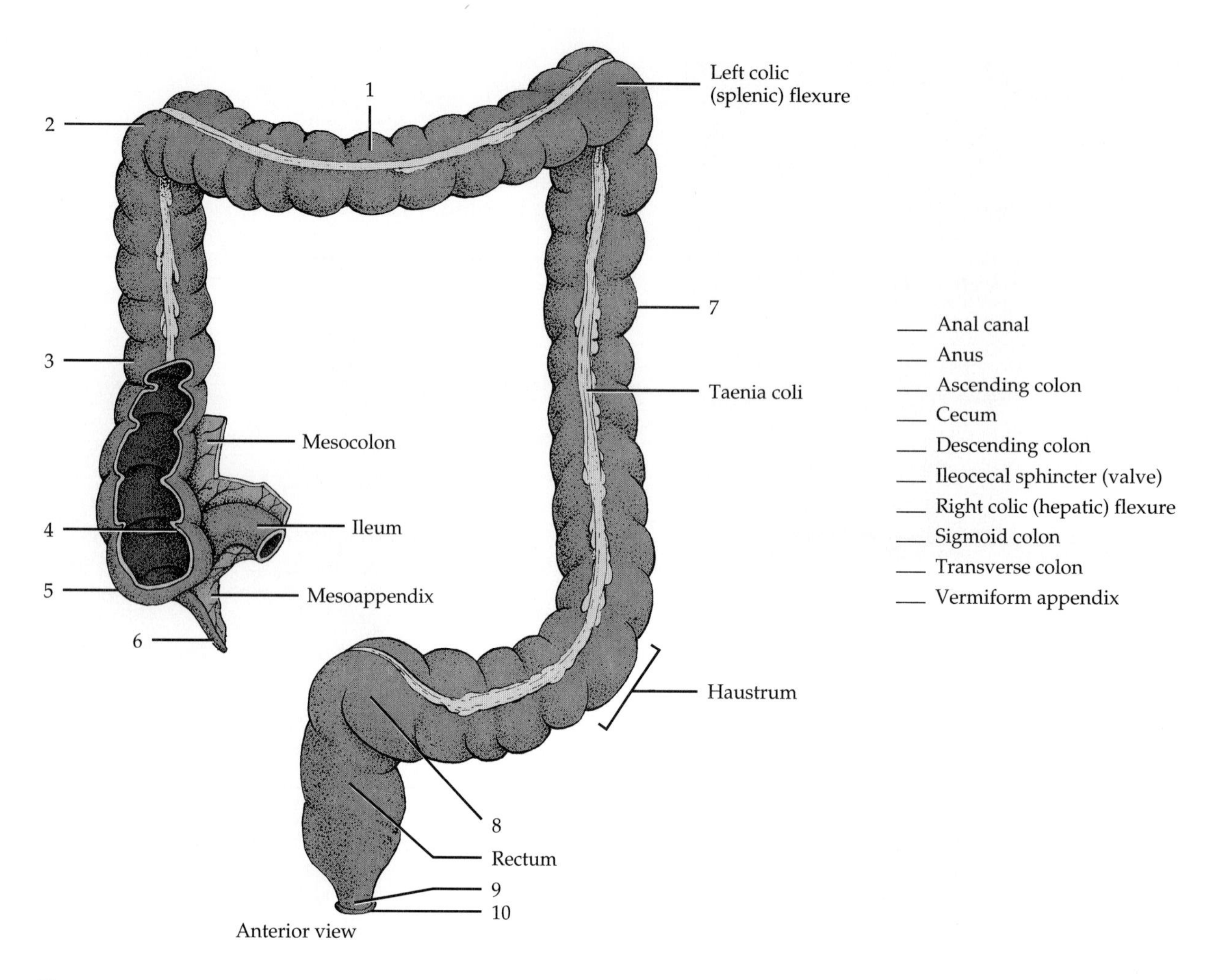

FIGURE 21.14 Large intestine.

layer and an inner circular layer of smooth muscle. However, the longitudinal layer is not continuous; it is broken up into three flat bands, the ***taeniae coli*** (TĒ-nē-a KŌ-lī; *taenia* = flat), which gather the colon into a series of pouches called ***haustra*** (HAWS-tra). Singular is ***haustrum*** (= shaped like a pouch) (see Figure 21.14). The ***serosa*** of the large intestine is visceral peritoneum.

Examine a prepared slide of the large intestine showing its tunics. Compare your observations to Figure 21.15.

C. DISSECTION OF CAT DIGESTIVE SYSTEM

CAUTION! *Please reread Section D, "Precautions Related to Dissection" at the beginning of the laboratory manual on page xi* ***before*** *you begin your dissection.*

PROCEDURE

1. Salivary Glands

1. The head of the cat contains five pairs of ***salivary glands.***
2. Working on the right side of the head of your specimen, locate the following glands (Figure 21.16 on page 422).

a. PAROTID GLAND

1. The largest of the salivary gland, the parotid is found in front of the ear over the masseter muscle.
2. The ***parotid duct*** passes over the masseter muscle and pierces the cheek opposite the last upper premolar tooth.
3. Look inside the cheek to identify the opening of the duct. Do not confuse the parotid duct with the branches of the anterior facial nerve. One branch is dorsal and one is ventral to the parotid duct.

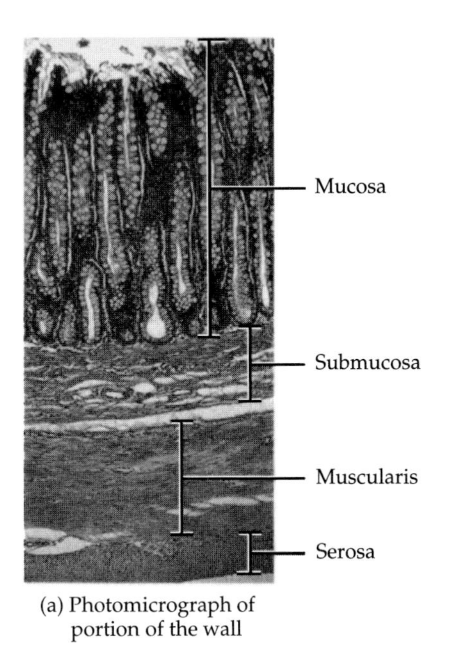

(a) Photomicrograph of portion of the wall

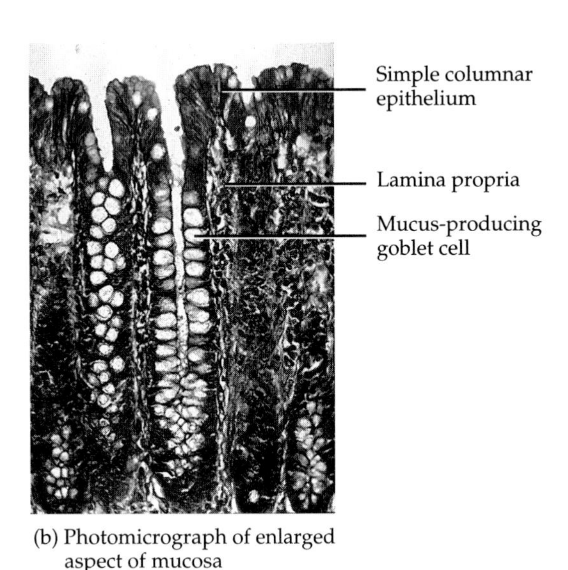

(b) Photomicrograph of enlarged aspect of mucosa

FIGURE 21.15 Histology of large intestine.

b. SUBMANDIBULAR GLAND

1. This gland is below the parotid at the angle of the jaw. The posterior facial vein passes over it.
2. Below the submandibular gland is a lymph node that you should remove. Lymph nodes have smooth surfaces, whereas salivary glands have lobulated surfaces.
3. Lift the anterior margin of the submandibular gland to find the ***submandibular duct.***
4. To trace the duct forward, dissect, transect, and reflect the digastric and mylohyoid muscles. The submandibular duct is joined by the duct from the sublingual gland, which is usually difficult to identify.
5. The ducts of the submandibular and the sublingual glands open into the floor of the mouth near the lower incisors.

c. SUBLINGUAL GLAND

This gland is cranial to the submandibular gland. Its duct joins that of the submandibular gland.

d. MOLAR GLAND

This is a very small mass located near the angle of the mouth between the masseter muscle and mandible. Dissection is not necessary.

e. INFRAORBITAL GLAND

This gland is within the floor of the orbit (it is not shown in Figure 21.16).

2. Mouth

1. Open the ***mouth*** by cutting through the muscles at each angle of the jaw with bone shears.
2. Press down on both sides of the mandible to expose the mouth.
3. Identify the following parts (Figure 21.17 on page 423).

a. LIPS

The ***lips*** are the folds of skin around the orifice of the mouth, lined internally by mucous membrane. They are attached to the mandible by a fold of mucous membrane at the midline.

b. CHEEKS

The ***cheeks*** are the lateral borders of the oral cavity.

c. VESTIBULE

The ***vestibule*** is the space between lips and teeth.

d. ORAL CAVITY

The ***oral cavity*** is the part of the mouth behind the teeth.

e. TEETH

1. The ***teeth*** of a cat may be represented by the dental formula $\frac{3\text{–}1\text{–}3\text{–}1}{3\text{–}1\text{–}2\text{–}1}$. The numbers represent, from left to right, the number of incisors, canines, premolars, and molars on each side. The upper row of numbers represents teeth in the upper jaw and the lower row represents teeth in the lower jaw.

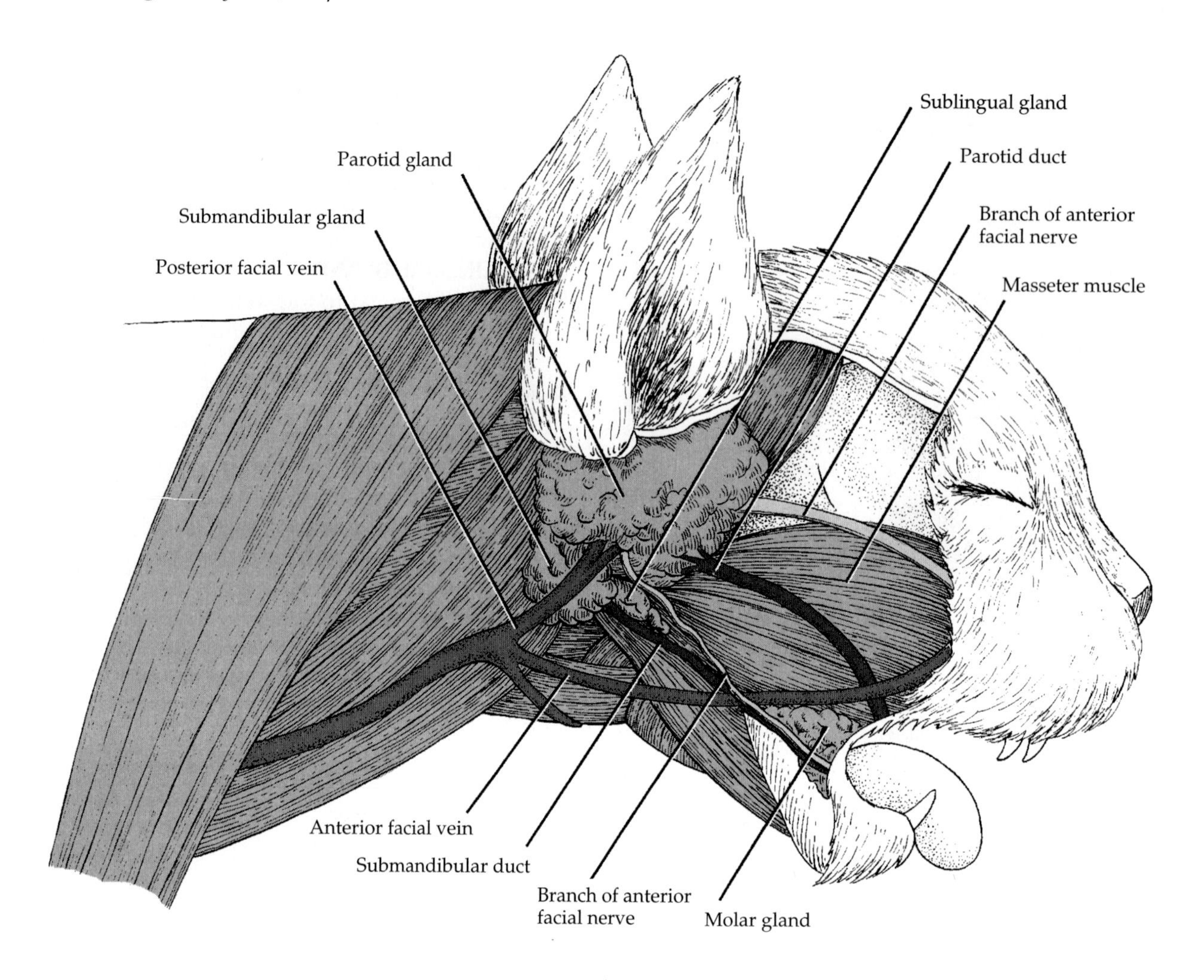

FIGURE 21.16 Salivary glands of cat.

2. Observe the teeth. How many teeth are in a complete permanent set?

f. TONGUE

1. The ***tongue*** forms the floor of the oral cavity.
2. Observe the ***lingual frenulum,*** a fold of mucous membrane that anchors the tongue to the floor of the mouth.
3. Now examine the dorsum (free surface) of the tongue and, using a hand lens, find the papillae.
4. The ***filiform papillae*** are located mostly in the front and middle of the tongue. They are pointed (spinelike) and are the most numerous of the papillae.
5. The ***fungiform papillae*** are small, mushroom-shaped structures between and behind the filiform papillae.
6. The ***circumvallate (vallate) papillae,*** about 12 in number, are large and rounded and are surrounded by a circular groove. They are found near the back of the tongue.
7. The ***foliate papillae,*** leaf-shaped and relatively few in number, are lateral to the vallate papillae. Taste buds are located between papillae.

g. HARD PALATE

1. This forms the cranial part of the palate. The bones of the hard palate are the palatine

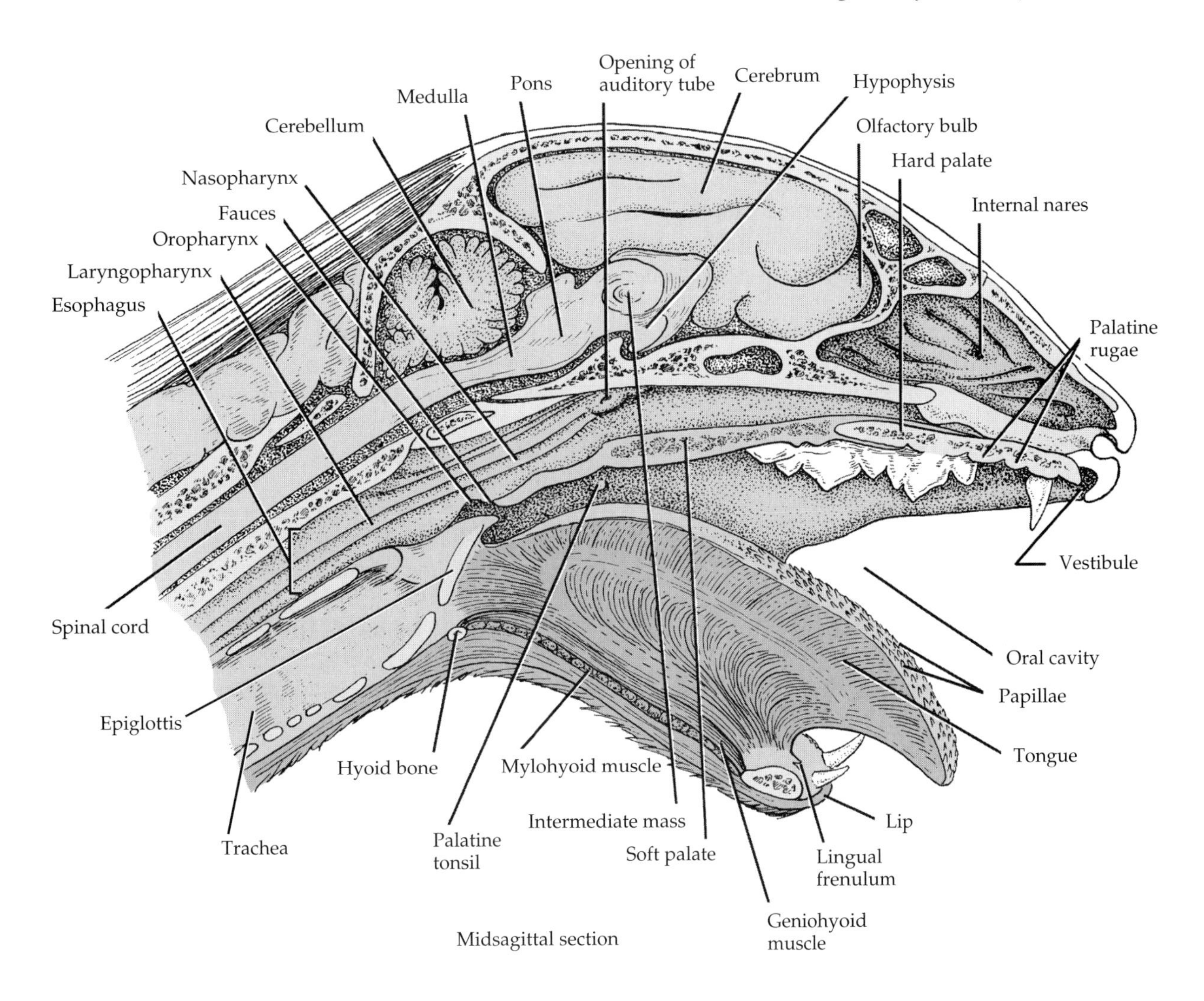

FIGURE 21.17 Head and neck of cat.

process of the premaxilla, the palatine process of the maxilla, and the palatine bone.

2. The mucosa of the hard palate contains transverse ridges called ***palatine rugae.***

h. SOFT PALATE

This forms the caudal portion of the palate and has no bony support.

i. PALATINE TONSILS

1. These are small rounded masses of lymphatic tissue in the lateral wall of the soft palate near the base of the tongue.
2. The tonsils are between lateral folds of tissue called the ***glossopalatine arches.*** The arches are best seen by pulling the tongue ventrally.

3. Pharynx

1. The ***pharynx*** (throat) is a common passageway for the digestive and respiratory systems.
2. Although the pharynx will not be dissected at this time, you can locate some of its important parts in Figure 21.17. These include the ***nasopharynx,*** the part of the pharynx above the soft palate; the ***auditory (Eustachian) tubes,*** a pair of slitlike openings in the laterodorsal walls of the nasopharynx; the ***internal nares (choanae),*** which enter the cranial portion of the nasopharynx; the ***oropharynx,*** the portion of the pharynx between the glossopalatine arches and free posterior margin of the soft palate (the ***fauces*** is the opening between the oral cavity and oropharynx); and the ***laryngopharynx,*** the part of the pharynx behind the larynx.

3. The laryngopharynx opens into the esophagus and larynx.

4. Esophagus

1. The *esophagus* is a muscular tube that begins at the termination of the laryngopharynx (Figure 21.17) It extends through the thoracic cavity, pierces the diaphragm, and terminates at the stomach.
2. The esophagus is dorsal to and a little left of the trachea, running posteriorly toward the diaphragm.
3. Just above the diaphragm, it lies dorsal to the heart and ventral to the aorta.

5. Abdominal Structures

a. PERITONEUM

1. The interior of the abdominal wall is lined by the *parietal peritoneum.*
2. The various abdominal organs are covered by *visceral peritoneum.*
3. Extensions of the peritoneum between the abdominal wall and viscera are termed *mesenteries, ligaments,* and *omenta.* Within them are blood vessels, lymphatics, and nerves that supply the viscera.
4. Other peritoneal extensions are the greater omentum, lesser omentum, and mesentery (Figure 21.18).

b. GREATER OMENTUM

1. This is a double sheet of peritoneum that attaches the greater curvature of the stomach to the dorsal body wall.
2. The greater omentum encloses the spleen and part of the pancreas and covers the transverse colon and a large part of the small intestine.
3. The cavity between the layers of the greater omentum is called the *omental bursa.* It contains considerable fat and is often entwined with the intestine.
4. The portion of the greater omentum between the stomach and spleen is called the *gastrosplenic ligament.*
5. Remove the greater omentum.

c. LESSER OMENTUM

1. This extends from the left lateral lobe of the liver to the stomach and the duodenum.
2. The portion between the stomach and the liver is called the *hepatogastric ligament,* and the portion between the duodenum and the liver is called the *hepatoduodenal ligament.*
3. Within the right lateral border of the lesser omentum are the common bile duct, hepatic artery, and portal vein.

d. MESENTERY

1. The *mesentery proper* suspends the small intestine from the dorsal body wall.
2. The *mesocolon* suspends the colon from the dorsal body wall.
3. Note the blood vessels and lymph nodes in the mesenteries.

e. LIVER

1. The *liver* is the largest abdominal organ. It is reddish brown and is located immediately caudal to the diaphragm (Figure 21.18).
2. If you pull the liver and diaphragm apart, you can see the central portion of the diaphragm is formed by a tendon into which its muscle fibers insert. This is the *central tendon* of the diaphragm.
3. The liver is dived into *right* and *left lobes.* Between these two main lobes is the *falciform ligament.*
4. The falciform ligament is continuous with the *coronary ligament,* which attaches the liver to the central tendon of the diaphragm.
5. At the free edge of the falciform ligament is a fibrous strand that represents the vestige of the umbilical vein. This strand is known as the *round ligament.*
6. Each main lobe of the liver is further subdivided into two lobes called the lateral and the medial lobes. Thus, there is a *left lateral lobe, left median lobe, right lateral lobe,* and *right median lobe.*
7. Caudal to the left lateral lobe is another small lobe, the *caudate lobe.*

f. GALLBLADDER

The *gallbladder* is a saclike structure located in a depression on the dorsal surface of the right median lobe of the liver (Figure 21.18).

g. STOMACH

1. The *stomach* lies to the left of the abdominal cavity (Figure 21.18). Its position can be noted by lifting the lobes of the liver.
2. Cut through the diaphragm to note the point at which the esophagus enters the stomach.
3. Identify the following parts of the stomach.
   a. *Cardiac region* Portion of the stomach adjacent to the esophagus.

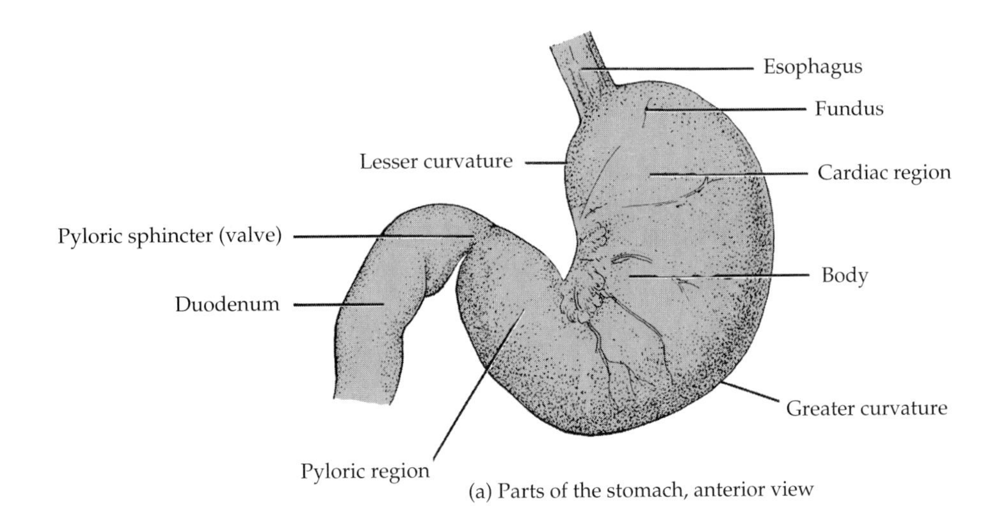

(a) Parts of the stomach, anterior view

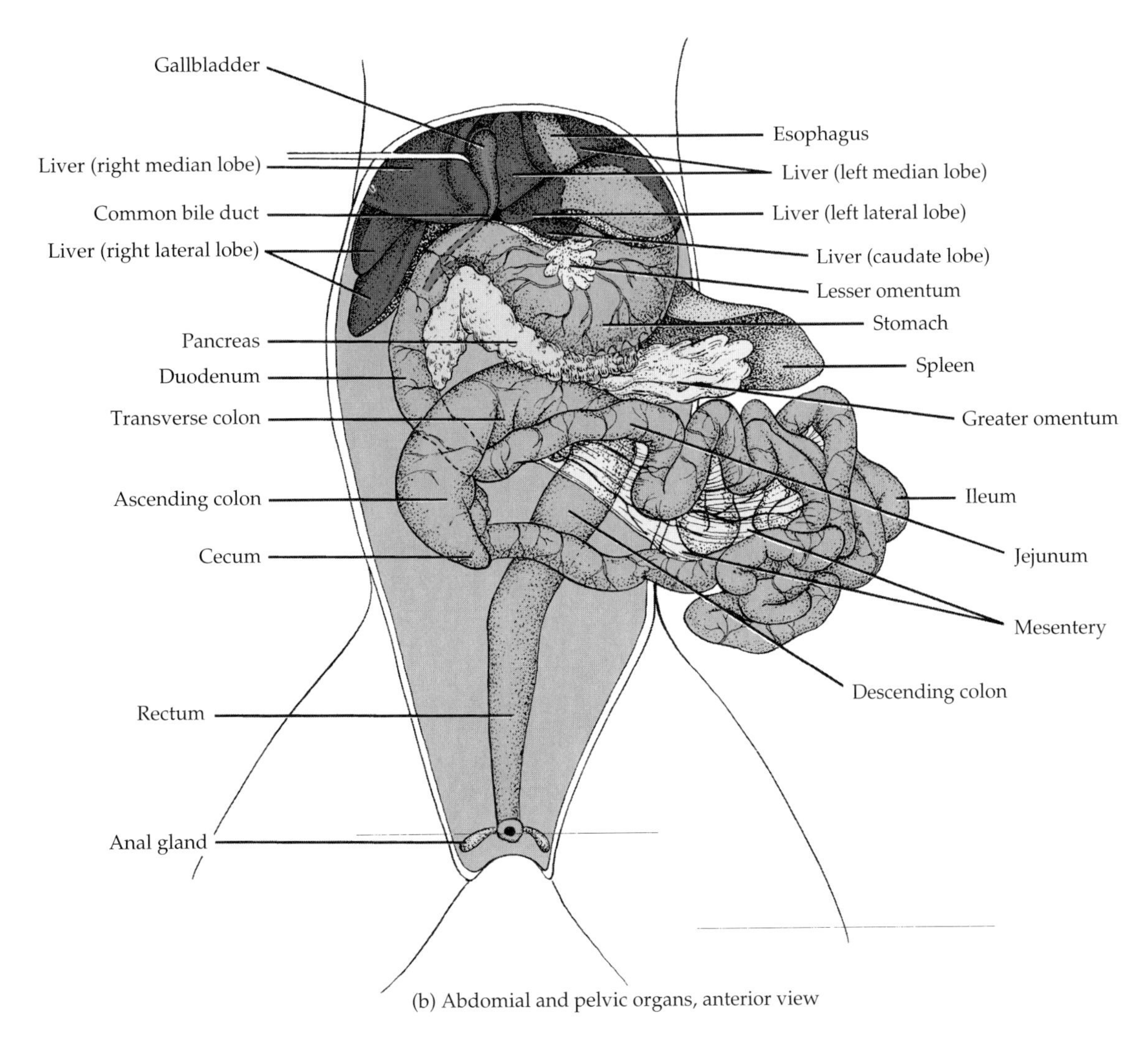

(b) Abdomial and pelvic organs, anterior view

FIGURE 21.18 Head and neck of cat.

b. ***Fundus*** Dome-shaped portion extending cranially to the left of the cardiac region.
c. ***Body*** Large section between the fundus and the pyloric region.
d. ***Pyloric region*** Narrow caudal portion that empties into the duodenum. Between the pyloric region and the duodenum is a valve called the ***pyloric sphincter (valve),*** which can be seen if the stomach is cut open. To see the pyloric sphincter, make a longitudinal incision from the body to the duodenum and then wash the contents out of the stomach. You can also see the longitudinal ***rugae*** if the stomach is cut open.
   i. ***Greater curvature*** The long, left, and caudal margin of the stomach.
   ii. ***Lesser curvature*** The short, right, and cranial margin of the stomach.

h. SPLEEN

The ***spleen*** (which is lymphatic tissue rather than part of the digestive system) is a reddish brown structure on the left side of the abdominal cavity near the greater curvature of the stomach (Figure 21.18).

i. SMALL INTESTINE

1. To examine the small intestine, reflect the greater omentum, which covers the transverse colon and most of the ***small intestine.***
2. The small intestine is divided into three major regions (Figure 21.18). The first region is the ***duodenum,*** which is U-shaped and about 6 in. long. The common bile duct and the pancreatic duct empty into the duodenum near the pyloric sphincter.
3. The enlargement of the duodenum where the ducts unite and enter the duodenum is called the ***hepatopancreatic ampulla*** (Figure 21.19).

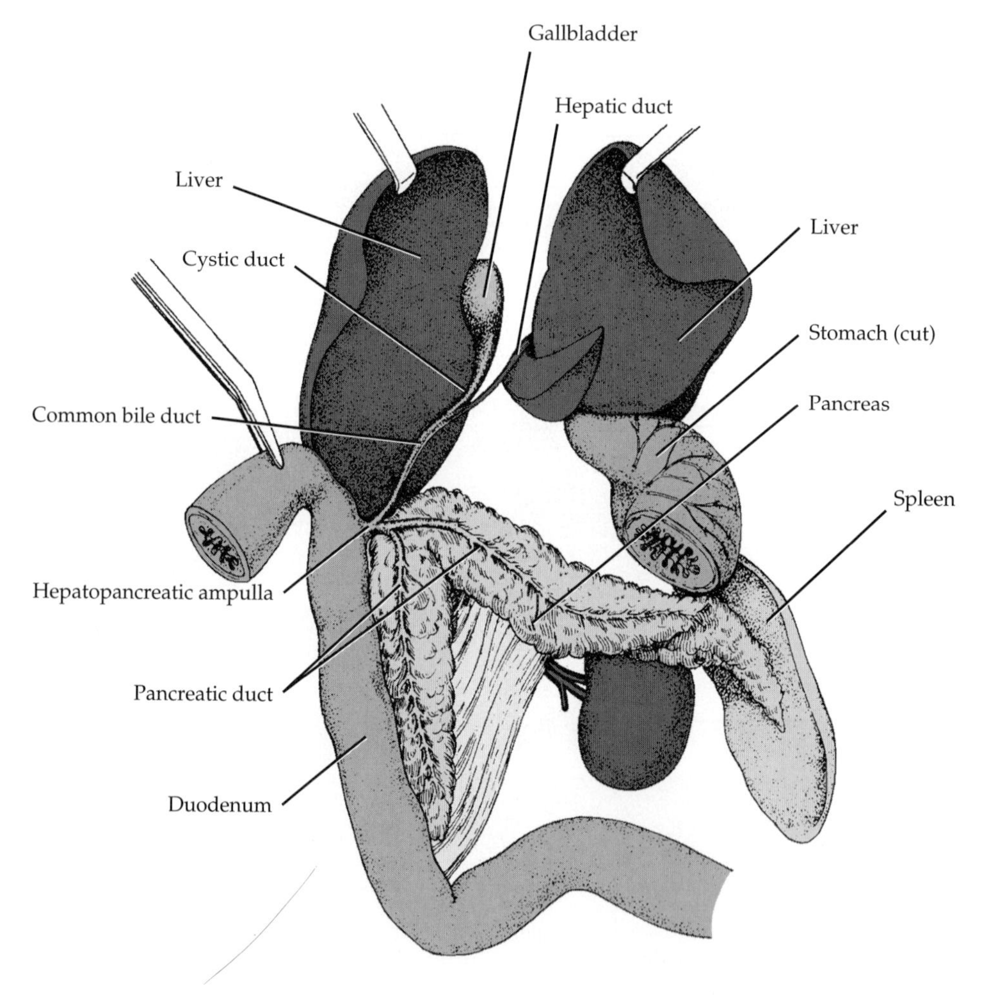

FIGURE 21.19 Relations of the liver, gallbladder, duodenum, and pancreas in cat.

4. The ***common bile duct*** is formed by the union of the ***hepatic ducts*** from the lobes of the liver and the ***cystic duct*** from the gallbladder.
5. The second region of the small intestine is the ***jejunum,*** which composes about the first half of its remaining length (see Figure 21.18).
6. The ***ileum,*** the third region of the small intestine, composes about the second half of its remaining length. There is no definite demarcation between the jejunum and ileum. Trace the ileum to its junction with the large intestine.
7. If you cut open the jejunum or ileum, you can note the velvety appearance of the mucosa. This appearance is due to villi, microscopic fingerlike projections that carry on absorption.
8. Remove any parasitic roundworms or tapeworms.

j. LARGE INTESTINE

1. The beginning of the ***large intestine*** consists of a short, blind pouch called the ***cecum*** (see Figure 21.18). In humans, the appendix is attached to the cecum. Cats have no appendix.
2. Cut open the wall of the cecum and colon opposite the entrance of the ileum.
3. Identify the ***ileocecal sphincter (valve)*** at the junction of the ileum and colon.
4. Cranial to the cecum is the ***ascending colon*** on the right side of the abdominal cavity. Here, it passes across the abdominal cavity as the ***transverse colon.***
5. The transverse colon continues as the ***descending colon,*** which runs caudally toward the midline. The cat has no sigmoid colon (which is the next part of the large intestine in humans).
6. Following the descending colon in the cat is the ***rectum,*** which descends into the pelvic cavity and opens to the exterior as the ***anus.***
7. Lateral to the anus are the ***anal (scent) glands.*** Their ducts empty into the rectum.

k. PANCREAS

1. The ***pancreas*** is a lobulated organ lying against the descending portion of the duodenum and across the body of the spleen (see Figure 21.18).
2. If you carefully tease away the pancreatic tissue, you can identify the ***pancreatic duct*** inside the pancreas. This duct is a white, threadlike structure that conveys pancreatic secretions to the duodenum.
3. The pancreatic duct unites with the common bile duct at the hepatopancreatic ampulla.

ANSWER THE LABORATORY REPORT QUESTIONS AT THE END OF THE EXERCISE.

LABORATORY REPORT QUESTIONS

Digestive System 21

Student ______________________ **Date** ______________________

Laboratory Section ______________________ **Score/Grade** ______________________

PART 1. Multiple Choice

_______ **1.** If an incision has to be made in the small intestine to remove an obstruction, the first layer of tissue to be cut is the (a) muscularis (b) mucosa (c) serosa (d) submucosa

_______ **2.** Mesentery, lesser omentum, and greater omentum are all directly associated with the (a) peritoneum (b) liver (c) esophagus (d) mucosa of the gastrointestinal tract

_______ **3.** Chemical digestion of carbohydrates is initiated in the (a) stomach (b) small intestine (c) mouth (d) large intestine

_______ **4.** A tumor of the villi and circular folds would interfere most directly with the body's ability to carry on (a) absorption (b) deglutition (c) mastication (d) peristalsis

_______ **5.** The main chemical activity of the stomach is to begin the digestion of (a) triglycerides (b) proteins (c) carbohydrates (d) all of the above

_______ **6.** Surgical cutting of the lingual frenulum would occur in which part of the body? (a) salivary glands (b) esophagus (c) nasal cavity (d) tongue

_______ **7.** To free the small intestine from the posterior abdominal wall, which of the following would have to be cut? (a) mesocolon (b) mesentery (c) lesser omentum (d) falciform ligament

_______ **8.** The cells of gastric glands that produce secretions directly involved in chemical digestion are the (a) mucous (b) parietal (c) chief (d) pancreatic islets (islets of Langerhans)

_______ **9.** An obstruction in the hepatopancreatic ampulla (ampulla of Vater) would affect the ability to transport (a) bile and pancreatic juice (b) gastric juice (c) salivary amylase (d) intestinal juice

_______ **10.** The terminal portion of the small intestine is known as the (a) duodenum (b) ileum (c) jejunum (d) pyloric sphincter (valve)

_______ **11.** The portion of the large intestine closest to the liver is the (a) right colic flexure (b) rectum (c) sigmoid colon (d) left colic flexure

_______ **12.** The lamina propria is found in which coat? (a) serosa (b) muscularis (c) submucosa (d) mucosa

_______ **13.** Which structure attaches the liver to the anterior abdominal wall and diaphragm? (a) lesser omentum (b) greater omentum (c) mesocolon (d) falciform ligament

_______ **14.** The opening between the oral cavity and pharynx is called the (a) vermilion border (b) fauces (c) vestibule (d) lingual frenulum

_______ **15.** All of the following are parts of a tooth *except* the (a) crown (b) root (c) cervix (d) papilla

_______ **16.** Cells of the liver that destroy worn-out white and red blood cells and bacteria are termed (a) hepatocytes (b) stellate reticuloendothelial (Kupffer) cells (c) alpha cells (d) beta cells

______ **17.** Bile is manufactured by which cells? (a) alpha (b) beta (c) hepatocytes (d) stellate reticuloendothelial (Kupffer)

______ **18.** Which part of the small intestine secretes the intestinal digestive enzymes? (a) intestinal glands (b) duodenal (Brunner's) glands (c) lacteals (d) microvilli

______ **19.** Structures that give the colon a puckered appearance are called (a) taenia coli (b) villi (c) rugae (d) haustra

PART 2. Completion

20. An acute inflammation of the serous membrane lining the abdominal cavity and covering the abdominal viscera is referred to as ________________.

21. The ________________ is a sphincter (valve) between the ileum and large intestine.

22. The portion of the small intestine that is attached to the stomach is the ________________.

23. The portion of the stomach closest to the esophagus is the ________________.

24. The ________________ forms the floor of the oral cavity and is composed of skeletal muscle covered with mucous membrane.

25. The convex lateral border of the stomach is called the ________________.

26. The three special structures found in the wall of the small intestine that increase its efficiency in absorbing nutrients are the villi, circular folds, and ________________.

27. The three pairs of salivary glands are the parotids, submandibulars, and ________________.

28. The small intestine is divided into three segments: duodenum, ileum, and ________________.

29. The large intestine is divided into four main regions: the cecum, colon, rectum, and ________________.

30. The enzyme that is present in saliva is called ________________.

31. The transition of the lips where the outer skin and inner mucous membrane meet is called the ________________.

32. The ________________ papillae are arranged in the form of an inverted V on the posterior surface of the tongue.

33. The portion of a tooth containing blood vessels, lymphatic vessels, and nerves is the ________________.

34. The teeth present in a permanent dentition, but not in a deciduous dentition, that replace the deciduous molars are the ________________.

35. The portion of the gastrointestinal tract that conveys food from the pharynx to the stomach is the ________________.

36. The inferior region of the stomach connected to the small intestine is the ________________.

37. The clusters of cells in the pancreas that secrete digestive enzymes are called ________________.

38. The caudate lobe, quadrate lobe, and central vein are all associated with the ________________.

39. The common bile duct is formed by the union of the common hepatic duct and ________________ duct.

40. The pear-shaped sac that stores bile is the ____________________.

41. ____________________ glands of the small intestine secrete an alkaline substance to protect the mucosa from excess acid.

42. The ____________________ attaches the large intestine to the posterior abdominal wall.

43. The ____________________ is the last 20 cm (7 to 8 in.) of the gastrointestinal tract.

44. A midline fold of mucous membrane that attaches the inner surface of each lip to its corresponding gum is the ____________________.

45. The bonelike substance that gives teeth their basic shape is called ____________________.

46. The ____________________ anchors teeth in position and helps to dissipate chewing forces.

47. The portion of the colon that terminates at the rectum is the ____________________ colon.

48. The palatine tonsils are between the palatoglossal and ____________________ arches.

49. The teeth closest to the midline are the ____________________.

50. The vermiform appendix is attached to the ____________________.

Urinary System

22

The *urinary system* functions to keep the body in homeostasis by controlling the composition and volume of the blood. The system accomplishes these functions by removing and restoring selected amounts of water and various solutes. The kidneys also excrete selected amounts of various wastes, assume a role in erythropoiesis by secreting erythropoietin, help control blood pH, help regulate blood pressure by secreting renin (which activates the renin-angiotensin pathway), participate in the synthesis of vitamin D, and perform gluconeogenesis (synthesis of glucose molecules) during periods of fasting or starvation.

The urinary system consists of two kidneys, two ureters, one urinary bladder, and a single urethra (see Figure 22.1). Other systems that help in waste elimination are the respiratory, integumentary, and digestive systems.

Using your textbook, charts, or models for reference, label Figure 22.1.

A. ORGANS OF URINARY SYSTEM

1. Kidneys

The paired *kidneys,* which resemble kidney beans in shape, are found just superior to the waist between the parietal peritoneum and the posterior wall of the abdomen. Because they are behind the peritoneal lining of the abdominal cavity, they are referred to as *retroperitoneal* (re'-trō-per-i-tō-NĒ-al; *retro* = behind). The kidneys are located between the levels of the last thoracic and third lumbar vertebrae, with the right kidney slightly lower than the left because of the position of the liver.

Three layers of tissue surround each kidney: the *renal* (*renalis* = kidney) *capsule, adipose capsule,* and *renal fascia.* They function to protect the kidney and hold it firmly in place.

Near the center of the kidney's concave border, which faces the vertebral column, is a vertical fissure called the *renal hilus,* through which the ureter leaves the kidney and through which blood vessels, lymphatic vessels, and nerves enter and exit the kidney. The hilus is the entrance to a cavity in the kidney called the *renal sinus.*

If a frontal section is made through a kidney, the following structures can be seen:

- **a.** *Renal cortex* (*cortex* = rind or bark) Superficial, narrow, reddish area.
- **b.** *Renal medulla* (*medulla* = inner portion) Deep, wide, reddish-brown area.
- **c.** *Renal (medullary) pyramids* Striated triangular structures, 8 to 18 in number, in the renal medulla. The bases of the renal pyramids face the renal cortex, and the apices, called *renal papillae,* are directed toward the center of the kidney. The term *renal lobe* is applied to a renal pyramid and its overlying area of renal cortex.
- **d.** *Renal columns* Cortical substance between renal pyramids.
- **e.** *Renal pelvis* Large cavity in the renal sinus, which is the enlarged proximal portion of the ureter.
- **f.** *Major calyces* (KĀ-li-sēz) Consist of 2 or 3 cuplike extensions of the renal pelvis.
- **g.** *Minor calyces* Consist of 8 to 18 cuplike extensions of the major calyces.

Within the renal cortex and renal pyramids of each kidney are more than 1,000,000 microscopic units called nephrons, the functional units of the kidneys (described shortly). As a result of their activity in regulating the volume and chemistry of the blood, they produce urine. Urine passes from the nephrons to the minor calyces, major calyces, renal pelvis, ureter, urinary bladder, and urethra.

Examine a specimen, model, or chart of the kidney and with the aid of your textbook, label Figure 22.2 on page 435.

2. Nephrons

Basically, a *nephron* (NEF-ron) consists of (1) a renal corpuscle (KOR-pus-sul; *corpus* = body; *cle* = tiny) where fluid is filtered and (2) a renal tubule into

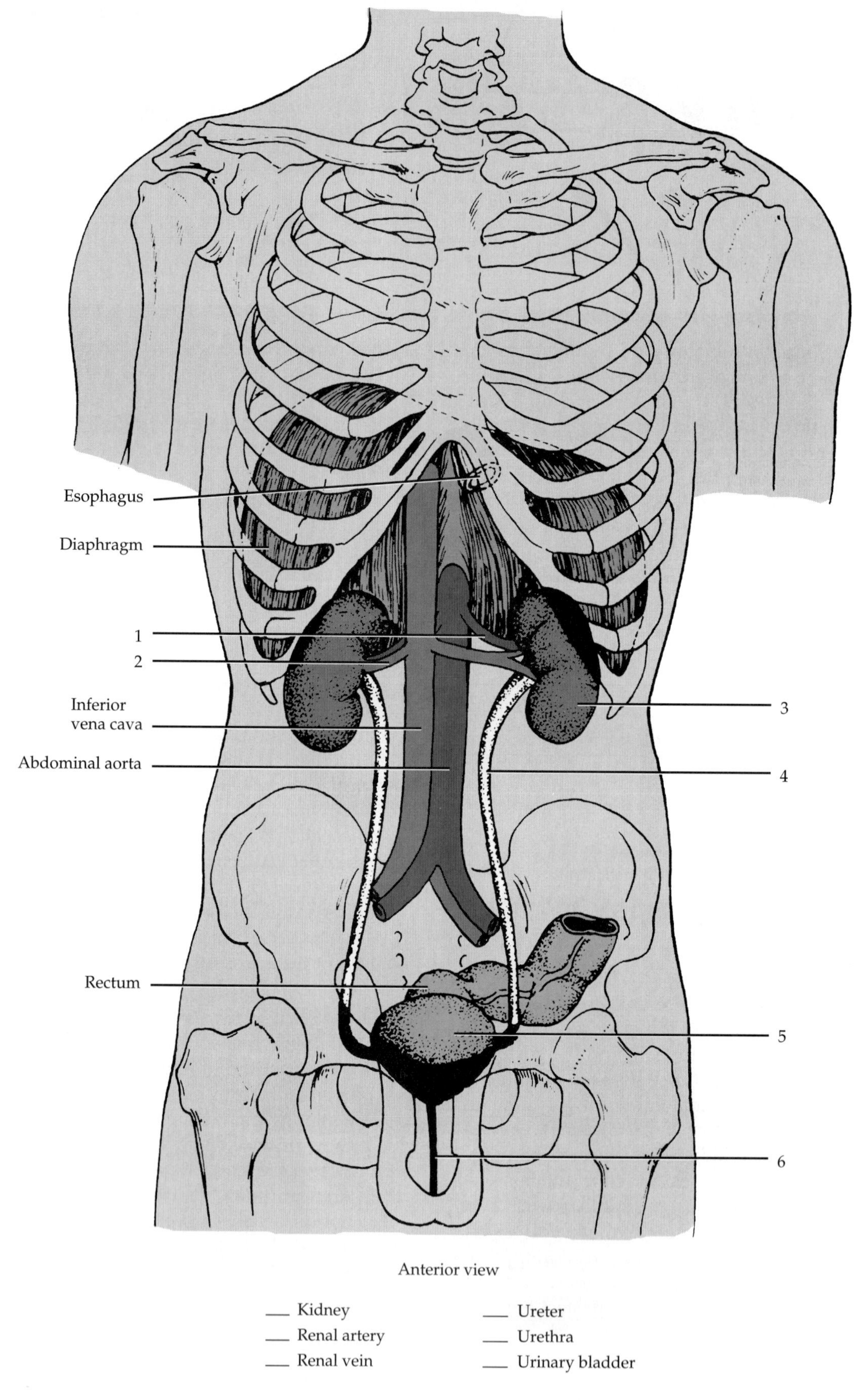

Anterior view

___ Kidney
___ Renal artery
___ Renal vein
___ Ureter
___ Urethra
___ Urinary bladder

FIGURE 22.1 Organs of male urinary system and associated structures.

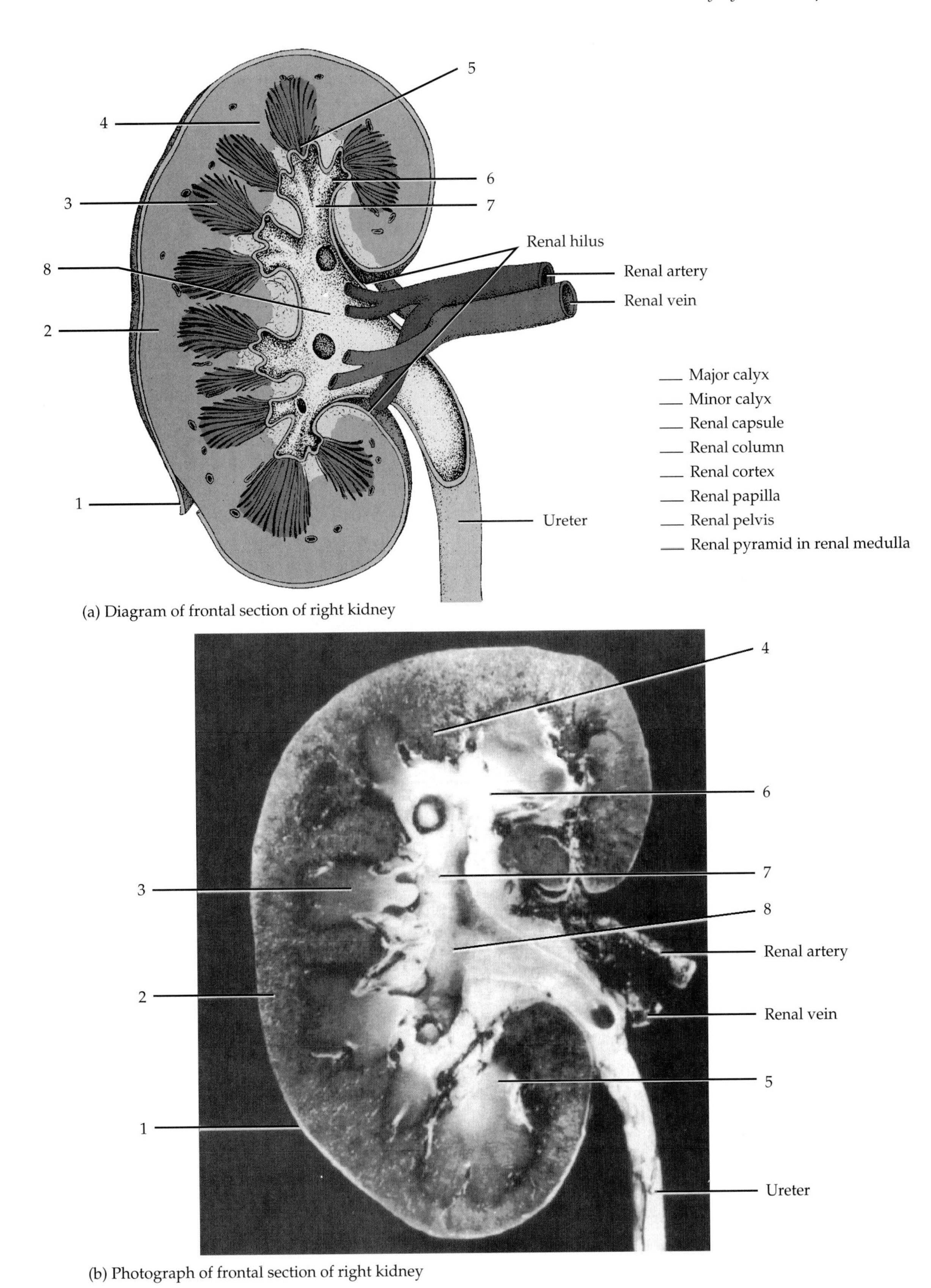

(a) Diagram of frontal section of right kidney

(b) Photograph of frontal section of right kidney

FIGURE 22.2 Kidney.

which the filtered fluid (filtrate) passes. A ***renal corpuscle*** has two components: a tuft (knot) of capillaries called a ***glomerulus*** (glō-MER-yoo-lus; *glomus* = ball; *ulus* = small) surrounded by a double-walled epithelial cup, called a ***glomerular (Bowman's) capsule,*** lying in the renal cortex of the kidney. The inner wall of the capsule, the ***visceral layer,*** consists of epithelial cells called ***podocytes*** and surrounds the glomerulus. A space called the ***capsular (Bowman's) space*** separates the visceral layer from the outer wall of the capsule, the ***parietal layer,*** which is composed of simple squamous epithelium.

The visceral layer of the glomerular (Bowman's) capsule and endothelium of the glomerulus form an ***endothelial-capsular (filtration) membrane,*** a very effective filter. Electron microscopy has determined that the membrane consists of the following components, given in the order in which substances are filtered (Figure 22.3).

a. ***Endothelial fenestrations (pores) of the glomerulus*** The single layer of endothelial cells has large fenestrations (pores) that prevent filtration of blood cells but allow all other components of blood plasma to pass through.
b. ***Basement membrane of the glomerulus*** This layer of extracellular material lies between the endothelium and the visceral layer of the glomerular capsule. It consists of fibrils in a glycoprotein matrix and prevents filtration of larger proteins.
c. ***Slit membranes between pedicels*** The specialized epithelial cells that cover the glomerular

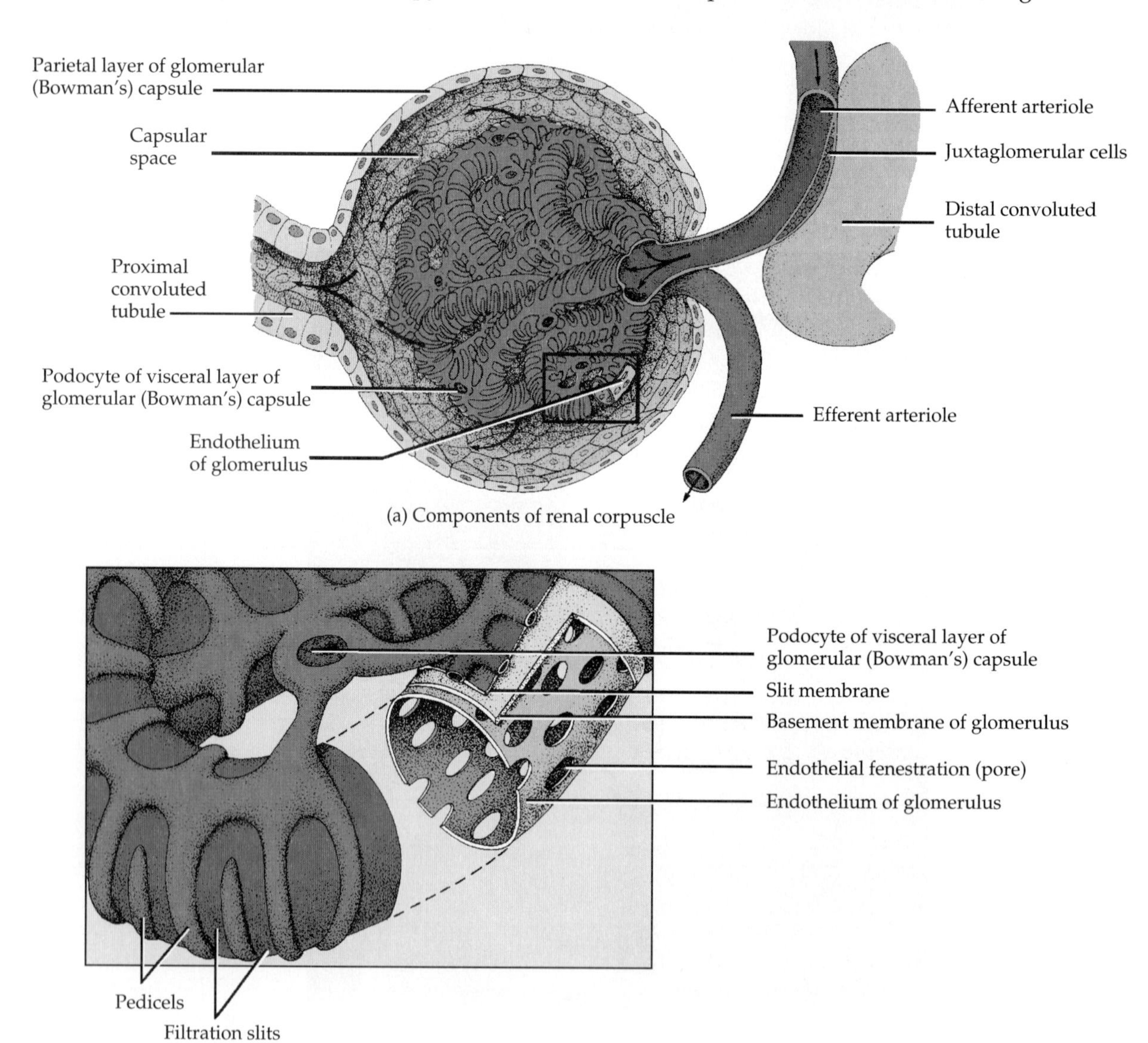

FIGURE 22.3 Endothelial-capsular membrane.

capillaries are called ***podocytes*** (*podos* = foot). Extending from each podocyte are thousands of footlike structures called ***pedicels*** (PED-i-sels; *pediculus* = little foot). The pedicels cover the basement membrane, except for spaces between them, which are called ***filtration slits.*** A thin membrane, the ***slit membrane***, extends across filtration slits and prevents filtration of medium-sized proteins.

The endothelial-capsular membrane filters blood passing through the kidney. Blood cells and large molecules, such as proteins, are retained by the filter and eventually are recycled into the blood. The filtered substances pass through the membrane and into the space between the parietal and visceral layers of the glomerular (Bowman's) capsule, and then enter the renal tubule (described shortly).

As the filtered fluid (filtrate) passes through the remaining parts of a nephron, substances are selectively added and removed. The end product of these activities is urine. Nephrons are frequently classified into two kinds. A ***cortical nephron*** usually has its glomerulus in the superficial renal cortex, and the remainder of the nephron penetrates only into the superficial renal medulla. A ***juxtamedullary nephron*** usually has its glomerulus close to the corticomedullary junction, and other parts of the nephron penetrate deeply into the renal medulla (see Figure 22.4). The following description of the remaining components of a nephron applies to juxtamedullary nephrons.

After the filtrate leaves the glomerular (Bowman's) capsule, it passes through the following parts of a renal tubule:

a. ***Proximal convoluted tubule (PCT)*** Coiled tubule in the renal cortex that originates at the glomerular (Bowman's) capsule; consists of simple cuboidal epithelium with microvilli.
b. ***Loop of Henle (nephron loop)*** U-shaped tubule that connects the proximal and distal convoluted tubules. It consists of a ***descending limb of the loop of Henle,*** an extension of the proximal convoluted tubule that dips down into the renal medulla and consists of simple squamous epithelium, and an ***ascending limb of the loop of Henle*** that ascends into the renal medulla and approaches the renal cortex and consists of simple squamous, cuboidal, and columnar epithelium; the ascending limb is wider in diameter than the descending limb.
c. ***Distal convoluted tubule (DCT)*** Coiled extension of ascending limb in the renal cortex; consists of simple cuboidal epithelium with fewer microvilli than in the proximal convoluted tubule.

Distal convoluted tubules terminate by merging with straight ***collecting ducts***. A ***renal lobule*** is a portion of a kidney within a renal lobe that includes a collecting duct and all the nephrons that drain into it. In the renal medulla, collecting ducts receive distal convoluted tubules from several nephrons, pass through the renal pyramids, and open at the renal papillae into minor calyces through about 30 large ***papillary ducts***. The processed filtrate, called urine, passes from the collecting ducts to papillary ducts, minor calyces, major calyces, renal pelvis, ureter, urinary bladder, and urethra.

With the aid of your textbook, label the parts of a nephron and associated structures in Figure 22.4.

Examine prepared slides of various components of nephrons and compare your observations to Figure 22.5 on page 439.

3. Blood and Nerve Supply

Nephrons are abundantly supplied with blood vessels, and the kidneys actually receive around 20 to 25% of the total cardiac output, or approximately 1200 ml, every minute. The blood supply originates in each kidney with the ***renal artery,*** which divides into many branches, eventually supplying the nephron and its complete tubule.

Before or immediately after entering the renal hilus, the renal artery divides into a larger anterior branch and a smaller posterior branch. From these branches, five ***segmental arteries*** originate. Each gives off several branches, the ***interlobar*** (*inter* = between; *lobar* = renal lobe) ***arteries,*** which pass between the renal lobes in the renal columns. At the bases of the pyramids, the interlobar arteries arch between the renal medulla and renal cortex and here are known as ***arcuate*** (*arcuatus* = shaped like a bow) ***arteries***. Branches of the arcuate arteries, called ***interlobular*** (*inter* = between; *lobular* = renal lobule) ***arteries,*** enter the renal cortex between lobules. ***Afferent*** (*ad* = toward; *ferre* = to carry) ***arterioles,*** branches of the interlobular arteries, are distributed to the ***glomeruli.*** Blood leaves the glomeruli via ***efferent*** (*efferers* = to bring out) ***arterioles.***

The next sequence of blood vessels depends on the type of nephron. Around convoluted tubules, efferent arterioles of cortical nephrons divide to form capillary networks called ***peritubular*** (*peri* = around) ***capillaries.*** Efferent arterioles of juxtamedullary nephrons also form peritubular capillaries and, in

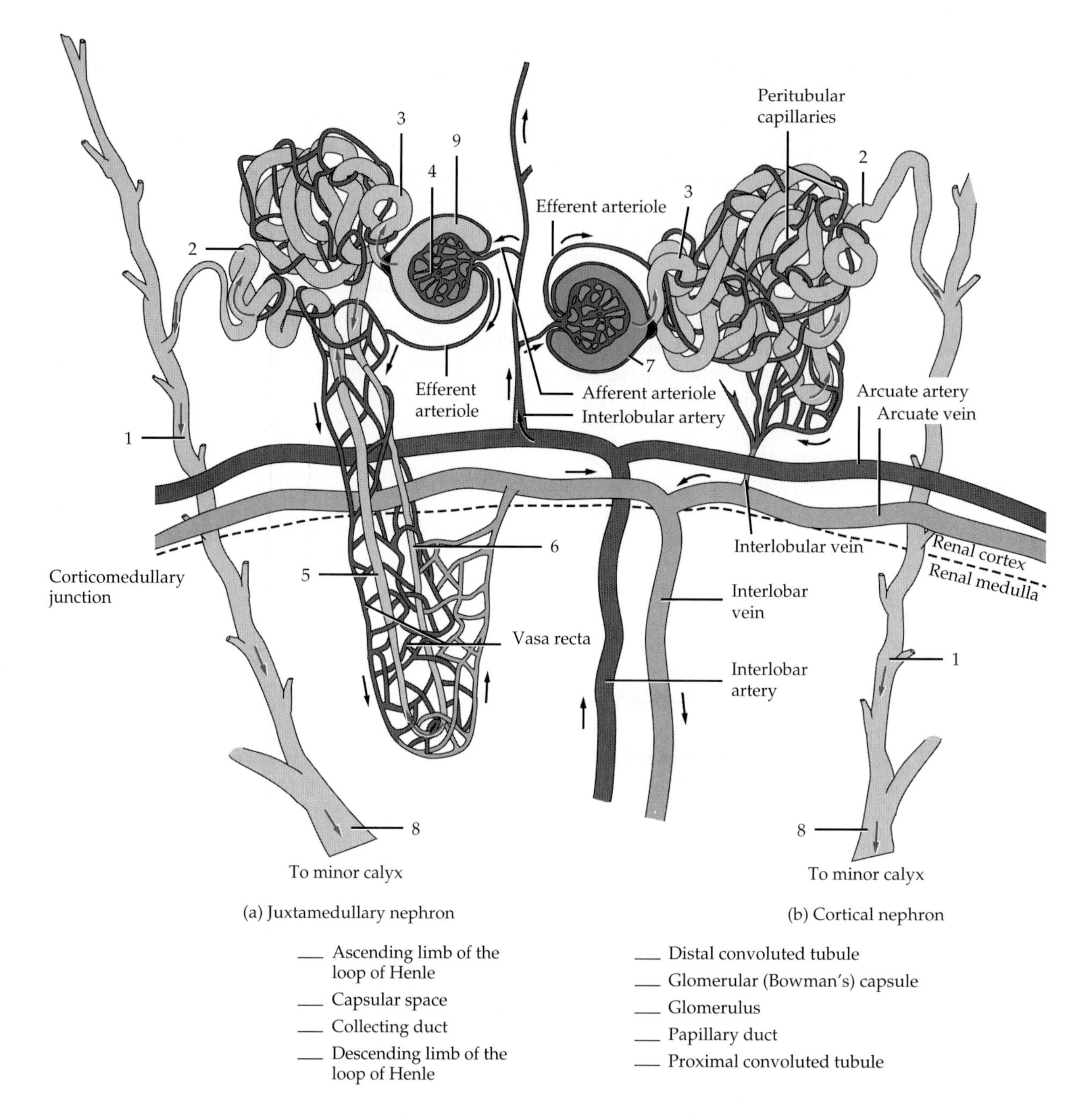

FIGURE 22.4 Nephrons. The arrows inside the nephrons indicate the direction in which the filtrate flows.

addition, form long loops of blood vessels around medullary structures called ***vasa recta*** (VĀ-sa REK-ta; *vasa* = vessels; *recta* = straight). Peritubular capillaries eventually reunite to form ***interlobular veins.*** Blood then drains into ***arcuate veins, interlobar veins,*** and ***segmental veins.*** Blood leaves the kidneys through the ***renal vein*** that exits at the hilus. (The vasa recta pass blood into the interlobular veins, arcuate veins, interlobar veins, and renal veins.)

In each nephron, the final portion of the ascending limb of the loop of Henle makes contact with the afferent arteriole serving its own renal corpuscle. The cells of the renal tubule in this region are tall and crowded together. Collectively, they are known as the ***macula densa*** (*macula* = spot; *densa* = dense). These cells monitor the Na^+ and Cl^- concentration of fluid in the tubule lumen. Next to the macula densa, the wall of the afferent arteriole (and sometimes efferent arteriole) contains modified smooth muscle fibers called ***juxtaglomerular (JG) cells.*** Together with the macula densa, they constitute the ***juxtaglomerular apparatus,*** or ***JGA.***

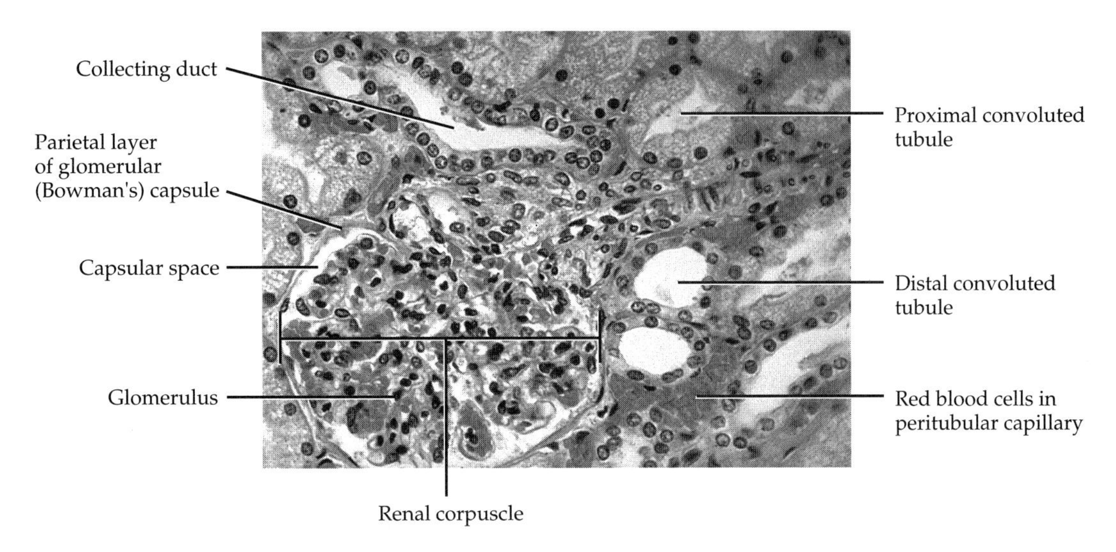

FIGURE 22.5 Histology of nephrons (250×).

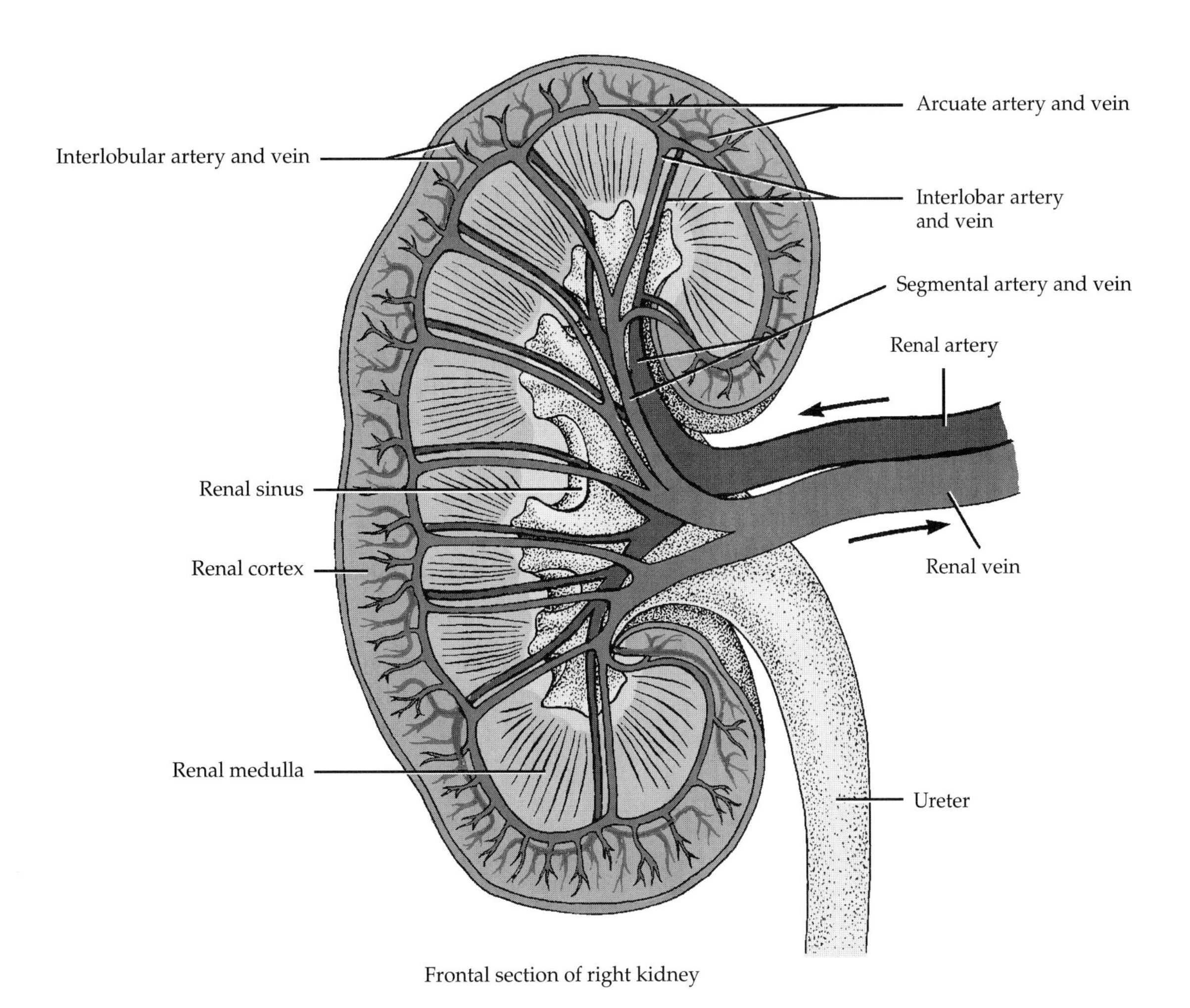

FIGURE 22.6 Macroscopic blood vessels of the kidney. Macroscopic and microscopic blood vessels are shown in Figure 22.4.

The JGA helps regulate arterial blood pressure and the rate of blood filtration by the kidneys. The distal convoluted tubule begins a short distance past the macula densa.

Using Figures 22.4 and 22.6 as guides, trace a drop of blood from its entrance into the renal artery to its exit through the renal vein. As you do so, name in sequence each blood vessel through which blood passes for both cortical and juxtaglomerular nephrons.

Label Figure 22.7.

The nerve supply to the kidneys comes from the ***renal plexus*** of the autonomic system. The nerves are vasomotor because they regulate the circulation of blood in the kidney by regulating the diameters of the arterioles.

4. Ureters

The body has two retroperitoneal ***ureters*** (YOO-re-ters), one for each kidney; each ureter is a continuation of the renal pelvis and runs to the urinary bladder (see Figure 22.1). Urine is carried through the ureters mostly by peristaltic contractions of the muscular layer of the ureters. Each ureter extends 25 to 30 cm (10 to 12 in.). At the base of the urinary bladder, the ureters turn medially and enter the posterior aspect of the urinary bladder.

Histologically, the ureters consist of an inner ***mucosa*** of transitional epithelium and an underlying lamina propria (connective tissue), a middle ***muscularis*** (inner longitudinal and outer circular smooth muscle), and an outer ***adventitia*** (areolar connective tissue).

Examine a prepared slide of the wall of the ureter showing its various layers. With the aid of your textbook, label Figure 22.8.

5. Urinary Bladder

The ***urinary bladder*** is a hollow muscular organ located in the pelvic cavity posterior to the pubic symphysis (see Figure 22.1). In the male, the bladder is directly anterior to the rectum; in the female, it is anterior to the vagina and inferior to the uterus.

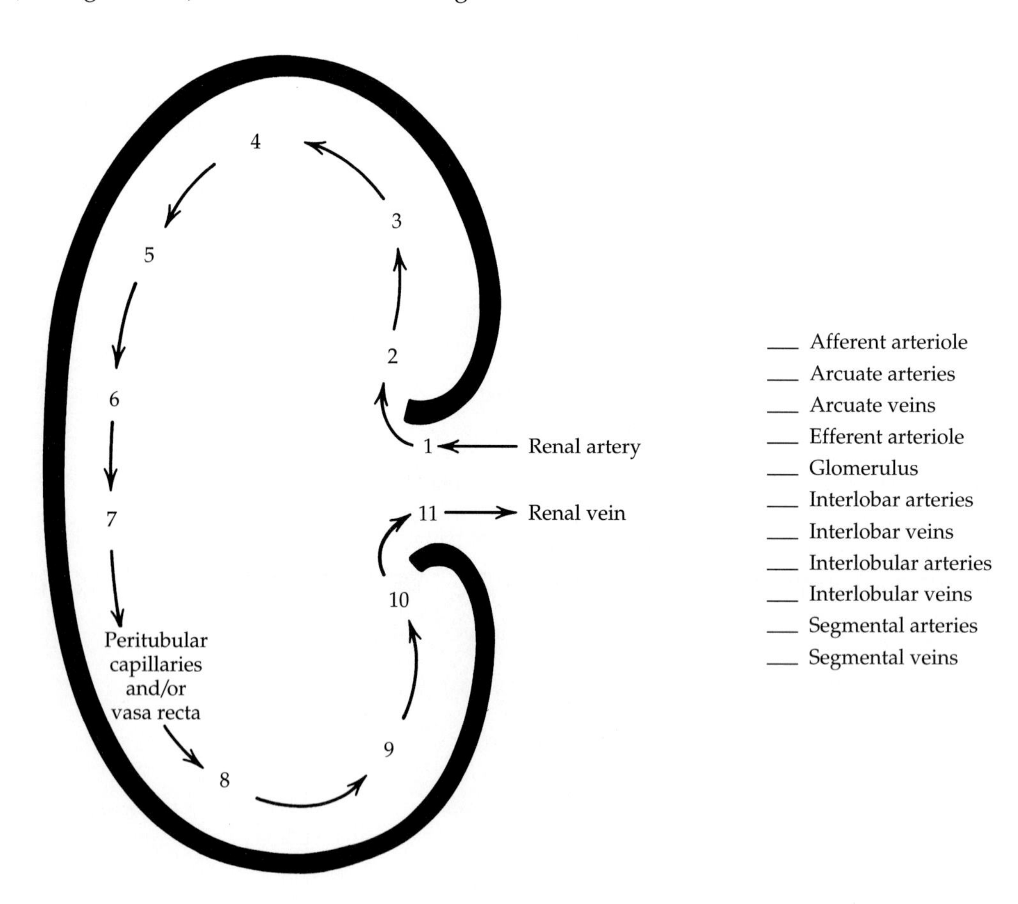

FIGURE 22.7 Blood supply of the right kidney. This view is designed to show the *sequence* of blood flow, not the anatomical location of blood vessels, which is shown in Figure 22.6.

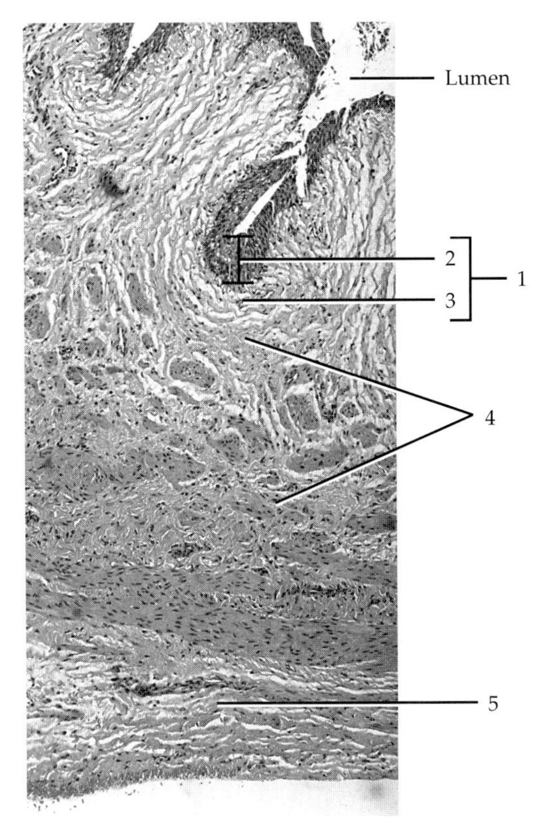

___ Adventitia
___ Lamina propria of mucosa
___ Mucosa
___ Muscularis
___ Transitional epithelium of mucosa

FIGURE 22.8 Histology of ureter.

At the base of the interior of the urinary bladder is the ***trigone*** (TRĪ-gōn; *trigonium* = triangle), a triangular area bounded by the opening into the urethra (internal urethral orifice) and ureteral openings into the bladder. The ***mucosa*** of the urinary bladder consists of transitional epithelium and lamina propria (connective tissue). Rugae (folds in the mucosa) are also present. The ***muscularis,*** also called the ***detrusor*** (de-TROO-ser; *detrudere* = to push down) ***muscle,*** consists of three layers of smooth muscle: inner longitudinal, middle circular, and outer longitudinal. In the region around the opening to the urethra, the circular muscle fibers form an ***internal urethral sphincter.*** Inferior to this is the ***external urethral sphincter*** which is composed of skeletal muscle and is a modification of the urogenital diaphragm. The superficial coat of the urinary bladder is the ***adventitia,*** a layer of areolar connective tissue that is continuous with that of the ureters. Over the superior surface of the urinary bladder, the adventitia is also covered with peritoneum and the two together constitute the ***serosa.***

Urine is expelled from the bladder by an act called ***micturition*** (mik'-too-RISH-un; *micturire* = to urinate), commonly known as urination, or voiding. The average capacity of the urinary bladder is 700 to 800 ml.

Using your textbook as a guide, label the external urethral sphincter, internal urethral sphincter, ureteral openings, and ureters in Figure 22.9.

Examine a prepared slide of the wall of the urinary bladder. With the aid of your textbook, label Figure 22.10 on page 443.

6. Urethra

The ***urethra*** is a small tube leading from the internal urethral orifice in the floor of the urinary bladder to the exterior of the body. In females, this tube is posterior to the pubic symphysis and is embedded in the anterior wall of the vagina; its length is approximately 3.8 cm (1½ in.). The opening of the urethra to the exterior, the ***external urethral orifice,*** is between the clitoris and vaginal orifice. In males, its length is around 20 cm (8 in.), and it follows a route different from that of the female. Immediately inferior to the urinary bladder, the urethra passes through the prostate gland (prostatic urethra), pierces the urogenital diaphragm (membranous urethra), and traverses the penis (spongy urethra). The urethra is the terminal portion of the urinary system, and serves as the passageway for discharging urine from the body. In addition, in the male, the urethra serves as the duct through which reproductive secretions are discharged from the body.

Label the urethra and urethral orifice in Figure 22.9.

B. DISSECTION OF CAT URINARY SYSTEM

PROCEDURE

CAUTION! *Please reread Section D, "Precautions Related to Dissection" at the beginning of the laboratory manual on page xi before you begin your dissection.*

1. Kidneys

1. The ***kidneys*** are situated on either side of the vertebral column at about the level of the third to fifth lumbar vertebrae. The right kidney is

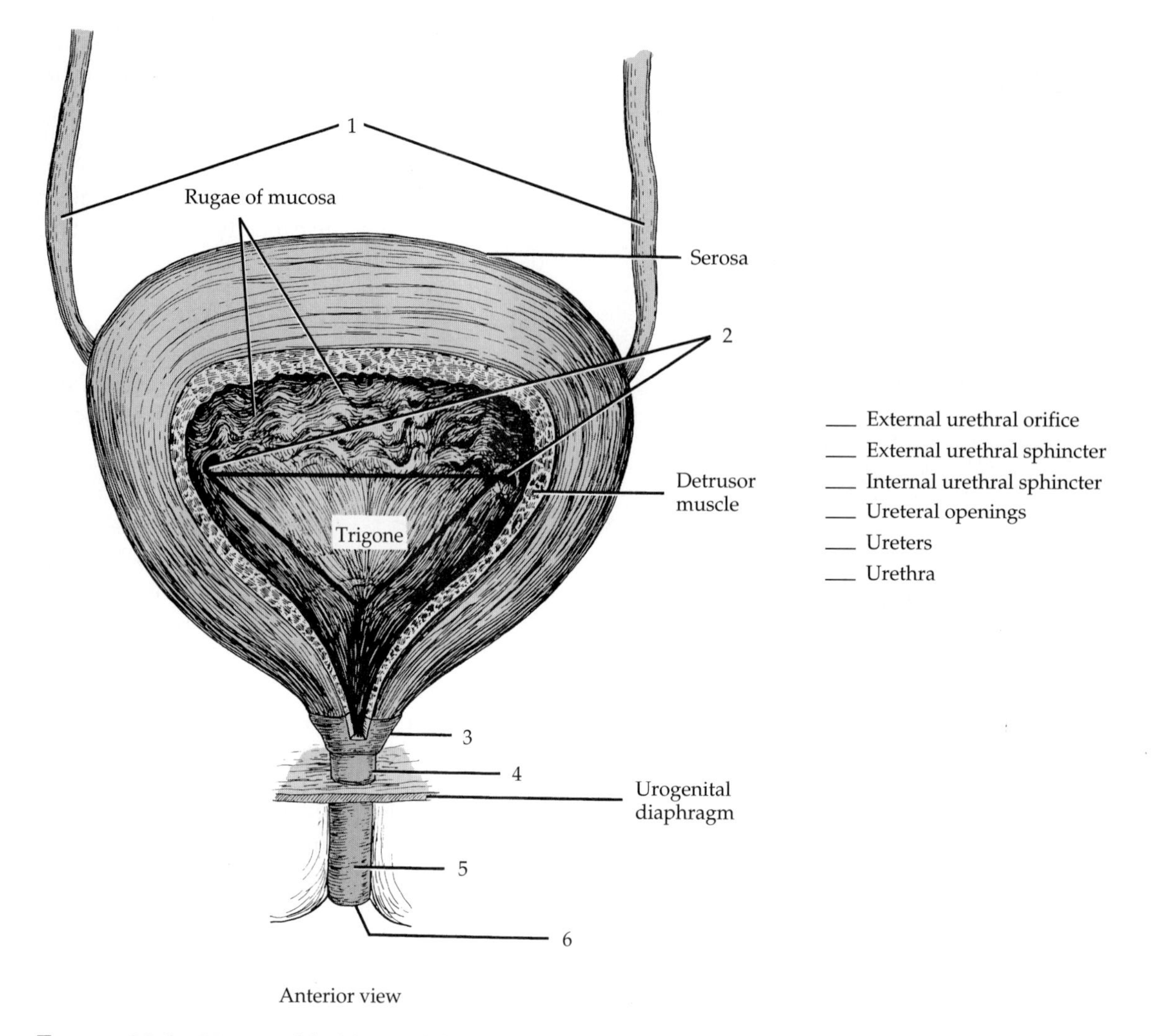

FIGURE 22.9 Urinary bladder and female urethra.

slightly higher than the left (Figure 22.11 on page 444).

2. Each is surrounded by a mass of fat called the ***adipose capsule,*** which should be removed.
3. Unlike other abdominal organs that are suspended by an extension of the peritoneum, the kidneys are covered by parietal peritoneum only on their ventral surfaces. Because of this, the kidneys are said to be ***retroperitoneal.***
4. Identify the ***adrenal (suprarenal) glands*** located cranial and medial to the kidneys. They are part of the endocrine system.
5. Note that the medial surface of each kidney contains a concave opening, the ***renal hilus,*** through which blood vessels and the ureter enter or leave the kidney.
6. Identify the ***renal artery*** branching off the aorta and entering the renal hilus and the ***renal vein*** leaving the renal hilus and joining the inferior vena cava. Also, identify the ureter caudal to the renal vein.
7. If a detailed dissection of the cat kidney is to be done, follow the directions of the sheep kidney dissection and identify the following structures of the cat that correlate with the sheep kidney.
   a. Renal hilus
   b. Renal pelvis
   c. Calyces
   d. Renal cortex
   e. Renal (medullary) pyramids
   f. Renal column
   g. Renal papillae

2. Ureters

1. The ***ureters,*** like the kidneys, are retroperitoneal.
2. Each begins as the renal pelvis in the renal sinus and passes caudally to the urinary bladder.

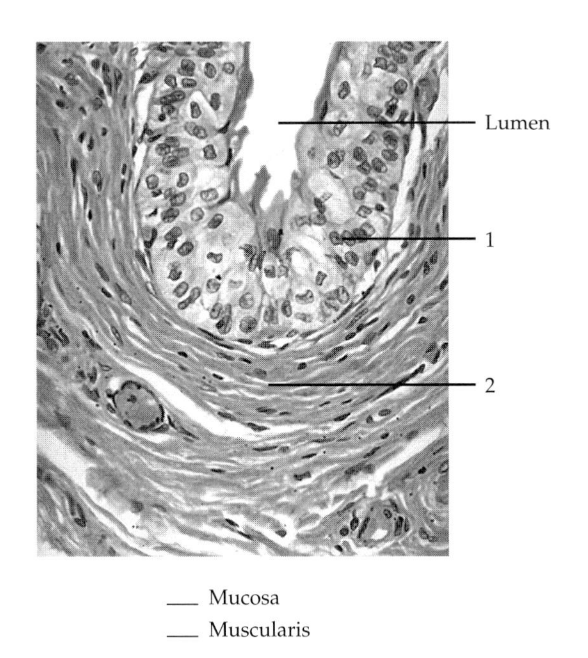

___ Mucosa
___ Muscularis

FIGURE 22.10 Histology of urinary bladder (400×).

3. The ureters pass posterior to the urinary bladder and open into its floor.

3. Urinary Bladder

1. The ***urinary bladder***, a pear-shaped musculomembranous sac located just cranial to the pubic symphysis (Figure 22.11), is also retroperitoneal.
2. The urinary bladder is attached to the abdominal wall by peritoneal folds termed ***ligaments.*** The ***ventral suspensory ligament*** extends from the ventral side of the bladder to the ***linea alba.*** The ***lateral ligaments***, one on either side, connect the sides of the bladder to the dorsal body wall. They contain considerable fat.
3. The broad, rounded, cranial portion of the bladder is called the ***fundus.***
4. The narrow, caudal, attached portion is referred to as the ***neck.***
5. If you have a male cat, identify the ***rectovesical pouch,*** the space between the urinary bladder and the rectum.
6. If you have a female cat, identify the ***vesicouterine*** pouch, the space between the urinary bladder and the uterus.
7. If you cut open the urinary bladder, you might be able to see the openings of the ureters into the bladder.

4. Urethra

1. The ***urethra*** is a duct that conducts urine from the neck of the urinary bladder to the exterior.
2. Dissection of the urethra will be delayed until the reproductive system is studied.

C. DISSECTION OF SHEEP (OR PIG) KIDNEY

The sheep kidney is very similar to both the human and cat kidney. You may use Figure 22.2 as a reference for this dissection.

CAUTION! *Please reread Section D, "Precautions Related to Dissection" at the beginning of the laboratory manual on page xi before you begin your dissection.*

PROCEDURE

1. Examine the intact kidney and notice the renal hilus and the fatty tissue that normally surrounds the kidney. Strip away the fat.
2. As you peel the fat off, look carefully for the ***adrenal (suprarenal) gland.*** This gland is usually found attached to the superior surface of the kidney, as it is in the human. Most preserved kidneys do not have this gland. If it is present, remove it, cut it in half, and note its distinct outer ***cortex*** and inner ***medulla.***
3. Look at the ***renal hilus,*** which is the concave area of the kidney. From here the ***ureter, renal artery,*** and ***renal vein*** enter and exit.
4. Differentiate these blood vessels by examining the thickness of their walls. Which vessel has the thicker wall?
5. With a sharp scalpel *carefully* make a frontal section through the kidney.
6. Identify the ***renal capsule*** as a thin, tough layer of connective tissue completely surrounding the kidney.
7. Immediately beneath this capsule is an outer light-colored area called the ***renal cortex.*** The inner dark-colored area is the ***renal medulla.***
8. The ***renal pelvis*** is the large chamber formed by the expansion of the ureter inside the kidney. This renal pelvis divides into many smaller areas called ***renal calyces,*** each of which has a dark tuft of kidney tissue called a ***renal pyramid.***
9. The bases of these pyramids face the cortical area. Their apices, called ***renal papillae,*** are directed toward the center of the kidney.

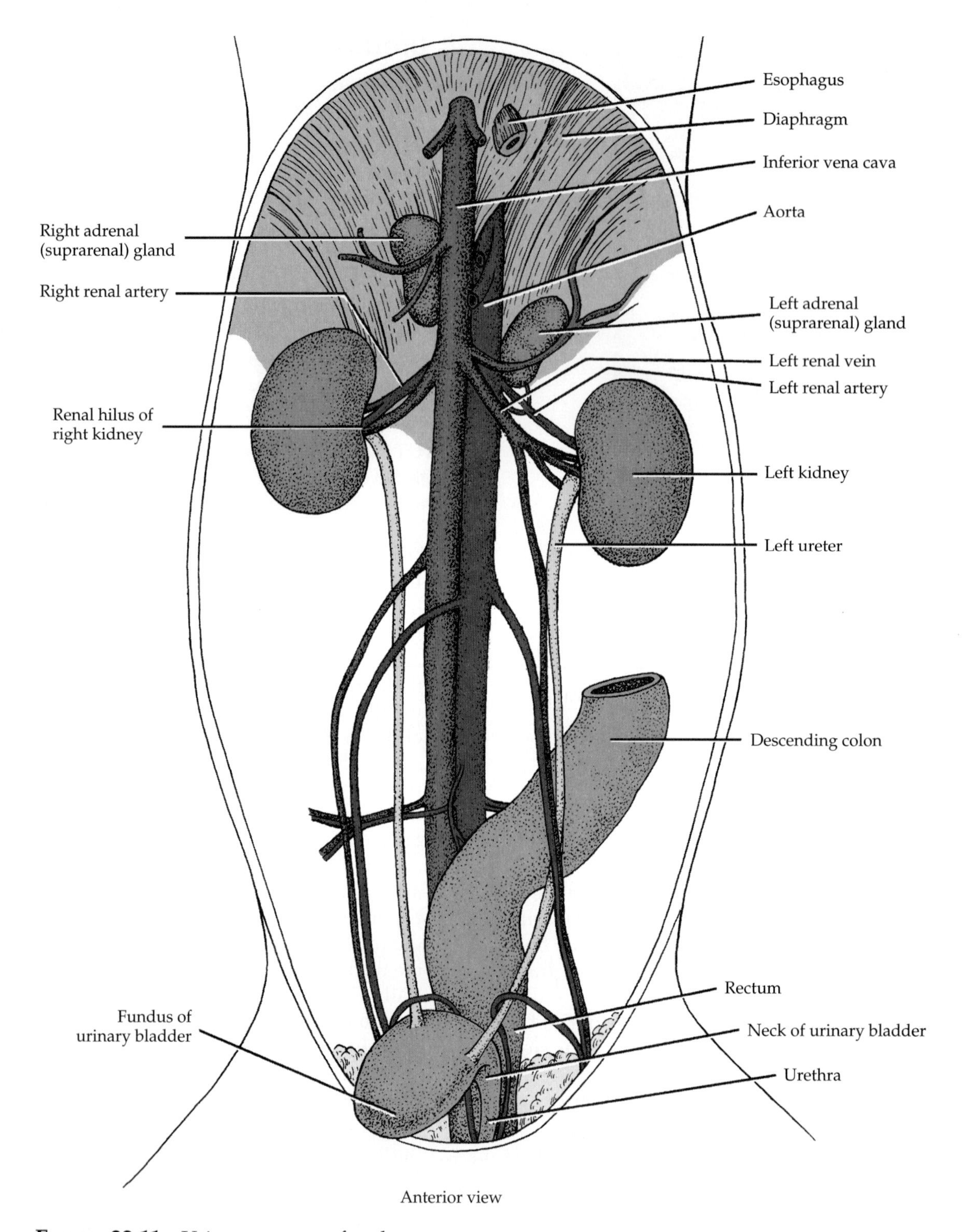

FIGURE 22.11 Urinary system of male cat.

10. The calyces collect urine from collecting ducts and drain it into the renal pelvis and out through the ureter.
11. The renal artery divides into several branches that pass between the renal pyramids. These vessels are small and delicate and may be too difficult to dissect and trace through the renal medulla.

ANSWER THE LABORATORY REPORT QUESTIONS AT THE END OF THE EXERCISE.

LABORATORY REPORT QUESTIONS

Urinary System 22

Student ______________________ **Date** ______________________

Laboratory Section ______________________ **Score/Grade** ______________________

PART 1. Multiple Choice

_______ **1.** Beginning at the deepest layer and moving toward the superficial layer, identify the order of tissue layers surrounding the kidney. (a) renal capsule, renal fascia, adipose capsule (b) renal fascia, adipose capsule, renal capsule (c) adipose capsule, renal capsule, renal fascia (d) renal capsule, adipose capsule, renal fascia

_______ **2.** The functional unit of the kidney is the (a) nephron (b) ureter (c) urethra (d) hilus

_______ **3.** Substances filtered by the kidney must pass through the endothelial-capsular membrane, which is composed of several parts. Which of the following choices lists the correct order of the parts as substances pass through the membrane? (a) epithelium of the visceral layer of the glomerular (Bowman's) capsule, endothelium of the glomerulus, basement membrane of the glomerulus (b) endothelium of the glomerulus, basement membrane of the glomerulus, epithelium of the visceral layer of the glomerular (Bowman's) capsule (c) basement membrane of the glomerulus, endothelium of the glomerulus, epithelium of the visceral layer of the glomerular (Bowman's) capsule (d) epithelium of the visceral layer of the glomerular (Bowman's) capsule, basement membrane of the glomerulus, endothelium of the glomerulus

_______ **4.** In the glomerular (Bowman's) capsule, the afferent arteriole divides into a capillary network called a(n) (a) glomerulus (b) interlobular artery (c) peritubular capillary (d) efferent arteriole

_______ **5.** Transport of urine from the renal pelvis into the urinary bladder is the function of the (a) urethra (b) calculi (c) casts (d) ureters

_______ **6.** The terminal portion of the urinary system is the (a) urethra (b) urinary bladder (c) ureter (d) nephron

_______ **7.** Damage to the renal medulla would interfere first with the functioning of which parts of a juxtamedullary nephron? (a) glomerular (Bowman's) capsule (b) distal convoluted tubule (c) collecting duct (d) proximal convoluted tubule

_______ **8.** An obstruction in the glomerulus would affect the flow of blood into the (a) renal artery (b) efferent arteriole (c) afferent arteriole (d) intralobular artery

_______ **9.** Urine that leaves the distal convoluted tubule passes through the following structures in which sequence? (a) collecting duct, hilus, calyces, ureter (b) collecting duct, calyces, pelvis, ureter (c) calyces, collecting duct, pelvis, ureter (d) calyces, hilus, pelvis, ureter

_______ **10.** The position of the kidneys posterior to the peritoneal lining of the abdominal cavity is described by the term (a) retroperitoneal (b) anteroperitoneal (c) ptosis (d) inferoperitoneal

_______ **11.** Of the following structures, the one to receive filtrate *last* as it passes through the nephron is the (a) proximal convoluted tubule (b) ascending limb of the loop of Henle (c) glomerulus (d) collecting duct

_______ **12.** Peristalsis of the ureter is a function of the (a) serosa (b) mucosa (c) submucosa (d) muscularis

_______ **13.** The trigone and the detrusor muscle are associated with the (a) kidney (b) urinary bladder (c) urethra (d) ureters

_______ **14.** The notch on the medial surface of the kidney through which blood vessels enter and exit is called the (a) renal medulla (b) major calyx (c) renal hilus (d) renal column

_______ **15.** Blood is drained from the kidneys by the (a) renal arteries (b) interlobar arteries (c) interlobular veins (d) renal veins

_______ **16.** The epithelium of the urinary bladder that permits distension is (a) stratified squamous (b) transitional (c) simple squamous (d) pseudostratified

PART 2. Completion

17. In addition to the urinary system, other systems that help eliminate wastes are the respiratory, integumentary, and _______________ systems.

18. The double-walled cup found in a nephron is called a(n) _______________.

19. The special capillary network found inside of this double-walled cup is the _______________.

20. The major blood vessel that enters each kidney is the _______________.

21. The nerve supply to the kidneys comes from the autonomic nervous system and is called the _______________.

22. Urine is expelled from the urinary bladder by an act called urination, voiding, or _______________.

23. The small tube in the urinary system that leads from the floor of the urinary bladder to the outside is the _______________.

24. The apices of renal pyramids are referred to as renal _______________.

25. The cortical substance between renal pyramids is called a renal _______________.

26. Cuplike extensions of the renal pelvis, usually two or three in number, are referred to as _______________.

27. Epithelial cells of the visceral layer of the glomerular (Bowman's) capsule are called _______________.

28. Distal convoluted tubules terminate by merging with _______________.

29. Long loops of blood vessels around the medullary structures of juxtamedullary nephrons are called _______________.

30. Which blood vessel comes next in the sequence? Interlobar artery, arcuate artery, interlobular artery, _______________.

Reproductive Systems

23

Reproduction is the process by which new individuals of a species are produced and the genetic material is passed from generation to generation. This maintains continuation of the species. Cell division in a multicellular organism is necessary for growth as well as repair and it involves passing of genetic material from parent cells to daughter cells. In somatic cell division a parent cell produces two identical daughter cells. This process is involved in replacing cells and growth. In reproductive cell division, sperm and egg cells are produced for continuity of the species.

The organs of the male and female reproductive systems may be grouped by function. (1) The testes and ovaries, also called ***gonads*** (*gonos* = seed), function in the production of gametes—sperm cells and ova, respectively. The gonads also secrete hormones. (2) The ***ducts*** of the reproductive systems transport, receive, and store gametes. (3) Still other reproductive organs, called ***accessory sex glands,*** produce materials that support gametes. (4) Finally, several ***supporting structures,*** including the penis, have various roles in reproduction. In this exercise, you will study the structure of the male and female reproductive organs and associated structures.

A. ORGANS OF MALE REPRODUCTIVE SYSTEM

The ***male reproductive system*** includes (1) the testes, or male gonads, which produce sperm and secrete hormones; (2) a system of ducts that transport, receive, or store sperm; (3) accessory glands, whose secretions contribute to semen; and (4) several supporting structures, including the penis.

1. Testes

The ***testes,*** or ***testicles,*** are paired oval glands that lie in the pelvic cavity for most of fetal life. They usually begin to enter the scrotum during the latter half of the seventh month of fetal development; full descent is not complete until just before birth. If the testes do not descend, the condition is called ***cryptorchidism*** (krip-TOR-ki-dizm; *kryptos* = hidden; *orchis* = testis). Untreated cryptorchidism on both sides results in sterility, because the cells that stimulate the initial development of sperm cells are destroyed by the higher temperature of the pelvic cavity. The chance of testicular cancer is 30 to 50% greater in cryptorchid testes.

Each testis is partially covered by a serous membrane called the ***tunica*** (*tunica* = sheath) ***vaginalis,*** which is derived from the peritoneum. Internal to the tunica vaginalis is a dense white fibrous capsule, the ***tunica albuginea*** (al'-byoo-JIN-ē-a; *albus* = white), which extends inward and divides the testis into a series of 200 to 300 internal compartments called ***lobules.*** Each lobule contains one to three tightly coiled ***seminiferous*** (*semen* = seed; *ferre* = to carry) ***tubules*** where sperm production ***(spermatogenesis)*** occurs.

Label the structures associated with the testes in Figure 23.1.

Spermatogenic cells are sperm-forming cells in various stages that undergo mitosis and differentiation to eventually produce sperm. Together with supporting cells, they line the seminiferous tubules. The most immature spermatogenic cells are called ***spermatogonia*** (sper'-ma-tō-GŌ-nē-a; *sperm* = seed; *gonium* = generation or offspring; singular is ***spermatogonium***). They lie next to the basement membrane. Toward the lumen of the tubule are layers of progressively more mature cells. In order of advancing maturity, these are ***primary spermatocytes*** (SPER-ma-tō-sīts), ***secondary spermatocytes,*** and ***spermatids.*** By the time a ***sperm cell,*** or ***spermatozoon*** (sper'-ma-tō-ZŌ-on; *zoon* = life; plural is ***sperm,*** or ***spermatozoa***), has nearly reached maturity, it is released into the lumen of the tubule and begins to move out of the rete testis.

Embedded among the spermatogenic cells in the tubules are large ***sustentacular*** (sus'-ten-TAK-yoo-lar;

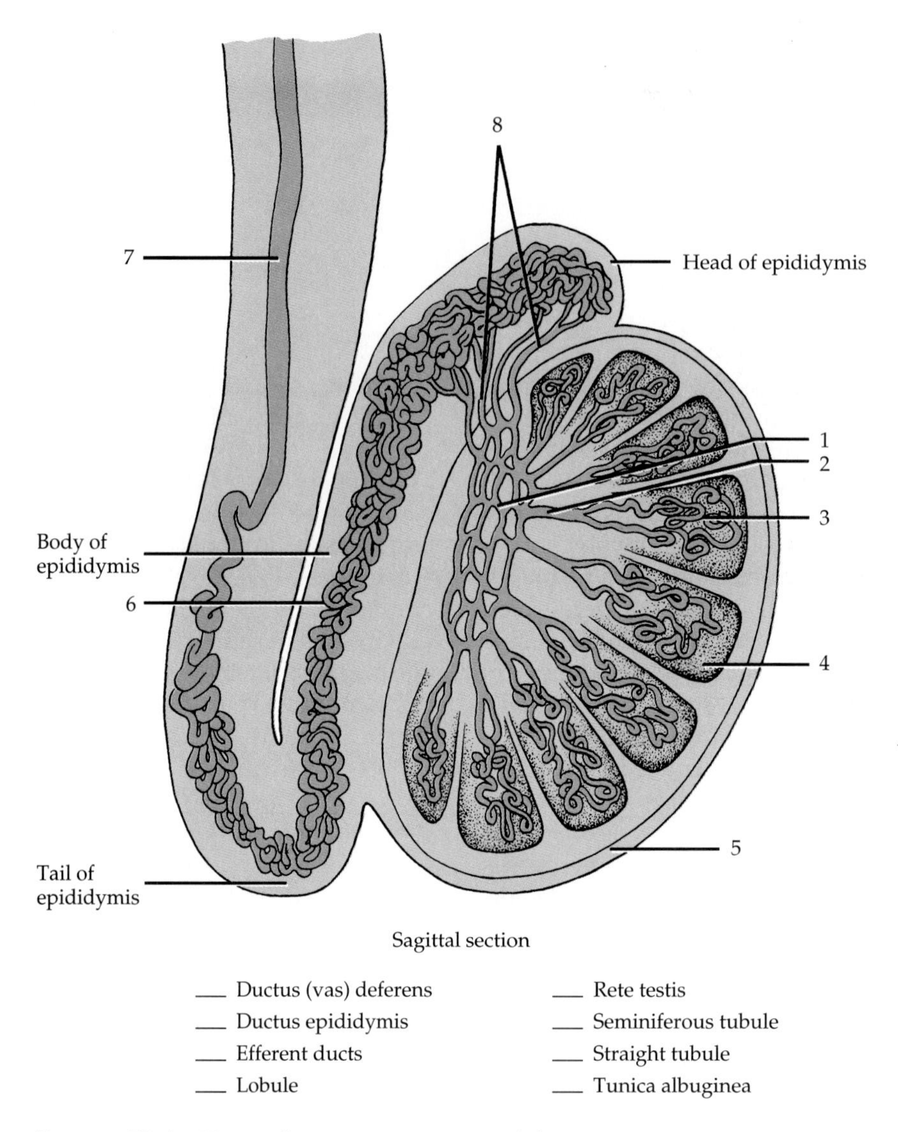

FIGURE 23.1 Testis showing its system of ducts.

sustentare = to support), or ***Sertoli, cells*** that extend from the basement membrane to the lumen of the tubule. Sustentacular cells support and protect developing spermatogenic cells; nourish spermatocytes, spermatids, and sperm; phagocytize excess spermatid cytoplasm as development proceeds; and regulate the effects of testosterone and follicle-stimulating hormone (FSH). Sustentacular cells also control movements of spermatogenic cells and the release of sperm into the lumen of the seminiferous tubules. They produce fluid for sperm transport and secrete the hormone inhibin, which helps regulate sperm production by inhibiting the secretion of FSH. In the spaces between adjacent seminiferous tubules are clusters of cells called ***interstitial endocrinocytes,*** or ***Leydig cells.*** These cells secrete testosterone, the most important androgen (male sex hormone). Because they produce both sperm and hormones, the testes are both exocrine and endocrine glands.

Using your textbook, charts, or models as reference, label Figure 23.2.

Sperm are produced at the rate of about 300 million per day. Once ejaculated, they usually live about 48 hr in the female reproductive tract. The parts of a sperm are as follows:

a. ***Head*** Contains the ***nucleus*** and ***acrosome*** (produces hyaluronic acid and proteinases to bring about penetration of secondary oocyte).
b. ***Midpiece*** Contains numerous mitochondria in which the energy for locomotion is generated.
c. ***Tail*** Typical flagellum used for locomotion.

With the aid of your textbook, label Figure 23.3 on page 450.

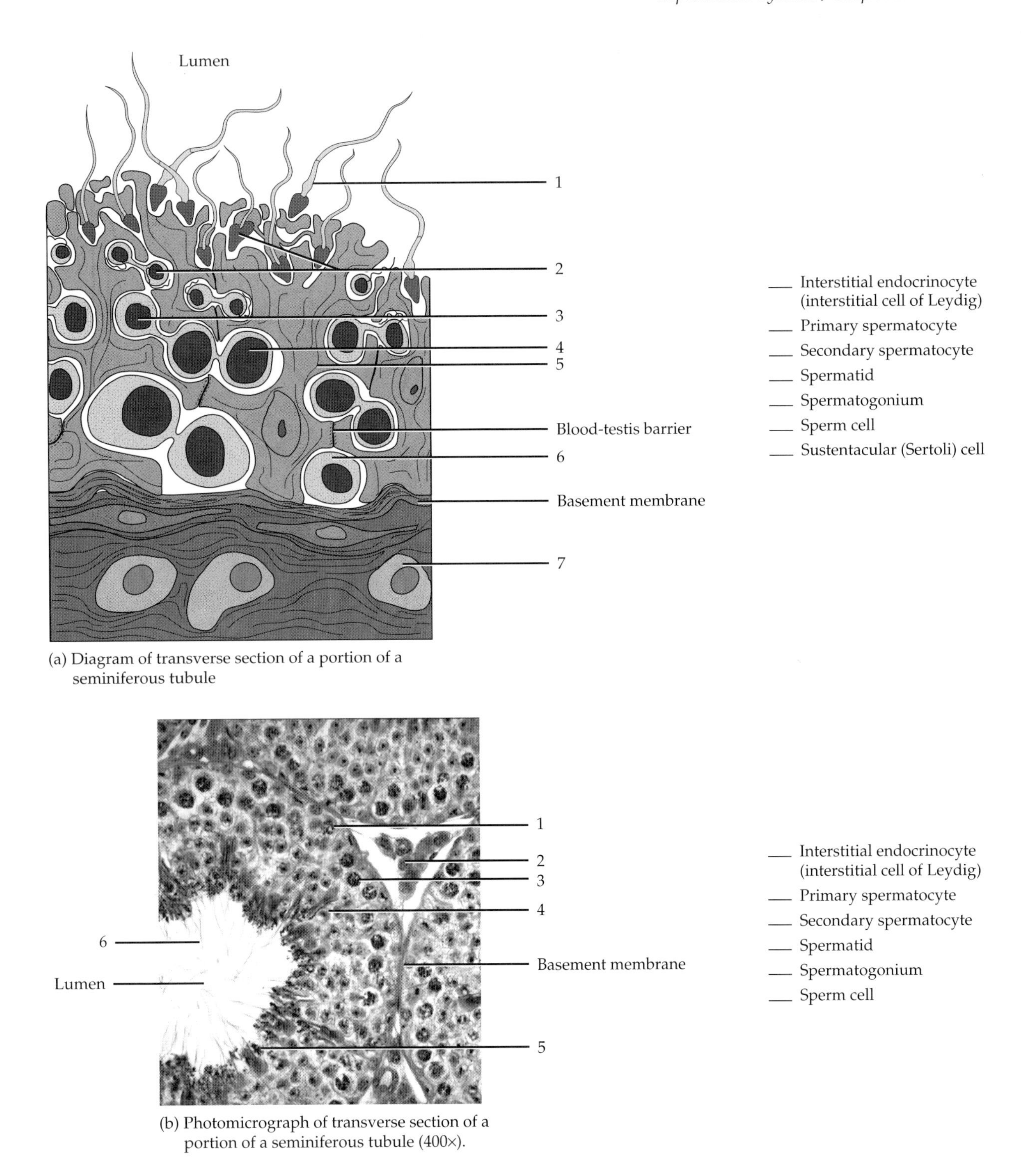

(a) Diagram of transverse section of a portion of a seminiferous tubule

(b) Photomicrograph of transverse section of a portion of a seminiferous tubule (400×).

FIGURE 23.2 Seminiferous tubules showing various stages of spermatogenesis.

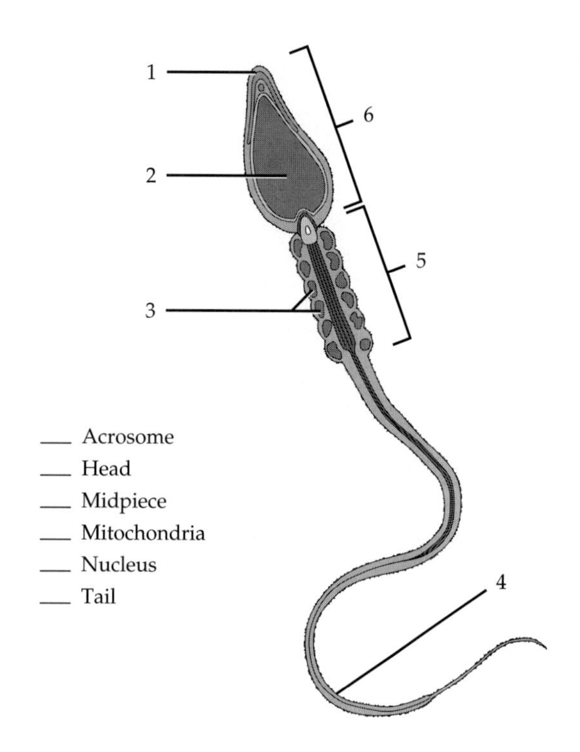

FIGURE 23.3 Parts of a sperm cell.

2. Ducts

As sperm cells mature, they are moved through seminiferous tubules into tubes called ***straight tubules,*** from which they are transported into a network of ducts, the ***rete*** (RĒ-tē; *rete* = network) ***testis.*** The sperm cells are next transported out of the testes through a series of coiled ***efferent ducts*** that empty into a single ***ductus epididymis*** (ep'-i-DID-i-mis; *epi* = above; *didymos* = testis). From there, they are passed into the ***ductus (vas) deferens,*** which ascends along the posterior border of the testis, penetrates the inguinal canal, enters the pelvic cavity, and loops over the side and down the posterior surface of the urinary bladder. The ductus (vas) deferens and duct from the seminal vesicle (gland) together form the ***ejaculatory*** (e-JAK-yoo-la-tō'-rē; *ejectus* = to throw out) ***duct,*** which propels the sperm cells into the ***urethra,*** the terminal duct of the system. The male urethra is divisible into (1) a *prostatic portion,* which passes through the prostate gland; (2) a *membranous portion,* which passes through the urogenital diaphragm; and (3) a *spongy (penile) portion,* which passes through the corpus spongiosum of the penis (see Figure 23.6 on page 452). The ***epididymis*** is a comma-shaped organ that is divisible into a head, body, and tail. The head is the superior portion that contains the efferent ducts; the body is the middle portion that contains the ductus epididymis; and the tail is the inferior portion in which the ductus epididymis continues as the ductus (vas) deferens.

Label the various ducts of the male reproductive system in Figure 23.1.

The ductus epididymis is lined with ***pseudostratified columnar epithelium.*** The free surfaces of the cells contain long, branching microvilli called ***stereocilia.*** The muscularis deep to the epithelium consists of smooth muscle. Functionally, the ductus epididymis is the site of sperm maturation (increased motility and fertility potential). They require between 10 and 14 days to complete their maturation—that is, to become capable of fertilizing a secondary oocyte. The ductus epididymis also stores sperm cells and propels them toward the urethra during emission by peristaltic contraction of its smooth muscle. Sperm cells may remain in storage in the ductus epididymis up to a month or more. After that, they are expelled from the epididymis or reabsorbed in the epididymis.

Obtain a prepared slide of the ductus epididymis showing its mucosa and muscularis. Compare your observations to Figure 23.4.

Histologically, the ***ductus (vas) deferens (seminal duct)*** is also lined with ***pseudostratified columnar epithelium*** and its muscularis consists of three layers of smooth muscle. Peristaltic contractions of the muscularis propel sperm cells toward the urethra during ejaculation. One method of sterilization in males, ***vasectomy,*** involves removal of a portion of each ductus (vas) deferens.

Obtain a prepared slide of the ductus (vas) deferens showing its mucosa and muscularis. Compare your observations to Figure 23.5.

3. Accessory Sex Glands

Whereas the ducts of the male reproductive system store or transport sperm, a series of ***accessory sex glands*** secrete most of the liquid portion of ***semen.*** Semen is a mixture of sperm cells and the secretions of the seminal vesicles, prostate gland, and bulbourethral glands.

The ***seminal*** (*seminalis* = pertaining to seed) ***vesicles*** (VES-i-kuls) are paired, convoluted, pouchlike structures posterior to and at the base of the urinary bladder anterior to the rectum. The

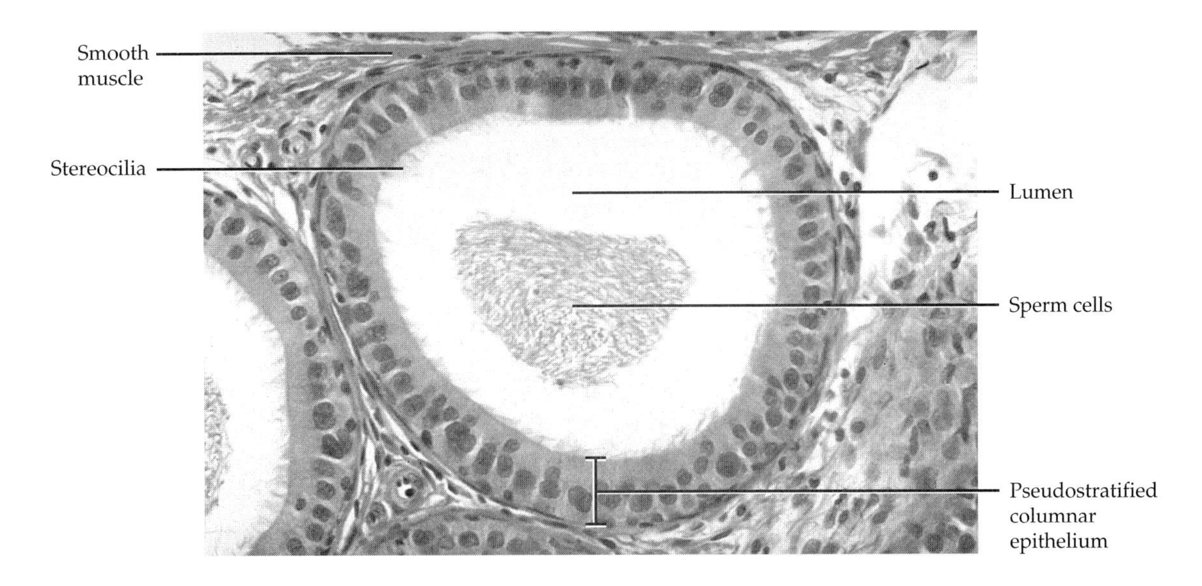

FIGURE 23.4 Histology of ductus epididymis.

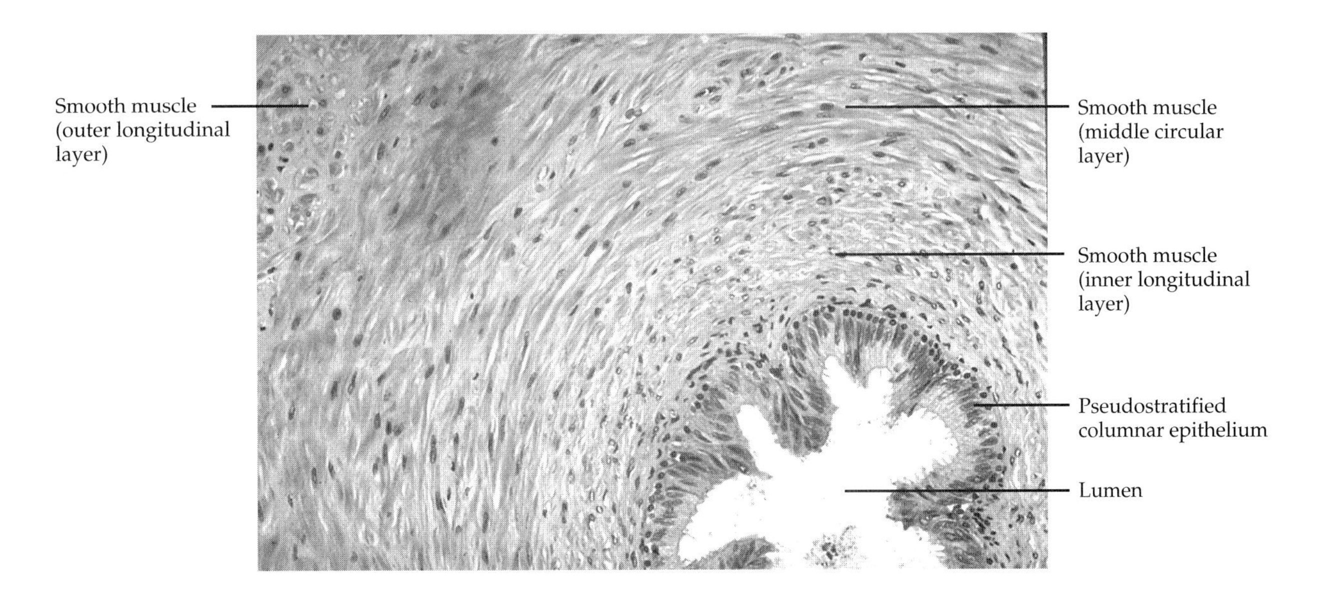

FIGURE 23.5 Histology of ductus (vas) deferens.

glands secrete the alkaline viscous component of semen into the ejaculatory duct. The seminal vesicles contribute about 60% of the volume of semen.

The ***prostate*** (PROS-tāt) ***gland,*** a single doughnut-shaped gland inferior to the urinary bladder, surrounds the prostatic urethra. The prostate secretes a slightly acidic fluid into the prostatic urethra. The prostatic secretion constitutes about 25% of the total semen produced.

The paired ***bulbourethral*** (bul'-bō-yoo-RĒ-thral), or ***Cowper's, glands,*** located inferior to the prostate on either side of the membranous urethra, are about the size of peas. They secrete an alkaline substance through ducts that open into the spongy (penile) urethra.

With the aid of your textbook, label the accessory glands and associated structures in Figure 23.6.

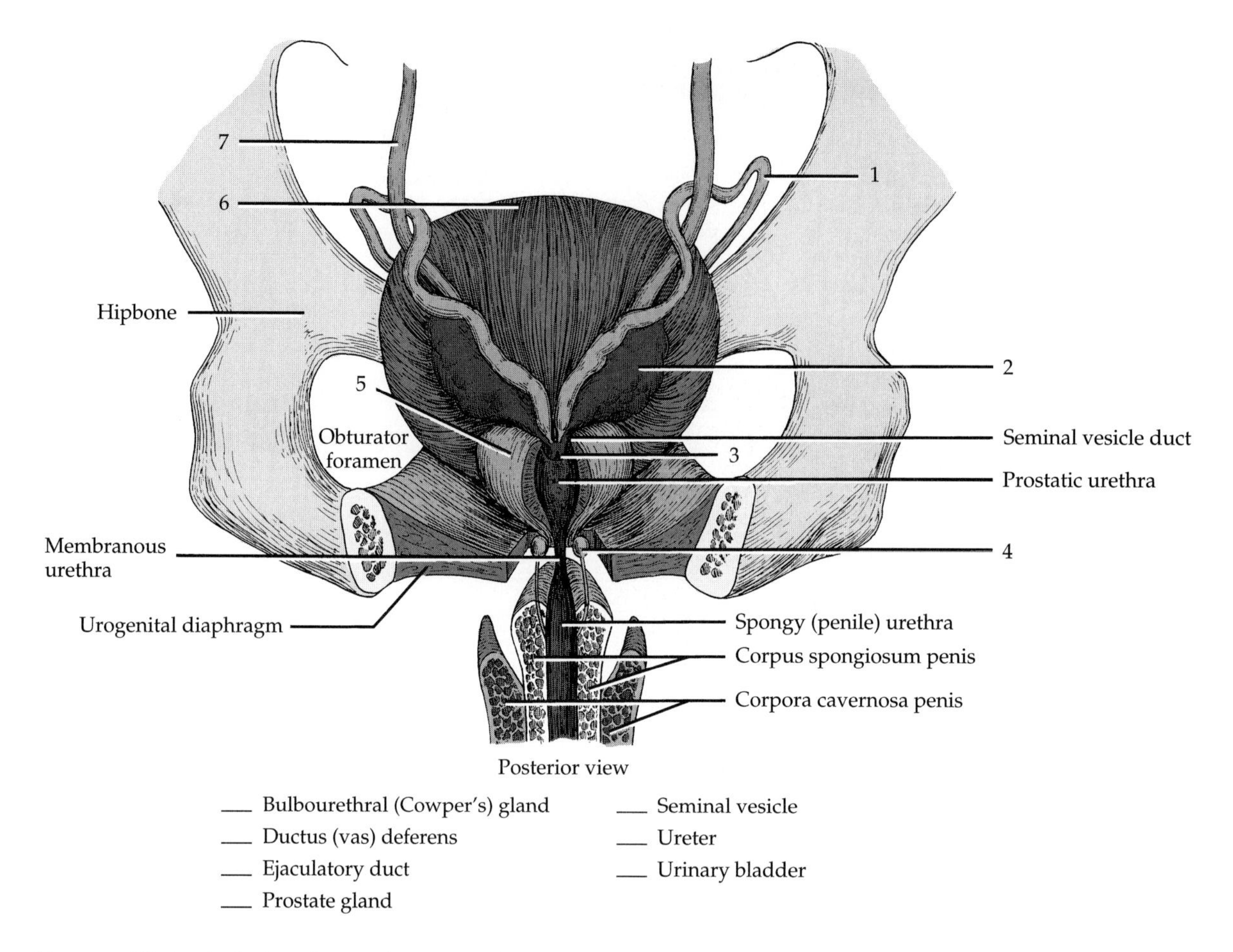

FIGURE 23.6 Relationships of some male reproductive organs.

4. Penis

The ***penis*** conveys urine to the exterior and introduces sperm cells into the vagina during copulation. Its principal parts are

a. ***Glans*** (*glandes* = acorn) ***penis*** Slightly enlarged distal end.
b. ***Corona*** Margin of glans penis.
c. ***Prepuce*** (PRĒ-pyoos) Foreskin; loosely fitting skin covering glans penis.
d. ***Corpora cavernosa*** (*corpus* = body; *caverna* = hollow) ***penis*** Two dorsolateral masses of erectile tissue.
e. ***Corpus spongiosum penis*** Midventral mass of erectile tissue that contains spongy (penile) urethra.
f. ***External urethral orifice*** Opening of spongy (penile) urethra to exterior.

With the aid of your textbook, label the parts of the penis in Figure 23.7.

Now that you have completed your study of the organs of the male reproductive system, label Figure 23.8 on page 454.

B. ORGANS OF FEMALE REPRODUCTIVE SYSTEM

The ***female reproductive system*** includes the female gonads (ovaries), which produce secondary oocytes; uterine (Fallopian) tubes, or oviducts, which transport secondary oocytes and fertilized ova to the uterus; vagina; external organs that compose the vulva; and the mammary glands.

1. Ovaries

The ***ovaries*** (*ovarium* = egg receptacle) are paired glands that resemble almonds in size and shape. Functionally, the ovaries produce secondary

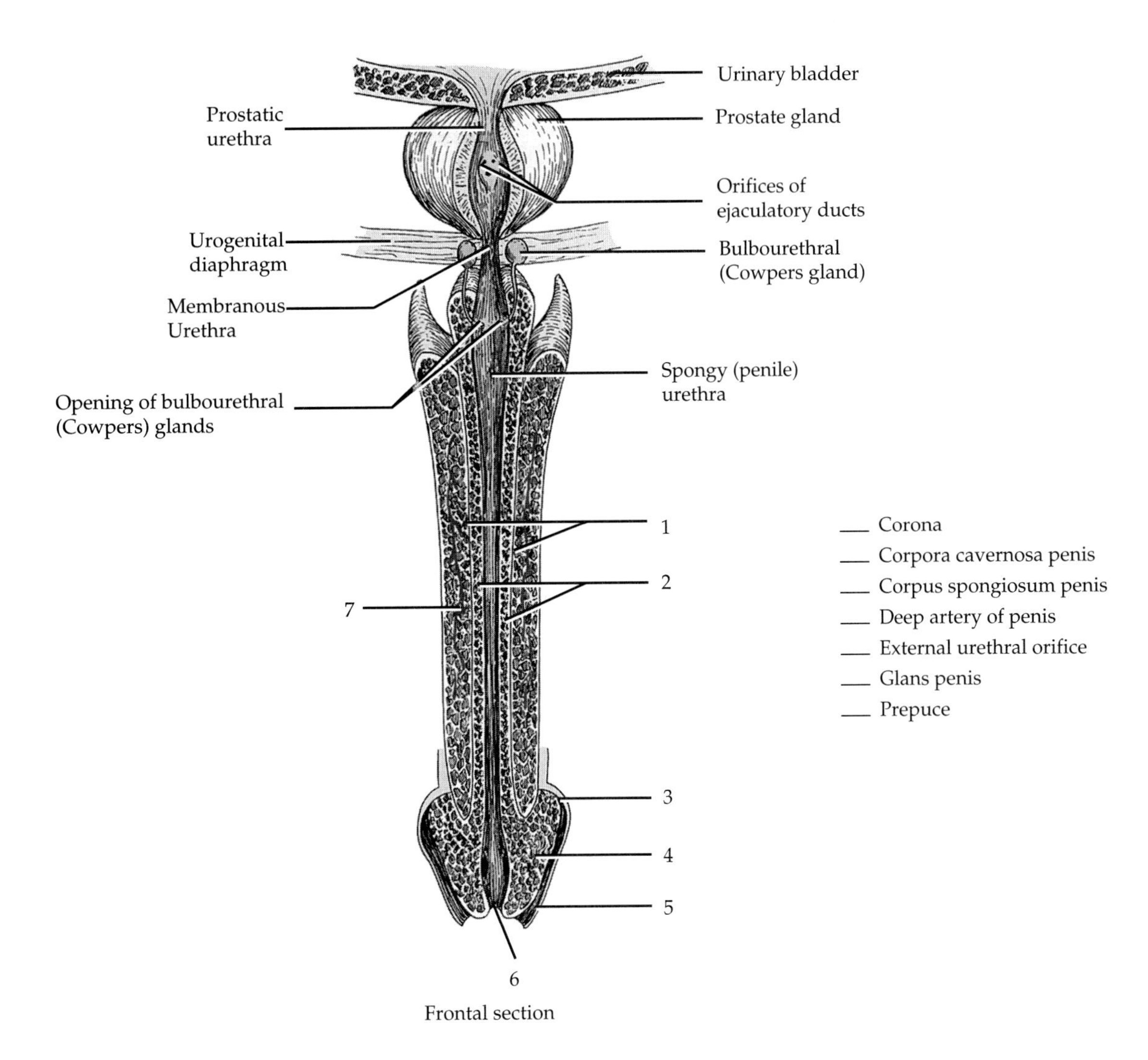

FIGURE 23.7 Internal structure of penis viewed from floor of penis.

oocytes, discharge them about once a month by a process called ovulation, and secrete female sex hormones (estrogens, progesterone, relaxin, and inhibin). The point of entrance for blood vessels and nerves is the ***hilus.*** The ovaries are positioned in the superior pelvic cavity, one on each side of the uterus, by a series of ligaments:

a. ***Mesovarium*** Double-layered fold of peritoneum that attaches ovaries to broad ligaments of uterus.
b. ***Ovarian ligament*** Anchors ovary to uterus.
c. ***Suspensory ligament*** Attaches ovary to pelvic wall.

With the aid of your textbook, label the ovarian ligaments in Figure 23.9 on page 455.

Histologically, the ovaries consist of the following parts:

1. ***Germinal epithelium*** A layer of simple epithelium (low cuboidal or squamous) that covers the surface of the ovary and is continuous with the mesothelium that covers the mesovarium. The term *germinal epithelium* is a misnomer since it does not give rise to oocytes, although at one time it was believed that it did.
2. ***Tunica albuginea*** A whitish capsule of dense, irregular connective tissue immediately deep to the germinal epithelium.
3. ***Ovarian cortex*** A region just deep to the tunica albuginea that consists of dense connective tissue and contains ovarian follicles (described shortly).
4. ***Ovarian medulla*** A region deep to the ovarian cortex that consists of loose connective tissue and contains blood vessels, lymphatics, and nerves.

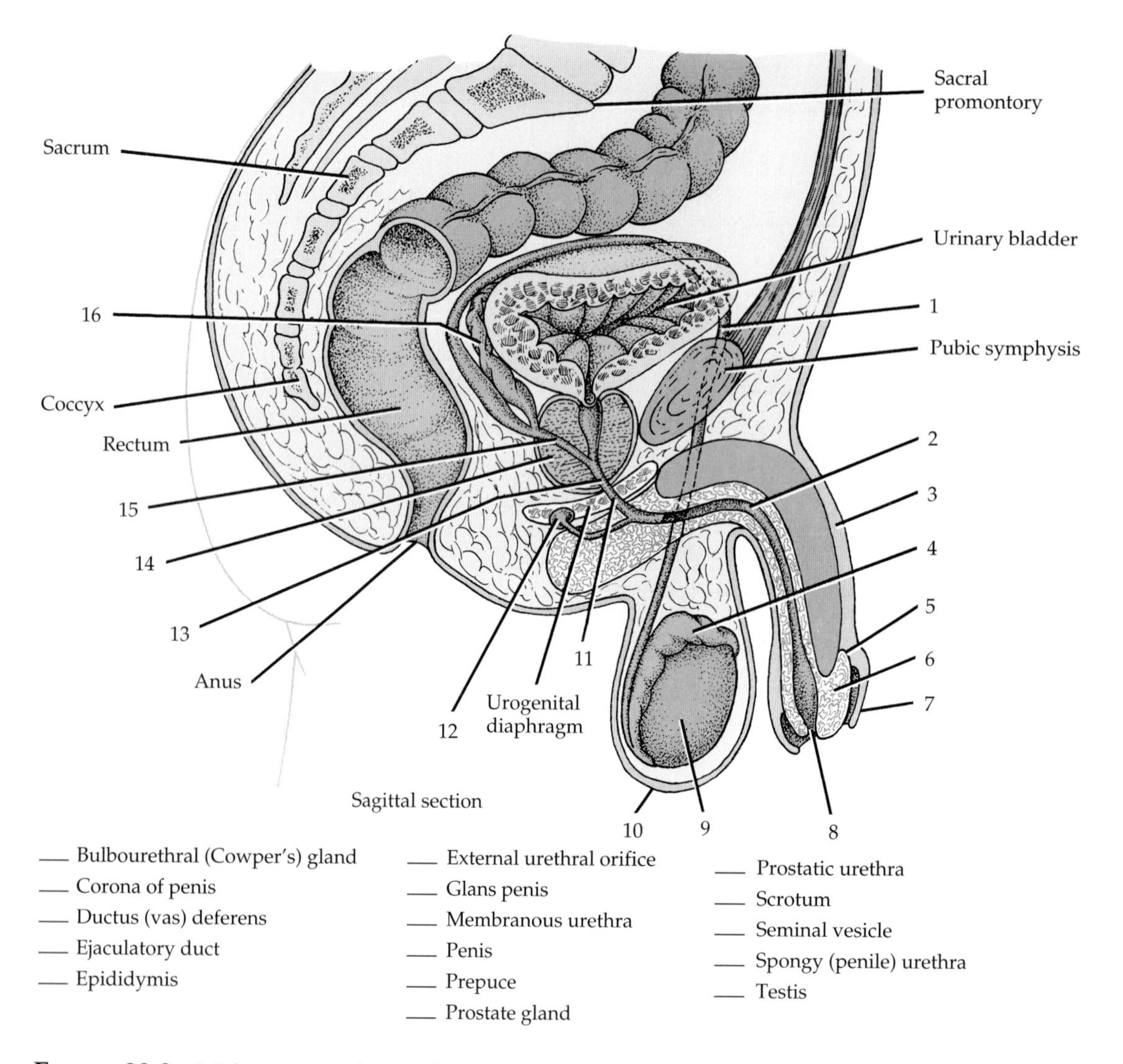

FIGURE 23.8 Male organs of reproduction and surrounding structures.

5. ***Ovarian follicles*** (*folliculus* = little bag) Lie in the cortex and consist of ***oocytes*** (immature ova) in various stages of development and their surrounding cells. When the surrounding cells form a single layer, they are called ***follicular cells.*** Later in development, when they form several layers, they are referred to as ***granulosa cells.*** The surrounding cells nourish the developing oocyte and begin to secrete estrogens as the follicle grows larger. Ovarian follicles undergo a series of changes prior to ovulation, progressing through several distinct stages. The most numerous and peripherally arranged follicles are termed ***primordial follicles.*** If a primordial follicle progresses to ovulation (release of a mature ovum), it will sequentially transform into a ***primary (preantral) follicle,*** then a ***secondary (antral) follicle,*** and finally a ***mature (Graafian) follicle.***
6. ***Mature (Graafian) follicle*** A large, fluid-filled follicle that soon will rupture and expel a secondary oocyte, a process called ***ovulation.***
7. ***Corpus luteum*** (= yellow body) Contains the remnants of an ovulated mature follicle. The corpus luteum produces progesterone, estrogens, relaxin, and inhibin until it degenerates and turns into fibrous tissue called a ***corpus albicans*** (= white body).

With the aid of your textbook, label the parts of an ovary in Figure 23.10 on page 456.

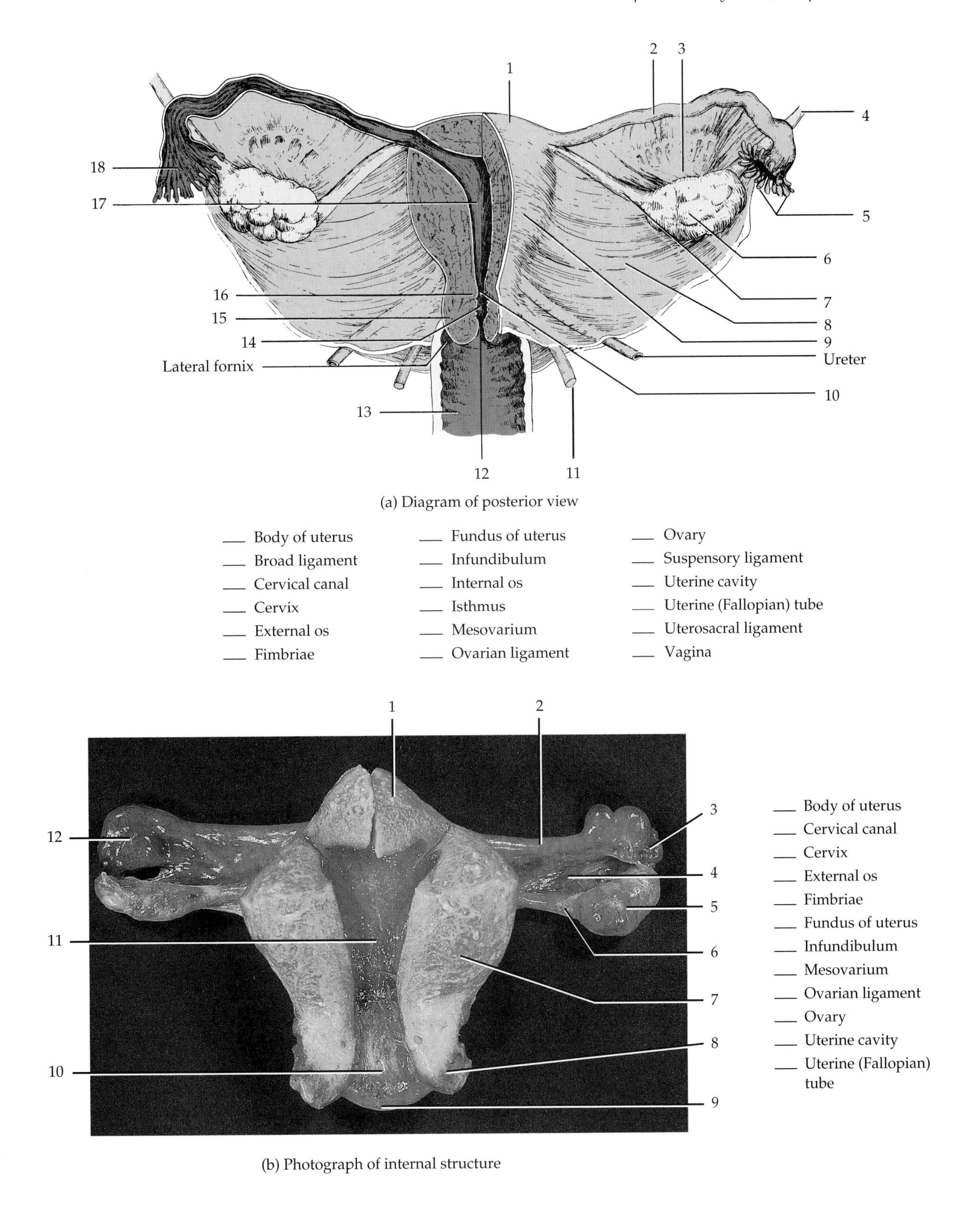

FIGURE 23.9 Uterus and associated female reproductive structures. The left side of figure (a) has been sectioned to show internal structures.

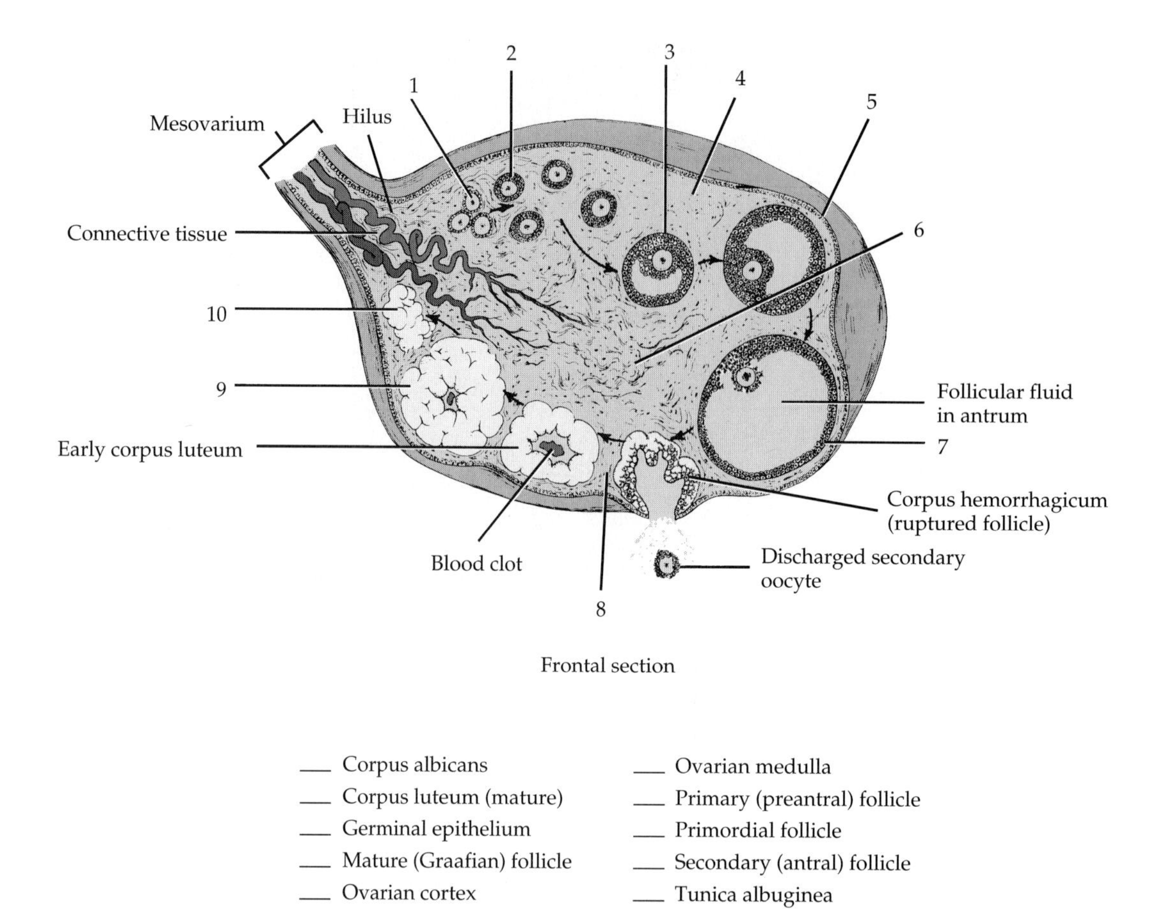

___ Corpus albicans
___ Corpus luteum (mature)
___ Germinal epithelium
___ Mature (Graafian) follicle
___ Ovarian cortex
___ Ovarian medulla
___ Primary (preantral) follicle
___ Primordial follicle
___ Secondary (antral) follicle
___ Tunica albuginea

FIGURE 23.10 Histology of ovary. Arrows indicate sequence of developmental stages that occur as part of ovarian cycle.

Obtain prepared slides of the ovary, examine them, and compare your observations to Figure 23.11.

2. Uterine (Fallopian) Tubes

The ***uterine (Fallopian) tubes,*** or ***oviducts,*** extend laterally from the uterus and transport secondary oocytes from the ovaries to the uterus. Fertilization normally occurs in the uterine tubes. The tubes are positioned between folds of the broad ligaments of the uterus. The funnel-shaped, open distal end of each uterine tube, called the ***infundibulum,*** is surrounded by a fringe of fingerlike projections called ***fimbriae*** (FIM-bre-ē; *fimbrae* = fringe). The ***ampulla*** (am-POOL-la) of the uterine tube is the widest, longest portion, constituting about two-thirds of its length. The ***isthmus*** (IS-mus) is the short, narrow, thick-walled portion that joins the uterus.

With the aid of your textbook, label the parts of the uterine tubes in Figure 23.9.

Histologically, the mucosa of the uterine tubes consists of ciliated columnar cells and secretory cells. The muscularis is composed of inner circular and outer longitudinal layers of smooth muscle. Wavelike contractions of the muscularis help move the ovum down into the uterus. The serosa is the outer covering.

Examine a prepared slide of the wall of the uterine tube, and compare your observations to Figure 23.12 on page 458.

3. Uterus

The ***uterus (womb)*** is the site of menstruation, implantation of a fertilized ovum, development of the fetus during pregnancy, and labor. Located between the urinary bladder and the rectum, the organ is shaped like an inverted pear. The uterus is subdivided into the following regions:

a. ***Fundus*** Dome-shaped portion superior to the uterine tubes.

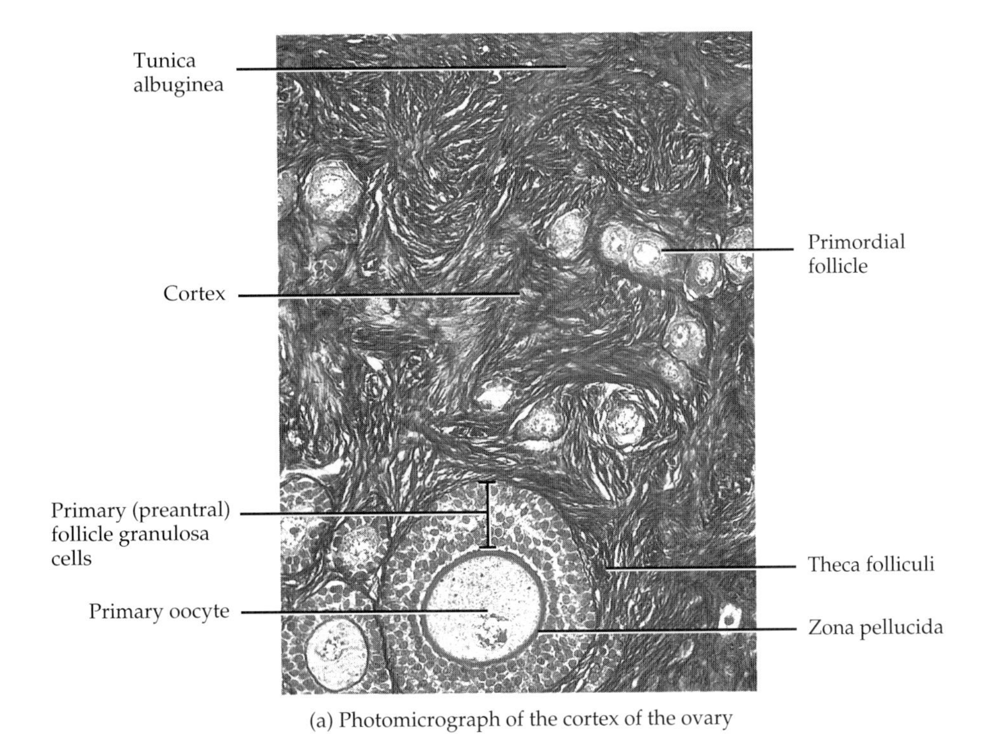

(a) Photomicrograph of the cortex of the ovary

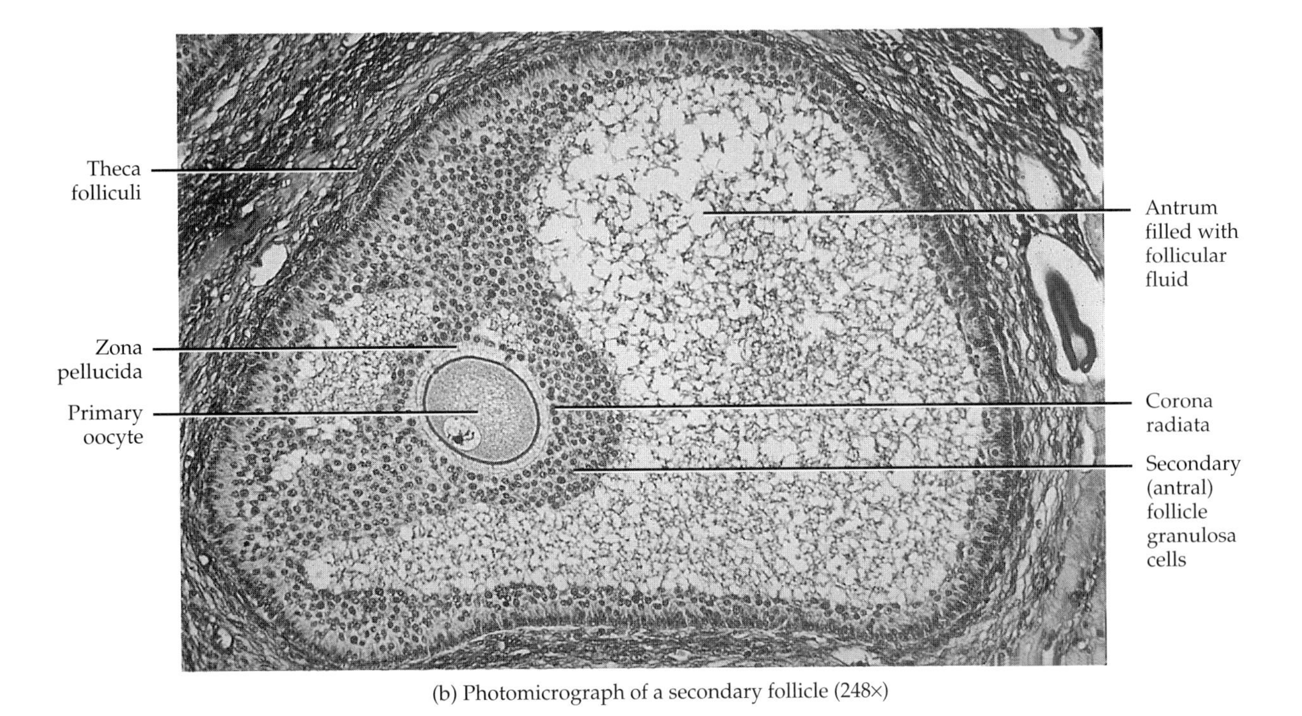

(b) Photomicrograph of a secondary follicle (248×)

FIGURE 23.11 Histology of ovary.

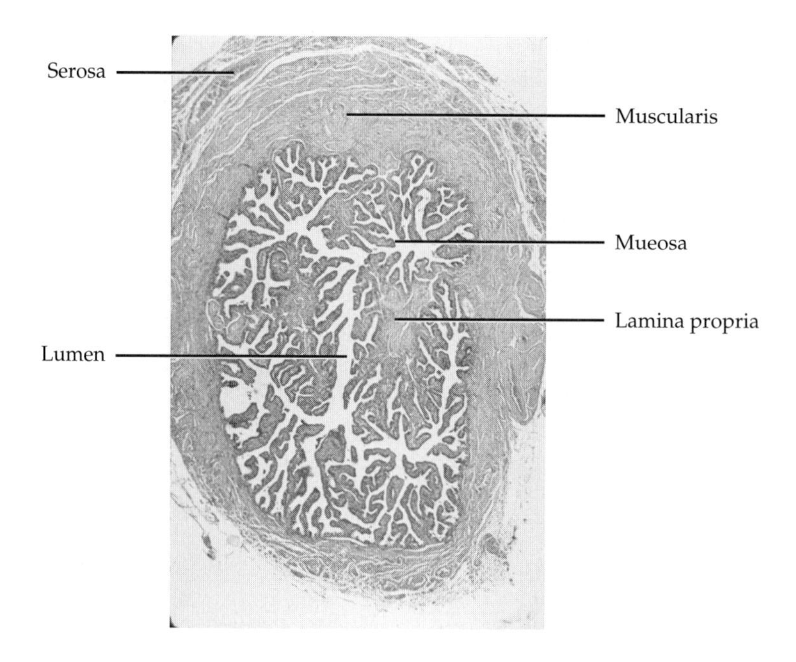

FIGURE 23.12 Histology of uterine (Fallopian) tube (3×).

b. ***Body*** Major, tapering portion.
c. ***Cervix*** Inferior narrow opening into vagina.
d. ***Isthmus*** (IS-mus) Constricted region between body and cervix.
e. ***Uterine cavity*** Interior of the body.
f. ***Cervical canal*** Interior of the cervix.
g. ***Internal os*** Site where cervical canal opens into uterine cavity.
h. ***External os*** Site where cervical canal opens into vagina.

Label these structures in Figure 23.9.

The uterus is maintained in position by the following ligaments:

a. ***Broad ligaments*** Double folds of parietal peritoneum that anchor the uterus to either side of the pelvic cavity.
b. ***Uterosacral ligaments*** Parietal peritoneal extensions that connect the uterus to the sacrum.
c. ***Cardinal (lateral cervical) ligaments*** Tissues containing smooth muscle, uterine blood vessels, and nerves. These ligaments extend below the bases of the broad ligaments between the pelvic wall and the cervix and vagina, and are the chief ligaments that maintain the position of the uterus, helping to keep it from dropping into the vagina.
d. ***Round ligaments*** Extend from uterus to external genitals (labia majora) between folds of broad ligaments.

Label the uterine ligaments in Figure 23.9.

Histologically, the uterus consists of three principal layers: endometrium, myometrium, and perimetrium (serosa). The inner ***endometrium*** (*endo* = within) is a mucous membrane that consists of simple columnar epithelium, an underlying endometrial stroma composed of connective tissue and endometrial glands. The endometrium consists of two layers: (1) ***stratum functionalis,*** the layer closer to the uterine cavity that is shed during menstruation; and (2) ***stratum basalis*** (ba-SAL-is), the permanent layer that produces a new stratum functionalis after menstruation. The middle ***myometrium*** (*myo* = muscle) forms the bulk of the uterine wall and consists of three layers of smooth muscle. During labor its coordinated contractions help to expel the fetus. The outer layer is the ***perimetrium*** (*peri* = around; *metron* = uterus), or ***serosa,*** part of the visceral peritoneum.

4. Vagina

A muscular, tubular organ lined with a mucous membrane, the ***vagina*** (*vagina* = sheath) is the passageway for menstrual flow, the receptacle for the penis during copulation, and the inferior portion of the birth canal. The vagina is situated between the urinary bladder and rectum and extends from the cervix of the uterus to the vestibule of the vulva. Recesses called ***fornices*** (FOR-ni-sēz'; *fornix* = arch or vault) surround the vaginal attachment to the cervix (see Figure 23.9) and make possible the use of contraceptive diaphragms. The opening of the vagina to the exterior, the ***vaginal orifice,*** may be bordered by a thin fold of vascularized membrane, the ***hymen*** (*hymen* = membrane).

Label the vagina in Figure 23.9.

Histologically, the mucosa of the vagina consists of nonkeratinized stratified squamous epithelium and connective tissue that lies in a series of transverse folds, the ***rugae.*** The muscularis is composed of an outer circular and an inner longitudinal layer of smooth muscle.

5. Vulva

The ***vulva*** (VUL-va; *volvere* = to wrap around), or ***pudendum*** (pyoo-DEN-dum), is a collective term for the external genitals of the female. It consists of the following parts:

a. ***Mons pubis*** (MONZ PŪ-bis) Elevation of adipose tissue over the pubic symphysis covered by skin and pubic hair.
b. ***Labia majora*** (LĀ-bē-a ma-JŌ-ra; *labium* = lip) Two longitudinal folds of skin that extend inferiorly and posteriorly from the mons pubis. Singular is ***labium majus.*** The folds, covered by pubic hair on their superior lateral surfaces, contain abundant adipose tissue and sebaceous (oil) and sudoriferous (sweat) glands.
c. ***Labia minora*** (MĪ-nō-ra) Two folds of mucous membrane medial to labia majora. Singular is ***labium minus.*** The folds have numerous sebaceous glands but few sudoriferous glands and no fat or pubic hair.
d. ***Clitoris*** (KLI-to-ris) Small cylindrical mass of erectile tissue and nerves at anterior junction of labia minora. The exposed portion is called the ***glans*** and the covering is called the ***prepuce*** (foreskin).
e. ***Vestibule*** Cleft between labia minora containing vaginal orifice, hymen (if present), external urethral orifice, and openings of several ducts.
f. ***Vaginal orifice*** Opening of vagina to exterior.
g. ***Hymen*** Thin fold of vascularized membrane that borders vaginal orifice.
h. ***External urethral orifice*** Opening of urethra to exterior.
i. ***Orifices of paraurethral (Skene's) glands*** Located on either side of external urethral orifice. The glands secrete mucus.
j. ***Orifices of ducts of greater vestibular*** (ves-TIB-yoo-lar), or ***Bartholin's, glands*** Located in a groove between hymen and labia minora. These glands produce a mucoid secretion that supplements lubrication during intercourse.
k. ***Orifices of ducts of lesser vestibular glands*** Microscopic orifices opening into vestibule.

Using your textbook as an aid, label the parts of the vulva in Figure 23.13.

6. Mammary Glands

The ***mammary*** (*mamma* = breast) ***glands*** are modified sweat glands that lie over the pectoralis major muscles and are attached to them by a layer of deep fascia. They consist of the following structures:

a. ***Lobes*** Around 15 to 20 compartments separated by adipose tissue.
b. ***Lobules*** Smaller compartments in lobes that contain clusters of milk-secreting glands called ***alveoli*** (*alveolus* = small cavity).
c. ***Secondary tubules*** Receive milk from alveoli.
d. ***Mammary ducts*** Receive milk from secondary tubules.
e. ***Lactiferous*** (*lact* = milk; *ferre* = to carry) ***sinuses*** Expanded distal portions of mammary ducts that store milk.
f. ***Lactiferous ducts*** Receive milk from lactiferous sinuses.
g. ***Nipple*** Projection on anterior surface of mammary gland that contains lactiferous ducts.
h. ***Areola*** (a-RĒ-ō-la; *areola* = small space) Circular pigmented skin around nipple.

With the aid of your textbook, label the parts of the mammary gland in Figure 23.14 on page 461.

Examine a prepared slide of alveoli of the mammary gland, and compare your observations to Figure 23.15 on page 462.

Now that you have completed your study of the organs of the female reproductive system, label Figure 23.16 on page 462.

7. Female Reproductive Cycle

During their reproductive years, nonpregnant females normally experience a cyclical sequence of changes in the ovaries and uterus. Each cycle takes about a month and involves both oogenesis and preparation of the uterus to receive a fertilized ovum. Hormones secreted by the hypothalamus, anterior pituitary gland, and ovaries control the

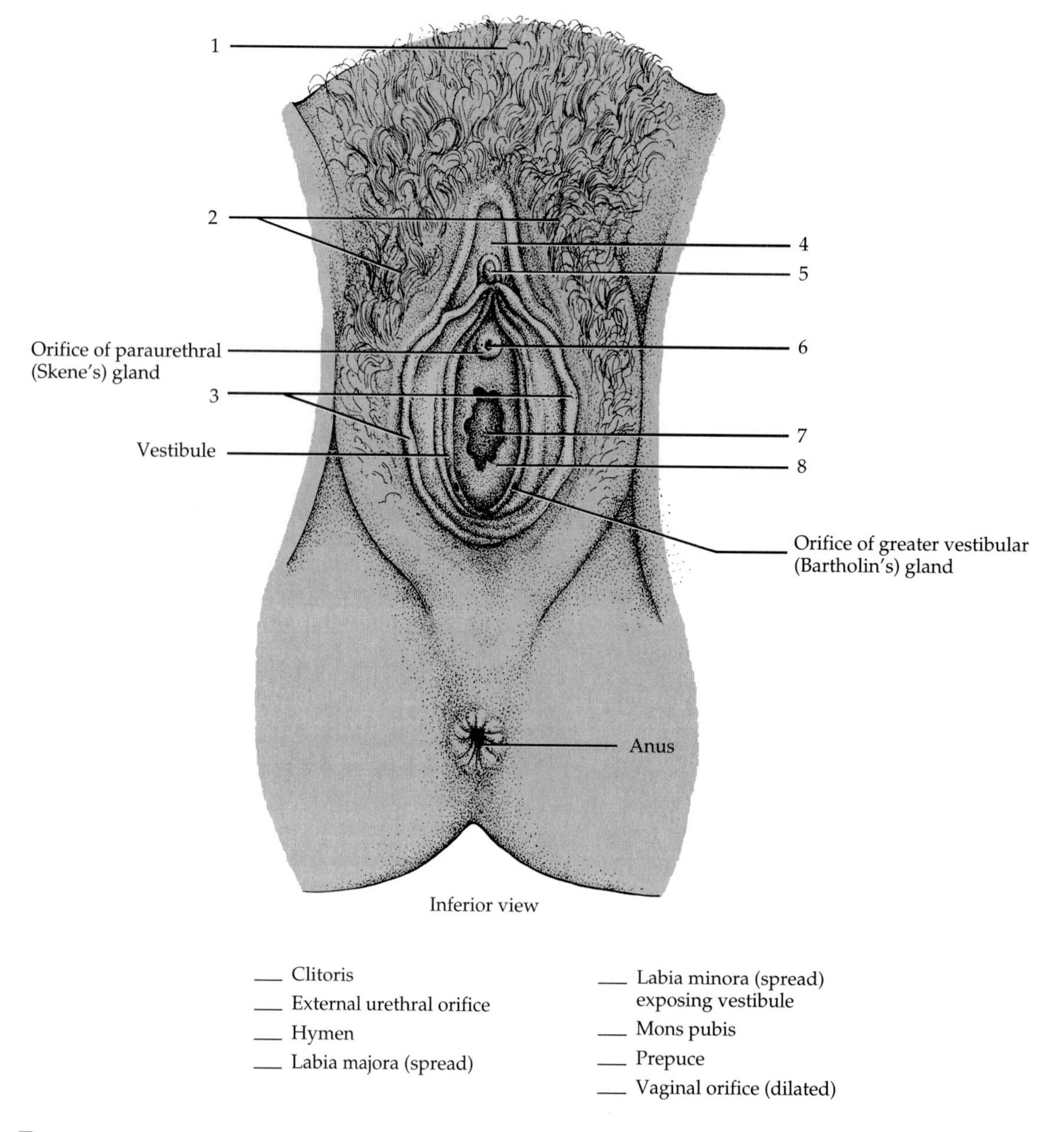

FIGURE 23.13 Vulva.

principal events. The ***ovarian cycle*** is a series of events associated with the maturation of an oocyte. The ***uterine (menstrual) cycle*** is a series of changes in the endometrium of the uterus. Each month, the endometrium is prepared for the arrival of a fertilized ovum that will develop in the uterus until birth. If fertilization does not occur, the stratum functionalis portion of the endometrium is shed. The general term ***female reproductive cycle*** includes the ovarian and uterine cycles, the hormonal changes that regulate them, and cyclical changes in the breasts and cervix.

a. HORMONAL REGULATION

The uterine cycle and ovarian cycle are controlled by gonadotropin releasing hormone (GnRH) from the hypothalamus. See Figure 23.17 on page 463. GnRH stimulates the release of follicle-stimulating hormone (FSH) and luteinizing hormone (LH) from the anterior pituitary gland. FSH stimulates the initial secretion of estrogens by the follicles. LH stimulates the further development of ovarian follicles and their full secretion of estrogens, brings about ovulation, and stimulates the production of estrogens, progesterone, relaxin, and inhibin by the corpus luteum.

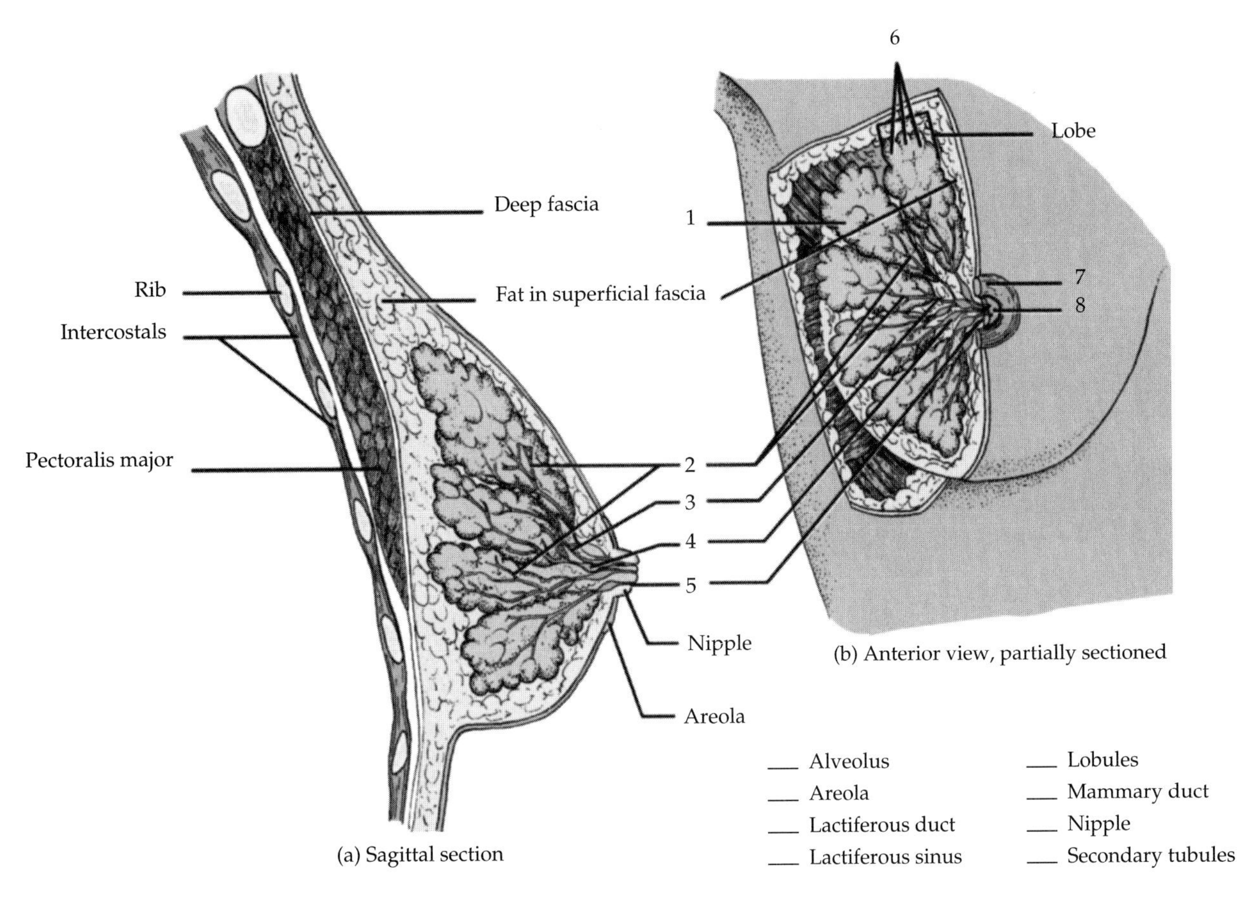

FIGURE 23.14 Mammary glands.

b. PHASES OF THE FEMALE REPRODUCTIVE CYCLE

The duration of the female reproductive cycle typically is 24–35 days. For this discussion, we shall assume a duration of 28 days, divided into three phases: the menstrual phase, preovulatory phase, and postovulatory phase (Figure 23.17).

(1) Menstrual Phase (Menstruation) The ***menstrual*** (MEN-stroo-al) ***phase,*** also called ***menstruation*** (men'-stroo-Ā-shun) or ***menses*** (*mensis* = month), lasts for roughly the first five days of the cycle. (By convention, the first day of menstruation marks the first day of a new cycle.)

Events in the Ovaries During the menstrual phase, 20 or so small secondary (antral) follicles, some in each ovary, begin to enlarge. Follicular fluid, secreted by the granulosa cells and oozing from blood capillaries, accumulates in the enlarging antrum while the oocyte remains near the edge of the follicle (see Figure 23.10).

Events in the Uterus Menstrual flow from the uterus consists of 50–150 ml of blood, tissue fluid, mucus, and epithelial cells derived from the endometrium. This discharge occurs because the declining level of estrogens and progesterone causes the uterine spiral arteries to constrict. As a result, the cells they supply become ischemic (deficient in blood) and start to die. Eventually, the entire stratum functionalis sloughs off. At this time the endometrium is very thin because only the stratum basalis remains. The menstrual flow passes from the uterine cavity to the cervix and through the vagina to the exterior.

(2) Preovulatory Phase The ***preovulatory phase,*** the second phase of the female reproductive cycle, is the time between menstruation and ovulation. The preovulatory phase of the cycle is more variable in length than the other phases and accounts for most of the difference when cycles are shorter or longer than 28 days. It lasts from days 6 to 13 in a 28-day cycle.

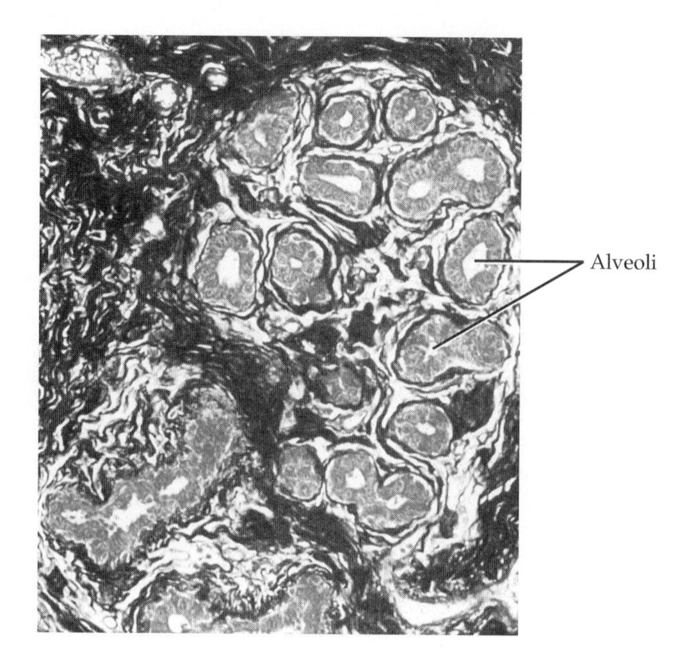

FIGURE 23.15 Histology of mammary gland showing alveoli (180×).

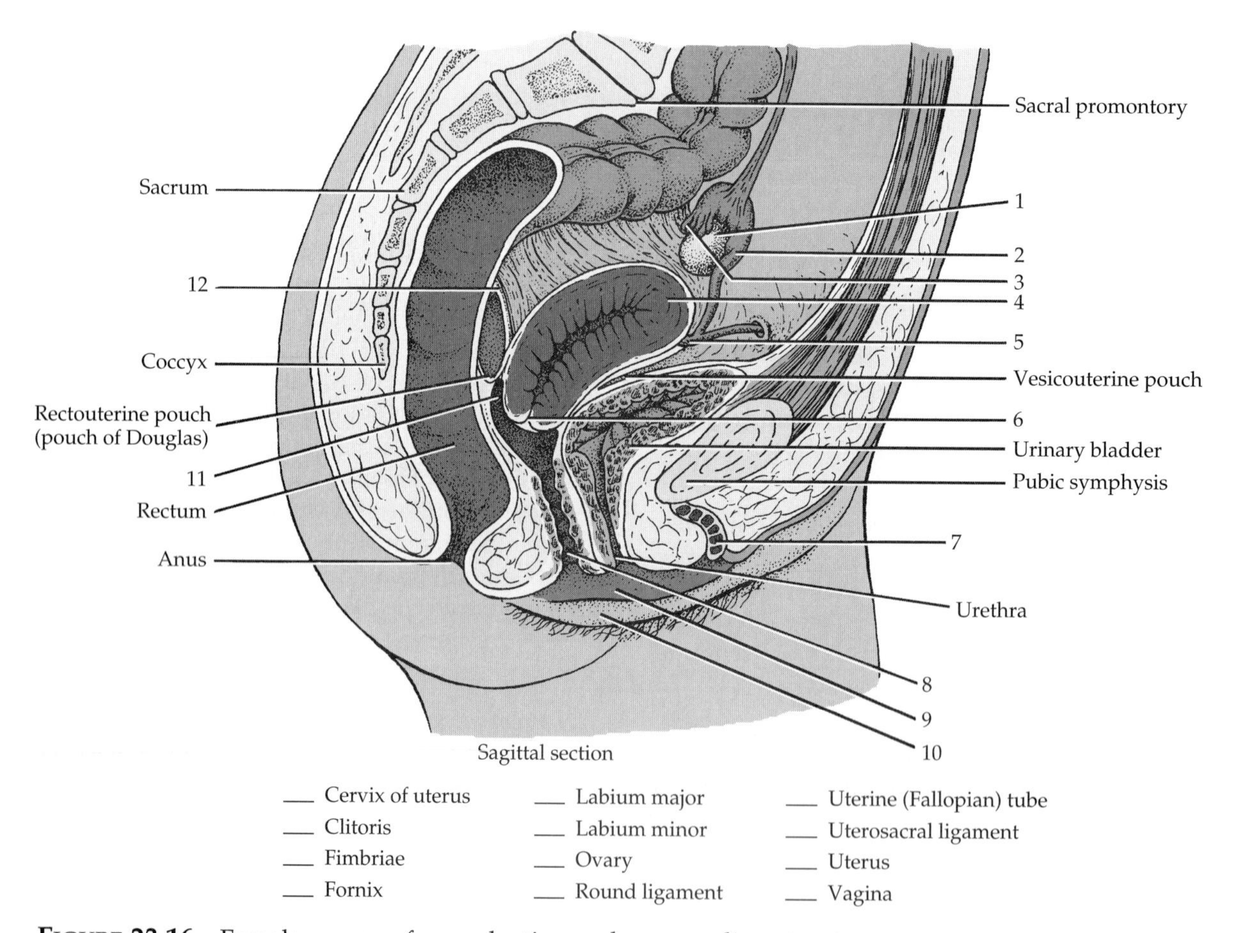

FIGURE 23.16 Female organs of reproduction and surrounding structures.

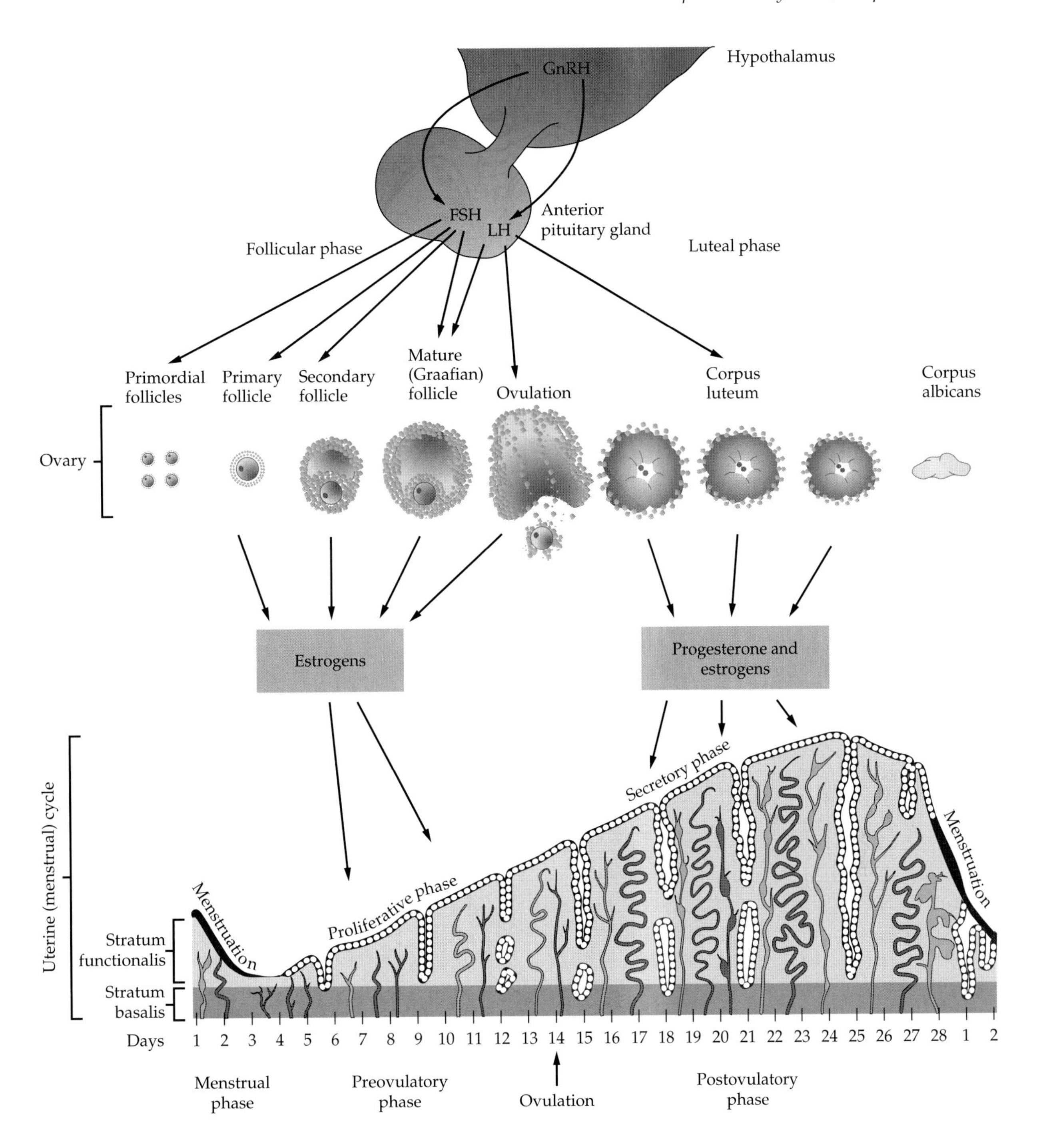

FIGURE 23.17 Menstrual cycle.

Events in the Ovaries Under the influence of FSH, the group of about 20 secondary follicles continues to grow and begins to secrete estrogens and inhibin. By about day 6, one follicle in one ovary has outgrown all the others and is called the ***dominant follicle.*** Estrogens and inhibin secreted by the dominant follicle decrease the secretion of FSH, which causes the other less well-developed follicles to stop growing and undergo atresia.

The one dominant follicle becomes the ***mature (Graafian) follicle*** that continues to enlarge until it is more than 20 mm in diameter and ready for ovulation (see Figure 23.10). This follicle forms a blisterlike bulge on the surface of the ovary. Fraternal (nonidentical) twins may result if two secondary follicles achieve dominance and both ovulate. During the final maturation process, the dominant follicle continues to increase its production of estrogens under the influence of an increasing level of LH. Estrogens are the primary ovarian hormones before ovulation, but small amounts of progesterone are produced by the mature follicle a day or two before ovulation.

With reference to the ovaries, the menstrual phase and preovulatory phase together are termed the ***follicular*** (fō-LIK-yoo-lar) ***phase*** because ovarian follicles are growing and developing.

Events in the Uterus Estrogens being liberated into the blood by growing follicles stimulate the repair of the endometrium. Cells of the stratum basalis undergo mitosis and produce a new stratum functionalis. As the endometrium thickens, the short, straight endometrial glands develop and the arterioles coil and lengthen as they penetrate the stratum functionalis. The thickness of the endometrium approximately doubles to about 4–6 mm. With reference to the uterus, the preovulatory phase is also termed the ***proliferative phase*** because the endometrium is proliferating.

(3) Ovulation The rupture of the mature (Graafian) follicle with release of the secondary oocyte into the pelvic cavity, called ***ovulation,*** usually occurs on day 14 in a 28-day cycle. During ovulation, the secondary oocyte remains surrounded by its zona pellucida and corona radiata. It generally takes a total of about 20 days (spanning the last 6 days of the previous cycle and the first 14 days of the current cycle) for a secondary follicle to develop into a fully mature follicle. During this time the developing ovum completes reduction division (meiosis I) and reaches metaphase of equatorial division (meiosis II). The *high* levels of estrogens during the last part of the preovulatory phase exert a *positive feedback* effect on both LH and GnRH and cause ovulation.

An over-the-counter home test that detects the LH surge associated with ovulation is available. The test predicts ovulation a day in advance. FSH also increases at this time, but not as dramatically as LH because FSH is stimulated only by the increase in GnRH. The positive feedback effect of estrogens on the hypothalamus and anterior pituitary gland does not occur if progesterone is present at the same time.

After ovulation, the mature follicle collapses and blood within it forms a clot due to minor bleeding during rupture and collapse of the follicle to become the ***corpus hemorrhagicum*** (*hemo* = blood; *rhegnynai* = to burst forth). (See Figure 23.10.) The clot is eventually absorbed by the remaining follicular cells. In time, the follicular cells enlarge, change character, and form the corpus luteum under the influence of LH. Stimulated by LH, the corpus luteum secretes progesterone, estrogens, relaxin, and inhibin.

(4) Postovulatory Phase The postovulatory phase of the female reproductive cycle is the most constant in duration and lasts for 14 days, from days 15 to 28 in a 28-day cycle. It represents the time between ovulation and the onset of the next menses. After ovulation, LH secretion stimulates the remnants of the mature follicle to develop into the corpus luteum. During its 2-week lifespan, the corpus luteum secretes increasing quantities of progesterone and some estrogens.

Events in One Ovary If the egg is fertilized and begins to divide, the corpus luteum persists past its normal 2-week lifespan. It is maintained by ***human chorionic*** (kō-rē-ON-ik) ***gonadotropin (hCG),*** a hormone produced by the chorion of the embryo as early as 8–12 days after fertilization. The chorion eventually develops into the placenta and the presence of hCG in maternal blood or urine is an indication of pregnancy. As the pregnancy progresses, the placenta itself begins to secrete estrogens to support pregnancy and progesterone to support pregnancy and breast development for lactation. Once the placenta begins its secretion, the role of the corpus luteum becomes minor. With reference to the ovaries, this phase of the cycle is also called the ***luteal phase.***

If hCG does not rescue the corpus luteum, after 2 weeks its secretions decline and it degenerates into a corpus albicans (see Figure 23.10). The lack of progesterone and estrogens due to degeneration of the corpus luteum then causes menstruation. In addition, the decreased levels of progesterone, estrogens, and inhibin promote the release of GnRH, FSH, and LH, which stimulate follicular growth, and a new ovarian cycle begins.

Events in the Uterus Progesterone produced by the corpus luteum is responsible for preparing the endometrium to receive a fertilized ovum. Preparatory activities include growth and coiling of the endometrial glands, which begin to secrete glycogen, vascularization of the superficial endometrium, thickening of the endometrium, and an increase in the amount of tissue fluid. These preparatory changes are maximal about 1 week after ovulation, corresponding to the time of possible arrival of a fertilized ovum. With reference to the uterus, this phase of the cycle is called the ***secretory phase*** because of the secretory activity of the endometrial glands.

Obtain microscope slides of the endometrium showing the menstrual, preovulatory, and postovulatory phases of the menstrual cycle. See if you can note the differences in thickness of the endometrium, distribution of blood vessels, and distribution and size of endometrial glands.

C. DISSECTION OF CAT-REPRODUCTIVE SYSTEMS

CAUTION! Please reread Section D, "Precautions Related to Dissection" at the beginning of the laboratory manual on page xi ***before*** *you begin your dissection.*

1. Male Reproductive System

The male reproductive system of the cat is shown in Figure 23.18.

PROCEDURE

a. SCROTUM

1. The ***scrotum,*** a double sac ventral to the anus that contains the testes, consists of skin, muscle, and connective tissue.
2. Internally, the scrotum is divided into two chambers by a ***median septum.*** A testis is found in each chamber.

b. TESTES

1. Carefully cut open the scrotum to examine the ***testes.***
2. The glistening covering of peritoneum around the testes is the ***tunica vaginalis.***

c. EPIDIDYMIDES

1. Cut away the tunica vaginalis.
2. Each ***epididymis*** is a long, coiled tube that receives spermatozoa from a testis. It is located anterolateral to its testis.
3. Each epididymis consists of an enlarged cranial end (***head***), a narrow middle portion (***body***), and an expanded caudal portion (***tail***).

d. DUCTUS (VAS) DEFERENS

1. Each ***ductus (vas) deferens*** is a continuation from the tail of the epididymis. It carries sperm cells from the epididymis through the inguinal canal to the urethra.
2. The ***inguinal canal*** is a short passageway through the abdominal musculature.
3. Passing through the inguinal canal with the ductus deferens are the testicular artery, vein, lymphatic vessel, and nerve. Collectively, these structures constitute the ***spermatic cord,*** which is covered by connective tissue.
4. In the abdomen, the ductus deferens separates from the spermatic cord, passes ventral to the ureter, and then medially to the dorsal surface of the urinary bladder. From there it enters the ***prostatic urethra,*** the portion of the urethra that passes through the prostate gland.
5. The pelvic cavity must be exposed by cutting through the pubic symphysis at the midline and opening the pelvic girdle.

e. PROSTATE GLAND

1. At the junction of the ductus (vas) deferens and the neck of the urinary bladder, locate the *prostate gland.*
2. The ducts of the prostate gland empty into the prostatic urethra.

f. BULBOURETHRAL GLANDS

1. These two glands are located on either side of the membranous urethra, the portion of the urethra between the prostatic urethra and the spongy urethra (the part passing through the penis).
2. The bulbourethral (Cowper's) glands are just dorsal to the penis.

g. PENIS

1. This is the external organ of copulation ventral to the scrotum.
2. The free (distal) end of the penis is enlarged and is referred to as the ***glans penis.*** The loose skin covering the penis is the ***foreskin,*** or ***prepuce.***
3. Cut open the prepuce to examine the glans penis. The opening of the penis to the exterior is the ***urogenital aperture.***
4. Internally the penis consists of two dorsally located cylinders and one ventrally located cylinder of erectile tissue. The dorsal masses are side by side and are known as the ***corpora cavernosa penis.*** Their proximal ends are known as the ***crura.*** Each crus is attached to the ischium on its own side by the ***ischiocavernosus muscle.*** The ventral mass of erectile tissue is

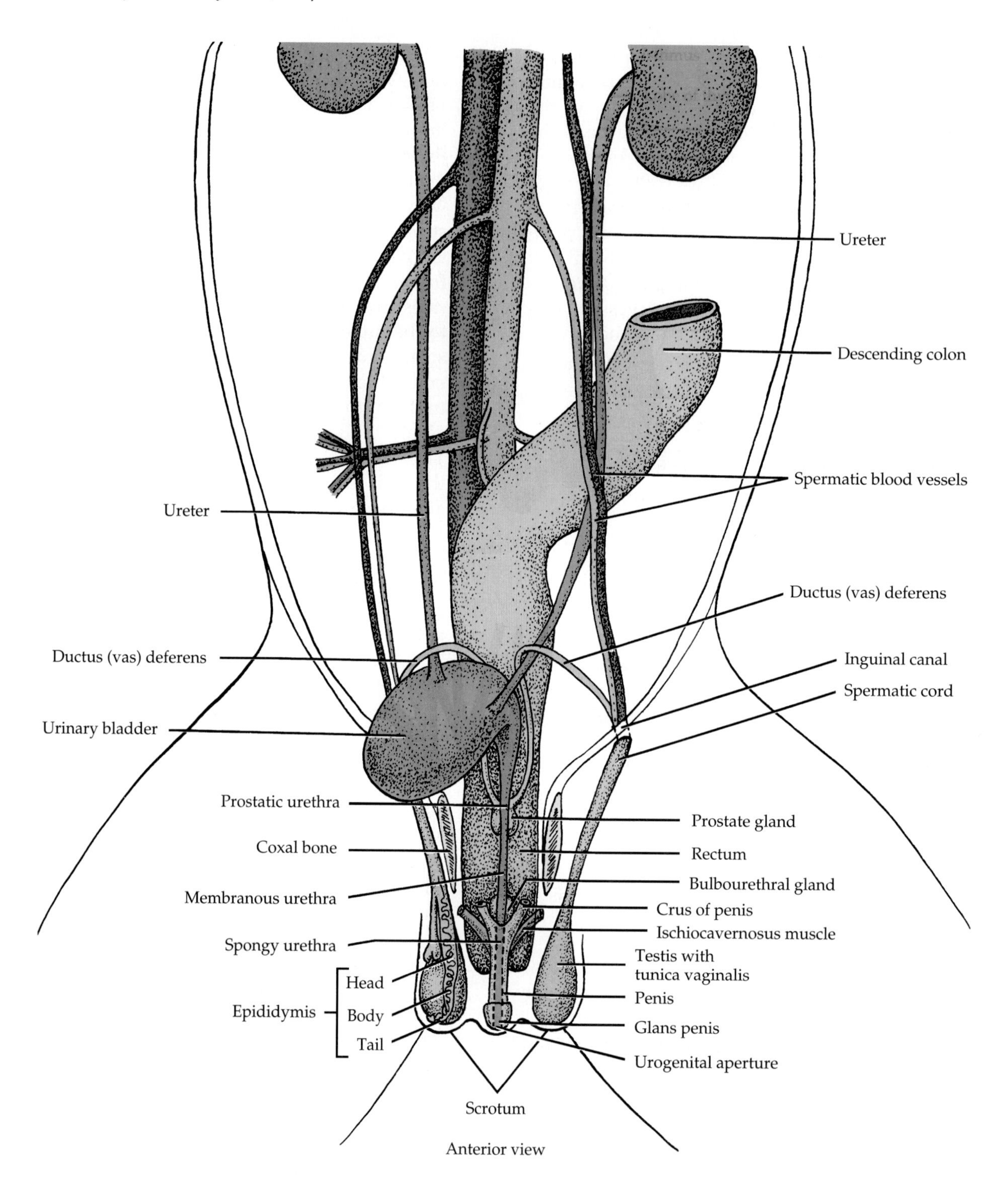

FIGURE 23.18 Male reproductive and urinary systems of cat.

called the ***corpus spongiosum penis.*** Through it passes the ***spongy (cavernosus) urethra.***

h. OS PENIS

1. This is the small bone embedded on one surface of the urethra in the glans penis.
2. The bone helps to stiffen this part of the penis.

i. ANAL GLANDS

1. These are a pair of round glands beneath the skin on either side of the rectum near the anus.
2. They produce an odoriferous secretion into the anus that is believed to be used for sexual attraction.

2. Female Reproductive System

The female reproductive system of the cat is shown in Figure 23.19.

PROCEDURE

a. OVARIES

1. The ***ovaries*** are a pair of small, oval organs located slightly caudal to the kidneys.
2. The lumpy appearance of the ovaries is due to the presence, within the ovaries, of ***vesicular ovarian (Graafian) follicles.*** These are small vesicles that contain ova and protrude to the ovarian surface.
3. Each ovary is suspended by the ***mesovarium,*** a peritoneal fold extending from the dorsal body wall to the ovary.

b. UTERINE (FALLOPIAN) TUBES

1. The paired ***uterine (Fallopian) tubes*** lie on the cranial surfaces of the ovaries.
2. The expanded end of the uterine tube, the ***infundibulum,*** has a fringed border of small fingerlike projections called ***fimbriae*** and also has an opening termed the ***ostium,*** which receives the ovum.
3. The peritoneum of the uterine tubes is called the ***mesosalpinx.*** It is continuous with the mesovarium.

c. UTERUS

1. The uterus is a Y-shaped structure consisting of two principal parts—uterine horns and body.
2. The ***uterine horns*** are the enlarged caudal communications of the uterine tubes. The embryos develop in the horns.
3. The ***body*** of the uterus is the caudal, median portion of the uterus. It is located between the urinary bladder and the rectum.

d. BROAD LIGAMENT

1. The uterine horns and body of the uterus are supported by a fold of peritoneum called the ***mesometrium.*** It extends from these structures to the lateral body walls.
2. Together, the mesometrium, mesosalpinx, and mesovarium are referred to as the ***broad ligament.***

e. ROUND LIGAMENT

1. The ***round ligament*** is a thin, fibrous band that extends from the dorsal body wall to the middle of each uterine horn.
2. The remainder of the female reproductive system can be seen only if the pelvic cavity is exposed. You will have to cut through the pelvic muscles and pubic symphysis at the midline.
3. Spread the thighs back.

f. VAGINA

1. Before examining the vagina, locate the ***urethra,*** a tube caudal to the urinary bladder.
2. Dorsal to the urethra is the ***vagina,*** a tube leading from the body of the uterus to the ***urogenital sinus (vestibule).*** This is a common passageway formed by the union of the urethra and vagina.

g. VULVA

1. The urogenital sinus opens to the exterior as the ***urogenital aperture,*** just ventral to the anus.
2. On either side of the urogenital aperture are folds of skin called the ***labia majora.***
3. In the ventral wall of the urogenital sinus is the ***clitoris,*** a mass of erectile tissue that is homologous to portions of the penis.
4. The urogenital aperture, labia majora, and clitoris constitute the ***vulva.***

D. DISSECTION OF FETUS-CONTAINING PIG UTERUS

Examination of the uterus of a pregnant pig reveals that the fetuses are equally spaced in the two uterine horns. Each fetus produces a local enlargement of the horn. The litter size normally ranges from 6 to 12. Your instructor may have you dissect the fetus-containing uterus of a pregnant pig or have

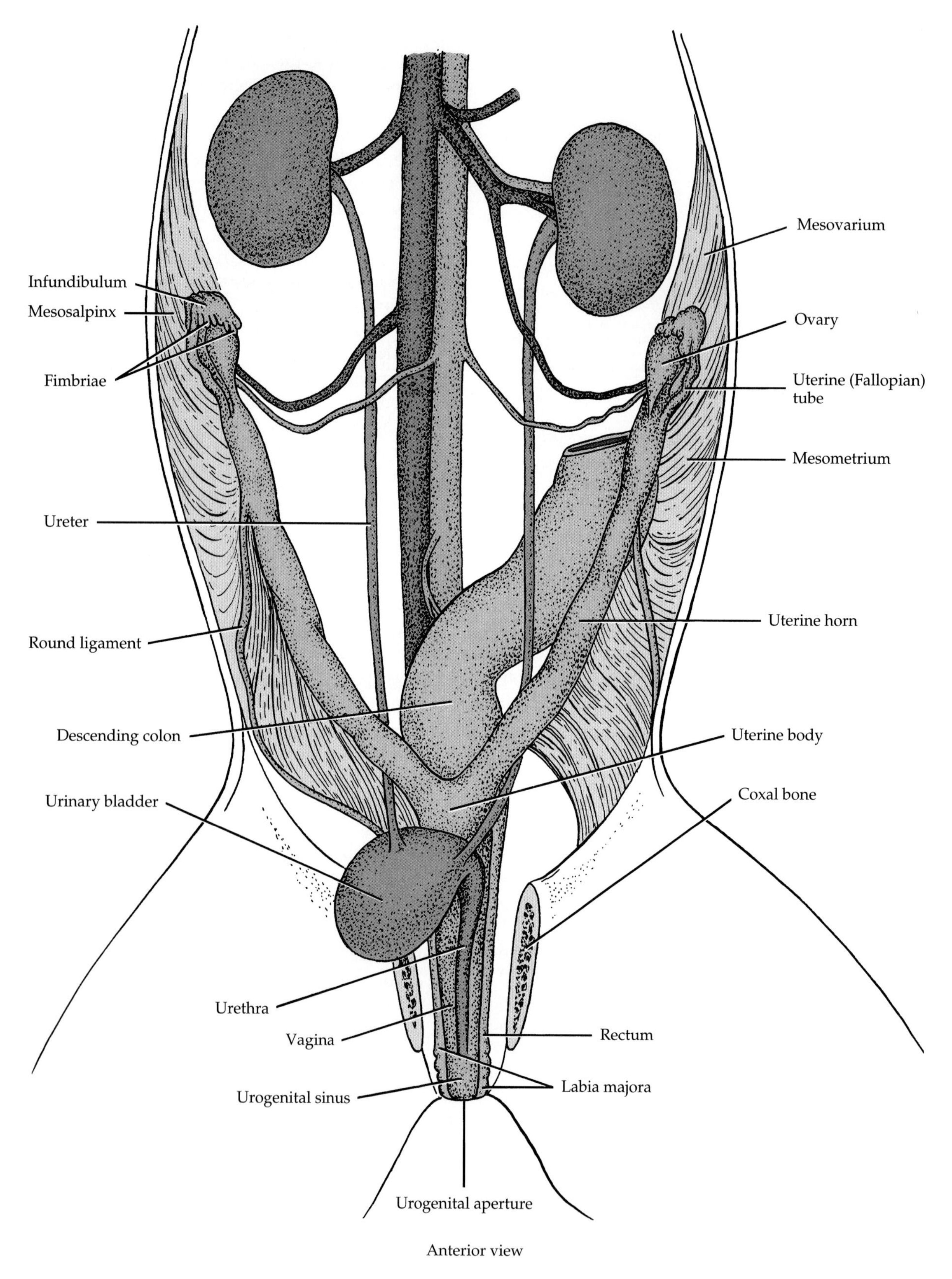

FIGURE 23.19 Female reproductive and urinary systems of cat.

one available as a demonstration (Figure 23.20). If you do a dissection, use the following directions. Also examine a chart or model of a human fetus and pregnant uterus if they are available.

CAUTION! *Please reread Section D, "Precautions Related to Dissection" at the beginning of the laboratory manual on page xi before you begin your dissection.*

PROCEDURE

1. Using a sharp scissors, cut open one of the enlargements of the horn and you will see that each fetus is enclosed together with an elongated, sausage-shaped ***chorionic vesicle.***
2. You will also notice many round bumps called ***areolae*** located over the chorionic surface.
3. The lining of the uterus together with the wall of the chorionic vesicle forms the ***placenta.***
4. Carefully cut open the chorionic vesicle, avoiding cutting or breaking the second sac lying within that surrounds the fetus itself.
5. The vesicle wall is the fusion of two extraembryonic membranes, the outer ***chorion*** (KOR-ē-on) and the inner ***allantois*** (a-LAN-tō-is). The allantois is the large sac growing out from the fetus, and the umbilical cord contains its stalk.
6. The ***umbilical blood vessels*** are seen in the allantoic wall spreading out in all directions and are also seen entering the ***umbilical cord.***
7. A thin-walled nonvascular ***amnion*** surrounds the fetus. This membrane is filled with ***amniotic fluid,*** which acts as a protective water cushion and prevents adherence of the fetus and membranes.

ANSWER THE LABORATORY REPORT QUESTIONS AT THE END OF THE EXERCISE.

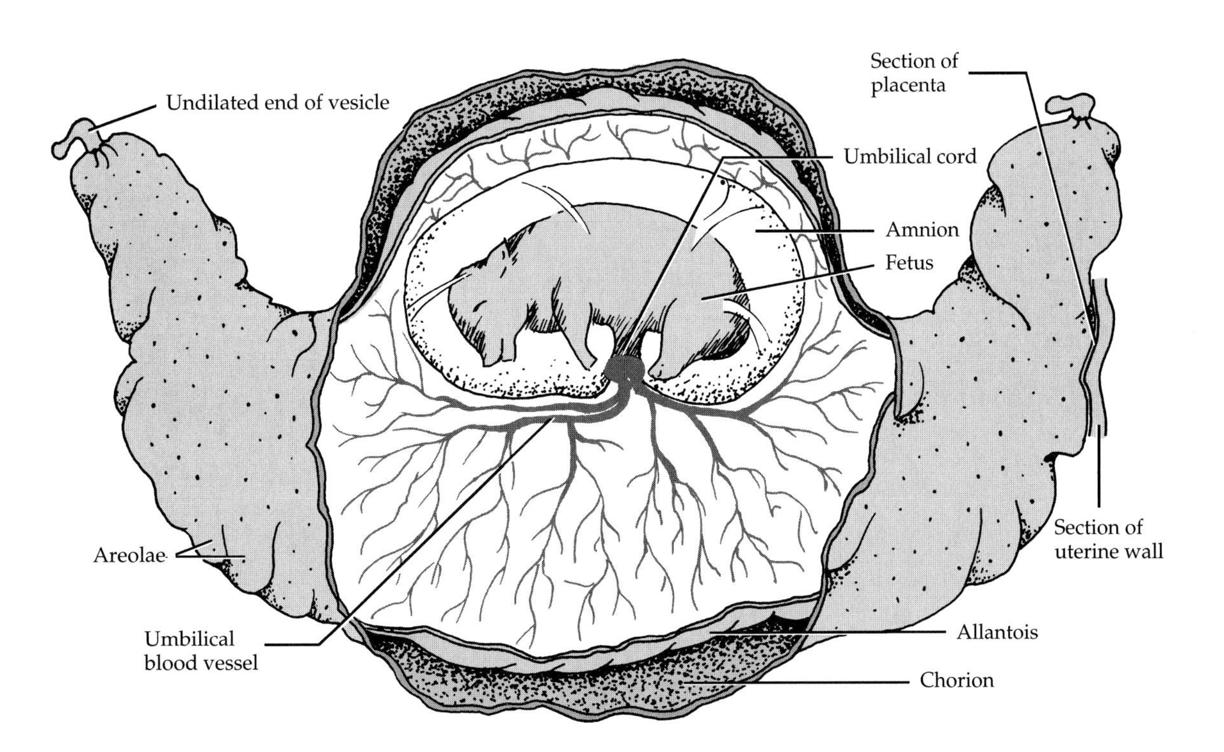

FIGURE 23.20 Fetal pig in opened chorionic vesicle.

LABORATORY REPORT QUESTIONS

Reproductive Systems 23

Student ______________________ Date ______________________

Laboratory Section ______________________ Score/Grade ______________________

PART 1. Multiple Choice

_______ **1.** Structures of the male reproductive system responsible for the production of sperm cells are the (a) efferent ducts (b) seminiferous tubules (c) seminal vesicles (d) rete testis

_______ **2.** The superior portion of the male urethra is encircled by the (a) epididymis (b) testes (c) prostate gland (d) seminal vesicles

_______ **3.** Cryptorchidism is a condition associated with the (a) prostate gland (b) testes (c) seminal vesicles (d) bulbourethral (Cowper's) glands

_______ **4.** Weakening of the suspensory ligament would directly affect the position of the (a) mammary glands (b) uterus (c) uterine (Fallopian) tubes (d) ovaries

_______ **5.** Organs in the female reproductive system responsible for transporting secondary oocytes and ova from the ovaries to the uterus are the (a) uterine (Fallopian) tubes (b) seminal vesicles (c) inguinal canals (d) none of the above

_______ **6.** The name of the process that is responsible for the actual production of sperm cells is called (a) cryptorchidism (b) oogenesis (c) spermatogenesis (d) spermatogonia

_______ **7.** Fertilization normally occurs in the (a) uterine (Fallopian) tubes (b) vagina (c) uterus (d) ovaries

_______ **8.** The portion of the uterus that assumes an active role during labor is the (a) serosa (b) endometrium (c) peritoneum (d) myometrium

_______ **9.** Stereocilia are associated with the (a) ductus (vas) deferens (b) oviduct (c) epididymis (d) rete testis

_______ **10.** The major portion of the volume of semen is contributed by the (a) bulbourethral (Cowper's) glands (b) testes (c) prostate gland (d) seminal vesicles

_______ **11.** The chief ligament supporting the uterus and keeping it from dropping into the vagina is the (a) cardinal ligament (b) round ligament (c) broad ligament (d) ovarian ligament

_______ **12.** Which sequence, from inside to outside, best represents the histology of the uterus? (a) stratum basalis, stratum functionalis, myometrium, perimetrium (b) myometrium, perimetrium, stratum functionalis, stratum basalis (c) stratum functionalis, stratum basalis, myometrium, perimetrium (d) stratum basalis, stratum functionalis, perimetrium, myometrium

_______ **13.** The white fibrous capsule that divides the testis into lobules is called the (a) dartos (b) raphe (c) tunica albuginea (d) germinal epithelium

_______ **14.** Which sequence best represents the course taken by sperm cells from their site of origin to the exterior? (a) seminiferous tubules, efferent ducts, epididymis, ductus (vas) deferens, ejaculatory duct, urethra (b) seminiferous tubules, efferent ducts, epididymis, ductus (vas) deferens, urethra, ejaculatory duct (c) seminiferous tubules, efferent ducts, ductus (vas) deferens, epididymis, ejaculatory duct, urethra (d) seminiferous tubules, epididymis, efferent ducts, ductus (vas) deferens, ejaculatory duct, urethra

_______ **15.** The ovaries are anchored to the uterus by the (a) ovarian ligament (b) broad ligament (c) suspensory ligament (d) mesovarium

_______ **16.** The terminal duct for the male reproductive system is the (a) urethra (b) ductus (vas) deferens (c) inguinal canal (d) ejaculatory duct

_______ **17.** The site of sperm cell maturation is the (a) ductus (vas) deferens (b) spermatic cord (c) epididymis (d) testes

_______ **18.** Which of the following is the site of menstruation, implantation of a fertilized ovum, development of the fetus during pregnancy, and labor? (a) uterus (b) uterine (Fallopian) tubes (c) vagina (d) cervix

_______ **19.** Glands lying over the pectoralis major muscles are the (a) lesser vestibular glands (b) adrenal (suprarenal) glands (c) mammary glands (d) greater vestibular (Bartholin's) glands

PART 2. Completion

20. Discharge of a secondary oocyte from the ovary about once each month is a process referred to as

_______________.

21. The inferior, narrow portion of the uterus that opens into the vagina is the _______________.

22. The clusters of milk-secreting cells of the mammary glands are referred to as

_______________.

23. The distal end of the penis is a slightly enlarged region called the _______________.

24. Covering the slightly enlarged region of the penis is a loosely fitting skin called the

_______________.

25. The circular pigmented area surrounding each nipple of the mammary glands is the

_______________.

26. After a secondary oocyte leaves the ovary, it enters the open, funnel-shaped distal end of the uterine (Fallopian) tube called the _______________.

27. The portion of a sperm cell that contains the nucleus and acrosome is the

_______________.

28. Vasectomy refers to removal of a portion of the _______________.

29. The mass of erectile tissue in the penis that contains the spongy urethra is the

_______________.

30. Both the mature (Graafian) follicle and _______________ of the ovary secrete hormones.

31. The superior dome-shaped portion of the uterus is called the _______________.

32. The _______________ anchor the uterus to either side of the pelvic cavity.

33. The passageway for menstrual flow and inferior portion of the birth canal is the

_______________.

34. Two longitudinal folds of skin that extend inferiorly and posteriorly from the mons pubis and are covered with pubic hair are the ______________________.

35. The ______________________ is a small mass of erectile tissue at the anterior junction of the labia minora.

36. The thin fold of vascularized membrane that borders the vaginal orifice is the ______________________.

37. Complete the following sequence for the passage of milk: alveoli, secondary tubules, lactiferous sinuses, ______________________, lactiferous ducts, nipple.

38. The layer of simple epithelium covering the free surface of the ovary is the ______________________.

39. The phase of the menstrual cycle between days 6 and 13 during which endometrial repair occurs is the ______________________ phase.

40. During menstruation, the stratum ______________________ of the endometrium is sloughed off.

41. The most immature spermatogenic cells are called ______________________.

42. The ______________________ contains the remnants of an ovulated mature follicle.

43. The hypothalamic hormone that controls the uterine and ovarian cycles is ______________________.

44. High levels of estrogens exert a positive feedback on LH and GnRH that cause ______________________.

Development

24

Development refers to the sequence of events starting with fertilization of a secondary oocyte and ending with the formation of a complete organism. Consideration will be given to how reproductive cells are produced and to a few developmental events associated with pregnancy.

A. SPERMATOGENESIS

The process by which the testes produce haploid (*n*) sperm cells involves several phases, including meiosis, and is called ***spermatogenesis*** (sper'-ma-tō-JEN-e-sis; *spermato* = sperm; *genesis* = to produce). In order to understand spermatogenesis, review the following concepts.

1. In sexual reproduction, a new organism is produced by the union and fusion of sex cells called ***gametes*** (*gameto* = to marry). Male gametes, produced in the testes, are called sperm cells, and female gametes, produced in the ovaries, are called oocytes.
2. The cell resulting from the union and fusion of gametes, called a ***zygote*** (*zygosis* = a joining), contains two full sets of chromosomes (DNA), one set from each parent. Through repeated mitotic cell divisions, a zygote develops into a new organism.
3. Gametes differ from all other body cells (somatic cells) in that they contain the ***haploid*** (one-half) ***chromosome number,*** symbolized as *n*. In humans, this number is 23, which composes a single set of chromosomes. The nucleus of a somatic cell contains the ***diploid chromosome number,*** symbolized as 2*n*. In humans, this number is 46, which composes two sets of paired chromosomes. One set of 23 chromosomes comes from the mother and the other set comes from the father.
4. In a diploid cell, two chromosomes that belong to a pair are called ***homologous*** (*homo* = same) ***chromosomes (homologues).*** In human diploid cells, the members of 22 of the 23 pairs of chromosomes are morphologically similar and are called ***autosomes.*** The other pair, termed X and Y chromosomes, are called the ***sex chromosomes*** because they determine one's gender. In the female, the homologous pair of sex chromosomes are two similar X chromosomes; in the male, the pair consists of an X and a Y chromosome.
5. If gametes were diploid (2*n*), like somatic cells, the zygote would contain twice the diploid number (4*n*), and with every succeeding generation the chromosome number would continue to double and normal development could not occur.
6. This continual doubling of the chromosome number does not occur because of ***meiosis*** (*meio* = less), a process by which gametes produced in the testes and ovaries receive the haploid chromosome number. Thus, when haploid (*n*) gametes fuse, the zygote contains the diploid chromosome number (2*n*) and can undergo normal development.

In humans, spermatogenesis takes about 74 days. The seminiferous tubules are lined with immature cells called ***spermatogonia*** (sper'-ma-tō-GŌ-nē-a; *sperm* = seed; *gonium* = generation or offspring), or sperm mother cells (see Figure 24.2 on page 477). Singular is ***spermatogonium.*** These cells develop from ***primordial*** (*primordialis* = primitive or early form) ***germ cells*** that arise from yolk sac endoderm and enter the testes early in development. In the embryonic testes, the primordial germ cells differentiate into spermatogonia but remain dormant until they begin to undergo mitotic proliferation at puberty. Spermatogonia contain the diploid (2*n*) chromosome number. Some spermatogonia remain relatively undifferentiated and capable of extensive mitotic division. Following division, some of the daughter cells remain undifferentiated and serve as a reservoir of precursor cells to prevent depletion of the stem cell population. Such cells remain near the basement membrane. The remainder of the daughter cells differentiate into spermatogonia that lose contact with the basement membrane of the seminiferous tubule, undergo certain developmental changes, and become known as ***primary spermatocytes*** (sper-MAT-ō-sītz'). Primary spermatocytes, like spermatogonia, are diploid (2*n*); that is, they have 46 chromosomes.

1. Reduction Division (Meiosis I)

Each primary spermatocyte enlarges before dividing. Then two nuclear divisions take place as part of meiosis. In the first, DNA is replicated and 46 chromosomes (each made up of two identical chromatids from the replicated DNA) form and move toward the equatorial plane of the cell. There they line up in homologous pairs so that there are 23 pairs of duplicated chromosomes in the center of the cell. This pairing of homologous chromosomes is called ***synapsis.*** The four chromatids of each homologous pair then become associated with each other to form a ***tetrad.*** In a tetrad, portions of one chromatid may be exchanged with portions of another. This process, called ***crossing over,*** permits an exchange of genes among maternal and paternal chromosomes (Figure 24.1) that results in the ***recombination*** of genes. Thus, the sperm cells eventually produced are genetically unlike each other and unlike the cell that produced them—one reason for the great variation among humans. Next, the meiotic spindle forms and the kinetochore microtubules organized by the centromeres extend toward the poles of the cell. As the pairs separate, one member of each pair migrates to opposite poles of the dividing cell. The random arrangement of chromosome pairs on the spindle is another reason for variation among humans. The cells formed by the first nuclear division (reduction division) are called ***secondary spermatocytes.*** Each cell has 23 chromosomes—the haploid number. Each chromosome of the secondary spermatocytes, however, is made up of two chromatids (two copies of the DNA) still attached by a centromere. Moreover, the genes of the chromosomes of secondary spermatocytes may be rearranged as a result of crossing over.

2. Equatorial Division (Meiosis II)

The second nuclear division of meiosis is ***equatorial division.*** There is no replication of DNA. The chromosomes (each composed of two chromatids) line up in single file along the equatorial plane, and the chromatids of each chromosome separate from each other. The cells formed from the equatorial division are called ***spermatids.*** Each contains half the original chromosome number, or 23 chromosomes, and is haploid. Each primary spermatocyte therefore produces four spermatids by meiosis (reduction division and equatorial division). Spermatids lie close to the lumen of the seminiferous tubule.

3. Spermiogenesis

The final stage of spermatogenesis, called ***spermiogenesis*** (sper'-mē-ō-JEN-e-sis), involves the maturation of spermatids into sperm cells. Each spermatid embeds in a sustentacular (Sertoli) cell and develops a head with an acrosome (enzyme-containing granule) and a flagellum (tail). Sustentacular cells extend from the basement membrane to the lumen of the seminiferous tubule, where they nourish the developing spermatids. Since there is no cell division in spermiogenesis, each spermatid develops into a single ***sperm cell (spermatozoon).*** The release of a sperm cell from a sustentacular cell is known as ***spermiation.***

Sperm cells enter the lumen of the seminiferous tubule and migrate to the ductus epididymis,

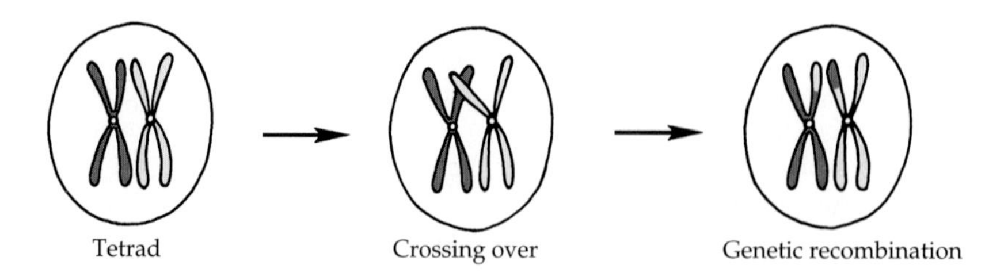

FIGURE 24.1 Crossing over within a tetrad resulting in genetic recombination.

where in 10 to 14 days they complete their maturation and become capable of fertilizing a secondary oocyte. Sperm cells are also stored in the ductus (vas) deferens. Here, they can retain their fertility for up to several months.

With the aid of your textbook, label Figure 24.2.

B. OOGENESIS

The formation of haploid (n) secondary oocytes in the ovary involves several phases, including meiosis, and is referred to as ***oogenesis*** (ō'-ō-JEN-e-sis; *oo* = egg; *genesis* = to produce). With some important

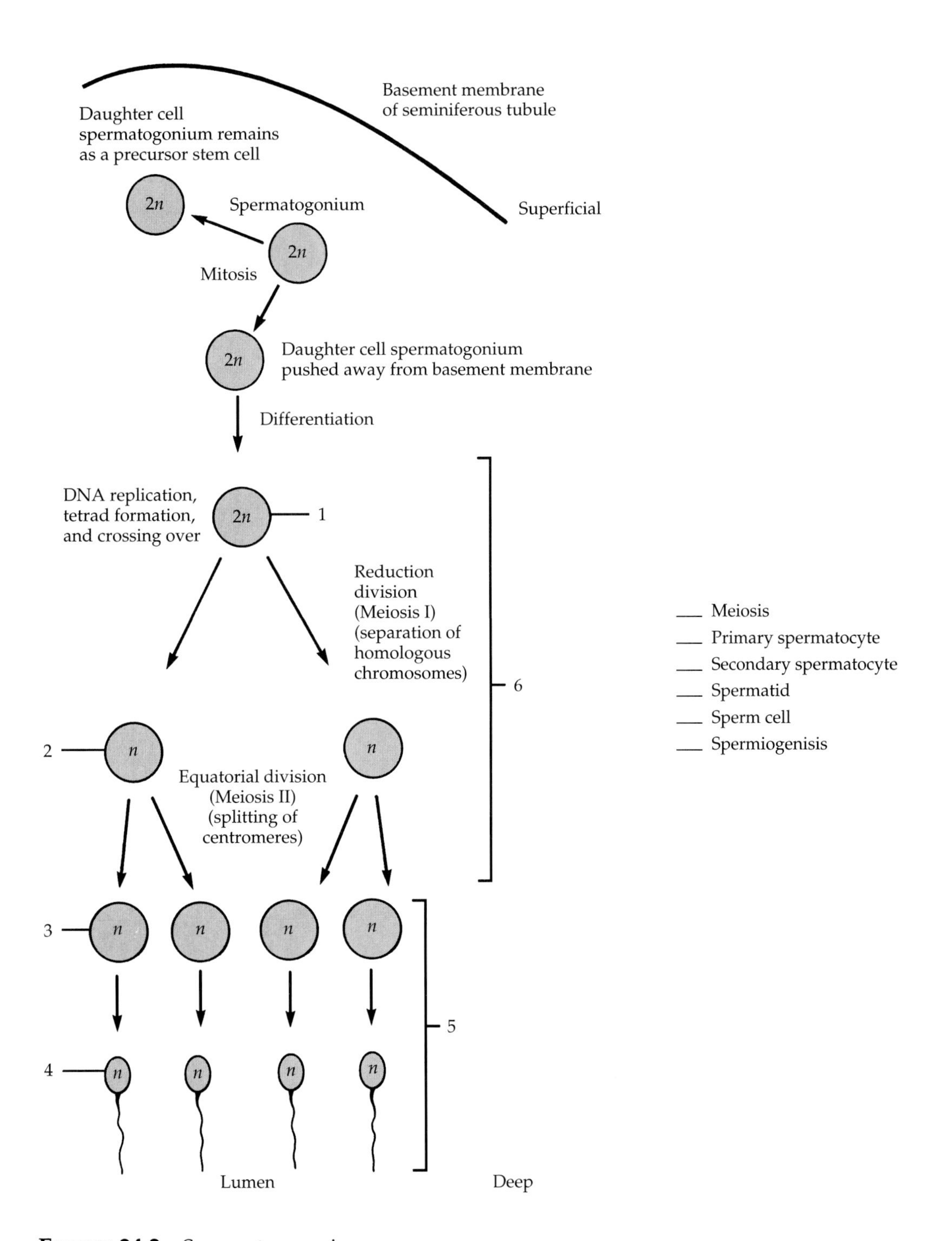

FIGURE 24.2 Spermatogenesis.

exceptions, oogenesis occurs in essentially the same manner as spermatogenesis.

1. Reduction Division (Meiosis I)

During early fetal development, primordial (primitive) germ cells migrate from the endoderm of the yolk sac to the ovaries. There, germ cells differentiate within the ovaries into ***oogonia*** (ō'-ō-GŌ-nē-a; singular is ***oogonium*** (ō'-ō-GŌ-nē-um). Oogonia are diploid ($2n$) cells that divide mitotically to produce millions of germ cells. Even before birth, many of these germ cells degenerate, a process known as ***atresia.*** A few develop into larger cells called ***primary*** (*primus* = first) ***oocytes*** (Ō'-ō-sītz) that enter the prophase of reduction division (meiosis I) during fetal development but do not complete it until puberty. At birth 200,000–2,000,000 oogonia and primary oocytes remain in each ovary. Of these, about 400 will mature and ovulate during a woman's reproductive lifetime; the remaining 99.98% undergo atresia.

Each primary oocyte is surrounded by a single layer of follicular cells, and the entire structure is called a ***primordial follicle*** (see Figure 23.10). Although the stimulating mechanism is unclear, a few primordial follicles start to grow, even during childhood. They become ***primary (preantral) follicles,*** which are surrounded first by one layer of cuboidal-shaped follicular cells and then by six to seven layers of cuboidal and low-columnar cells called ***granulosa cells.*** As a follicle grows, it forms a clear glycoprotein layer, called the ***zona pellucida*** (pe-LOO-si-da) between the oocyte and the granulosa cells. The innermost layer of granulosa cells becomes firmly attached to the zona pellucida and is called the ***corona radiata*** (*corona* = crown; *radiata* = radiation). The outermost granulosa cells rest on a basement membrane that separates them from the surrounding ovarian stroma. This outer region is called the ***theca folliculi.*** As the primary follicle continues to grow, the theca differentiates into two layers: (1) the ***theca interna,*** a vascularized internal layer of secretory cells, and (2) the ***theca externa,*** an outer layer of connective tissue cells. The granulosa cells begin to secrete follicular fluid, which builds up in a cavity called the ***antrum*** in the center of the follicle. The follicle is now termed a ***secondary (antral) follicle.*** During early childhood, primordial and developing follicles continue to undergo atresia.

After puberty, under the influence of the gonadotropin hormones secreted by the anterior pituitary gland, each month meiosis resumes in one secondary follicle. The diploid primary oocyte completes reduction division (meiosis I) and two haploid cells of unequal size, both with 23 chromosomes (n) of two chromatids each, are produced. The follicle in which these events are taking place, termed the ***mature (Graafian) follicle*** (also called a ***vesicular ovarian follicle***), will soon rupture and release its oocyte.

The smaller cell produced by meiosis I, called the ***first polar body,*** is essentially a packet of discarded nuclear material. The larger cell, known as the ***secondary oocyte,*** receives most of the cytoplasm. Once a secondary oocyte is formed, it proceeds to the metaphase of equatorial division (meiosis II) and then stops at this stage.

2. Equatorial Division (Meiosis II)

At ovulation, usually one secondary oocyte (with the first polar body and corona radiata) is expelled into the pelvic cavity. Normally, the cells are swept into the uterine (Fallopian) tube. If fertilization does not occur, the oocyte and other cells degenerate. If sperm cells are present in the uterine tube and one penetrates the secondary oocyte (fertilization), however, equatorial division (meiosis II) resumes. The secondary oocyte splits into two haploid (n) cells of unequal size. The larger cell is the ***ovum,*** or mature egg; the smaller one is the ***second polar body.*** The nuclei of the sperm cell and the ovum then unite, forming a diploid ($2n$) ***zygote.*** The first polar body may also undergo another division to produce two polar bodies. If it does, the primary oocyte ultimately gives rise to a single haploid (n) ovum and three haploid (n) polar bodies, which all degenerate. Thus an oogonium gives rise to a single gamete (ovum), whereas a spermatogonium produces four gametes (sperm).

With the aid of your textbook, label Figure 24.3.

C. EMBRYONIC PERIOD

The ***embryonic period*** is the first two months of development, and the developing human is called an ***embryo.***

1. Fertilization

During ***fertilization*** (fer'-til-i-ZĀ-shun; *fertilis* = reproductive) the genetic material from the sperm cell and secondary oocyte merges into a single nucleus (Figure 24.4a). Of the 300–500 million sperm introduced into the vagina, less than 1%

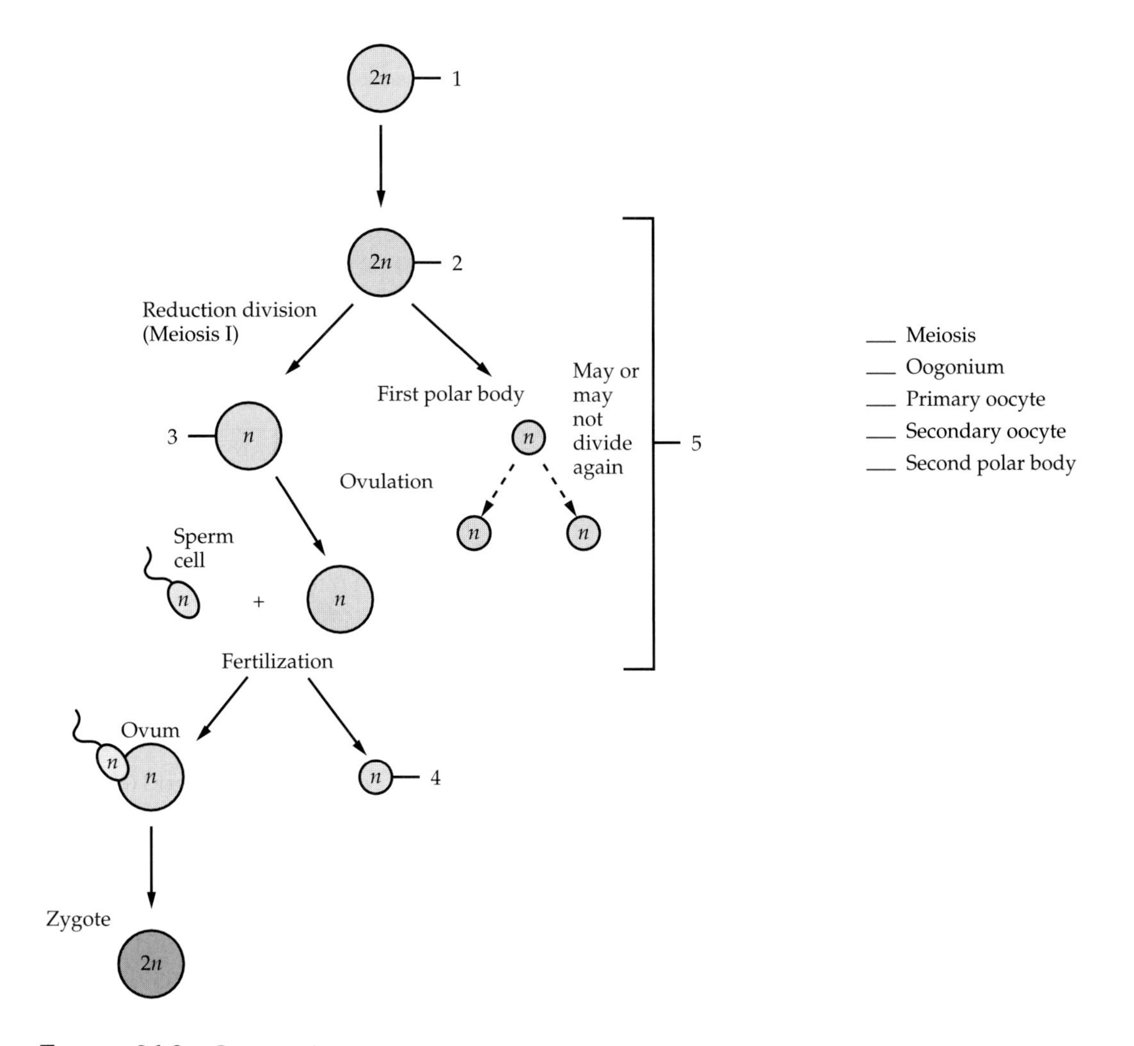

FIGURE 24.3 Oogenesis.

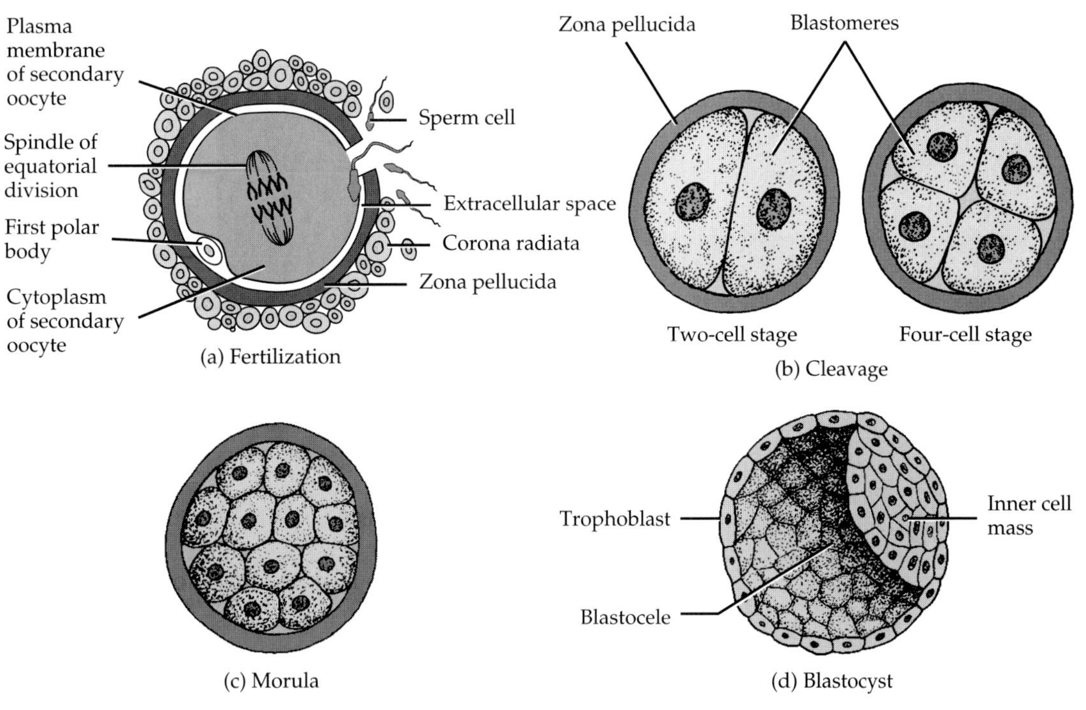

FIGURE 24.4 Fertilization.

reach the secondary oocyte. Fertilization normally occurs in the uterine (Fallopian) tube about 12–24 hours after ovulation. Since ejaculated sperm cells remain viable for about 48 hours and an oocyte is viable for about 24 hours after ovulation, there typically is a 3-day window during which pregnancy can occur—from 2 days before to 1 day after ovulation. Peristaltic contractions and the action of cilia transport the oocyte through the uterine tube. Sperm cells swim up the female tract by whiplike movements of their tails (flagella). The acrosome of sperm cells produces an enzyme called ***acrosin*** that stimulates sperm cell motility and migration within the female reproductive tract. Also, muscular contractions of the uterus, stimulated by prostaglandins in semen, probably aid sperm cell movement toward the uterine tube. Finally, the oocyte is thought to secrete a chemical substance that attracts sperm cells.

Besides contributing to sperm cell movement, the female reproductive tract also confers on sperm cells the capacity to fertilize a secondary oocyte. Although sperm cells undergo maturation in the epididymus, they are still not able to fertilize an oocyte until they have been in the female reproductive tract for several hours.

Capacitation (ka-pas'-i-TĀ-shun) refers to the functional changes that sperm cells undergo in the female reproductive tract that allow them to fertilize a secondary oocyte. During this process, the membrane around the acrosome becomes fragile so that several destructive enzymes—hyaluronidase, acrosin, and neuraminidase—are secreted by the acrosomes. It requires the collective action of many sperm cells to have just one penetrate the secondary oocyte. The enzymes help penetrate the corona radiata and zona pellucida around the oocyte. Sperm cells bind to receptors in the zona pellucida. Normally only one sperm cell penetrates and enters a secondary oocyte. This event is called ***syngamy*** (*syn* = together; *gamos* = marriage). Syngamy causes depolarization, which triggers the release of calcium ions inside the cell. Calcium ions stimulate the release of granules by the oocyte that, in turn, promote changes in the zona pellucida to block entry of other sperm cells. This prevents ***polyspermy,*** fertilization by more than one sperm cell. Once a sperm cell has entered a secondary oocyte, the oocyte completes equatorial division (meiosis II). It divides into a larger ovum (mature egg) and a smaller second polar body that fragments and disintegrates (see Figure 24.3).

When a sperm cell has entered a secondary oocyte, the tail is shed and the nucleus in the head develops into a structure called the ***male pronucleus.*** The nucleus of the secondary oocyte develops into a ***female pronucleus.*** After the pronuclei are formed, they fuse to produce a ***segmentation nucleus.*** The segmentation nucleus is diploid since it contains 23 chromosomes (n) from the male pronucleus and 23 chromosomes (n) from the female pronucleus. Thus the fusion of the haploid (n) pronuclei restores the diploid number ($2n$). The fertilized ovum, consisting of a segmentation nucleus, cytoplasm, and zona pellucida, is called a ***zygote*** (ZĪ-gōt; *zygosis* = a joining).

2. Formation of the Morula

After fertilization, rapid mitotic cell divisions of the zygote take place. These early divisions of the zygote are called ***cleavage*** (Figure 24.4b). Although cleavage increases the number of cells, it does not increase the size of the embryo, which is still contained within the zona pellucida.

The first cleavage begins about 24 hours after fertilization and is completed about 30 hours after fertilization, and each succeeding division takes slightly less time. By the second day after fertilization, the second cleavage is completed. By the end of the third day, there are 16 cells. The progressively smaller cells produced by cleavage are called ***blastomeres*** (BLAS-tō-mērz; *blast* = germ, sprout; *meros* = part). Successive cleavages produce a solid sphere of cells, still surrounded by the zona pellucida, called the ***morula*** (MOR-yoo-la; *morula* = mulberry) (Figure 24.4c). A few days after fertilization, the morula is about the same size as the original zygote.

3. Development of the Blastocyst

By the end of the fourth day, the number of cells in the morula increases and it continues to move through the uterine (Fallopian) tube toward the uterine cavity. At 4½–5 days, the dense cluster of cells has developed into a hollow ball of cells and enters the uterine cavity; it is now called a ***blastocyst*** (*kystis* = bag) (Figure 24.4d).

The blastocyst has an outer covering of cells called the ***trophoblast*** (TRŌ-fō-blast; *troph* = nourish), an ***inner cell mass (embryoblast),*** and an internal fluid-filled cavity called the ***blastocele*** (BLAS-tō-sēl; *koilos* = hollow). The trophoblast ultimately forms part of the membranes composing the fetal portion of the placenta; part of the inner cell mass develops into the embryo.

PROCEDURE

1. Obtain prepared slides of the embryonic development of the sea urchin. First try to find a zygote. This will appear as a single cell surrounded by an inner fertilization membrane and an outer, jellylike membrane. Draw a zygote in the space provided.
2. Now find several cleavage stages. See if you can isolate two-cell, four-cell, eight-cell, and sixteen-cell stages. Draw the various stages in the spaces provided.
3. Try to find a blastula (called a blastocyst in humans), a hollow ball of cells with a lighter center due to the presence of the blastocele. Draw a blastula in the space provided.

Zygote

Two-cell stage

Four-cell stage

Eight-cell stage

Sixteen-cell stage

Blastula

4. Implantation

The blastocyst remains free within the cavity of the uterus for a short period of time before it attaches to the uterine wall. During this time, the zona pellucida disintegrates and the blastocyst enlarges. The blastocyst receives nourishment from glycogen-rich secretions of endometrial (uterine) glands, sometimes called uterine milk. About 6 days after fertilization the blastocyst attaches to the endometrium, a process called ***implantation*** (Figure 24.5). At this time, the endometrium is in its secretory phase.

As the blastocyst implants, usually on the posterior wall of the fundus or body of the uterus, it is oriented so that the inner cell mass is toward the endometrium. The trophoblast develops two layers in the region of contact between the blastocyst and endometrium. These layers are a **syncytiotrophoblast** (sin-sī t'-ē-ō-TRŌF-ō-blast; *syn* = joined; *cyto* = cell) that contains no cell boundaries and a **cytotrophoblast** (sī '-tō-TRŌF-ō-blast) between the inner cell mass and syncytiotrophoblast that is composed of distinct cells (Figure 24.5c). These two layers of trophoblast become part of the chorion (one of the fetal membranes) as they undergo further growth. During implantation, the syncytiotrophoblast secretes enzymes that enable the blastocyst to penetrate the uterine lining. The enzymes digest and liquefy the endometrial cells. The fluid and nutrients further nourish the burrowing blastocyst for about a week after implantation. Eventually, the blastocyst becomes buried in the endometrium. The trophoblast also secretes human chorionic gonadotropin (hCG) that rescues the corpus luteum from degeneration and sustains its secretion of progesterone and estrogens.

5. Primary Germ Layers

After implantation, the first major event of the embryonic period occurs. The inner cell mass of the blastocyst begins to differentiate into the three ***primary germ layers:*** ectoderm, endoderm, and mesoderm. These are the major embryonic tissues from which all tissues and organs of the body will develop. The process by which the two-layered inner cell mass is converted into a structure composed of the primary germ layers is called ***gastrulation*** (gas'-troo-LĀ-shun; *gastrula* = little belly).

Within eight days after fertilization, the cells of the inner cytotrophoblast proliferate and form the amnion (a fetal membrane) and a space, the

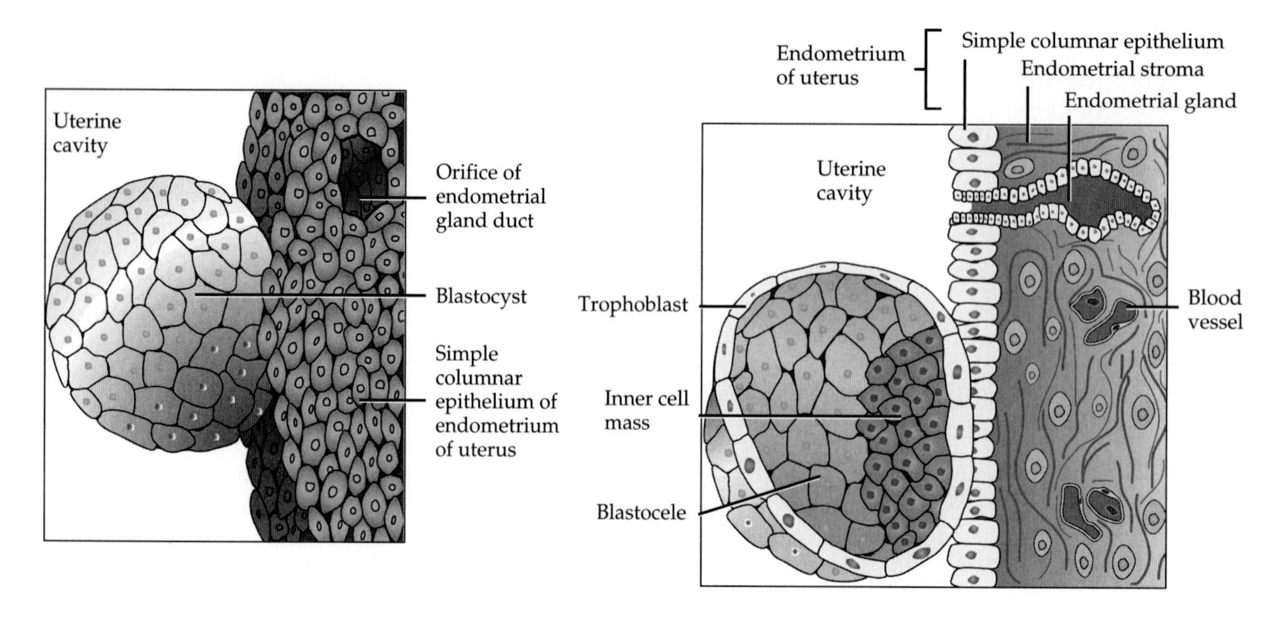

(a) External view, about 5 days after fertilization

(b) Internal view, about 6 days after fertilization

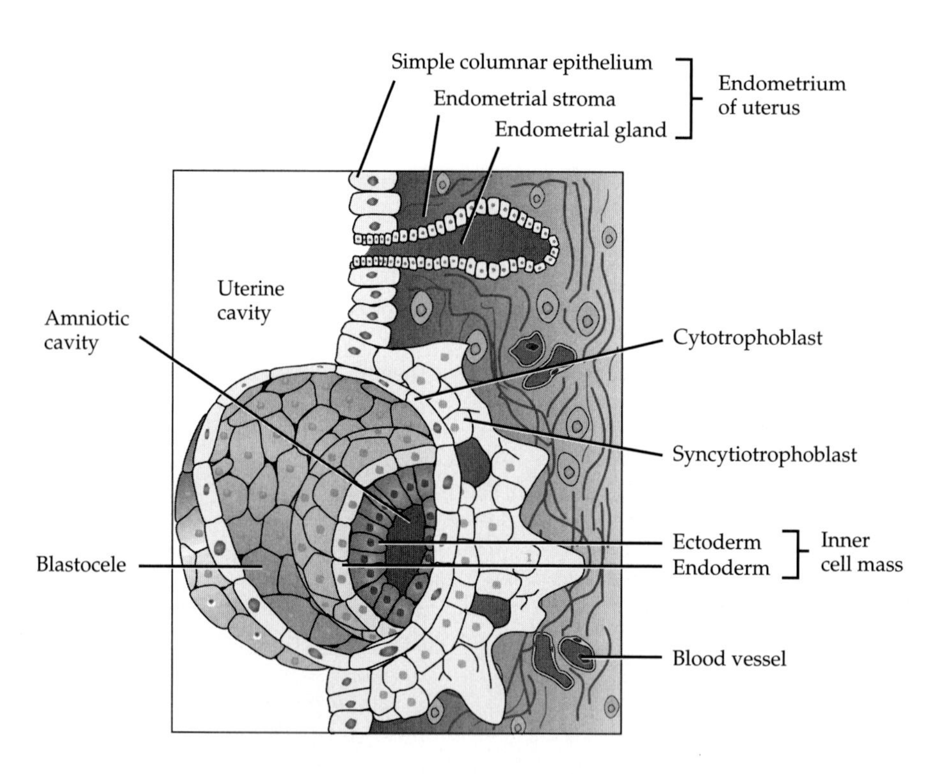

(c) Internal view, about 7 days after fertilization

FIGURE 24.5 Implantation.

amniotic (am'-nē-OT-ik; *amnion* = lamb) or ***amniotic cavity,*** over the inner cell mass. The layer of cells of the inner cell mass that is closer to the amniotic cavity develops into the ***ectoderm*** (*ecto* = outside; *derm* = skin). The layer of the inner cell mass that borders the blastocele develops into the ***endoderm*** (*endo* = inside). As the amniotic cavity forms, the inner cell mass at this stage is called the ***embryonic disc.*** It will form the embryo. At this stage, the embryonic disc contains ectodermal and endodermal cells; the mesodermal cells are scattered external to the disc.

About the 12th day after fertilization, striking changes appear (Figure 24.6a). The cells of

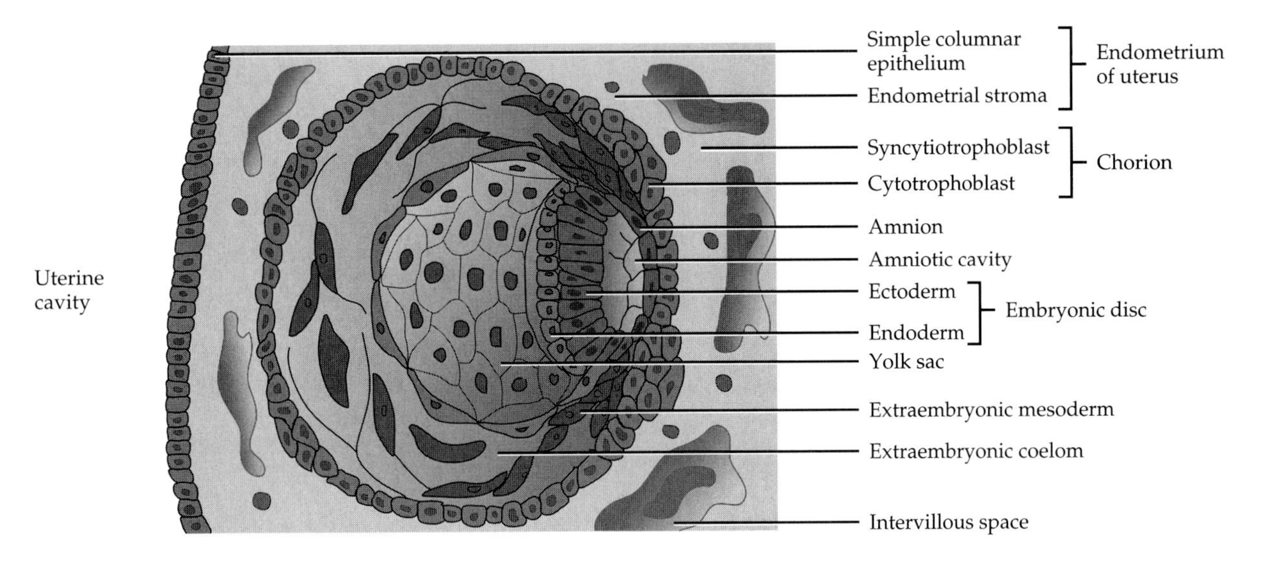

(a) Internal view, about 12 days after fertilization

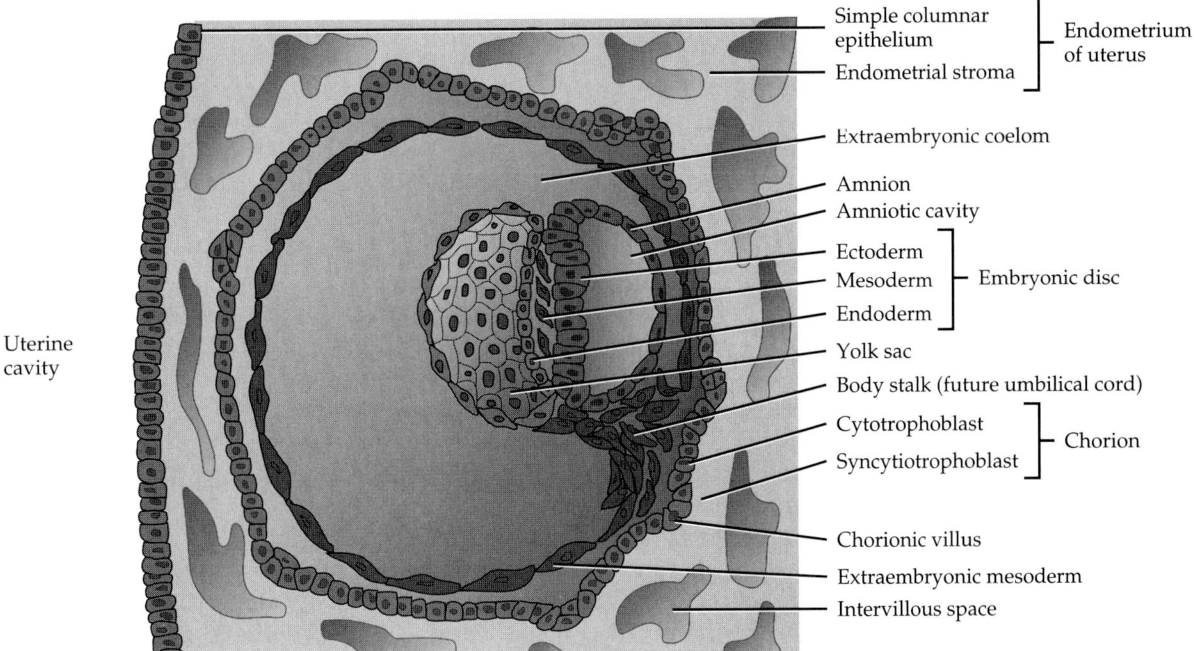

(b) Internal view, about 14 days after fertilization

FIGURE 24.6 Formation of the primary germ layers and associated structures.

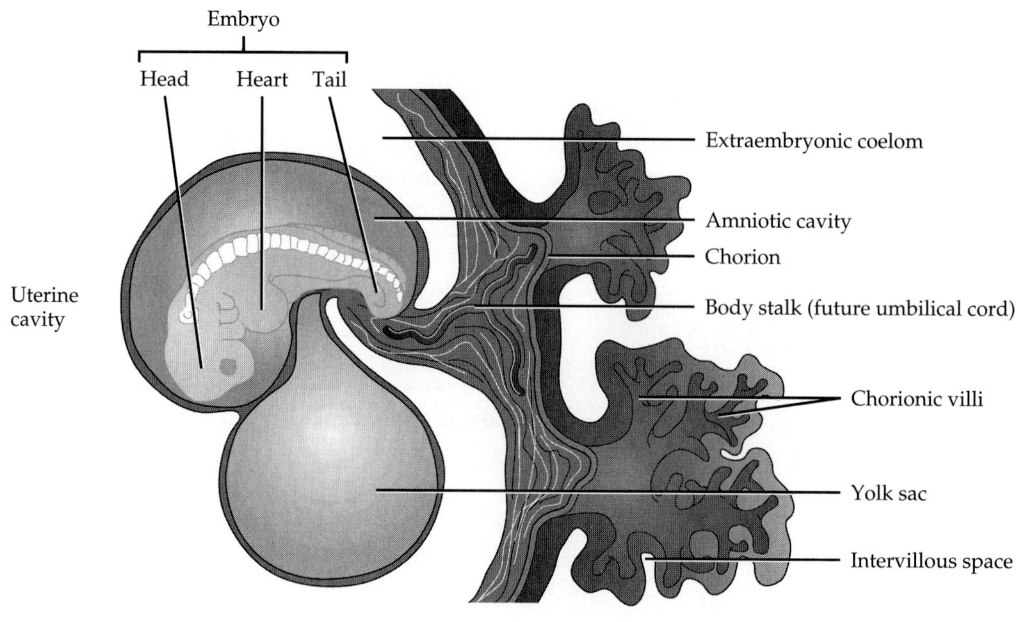

Figure 24.6 ***(Continued)*** Formation of the primary germ layers and associated structures.

the endodermal layer have been dividing rapidly, so that groups of them now extend around in a circle, forming the yolk sac, another fetal membrane (described shortly). Cells of the cytotrophoblast give rise to a loose connective tissue, the ***extraembryonic mesoderm*** (*meso* = middle). This completely fills the space between the cytotrophoblast and yolk sac. Soon large spaces develop in the extraembryonic mesoderm and come together to form a single, larger cavity called the ***extraembryonic coelom*** (SĒ-lōm; *koiloma* = cavity), the future ventral body cavity.

About the 14th day, the cells of the embryonic disc differentiate into three distinct layers: the ectoderm, the mesoderm (intraembryonic), and the endoderm (Figure 24.6b). As the embryo develops (Figure 24.6c), the endoderm becomes the epithelial lining of most of the gastrointestinal tract, urinary bladder, gallbladder, liver, pharynx, larynx, trachea, bronchi, lungs, vagina, urethra, and thyroid, parathyroid, and thymus glands, among other structures. The mesoderm develops into muscle; cartilage, bone and other connective tissues; red bone marrow, lymphoid tissue, endothelium of blood and lymphatic vessels, gonads, dermis of the skin, and other structures. The ectoderm develops into the entire nervous system, epidermis of skin, epidermal derivatives of the skin, and portions of the eye and other sense organs.

6. Embryonic Membranes

A second major event that occurs during the embryonic period is the formation of the ***embryonic (extraembryonic) membranes*** (Figure 24.7). These membranes lie outside the embryo and protect and nourish the embryo and, later, the fetus. The membranes are the yolk sac, amnion, chorion, and allantois.

In many species, the ***yolk sac*** is a membrane that is the primary source of blood vessels that transport nutrients to the embryo. A human embryo receives nutrients from the endometrium, however; the yolk sac remains small and functions as an early site of blood formation. The yolk sac also contains cells that migrate into the gonads and differentiate into the primitive germ cells (spermatogonia and oogonia).

The ***amnion*** is a thin, protective membrane that forms by the eighth day after fertilization and initially overlies the embryonic disc. As the embryo grows, the amnion comes to entirely surround the embryo, creating a cavity that becomes filled with ***amniotic fluid.*** Most amniotic fluid is initially derived from a filtrate of maternal blood. Later, the fetus makes daily contributions to the fluid by

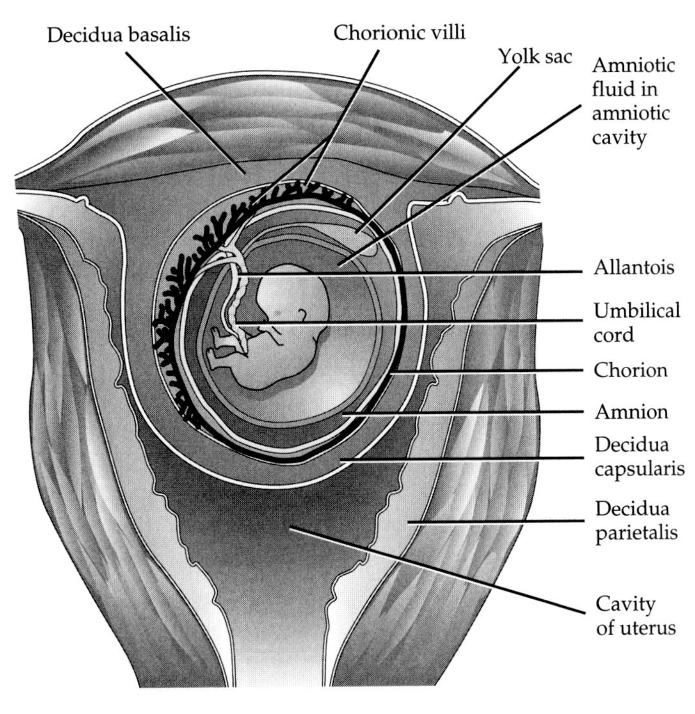

FIGURE 24.7 Embryonic membranes.

excreting urine into the amniotic cavity. Amniotic fluid serves as a shock absorber for the fetus, helps regulate fetal body temperature, and prevents adhesions between the skin of the fetus and surrounding tissues. Embryonic cells are sloughed off into amniotic fluid; they can be examined in the procedure called ***amniocentesis*** (am'-nē-ō-sen-TĒ-sis). The amnion usually ruptures just before birth and with its fluid constitutes the "bag of waters."

The ***chorion*** (KŌ-rē-on) is derived from the trophoblast of the blastocyst and the mesoderm that lines the trophoblast. It surrounds the embryo and, later, the fetus. Eventually, the chorion becomes the principal embryonic part of the placenta, the structure for exchange of materials between the mother and fetus. It also produces human chorionic gonadotropin (hCG). The amnion, which also surrounds the fetus, eventually fuses to the inner layer of the chorion.

The ***allantois*** (a-LAN-tō-is; *allas* = sausage) is a small, vascularized outpouching of the hindgut. It serves as an early site of blood formation. Later its blood vessels serve as the umbilical connection in the placenta between mother and fetus. This connection is the umbilical cord.

7. Placenta and Umbilical Cord

Development of the ***placenta*** (pla-SEN-ta; *placenta* = flat cake), the third major event of the embryonic period, is accomplished by the third month of pregnancy. The placenta has the shape of a flat cake when fully developed and is formed by the chorion of the embryo and a portion of the endometrium (decidua basalis) of the mother (Figure 24.7). Functionally, the placenta allows oxygen and nutrients to diffuse into fetal blood from maternal blood. Simultaneously, carbon dioxide and wastes diffuse from fetal blood into maternal blood at the placenta.

The placenta also is a protective barrier since most microorganisms cannot cross it. However, certain viruses, such as those that cause AIDS, German measles, chickenpox, measles, encephalitis, and poliomyelitis, may pass through the placenta. The placenta also stores nutrients such as carbohydrates, proteins, calcium, and iron, which are released into fetal circulation as required. Finally, the placenta produces several hormones that are necessary to maintain pregnancy. Almost all drugs, including alcohol, and many substances that can cause birth defects pass freely through the placenta.

If implantation occurs, a portion of the endometrium becomes modified and is known as the ***decidua*** (dē-SID-yoo-a; *deciduus* = falling off). The decidua includes all but the stratum basalis layer of the endometrium and separates from the endometrium after the fetus is delivered. Different regions of the decidua, which are all areas of the stratum functionalis, are named based on their positions relative to the site of the implanted, fertilized ovum (Figure 24.7). The ***decidua basalis*** is the portion of the endometrium between the chorion and the stratum basalis of the uterus. It becomes the maternal part of the placenta. The ***decidua capsularis*** is the portion of the endometrium that covers the embryo and is located between the embryo and the uterine cavity. The ***decidua parietalis*** (pa-rī-e-TAL-is) is the remaining modified endometrium that lines the noninvolved areas of the entire pregnant uterus. As the embryo and later the fetus enlarge, the decidua capsularis bulges into the uterine cavity and initially fuses with the decidua parietalis, thus obliterating the uterine cavity. By about 27 weeks, the decidua capsularis degenerates and disappears.

During embryonic life, fingerlike projections of the chorion, called ***chorionic villi*** (kō'-rē-ON-ik VIL-ī), grow into the decidua basalis of the endometrium (Figure 24.7). These will contain fetal blood vessels of the allantois. They continue growing until they are bathed in maternal blood sinuses called ***intervillous*** (in-ter-VIL-us) ***spaces.*** Thus maternal and fetal blood vessels are brought into proximity. It should be noted, however, that maternal and fetal blood do not normally mix. Rather, oxygen and nutrients in the blood of the mother's intervillous spaces diffuse across the cell membranes into the capillaries of the villi while waste products diffuse in the opposite direction. From the capillaries of the villi, nutrients and oxygen enter the fetus through the umbilical vein. Wastes leave the fetus through the umbilical arteries, pass into the capillaries of the villi, and diffuse into the maternal blood.

The ***umbilical*** (um-BIL-i-kul) ***cord*** is a vascular connection between mother and fetus. It consists of two umbilical arteries that carry deoxygenated fetal blood to the placenta, one umbilical vein that carries oxygenated blood into the fetus, and supporting mucous connective tissue called Wharton's jelly from the allantois. The entire umbilical cord is surrounded by a layer of amnion (Figure 24.7).

At delivery, the placenta detaches from the uterus and is termed the ***afterbirth.*** At this time, the umbilical cord is severed, leaving the baby on its own. The small portion (about an inch) of the cord that remains still attached to the infant begins to wither and falls off, usually within 12–15 days after birth. The area where the cord was attached becomes covered by a thin layer of skin and scar tissue forms. The scar is the ***umbilicus (navel).***

D. FETAL PERIOD

During the ***fetal period,*** the months of development after the second month, all the organs of the body grow rapidly from the original primary germ layers, and the organism takes on a human appearance. During this time, the developing human is called a ***fetus.*** Some of the principal changes associated with fetal growth are summarized in Table 24.1.

ANSWER THE LABORATORY REPORT QUESTIONS AT THE END OF THE EXERCISE.

TABLE 24.1
Changes Associated with Embryonic and Fetal Growth

End of month	Approximate size and weight	Representative changes
1	0.6 cm ($\frac{3}{16}$ in.)	Eyes, nose, and ears not yet visible. Vertebral column and vertebral canal form. Small buds that will develop into limbs form. Heart forms and starts beating. Body systems begin to form. The central nervous system appears at the start of the third week.
2	3 cm ($1\frac{1}{4}$ in.) 1 g ($\frac{1}{30}$ oz)	Eyes far apart, eyelids fused, nose flat. Ossification begins. Limbs become distinct and digits are well formed. Major blood vessels form. Many internal organs continue to develop.
3	$7\frac{1}{2}$ cm (3 in.) 30 g (1 oz)	Eyes almost fully developed but eyelids still fused, nose develops bridge, and external ears are present. Ossification continues. Limbs are fully formed and nails develop. Heartbeat can be detected. Urine starts to form. Fetus begins to move, but it cannot be felt by mother. Body systems continue to develop.
4	18 cm ($6\frac{1}{2}$–7 in.) 100 g (4 oz)	Head large in proportion to rest of body. Face takes on human features and hair appears on head. Skin bright pink. Many bones ossified, and joints begin to form. Rapid development of body systems.
5	25–30 cm (10–12 in.) 200–450 g ($\frac{1}{2}$–1 lb)	Head less disproportionate to rest of body. Fine hair (lanugo) covers body. Skin still bright pink. Brown fat forms and is the site of heat production. Fetal movements commonly felt by mother (quickening). Rapid development of body systems.
6	27–35 cm (11–14 in.) 550–800 g ($1\frac{1}{4}$–$1\frac{1}{2}$ lb)	Head becomes even less disproportionate to rest of body. Eyelids separate and eyelashes form. Substantial weight gain. Skin wrinkled and pink. Type II alveolar cells begin to produce surfactant.
7	32–42 cm (13–17 in.) 1100–1350 g ($2\frac{1}{2}$–3 lb)	Head and body more proportionate. Skin wrinkled and pink. Seven-month fetus (premature baby) is capable of survival. Fetus assumes an upside-down position. Testes descend into scrotum.
8	41–45 cm ($16\frac{1}{2}$–18 in.) 2000–2300 g ($4\frac{1}{2}$–5 lb)	Subcutaneous fat deposited. Skin less wrinkled. Chances of survival much greater at end of eighth month.
9	50 cm (20 in.) 3200–3400 g (7–$7\frac{1}{2}$ lb)	Additional subcutaneous fat accumulates. Lanugo shed. Nails extend to tips of fingers and maybe even beyond.

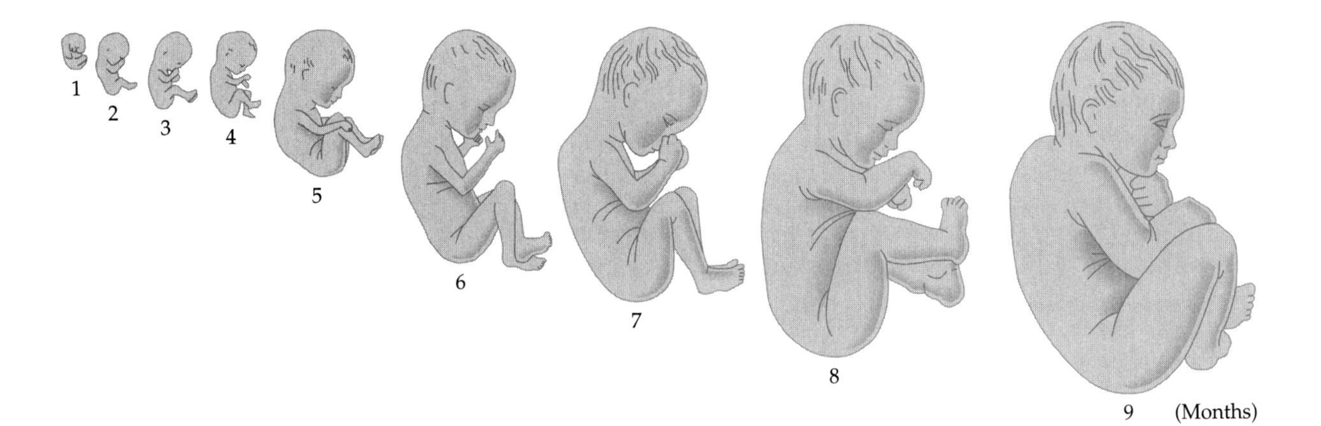

LABORATORY REPORT QUESTIONS

Development 24

Student ______________________ Date ______________________

Laboratory Section ______________________ Score/Grade ______________________

PART 1. Multiple Choice

_______ 1. The basic difference between spermatogenesis and oogenesis is that (a) two more polar bodies are produced in spermatogenesis (b) the secondary oocyte contains the haploid chromosome number, whereas the mature sperm cell contains the diploid number (c) in oogenesis, one secondary oocyte is produced, and in spermatogenesis four mature sperm cells are produced (d) both mitosis and meiosis occur in spermatogenesis, but only meiosis occurs in oogenesis

_______ 2. The union of a sperm cell nucleus and a secondary oocyte nucleus resulting in formation of a zygote is referred to as (a) implantation (b) fertilization (c) gestation (d) parturition

_______ 3. The most advanced stage of development for these stages is the (a) morula (b) zygote (c) ovum (d) blastocyst

_______ 4. Damage to the mesoderm during embryological development would directly affect the formation of (a) muscle tissue (b) the nervous system (c) the epidermis of the skin (d) hair, nails, and skin glands

_______ 5. The placenta, the organ of exchange between mother and fetus, is formed by union of the endometrium with the (a) yolk sac (b) amnion (c) chorion (d) umbilicus

_______ 6. One oogonium produces (a) one ovum and three polar bodies (b) two ova and two polar bodies (c) three ova and one polar body (d) four ova

_______ 7. Implantation is defined as (a) attachment of the blastocyst to the uterine (Fallopian) tube (b) attachment of the blastocyst to the endometrium (c) attachment of the embryo to the endometrium (d) attachment of the morula to the endometrium

_______ 8. Epithelium lining most of the gastrointestinal tract and a number of other organs is derived from (a) ectoderm (b) mesoderm (c) endoderm (d) mesophyll

_______ 9. The nervous system is derived from the (a) ectoderm (b) mesoderm (c) endoderm (d) mesophyll

_______ 10. Which of the following is *not* an embryonic membrane? (a) amnion (b) placenta (c) chorion (d) allantois

PART 2. Completion

11. A normal human sperm cell, as a result of meiosis, contains ______________________ chromosomes.
12. The process that permits an exchange of genes resulting in their recombination and a part of the variation among humans is called ______________________.

13. The result of meiosis in spermatogenesis is that each primary spermatocyte produces four

_____________________.

14. The stage of spermatogenesis that results in maturation of spermatids into sperm cells is called

_____________________.

15. The afterbirth expelled in the final stage of delivery is the _____________________.

16. After the second month, the developing human is referred to as a(n) _____________________.

17. Embryonic tissues from which all tissues and organs of the body develop are called the

_____________________.

18. The cells of the inner cell mass divide to form two cavities: amniotic cavity and

_____________________.

19. Somatic cells that contain two sets of chromosomes are referred to as _____________________.

20. At the end of the _____________________ month of development, a heartbeat can be detected.

21. In oogenesis, primordial follicles develop into _____________________ follicles.

22. The clear, glycoprotein layer between the oocyte and granulosa cells is called the

_____________________.

23. _____________________ refers to the functional changes that sperm cells undergo in the female reproductive tract that allow them to fertilize a secondary oocyte.

24. The _____________________ of a blastocyst develops into an embryo.

25. The decidua _____________________ is the portion of the endometrium between the chorion and stratum basalis of the uterus.

26. At the end of the _____________________ month of development, the testes descend into the scrotum.

Some Important Units of Measurement

English Units of Measurement

Fundamental or Derived Unit	Units and Equivalents
Length	12 inches (in.) = 1 foot (ft) = 0.333 yard (yd) 3 ft = 1 yd 1760 yd = 1 mile (mi) 5280 ft = 1 mi
Mass	1 ounce (oz) = 28.35 grams (g); 1 g = 0.0353 oz 1 pound (lb) = 453 g = 16 oz; 1 kilogram (kg) = 2.205 lb 1 ton = 2000 lb = 907 kg
Time	1 second (sec) = 1/86,400 of a mean solar day 1 minute (min) = 60 sec 1 hour (hr) = 60 min = 3600 sec 1 day = 24 hr = 1440 min = 86,400 sec
Volume	1 fluid dram (fl dr) = 0.125 fluid ounce (fl oz) 1 fl oz = 8 fl dr = 0.0625 quart (qt) = 0.008 gallon (gal) 1 qt = 256 fl dr = 32 fl oz = 2 pints (pt) = 0.25 gal 1 gal = 4 qt = 128 fl oz = 1024 fl dr

Metric Units of Length and Some English Equivalents

Metric Unit	Meaning of Prefix	Metric Equivalent	English Equivalent
1 kilometer (km)	kilo = 1000	1000 m	3280.84 ft or 0.62 mi; 1 mi = 1.61 km
1 hectometer (hm)	hecto = 100	100 m	328 ft
1 dekameter (dam)	deka = 10	10 m	32.8 ft
1 meter (m)	Standard unit of length		39.37 in. or 3.28 ft or 1.09 yd
1 decimeter (dm)	deci = $\frac{1}{10}$	0.1 m	3.94 in.
1 centimeter (cm)	centi = $\frac{1}{100}$	0.01 m	0.394 in.; 1 in. = 2.54 cm
1 millimeter (mm)	milli = $\frac{1}{1000}$	0.001 m = $\frac{1}{10}$ cm	0.0394 in.
1 micrometer (μm) [formerly micron (μ)]	micro = $\frac{1}{1,000,000}$	0.0000001 m = $\frac{1}{10,000}$ cm	3.94×10^{-5} in.
1 nanometer (nm) [formerly millimicrons (mμ)]	nano = $\frac{1}{1,000,000,000}$	0.000000001 m = $\frac{1}{10,000,000}$ cm	3.94×10^{-8} in.

Temperature

Unit	K	°F	°C
1 degree Kelvin (K)	1	$\frac{9}{5}$(K) – 459.7	K + 273.16*
1 degree Fahrenheit (°F)	$\frac{5}{9}$(°F) + 255.4	1	$\frac{5}{9}$(°F – 32)
1 degree Celsius (°C)	°C – 273	$\frac{9}{5}$(°C) + 32	1

* Absolute zero (K) = –273.16°C

Volume

Unit	ml	cm^3	qt	oz
1 milliliter (ml)	1	1	1.06×10^{-3}	3.392×10^{-2}
1 cubic centimeter (cm^3)	1	1	1.06×10^{-3}	3.392×10^{-2}
1 quart (qt)	943	943	1	32
1 fluid ounce (fl oz)	29.5	29.5	3.125×10^{-2}	1

Periodic Table of the Elements

APPENDIX **B**

KEY

6 ← Atomic Number
C ← Symbol
12.01 ← Atomic Weight
Carbon ← Name

1 H 1.0080 Hydrogen																	2 He 4.003 Helium
3 Li 6.940 Lithium	4 Be 9.013 Berilium											5 B 10.82 Boron	6 C 12.011 Carbon	7 N 14.008 Nitrogen	8 O 16.000 Oxygen	9 F 19.00 Fluorine	10 Ne 20.183 Neon
11 Na 22.991 Sodium	12 Mg 24.32 Mag-nesium											13 Al 26.98 Alminum	14 Si 28.09 Silicon	15 P 30.975 Phos-phorus	16 S 32.066 Sulfur	17 Cl 35.457 Chlorine	18 Ar 39.944 Argon
19 K 39.100 Potas-sium	20 Ca 40.08 Calcium	21 Sc 44.96 Scandium	22 Ti 47.90 Titanium	23 V 50.95 Vanadium	24 Cr 52.01 Chromium	25 Mn 54.94 Manganese	26 Fe 55.85 Iron	27 Co 58.94 Cobalt	28 Ni 58.71 Nickel	29 Cu 63.54 Copper	30 Zn 65.38 Zinc	31 Ga 69.72 Gallium	32 Ge 72.60 German-ium	33 As 74.91 Arsenic	34 Se 78.96 Selenium	35 Br 79.916 Bromine	36 Kr 83.80 Krypton
37 Rb 85.48 Rubidium	38 Sr 87.63 Strontium	39 Y 88.92 Yttrium	40 Zr 91.22 Zirconium	41 Nb 92.91 Niobium	42 Mo 95.95 Molyb-denum	43 Tc (99) Tech-netium	44 Ru 101.1 Ruthenium	45 Rh 102.91 Rhodium	46 Pd 106.4 Palladium	47 Ag 107.880 Silver	48 Cd 112.41 Cadmium	49 In 114.82 Indium	50 Sn 118.70 Tin	51 Sb 121.76 Antimony	52 Te 127.61 Tellurium	53 I 126.91 Iodine	54 Xe 131.30 Xenon
55 Cs 132.91 Cesium	56 Ba 137.36 Barium	57 La 138.92 Lanthanum	72 Hf 178.50 Hafnium	73 Ta 180.95 Tantalum	74 W 183.86 Wolfram	75 Re 186.22 Rhenium	76 Os 190.2 Osmium	77 Ir 192.2 Iridium	78 Pt 195.09 Platinum	79 Au 197.0 Gold	80 Hg 200.61 Mercury	81 Tl 204.39 Thallium	82 Pb 207.21 Lead	83 Bi 209.00 Bismuth	84 Po (210) Polonium	85 At (210) Astatine	86 Rn (222) Radon
87 Fr (223) Francium	88 Ra (226) Radium	89 Ac (227) Actinium	104 Unq (261) Unnil-quadium	105 Unp (262) Unnil-pentium	106 Unh (263) Unnil-hexium	107 Uns (262) Unnil-septium	108 Uno (265) Unnil-octium	109 Une (267) Unnil-ennium									
				58 Ce 140.13 Cerium	59 Pr 140.92 Praseo-dymium	60 Nd 144.27 Neo-dymium	61 Pm (147) Pro-methium	62 Sm 150.35 Samarium	63 Eu 152.0 Europium	64 Gd 157.26 Gado-linium	65 Tb 158.93 Terbium	66 Dy 162.51 Dys-prosium	67 Ho 164.94 Holmium	68 Er 167.27 Erbium	69 Tm 168.94 Thulium	70 Yb 173.04 Ytterbium	71 Lu 174.99 Lutetium
				90 Th (232) Thorium	91 Pa (231) Protactin-ium	92 U 238.07 Uranium	93 Np (237) Neptunium	94 Pu (242) Plutonium	95 Am (243) Americium	96 Cm (247) Curium	97 Bk (249) Berkelium	98 Cf (251) Californ-ium	99 Es (254) Einstein-ium	100 Fm (253) Fermium	101 Md (256) Mendelev-ium	102 No (253) Nobelium	103 Lw 257 Lawrenc-ium

Eponyms Used in This Laboratory Manual

APPENDIX C

Eponym	Current Terminology
Achilles tendon	calcaneal tendon
Adam's apple	thyroid cartilage
ampulla of Vater (VA-ter)	hepatopancreatic ampulla
Bartholin's (BAR-tō-linz) gland	greater vestibular gland
Billroth's (BIL-rōtz) cord	splenic cord
Bowman's (BŌ-manz) capsule	glomerular capsule
Bowman's (BŌ-manz) gland	olfactory gland
Broca's (BRŌ-kaz) area	motor speech area
Brunner's (BRUN-erz) gland	duodenal gland
bundle of His (HISS)	atrioventricular (AV) bundle
canal of Schlemm (SHLEM)	scleral venous sinus
circle of Willis (WIL-is)	cerebral arterial circle
Cooper's (KOO-perz) ligament	suspensory ligament of the breast
Cowper's (KOW-perz) gland	bulbourethral gland
crypt of Lieberkühn (LĒ-ber-kyoon)	intestinal gland
duct of Rivinus (ri-VĒ-nus)	lesser sublingual duct
duct of Santorini (san'-tō-RĒ-nē)	accessory duct
duct of Wirsung (VĒR-sung)	pancreatic duct
end organ of Ruffini (roo-FĒ-nē)	type-II cutaneous mechanoreceptor
Eustachian (yoo-STĀ-kē-an) tube	auditory tube
Fallopian (fal-LŌ-pē-an) tube	uterine tube
gland of Zeis (ZĪS)	sebaceous ciliary gland
Golgi (GOL-jē) tendon organ	tendon organ
Graafian (GRAF-ē-an) follicle	mature follicle
Hassall's (HAS-alz) corpuscle	thymic corpuscle
Haversian (ha-VĒR-shun) canal	central canal
Haversian (ha-VĒR-shun) system	osteon
interstitial cell of Leydig (LĪ-dig)	interstitial endocrinocyte
islet of Langerhans (LANG-er-hanz)	pancreatic islet
Kupffer's (KOOP-ferz) cell	stellate reticuloendothelial cell
loop of Henle (HEN-lē)	loop of the nephron
Malpighian (mal-PIG-ē-an) corpuscle	splenic nodule
Meibomian (mī-BŌ-mē-an) gland	tarsal gland
Meissner's (MĪS-nerz) corpuscle	corpuscle of touch
Merkel's (MER-kelz) disc	tactile disc
Müller's (MIL-erz) duct	paramesonephric duct
Nissl (NISS-l) bodies	chromatophilic substance
node of Ranvier (ron-VĒ-ā)	neurofibral node
organ of Corti (KOR-tē)	spiral organ
Pacinian (pa-SIN-ē-an) corpuscle	lamellated corpuscle
Peyer's (PI-erz) patches	aggregated lymphatic follicles
plexus of Auerbach (OW-er-bak)	myenteric plexus
plexus of Meissner (MĪS-ner)	submucous plexus
pouch of Douglas	rectouterine pouch
Purkinje (pur-KIN-jē) fiber	conduction myofiber

Eponym	Current Terminology
Rathke's (RATH-kēz) pouch	hypophyseal pouch
Schwann (SCHVON) cell	neurolemmocyte
Sertoli (ser-TŌ-lē) cell	sustentacular cell
Skene's (SKĒNZ) gland	paraurethral gland
sphincter of Oddi (OD-dē)	sphincter of the hepatopancreatic ampulla
Stensen's (STEN-senz) duct	parotid duct
Volkmann's (FŌLK-manz) canal	perforating canal
Wharton's (HWAR-tunz) duct	submandibular duct
Wharton's (HWAR-tunz) jelly	mucous connective tissue
Wormian (WER-mē-an) bone	sutural bone

Figure Credits

1.1 Courtesy of Olympus America, Inc.
3.5 (a)–(f) © Carolina Biological Supply/Photo-Take NYC.
4.1 (a) Biophoto Associates/Photo Researchers, Inc. (b) M. I. Walker/Photo Researchers, Inc. (c) Biophoto Associates/Photo Researchers, Inc. (d) G.W. Willis/Biological Photo Service (e) David M. Phillips/Visuals Unlimited (f), (g) Ed Reschke.
4.3 (a) (b) Biophoto Associates/Photo Researchers, Inc. (c) Robert Brons/Biological Photo Service (d) Biophoto Associates/Photo Researchers, Inc. (e) Ed Reschke (f) Bruce Iverson/Visuals Unlimited (g) Fred Hossler/Visuals Unlimited (h) Frederick C. Skvara (i) Chuck Brown/Photo Researchers, Inc.
5.2 Dr. Richard Kessel and Dr. Randy H. Kardon.
6.2 (d) Manfred Kage/Peter Arnold, Inc.
6.3 Biophoto Associates/Photo Researchers, Inc.
8.2 (a), (b), (c), Copyright © 1983 by Gerard J. Tortora, Courtesy of Matt Iacobino and Lynne Borghesi (d) Courtesy of Evan J. Colella, (f), (g), (h) Courtesy of Matt Iacobino (i) © 1991 Evan J. Colella, (e), (j) Copyright © 1997, Courtesy of Lynne Borghesi.
8.4 (c) Robert A. Chase, M.D.
9.1 Biophoto Associates/Photo Researchers, Inc.
9.3 Ed Reschke.
9.4 Biophoto Associates/Photo Researchers, Inc.
11.2–11.8 Biomedical Graphics Department, University of Minnesota Hospitals.
11.9 (a) Biomedical Graphics Department, University of Minnesota Hospitals (b) O. Richard Johnson.
11.10–11.12 Biomedical Graphics Department, University of Minnesota Hospitals.
12.1 (c) © Carolina Biological Supply/Photo-Take NYC.
13.10 (b) Martin M. Rotker/Science Source/Photo Researchers, Inc.
13.14 (b) Fred Hossler/Visuals Unlimited.
13.17 (b) Martin M. Rotker.
14.5 Visuals Unlimited.
14.6 (b) By Andrew Kuntzman, Wright State University.
14.9 Frank Awbrey/Visuals Unlimited.
14.17 Copyright © 1983 by Gerard J. Tortora, Courtesy of James Borghesi.
15.2 Bruce Iverson.
15.3 Lester V. Bergman & Associates, Inc.
15.4 © G. W. Willis/Biological Photo Service.
15.5 R. Calentine/Visuals Unlimited.
15.6 Ed Reschke/Peter Arnold, Inc.
16.1 (b) Ed Reschke.
17.2 (c) Courtesy of Chihiro Yokochi, MD and Johannes W. Rohen, MD from the book *Photographic Anatomy of the Human Body,* Igaku-Shoin Medical Publishers, New York, NY; 1978.
17.3 (b) John Eads.
18.1 (a) © Martin Rotker/PhotoTake NYC (b) © Carolina Biological Supply/PhotoTake NYC (c) © CNRI / PhotoTake NYC.
20.3 Copyright © 1997, Gerard J. Tortora, Courtesy of Lynne Borghesi.
20.5 Courtesy of Chihiro Yokochi, MD and Johannes W. Rohen, MD from the book *Photographic Anatomy of the Human Body,* Igaku-Shoin Medical Publishers, New York, NY; 1978.
20.7 © Carolina Biological Supply/PhotoTake NYC.
20.10 (a) Reproduced by permission from R. G. Kessel and R. H. Kardon, *Tissues and Organs: A Text-Atlas of Scanning Electron Microscopy*, W. H. Freeman, 1979 (b) G. W. Willis, MD/Biological Photo Service.
21.4 G. W. Willis/Biological Photo Service.
21.7 © Carolina Biological Supply/PhotoTake NYC.
21.8 (b) Michael Ballo.
21.9 James R. Smail and Russell Whitehead, Macalester College.
21.11 James R. Smail and Russell Whitehead, Macalester College.
21.13 (a) Biophoto Associates/Photo Researchers, Inc. (b), (c) Don W. Fawcett/Visuals Unlimited.
21.15 (a) Bruce Iverson/Visuals Unlimited. (b) Cabisco/Visuals Unlimited.

22.2 (b) Ed Reschke/Peter Arnold, Inc.
22.5 Dr. Andrew Kuntzman.
22.8 James R. Smail and Russell Whitehead, Macalester College.
22.10 Bruce Iverson/Visuals Unlimited. Inc.
23.2 (b) By Andrew Kuntzman, Wright State University.
23.4 By James R. Smail and Russell A. Whitehead, Macalester College.
23.5 Dr. Andrew Kuntzman
23.11 (a) By James R. Smail and Russell A. Whitehead, Macalester College (b) © Dr. Mary Notter/PhotoTake NYC.
23.12 Don W. Fawcett/Visuals Unlimited.
23.15 By James R. Smail and Russell A. Whitehead, Macalester College.

Index